Classical
Mechanics
and Relativity

Other World Scientific texts by the author:

Introduction to Quantum Mechanics: Schrödinger Equation and Path Integral, 2006

Electrodynamics: An Introduction including Quantum Effects, 2004

and with A. Wiedemann:

Supersymmetry, An Introduction with Conceptual and Calculational Details, 1987

Classical
Mechanics
and Relativity

Harald J W Müller-Kirsten

University of Kaiserslautern, Germany

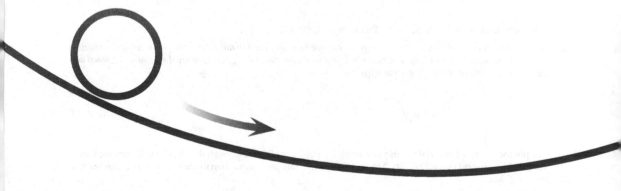

 **World Scientific**

NEW JERSEY · LONDON · SINGAPORE · BEIJING · SHANGHAI · HONG KONG · TAIPEI · CHENNAI

Published by

World Scientific Publishing Co. Pte. Ltd.

5 Toh Tuck Link, Singapore 596224

USA office: 27 Warren Street, Suite 401-402, Hackensack, NJ 07601

UK office: 57 Shelton Street, Covent Garden, London WC2H 9HE

Library of Congress Cataloging-in-Publication Data
Müller-Kirsten, H. J. W.
    Classical mechanics and relativity / Harald J. W. Müller-Kirsten.
        p. cm.
    Includes bibliographical references and index.
    ISBN-13: 978-981-283-251-1 (hardcover : alk. paper)
    ISBN-10: 981-283-251-3 (hardcover: alk. paper)
    ISBN-13: 978-981-283-252-8 (softcover: alk. paper)
    ISBN-10: 981-283-252-1 (softcover: alk. paper)
    1. Relativistic mechanics. 2. Lagrange equations. 3. Hamiltonian operator.
4. Einstein field equations.   I. Title.
    QA808.5.M85 2008
    530.12--dc22

                                            2008029186

British Library Cataloguing-in-Publication Data
A catalogue record for this book is available from the British Library.

Printed in Singapore.

Contents

Preface   xi

1   **Introduction**   1
    1.1   Introduction   1

2   **Recapitulation of Newtonian Mechanics**   3
    2.1   Introductory Remarks   3
    2.2   Recapitulation of Newton's Laws   4
    2.3   Further Definitions and Rotational Motion   5
    2.4   Conservative Forces   7
    2.5   Mechanics of a System of Particles   9
    2.6   Newton's Law of Gravitation   14
    2.7   Miscellaneous Examples   15

3   **The Lagrange Formalism**   29
    3.1   Introductory Remarks   29
    3.2   The Generalized Coordinates   30
    3.3   The Principle of Virtual Work   32
    3.4   D'Alembert's Principle, Lagrange Equations   37
    3.5   Hamilton's Variational Principle, and Euler–Lagrange Equations   42
        3.5.1   Hamilton's variational principle   42
        3.5.2   Hamilton's principle for conservative systems   42
        3.5.3   Hamilton's principle for holonomic systems   46
        3.5.4   Hamilton's principle for nonholonomic systems   48
        3.5.5   The general procedure   50
    3.6   Symmetry Properties and Conservation Laws   51
    3.7   Miscellaneous Examples   53

4   **The Canonical or Hamilton Formalism**   77
    4.1   Introductory Remarks   77
    4.2   Hamilton's Equations of Motion   78

4.3	Physical Significance of the Hamilton Function	82
4.4	Variational Principle for Hamilton's Equations	87
4.5	Transformation of Canonical Coordinates	88
4.6	Lagrange and Poisson Brackets	91
	4.6.1 The fundamental Lagrange and Poisson brackets	91
	4.6.2 Connection between Lagrange and Poisson brackets	92
4.7	The Poisson Algebra and its Significance	97
4.8	Miscellaneous Examples	100

5 Symmetries and Transformations   **103**
5.1	Introductory Remarks	103
5.2	Symmetries	103
5.3	The Galilei Transformation	104
5.4	Rotation and Rotation Group	109
	5.4.1 Group property of coordinate transformations	109
	5.4.2 The group concept	110
	5.4.3 The orthogonal group $O(n)$	112
	5.4.4 The groups $O(2)$ and $SO(2)$	114
	5.4.5 The groups $O(3)$ and $SO(3) =: \{R\}$	116
	5.4.6 The unitary groups $U(n)$ and $SU(n)$	119
	5.4.7 The infinitesimal rotation of a vector	127
5.5	Rotating Reference Frames	130
5.6	Definition of Scalars, Vectors, Tensors	134
5.7	The Theorem of E. Noether	138
5.8	Canonical Transformations	140
	5.8.1 Generators of canonical transformations	141
	5.8.2 Invariance of Poisson brackets	147
5.9	Conserved Quantities	152
	5.9.1 Infinitesimal canonical transformations	154
	5.9.2 Infinitesimal transformations and Poisson brackets	155
	5.9.3 Angular momenta and Poisson brackets	158
5.10	Miscellaneous Examples	160

6 Looking Beyond Classical Mechanics   **167**
6.1	Introductory Remarks	167
6.2	Aspects of Classical Statistics	168
	6.2.1 Classical probabilities	168
	6.2.2 The Liouville equation	175
	6.2.3 Probable values of observables	179
6.3	Spacetime Formulations	183
	6.3.1 Spacetime (Lorentz) transformations	184

      6.3.2   The Poincaré group 186

      6.3.3   Derivatives 189

  6.4 From Particles to Fields 190

      6.4.1   Euler–Lagrange equations 190

      6.4.2   The Noether theorem 192

      6.4.3   Curved spacetime 194

  6.5 Miscellaneous Examples 197

7 Two-Body Central Forces   199

  7.1 Introductory Remarks 199

  7.2 Equations of Motion 200

  7.3 Solution of the Equations 204

  7.4 Differential Equation of the Orbit 208

  7.5 The Kepler Problem 210

  7.6 Tangential Equations of Orbits 216

  7.7 Maxima and Minima of Velocities 224

  7.8 Same Orbit, Different Forces 226

  7.9 Period 227

  7.10 Perihelion Precession of Mercury 228

  7.11 Stability of Circular Orbits 233

  7.12 Scattering in Central Force Fields 235

  7.13 Miscellaneous Examples 241

8 Rigid Body Dynamics   257

  8.1 Introductory Remarks 257

  8.2 Moments of Inertia 258

  8.3 Diagonalization and Principal Axes 265

      8.3.1   The ellipsoid of inertia 265

      8.3.2   Transformation to principal axes 268

  8.4 The Equations of Motion 275

  8.5 Miscellaneous Examples I 279

  8.6 Force-free Motion 293

  8.7 The Spinning Top in the Gravitational Field 302

  8.8 Motion Relative to Rotations: Centrifugal and Coriolis Forces 311

  8.9 Miscellaneous Examples II 321

9 Small Oscillations and Stability   333

  9.1 Introductory Remarks 333

  9.2 Resonance Frequencies and Normal Modes 333

  9.3 Stability 342

  9.4 Miscellaneous Examples 348

10 Motivation of the Theory of Relativity   **353**
   10.1 Introductory Remarks 353
   10.2 The Weak Equivalence Principle 354
   10.3 Inertial Frames 358
   10.4 The Strong Principle of Equivalence 359
   10.5 The Fundamental Postulate 363
   10.6 Curvature 365
   10.7 Miscellaneous Examples 376

11 A Simple Look at Phenomenological Consequences   **385**
   11.1 Introductory Remarks 385
   11.2 Results of the Special Theory Summarized 386
   11.3 Main Tests of General Relativity 387
     11.3.1 The gravitational redshift 387
     11.3.2 The gravitational deflection of light 387
     11.3.3 The precession of the planet Mercury's perihelion . . . 388

12 Aspects of Special Relativity   **391**
   12.1 Introductory Remarks 391
   12.2 Basics and Physical Motivation of the Lorentz Transformation 392
   12.3 Active and Passive Transformations 395
   12.4 Proper Time and Light Cones 400
   12.5 Lorentz Indices and Transformations 406
     12.5.1 Contravariant vectors and covariant vectors 407
     12.5.2 Tensors 409
   12.6 Lorentz Boosts in Electrodynamics 410
   12.7 Curvature due to Lorentz Contraction 411
   12.8 Covariantization of Newton's Equation of a Charged Particle 412
   12.9 The Tangent Vector 417
   12.10 Miscellaneous Examples 419

13 Equation of Motion of a Particle in a Gravitational Field   **433**
   13.1 Introductory Remarks 433
   13.2 Equation of Motion 433
   13.3 Reduction to Newton's Equation 437
   13.4 Rotation Observed from an Inertial Frame 440
   13.5 The Redshift 443

14 Tensor Calculus for Riemann Spaces   **445**
   14.1 Introductory Remarks 445
   14.2 Tensors 446

14.3 Symmetric and Antisymmetric Tensors 449
14.4 Definition of Other Important Quantities 450
 14.4.1 Transformation of the metric tensor 450
 14.4.2 Pseudo-tensors and duals 450
 14.4.3 Volume forms 452
14.5 Covariant Derivatives by the Method of Parallel Transport of
 a Vector 455
14.6 Metric Affinity and Christoffel Symbols 462
14.7 Raising and Lowering of Indices 464
14.8 Rewriting Co- and Contravariant Derivatives 465
14.9 Covariant Divergence, Rotation etc. 466

15 Einstein's Equation of the Gravitational Field   **471**
15.1 Introductory Remarks 471
15.2 The Riemann Curvature Tensor 471
15.3 Bianchi Identities and Ricci–Einstein Tensor 475
15.4 The Energy–Momentum Tensor 481
 15.4.1 The energy–momentum tensor in electrodynamics . . 481
 15.4.2 The general case 485
15.5 Einstein's Equation of the Gravitational Field 487
15.6 Newton's Potential from Einstein's Equation 491

16 The Schwarzschild Solution   **497**
16.1 Introductory Remarks 497
16.2 The Spherical Solution Outside the Source 499
16.3 The Schwarzschild Solution for $\Lambda = 0$ 501
16.4 The Schwarzschild Solution for $\Lambda \neq 0$ 507
16.5 The Relativistic Kepler Problem 511
16.6 The Light Ray in the Schwarzschild Field 516

Appendix A: Schwarzschild Orbit Solution   **525**
 A.1 Introductory remarks 525
 A.2 The elliptic integral 526
 A.3 Evaluating the elliptic integral 528

Appendix B: Reissner–Nordstrom Metric   **531**
 B.1 Introductory remarks 531
 B.2 The metric 531
 B.3 The energy-momentum tensor 535
 B.4 The energy-momentum tensor for an electrostatic field 537
 B.5 Christoffel symbols and Riemann tensor 539

B.6   The Einstein equation 542
B.7   Evaluating the electrostatic and gravitational fields . . 544

Bibliography   **549**

Index   **553**

Preface

Classical Mechanics is the most basic subject of any degree course in physics, and the subject is now so well established that its main building blocks, the Lagrangian and Hamiltonian methods, provide it with a universal character which forms the basis of other fundamental areas of physics, such as electrodynamics, quantum mechanics and statistical physics. A thorough understanding of classical mechanics is therefore a far-reaching help for the understanding also of these other areas. Classical mechanics became a theoretical subject with Newton's laws of motion, and attained its universality with the replacement of these laws by Hamilton's action principle as the basic axiom. This principle is of such fundamentality that today almost any model theory proposed in particle physics is first formulated in terms of such an action.

It is common knowledge today that Einstein's theory of relativity, in both its Special and General forms, extended Newton's mechanics to spectacularly precise confirmations by experiment, the Special Theory, for instance, in electrodynamics, and the General Theory in celestial mechanics. Einstein's mechanics is widely regarded as unusually complex and is popularly considered difficult to grasp. The theory of relativity is therefore rarely presented along with ordinary classical mechanics. In some sense this is a pity since the student of classical mechanics — and the latter includes always Newton's gravitational law and planetary motion — knows very well that there is more to it, that Einstein's relativity provides corrections, but with no reference to this he is supported in his belief that this is a theory beyond his understanding. This does not have to be the case, and in fact can be expected to change, since *e.g.* texts on space science assume already a first acquaintance with general relativity. Therefore we make an attempt here to direct the reader's view from as early a stage as possible to the ultimate theory of Einstein, so that even if the reader's prime interest is focussed on classical mechanics, he can turn the pages and explore the steps to this ultimate theory. Thus, an example in Lagrangian dynamics serves to derive the full Kepler problem from the Schwarzschild metric; the Kepler problem is then treated in full in the context of central forces, and later Newton's gravitational force and the Schwarzschild metric are obtained in Einstein's theory of relativity. An essential complementary aspect is the stepwise generalization of rotations to spacetime and general coordinate transformations, much of which can be grasped by analogy.

Apart from the inclusion of an introduction to relativity, the text covers most of the topics usually treated in standard texts on classical mechanics, such as for instance, the very wellknown book of H. Goldstein, and as in this and in texts of other authors, such as T. W. B. Kibble and F. H. Berkshire, frequent reference is made to electrodynamics, especially in examples. Our presentation here is in some respects more detailed and includes a larger number of worked problems, including a number of problems given as unsolved in the text of H. Goldstein. As in the author's earlier texts on electrodynamics and quantum mechanics, phrases like "it is easy to see" or "it is easy to show" are consistently avoided since if true the explanations can readily be provided by an author, and a condensed form which leaves many derivations to the reader seems inappropriate for the companion text of a basic course. In addition, for the benefit of readers who are new to the field, some vital points are occassionally repeated or re-emphasized. Thus, as a rule, besides detailed explanations, all derivations are given, assuming that it is easier for a reader to skip or run through given calculations than to reproduce these himself at numerous points of the text, which in any case leaves the reader a lot of other points to struggle with for himself.* A proper understanding requires a thorough understanding of each and every step, a mental struggle being in any case unavoidable. As the philosopher A. Schopenhauer remarked: The process of cognition is the path from a paradox to triviality. Once something has been thoroughly understood, it appears trivial.

Harald J. W. Müller–Kirsten

*Throughout the text names are cited without first names, *e.g.* Einstein, and only these are named in the index. Names in footnotes, usually with reference to literature, are given with initials of first names, but are not cited in the index.

Chapter 1

Introduction

1.1 Introduction

Mechanics is the theory of motion of material bodies. In order to distinguish this theory from other theories, particularly quantum mechanics, one also refers to it as Classical Mechanics. This latter specialization is also meant to include the Special Theory of Relativity. We assume here familiarity with fundamental physical concepts such as space, time, mass and force. We begin with the mechanics of a single pointlike mass called a particle.

A course in Classical Mechanics is the first and basic course in theoretical physics. Apart from covering the mechanics of material bodies, the course serves also as an essential prerequisite for subsequent courses in electrodynamics, quantum mechanics and statistical mechanics. Although we recapitulate first the laws postulated by Newton, we shall then proceed very differently. As in any other modern text, the basis of the mechanics we develop here will be the postulate of extremization of the so-called action integral, and we shall convince ourselves that this principle reproduces the totality of Newtonian mechanics. Moreover, we shall see (to some extent) that the principle is of much wider generality, so that its application can be found in many other branches of physics. The Newton equations of motion are obtained from this extremization principle as Euler–Lagrange equations, and the formalism as such is known as the Lagrange formalism. We shall consider important applications. The vital continuation of this formalism leads to the canonical Hamilton formalism, which serves as a basis for the transition to quantum mechanics. It is evident that for this reason alone these formalisms are already of fundamental significance. In fact, with the derivation of the so-called Poisson algebra in the context of Hamilton's mechanics, one is approaching quantum mechanics as closely, as is possible within classical mechanics (there is no derivation, again one has to start from some postulates). More complicated problems are those involving constraints, and we shall be concerned with some here. A method parallel to that of Poisson brackets but naturally more complicated can be developed to deal with constraints in quantum mechanics. Thus both, the Lagrange and the Hamilton formalisms, play a fundamental role not only in mechanics but also in subsequent courses such as electrodynamics and quantum mechanics. In addition they permit a deep insight into the structure of mechanics, into relationships between symmetry properties and conservation laws, and other basic properties. In the following chapter we begin with a very brief recapitulation of Newton's mechanics mainly for the purpose of making the difference in the postulates underlying Newton's approach and those underlying the modern extremization approach as clear as possible. In this recapitulation of Newton's mechanics, we also emphasize in particular that this does not require a special postulate for rotational motion. In addition, we consider immediately many particle systems, and thereby introduce also the concept of constraints which is required for the subsequent consideration of rigid bodies and their dynamics. Our brief recapitulation of Newtonian mechanics serves also our other purpose here, namely to uncover its connection with relativity, *i.e.* Einstein's mechanics. On the way we make brief and elementary forays into electrodynamics, group theory and statistical physics. The large number of worked problems scattered throughout the text is meant to illustrate specific aspects, and more than that to demonstrate the wide spectrum of relevance of classical mechanics in physics, reaching from atomic physics, through phenomena of daily life to those of celestial dimensions. Thus we aim at a presentation here which spans the entire domain of mechanics from Newton to Einstein, which is one of the most fascinating domains of the natural sciences.

Chapter 2

Recapitulation of Newtonian Mechanics

2.1 Introductory Remarks

Although Newton's approach to mechanics is today obsolete, it is helpful for an understanding of modern reasoning to be fully aware of the difference in the basic principles of both approaches. One can then gain a deeper appreciation of the more general modern approach — leading also to the neighbourhood of quantum mechanics — without losing anything of Newton's mechanics.* Thus we first recapitulate Newton's laws for particles and systems of particles, and distinguish clearly between linear and rotational motion. Since this is basically school mechanics we refrain from presenting numerous illustrative examples, though it would be wrong to conclude that all such examples would be simple or even trivial. Many books exist with hundreds of worked or unworked examples, of which some can be quite tricky or easier to handle with the Lagrangian formalism.† Some examples which usually require second thoughts are, for instance, those with varying mass, as in the case of rockets, chains slipping off a table and similar cases. Thus we include some examples at the end of this chapter, also for purposes of illustration. The power of the modern Lagrangian and Hamiltonian methods results from their suitability to treat much more general situations as will become evident in later chapters.

---

*The reader interested in a fascinating discussion of the development of the concept of spacetime from Aristotle to Galilei, Newton and Einstein is recommended to read R. Penrose [38], Chapter 17, pp. 383 – 411. This book will be referred to at various points of the text. The author is indebted to Andrew Y.T. Chan, then an Editor of World Scientific, for bringing this book to his attention.

†Books with hundreds of examples (solved and/or unsolved) are, for instance, those of T. W. B. Kibble and F. H. Berkshire [25] and *The Physics Coaching Class* [48].

2.2   Recapitulation of Newton's Laws

In *Newton's formulation* of classical mechanics, his wellknown three laws of motion are the basic postulates or axioms from which everything else is deduced. We recall these first and comment thereafter on certain terms used:

1. *The law of inertia*: Every body (of mass $\neq 0$) continues in its state of rest or of uniform motion in a straight line, unless compelled by some external force to change that state.

2. *The equation of motion*: The time rate of change of motion (i.e. of the momentum) is proportional to the applied force, and takes place in the direction of this force.

3. *Actio = reactio*: To every action there is an equal and opposite reaction.

The first law may sound trivial in the way we all became acquainted with this at school. However, the phrase "state of rest or of uniform motion" (constant velocity) reveals already relativity, *i.e.* Galilean relativity: There is nothing to distinguish the physics of the so-called state of rest from that of uniform motion, since there is no such notion as a fixed point in space.

The word *"inertia"* (from 'inert' meaning sluggish, slow) means with nonzero mass but without inherent action or motion, and hence implies what the first law spells out.[‡] The word *"uniform"* means with constant velocity, and also that *e.g.* a bottle with a cork stuck on moves in this form with the same magnitude and direction, and not *e.g.* sometimes with a cork flying somewhere nearby. In the second law the phrase "time rate of change" means the derivative with respect to time t, *i.e.* d/dt, and the word "motion" is to be understood as linear momentum $\mathbf{p}$. Thus for a single particle under the influence of an applied force $\mathbf{F}$ the second law implies

$$\frac{d\mathbf{p}}{dt} = \mathbf{F}. \tag{2.1}$$

[‡]We shall define later an "inertial reference frame" as an unaccelerated one defined by mass × velocity = const., *i.e.* velocity = const./mass, so that the mass cannot be zero (hence the name "inertial frame", and later the name "moment of inertia" meaning, effectively, moment of mass). The larger the mass is, the sluggisher, more "inert" is the motion, or "the resistance to acceleration", as R. Penrose says [38], p. 392. In Newton's nonrelativistic mechanics such motion is conveniently considered with respect to an assumed fixed origin in space, or correspondingly to the position of an extremely massive masspoint like that of a fixed star or the sun. Accelerated systems are the subject of the General Theory of Relativity. Since acceleration is there argued to be indistinguishable from gravity, inertial frames are those far away from any influence of gravity. An inertial frame of reference is therefore also described as a non-accelerated frame of reference, and in what is called *"free fall"* ("free" meaning the right hand side of the geodesic equation of Sec. 13.2 is zero) all bodies move inertially whether or not gravity is present, *i.e.* at least locally (*cf.* also Secs. 10.2, 10.4).

Introducing the position vector $\mathbf{r}$, the particle mass m, and the particle's velocity $\mathbf{v}$, then here (*cf.* Fig. 2.1)

$$\mathbf{p} = m\mathbf{v} = m\frac{d\mathbf{r}}{dt} = m\lim_{\triangle t \to 0}\frac{\mathbf{r}_2 - \mathbf{r}_1}{\triangle t} = m\lim_{\triangle t \to 0}\frac{\triangle \mathbf{s}}{\triangle t}, \qquad (2.2)$$

so that Eq. (2.1) becomes $\mathbf{F} = d(m\mathbf{v})/dt$.

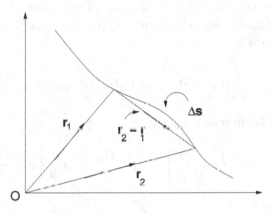

Fig. 2.1 Path element $\triangle$s.

For the reason explained above the mass m here in Newton's equation is called "*inertial mass*". In Sec. 2.6 we shall encounter the term "*gravitational mass*", and in Chapter 10 we shall see that these are equal, this being (as explained there) the observation of Galilei.

2.3   Further Definitions and Rotational Motion

The *acceleration* of the particle of mass m is defined as

$$\mathbf{a} = \frac{d^2\mathbf{r}}{dt^2}. \qquad (2.3)$$

The particle's *angular momentum* $\mathbf{L}$ around a point 0, and it's *torque* $\mathbf{N}$ around 0 are respectively defined as the moments

$$\mathbf{L} = \mathbf{r} \times \mathbf{p}, \quad \mathbf{N} = \mathbf{r} \times \mathbf{F}. \qquad (2.4)$$

Thus the angular momentum is the moment of momentum. We also observe the correspondence

$$\mathbf{p} \leftrightarrow \mathbf{L}, \quad \mathbf{F} \leftrightarrow \mathbf{N}$$

between linear motion and rotational motion. This correspondence becomes even more conspicuous by the following relation which we verify below:

$$\mathbf{N} = \frac{d\mathbf{L}}{dt}. \tag{2.5}$$

In verifying this relation we gain some experience in dealing with vectors and their cross products. The mass m is a constant in time. Here in Newtonian mechanics we also consider the orientation of the axes of our Cartesian coordinate system (characterized by unit vectors $\mathbf{e}_x, \mathbf{e}_y, \mathbf{e}_z$) as not changing in the course of time, *i.e.*

$$\frac{d}{dt}(\mathbf{e}_i) = 0, \quad i = 1, 2, 3 \ \ (\text{or } x, y, z).$$

The definition of $\mathbf{L}$ therefore implies

$$\frac{d\mathbf{L}}{dt} = m\frac{d}{dt}\left\{\mathbf{r} \times \frac{d\mathbf{r}}{dt}\right\}, \quad \text{or} \quad \frac{dL_i}{dt} = m\frac{d}{dt}\left\{\mathbf{r} \times \frac{d\mathbf{r}}{dt}\right\}_i. \tag{2.6}$$

For a vector $\boldsymbol{\omega}$ we have

$$\left(\frac{d\boldsymbol{\omega}}{dt}\right)_i = \left(\frac{d\omega_x}{dt}\mathbf{e}_x + \frac{d\omega_y}{dt}\mathbf{e}_y + \frac{d\omega_z}{dt}\mathbf{e}_z\right)_i = \frac{d}{dt}(\boldsymbol{\omega})_i.$$

A cross product as in the definition of $\mathbf{L}$ can be expressed in terms of the so-called *Levi–Civita symbol* which is defined as follows:

$$\epsilon_{ijk} = \begin{cases} +1, & \text{if } i, j, k \text{ a cyclic permutation of } 1, 2, 3, \\ -1, & \text{if } i, j, k, \text{ an anticyclic permutation of } 1, 2, 3, \\ 0, & \text{if two of } i, j, k \text{ are equal}. \end{cases} \tag{2.7}$$

We then have

$$(\mathbf{A} \times \mathbf{B})_i = \sum_{j,k=1}^{3} \epsilon_{ijk} A_j B_k. \tag{2.8}$$

Thus we have in the above case of Eq. (2.6)

$$\begin{aligned} \frac{dL_i}{dt} &= m\frac{d}{dt}\left(\mathbf{r} \times \frac{d\mathbf{r}}{dt}\right)_i = m\frac{d}{dt}\sum_{j,k=1}^{3} \epsilon_{ijk}(\mathbf{r})_j\left(\frac{d\mathbf{r}}{dt}\right)_k \\ &= m\sum_{j,k} \epsilon_{ijk}\left[\left(\frac{d\mathbf{r}}{dt}\right)_j\left(\frac{d\mathbf{r}}{dt}\right)_k + (\mathbf{r})_j\left(\frac{d^2\mathbf{r}}{dt^2}\right)_k\right]. \end{aligned} \tag{2.9}$$

But for a general three-vector $\boldsymbol{\alpha}$ (with half the original expression plus half that with j, k renamed k, j):

$$\sum_{j,k=1}^{3} \epsilon_{ijk}\alpha_j\alpha_k = \frac{1}{2}\sum_{j,k=1}^{3} \alpha_j\alpha_k(\epsilon_{ijk} + \epsilon_{ikj}) = 0. \qquad (2.10)$$

This result follows from the antisymmetry of the Levi–Civita symbol. Hence the first term on the right hand side of Eq. (2.9) vanishes, and we obtain:

$$\frac{dL_i}{dt} = m\sum_{j,k} \epsilon_{ijk}(\mathbf{r})_j(\ddot{\mathbf{r}})_k, \quad i.e. \quad \frac{d\mathbf{L}}{dt} = \mathbf{r} \times \mathbf{F} = \mathbf{N}, \qquad (2.11)$$

where $\mathbf{F} = m\ddot{\mathbf{r}}$. The moment of the force $\mathbf{F}$ is called *torque*. It is this which causes a turning effect. We thus arrive at the analogues of Newton's laws for rotation[§] (however, derived from these by considering effectively infinitesimal steps and summing these):

1. If a body is rotating about a given axis through its centre of mass, it will continue to do so with constant angular velocity unless compelled by some external torque to change that state.

2. The time rate of change of angular momentum is proportional to the external torque and takes place about the axis about which the torque is applied and in the same sense.

3. To every couple there is an equal and opposite couple (two equal unlike parallel forces constituting a *couple*).

2.4 Conservative Forces

We now define the *work* W_{12} done by the external force $\mathbf{F}$ applied to a particle in moving this from a point 1 to a point 2 by the expression

$$W_{12} = \int_1^2 \mathbf{F} \cdot d\mathbf{s}. \qquad (2.12a)$$

For a constant mass m we have:

$$\int_1^2 \mathbf{F} \cdot d\mathbf{s} = m \int \frac{d\mathbf{v}}{dt} \cdot (\mathbf{v}dt) = T_2 - T_1, \quad T_i = \frac{1}{2}mv_i^2. \qquad (2.12b)$$

[§]See *e.g.* E. H. Booth and P. M. Nicol [6], p. 73.

Thus the work $W_{12} = T_2 - T_1$ is equal to the change in kinetic energy. The work can here, for instance, be that performed by the force called *weight mg* in moving the mass m from a point 1 to a point 2. If the force or, as one says, the *force field*, is such that the integral taken around a closed path vanishes, *i.e.*

$$\oint \mathbf{F} \cdot d\mathbf{s} = 0, \tag{2.13}$$

the force (and hence the system) is described as *conservative*. We emphasize: The integral in Eq. (2.12b) does not only depend on the endpoints 1 and 2, but also on the path from point 1 to point 2 ($d\mathbf{s}$ is an element of this path). We now recall *Stokes's theorem*:

$$\oint \mathbf{F} \cdot d\mathbf{s} = \int_{\mathcal{F}} \text{curl } \mathbf{F} \cdot d\mathbf{f} = \int_{\mathcal{F}} \mathbf{n} df \cdot \text{curl } \mathbf{F}, \tag{2.14}$$

where $\mathbf{n}$ is a unit vector perpendicular out of the plane of f and $\mathcal{F}$ the entire area integrated over, so that for conservative forces, *i.e.* those satisfying Eq. (2.13),

$$\text{curl } \mathbf{F} = 0, \quad i.e. \quad \boldsymbol{\nabla} \times \mathbf{F} = 0. \tag{2.15}$$

Since always curl grad of something vanishes (as one can verify by explicit evaluation), it follows that the conservative force can be expressed as a gradient, *i.e.*

$$\mathbf{F} = -\boldsymbol{\nabla} V, \tag{2.16}$$

and the scalar quantity $V(\mathbf{r})$ is described as *potential*. If we use for curl $\mathbf{F}$ the representation

$$\text{curl } \mathbf{F} = \begin{vmatrix} \mathbf{e}_x & \mathbf{e}_y & \mathbf{e}_z \\ \frac{\partial}{\partial x} & \frac{\partial}{\partial y} & \frac{\partial}{\partial z} \\ F_x & F_y & F_z \end{vmatrix}, \tag{2.17}$$

the vanishing of curl $\mathbf{F}$ with $\mathbf{F} = -\boldsymbol{\nabla} V$ follows from the theorem, that a determinant vanishes if the elements of two rows are identical. The potential $V(\mathbf{r})$ is a scalar quantity (for the definition see Sec. 5.6) and is also described as *potential energy*. Thus for a conservative system

$$W_{12} = \int_1^2 \mathbf{F} \cdot d\mathbf{s} = -\int_1^2 \boldsymbol{\nabla} V \cdot \mathbf{s} = V_1 - V_2. \tag{2.18}$$

Since also $W_{12} = T_2 - T_1$, *cf.* Eqs. (2.12a), (2.12b), we have $V_1 - V_2 = T_2 - T_1$, or

$$T_1 + V_1 = T_2 + V_2 = \cdots = \text{const.} \tag{2.19}$$

This relation expresses the *conservation of the total energy*. We can also reverse the argument. Let

$$\int_A^B \nabla V \cdot d\mathbf{s} = V_B - V_A \qquad (2.20)$$

be a potential difference with V single-valued and continuous. Then

$$\oint \nabla V \cdot d\mathbf{s} = \left(\int_A^B + \int_B^A \right) \nabla V \cdot d\mathbf{s} = (V_B - V_A) + (V_A - V_B) = 0, \quad (2.21)$$

i.e. $\oint \nabla V \cdot d\mathbf{s} = 0$ if V is single-valued and continuous. Thus for $\mathbf{F} = -\nabla V$ then $\oint \mathbf{F} \cdot d\mathbf{s} = 0$. Hence also:

$$\oint \mathbf{F} \cdot d\mathbf{s} \stackrel{\text{Stokes}}{=} \oint_{\mathcal{F}} \text{curl}\, \mathbf{F} \cdot d\mathbf{f} = 0, \quad \text{or} \quad \text{curl}\mathbf{F} = 0.$$

Finally we observe from Eq. (2.1) the *conservation of the linear momentum* when the total force $\mathbf{F}$ vanishes, *i.e.*

$$\mathbf{p} = \text{const.} \quad \text{for} \quad \mathbf{F} = 0, \qquad (2.22)$$

and from Eq. (2.5) the *conservation of angular momentum* when the total torque $\mathbf{N}$ vanishes, *i.e.*

$$\mathbf{L} = \text{const.} \quad \text{for} \quad \mathbf{N} = 0. \qquad (2.23)$$

2.5 Mechanics of a System of Particles

In considering a system of several particles, we distinguish between *external forces* originating from some source outside the particle system, and *internal forces*, that the particles exert on each other. Let $\mathbf{F}_i^{(e)}$ be the external force acting on particle 'i' (superscript (e) for 'external'), and $\mathbf{F}_{ji}$ the (internal) force that particle 'j' exerts on particle 'i'. Newton's second law applied to particle 'i' then implies the equation:

$$\mathbf{F}_i^{(e)} + \sum_j \mathbf{F}_{ji} = \frac{d}{dt}\mathbf{p}_i \equiv \dot{\mathbf{p}}_i, \qquad (2.24)$$

where $\mathbf{p}_i$ is, of course, the momentum of particle 'i'. Newton's third law yields the equations

$$\mathbf{F}_{ij} = -\mathbf{F}_{ji}, \quad \mathbf{F}_{ii} = 0. \qquad (2.25)$$

It follows that

$$\sum_{i,j} \mathbf{F}_{ji} = \frac{1}{2} \sum_{i,j} (F_{ji} + F_{ij}) = 0.$$

Summing the external forces acting on particle 'i' by summing over index 'i', we obtain

$$\sum_i \mathbf{F}_i^{(e)} = \sum_i \dot{\mathbf{p}}_i = \sum_i m_i \frac{d^2 \mathbf{r}_i}{dt^2}. \tag{2.26}$$

We now define the coordinate of a point called *centre of mass* of the system by the vector

$$\mathbf{R} = \frac{\sum_i m_i \mathbf{r}_i}{\sum_i m_i}, \qquad M = \sum_i m_i, \tag{2.27}$$

where M is the total mass of the system. The centre of mass is a so-called *collective coordinate* of the system, since it describes the collective position of the system. Later we shall use Eq. (2.27) in the form

$$\sum_i m_i \mathbf{r}_i - \sum_i \mathbf{R} m_i = 0. \tag{2.28}$$

It follows that we can rewrite Eq. (2.26) as

$$\mathbf{F}^{(e)} \equiv \sum_i \mathbf{F}_i^{(e)} = \sum_i m_i \frac{d^2 \mathbf{R}}{dt^2} = M \frac{d^2 \mathbf{R}}{dt^2}. \tag{2.29}$$

This equation says that the mass centre moves as if the entire external force acts on the total mass of the system. An example is an exploding grenade. The mass centre of its pieces moves as if the grenade were still a single piece.

For the total linear momentum $\mathbf{P}$ of the system we obtain with Eq. (2.28)

$$\mathbf{P} = \sum_i m_i \frac{d\mathbf{r}_i}{dt} = M \frac{d\mathbf{R}}{dt}. \tag{2.30}$$

It follows that if $\mathbf{F}^{(e)} = \sum_i F_i^{(e)} = 0$, i.e. $M d^2 \mathbf{R}/dt^2 = 0$,

$$\mathbf{P} = M \frac{d\mathbf{R}}{dt} = \text{const.} \tag{2.31}$$

This means, the total linear momentum is constant or, as one says, conserved.

Consider next the total torque $\mathbf{N}$ about a point O:

$$\mathbf{N} = \frac{d\mathbf{L}}{dt} = \sum_i \mathbf{r}_i \times \mathbf{F}_i = \sum_i \mathbf{r}_i \times \mathbf{F}_i^{(e)} + \sum_{i,j} \mathbf{r}_i \times \mathbf{F}_{ij}. \tag{2.32}$$

Let the external torque be the following and zero, *i.e.*

$$\mathbf{N}^{(e)} = \sum_i \mathbf{r}_i \times \mathbf{F}_i^{(e)} = 0. \tag{2.33}$$

Then

$$\begin{aligned}
\frac{d\mathbf{L}}{dt} &= \sum_{i,j} \mathbf{r}_i \times \mathbf{F}_{ij} = \frac{1}{2} \sum_{i,j} [\mathbf{r}_i \times \mathbf{F}_{ij} + \mathbf{r}_j \times \mathbf{F}_{ji}] \\
&= \frac{1}{2} \sum_{i,j} (\mathbf{r}_i - \mathbf{r}_j) \times \mathbf{F}_{ij} = \frac{1}{2} \sum_{i,j} \mathbf{r}_{ij} \times \mathbf{F}_{ij} = 0, \tag{2.34}
\end{aligned}$$

since by definition the force $\mathbf{F}_{ij}$ is parallel to $\mathbf{r}_{ij}$ (*cf.* Fig. 2.1). It follows that if, as here assumed, the external torque $\mathbf{N}^{(e)}$ vanishes, the total angular momentum $\mathbf{L}$ is conserved, *i.e.* constant in time.

Equation (2.30) gives the *total linear momentum* in terms of centre of mass quantities. What is the corresponding expression for the *total angular momentum*? We have

$$\mathbf{L} = \sum_i \mathbf{r}_i \times \mathbf{p}_i. \tag{2.35}$$

We set as in Fig. 2.2

$$\mathbf{r}'_i = \mathbf{r}_i - \mathbf{R}, \quad \text{and} \quad \dot{r}_i - v_i, \ \dot{r}'_i = v'_i, \tag{2.36}$$

and hence

$$\mathbf{v}'_i = \mathbf{v}_i - \mathbf{v}, \quad \mathbf{v} = \dot{\mathbf{R}}. \tag{2.37}$$

It follows that we can rewrite the expression (2.35) of the angular momentum as

$$\begin{aligned}
\mathbf{L} &= \sum_i m_i \mathbf{r}_i \times \mathbf{v}_i = \sum_i m_i (\mathbf{r}'_i + \mathbf{R}) \times (\mathbf{v}'_i + \mathbf{v}) \\
&= \sum_i m_i \mathbf{R} \times \mathbf{v} + \sum_i m_i \mathbf{r}'_i \times \mathbf{v}'_i \\
&\quad + \sum_i m_i \mathbf{r}'_i \times \mathbf{v} + \sum_i m_i \mathbf{R} \times \mathbf{v}'_i. \tag{2.38a}
\end{aligned}$$

The sum of the last two terms here is

$$\sum_i m_i (\mathbf{r}'_i \times \mathbf{v} + \mathbf{R} \times \mathbf{v}'_i) = \sum_i m_i \mathbf{r}'_i \times \mathbf{v} + \mathbf{R} \times \frac{d}{dt} \sum_i m_i \mathbf{r}'_i = 0 + 0,$$

since

$$\sum_i m_i \mathbf{r}'_i = \sum_i m_i \mathbf{r}_i - \sum_i m_i \mathbf{R} \overset{(2.28)}{=} 0.$$

It follows that

$$\mathbf{L} = \mathbf{R} \times M\mathbf{v} + \sum_i \mathbf{r}_i' \times \mathbf{p}_i'. \tag{2.38b}$$

This means, the total angular momentum of the system is equal to the angular momentum of the system concentrated in the centre of mass plus the angular momentum of the motion around the mass centre. In an analogous way one obtains for the *total kinetic energy* of the system

$$T = \frac{1}{2}Mv^2 + \frac{1}{2}\sum_i m_i v_i'^2. \tag{2.39}$$

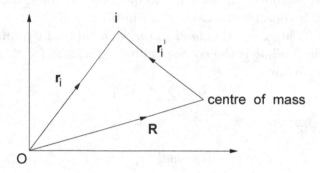

Fig. 2.2 Centre of mass.

Considering now the *work* W_{12} performed by all forces when the system, originally in configuration 1, has reached a configuration 2, we have

$$W_{12} = \int_1^2 \sum_i \mathbf{F}_i^{(e)} \cdot d\mathbf{s}_i + \int_1^2 \sum_{j.i} \mathbf{F}_{ji} \cdot d\mathbf{s}_i. \tag{2.40}$$

If the internal as well as the external forces are conservative, *i.e.* if these can be derived from potentials V_{ij} and V_i respectively, we have

$$\mathbf{F}_{ji} = -\boldsymbol{\nabla}_i V_{ij} = +\boldsymbol{\nabla}_j V_{ij} = -\mathbf{F}_{ij}, \quad \mathbf{F}_i^{(e)} = -\boldsymbol{\nabla}V_i, \tag{2.41}$$

where

$$V_{ij} = V_{ij}(|\mathbf{r}_i - \mathbf{r}_j|), \quad \text{so that} \quad \mathbf{F}_{ij} = -\mathbf{F}_{ji},$$

and

$$\boldsymbol{\nabla}_i V_{ij}(|\mathbf{r}_i - \mathbf{r}_j|) = \mathbf{r}_{ij}f, \quad f \text{ a scalar function}, \quad \mathbf{r}_{ij} \equiv \mathbf{r}_i - \mathbf{r}_j.$$

The second term on the right hand side of Eq. (2.40) can therefore be written:

$$\sum_{j,i} \int_1^2 \mathbf{F}_{ji} \cdot d\mathbf{s}_i = \frac{1}{2} \sum_{j,i} \int_1^2 \{ \mathbf{F}_{ji} \cdot d\mathbf{s}_i + \mathbf{F}_{ij} \cdot d\mathbf{s}_j \}$$

$$= \frac{1}{2} \sum_{j,i,i\neq j} \int_1^2 \{ -\boldsymbol{\nabla}_i V_{ij} \cdot d\mathbf{s}_i - \boldsymbol{\nabla}_j V_{ij} \cdot d\mathbf{s}_j \}. \quad (2.42)$$

However,

$$\boldsymbol{\nabla}_i V_{ij} = \boldsymbol{\nabla}_{ij} V_{ij} = -\boldsymbol{\nabla}_j V_{ij} \quad \text{and} \quad d\mathbf{s}_i - d\mathbf{s}_j = d\mathbf{r}_i - d\mathbf{r}_j = d\mathbf{r}_{ij},$$

so that

$$\sum_{i,i} \int_1^2 \mathbf{F}_{ji} \cdot d\mathbf{s}_i = -\frac{1}{2} \sum_{i,j,i\neq j} \int_1^2 \boldsymbol{\nabla}_{ij} V_{ij} \cdot d\mathbf{r}_{ij} = -\frac{1}{2} \sum_{i,j,i\neq j} V_{ij} \Big|_1^2.$$

For the work W_{12} we obtain therefore

$$W_{12} = -V \Big|_1^2, \quad V = \sum_i V_i + \frac{1}{2} \sum_{i,j,i\neq j} V_{ij}. \quad (2.43)$$

The quantity V defines the *total potential energy of the system*. We see that the total energy $T+V$ is again conserved, since, as we saw above, W_{12} is also equal to $T_2 - T_1$, the difference of the kinetic energy in the two onfigurations.

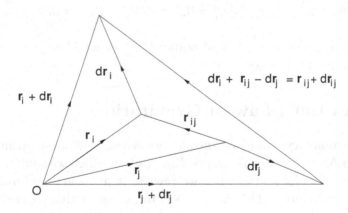

Fig. 2.3 Definition of $d\mathbf{r}_{ij}$.

The internal potential energy, *i.e.* $\sum_{i,j,i\neq j} V_{ij}$, remains constant only for specific bodies which are therefore known as *"rigid bodies"*. Thus the internal forces of a rigid body do not lead to changes in the configuration of the body, *i.e.* they do not perform work and can therefore be ignored in the discussion of the motion of the entire body. We look at this more closely, *i.e.* that the work performed by the constraining forces in a rigid body vanishes. We have as definition of $d\mathbf{r}_{ij}$ (see Fig.2.3):

$$d\mathbf{r}_i + \mathbf{r}_{ij} - d\mathbf{r}_j \equiv \mathbf{r}_{ij} + d\mathbf{r}_{ij} \quad \text{or} \quad d\mathbf{r}_{ij} = (\mathbf{r}_{ij} + d\mathbf{r}_{ij}) - \mathbf{r}_{ij} = d\mathbf{r}_i - d\mathbf{r}_j. \quad (2.44)$$

Thus $d\mathbf{r}_{ij}$ is not parallel to $\mathbf{r}_{ij}$. If

$$\mathbf{F}_{ij} = (\mathbf{r}_i - \mathbf{r}_j)f = \mathbf{r}_{ij}f, \qquad (2.45)$$

(*i.e.* the direction of $\mathbf{F}_{ij}$ is that of $\mathbf{r}_{ij}$ or of $-\mathbf{r}_{ij}$), then the work done by the constraining force $\mathbf{F}_{ij}$ is:

$$\mathbf{F}_{ij} \cdot d\mathbf{r}_{ij} = \mathbf{F}_{ij} \cdot (d\mathbf{r}_i - d\mathbf{r}_j). \qquad (2.46)$$

This force vanishes when either

1. $d\mathbf{r}_i = 0, d\mathbf{r}_j = 0$, *i.e.* no displacements inside the rigid body, or

2. $d\mathbf{r}_i - d\mathbf{r}_j$ perpendicular to $\mathbf{F}_{ij}$, which also applies in the case of a rigid body, since there the distance between two points remains constant, *i.e.*

$$(\mathbf{r}_i - \mathbf{r}_j)^2 - c_{ij}^2 = 0, \quad c_{ij} = \text{const.,}$$

   and hence

$$\mathbf{r}_{ij} \cdot d\mathbf{r}_{ij} = 0, \quad i.e. \quad \mathbf{F}_{ij} \cdot d\mathbf{r}_{ij} = 0.$$

It follows that in the case of rigid bodies the (see also below) *"virtual work"* done by the internal forces is zero.

2.6   Newton's Law of Gravitation

The above summary would be incomplete without at least an introductory reference to *Newton's law of universal gravitation* which we shall be concerned with at various points later; also, we rewrite it in various different forms as the context demands. This law of Newton's states that between any two particles of matter (note: particles, *i.e.* mass points which have no spatial extension) there acts a force F of attraction which is proportional to the two masses m_1, m_2 and is inversely proportional to the square of their separation r, *i.e.* (*cf.* Eq. (7.35))

$$F = -G\frac{m_1 m_2}{r^2}.$$

Here G is the *universal gravitational constant* with $G = 6.7 \times 10^{-8} \mathrm{g}^{-1} \mathrm{cm}^3 \mathrm{s}^{-2}$. The value of G has been determined by diverse experimental methods. The radius r of the Earth is determined *e.g.* astronomically from the curvature of the Earth, so that the mass of the Earth may be determined. If $m_2 = M =$ mass of the Earth, the mass m_1 of an object on Earth or thereabouts is called the *"gravitational mass"* of this object. As remarked in Sec. 2.2 we shall see in Chapter 10 that this gravitational mass is equal to the *inertial mass*, as observed by Galilei.[¶] For why should these masses be equal, as one may be inclined to assume at first sight? Some authors[‖] employ the additinal distinction between "active gravitational mass", *i.e.* one that gives rise to a gravitational field, and "passive gravitational mass", which is a quantity acted upon by a gravitational field. In Newtonian mechanics the equivalence of these follows from the equation for the gravitational potential ϕ, namely $\triangle\phi = 4\pi\rho(\mathbf{r})$, where ρ is the density of inertial mass, and $\mathbf{F} = -\boldsymbol{\nabla}\phi(\mathbf{r}), \phi = -Gm_1m_2/r$. Newton's potential will be derived from Einstein's equation in Sec. 15.6.

2.7 Miscellaneous Examples

As stated above, we restrict ourselves here to a few miscellaneous examples illustrating a number of directions of applications. It will be seen that these problems are not always as trivial as might be supposed. In particular Examples 2.5 to 2.8 deal with Newton's law of gravitation which will play a dominant role in later chapters.

Example 2.1: The rotating umbrella
A vertically held umbrella with horizontal rim is rotated about its axis n times in t_0 seconds and thereby scatters drops of water tangentially away from its (horizontal) rim. The rim is the circumference of a circle of diameter of $2R$ meters, and it is h meters above the ground. What is the radius of the circle formed by the pointlike drops on the ground?

Solution: The angular velocity of the umbrella is $\omega = 2\pi n/t_0$ radian/s. Hence the horizontal velocity of a drop is $\omega R = 2\pi n R/t_0$ meters/s. The time t it takes the drop to reach the ground is equal to the time it requires to fall through the distance of h meters (with initial velocity zero). Hence we have

$$\ddot{x} = g \text{ meters/s}^2, \quad x = \frac{1}{2}gt^2 = h \text{ meters}, \quad \text{so that} \quad t = \left(\frac{2h}{g}\right)^{1/2} \text{ meters}.$$

In this interval of time the drops move horizontally and tangentially to the rim of the umbrella through a distance

$$d = \frac{2\pi n R}{t_0} \left(\frac{2h}{g}\right)^{1/2} \text{ meters}.$$

[¶]The reader is again recommended to read the discussion of this topic given by R. Penrose [38], pp. 390 – 399.

[‖]C. W. Misner and P. Putnam [32].

It follows — *cf.* Fig. 2.4 — that the drops form a circle on the ground which has the radius

$$\sqrt{R^2 + d^2} = R\left[1 + \frac{8\pi^2 n^2 h}{g t_0^2}\right]^{1/2} \text{ meters.}$$

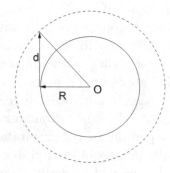

Fig. 2.4   View on the umbrella from above.

Example 2.2: The ball hopping on an inclined plane

A pointlike ball falls on an infinitely long, smooth, inclined plane with inclination angle α. Examine the motion of the ball. After how many jumps does the ball begin to roll, if at all? Show that the trajectories of the hopping motion have the shape of parabolas.

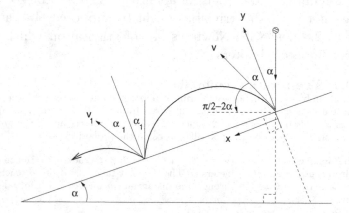

Fig. 2.5   The ball hopping on the inclined plane.

Solution: According to the law of conservation of momentum the ball with incident angle α to the vertical to the smooth, inclined plane with the same angle of inclination, is reflected through the same angle, as indicated in Fig. 2.5. With no friction, the velocity of the ball remains the same as that of its incidence, v. We choose the x-axis along the plane as in Fig. 2.5, and the y-axis perpendicular to the plane. Then we have for the components of the velocity v along the x and y directions in the *initial reflection,* and for the components of the acceleration due to gravity g there (note that g is always vertically directed) :

$$v_x = v\sin\alpha, \quad v_y = v\cos\alpha, \quad g_x = g\sin\alpha, \quad g_y = -g\cos\alpha.$$

Let x be the distance travelled by the ball in t seconds after its first impact on the plane. Then the *velocities of reflection* are

$$\dot{x} = v\sin\alpha + gt\sin\alpha, \qquad \dot{y} = v\cos\alpha - gt\cos\alpha. \tag{2.47}$$

By integration we obtain (with appropriate initial conditions)

$$x = vt\sin\alpha + g\frac{t^2}{2}\sin\alpha, \qquad y = vt\cos\alpha - g\frac{t^2}{2}\cos\alpha. \tag{2.48}$$

At the next (first) point of impact at time $t = t_1$, we have

$$y = 0, \quad \text{and} \quad \therefore t_1 = \frac{2v}{g}. \tag{2.49}$$

The corresponding distance along the plane is (inserting $t = t_1$)

$$x_1 = \frac{2v}{g}v\sin\alpha + \frac{1}{2}g\sin\alpha\left(\frac{2v}{g}\right)^2 = \frac{1}{g}4v^2\sin\alpha. \tag{2.50}$$

We compute the angle α_1 which the *velocity of impact* v_1 makes with the vertical to the inclined plane from the components of v_1. First, however, we have from Eq. (2.47) with the negative of $\dot{y}$ (for impact instead of reflection)

$$\dot{x}(t_1) \equiv v_{1x} = v\sin\alpha + g\frac{2v}{g}\sin\alpha = 3v\sin\alpha, \quad -\dot{y}(t_1) \equiv v_{1y} = -v\cos\alpha + g\frac{2v}{g}\cos\alpha = v\cos\alpha, \tag{2.51}$$

and

$$v_1 = v\sqrt{\cos^2\alpha + 9\sin^2\alpha} = v\sqrt{1 + 8\sin^2\alpha}. \tag{2.52}$$

Hence

$$\alpha_1 = \tan^{-1}\left(\frac{v_{1x}}{v_{1y}}\right) = \tan^{-1}(3\tan\alpha). \tag{2.53}$$

The ball is now *reflected* with velocity v_1 in the direction of angle α_1. In this case we obtain parallel to Eq. (2.47) with $v \to v_1, \alpha \to \alpha_1$ but the acceleration due to gravity g directed vertically as before, so that its components along and perpendicular to the plane remain unchanged:

$$\dot{x} = v_1\sin\alpha_1 + gt\sin\alpha, \qquad \dot{y} = v_1\cos\alpha_1 - gt\cos\alpha \tag{2.54}$$

($\sin\alpha_1 = v_{1x}/v_1, \cos\alpha_1 = v_{1y}/v_1$). By integration we obtain

$$x = v_1t\sin\alpha_1 + g\frac{t^2}{2}\sin\alpha + \text{const.} \to x_2, \qquad y = v_1t\cos\alpha_1 - g\frac{t^2}{2}\cos\alpha + \text{const.} \to y_2. \tag{2.55}$$

Let the point of impact (above x_1, y_1) be the origin $x = 0, y = 0$ of a new coordinate frame at t=0. Then const. $= 0$, and x, y here become x_2, y_2, and from $y_2 = 0$ at time t_2 we obtain

$$t_2 = \frac{2v_1\cos\alpha_1}{g\cos\alpha} = \frac{2v_{1y}}{g\cos\alpha} \overset{(2.51)}{\equiv} \frac{2v\cos\alpha}{g\cos\alpha} = \frac{2v}{g} = t_1. \tag{2.56}$$

Thus for every hop the ball requires the same length of time. Moreover:

$$x_2 = v_1\frac{2v}{g}\sin\alpha_1 + \frac{g}{2}\left(\frac{2v}{g}\right)^2\sin\alpha - \frac{2v}{g}(v_1\sin\alpha_1 + v\sin\alpha) = \frac{2v}{g}(v_{1x} + v_x).$$

Does the height which the ball falls through increase or decrease? Since (using Eq. (2.51)) $v_1 \sin \alpha_1 = v_{1x} = 3v \sin \alpha$, we have

$$x_2 = \frac{2v}{g}(v_1 \sin \alpha_1 + v \sin \alpha) = \frac{8v^2}{g} \sin \alpha > \frac{4v^2 \sin \alpha}{g} = x_1.$$

Thus the height the ball has fallen through from the first to the second impact, $x_2 \sin \alpha$, is larger than the height fallen through from its initial to the first impact, $x_1 \sin \alpha$. We obtain the angle α_2, which the velocity of impact v_2 makes with the perpendicular to the plane, from the components of v_2:

$$\dot{x}(t_2) \equiv v_{2x} \overset{(2.54)}{=} \overbrace{v_1 \sin \alpha_1}^{v_{1x}} + g\frac{2v}{g} \sin \alpha \overset{(2.51)}{=} 3v \sin \alpha + 2v \sin \alpha = 5v \sin \alpha,$$

$$-\dot{y}(t_2) \equiv v_{2y} \overset{(2.54)}{=} -\overbrace{v_1 \cos \alpha_1}^{v_{1y}} + g\frac{2v}{g} \cos \alpha \overset{(2.51)}{=} -v \cos \alpha + 2v \cos \alpha = v \cos \alpha,$$

$$v_2 = \quad = \sqrt{v_{2x}^2 + v_{2y}^2} = \sqrt{\cos^2 \alpha + 25 \sin^2 \alpha} = v\sqrt{1 + 24 \sin^2 \alpha}.$$

It follows that

$$\alpha_2 = \tan^{-1}\left(\frac{v_{2x}}{v_{2y}}\right) = \tan^{-1}(5\tan\alpha), \quad i.e. \quad \tan\alpha_2 = 5\tan\alpha = \frac{5}{3}\tan\alpha_1, \quad i.e. \quad \alpha_2 > \alpha_1.$$

We observe that the direction of reflection becomes increasingly flatter ($\alpha_2 > \alpha_1$). This effect continues indefinitely on the infinitely long, smooth, inclined plane. The y-component of the velocity of impact is the constant value $v \cos \alpha$; however, its x-component increases. Hence $\tan\alpha_i$ also increases. We obtain the angle of reflection after n reflections, α_n, from the relations:

$$\tan\alpha_1 = 3\tan\alpha, \quad \tan\alpha_2 = 5\tan\alpha, \quad \tan\alpha_3 = 7\tan\alpha, \ldots,$$

so that $\tan\alpha_n = (2n+1)\tan\alpha$. In order that

$$y_{n+1} = v_n t \cos \alpha_n - g\frac{t^2}{2} \cos \alpha = 0, \quad \text{we have always} \quad t = \frac{2v}{g}.$$

The ball continues to jump until $\alpha_n = \pi/2$, so that $\cos \alpha_n = 0$. After that y_{n+1} is always zero or negative.

We now show that the trajectory of every hop is a parabola. We achieve this by eliminating t from x and y. From Eqs. (2.48) we obtain

$$t = \frac{-v \sin \alpha \pm \sqrt{v^2 \sin^2 \alpha + (4gx \sin \alpha)/2}}{(2g \sin \alpha)/2}, \quad t = \frac{-v \cos \alpha \pm \sqrt{v^2 \cos^2 \alpha - (4gx \cos \alpha)/2}}{-(2g \cos \alpha)/2},$$

$$\therefore \quad -2v \pm \sqrt{v^2 + \frac{2gx}{\sin \alpha}} = \pm\sqrt{v^2 - \frac{2gy}{\cos \alpha}}.$$

Squaring both sides of the last equation and then a second time, we obtain

$$\left[2v^2 + g\left(\frac{x}{\sin \alpha} + \frac{y}{\cos \alpha}\right)\right]^2 = 4v^2\left(v^2 + \frac{2gx}{\sin \alpha}\right), \quad g^2\left(\frac{x}{\sin \alpha} + \frac{y}{\cos \alpha}\right)^2 + 4gv^2\left(-\frac{x}{\sin \alpha} + \frac{y}{\cos \alpha}\right) = 0,$$

which assumes the standard form of the equation of a parabola with the transformation

$$X = \frac{x}{\sin \alpha} + \frac{y}{\cos \alpha}, \quad Y = \frac{x}{\sin \alpha} - \frac{y}{\cos \alpha} \quad i.e. \quad X^2 = \frac{4v^2}{g}Y.$$

Example 2.3: The velocity of escape from the Earth

Calculate the smallest velocity, known as the *escape velocity*, with which a particle has to be shot vertically upwards in order to escape from the Earth. Ignore friction of the atmosphere and effects resulting from the rotation of the Earth and the moon. (This is a wellknown and simple problem. However, the result will re-appear in the Schwarzschild solution of Einstein's equation, as we shall see in Chapter 16, *e.g.* Eq. (16.58b), and in the trick calculation of Sec. 11.3).

Solution: Let M be the mass of the Earth and m that of the particle. Let R be the radius of the Earth and v the velocity of the particle at the height x above its point of launching into the air. Then according to both of Newton's laws:

$$m\ddot{x} = m\frac{d}{dx}\left(\frac{1}{2}\dot{x}^2\right) = -G\frac{mM}{(x+R)^2}, \qquad (2.57)$$

where G is Newton's gravitational constant $G = 6.67 \times 10^{-8}$ cm^3 g^{-1}s^{-2}. Integrating the equation we obtain

$$\frac{1}{2}\dot{x}^2 = -GM\int \frac{dx}{(x+R)^2} = \frac{GM}{x+R} + \text{const.}$$

With $\dot{x} = 0$ at $x = \infty$, the constant vanishes. Hence the velocity v at $x = 0$ is given by $v^2 = 2GM/R$. For a particle of unit mass at the surface of the Earth we have $g = G \times (1 \times M)/R^2$, so that $GM/R = gR$ and $v^2 = 2gR$. With $R = 6370$ km, $g = 980$ cm s^{-2}, we obtain $v^2 = 2 \times 980 \times 6370 \times 10^5$ cm^2 s^{-2}, $v \simeq 11.2$ km s^{-1}. The velocity of escape $v_{esc} = \sqrt{2GM/R}$ obtained here is a characteristic speed in Newton's theory of gravity, and is related to the result of relativity that light from M cannot reach a distant observer when $v_{esc} > c$, where c is the velocity of light.[**]

Example 2.4: The rocket fired vertically upwards

Show that the equation of motion of a rocket of initial mass m_0 (no constant mass!), which is fired vertically upwards in a uniform gravitational field of negligible air resistance is

$$m\frac{dv}{dt} = -v'\frac{dm}{dt} - mg, \qquad (2.58)$$

where m is the mass of the rocket at time t and v' the *forward velocity* of the expelled gas *relative* to that of the rocket ($v > 0, v' < 0$). Integrate this equation and determine v as a function of m, assuming that the loss of mass is proportional to time. Also show that for a rocket starting from rest with $v' = -2070$ m s^{-1} and a loss of mass of $1/60$ of the initial mass per second, and which is to attain the velocity of escape from the Earth, the ratio of loss of mass of fuel relative to that of the empty rocket must be almost 300.

Solution: In the time interval dt the change of momentum p of the rocket is[††]

$$dp = (m + dm)(v + dv) - mv - dm(v' + v) = mdv - v'dm,$$

i.e. we have (observe that dm is negative, $\int_{m_0}^{m} dm = m - m_0 < 0$)

$$\frac{dp}{dt} = m\frac{dv}{dt} - v'\frac{dm}{dt}. \qquad (2.59)$$

[**]See also D. Raine and E. Thomas [40], p. 1.

[††]We emphasize: v' is defined as *relative* to v. If dm had an *absolute forward velocity* v', Eq. (2.59) would be (this is the equation given *e.g.* by K. E. Bullen [8], p. 79)

$$\frac{dp}{dt} = m\frac{dv}{dt} - (v' - v)\frac{dm}{dt} = \frac{d(mv)}{dt} - v'\frac{dm}{dt}.$$

A similar problem is given in *The Physics Coaching Class* [48], problem 1138, p. 228.

Newton's equation of motion, *i.e.* $dp/dt = -mg$, therefore implies

$$m\frac{dv}{dt} = v'\frac{dm}{dt} - mg, \quad \text{or} \quad dv = v'\frac{dm}{m} - g\,dt, \tag{2.60}$$

where also $m = m(t)$. Integrating with v_0, m_0 the initial values of v, m, we obtain:

$$v = v_0 + v'\ln\frac{m}{m_0} - gt. \tag{2.61}$$

Since the loss of mass is proportional to time t, we have $m_0 - m = \mu t$, $\mu = $ const., and the equation becomes

$$v = v_0 + v'\ln\frac{m}{m_0} + g\frac{m - m_0}{\mu}. \tag{2.62}$$

We are given: $v_0 = 0$, $v' = -2070$ m s^{-1}, $v = $ velocity of escape $= 11200$ m s^{-1} (see Example 2.3), $g = 9.81$ m s^{-2}, $\mu = m_0/60$. In Eq. (2.62) we have $m \ll m_0$, if we assume that the initial mass of the rocket is overwhelmingly that of the fuel. Then $m - m_0 \approx -m_0$, and with the numerical values: $11200 = -2070\ln m/m_0 - 9.81 \times 60$ m/s^{-1} From this we obtain: $m_0/m = e^{5.7} = 299 \simeq 300$.

Example 2.5: Equilibrium at and approximate limit of atmosphere

A body of mass m and originally at rest falls from a vertical height h_0 on the surface of the Earth. Ignoring atmospheric effects calculate the velocity of the body taking into account the difference in the acceleration due to gravity at the height and at the surface of the Earth, assuming the latter is a sphere of radius r. How does the velocity differ in the two cases $h_0/r \ll 1$ and $h_0/r \gg 1$? The rotation of the Earth also imposes on the body a centrifugal force. The angular velocity of the Earth is 0.00007292 radian/s and $r = 6370284$ meters at the equator, and $g = 980$ cm/s^{-2}. What is the distance h above the equator at which the body is in a state of equilbrium?

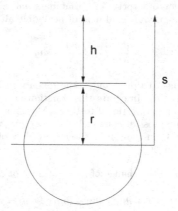

Fig. 2.6 The particle at height h above the surface of the Earth.

Solution: Let g be the acceleration due to gravity at the surface of the Earth, and g' that at a height h above, as indicated in Fig. 2.6. Then, according to Newton's law of gravitation, and with G the gravitational constant and M the mass of the Earth (and dividing out m from Newton's equation of motion and using s as in Fig. 2.6):

$$g = G\frac{M}{r^2}, \quad g' = G\frac{M}{(r+h)^2}, \quad \text{so that} \quad \frac{g}{g'} = \frac{(r+h)^2}{r^2} \equiv \frac{s^2}{r^2}, \quad g' = \frac{gr^2}{s^2}.$$

But

$$g' = -\frac{d^2s}{dt^2} = -\frac{1}{2}\frac{d}{ds}\left(\frac{ds}{dt}\right)^2, \quad \therefore \frac{d^2s}{dt^2} = -\frac{gr^2}{s^2} = \frac{1}{2}\frac{d}{ds}\left(\frac{ds}{dt}\right)^2, \quad \text{and} \quad \left(\frac{ds}{dt}\right)^2 = \frac{2gr^2}{s} + \text{const.}$$

At time $t = 0$ we have $(ds/dt)_{s=s_0} = 0$. Hence const. $= -2gr^2/s_0$, i.e.

$$v^2 = \left(\frac{ds}{dt}\right)^2 = 2gr^2\left(\frac{1}{s} - \frac{1}{s_0}\right) = 2gr^2\left(\frac{1}{r+h} - \frac{1}{s_0}\right).$$

At the surface of the Earth, where $h = 0$, the velocity is v_0 and is given by

$$v_0^2 = 2gr^2\left(\frac{1}{r} - \frac{1}{s_0}\right), \quad \text{or} \quad v_0^2 = 2gr^2\left(\frac{1}{r} - \frac{1}{r+h_0}\right),$$

where $s_0 = r + h_0$. For $h_0 \ll r$ the latter expression can be approximated by $v_0^2 = 2gh_0$, and for $h_0 \gg r$ we have $v'^2 = 2gr$. This velocity v' can also be interpreted as that particular velocity with which a body must be shot vertically upwards in order not to return to the Earth (in other words, in order to travel an infinitely long distance, $h_0 \to \infty$). This velocity is the *velocity of escape*, also considered in Example 2.3. For the body of mass m the centrifugal force of the Earth is at the surface (note that $\omega^2 r = 3.4\text{cm/s}^2$, as shown later, see Eq. (8.128)):

$$m\omega^2 r = \frac{\omega^2 r}{g}mg = \frac{mg}{289} \text{ rad}^2\text{kg meter s}^{-2},$$

At the distance $h = s - r$ above the surface, the outwardly directed centrifugal force is $F = (mg/289)(s/r)$, where the inwardly directed attractive force of the Earth is $F = mg'$, $g' = gr^2/s^2$. Thus for equilibrium we have

$$\frac{mg}{289}\frac{s}{r} = m\frac{gr^2}{s^2}, \quad \therefore \ s^3 = 289r^3, \quad s = 6.6r.$$

The requested distance is therefore roughly $5.6r$. Very roughly this is the top limit of the atmosphere.

Example 2.6: The spherical oil field

(a) A large sphere of radius R_0 consists of a material of constant density ρ_0 except for a spherical insert of radius R_1 and density $\rho_1 < \rho_0$ whose centre is located at a depth t beneath the surface of the larger sphere. Calculate the dependence of the vertical and horizontal components of the force of gravity at a point P on the surface of the larger sphere as a function of the angle ϕ indicated in Fig. 2.7. (b) Precision measurements of ordinary gravimeters are based on the deflection of a horizontally adjusted quartz fibre. These gravimeters have the sensitivity to detect a gravitational acceleration of approximately 10^{-4} cm/s^2. Considering the highly idealized case of a spherical oil field at a depth of $t = 500$ meters in the crust of the Earth and densities $\rho_0 = 5.5, \rho_1 = 1$ g/cm^3, calculate what the diameter of the spherical oil field would have to be in order to be gravimetrically observable.

Solution: (a) We assume the geometry as shown in Fig. 2.7. There O_1 is the origin of the small spherical insert, and $s = O_1P$, as indicated there. Then from triangles O_1OP, O_1NP:

$$s^2 = R_0^2 + (R_0 - t)^2 - 2R_0(R_0 - t)\cos\phi,$$
$$\cos\theta = \frac{R_0 - (R_0 - t)\cos\phi}{s} = \sin\left(\frac{\pi}{2} - \theta\right), \quad \sin\theta = \frac{(R_0 - t)\sin\phi}{s}. \tag{2.63}$$

From Newton's law of gravitation, $\mathbf{F} = Gm_1m_2\mathbf{r}/r^3$, for the attractive force $\mathbf{F}$ between two masses m_1, m_2 at a separation r, we obtain for the force acting on a unit mass at a point P as indicated in Fig. 2.7:

$$\frac{\mathbf{F}}{\text{unit mass} \times G} = \frac{\rho_0(4/3)\pi R_0^3}{R_0^2}\frac{\mathbf{y}}{y} + \frac{(\rho_1 - \rho_0)(4/3)\pi R_1^3}{s^2}\frac{\mathbf{s}}{s},$$

so that ($s_x = (R_0 - t)\sin\phi$, $s_y = R_0 - (R_0 - t)\cos\phi$ with x and y axes as at P in Fig. 2.7)

$$\frac{F_x}{1 \times G} = \frac{(\rho_1 - \rho_0)(4/3)\pi R_1^3 (R_0 - t)\sin\phi}{s^2} \frac{}{s},$$

$$\frac{F_y}{1 \times G} = \frac{\rho(4/3)\pi R_0^3}{R_0^2} + \frac{(\rho_1 - \rho_0)(4/3)\pi R_1^3}{s^2} \frac{R_0 - (R_0 - t)\cos\phi}{s}. \tag{2.64}$$

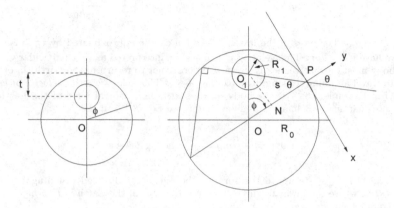

Fig. 2.7 The spherical oil field in the Earth.

With the expressions of Eq. (2.63) these become

$$\frac{F_x}{1 \times G} = \frac{(\rho_1 - \rho_0)(4/3)\pi R_1^3 (R_0 - t)\sin\phi}{\{R_0^2 + (R_0 - t)^2 - 2R_0(R_0 - t)\cos\phi\}^{3/2}},$$

$$\frac{F_y}{1 \times G} = \frac{\rho_0(4/3)\pi R_0^3}{R_0^2} + \frac{(\rho_1 - \rho_0)(4/3)\pi R_1^3\{R_0 - (R_0 - t)\cos\phi\}}{\{R_0^2 + (R_0 - t)^2 - 2R_0(R_0 - t)\cos\phi\}^{3/2}}. \tag{2.65}$$

For $R_0 \gg t$ we have approximately:

$$\frac{F_x}{1 \times G} = \frac{(\rho_1 - \rho_0)(4/3)\pi R_1^3 R_0 \sin\phi}{R_0^3 2^{3/2}(1 - \cos\phi)^{3/2}},$$

$$\frac{F_y}{1 \times G} = \frac{\rho_0(4/3)\pi R_0^3}{R_0^2} + \frac{(\rho_1 - \rho_0)(4/3)\pi R_1^3 R_0(1 - \cos\phi)}{R_0^3 2^{3/2}(1 - \cos\phi)^{3/2}}. \tag{2.66}$$

One should note that only in this approximation (but not exactly) is F_y/G at $\phi = 0$ equal to $\rho_0(4/3)\pi R_0$.

(b) We go to $\phi = 0$, *i.e.* directly to the oil field. Then, according to Eq. (2.66) $F_x = 0$, and from Eq. (2.65) we obtain

$$\frac{F_y}{G} = \frac{4}{3}\pi R_0 \rho_0 + \frac{(\rho_1 - \rho_0)(4/3)\pi R_1^3}{t^2}. \tag{2.67}$$

For $F_y \equiv g_y \times$ unit mass we have

$$g_y = G\frac{4}{3}\pi R_0 \rho_0 + \frac{G(\rho_1 - \rho_0)(4/3)\pi R_1^3}{t^2}.$$

The second term originates from the difference $\rho_1 \neq \rho_0$; hence it is responsible for the deviation from the first term. Hence

$$\Delta g_y = \frac{G(\rho_1 - \rho_0)(4/3)\pi R_1^3}{t^2}. \tag{2.68}$$

For $\triangle g_y = -10^{-4}\,\mathrm{cm/s^2}, G^2 = 6.6732 \times 10^{-8}\,\mathrm{cm^3 g^{-1} s^{-2}}, \rho_0 = 5.5\,\mathrm{g\,cm^{-3}}, \rho_1 = 1.0\,\mathrm{g\,cm^{-3}}, R_0 = 6370\,\mathrm{km}, t = 500$ meters, we obtain $R_1^3 = 0.2 \times 10^{18}\,\mathrm{meter^3}, R_1 = 58.35$ meters.

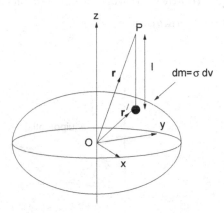

Fig. 2.8 The potential of the nonspherical Earth.

Example 2.7: Gravitational potential of the nonspherical, oblate Earth

Consider a solid oblate ellipsoid like the Earth of mass M_E and of constant mass density σ, which is rotationally symmetric about the z-axis, the equatorial plane being the (x,y)-plane, with the origin at the geometrical centre, and R the radius of the circular plane at the equator. Establish the gravitational potential $U(\mathbf{r})$ at a point P with coordinates $\mathbf{r} = (x,y,z)$ outside the ellipsoid and show that for $R \ll r$ this can be expressed in the following form (where $\mu = GM_E$):

$$
U(\mathbf{r}) = -\frac{\mu}{r}\left[1 - \left(\frac{R}{r}\right)^2 P_2(\sin\theta) J_2 + \cdots\right],
$$

$$
J_2 = -\frac{\sigma}{M}\int_{\text{ellipsoid}} \left(\frac{r'}{R}\right)^2 P_2(\sin\theta')d\mathbf{r}', \quad P_2(x) = -\frac{1}{2}(1 - 3x^2). \tag{2.69}
$$

What is the physical significance of J_2 (re-expressed in terms of Cartesian coordinates)? Finally show that the equipotential surfaces $U(\mathbf{r}) = \text{const.}$ have the shape of rotational ellipsoids. The Cartesian and polar equations of an ellipsoid, the latter with pole at the centre and coordinates ρ, ϕ, θ (see transformation below), are given by

$$
1 = \frac{x^2}{a^2} + \frac{y^2}{b^2} + \frac{z^2}{c^2}, \quad \text{and} \quad \frac{1}{\rho^2} = \frac{\cos^2\theta\cos^2\phi}{a^2} + \frac{\cos^2\theta\sin^2\phi}{b^2} + \frac{\sin^2\theta}{c^2}.
$$

In the case of an ellipsoid which is rotationally symmetric about the z-axis (*i.e.* whose equipotentials are independent of ϕ) as depicted in Fig. 2.9 we can choose $\phi = 0$ and obtain as in Fig. 2.9 the cross sectional ellipse

$$
\frac{1}{\rho^2} = \frac{\cos^2\theta}{a^2} + \frac{\sin^2\theta}{c^2}.
$$

Solution: We can choose the spherical polar coordinates of a point P (*cf.* Fig. 2.8) as $x = r\cos\theta\cos\phi, y = r\cos\theta\sin\phi, z = r\sin\theta, 0 \le \phi \le 2\pi, 0 \le \theta \le \pi$. The potential $U(\mathbf{r})$ at a point $\mathbf{r}$ outside the ellipsoid due to an infinitesimal mass element dm located at $\mathbf{r}'$ is

$$
U(\mathbf{r}) = -G\int_{\text{ellipsoid}} \frac{dm}{l}, \quad dm = \sigma dv = \sigma r'^2 dr' \cos\theta' d\theta' d\phi', \quad l^2 = r^2 + r'^2 - 2rr'\cos\chi. \tag{2.70}
$$

We also have (*cf.* x, y, z)

$$
\begin{aligned}
\cos \chi &= \frac{\mathbf{r} \cdot \mathbf{r}'}{rr'} = \cos \theta \cos \theta' \cos \phi \cos \phi' + \cos \theta \cos \theta' \sin \phi \sin \phi' + \sin \theta \sin \theta' \\
&= \sin \theta \sin \theta' + \cos \theta \cos \theta' \cos(\phi - \phi').
\end{aligned} \tag{2.71}
$$

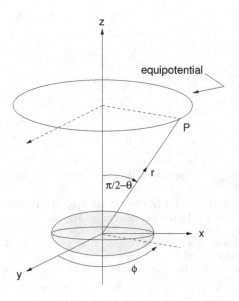

Fig. 2.9 Rotational symmetry about the z-axis.

With the following expressions which are known as *Legendre polynomials*,

$$
P_1(\cos \chi) = \cos \chi, \quad P_2(\cos \chi) = -\frac{1}{2}(1 - 3\cos^2 \chi), \quad P_3(\cos \chi) = -\frac{1}{2}\cos \chi (3 - 5\cos^2 \chi), \quad \text{etc.,} \tag{2.72}
$$

we have

$$
\begin{aligned}
\frac{1}{l} &= (r^2 + r'^2 - 2rr' \cos \chi)^{-1/2} = \frac{1}{r}\left(1 + \frac{r'^2}{r^2} - 2\frac{r'}{r}\cos \chi\right)^{-1/2} \\
&= \frac{1}{r}\left[1 - \frac{1}{2}\left(\frac{r'^2}{r^2} - 2\frac{r'}{r}\cos \chi\right) + \frac{3}{8}\left(\frac{r'^2}{r^2} - 2\frac{r'}{r}\cos \chi\right)^2 - \cdots\right] \\
&= \frac{1}{r}\left[1 + \frac{r'}{r}\cos \chi - \frac{1}{2}\frac{r'^2}{r^2}(1 - 3\cos^2 \chi) + \cdots\right] = \frac{1}{r}\left[1 + \frac{r'}{r}P_1 + \frac{r'^2}{r^2}P_2 + \cdots\right]. \tag{2.73}
\end{aligned}
$$

Here $\int dm = M_E$ = mass of the Earth. We assume that the Earth is symmetric about its rotational axis. In Eq. (2.73) we consider terms up to and including the term in P_2. Then we have to evaluate some integrals. We begin with the integral containing P_1 and use the results:

$$
\int_0^\pi \sin \theta' \cos \theta' \, d\theta' = \int_0^\pi \sin \theta' \, d(\sin \theta') = \left[\frac{\sin^2 \theta'}{2}\right]_0^\pi = 0,
$$

$$
\int_0^{2\pi} \cos(\phi - \phi') \, d\phi' = [\sin(\phi - \phi')]_0^{2\pi} = \sin(\phi - 2\pi) - \sin \phi = 0.
$$

Then:

$$\int dm P_1(\cos\chi) \;=\; \sigma \int r'^2 dr' \cos\theta' d\theta' d\phi' [\sin\theta\sin\theta' + \cos\theta\cos\theta'\cos(\phi-\phi')] = 0,$$

$$\int dm P_2(\cos\chi) \;=\; \int dm\left[\frac{3}{2}\cos^2\chi - \frac{1}{2}\right] = \int dm\left[\frac{3}{2}\sin^2\theta\sin^2\theta' + \frac{3}{2}\cos^2\theta\cos^2\theta'\cos^2(\phi-\phi')\right.$$
$$\left. + \frac{3}{4}\sin 2\theta\sin 2\theta'\cos(\phi-\phi') - \frac{1}{2}\right].$$

In the last expression we replace $\cos^2(\phi-\phi')$ by $[1+\cos 2(\phi-\phi')]/2$. The terms in $\cos n(\phi-\phi')$ do not contribute because

$$\int_0^{2\pi}\cos n(\phi-\phi')d\phi' = \left[\frac{\sin n(\phi-\phi')}{-n}\right]_0^{2\pi} = -\frac{1}{n}[\sin n\phi - \sin n\phi] = 0.$$

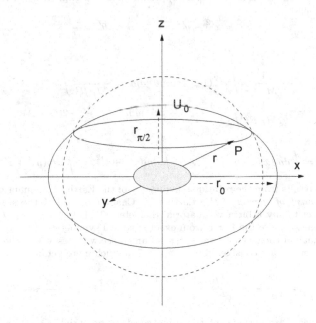

Fig. 2.10 The equipotential and its envelope in the dipole approximation.

We now have

$$\int dm P_2(\cos\chi)r'^2 \;=\; \int dm\left[\frac{3}{2}\sin^2\theta\sin^2\theta' + \frac{3}{4}\cos^2\theta\cos^2\theta' - \frac{1}{2}\right]r'^2$$
$$=\; \int dm\left[\frac{3}{2}\sin^2\theta\sin^2\theta' + \frac{3}{4}(1-\sin^2\theta)(1-\sin^2\theta') - \frac{1}{2}\right]r'^2$$
$$=\; \int r'^2 dm\left[\frac{9}{4}\sin^2\theta\sin^2\theta' - \frac{3}{4}(\sin^2\theta+\sin^2\theta') + \frac{1}{4}\right]$$
$$=\; \int r'^2 dm\left(\frac{3}{2}\sin^2\theta - \frac{1}{2}\right)\left(\frac{3}{2}\sin^2\theta' - \frac{1}{2}\right).$$

It follows that, with $GM_E = \mu$,

$$
\begin{aligned}
U(\mathbf{r}) &= -\frac{\mu}{r}\left[1 + \frac{1}{M_E}\int \frac{r'^2}{r^2}\left(\frac{3}{2}\sin^2\theta - \frac{1}{2}\right)\left(\frac{3}{2}\sin^2\theta' - \frac{1}{2}\right)dm + \cdots\right] \\
&= -\frac{\mu}{r}\left[1 - \left(\frac{R}{r}\right)^2 P_2(\sin\theta)J_2 + \cdots\right],
\end{aligned}
\tag{2.74}
$$

where, with R the equatorial radius of the Earth,

$$
J_2 = -\frac{1}{M_E}\int_{\text{Earth}}\left(\frac{r'}{R}\right)^2 P_2(\sin\theta')dm.
\tag{2.75}
$$

The result (2.74) contains a deviation of the equipotential surfaces $U = \text{const.}$ from pure spherical symmetry, whereas J_2 describes the nonspherical deformation of the Earth; this term, proportional to $1/r^3$, is known as the *dipole* contribution, and is more familiar in electrodynamics.[‡‡] We also note that the potential is independent of the angle ϕ. The reason for this is the rotational symmetry of the potential about the z-axis, as illustrated in Fig. 2.10.

We now restrict ourselves to the dipole approximation and express J_2 in terms of Cartesian coordinates. We have:

$$
r'^2 = x'^2 + y'^2 + z'^2, \qquad \sin\theta' = \frac{z'}{r'},
$$

so that

$$
\begin{aligned}
J_2 &= \frac{1}{2M_E R^2}\int r'^2(1 - 3\sin^2\theta')dm = \frac{1}{2M_E R^2}\int (r'^2 - 3z'^2)dm \\
&= \frac{1}{2M_E R^2}\int (x'^2 + y'^2 - 2z'^2)dm = \frac{1}{2M_E R^2}2(C - A),
\end{aligned}
\tag{2.76}
$$

where

$$
C = \int dm(x'^2 + y'^2), \qquad A = \int dm(x'^2 + z'^2) = \int dm(y'^2 + z'^2).
$$

The last equality results from the rotational symmetry of the Earth. The quantity C is, of course, the *principal moment of inertia* of the Earth (*cf.* Chapter 8) and A the smallest moment of inertia with respect to any arbitrary equatorial diameter. Thus the quantity J_2 is a measure of the deviation of the shape of the Earth from exact spherical symmetry.

The polar equation (polar coordinates ρ, θ) of an ellipse with respect to the centre as pole is, as we saw (these aspects are considered in Chapter 7 in much more detail),[*]

$$
\frac{1}{\rho^2} = \frac{\cos^2\theta}{a^2} + \frac{\sin^2\theta}{c^2}, \qquad \text{or} \qquad \frac{1}{\gamma^2} \equiv \frac{a^2}{\rho^2} = \left[1 + \frac{a^2 - c^2}{c^2}\sin^2\theta\right].
\tag{2.77}
$$

Equipotential surfaces are surfaces of the same constant potential, *i.e.* $U(\mathbf{r}) = U_0 = \text{const.}$ Inserting this into Eq. (2.74), we obtain for these surfaces (note that the factor 3 originates from $P_2(\sin\theta)$):

$$
-U_0 = \frac{\mu}{r} + \frac{\beta}{r^3}(1 - 3\sin^2\theta), \qquad \beta = \frac{1}{2}\mu J_2 R^2.
\tag{2.78}
$$

In order to be able to compare this equation with Eq. (2.77), we have to rewrite it in a form which has the factor $(1 + 3\sin^2\theta)$ on the right hand side. Thus we obtain

$$
\frac{\mu}{r} + \frac{2\beta}{r^3} + U_0 = \frac{\beta}{r^3}(1 + 3\sin^2\theta), \qquad \text{or} \qquad 2 + r^2\frac{\mu + rU_0}{\beta} = (1 + 3\sin^2\theta).
\tag{2.79}
$$

[‡‡]See *e.g.* H. J. W. Müller–Kirsten [34], pp. 78 –82.

[*]For an analogous case see Example 10.3 on the tidal effect.

Comparing with Eq. (2.77) we obtain

$$\frac{1}{\gamma^2} \equiv \frac{a^2}{\rho^2} = 2 + r^2 \frac{\mu + rU_0}{\beta}, \quad a^2 = 4c^2. \tag{2.80}$$

We can see the behaviour of the equipotentials with respect to the circle $-U_0 = \mu/r$ by looking at the points $\theta = 0, \pi$ and $\theta = \pm\pi/2$. In these cases we have:

$$\left.\begin{array}{ll} r \to r_0, & \theta = 0, \pi: \quad -U_0 = \frac{\mu}{r_0} + \frac{\beta}{r_0^3}, \\[2mm] r \to r_{\pi/2}, & \theta = \pm\frac{\pi}{2}: \quad -U_0 = \frac{\mu}{r_{\pi/2}} - \frac{2\beta}{r_{\pi/2}^3}, \end{array}\right\} \quad \therefore \ r_0 > r_{\pi/2}. \tag{2.81}$$

Thus $r_0 > r_{\pi/2}$ as indicated in Fig. 2.10.

This example demonstrates that the potential of the nonspherical Earth is given by a series of the following form:[†]

$$U(\mathbf{r}) = -\frac{\mu}{r}\left[1 - \sum_{n=2}^{\infty} \left(\frac{R}{r}\right)^n P_n(\sin\theta) J_n\right]. \tag{2.82}$$

Example 2.8: A particle going back and forth in a tunnel under gravity

A chord–like tunnel of length $AB = 2R\cos\theta$ is driven through the Earth of radius R from a point A on the surface to a point B on the surface. The particle of mass m is inserted at A with zero velocity. Determine the period of oscillation of the particle between points A and B.[‡]

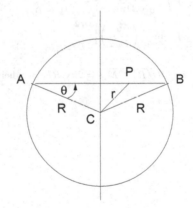

Fig. 2.11 The tunnel AB through the Earth of radius R.

Solution: The potential of the particle at point P of the tunnel in Fig. 2.11, a distance r from the centre of the Earth and distance d from A, is

$$V(r) = -G\frac{mM_E}{r}, \quad r^2 = R^2 + d^2 - 2Rd\cos\theta, \tag{2.83}$$

where M_E is the mass of the Earth, and θ is the angle shown in Fig. 2.11. Hence

$$V(r) = -GmM_E[R^2 + d^2 - 2Rd\cos\theta]^{-1/2} \simeq -G\frac{mM_E}{R}\left[1 - \frac{d^2 - 2Rd\cos\theta}{2R^2}\right]. \tag{2.84}$$

[†]For a somewhat similar problem see *The Physics Coaching Class* [48], problem 1149, p. 245.
[‡]For an analogous problem see *The Physics Coaching Class* [48], problem 1030, p. 38.

Taking the potential at the surface of the Earth as zero, the potential energy of the particle is

$$V(r) = G\frac{mM_E}{2R^3}[d^2 - 2Rd\cos\theta] = G\frac{mM_E}{2R^3}[r^2 - R^2]. \tag{2.85}$$

Thus at the surface with $r = R$ this potential energy vanishes.

The total energy of the particle which starts off from rest at point A is therefore

$$E = \frac{1}{2}mv^2 + G\frac{mM_E}{2R^3}[r^2 - R^2] = 0. \tag{2.86}$$

Setting $v = dx/dt, x = AP$ in Fig. 2.11, we have $r^2 = R^2 + x^2 - 2xR\cos\theta$, so that

$$v^2 = G\frac{M_E}{R^3}[2xR\cos\theta - x^2], \tag{2.87}$$

and therefore we obtain for the time it takes the particle to travel from A to B:

$$t_{AB} = \int_A^B dt = \int_{x=0}^{AB} dx\sqrt{\frac{R^3}{GM_E}}\frac{1}{\sqrt{2xR\cos\theta - x^2}}. \tag{2.88}$$

With $D = 2R\cos\theta$ and $f = \sqrt{R^3/GM_E}$ and $x \to y + D/2$, this is

$$
\begin{aligned}
t_{AB} &= f\int_{x=0}^{D}\frac{dx}{\sqrt{Dx - x^2}} = f\int_{y=-D/2}^{D/2}\frac{dy}{\sqrt{D^2/4 - y^2}} \\
&= f\left[\sin^{-1}\left(\frac{y}{D/2}\right)\right]_{-D/2}^{D/2} = \pi f. \tag{2.89}
\end{aligned}
$$

Thus the period $T = 2t_{AB}$ is $2\pi\sqrt{R}\sqrt{R^2/GM_E} = 2\pi\sqrt{R}/\sqrt{g}$, where g is the acceleration due to gravity.

Chapter 3

The Lagrange Formalism

3.1   Introductory Remarks

We have already remarked that compared with the Newtonian approach, the
Lagrangian formalism has the power of permitting the treatment of much
more general situations. Thus Maxwell's electrodynamics, Einstein's general
theory of relativity and practically any other field theory can be formulated in
this way. It is worth to recall some comments on the formalism by the Oxford
mathematician R. Penrose, although after these comments he expresses some
uneasiness about the fundamentality of the Lagrangian method. In his highly
recommendable book "*The Road to Reality*" he has a section with title "*How
Lagrangians drive modern theory*" and says:* "Lagrangian theory (as well as
Hamiltonian theory) has a highly influential role in modern physics, there
being many remarkable uses to which it can be put. ... In modern attempts
at fundamental physics, when some suggested new theory is put forward, it is
almost invariably given in the form of some Lagrangian functional. This has
many advantages, such as the fact that there is a greater chance (but not an
absolute certainty) of the resulting theory having required consistency and
invariance properties, and that some form of 'Newton's third law' is implicit
(in the sense that if two fields interact then the interaction is mutual: if one
acts upon the other then the other acts equally back on the one). Moreover
Lagrangians have the pleasant property that, if a new field is introduced, then
its contributions can usually simply be added to the Lagrangian that one had
before, with any required interaction terms added also. More importantly,
perhaps, there is a direct route to the formation of a quantum theory...".
In the present chapter we deal only with the Lagrangian formalism; the
Hamiltonian formalism which is built upon this will be dealt with in the

---

*R. Penrose [38], pp. 489 – 492.

29

next chapter.

The important steps of the Lagrange formalism are the definition of generalized coordinates, then the formulation of Hamilton's variational principle and subsequently the derivation of the Euler–Lagrange equation (or several) by extremization of a postulated action integral. It is this *Euler–Lagrange equation* in mechanics which replaces (and is identical with) *Newton's equation*. The basic formalism and procedure, however, is such that in corresponding contexts of *e.g.* electrodynamics or relativity the basic equations there (*i.e.* Maxwell or Einstein) can be derived in an analogous fashion, though naturally with more involved mathematics. We illustrate the vast diversity of applications by numerous examples.

3.2   The Generalized Coordinates

The first step in the formulation of Lagrangian dynamics is the definition of an N-dimensional coordinate space called *configuration space C*, spanned by coordinates $q_i, i = 1, 2, \ldots, N$, called *generalized coordinates*. This space is, in general, not identical with the physical space of our physical experiences. One reason for introducing the space of the set $\{q_i\}$ — as will be seen — is that not every force in mechanics can be given explicitly, another reason is that if symmetry properties of a situation are exploited the considerations permit substantial simplifications. One describes the independent coordinates of a system as its "*degrees of freedom*"; here "free" refers to the absence of constraints. A system of N particles (pointlike masses $m_i, i = 1, 2, \ldots, N$) in 3-dimensional Euclidean space, which is not subjected to any additional conditions called "*constraints*", has $3N$ independent coordinates or degrees of freedom. If the system is subjected to some constraint, like a condition that forces the system along a particular path, then this constraint is an equation in the $3N$ coordinates, and this equation then allows one to express one coordinate in terms of the other $3N - 1$ coordinates which reduces the number of independent coordinates by one. In such cases it makes sense to formulate a problem in terms of the reduced number of coordinates, and to select these in a suitable way, *e.g.* by exploiting some symmetry of the system. The coordinates chosen in such a way are described as "*generalized coordinates*". Constraints in the form of equations like

$$f(\mathbf{r}_1, \mathbf{r}_2, \ldots, t) = 0,$$

are said to be "*holonomic* constraints". Thus these are those constraints that can be used to eliminate some variables so that the remaining ones are free of constraints. An example is provided by the rigid body with constraints given by the equations

$$(\mathbf{r}_i - \mathbf{r}_j)^2 = c_{ij}^2 = \text{const.}$$

Another example is a particle constrained to move along a curve given by $xy = a = \text{const.}$ Constraints which cannot be expressed in such an explicit or 'integrated' form, are described as "*nonholonomic*". An example is the constraint provided by the walls of a flask on the particles of the gas it contains. Another example is the constraint which forces a particle to move only on the surface of a sphere in the gravitational field; in this case the particle will roll along some part of the surface of the sphere, but will fall off from other parts. For instance, the following constraint is non-holonomic:

$$f(\dot{\mathbf{r}}_1, \dot{\mathbf{r}}_2) = 0,$$

since this is a condition on the velocities and can be expressed in an integrated form only after solution of the problem. A simple way to remember the difference between the two types of constraints is to keep in mind the following examples:

- A *holonomic constraint* is given by the equation

$$s = (x^2 + y^2)^{1/2} = a\phi,$$

- but a *nonholonomic constraint* by the nonintegrated form

$$ds = ad\phi, \quad v = \frac{ds}{dt} = a\dot{\phi}.$$

It is true that the second follows from the first, but the reverse is not true, since this requires knowledge of the path.

If a system is subjected to constraints, these constraints imply in a different way that there are forces acting on the system which it is difficult, if not impossible, to describe as such in detail, but whose effect on the dynamics of the system is known, as, for instance, in the case of a bead sliding along a curved wire in the gravitational field of the Earth. It would be very difficult to determine the forces acting at every single point of the wire, but it is much easier to exploit the mathematical equation representing the shape of its curve.

If a system with $3N$ degrees of freedom is constrained by k holonomic constraints expressible as k equations, then the system has $3N - k$ degrees of freedom and hence $3N - k$ independent, generalized coordinates

$q_1, q_2, q_3, \ldots, q_{3N-k}$. These generalized coordinates are related to the original $3N$ coordinates $\mathbf{r}_1, \mathbf{r}_2, \mathbf{r}_3, \ldots, \mathbf{r}_N$ by the set of equations (*cf.* Fig. 3.1):

$$\mathbf{r}_1 = \mathbf{r}_1(q_1, q_2, q_3, \ldots, q_{3N-k}, t), \ldots, \mathbf{r}_N = \mathbf{r}_N(q_1, q_2, q_3, \ldots, q_{3N-k}, t).$$

Time dependent constraints are said to be "*rheonomic*" and time independent constraints "*scleronomic*". Thus in the case of a particle which is constrained to move on the surface of a sphere, the two generalized coordinates are obviously the two angles describing its longitude and latitude on the sphere.

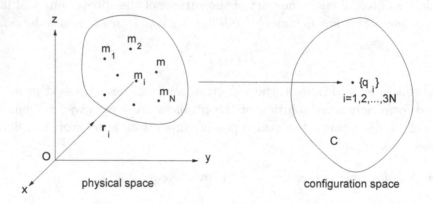

Fig. 3.1 Transition from physical space to configuration space C.

Example 3.1: In scleronomic cases the kinetic energy is a quadratic form
Show that in the case of a system of N particles of mass m_i, velocity $v_i, i = 1, 2, \ldots, N$, and scleronomic constraints, the kinetic energy T is a quadratic form.

Solution: We have

$$T = \frac{1}{2} \sum_i m_i v_i^2 = \frac{1}{2} \sum_i m_i \left(\sum_j \frac{\partial \mathbf{r}_i}{\partial q_j} \dot{q}_j + \frac{\partial \mathbf{r}_i}{\partial t} \right)^2 = a + \sum_j a_j \dot{q}_j + \sum_{j,k} a_{jk} \dot{q}_j \dot{q}_k,$$

where

$$a = \frac{1}{2} \sum_i m_i \left(\frac{\partial \mathbf{r}_i}{\partial t} \right)^2, \quad a_i = \sum_j m_j \frac{\partial \mathbf{r}_j}{\partial t} \cdot \frac{\partial \mathbf{r}_j}{\partial q_i}, \quad a_{jk} = \frac{1}{2} \sum_i m_i \frac{\partial \mathbf{r}_i}{\partial q_j} \cdot \frac{\partial \mathbf{r}_i}{\partial q_k}.$$

It follows that for cases in which $(\partial \mathbf{r}_i/\partial t) = 0$, the coefficients a and a_j vanish.

3.3   The Principle of Virtual Work

The principle named in the title of this section is unrelated to the Lagrange formalism. Nonetheless, it is convenient to introduce this here, since its applications require a clear distinction between applied forces and those of constraints, which we introduced above.

One describes as an infinitesimal *virtual displacement* of a system a displacement of the *configuration of the system* as a result of some arbitrary infinitesimal changes of the coordinates $\delta \mathbf{r}_i$, which are compatible with the forces and constraints of the system at any given instant of time; the word "virtual" implies that there is no shift in time t.[†] Assuming the system to be in a state of equilibrium, so that the total force acting on every particle "i" vanishes, we have $\mathbf{F}_i = 0$. This state of the system is described as that of *statics*. It follows that with this condition, also the work performed by the forces $\mathbf{F}_i$ in displacing every coordinate $\mathbf{r}_i$ by an amount $\delta \mathbf{r}_i$ vanishes, *i.e.*

$$\sum_i \mathbf{F}_i \cdot \delta \mathbf{r}_i = 0. \tag{3.1}$$

We now set

$$\mathbf{F}_i = \mathbf{F}_i^{(a)} + \mathbf{f}_i, \tag{3.2}$$

where

$$\mathbf{F}_i^{(a)} = \text{applied force} \quad \text{and} \quad \mathbf{f}_i = \text{constraining force}.$$

Next we restrict ourselves further to only such systems, for which the total virtual work of the constraining forces vanishes, *i.e.* we assume

$$\sum_i \mathbf{f}_i \cdot \delta \mathbf{r}_i = 0. \tag{3.3}$$

Here $\delta \mathbf{r}_i$ perpendicular to $\mathbf{f}_i$ is a possibility, but not necessary. The condition (3.3) applies, for instance, in the case of particles, which are forced to move on a plane (the displacements are within the plane, but the constraining forces, which keep the particles on this plane, are perpendicular to this plane). The condition applies also, as shown above, for rigid bodies, and for many other cases, and in general for statics. The condition does not apply, for instance, when nonnegligible frictional forces are present. We can summarize therefore, that for systems in equilibrium (meaning the total force acting on each particle vanishes) whose constraining forces perform no virtual work (*i.e.* those with $\sum_i \mathbf{f}_i \cdot \delta \mathbf{r}_i = 0$), the principle of virtual work states that

$$\sum_i \mathbf{F}_i^{(a)} \cdot \delta \mathbf{r}_i = 0. \tag{3.4}$$

This means the virtual (*i.e.* with $\delta t = 0$) work of the applied forces is zero. This is the statement of the principle of virtual work. One should

---

[†]Strictly speaking of higher order than δt. If a velocity dx/dt is zero when $t = t_0$, where x is a differentiable function of t, the increment δx in x when t changes from t_0 to $t_0 + \delta t$, is of higher order than δt, *i.e.* $\delta x = x(t) - x(t_0) = (\delta t)^2 x''(t_0)/2 + \cdots$. For discussion see K. E. Bullen [8], p. 285.

note that in view of the assumed presence of constraints, the infinitesimal displacements $\delta \mathbf{r}_i$ are not independent. It is the generalized coordinates q_i which are independent of each other.

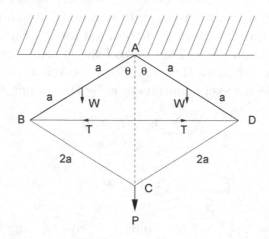

Fig. 3.2 The rhombus rod structure hanging from A.

Example 3.2: The hanging rhombus

Four rods, each of length $2a$, are arranged to form a rhombus $ABCD$ with a light (*i.e.* weightless) rod between the joints at D and B. By attaching the rhombus at A to a ceiling, as shown in Fig. 3.2, the system can hang freely in the gravitational field of the Earth. The rods BC and CD are also weightless, whereas each of the other two rods AB, AD has a weight W. At the point C a weight P is attached, the rhombus maintaining its shape in view of the weightless rod BD. Calculate the tension T in BD.

Solution: We assume the system undergoes a small virtual displacement along BD. With the principle of virtual work, *i.e.* Eq. (3.4), we then obtain

$$
\begin{aligned}
0 &= T\delta(DB) + 2W\frac{\delta(AD\cos\theta)}{2} + P\delta[(AB + BC)\cos\theta], \\
0 &= Td(4a\sin\theta) + 2Wd(a\cos\theta) + Pd(4a\cos\theta), \\
0 &= T(4a\cos\theta)d\theta + 2W(-a\sin\theta)d\theta + P(-4a\sin\theta)d\theta.
\end{aligned}
\tag{3.5}
$$

It follows that

$$
T = \frac{1}{2}(W + 2P)\tan\theta.
\tag{3.6}
$$

Example 3.3: A heavy rope resting on a sphere, another on a cone

A uniform, heavy, endless rope of weight W rests in equilibrium on a smooth solid sphere and assumes on this sphere the position of a horizontal circle. The radius of the rope subtends an angle θ at the centre of the sphere as indicated in Fig.3.3. Show that the tension T in the rope is given by $(W/2\pi)\tan\theta$. What is it in the case of the cone shown in Fig. 3.3?

Solution: Of the three spherical coordinates r, θ, ϕ the angle ϕ plays no role in view of the circular symmetry of the rope. Moreover $r = a$ is a constraint. Thus there remains only the one independent variable θ. Three forces act at every point P of the rope: The weight of the masspoint (of the rope there), the tension T in the rope, and the reaction N (of the sphere) which keeps the rope in place. We now assume an infinitesimal extension dl of the length l of the rope orthogonal to the constraining force N. The result is a small downward displacement of the rope. The distance $OP = a$ remains unchanged, which means that the virtual work of the constraining force N is zero. As a result only the tension T and the weight W appear in the equation of the principle of virtual work. Wit l the length of the rope, and W its weight, we have $l = 2\pi PQ = 2\pi a \sin\theta$ and $dl = 2\pi a \cos\theta d\theta$. The principle of virtual work then yields the relation

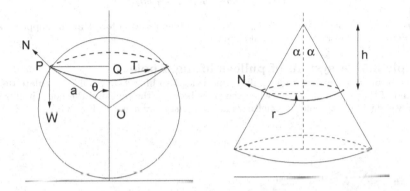

Fig. 3.3 A heavy rope resting on a sphere, another on a cone.

$$Tdl + Wd(OQ) = 0, \quad \text{and} \quad \therefore\ T2\pi a\cos\theta d\theta + Wd(a\cos\theta) = 0.$$

Thus $T = W\tan\theta/2\pi$. With analogous calculations for the cone we have, since $\tan\alpha = r/h$,

$$Td(2\pi r) + Wdh = 0, \qquad Td(2\pi h\tan\alpha) + Wdh = 0,$$

$$\therefore\ T = -\frac{W}{2\pi\tan\alpha} = -\frac{Wh}{2\pi r}.$$

Example 3.4: A bridge structure consisting of beams and struts

Use the principle of virtual work to determine the force F_u in a strut at the bottom of the bridge structure shown in Fig. 3.4. A vertical force F acts at each of the three lower joints.

Solution: We cut one of the struts of length $2h$ at the bottom of the bridge structure of height h' as indicated in Fig. 3.4. Then we allow an infinitesimal rotation of the left hand part shown hatched in Fig. 3.4 about the point A through an angle $\delta\alpha$, with an appropriate rotation of the right hand part about B through an angle $\delta\beta$. The infinitesimal displacement of the point C

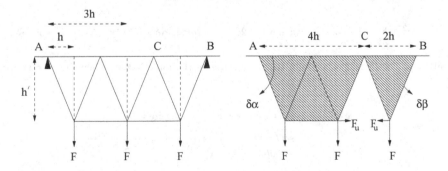

Fig. 3.4 A bridge structure consisting of beams and struts with vertical forces F.

implies the equation $4h\delta\alpha = 2h\delta\beta$, $i.e.$ $\delta\beta = 2\delta\alpha$. The associated virtual work done is therefore

$$\delta W = Fh\delta\alpha + F3h\delta\alpha - F_u h'\delta\alpha + Fh\delta\beta - F_u h'\delta\beta = 0,$$

$i.e.$

$$4hF\delta\alpha + Fh\delta\beta - F_u h'(\delta\alpha + \delta\beta) = 0,$$

and hence $6hF\delta\alpha = F_u h'3\delta\alpha$. Thus $F_u = 2(h/h')F$. The reaction forces at the supports A and B do not perform any work in the virtual displacement.

Example 3.5: A system of pulleys lifting a pole

A system of pulleys consists of four pulleys and is used to lift a heavy pole. Calculate the force F at the end of the system of pulleys when the angle between the pole and the horizontal is α (given also the weight G of the pole, its length $2s$ and the distances a, b, c in Fig. 3.5.

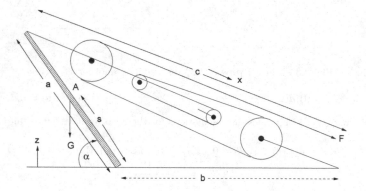

Fig. 3.5 A system of pulleys lifting a pole.

Solution: Let δW be the virtual work of the two external forces G and F. Tilting the pole through an infinitesimal angle $\delta\alpha$ implies infinitesimal displacements in x and z directions, as indicated in Fig. 3.5, and we have (with arguments as in the preceding examples)

$$\delta W = -G\delta z_A + F\delta x = 0.$$

Since $z_A = s\sin\alpha$, we have $\delta z_A = s\cos\alpha\delta\alpha$. In the case of the system with four pulleys (five pieces of belt along x) we have $\delta x = -5\delta c$. Using the cosine theorem

$$c^2 = a^2 + b^2 + 2ab\cos\alpha, \quad \text{we have} \quad 2c\delta c = -2ab\sin\alpha\delta\alpha, \quad \delta x = -5\delta c = \frac{5ab}{c}\sin\alpha\delta\alpha.$$

For δW and the force F we obtain therefore (observe the force is zero when the pole is upright, $\alpha = \pi/2$)

$$\delta W = \left[-Gs\cos\alpha + F\frac{5ab}{c}\sin\alpha \right]\delta\alpha = 0,$$

$$\therefore \quad F = \frac{scG\cot\alpha}{5ab} = \frac{sG\cot\alpha}{5ab}\sqrt{a^2 + b^2 + 2ab\cos\alpha}.$$

3.4 D'Alembert's Principle, Lagrange Equations

The principle we now consider briefly is of interest here only in as far as it yields already at this stage the form of the basic *Euler–Lagrange equations*, that we shall subsequently investigate in much more detail. This is therefore the sole reason for our consideration of D'Alembert's principle here.[‡]

As we saw above, the principle of virtual work as enunciated there applies only to statics. However, we also want to be able to handle dynamical systems, and this can be achieved by *formally* reducing dynamics to statics as follows. To this end we rewrite the equation of motion $\dot{\mathbf{p}}_i = \mathbf{F}_i$ in the form

$$\mathbf{F}_i - \dot{\mathbf{p}}_i = 0. \tag{3.7}$$

One now interprets $-\dot{\mathbf{p}}_i$ as an oppositely directed effective force. We then have instead of Eq. (3.1) the equation

$$\sum_i (\mathbf{F}_i - \dot{\mathbf{p}}_i) \cdot \delta\mathbf{r}_i = 0, \tag{3.8}$$

and hence

$$\sum_i (\mathbf{F}_i^{(a)} + \mathbf{f}_i - \dot{\mathbf{p}}_i) \cdot \delta\mathbf{r}_i = 0. \tag{3.9}$$

The statement of this equation is now, that in the case of a virtual ($\delta t = 0$) displacement of the particle system (which moves in accordance with Newton's laws), the work performed vanishes. Considering again only systems for which the total virtual work of the constraining forces vanishes, *i.e.* those for which $\sum_i \mathbf{f}_i \cdot \delta\mathbf{r}_i = 0$, *i.e.* considering only nondeformable systems (a sphere remains a sphere and does not become an ellipsoid), then

$$\sum_i (\mathbf{F}_i^{(a)} - \dot{\mathbf{p}}_i) \cdot \delta\mathbf{r}_i = 0. \tag{3.10}$$

This equation is known as *D'Alembert's principle*.

We recall that the coordinates $\mathbf{r}_i$ are connected by the constraints. Thus in general, if the displacements $\delta\mathbf{r}_i$ are not independent, we cannot equate their coefficients to zero. However, we can do this in the case of equations of the form

$$\sum_i (\ldots)\delta q_i = 0,$$

[‡]Although strictly speaking the equations obtained with D'Alembert's principle are called Lagrange equations, we shall simply call all these equations Euler–Lagrange equations, since this is what they are eventually.

where the coordinates q_i are independent generalized coordinates. We therefore wish to re-express Eq. (3.10) in terms of these coordinates. We have

$$\mathbf{r}_i \equiv \mathbf{r}_i(q_1, q_2, q_3, \ldots, q_n, t),$$

and hence

$$\mathbf{v}_i \equiv \dot{\mathbf{r}}_i = \sum_j \frac{\partial \mathbf{r}_i}{\partial q_j} \dot{q}_j + \frac{\partial \mathbf{r}_i}{\partial t} \quad \text{and} \quad \frac{\partial \mathbf{v}_i}{\partial \dot{q}_i} = \frac{\partial \mathbf{r}_i}{\partial q_j}, \tag{3.11}$$

and generally

$$\delta \mathbf{r}_i = \sum_j \frac{\partial \mathbf{r}_i}{\partial q_j} \delta q_j + \frac{\partial \mathbf{r}_i}{\partial t} \delta t, \quad (\delta \mathbf{r}_i)_{\text{virtual}} = \sum_j \frac{\partial \mathbf{r}_i}{\partial q_j} \delta q_j, \quad i.e. \; \delta t = 0. \tag{3.12}$$

We now have

$$\sum_i \mathbf{F}_i^{(a)} \cdot \delta \mathbf{r}_i = \sum_{i,j} \mathbf{F}_i^{(a)} \cdot \frac{\partial \mathbf{r}_i}{\partial q_j} \delta q_j \equiv \sum_j Q_j \delta q_j, \tag{3.13}$$

where

$$Q_j = \sum_i \mathbf{F}_i^{(a)} \cdot \frac{\partial \mathbf{r}_i}{\partial q_j}. \tag{3.14}$$

The quantities Q_j are called the components of the "*generalized force*". We now deal with the second term in Eq. (3.10):

$$\sum_i \dot{\mathbf{p}}_i \cdot \delta \mathbf{r}_i = \sum_i m_i \ddot{\mathbf{r}}_i \cdot \delta \mathbf{r}_i = \sum_{i,j} m_i \ddot{\mathbf{r}}_i \cdot \frac{\partial \mathbf{r}_i}{\partial q_j} \delta q_j$$

$$= \sum_j \delta q_j \sum_i \left\{ \frac{d}{dt} \left(m_i \dot{\mathbf{r}}_i \cdot \frac{\partial \mathbf{r}_i}{\partial q_j} \right) - m_i \dot{\mathbf{r}}_i \cdot \frac{d}{dt} \left(\frac{\partial \mathbf{r}_i}{\partial q_j} \right) \right\}. \tag{3.15}$$

But

$$\frac{d}{dt} \left(\frac{\partial \mathbf{r}_i}{\partial q_j} \right) = \sum_k \frac{\partial^2 \mathbf{r}_i}{\partial q_k \partial q_j} \frac{\partial q_k}{\partial t} + \frac{\partial^2 \mathbf{r}_i}{\partial t \partial q_j} \tag{3.16}$$

and

$$\frac{\partial}{\partial q_j} \left(\frac{d\mathbf{r}_i}{dt} \right) = \frac{\partial}{\partial q_j} \left(\frac{\partial \mathbf{r}_i}{\partial q_k} \frac{\partial q_k}{\partial t} + \frac{\partial \mathbf{r}_i}{\partial t} \right) \overset{(3.16)}{=} \frac{d}{dt} \left(\frac{\partial \mathbf{r}_i}{\partial q_j} \right) + \frac{\partial \mathbf{r}_i}{\partial q_k} \overbrace{\frac{\partial^2 q_k}{\partial q_j \partial t}}^{\frac{\partial}{\partial t} \frac{\partial q_k}{\partial q_j} = \frac{\partial}{\partial t} \delta_{kj} = 0}$$

$$= \frac{d}{dt} \left(\frac{\partial \mathbf{r}_i}{\partial q_j} \right). \tag{3.17}$$

Moreover, from Eq. (3.11) we obtain

$$\frac{\partial \mathbf{v}_i}{\partial \dot{q}_j} = \frac{\partial \mathbf{r}_i}{\partial q_j}. \tag{3.18}$$

Hence finally we obtain from Eq. (3.15), inserting (3.18) in the first term and (3.17) in the second,

$$\sum_i \dot{\mathbf{p}}_i \cdot \delta \mathbf{r}_i = \sum_j \delta q_j \sum_i \left\{ \frac{d}{dt}\left(m_i \mathbf{v}_i \cdot \frac{\partial \mathbf{v}_i}{\partial \dot{q}_j}\right) - m_i \mathbf{v}_i \frac{\partial \mathbf{v}_i}{\partial q_j} \right\}$$

$$= \sum_j \delta q_j \left\{ \frac{d}{dt}\frac{\partial}{\partial \dot{q}_j}(T) - \frac{\partial}{\partial q_j}(T) \right\}, \tag{3.19}$$

where

$$T = \sum_i \frac{1}{2} m_i \mathbf{v}_i^2. \tag{3.20}$$

The D'Alembert principle, *i.e.*

$$\sum_i (\mathbf{F}_i^{(a)} - \dot{\mathbf{p}}_i) \cdot \delta \mathbf{r}_i = -\sum_i \mathbf{f}_i \cdot \delta \mathbf{r}_i = 0$$

now implies

$$-\sum_j \delta q_j \left[\left\{ \frac{d}{dt}\frac{\partial T}{\partial \dot{q}_j} - \frac{\partial T}{\partial q_j} \right\} - Q_j \right] = 0. \tag{3.21}$$

In the case of holonomic constraints the generalized coordinates q_j are independent, and we obtain

$$\frac{d}{dt}\left(\frac{\partial T}{\partial \dot{q}_j}\right) - \frac{\partial T}{\partial q_j} = Q_j. \tag{3.22}$$

These equations for $j = 1, 2, 3, \ldots, n$ are called *Lagrangian equations*. If the applied forces $\mathbf{F}_i^{(a)}$ are conservative, *i.e.* when

$$\mathbf{F}_i^{(a)} = -\boldsymbol{\nabla}_i V \equiv -\frac{\partial V}{\partial \mathbf{r}_i}, \tag{3.23}$$

then

$$Q_j = \sum_i \mathbf{F}_i^{(a)} \cdot \frac{\partial \mathbf{r}_i}{\partial q_j} = -\sum_i \boldsymbol{\nabla}_i V \cdot \frac{\partial \mathbf{r}_i}{\partial q_j} = -\frac{\partial V(\mathbf{r}_1, \ldots, \mathbf{r}_n)}{\partial q_j}, \tag{3.24}$$

as long as the potential V is only a function of the position coordinates, but not of the velocities. However, something analogous applies in the case of some special velocity dependent potentials with forces Q_j defined by

$$Q_j = -\frac{\partial V}{\partial q_j} + \frac{d}{dt}\left(\frac{\partial V}{\partial \dot{q}_j}\right). \tag{3.25}$$

Hence the Lagrangian equations can be written

$$\frac{d}{dt}\left(\frac{\partial T}{\partial \dot{q}_j}\right) - \frac{\partial(T - V)}{\partial q_j} = 0,$$

or ($\partial V/\partial \dot{q}_j$ must not necessarily be zero)

$$\frac{d}{dt}\left(\frac{\partial L}{\partial \dot{q}_j}\right) - \frac{\partial L}{\partial q_j} = 0, \qquad L = T - V. \tag{3.26}$$

The functional $L(q_j, \dot{q}_j)$ is known as *Lagrangian*. More generally, if some but not all forces are conservative, the equations can be written in the form

$$\frac{d}{dt}\left(\frac{\partial L}{\partial \dot{q}_j}\right) - \frac{\partial L}{\partial q_j} = Q_j. \tag{3.27}$$

Here L contains the potential of the conservative forces, and Q_j represents those forces which do not result from a potential, *e.g.* (dissipative) friction forces. One should note that the generalized forces Q_j do not necessarily have the usual dimension of a force. If, for instance, q_j is an angle, then Q_j has the dimension of a torque. The *generalized forces* can be obtained as follows (see also Example 3.6). We assume that a coordinate q_j is shifted by an amount δq_j, but that all other coordinates contained in L are kept fixed. One then calculates (somehow) the work δW_{q_j} of the external forces (*e.g.* of the gravitational force). Then Q_j follows from

$$\delta W_{q_j} = Q_j \delta q_j. \tag{3.28}$$

Example 3.6: The pendulum bob attached to a rubber band

A mass m is attached to an extendable string of unstreched length r_0, and is allowed to swing as a pendulum in a vertical plane in the gravitational field. Calculate the generalized forces for this case.

Solution: The planar motion suggests the use of two-dimensional polar coordinates r and θ, as shown in Fig. 3.6. Thus there are two degrees of freedom and the kinetic energy T of the mass is given by

$$T = \frac{1}{2}m(\dot{r}^2 + r^2\dot{\theta}^2). \tag{3.29}$$

It follows that

$$\frac{\partial T}{\partial r} = mr\dot{\theta}^2, \quad \frac{\partial T}{\partial \dot{r}} = m\dot{r}, \quad \frac{\partial T}{\partial \theta} = 0, \quad \frac{\partial T}{\partial \dot{\theta}} = mr^2\dot{\theta}.$$

With these derivatives the Lagrange equations (3.27) for $q_j = r, \theta$ are seen to be

$$m\ddot{r} - mr\dot{\theta}^2 = Q_r, \qquad \frac{d}{dt}(mr^2\dot{\theta}) = Q_\theta. \tag{3.30}$$

We have to calculate the generalized forces Q_r, Q_θ. We begin with

$$Q_r \overset{(3.14)}{=} F_x \frac{\partial x}{\partial r} + F_y \frac{\partial y}{\partial r} + F_z \frac{\partial z}{\partial r}. \tag{3.31}$$

The forces acting on the pendulum bob are the gravitational force and the force in the elastic

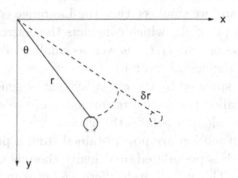

Fig. 3.6 The pendulum bob attached to a rubber band.

string, the latter being that obtained from *Hooke's law* with constant k. Then, *cf.* Fig. 3.6,

$$F_x = -k(r - r_0)\sin\theta, \qquad F_y = mg - k(r - r_0)\cos\theta, \qquad F_z = 0. \tag{3.32}$$

Here r_0 is the unextended length of the taut string. Since,

$$x = r\sin\theta, \quad y = r\cos\theta, \quad i.e. \quad \frac{\partial x}{\partial r} = \sin\theta, \quad \frac{\partial y}{\partial r} = \cos\theta,$$

we obtain

$$Q_r = -k(r - r_0)\sin^2\theta + [mg - k(r - r_0)\cos\theta]\cos\theta = -k(r - r_0) + mg\cos\theta. \tag{3.33}$$

Similarly one obtains

$$Q_\theta = F_x \frac{\partial x}{\partial \theta} + F_y \frac{\partial y}{\partial \theta} = -mgr\sin\theta. \tag{3.34}$$

With these results the equations of motion of the pendulum have been completely determined:

$$m\ddot{r} - mr\dot{\theta}^2 = -k(r - r_0) + mg\cos\theta, \qquad \frac{d}{dt}(mr^2\dot{\theta}) = -mgr\sin\theta. \tag{3.35}$$

Of course, in the present case, we could also have written down the potential V, *i.e.*

$$V = -mgr\cos\theta + \frac{1}{2}k(r - r_0)^2 = -mgy + \frac{1}{2}k\{(x - x_0)^2 + (y - y_0)^2\},$$

so that

$$F_x = -\frac{\partial V}{\partial x} = -k(x - x_0) = -k(r - r_0)\sin\theta, \quad F_y = -\frac{\partial V}{\partial y} = -k(y - y_0) + mg = mg - k(r - r_0)\cos\theta.$$

Then

$$L = T - V = \frac{1}{2}m(\dot{r}^2 + r^2\dot{\theta}^2) + mgr\cos\theta - \frac{1}{2}k(r - r_0)^2. \tag{3.36}$$

3.5   Hamilton's Variational Principle, and Euler–Lagrange Equations

3.5.1   Hamilton's variational principle

The two principles considered previously, *i.e.* the principle of virtual work and D'Alembert's principle are *differential principles*, which are based on infinitesimal, virtual displacements of a system. We now proceed to consider *integral principles*, and we shall see that the Lagrange equations can also be derived from such a principle, which considers the entire motion of a system between two instants of time t_1, t_2, as well as small, virtual deviations of the entire motion from the actual motion.

The hyperspace spanned by the entire system of generalized coordinates $q_1, q_2, q_3, \ldots, q_n$ is called the *configuration space* of the system. The motion of the system ensues along a *path* in this space.

The equations of motion are now obtained from a principle, *i.e.* *Hamilton's principle*, which is postulated, and hence there is no reference to Newton's laws anymore. This is the vital difference between Hamilton's mechanics and that of Newton.

3.5.2   Hamilton's principle for conservative systems

We consider first conservative systems. One postulates: The physical system moves from some initial configuration at time t_1 to a final configuration at time $t_2 > t_1$, such that the following integral, described as representing the *action* of the system, and irrespective of all possible paths of reaching the final configuration from that of the initial configuration,

$$I := \int_{t_1}^{t_2} L(q_i, \dot{q}_i, t)dt, \qquad L = T - V, \tag{3.37}$$

is an *extremum*, *i.e.* a minimum or maximum. Expressed as a *variational principle*, the postulate implies:

$$\delta I = \delta \int_{t_1}^{t_2} L(q_i, \dot{q}_i, t)dt = 0 \tag{3.38}$$

We shall obtain the Euler–Lagrange equations of motion from this condition, which is therefore the *fundamental postulate of the mechanics of conservative systems*.[§]

[§]The reverse is also possible, *i.e.* the derivation of Hamilton's principle from the Euler–Lagrange equations.

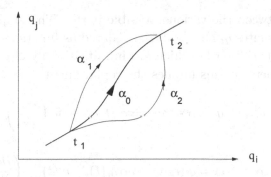

Fig. 3.7 The extremized path.

For a better understanding we repeat: Every function or functional L called Lagrange function or Lagrangian defines a theory. We could consider Lagrangians depending on arbitrarily high derivatives of the generalized coordinates (as functions of time), and each case would define a theory (see *e.g.* Example 3.8 below). However, we reproduce the equations of Newtonian mechanics only by permitting derivatives of the first order. The Lagrangian itself is not derived — in each case the chosen form of the Lagrangian is part of the postulate. The claim is the consequence of the postulate, and this means that the system defined by the Lagrangian chooses such a path in *configuration space* (the space spanned by the generalized coordinates) that the integral I, called action or action integral, is extremized (no statement regarding maximum or minimum).

Above we defined the concept of configuration space. A different space called '*phase space*' is the space used in Hamiltonian mechanics (the topic of the next chapter); this is the hyperspace spanned by both the generalized coordinates q_i and the generalized velocities $\dot{q}_i$, or better, as we shall see the corresponding momenta p_i.

We proceed to the explicit derivation of the Lagrange equations from *Hamilton's principle*. We write the *variational principle*:

$$\delta I = \delta \int_{t_1}^{t_2} L(q_1(t), q_2(t), \ldots, \dot{q}_1(t), \dot{q}_2(t), \ldots; t) dt = 0. \qquad (3.39)$$

We now introduce a parameter α whose values enumerate the various paths of the set of possible paths from a point with coordinates $q_i(t_1)$ at an initial time t_1 to a point with coordinates $q_i(t_2)$ at a later or final time $t_2 > t_1$, as indicated in Fig. 3.7. Only a specific path, say that with $\alpha = \alpha_0$, will extremize the action integral I. A path neighbouring this extremal path is characterized by an infinitesimally different value of the parameter α, *i.e.* by $\alpha_0 + \delta\alpha$. We consider the action I as a function also of this parameter which distinguishes between the various possible paths. This implies that also the generalized coordintes q_i have to be considered as functions of the parameter α, *i.e.* as $q_i(t, \alpha)$. For paths which are infinitessimally close to the extremal path with parameter α this implies the replacement

$$q_i \longrightarrow q_i(t, \alpha + \delta\alpha) = q_i(t, \alpha) + \delta\alpha\left(\frac{\partial q_i}{\partial \alpha}\right),$$

or

$$q_i(t, \alpha + \delta\alpha) = q_i(t, \alpha) + \delta\alpha\eta_i(t), \quad \eta_i(t) = \left(\frac{\partial q_i}{\partial \alpha}\right). \tag{3.40}$$

All possible paths are required to originate from one and the same point defined by $q_i(t_1)$ and to terminate at one and the same point $q_i(t_2), t_2 > t_1$. Thus at $t = t_1$ and at $t = t_2$ we must have

$$\eta_i(t_1) = 0 = \eta_i(t_2), \tag{3.41}$$

since for any arbitrary values of α:

$$\text{at} \quad t = t_1, t_2: \quad q_i(t, \alpha) = q_i(t).$$

Thus the action integral I is also a function of the parameter α, but through the generalized coordinates q_i.

We proceed to calculate the variation of I, *i.e.* the difference between the values of I for two infinitesimally close paths. In performing the calculation we consider again virtual displacements, *i.e.* the parameter of time t is not varied (variational principles which consider real displacements and therefore involve also the variation of time, are generally described as principles of *'least action'*). The Lagrange function is a functional of $q_i, \dot{q}_i$ and possibly of t, as postulated in Hamilton's principle for *mechanics* (in an abstract consideration one could, of course, include higher order time derivatives). Thus we have (with $\alpha = \alpha_0 + \delta\alpha$)

$$\begin{aligned} \delta I &= I(\alpha_0 + \delta\alpha) - I(\alpha_0) = \left(\frac{\partial I}{\partial \alpha}\right)_{\alpha=\alpha_0} \delta\alpha \\ &= \int_{t_1}^{t_2} \sum_i \left(\frac{\partial L}{\partial q_i}\frac{\partial q_i}{\partial \alpha}\delta\alpha + \frac{\partial L}{\partial \dot{q}_i}\frac{\partial \dot{q}_i}{\partial \alpha}\delta\alpha\right) dt \\ &= \int_{t_1}^{t_2} \sum_i \left(\frac{\partial L}{\partial q_i}\delta q_i + \frac{\partial L}{\partial \dot{q}_i}\delta\dot{q}_i\right) dt, \end{aligned} \tag{3.42}$$

where

$$\delta q_i = \frac{\partial q_i}{\partial \alpha}\delta\alpha, \quad \delta\dot{q}_i = \frac{\partial \dot{q}_i}{\partial \alpha}\delta\alpha. \tag{3.43}$$

One should note that $q_i = q_i(\alpha, t)$. By not considering variations in t, *i.e.* taking $\delta t = 0$, we consider again virtual displacements. Correspondingly we have

$$\delta \dot{q}_i = \delta \frac{dq_i(\alpha, t)}{dt} = \frac{d}{dt}\left(\frac{\partial q_i(\alpha, t)}{\partial \alpha}\right)\delta\alpha = \frac{d}{dt}(\delta q_i), \qquad (3.44)$$

since the variation δ defined by Eq. (3.43) means only the variation with respect to α. Also

$$\delta q_i = \frac{\partial q_i}{\partial \alpha}\delta\alpha = \eta_i(t)\delta\alpha = 0 \quad \text{for} \quad t = t_1, t_2, \qquad (3.45)$$

since $\eta_i(t_1) = 0 = \eta_i(t_2)$. With partial integration we obtain

$$\int_{t_1}^{t_2} \frac{\partial L}{\partial \dot{q}_i}\delta\dot{q}_i dt = \int_{t_1}^{t_2} \frac{\partial L}{\partial \dot{q}_i}\frac{d}{dt}(\delta q_i)dt = \left[\frac{\partial L}{\partial \dot{q}_i}\delta q_i\right]_{t_1}^{t_2} - \int_{t_1}^{t_2} \frac{d}{dt}\left(\frac{\partial L}{\partial \dot{q}_i}\right)\delta q_i dt$$

$$= 0 - \int_{t_1}^{t_2} \frac{d}{dt}\left(\frac{\partial L}{\partial \dot{q}_i}\right)\delta q_i dt, \qquad (3.46)$$

since $\delta q_i = 0$ for $t = t_1, t_2$. Inserting the result into δI, we obtain

$$\delta I = \int_{t_1}^{t_2} \sum_i \left[\frac{\partial L}{\partial q_i} - \frac{d}{dt}\left(\frac{\partial L}{\partial \dot{q}_i}\right)\right]\delta q_i dt. \qquad (3.47)$$

Here

$$\delta q_i = \frac{\partial q_i}{\partial \alpha}\delta\alpha = \eta_i(t)\delta\alpha.$$

Since for holonomic constraints the increments δq_i and hence η_i are independent (*i.e.* after elimination of superfluous degrees of freedom), the condition $\delta I = 0$ is satisfied provided for every value of i:

$$\frac{\partial L}{\partial q_i} - \frac{d}{dt}\left(\frac{\partial L}{\partial \dot{q}_i}\right) = 0. \qquad (3.48)$$

We have thus obtained the *Euler–Lagrange equations* of conservative systems from the Hamilton principle. In the more general form, these equations are known as *Euler–Lagrange equations*.

Example 3.7: Extremal distance separating two points in a plane

The line element ds in the (x, y)-plane is $ds = \sqrt{dx^2 + dy^2}$. Show that the extremal (shortest) distance connecting points 1 amd 2 is given by a straight line, *i.e.* $y = mx + c$, where m and c are constants.

Solution: We have $I = \int_1^2 ds = \int_{x_1}^{x_2} \sqrt{1 + (dy/dx)^2}dx$. Thus we identify the Lagrangian with $L(y, \dot{y}; x) = \sqrt{1 + \dot{y}^2}$ and the Euler–Lagrange equation is

$$\frac{d}{dx}\left(\frac{\partial L}{\partial \dot{y}}\right) - \frac{\partial L}{\partial y} = 0, \quad \text{i.e.} \quad \frac{\partial L}{\partial \dot{y}} - \frac{\dot{y}}{\sqrt{1 + \dot{y}^2}} = \text{const}.$$

Thus $\dot{y} = \text{const.}, y = mx + c$, where m and c are constants. This demonstrates only that this straight line is an extremal solution. One would still have to show — by going to the second variation — that this is actually a minimum; maxima and minima are treated in Example 9.6.

Example 3.8: Euler's differential equation

Considering the functional $L(q, \dot{q}, \ddot{q}; t)$ (*i.e.* involving also a second order derivative), derive the Euler equation resulting from the extremum condition $\delta I = 0, I = \int L dt$.¶

Solution: We use a simplified notation and do not repeat the detailed discussion provided above. Thus we assume that δq and its derivatives are continuous up to fourth order and vanish at the integration limits $t = t_1, t_2$. Then, with appropriate partial integrations,

$$\delta I[q] = \int_{t_1}^{t_2} dt \left[\frac{\partial L}{\partial q} \delta q + \frac{\partial L}{\partial \dot{q}} \delta \dot{q} + \frac{\partial L}{\partial \ddot{q}} \delta \ddot{q} \right] = \int_{t_1}^{t_2} dt \left[\frac{\partial L}{\partial q} \delta q - \frac{d}{dt} \left(\frac{\partial L}{\partial \dot{q}} \right) \delta q - \frac{d}{dt} \left(\frac{\partial L}{\partial \ddot{q}} \right) \delta \dot{q} \right]$$

$$= \int_{t_1}^{t_2} dt \left[\frac{\partial L}{\partial q} \delta q - \frac{d}{dt} \left(\frac{\partial L}{\partial \dot{q}} \right) \delta q + \frac{d^2}{dt^2} \left(\frac{\partial L}{\partial \ddot{q}} \right) \delta q \right].$$

This expression vanishes if *Euler's equation* holds,

$$\frac{\partial L}{\partial q} - \frac{d}{dt} \left(\frac{\partial L}{\partial \dot{q}} \right) + \frac{d^2}{dt^2} \left(\frac{\partial L}{\partial \ddot{q}} \right) = 0.$$

As an example consider $L(q, \dot{q}, \ddot{q}; t) = \ddot{q}^2 - 2f(t)q$. Euler's equation is $d^4 q/dt^4 - f(t) = 0$.

3.5.3   Hamilton's principle for holonomic systems

We now consider Hamilton's principle for more general cases, in particular for nonconservative and nonholonomic systems. The principle in these cases is the *extended Hamilton principle*:

$$\delta I = \delta \int_{t_1}^{t_2} (T + W) dt = 0 \tag{3.49}$$

with *fixed endpoints* as before. In this expression (see Eq. (3.28))

$$\delta W = \sum_i \mathbf{F}_i \cdot \delta \mathbf{r}_i = \sum_j Q_j \delta q_j, \tag{3.50}$$

$\mathbf{F}_i \equiv \mathbf{F}_i^{(a)}$ as in Eq. (3.13). Since there is no variation in time, the variations $\delta \mathbf{r}_i, \delta q_j$ are virtual. Returning briefly to the case of conservative systems, we show that $\delta I = \delta \int L dt$. If the forces Q_j can be derived from a potential V as described above (*cf.* Eqs. (3.23), (3.25)), we have

$$Q_j = -\frac{\partial V}{\partial q_j},$$

¶R. Courant [12], Vol. II, pp. 512 – 514.

or more generally, if V is allowed to be velocity-dependent,

$$Q_j = -\frac{\partial V}{\partial q_j} + \frac{d}{dt}\frac{\partial V}{\partial \dot{q}_j}, \tag{3.51}$$

and then obtain the case considered above, *i.e.*

$$
\begin{aligned}
\delta I &= \delta \int_{t_1}^{t_2} T\, dt - \int_{t_1}^{t_2} \sum_j \left(\frac{\partial V}{\partial q_j}\delta q_j - \frac{d}{dt}\frac{\partial V}{\partial \dot{q}_j}\delta q_j \right) dt \\
&= \delta \int_{t_1}^{t_2} T\, dt - \int_{t_1}^{t_2} \sum_j \left(\frac{\partial V}{\partial q_j}\delta q_j + \frac{\partial V}{\partial \dot{q}_j}\delta \dot{q}_j \right) dt \\
&= \delta \int_{t_1}^{t_2} (T - V)\, dt = \delta \int_{t_1}^{t_2} L\, dt,
\end{aligned}
$$

after the partial integration:

$$
\begin{aligned}
\int_{t_1}^{t_2} \frac{d}{dt}\left(\frac{\partial V}{\partial \dot{q}_j} \right) \delta q_j\, dt &= \sum_j \left[\delta q_j \frac{\partial V}{\partial \dot{q}_j} \right]_{t_1}^{t_2} - \sum_j \int_{t_1}^{t_2} \delta \dot{q}_j \frac{\partial V}{\partial \dot{q}_j}\, dt \\
&= 0 - \sum_j \int_{t_1}^{t_2} \delta \dot{q}_j \frac{\partial V}{\partial \dot{q}_j}\, dt,
\end{aligned}
$$

since $\delta q_j = 0$ for $t = t_1, t_2$. Hence $\delta I = \delta \int L\, dt$ for conservative systems. We now return to *the general case*. Since $T = T(q_j, \dot{q}_j)$, we have

$$\delta \int T\, dt = \int \sum_j \left[\frac{\partial T}{\partial q_j} - \frac{d}{dt}\frac{\partial T}{\partial \dot{q}_j} \right] \delta q_j\, dt, \tag{3.52}$$

and hence the extended Hamilton principle $\delta I = \delta \int (T + W)\, dt = 0$ implies

$$\int_{t_1}^{t_2} \sum_j \left(\frac{\partial T}{\partial q_j} - \frac{d}{dt}\frac{\partial T}{\partial \dot{q}_j} + Q_j \right) \delta q_j\, dt = 0. \tag{3.53}$$

In the case of holonomic constraints the coordinates q_j and hence their variations δq_j are again independent and we obtain

$$\frac{d}{dt}\left(\frac{\partial T}{\partial \dot{q}_j} \right) - \frac{\partial T}{\partial q_j} = Q_j. \tag{3.54}$$

These are the Lagrange equations or *Euler–Lagrange equations* in the case of external forces that cannot be derived from a potential.

3.5.4   Hamilton's principle for nonholonomic systems

We next consider the further extension of Hamilton's principle to nonholonomic systems. We recall that — as just demonstrated — in the case of holonomic constraints the generalized coordinates q_j are independent of one another, and this condition was implemented only in the very last step of the derivation of the Euler–Lagrange equations. Nonholonomic constraints can no longer be cast into a form like $f(q_1, q_2, \ldots, q_n, q', t) = 0$, where q' now represents the degree of freedom to be eliminated. In the case of nonholonomic constraints the q_j are not independent. Here we consider a specific type of nonholonomic constraints, *i.e.* those which can be expressed in terms of differentials dq_j as$^{\|}$

$$\sum_{k=1}^{n} a_{lk} dq_k + a_{lt} dt = 0 \qquad (3.55)$$

(m equations, $l = 1, 2, \ldots, m$). However, since in Hamilton's principle we are concerned with virtual displacements of the system, and not with real ones for which time t is also varied, the constraints are

$$\sum_{k=1}^{n} a_{lk} \delta q_k = 0. \qquad (3.56)$$

We thus have in addition for m undetermined constants λ_l:

$$\sum_l \lambda_l \sum_k a_{lk} \delta q_k = 0 \quad \text{and} \quad \int_{t_1}^{t_2} \sum_{k,l} \lambda_l a_{lk} \delta q_k dt = 0. \qquad (3.57)$$

The (as yet) undetermined constants λ_l are known as *Lagrange multipliers*. Adding this last relation to the relation of Eq. (3.47) (there obtained for conservative systems), we obtain

$$\int_{t_1}^{t_2} dt \sum_{k=1}^{n} \left[\frac{\partial L}{\partial q_k} - \frac{d}{dt} \left(\frac{\partial L}{\partial \dot{q}_k} \right) + \sum_{l=1}^{m} \lambda_l a_{lk} \right] \delta q_k = 0. \qquad (3.58)$$

Here the variations $\delta q_k, k = 1, 2, \ldots, n$, are *not independent*, but are connected as in Eq. (3.56) for $l = 1, 2, \ldots, m$.

We have n $(n > m)$ variations δq_k with m linking constraints. Thus $n-m$ of the variations are independent. We now choose the Lagrange multipliers λ_l such that for

$$k = n - m + 1, n - m + 2, \ldots, n$$

$^{\|}$For integrability the coefficients a_{lk} cannot assume arbitrary values, but must have the form given below in Eq. (3.62).

the following equations hold:

$$\frac{\partial L}{\partial q_k} - \frac{d}{dt}\frac{\partial L}{\partial \dot{q}_k} + \sum_{l=1}^{m} \lambda_l a_{lk} = 0. \tag{3.59}$$

Inserting these equations into Eq. (3.58), we obtain

$$\int_{t_1}^{t_2} dt \sum_{k=1}^{n-m} \left[\frac{\partial L}{\partial q_k} - \frac{d}{dt}\left(\frac{\partial L}{\partial \dot{q}_k}\right) + \sum_{l=1}^{m} \lambda_l a_{lk} \right] \delta q_k = 0. \tag{3.60}$$

This equation involves only the $n - m$ independent variations δq_k. Hence their coefficients vanish and we obtain

$$\frac{\partial L}{\partial q_k} - \frac{d}{dt}\left(\frac{\partial L}{\partial \dot{q}_k}\right) + \sum_{l=1}^{m} \lambda_l a_{lk} = 0 \tag{3.61}$$

for $k = 1, 2, \ldots, n-m$, but together with Eq. (3.59) for $k = 1, 2, \ldots, n$. These equations involve $n + m$ unknowns, i.e. n generalized coordinates q_k and m Lagrange multipliers λ_l. The additional equations required for the solution of the problem are the original constraints, but in differential form

$$\sum_{k=1}^{n} a_{lk}\dot{q}_k + a_{lt} = 0. \tag{3.62}$$

Next we ask: What is the physical meaning of the Lagrange multipliers λ_l? Comparing Eq. (3.61) with Eq. (3.27), i.e.

$$\frac{\partial L}{\partial q_k} - \frac{d}{dt}\left(\frac{\partial L}{\partial \dot{q}_k}\right) + Q'_k = 0, \tag{3.63}$$

where Q'_k are external forces replacing the constraining forces, we conclude that

$$Q'_k = \sum_{l=1}^{m} \lambda_l a_{lk}. \tag{3.64}$$

This means the Lagrange multipliers determine the generalized constraining forces Q'_k.

We note finally that Eq. (3.55) does not represent the most general type of nonholonomic constraints; for instance constraints in the form of inequalities are excluded. However, constraints of the type of Eq. (3.55) include also holonomic ones, since with

$$f(q_1, q_2, \ldots, q_n, t) = 0,$$

also

$$\sum_k \frac{\partial f}{\partial q_k} dq_k + \frac{\partial f}{\partial t} dt = 0 \qquad (3.65)$$

is given. This form of this equation is identical with that of Eq. (3.55), provided the coefficients have the derivative form given here, and the entire expression is then integrable.

3.5.5 The general procedure

Equations (3.58) and (3.60) reveal the procedure to be followed in general: We add to the Lagrange function L every constraint f_l multiplied by a Lagrange multiplier λ_l, and thereby obtain a new Lagrange function L':

$$L' = L(q_i, \dot{q}_i, t) + \sum_l \lambda_l f_l(q_i, t) \equiv L'(q_i, \dot{q}_i, \lambda_l). \qquad (3.66)$$

We now consider the Lagrange multipliers as new variables, that we treat exactly like the original generalized coordinates, so that we obtain additional Lagrange equations for these, *i.e.*

$$\frac{d}{dt}\left(\frac{\partial L'}{\partial \dot{q}_i}\right) - \frac{\partial L'}{\partial q_i} = 0, \qquad \frac{d}{dt}\left(\frac{\partial L'}{\partial \dot{\lambda}_l}\right) - \frac{\partial L'}{\partial \lambda_l} = 0. \qquad (3.67)$$

The first equations imply

$$\frac{d}{dt}\left(\frac{\partial L}{\partial \dot{q}_i}\right) - \frac{\partial L}{\partial q_i} - \sum_l \lambda_l \frac{\partial f_l}{\partial q_i} = 0, \qquad (3.68a)$$

and the second the constraints

$$f_l(q_i, t) = 0. \qquad (3.68b)$$

The latter equations agree with those above if we identify $f_l(q_i, t)$ with $\sum_k a_{lk} q_k$.

Example 3.9: A particle constrained to a circular orbit

Establish the Lagrangian of a particle of mass m which is forced to move on a planar circular orbit.

Solution: Assuming a radius $r = a$ of the orbit, we can write the Lagrangian either as

$$L(x, y, \lambda; \dot{x}, \dot{y}, \dot{\lambda}) = \frac{1}{2}m(\dot{x}^2 + \dot{y}^2) - \lambda(\sqrt{x^2 + y^2} - a),$$

or in the more convenient form (exploiting the symmetry)

$$L(r, \theta, \lambda; \dot{r}, \dot{\theta}, \dot{\lambda}) = \frac{1}{2}m(\dot{r}^2 + r^2\dot{\theta}^2) - \lambda(r - a).$$

One can now derive the equations of motion from

$$\frac{d}{dt}\left(\frac{\partial L}{\partial \dot{q}_i}\right) - \frac{\partial L}{\partial q_i} = 0 \quad \text{with} \quad q_i = r, \theta, \lambda.$$

One obtains:

$$m\ddot{r} - mr\dot{\theta}^2 + \lambda = 0, \quad \frac{d}{dt}(mr^2\dot{\theta}) = 0, \quad r - a = 0.$$

Thus one obtains, since $r = a = $ const., $ma^2\dot{\theta} = l = $ const. (angular momentum) and $\lambda = ma\dot{\theta}^2$ (centripetal force).

Example 3.10: A particle constrained to free motion on a sphere
Establish the Lagrangian of a particle of mass m which is constrained to move freely on the surface of a sphere of radius 1.

Solution: Assuming the radius $r = 1$ of the sphere, we can write the Lagrangian (in quadratic terms summation over $i = 1, 2, 3$ is understood)

$$L(x_i, \dot{x}_i; \lambda, \dot{\lambda}) = \frac{1}{2}m\dot{x}_i^2 + \lambda(x_i^2 - 1). \tag{3.69}$$

The equations of motion are

$$\frac{d}{dt}\left(\frac{\partial L}{\partial \dot{x}_i}\right) - \frac{\partial L}{\partial x_i} = 0, \quad \frac{d}{dt}\left(\frac{\partial L}{\partial \dot{\lambda}}\right) - \frac{\partial L}{\partial \lambda} = 0,$$

i.e.

$$m\ddot{x}_i - 2\lambda x_i = 0, \quad x_i^2 - 1 = 0. \tag{3.70}$$

Taking twice the derivative d/dt of the second of these equations, we obtain in vector notation

$$\mathbf{x} \cdot \dot{\mathbf{x}} = 0, \quad \dot{\mathbf{x}}^2 + \mathbf{x} \cdot \ddot{\mathbf{x}} = 0. \tag{3.71}$$

Scalar multiplication of the first of Eqs. (3.70) by $\mathbf{x}$ yields (since $\mathbf{x}^2 = 1$)

$$m\mathbf{x} \cdot \ddot{\mathbf{x}} - 2\lambda \mathbf{x}^2 = 0, \quad i.e. \quad -m\dot{\mathbf{x}}^2 - 2\lambda = 0,$$

i.e.

$$\lambda = -\frac{1}{2}m\dot{\mathbf{x}}^2, \quad \text{and hence} \quad m\ddot{\mathbf{x}} = -m\dot{\mathbf{x}}^2\mathbf{x}. \tag{3.72}$$

We see therefore that the expression on the right implies a velocity-dependent potential.

3.6 Symmetry Properties and Conservation Laws

The quantity

$$p_j \equiv \frac{\partial L}{\partial \dot{q}_j} \tag{3.73}$$

is called *conjugate momentum* and, in Hamilton's canonical formulation, *canonical momentum*. If the Lagrangian L does not contain a certain coordinate q_j (but possibly $\dot{q}_j$), this coordinate is described as a "*cyclic coordinate*". In the case of such a cyclic q_j the Euler–Lagrange equation

$$\frac{d}{dt}\left(\frac{\partial L}{\partial \dot{q}_j}\right) - \frac{\partial L}{\partial q_j} = 0 \tag{3.74}$$

reduces to

$$\frac{d}{dt}\left(\frac{\partial L}{\partial \dot{q}_j}\right) = 0, \quad i.e. \quad \frac{dp_j}{dt} = 0, \quad \text{or} \quad p_j = \text{const.} \qquad (3.75)$$

We thus obtain a *conservation law*: The conjugate momentum associated with a cyclic coordinate is conserved (*i.e.* a constant in time). Since in many cases $L = T - V$, Eq. (3.74) implies for *conservative systems*, for which V does not depend on velocities,

$$\frac{d}{dt}\left(\frac{\partial L}{\partial \dot{q}_j}\right) + \frac{\partial V}{\partial q_j} = \frac{\partial T}{\partial q_j} = 0, \quad i.e. \quad \dot{p}_j = -\frac{\partial V}{\partial q_j} \equiv Q_j. \qquad (3.76)$$

This equation expresses simply what *Newton's second law* says. If the coordinate q_j is cyclic, we obtain once again

$$\frac{\partial L}{\partial q_j} = -\frac{\partial V}{\partial q_j} = 0, \quad \dot{p}_j = 0, \quad p_j = \text{const.} \qquad (3.77)$$

This is the wellknown law of *conservation of momentum*. Recall once again the condition leading to this conservation law: q_j has to be cyclic, *i.e.* the Lagrangian L must not contain q_j, *i.e.* if we replace q_j by $q_j + \delta q_j$, *i.e.* if we subject the system to an infinitesimal displacement (translation) along q_j, the Lagrangian L remains unchanged, or — as one says — L remains invariant under this transformation. We see therefore, that the law of conservation of (linear) momentum follows from the invariance of the Lagrange function under translations. If q_j is a variable of length (*i.e.* x, y or z), the associated momentum is a linear momentum; if q_j is an angular variable, then the associated momentum is the corresponding angular momentum, and one refers not to a translation but to a rotation, *i.e.* if L remains unchanged under an infinitesimal rotation, this implies that the corresponding angle does not appear in L and the corresponding conjugate momentum is an angular momentum which is conserved, *i.e.* a constant of motion. One can imagine (we shall not prove this in detail) that every *finite* translation or rotation can be considered as a succession of infinitesimal translations or rotations. For this reason it suffices, to consider infinitesimal displacements or transformations when a conservation law is to be established. The function

$$H = T + V \qquad (3.78)$$

is known as the *Hamilton function*. In specific cases, *i.e.* for conservative systems and if the constraints are independent of time (so that $\partial L / \partial t = 0$) this function H is equal to the *total energy*. (Recall that if we had a holonomic constraint like for instance $5x + 3y - t = 0$, then after elimination of x with this, the Lagrange function would contain an explicit time dependence). The total energy is a conserved quantity, which follows from the invariance of L under shifts (*i.e.* translations) in time. These aspects are illustrated in explicit examples below.

3.7 Miscellaneous Examples

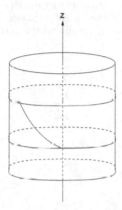

Fig. 3.8 The helix as shortest path connecting two points on a cylinder.

Example 3.11: Shortest line connecting two points on a cylinder

Show that the shortest line connecting two arbitrary points on the surface of a circular cylinder is of the form $r\theta = az + b$, where a and b are constants.

Solution: The cylinder has a circular cross section as indicated in Fig. 3.8. Hence we use cylindrical coordinates r, θ, z, with z measured along the axis of the cylinder. The length of an infinitesimal line element on the surface of the cylinder is thus given by $ds = \sqrt{dz^2 + r^2 d\theta^2}$. The length s of a line connecting points p_1, p_2 on the surface is then given by the integral

$$s = \int_{p_1}^{p_2} ds = \int_{p_1}^{p_2} \sqrt{1 + r^2 \left(\frac{d\theta}{dz}\right)^2}\, dz \equiv \int_{p_1}^{p_2} dz\, L(\theta(z), d\theta(z)/dz).$$

As we know from Hamilton's principle, an extremum of this integral is determined by the solution of the Euler–Lagrange equation

$$\frac{d}{dz}\left(\frac{\partial L}{\partial \dot{\theta}}\right) - \frac{\partial L}{\partial \theta} = 0, \quad \text{with} \quad L = [1 + r^2 \dot{\theta}^2]^{1/2}, \quad \dot{\theta} = \frac{d\theta}{dz}.$$

Since $\partial L/\partial \theta = 0$, we have $\partial L/\partial \dot{\theta} = $ const. Thus

$$(1 + r^2 \dot{\theta}^2)^{-1/2} r^2 \dot{\theta} = \text{const.} = c_1 \quad \text{or} \quad r^4 \dot{\theta}^2 = c_1^2(1 + r^2 \dot{\theta}^2).$$

Since $r = $ const. this implies $\dot{\theta} = $ const. Hence $r\theta = c_0 z + c$, where c_0, c are constants.

With $\theta = 0$ at $z = 0$ and $\theta = \theta_1$ at $z = z_1$, the constants become $c = 0, c_0 = r\theta_1/z_1$. The linear relation between θ and z is then $r\theta = (r\theta_1/z_1)z$. We observe that this is a linear relation like the equation of a straight line in Cartesian coordinates, say like $y = mx + c$. Considering the area $\triangle$ of a triangle on the cylinder, we have[*]

$$\triangle = \int_0^{\triangle} d(\text{area}) = \int dzr\theta = \int dz \frac{r\theta_1}{z_1} z = r\frac{\theta_1}{z_1} \int_0^{z_1} zdz = \frac{1}{2}r\theta_1 z_1.$$

One of the angles of this triangle was chosen to be $\pi/2$. What is the sum of the other two angles? Is this a trivial question? The linear relation $r\theta = (r\theta_1/z_1)z$ implies that these angles ϕ_1, ϕ_2 are given by $\tan\phi_1 = dz/dr\theta = z_1/r\theta_1, \tan\phi_2 = dr\theta/dz = r\theta_1/z_1 = (\tan\phi_1)^{-1}$. With these expressions we obtain

$$\tan(\phi_1 + \phi_2) = \frac{\tan\phi_1 + \tan\phi_2}{1 - \tan\phi_1\tan\phi_2} = \infty, \quad \therefore \ \phi_1 + \phi_2 = \frac{\pi}{2}.$$

Thus the sum of all three angles of the triangle is π, as for a triangle in a plane. Thus the surface of a circular cylinder is planar in the sense that its curvature (*cf.* Sec. 10.6) is zero. This follows from the linearity of the equation above. In the case of a spherical surface the equation of the geodesic is nonlinear in the two angles, and hence the sum of the three angles of a spherical triangle is not π (*cf.* Sec. 10.6).

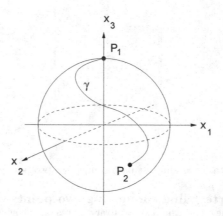

Fig. 3.9 Line connecting 2 points on a sphere.

Example 3.12: Shortest line connecting two points on a sphere
Show that the shortest line connecting two points on a sphere is the arc of a great circle.[†]

Solution: Without loss of generality we assume a sphere of unit radius and two points P_1, P_2 on the sphere. We choose a Cartesian reference frame with origin at the centre and axes x_1, x_2, x_3 as indicated in Fig. 3.9, and select the coordinates of the two points as

$$P_1 = (0, 0, 1), \quad P_2 = (\xi_1, \xi_2, \xi_3) \neq P_1, \quad \sum_{i=1}^{3} \xi_i^2 = 1.$$

The length l of a measurable curve $\gamma(s) = (x_1(s), x_2(s), x_3(s))$ connecting $P_1 = \gamma(0)$ with $P_2 \equiv (x_1(\xi) = \xi_1, x_2(\xi) = \xi_2, x_3(\xi) = \xi_3) = \gamma(\xi)$ is given by

$$l = \int_0^{\xi} \left\| \frac{d\gamma(s)}{ds} \right\| ds. \tag{3.79}$$

---

[*]A triangle here, as also a spherical triangle on a sphere (Chapter 10), is always understood as one whose sides have extremized (*i.e.* shortest) lengths.

[†]The author received the rather rigorous proof given below many years ago from a participant in the course, and does not know the original source. One could also proceed as in Examples 3.7 or 3.11.

As a curve on the surface of the unit sphere, we have the subsidiary condition

$$g(x_1, x_2, x_3) = x_1^2 + x_2^2 + x_3^2 - 1 = 0.$$

We introduce a *Lagrange multiplier* $\lambda(s)$ and extremize the integral, in which the dot now represents differentiation with respect to s,

$$I = \int_0^\xi [\| \dot{\gamma}(s) \| + \lambda(s)g(x_i(s))]ds, \quad \dot{\gamma}(s) = (\dot{x}_1(s), \dot{x}_2(s), \dot{x}_3(s)), \quad \| \dot{\gamma}(s) \| = \sqrt{\sum_j \dot{x}_j^2(s)} \equiv v. \tag{3.80}$$

The variation $\delta I = 0$ yields the following Euler–Lagrange equations:

$$\frac{\partial L}{\partial \lambda} = 0 \Rightarrow g = 0, \quad \frac{d}{ds}\left(\frac{\partial L}{\partial \dot{x}_i}\right) - \frac{\partial L}{\partial x_i} = 0, \quad i = 1, 2, 3, \tag{3.81}$$

where

$$\frac{\partial L}{\partial \dot{x}_i} = \frac{\partial}{\partial \dot{x}_i}\sqrt{\sum_j \dot{x}_j^2(s)} = \frac{\dot{x}_i(s)}{\sqrt{\sum_j \dot{x}_j^2(s)}} = \frac{\dot{x}_i(s)}{v}, \quad \frac{\partial L}{\partial x_i} = \lambda\frac{\partial g}{\partial x_i} = 2\lambda x_i(s).$$

Next we reparametrize $\gamma(s)$ in such a way, that γ is parametrized by its own arc-length, *i.e.* we write (with $l \leftrightarrow s$ bijective)

$$l(s) = \int_0^s \| \dot{\gamma}(\sigma) \| \, d\sigma, \quad \frac{dl}{ds} = \frac{d}{ds}\int_0^s \| \dot{\gamma}(\sigma) \| \, d\sigma = \| \dot{\gamma}(s) \| = v, \quad \frac{dl}{ds} = v, \tag{3.82}$$

and

$$dl = \frac{dl}{ds}ds \longrightarrow \frac{d}{ds} = \frac{dl}{ds}\frac{d}{dl} = v\frac{d}{dl}.$$

We denote unparametrized quantities by an overbar, *i.e.* $\overline{x}_i = x_i(s(l))$ (thus $\overline{x}_i$ is considered as a function of l, and x_i as a function of s). With dashes denoting differentiation with respect to l, we have

$$\overline{x}_i' = \frac{d}{dl}x_i(s(l)) = \frac{dx_i}{ds}\frac{ds}{dl} = \frac{\dot{x}_i(s)}{v}. \tag{3.83}$$

The Euler–Lagrange equation (3.81) now becomes (beginning with $\partial L/\partial \dot{x}_i$)

$$2\overline{\lambda}(l)\overline{x}_i = v\frac{d}{dl}\overline{x}_i' = v\overline{x}_i'', \quad i.e. \quad \overline{x}_i'' = \mu\overline{x}_i, \quad \mu = \frac{2\overline{\lambda}}{v}. \tag{3.84}$$

Equation (3.81) yields $g(\overline{x}_i) = 0, \sum_i \overline{x}_i^2 = 1$, *i.e.*

$$\frac{d^2}{dl^2}\left(\sum_i \overline{x}_i^2\right) = \frac{d}{dl}\left(2\sum_i \overline{x}_i\overline{x}_i'\right) = 2\sum_i (\overline{x}_i'^2 + \overline{x}_i\overline{x}_i'') = 0. \tag{3.85}$$

However (*cf.* Eq. (3.83)),

$$\sum_i \overline{x}_i'^2 = \sum_i \left(\frac{\dot{x}_i}{v}\right)^2 = \frac{1}{v^2}\sum_i \dot{x}_i^2 = \frac{v^2}{v^2} = 1,$$

so that $\sum_i \overline{x}_i\overline{x}_i'' = -1$. From Eq. (3.84) we obtain $\mu\sum_i \overline{x}_i^2 = -1$, which with $\sum_i \overline{x}_i^2 = 1$ implies $\mu = -1$. Hence we have to solve

$$\overline{x}_i'' \stackrel{(3.84)}{=} -\overline{x}_i, \quad \text{implying} \quad \overline{x}_i(l) = A_i\cos l + B_i\sin l, \quad \overline{x}_i'l = -A_i\sin l + B_i\cos l. \tag{3.86}$$

The solution has to satisfy the following conditions: (a) $\sum_i \bar{x}_i^2 = 1$, *i.e.*

$$\left(\sum_i A_i^2\right)\cos^2 l + \left(2\sum_i A_i B_i\right)\cos l \sin l + \left(\sum_i B_i^2\right)\sin^2 l = 1. \qquad (3.87)$$

(b) The next condition to be satisfied is that $\bar{\gamma}(l)$ is parametrized by its own length, *i.e.*

$$l = \int_0^l \| \bar{\gamma}'(\nu) \| \, d\nu \quad \Longrightarrow \quad \| \bar{\gamma}' \| = 1, \quad \therefore \sum_i \bar{x}_i'^2 = 1,$$

$$\left(\sum_i A_i^2\right)\sin^2 l - \left(2\sum_i A_i B_i\right)\cos l \sin l + \left(\sum_i B_i^2\right)^2\cos^2 l = 1. \qquad (3.88)$$

Now Eqs. (3.87) and (3.88) imply

$$\sum_i A_i^2 + \sum_i B_i^2 = 2. \qquad (3.89)$$

(c) Next we have to take into account the initial condition. The trajectory is to start at P_1 where $l = 0$, *i.e.*

$$\bar{x}_1(0) = 0 = A_1, \quad \bar{x}_2(0) = 0 = A_2, \quad \bar{x}_3(0) = 1 = A_3. \qquad (3.90)$$

Equation (3.89) implies for these values: $\sum_i B_i^2 = 1$. In addition Eq. (3.87) implies $\cos^2 l + 2B_3 \cos l \sin l + \sin^2 l = 1$, so that $B_3 = 0$. In view of the relation $B_1^2 + B_2^2 = 1$, only one free parameter α remains with $B_1 = \cos\alpha, B_2 = \sin\alpha$. Thus the calculated trajectory on the sphere is given by

$$\bar{\gamma}(l) = (\cos\alpha \sin l, \sin\alpha \sin l, \cos l). \qquad (3.91)$$

For fixed α the function $\bar{\gamma}$ describes a great circle through the point P_1 with the plane of x_1, x_2 rotated about the x_3-axis through the angle α. This angle α is determined by the condition that $\bar{\gamma}$ has to pass through the point $P_2 = (\xi_1, \xi_2, \xi_3)$. One obtains α and l by expressing ξ_1, ξ_2, ξ_3 in the "polar coordinates" α and l, *i.e.* $\xi_1 = \cos\alpha \sin l, \xi_2 = \sin\alpha \sin l, \xi_3 = \cos l$. The "spherical coordinate" l is the spherical distance from P_1 to P_2, and hence the arc $\bar{\gamma}$ of the great circle connecting P_1 and P_2. If $P_2 = (0, 0, -1)$, α is indeterminate. Every circle connecting opposite points is a great circle. The solution (3.91) is unique for $l \leq \pi$ (the maximum separation of two points on the sphere). If $l \geq \pi$, there is a second solution $(\sin\alpha \to -\sin\alpha, \cos\alpha \to -\cos\alpha)$ which describes the complementary great circle.

Example 3.13: Kepler problem from solution of Einstein's equation
In Einstein's theory of relativity the so-called *Schwarzschild solution* of the Einstein equation leads to the line element ds with*

$$(ds)^2 = \left(1 - \frac{2l}{r}\right)c^2 (dt)^2 - \frac{(dr)^2}{(1 - 2l/r)} - r^2[(d\theta)^2 + \sin^2\theta(d\phi)^2], \qquad (3.92)$$

where r, θ, ϕ are polar coordinates and t stands for time, c is the velocity of light, and $l = $ const. Equation (3.92) will be derived in Chapter 16 (*cf.* Eq. (16.100)), and l will be identified as $l = GM/c^2$, where G is Newton's gravitational constant and M the mass of the nonrotating spherical mass which is the source of the gravitational field. Derive the Euler–Lagrange equations resulting from the variation $\delta \int ds = 0$, show that $\theta = \pi/2$ is a solution and that the motion therefore is planar, and exploiting the fact that $(ds)^2/(ds)^2 = 1$, and eliminating θ and ϕ obtain the equation in $u = 1/r$. What is the meaning of the equation?

Solution: In the following derivatives with respect to s are denoted by an overdot.

---

*Note that the usual notation in the literature is not $(ds)^2$ etc. as given here for clarity, but ds^2 etc.

We define a Lagrangian $L = 1$ (numerical value) as

$$L(t, \theta, \phi, r, \dot{t}, \dot{\theta}, \dot{\phi}, \dot{r}) = \left(1 - \frac{2l}{r}\right)c^2\dot{t}^2 - \frac{\dot{r}^2}{(1 - 2l/r)} - r^2[\dot{\theta}^2 + \sin^2\theta\dot{\phi}^2], \qquad (3.93)$$

and consider $\delta I = \delta \int ds L = 0$. We consider one after the other the Euler–Lagrange equations in t, θ, ϕ, r. Thus considering $t(s)$ we obtain

$$\frac{\partial L}{\partial \dot{t}} = 2\left(1 - \frac{2l}{r}\right)c^2\dot{t}, \qquad \therefore \frac{d}{ds}\left[\left(1 - \frac{2l}{r}\right)\dot{t}\right] = 0. \qquad (3.94)$$

In the case of θ we obtain:

$$\frac{\partial L}{\partial \dot{\theta}} = -2r^2\dot{\theta}, \qquad \frac{d}{ds}\left(\frac{\partial L}{\partial \dot{\theta}}\right) = -2r^2\ddot{\theta} - 4r\dot{\theta}\dot{r}, \qquad \frac{\partial L}{\partial \theta} = -2r^2\sin\theta\cos\theta\dot{\phi}^2,$$

$$i.e. \quad -2r^2\ddot{\theta} - 4r\dot{\theta}\dot{r} + 2r^2\sin\theta\cos\theta\dot{\phi}^2 = 0, \qquad \therefore \frac{d}{ds}(r^2\dot{\theta}) = r^2\sin\theta\cos\theta\dot{\phi}^2. \qquad (3.95)$$

We observe already here that $\theta = \pi/2$ solves this equation. In the case of ϕ we obtain

$$\frac{\partial L}{\partial \dot{\phi}} = -2r^2\sin^2\theta\dot{\phi}, \qquad \therefore \frac{d}{ds}(r^2\sin^2\theta\dot{\phi}) = 0. \qquad (3.96)$$

Finally we come to the case of r. We obtain:

$$\frac{\partial L}{\partial \dot{r}} = -2\frac{\dot{r}}{(1 - 2l/r)}, \qquad \frac{\partial L}{\partial r} = \frac{2l}{r^2}c^2\dot{t}^2 + \frac{2l\dot{r}^2}{r^2(1 - 2l/r)^2} - 2r(\dot{\theta}^2 + \sin^2\theta\dot{\phi}^2),$$

$$\therefore \frac{d}{ds}\left[-\frac{2\dot{r}}{(1 - 2l/r)}\right] + 2r(\dot{\theta}^2 + \sin^2\theta\dot{\phi}^2) - \frac{2l}{r^2}c^2\dot{t}^2 - \frac{2l\dot{r}^2}{r^2(1 - 2l/r)^2} = 0. \qquad (3.97)$$

In the following we use instead of this equation the line element, *i.e.* for $\theta = \pi/2$ the equation:

$$1 = \left(1 - \frac{2l}{r}\right)c^2\dot{t}^2 - \frac{\dot{r}^2}{(1 - 2l/r)} - r^2\dot{\phi}^2. \qquad (3.98)$$

From Eq. (3.96) we obtain $r^2\dot{\phi} = \text{const.} = hc$ (note that hc has the meaning of angular momentum per unit mass and is therefore a conserved quantity), and from Eq. (3.94): $(1 - 2l/r)\dot{t} = \text{const.} = kc$ (this is the second constant we expect, and, in fact, kc is the relativistic energy per unit mass which, however, is not quite so easy to see,[†] but note it is related to the zero component of momentum $\dot{t}$). Inserting these into Eq. (3.98) we obtain

$$1 = \left(1 - \frac{2l}{r}\right)^{-1}[c^2k^2 - \dot{r}^2] - \frac{h^2}{r^2}.$$

We now use the following replacements (note carefully that the prime denotes differentiation with respect to the angle ϕ, whereas the dot denotes differentiation with respect to s):

$$r' = \frac{\dot{r}}{\dot{\phi}} = \frac{dr}{d\phi}, \qquad \dot{r} = r'\dot{\phi} = r'\frac{h}{r^2}, \qquad \therefore \left(1 - \frac{2l}{r}\right) = c^2k^2 - \frac{h^2 r'^2}{r^4} - \frac{h^2}{r^2}\left(1 - \frac{2l}{r}\right).$$

Next we set $u = 1/r, r = 1/u$. Then $r' = dr/d\phi = (dr/du)(du/d\phi) = -u'/u^2, u' = du/d\phi$. Hence

$$\dot{r} = \frac{r'h}{r^2} = -u'h,$$

[†]See D. Raine and E. Thomas [40], p. 20.

and

$$(1 - 2lu) = c^2 k^2 - h^2 u'^2 - h^2 u^2 (1 - 2lu), \quad i.e. \quad u'^2 = \frac{c^2 k^2 - 1}{h^2} + \frac{2l}{h^2} u - u^2 + 2lu^3.$$

Differentiating once again we obtain:

$$2u'u'' = \frac{2l}{h^2} u' - 2uu' + 6lu^2 u', \quad \therefore \quad u'\left(u'' - \frac{l}{h^2} + u - 3lu^2\right) = 0.$$

One solution of this equation is $u' = 0$, *i.e.* $r = $ const. implying a circular orbit. The other solution is given by the equation

$$u'' + u = \frac{l}{h^2} + 3lu^2. \tag{3.99}$$

We shall see in Chapter 7 that this is the equation which describes the classical Kepler problem, *cf.* Eq. (7.93), the equation without the term in u^2 being that obtained from Newton's gravitational force, whereas the term in u^2 is the correction given by Einstein's general theory of relativity, as we shall see in Chapter 16.

Example 3.14: The simple pendulum
Establish the Euler–Lagrange equation of motion of a simple pendulum[‡] consisting of a mass point attached to a weightless rod. Then obtain the period of the pendulum whose deflections θ are too large to allow the approximation $\sin\theta \simeq \theta$. Assume that the angular velocity $d\theta/dt$ vanishes at a position $\theta = \alpha$. (The answer involves an elliptic integral).

Solution: We let m be the mass of the pendulum bob, l the length of the rod, and θ its angular deflection from the vertical at time t. The Lagrange function is[§]

$$L = \text{kinetic energy } T - \text{potential energy } V = \frac{1}{2}ml^2\dot\theta^2 - (-mgl\cos\theta). \tag{3.100}$$

The equation of motion is obtained accordingly (with $q_j = \theta$) as

$$\frac{d}{dt}\left(\frac{\partial L}{\partial \dot q_j}\right) - \frac{\partial L}{\partial q_j} = 0, \quad ml^2\ddot\theta + mgl\sin\theta = 0, \quad \ddot\theta = -\frac{g}{l}\sin\theta. \tag{3.101}$$

With $\sin\theta \simeq \theta$ (*i.e.* θ small) this equation has the form of that of the well known harmonic oscillator with angular frequency $\omega = \sqrt{g/l}$. We now assume that this approximation is not permissible. Using the relation

$$\ddot\theta = \frac{d}{d\theta}\left(\frac{1}{2}\dot\theta^2\right) \quad \left[= \frac{d}{dt}\left(\frac{1}{2}\dot\theta^2\right)\frac{1}{\dot\theta}\right],$$

the equation becomes (with $\theta = \alpha$ when $\dot\theta = 0$)

$$\dot\theta^2 = \frac{2g}{l}(\cos\theta - \cos\alpha). \tag{3.102}$$

A further integration yields, assuming $\theta = 0$ at time $t = 0$ and $0 \le \theta_1 \le \alpha$,

$$t = \sqrt{\frac{l}{2g}} \int_0^{\theta_1} \frac{d\theta}{\sqrt{\cos\theta - \cos\alpha}}.$$

[‡]The word 'simple' actually refers to 'simple harmonic motion', and hence to the case of $\sin\theta$ approximated by θ.

[§]With $q = h\sin\theta, h \equiv l$, the Lagrange function is $L = mh^2\dot q^2/2(h^2 - q^2) + mg\sqrt{h^2 - q^2}$. This is the form given by R. Penrose [38], p. 480. Thus if we link $(h^2 - q^2)$ with the mass m, this may be looked at as the case of an effective q-dependent mass.

Using the relation $\cos\theta = 1 - 2\sin^2(\theta/2)$, the integral can be written

$$t = \sqrt{\frac{l}{2g}} \int_0^{\theta_1} \frac{d\theta}{\sqrt{2(\sin^2\alpha/2 - \sin^2\theta/2)}}.$$

We define an angle ϕ by the relation

$$\sin\frac{\theta}{2} = \sin\frac{\alpha}{2}\sin\phi, \quad \frac{1}{2}d\theta\cos\frac{\theta}{2} - \sin\frac{\alpha}{2}\cos\phi d\phi.$$

Now,

$$d\theta = \frac{2\sin(\alpha/2)\cos\phi d\phi}{\cos(\theta/2)} = \frac{2\sin(\alpha/2)\cos\phi d\phi}{\sqrt{1-\sin^2(\alpha/2)\sin^2\phi}}.$$

It then follows that

$$t = \sqrt{\frac{l}{2g}} \int_0^{\phi_1} \frac{2\sin(\alpha/2)\cos\phi d\phi}{\sqrt{2(\sin^2(\alpha/2) - \sin^2(\alpha/2)\sin^2\phi)}} \frac{1}{\sqrt{1-\sin^2(\alpha/2)\sin^2\phi}}$$

$$= \sqrt{\frac{l}{g}} \int_0^{\phi_1} \frac{d\phi}{\sqrt{1-\sin^2(\alpha/2)\sin^2\phi}}.$$

Let t be the time taken for one quarter of a revolution of the pendulum. Then $\theta_1 = \alpha$, and hence (since $\sin(\theta/2) = \sin(\alpha/2)\sin\phi$) we have $\phi = \pi/2$. The total period of the pendulum is therefore

$$T = 4\sqrt{\frac{l}{g}} \int_0^{\pi/2} \frac{d\phi}{\sqrt{1-k^2\sin^2\phi}} \equiv 4\sqrt{\frac{l}{g}}K(k^2), \quad \text{with} \quad k^2 \equiv \sin^2(\alpha/2),\ 0 \le k^2 \le 1. \quad (3.103)$$

The integral is an elliptic integral, and the function $K(k^2)$ is known as the *complete elliptic integral of the first kind*, and the parameter k is called *elliptic modulus*.[¶] We can expand the denominator of the integrand in rising powers of k^2, and then evaluate the resulting integrals with the help of the formula:

$$\int_0^{\pi/2} \sin^n\theta d\theta = \int_0^{\pi/2} \cos^n\theta d\theta = \frac{(n-1)(n-3)\cdots 2\ \text{or}\ 1}{n(n-2)\cdots 2\ \text{or}\ 1}.$$

Then

$$T = 4\sqrt{\frac{l}{g}} \int_0^{\pi/2} d\phi\left(1 + \frac{1}{2}k^2\sin^2\phi + \frac{3}{8}k^4\sin^2\phi + \cdots\right) = 2\pi\sqrt{\frac{l}{g}}\left(1 + \frac{1}{4}k^2 + \frac{9}{64}k^4 + \cdots\right).$$

Thus the period of this "ordinary" pendulum is not strictly independent of the amplitude (*i.e.* α). This observation led Huygens, in his efforts to construct accurate clocks, to seek a mechanism which eliminates this defect. In his search Huygens then recognized the cycloid as the required trajectory of oscillations. See Example 3.15 on the cycloid and Example 3.21 on the cycloidal pendulum.

Example 3.15: The cycloid, path of shortest time under gravity

Determine the curve connecting two points 1 and 2, along which a particle falls in the shortest time under gravity. The particle is to fall from rest and falls from a higher to a lower point. Show that the desired curve is a *cycloid* with a *cusp* at its initial point. Show also that there is a cusp if the particle is given an initial kinetic energy $mv_0^2/2$ given by $v_0^2 = 2gz$.[‖]

[¶] See, for instance, L. M. Milne–Thomson [30], p. 38. The treatment of the pendulum in terms of elliptic functions was given by L. M. Milne–Thomson [31]. The reader who is interested in the complete treatment of this problem in terms of elliptic functions can find this also in the *Handbook of Elliptic Integrals* of P. F. Byrd and M. D. Friedman [9], p. 7.

[‖] This is problem 3 of Chapter II of H. Goldstein [19], given there only with some introductory comments in the text. Valuable discussion of the cycloid and how it arises can be found in the treatise of R. Courant [12], Vol. 1, p. 261.

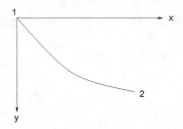

Fig. 3.10 A particle falling from 1 to 2.

Solution: The particle of mass m is to move from a position 1 to a position 2 with an initial velocity $v_0 = 0$, as indicated in Fig. 3.10. If the particle moves in the direction of the gravitational force, it is said to 'fall', if it is moved in the opposite direction, it is said to be 'lifted' (thus if a particle is allowed to fall from a certain point and is lifted back to the same point, the total work done is zero). We choose the origin O of coordinates such that at position 1 both the potential energy V and the kinetic energy T are zero. Then the total energy (of this conservative system with total energy $T + V$) is zero. During its motion, we have (note: weight mg is a conservative force and $mg \times$ distance is the work done by this force, and here the negative of this work is called potential)

$$V = -mgy, \quad T = \frac{1}{2}mv^2.$$

Since $T + V = 0$, it follows that $mgy = mv^2/2$, or $v = \sqrt{2gy}$. The time needed by the particle to move through the infinitesimal distance $ds = \sqrt{dx^2 + dy^2}$ is $dt = ds/v$. Hence we obtain

$$t_{12} \equiv \int_1^2 dt = \int_1^2 \frac{ds}{v} = \int_1^2 \frac{\sqrt{dx^2 + dy^2}}{\sqrt{2gy}} = \int_1^2 \frac{1}{\sqrt{2gy}} \sqrt{1 + \left(\frac{dx}{dy}\right)^2}\, dy. \qquad (3.104)$$

We set $\dot{x} = dx/dy$ and

$$f(x, \dot{x}, y) = \left[\frac{1 + \dot{x}^2}{2gy}\right]^{1/2}. \qquad (3.105)$$

Note that we could proceed in another way by extracting dx from the square root and then integrating over x; we return to this point below. As shown in the preceding text, an extremum of the integral is obtained with the Euler–Lagrange equation

$$\frac{d}{dy}\left(\frac{\partial f}{\partial \dot{x}}\right) - \frac{\partial f}{\partial x} = 0.$$

From Eq. (3.105) we obtain therefore for the path of shortest time

$$\frac{\partial f}{\partial x} = 0, \quad \text{implying} \quad \frac{d}{dy}\left(\frac{\partial f}{\partial \dot{x}}\right) = 0, \quad \frac{\partial f}{\partial \dot{x}} = \frac{\dot{x}}{\sqrt{2gy(1 + \dot{x}^2)}} = \text{const.} \equiv \frac{1}{2\alpha}. \qquad (3.106)$$

It follows that, setting $g/2\alpha^2 = 1/2\beta$,

$$\dot{x}^2 = \frac{2g}{4\alpha^2}y(1 + \dot{x}^2), \quad \dot{x}^2 - \frac{1}{2\beta}y\dot{x}^2 = \frac{1}{2\beta}y, \quad \dot{x}^2 = \frac{y}{2\beta(1 - y/2\beta)}.$$

Thus we obtain (putting the $\pm$ sign of the square root on the left hand side)

$$\pm\dot{x} = \sqrt{\frac{y}{2\beta - y}} = \frac{y}{\sqrt{2\beta y - y^2}}, \quad \pm x = \int \frac{y\, dy}{\sqrt{2\beta y - y^2}}. \qquad (3.107)$$

The integral can be evaluated with the help of Tables of Integrals** to give the equation of the requested curve

$$x = -\sqrt{y(2\beta - y)} + 2\beta \sin^{-1}(\sqrt{y/2\beta}).$$

The same integral (3.107) for x is obtained, if instead of Eqs. (3.104), (3.105), we had written $dy/dx \equiv \dot{y}$ and

$$l_{12} = \int_1^2 \frac{1}{\sqrt{2gy}}\sqrt{1+\dot{y}^2}dx, \quad g(y,\dot{y}) = \left[\frac{1+\dot{y}^2}{2gy}\right]^{1/2} \quad \text{with} \quad \frac{d}{dx}\left(\frac{\partial g}{\partial \dot{y}}\right) - \frac{\partial g}{\partial y} = 0.$$

Here the algebra is a little more involved. We have

$$\frac{\partial g}{\partial y} = -\frac{1}{2y^{3/2}}\sqrt{\frac{1+\dot{y}^2}{2g}}, \quad \frac{d}{dx}\left(\frac{\dot{y}}{\sqrt{2gy(1+\dot{y}^2)}}\right) + \frac{1}{2y^{3/2}}\sqrt{\frac{1+\dot{y}^2}{2g}} = 0.$$

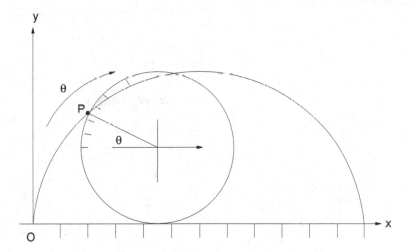

Fig. 3.11 The cycloid traced by point P on the rolling circle.

Differentiation and rearrangement of terms then yields (we omit these steps here)

$$\ddot{y} + \frac{1}{2y}(1+\dot{y}^2) = 0, \quad i.e. \quad \ddot{y} = \frac{d}{dy}\left(\frac{1}{2}\dot{y}^2\right) = -\frac{1}{2y}(1+\dot{y}^2),$$

and on integration

$$\int \frac{d\dot{y}^2}{1+\dot{y}^2} = -\int \frac{dy}{y}, \quad \ln(1+\dot{y}^2) \propto -\ln y, \quad \dot{y}^2 = \frac{\text{const.}}{y} - 1,$$

from which with const. $= 2\beta$ we obtain the result (3.107):

$$\pm x = \int \frac{\sqrt{y}dy}{\sqrt{2\beta - y}} = \int \frac{ydy}{\sqrt{2\beta y - y^2}}.$$

**H. B. Dwight [13], formula 322.01, p. 63.

We now set $y = \beta(1 \pm \cos\theta)$, $dy = \mp\beta\sin\theta d\theta$. Then

$$
\begin{aligned}
\pm x &= \int \frac{\mp\beta^2(1\pm\cos\theta)\sin\theta d\theta}{\sqrt{2\beta^2(1\pm\cos\theta) - \beta^2(1\pm\cos\theta)^2}} = \mp\int \frac{\beta^2(1\pm\cos\theta)\sin\theta d\theta}{\sqrt{\beta^2 - \beta^2\cos^2\theta}} \\
&= \mp\int \frac{\beta(1\pm\cos\theta)\sin\theta d\theta}{\sqrt{1-\cos^2\theta}} = \mp\beta(\theta\pm\sin\theta) + \text{const.}
\end{aligned}
$$

For $x = 0$ when $\theta = 0$, the constant vanishes. The equations

$$
\pm x_\pm = \mp\beta(\theta\pm\sin\theta), \quad y_\pm = \beta(1\pm\cos\theta), \tag{3.108}
$$

are the parameter-θ equations of the curve known as a *cycloid*. This is the curve or trajectory traced by a point fixed on the surface of a circle as the circle rolls along a straight line. This is illustrated in Fig. 3.11 which shows the downward curve of Fig. 3.10 rotated about the x-axis. At $\theta = 0, \pm 2\pi, \ldots$, $y_- = 0$ (y_+ displaced by 2β) and therefore there $\dot{x}^2 = 0$ (*cf.* Eq. (3.107)) and this means, there $\dot{x} = dx/dy = 0$, *i.e.* gradient $dy/dx = \infty$. Hence this cycloid meets the x-axis at right angles forming a cusp.[††]

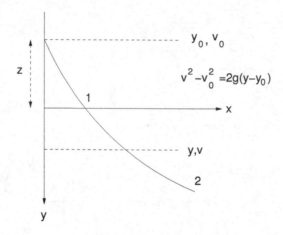

Fig. 3.12 The distance z.

We now assume the particle is given an initial velocity v_0 at a height y_0. The law of conservation of energy E then demands that (*cf.* Fig. 3.12)

$$
E = \frac{1}{2}mv^2 - mgy = \frac{1}{2}mv_0^2 - mgy_0.
$$

Then

$$
v = \sqrt{\frac{2}{m}(E + mgy)} = \sqrt{2g(y - y_0) + v_0^2}.
$$

The time the particle now requires to move through a distance ds is (*cf.* Fig. 3.12)

$$
dt = \frac{ds}{v} = \frac{\sqrt{dx^2 + dy^2}}{v} = \frac{\sqrt{1 + \dot{x}^2}}{\sqrt{2g(y - y_0) + v_0^2}} dy.
$$

[††]For discussion see R. Courant [12], Vol. 1, pp. 260 - 261, 287, 302 - 304.

Proceeding as before we have (this differs from Eq. (3.105) in having $y_0 \neq 0, v_0 \neq 0$)

$$f(x, \dot{x}, y) = \left[\frac{1 + \dot{x}^2}{2g(y - y_0) + v_0^2} \right]^{1/2},$$

and (*cf.* Eq. (3.107))

$$\dot{x}^2 = \frac{[2g(y - y_0) + v_0^2](1 + \dot{x}^2)}{4\alpha^2} \quad \text{or} \quad \dot{x}^2 = \frac{2g(y - y_0) + v_0^2}{4\alpha^2 - [2g(y - y_0) + v_0^2]}.$$

This expression vanishes at y given by $y = y_0 - v_0^2/2g$, *i.e.* at y_0 for $v_0 = 0$.

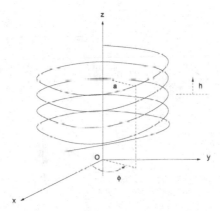

Fig. 3.13 A bead sliding down a frictionless spring.

Example 3.16: A bead sliding down a frictionless spring

A bead of mass m is under the influence of gravity sliding down a frictionless vertically held spring of radius a and element separation h. Establish the equations of motion using cylindrical coordinates and examine their solution.

Solution: In cylindrical coordinates r, ϕ, z the kinetic energy T of the bead is $T = (m/2)(\dot{r}^2 + r^2\dot{\phi}^2 + \dot{z}^2)$, and its potential energy $V = -mgz$, as in Fig. 3.13. The constraints are $r - a = 0, z - b\phi = 0$, where $b = \text{const}$. Since $\delta z = h$ for $\delta \phi = 2\pi$, we have $b = h/2\pi$. Thus we have $z - h\phi/2\pi = 0$. With the help of this constraint we can eliminate ϕ from T and obtain the Lagrangian:

$$L = T - V = \frac{1}{2}m\left[\dot{r}^2 + \dot{z}^2\left(1 + \frac{4\pi^2}{h^2}a^2\right) \right] + mgz.$$

Only two of the three coordinates remain and we obtain for r and z the equations:

$$\frac{d}{dt}(m\dot{r}) = 0, \quad \frac{d}{dt}\left[m\dot{z}\left(1 + \frac{4\pi^2}{h^2}a^2\right) \right] - mg = 0.$$

It follows that $\dot{r} = \text{const.}$, *i.e.* r is a cyclic variable. Since $r = a$, $\dot{r} = 0$, *i.e.* const. $= 0$. From the second equation we obtain

$$m\ddot{z}\left(1 + \frac{4\pi^2}{h^2}a^2\right) = mg, \quad \text{or} \quad \dot{z} = \frac{gt}{1 + 4\pi^2 a^2/h^2} + \text{const.}$$

With the initial condition $\dot{z} = 0, z = 0$ at $t = 0$, we obtain

$$z = \frac{gt^2}{2[1 + 4\pi^2 a^2/h^2]}.$$

Example 3.17: The spherical pendulum

Consider the equations of motion of a spherical pendulum* and integrate these in dominant approximation for (a) small deflections, and (b) for small deviations α of the pendulum bob from the circular orbit $\theta = \theta_0 = \text{const.}$ which is possible at a given angular momentum.

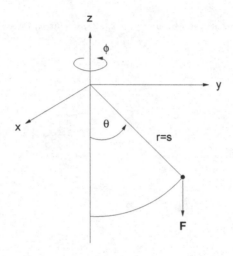

Fig. 3.14   The spherical pendulum.

Solution: The spherical coordinates are r, θ, ϕ. The constraint here is $r = s = \text{const.}$ We have, *cf.* Fig. 3.14,

$$\mathbf{r} = (x, y, -z) = (r \sin\theta \cos\phi, r \sin\theta \sin\phi, -r \cos\theta).$$

The force acting is the gravitational force $\mathbf{F} = -mg\mathbf{z}/z$. The generalized forces given by Eq. (3.14) are therefore given by

$$Q_r = \mathbf{F} \cdot \frac{\partial \mathbf{r}}{\partial r} = mg\cos\theta, \quad Q_\theta = \mathbf{F} \cdot \frac{\partial \mathbf{r}}{\partial \theta} = -mgr\sin\theta, \quad Q_\phi = \mathbf{F} \cdot \frac{\partial \mathbf{r}}{\partial \phi} = 0. \tag{3.109a}$$

From $Q_\phi = 0$ we deduce that $l = l_z = \text{const.}$ (principle of angular momentum), where l_z is the z-component of the angular momentum $\mathbf{l}$. Differentiating the coordinate vector $\mathbf{r}$ we obtain

$$\mathbf{v} = \dot{\mathbf{r}} = (-s\dot{\phi}\sin\theta\sin\phi + s\dot{\theta}\cos\theta\cos\phi, \; s\dot{\theta}\cos\theta\sin\phi + s\dot{\phi}\sin\theta\cos\phi, \; s\dot{\theta}\sin\theta),$$

so that

$$
\begin{aligned}
l &= m(\mathbf{r} \times \mathbf{v})_z = m(xv_y - yv_x) \\
&= m[s\sin\theta\cos\phi(s\dot{\theta}\cos\theta\sin\phi + s\dot{\phi}\sin\theta\cos\phi) \\
&\quad - s\sin\theta\sin\phi(-s\dot{\phi}\sin\theta\sin\phi + s\dot{\theta}\cos\theta\cos\phi)] \\
&= ms^2\sin^2\theta\dot{\phi}, \quad \text{so that} \quad \dot{\phi} = \frac{l}{ms^2\sin^2\theta}. \tag{3.109b}
\end{aligned}
$$

---

*A mass point attached to a rigid massless rod and constrained to move on a spherical surface; *cf.* problem 7, Chapter I of H. Goldstein [19]. See also *The Physics Coaching Class* [48], problem 2014, p. 489.

We obtain the second equation from the Euler–Lagrange equation. Since

$$V = mgs(1 - \cos\theta) \quad \text{and} \quad T = \frac{1}{2}m\mathbf{v}^2 = \frac{1}{2}ms^2\dot{\theta}^2 + \frac{1}{2}ms^2\sin^2\theta\dot{\phi}^2,$$

we have

$$L = T - V = \frac{1}{2}ms^2\dot{\theta}^2 + \frac{1}{2}ms^2\sin^2\theta\dot{\phi}^2 - mgs(1 - \cos\theta), \tag{3.110}$$

and hence

$$\frac{\partial L}{\partial\dot{\theta}} = ms\dot{\theta}s^2, \quad \frac{d}{dt}\left(\frac{\partial L}{\partial\dot{\theta}}\right) = ms^2\ddot{\theta}, \quad \frac{\partial L}{\partial\theta} = ms^2\sin\theta\cos\theta\dot{\phi}^2 - mgs\sin\theta.$$

The Euler–Lagrange equation

$$\frac{d}{dt}\left(\frac{\partial L}{\partial\dot{\theta}}\right) - \frac{\partial L}{\partial\theta} = 0 \quad \text{implies} \quad \ddot{\theta} - \frac{l^2\cos\theta}{m^2s^4\sin^3\theta} + \frac{g}{s}\sin\theta = 0. \tag{3.111}$$

(a) We now consider the case of θ small with $\sin\theta \approx \theta$ and $\cos\theta \approx 1 - (\theta^2/2)$. With these approximations Eq. (3.111) becomes

$$\ddot{\theta} - \frac{l^2}{m^2s^4}\frac{1}{\theta^3} + \frac{g}{s}\theta = 0. \tag{3.112}$$

We set $u = \dot{\theta}^2/2$. Then

$$\ddot{\theta} - \frac{du}{d\theta} = \frac{l^2}{m^2s^4}\frac{1}{\theta^3} - \frac{g}{s}\theta.$$

Integration yields

$$u + c_1 = -\frac{1}{2}\frac{l^2}{m^2s^4}\frac{1}{\theta^2} - \frac{1}{2}\frac{g}{s}\theta^2,$$

where c_1 is a constant. The equation can also be written

$$\dot{\theta}^2 + c = -\frac{l^2}{m^2s^4}\frac{1}{\theta^2} - \frac{g}{s}\theta^2, \quad \text{where} \quad c = 2c_1. \tag{3.113}$$

We now let $\dot{\theta}(t = 0) = 0$ and $\theta(t = 0) = \theta_0$. Then

$$c = -\frac{l^2}{m^2s^4}\frac{1}{\theta_0^2} - \frac{g}{s}\theta_0^2 \quad \text{and} \quad \left(\frac{d\theta}{dt}\right)^2 = -\frac{l^2}{m^2s^4}\left(\frac{1}{\theta^2} - \frac{1}{\theta_0^2}\right) - \frac{g}{s}(\theta^2 - \theta_0^2). \tag{3.114}$$

For more transparency we continue in the following with c. From Eq. (3.113) we obtain

$$\dot{\theta} = \left[-\frac{l^2}{m^2s^4}\frac{1}{\theta^2} - \frac{g}{s}\theta^2 - c\right]^{1/2},$$

$$\int dt + c' = \int \frac{d\theta}{\sqrt{-l^2/m^2s^4\theta^2 - g\theta^2/s - c}} = \int \frac{\theta d\theta}{\sqrt{g/s}\sqrt{-l^2/m^2s^3g - \theta^4 - sc\theta^2/g}}.$$

We set $v = \theta^2, dv = 2\theta d\theta$. Then, with c' a constant and using Tables of Integrals,[†]

$$t + c' = \frac{1}{2}\sqrt{\frac{s}{g}}\int \frac{dv}{\sqrt{-l^2/m^2s^3g - v^2 - scv/g}} = -\frac{1}{2}\sqrt{\frac{s}{g}}\sin^{-1}\left[\frac{-2v + (-cs/g)}{\sqrt{-4l^2/m^2s^3g + c^2s^2/g^2}}\right].$$

$$\tag{3.115}$$

[†]H. B. Dwight [13], formula 380.001, p. 70.

Since at $t = 0, \theta(t = 0) = \theta_0$, we obtain

$$c' - \frac{1}{2}\sqrt{\frac{s}{g}} \sin^{-1}\left[\frac{2\theta_0^2 + cs/g}{\sqrt{c^2 s^2/g^2 - 4l^2/m^2 s^3 g}}\right]. \qquad (3.116)$$

In the following we continue with c'. We have, from Eq. (3.115),

$$\sin\left(2\sqrt{\frac{g}{s}}(t + c')\right) = \frac{2v + cs/g}{\sqrt{-4l^2/m^2 s^3 g + c^2 s^2/g^2}},$$

$$v = \frac{1}{2}\left[\sqrt{-\frac{4l^2}{m^2 s^3 g} + \frac{c^2 s^2}{g^2}}\sin\left(2\sqrt{\frac{g}{s}}(t + c')\right) - \frac{cs}{g}\right].$$

Since $v = \theta^2$, we have $\theta^2 = A\sin(Bt + \tilde{c}) + c^*, c^* > 0$, and

$$\theta(t) = \sqrt{A\sin(Bt + \tilde{c}) + c^*}. \qquad (3.117)$$

With Eq. (3.109b), *i.e.* $\dot{\phi} = l/ms^2 \sin^2\theta$, we obtain therefore

$$\dot{\phi} = \frac{l}{ms^2 \sin^2(\sqrt{A\sin(Bt + \tilde{c}) + c^*})}. \qquad (3.118)$$

From this equation we obtain by integration with respect to t the solution $\phi(t)$.
(b) The case $\theta = \theta_0 + \alpha, \theta_0 = $ const. The starting point is Eq. (3.111). In the present case we have

$$\sin\theta = \sin(\theta_0 + \alpha) \approx \sin\theta_0 + \alpha\cos\theta_0, \quad \cos\theta = \cos(\theta_0 + \alpha) \approx \cos\theta_0 - \alpha\sin\theta_0.$$

For a given $l = l_0 = $ const. and α small we can now approximate Eq. (3.111) by

$$\ddot{\alpha} - \frac{l_0^2(\cos\theta_0 - \alpha\sin\theta_0)}{m^2 s^4 (\sin\theta_0 + \alpha\cos\theta_0)^3} + \frac{g}{s}(\sin\theta_0 + \alpha\cos\theta_0) = 0,$$

or

$$\ddot{\alpha} - \frac{l_0^2(\cos\theta_0 - \alpha\sin\theta_0)}{m^2 s^4 \sin^3\theta_0}(1 - 3\alpha\cot\theta_0) + \frac{g}{s}(\sin\theta_0 + \alpha\cos\theta_0) \simeq 0,$$

$$\ddot{\alpha} + n^2\alpha + \alpha_0, \quad \alpha_0 = \frac{g}{s}\sin\theta_0 - \frac{l_0^2\cos\theta_0}{m^2 s^4 \sin^3\theta_0},$$

$$n^2 = \frac{g}{s}\cos\theta_0 + \frac{l_0^2}{m^2 s^4 \sin^2\theta_0} + \frac{3l_0^2\cot^2\theta_0}{m^2 s^4 \sin^2\theta_0}. \qquad (3.119)$$

Setting $R = n^2\alpha + \alpha_0$, so that $\ddot{R} = n^2\ddot{\alpha}$, the equation becomes

$$\ddot{R} + n^2 R = 0 \quad \text{with solution} \quad R = A\cos(nt + \gamma). \qquad (3.120)$$

The constants A, γ are fixed by boundary conditions. Then since $\alpha = (R - \alpha_0)/n^2$, *i.e.*

$$\alpha = \frac{A}{n^2}\cos(nt + \gamma) - \frac{\alpha_0}{n^2},$$

where α_0 and n are given by expressions (3.119). Thus we obtain for $\theta(t)$ the result:

$$\theta(t) = \theta_0 + \alpha(t) = \theta_0 + \frac{A}{n^2}\cos(nt + \gamma) - \frac{\alpha_0}{n^2}.$$

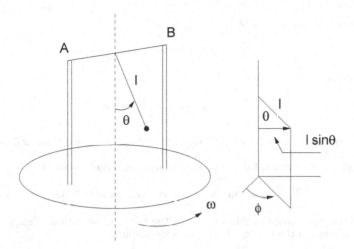

Fig. 3.15 The centrifugal pendulum.

Example 3.18: The centrifugal pendulum

The so-called *"centrifugal pendulum"* sketched in Fig. 3.15 rotates with constant angular velocity ω about the vertical axis. The mass m is attached to a weightless rod of length l which can oscillate from a support fixed on a rotating disc. Consider the pendulum as a spherical pendulum subjected to the time-dependent constraint $\phi = \omega t$. What is the equation of motion? What are the implications of the law of conservation of angular momentum?

Solution: We establish first the Lagrange function L. With the oscillation angle θ and rotation angle ϕ as defined in Fig. 3.15, the kinetic energy T and the potential energy V of the pendulum are given by

$$T = \frac{1}{2}m[(l\dot{\phi}\sin\theta)^2 + (l\dot{\theta})^2] \quad \text{and} \quad V = mgl(1 - \cos\theta),$$

so that

$$L = T - V = \frac{1}{2}m[(l\dot{\phi}\sin\theta)^2 + (l\dot{\theta})^2] - mgl(1 - \cos\theta). \tag{3.121}$$

The constraint is $\phi = \omega t$, i.e. $\dot{\phi} - \omega = 0$. This is a holonomic constraint, since integration of the latter relation reproduces the former. The constraint is built into the theory of L with the help of a *Lagrange multiplier*. In general one has

$$\frac{\partial L}{\partial q_k} - \frac{d}{dt}\left(\frac{\partial L}{\partial \dot{q}_k}\right) + \sum_l \lambda_l a_{lk} = 0 \quad \text{with} \quad \sum_k a_{lk}\dot{q}_k + a_{lt} = 0.$$

The latter relation corresponds to the constraint $\dot{\phi} - \omega = 0$ with $l = 1, k = 1, q_1 = \phi, a_{11} = 1, a_{1t} = -\omega$. We obtain therefore with $q_1 = \phi, q_2 = \theta$ the equations

$$\frac{\partial L}{\partial \phi} - \frac{d}{dt}\left(\frac{\partial L}{\partial \dot{\phi}}\right) + \lambda_1 a_{11} = 0, \quad \frac{\partial L}{\partial \theta} - \frac{d}{dt}\left(\frac{\partial L}{\partial \dot{\theta}}\right) + \lambda_1 a_{12} = 0, \quad \text{where } a_{12} = 0.$$

With Eq. (3.121) these equations become explicitly, first the equation in ϕ:

$$-\frac{d}{dt}[ml^2\dot{\phi}\sin^2\theta] + \lambda_1 = 0, \quad i.e. \quad ml^2\sin^2\theta\ddot{\phi} + 2ml^2\sin\theta\cos\theta\dot{\phi}\dot{\theta} = \lambda_1, \tag{3.122}$$

and then the equation in θ:

$$ml^2\dot{\phi}^2\sin\theta\cos\theta - mgl\sin\theta - \frac{d}{dt}[ml^2\dot{\theta}] = 0, \quad i.e. \quad ml^2\ddot{\theta} = ml^2\sin\theta\cos\theta\dot{\phi}^2 - mgl\sin\theta. \tag{3.123}$$

For $\dot{\phi} = \omega$ we obtain from Eq. (3.122)

$$\lambda_1 = 2ml^2\omega\dot{\theta}\sin\theta\cos\theta.$$

The equation of motion is therefore

$$\ddot{\theta} - \omega^2\sin\theta\cos\theta + \frac{g}{l}\sin\theta = 0. \tag{3.124}$$

This equation can be integrated with the help of the relation $\ddot{\theta} = d(\dot{\theta}^2/2)/d\theta$ and usually leads to an elliptic integral.

The angular momentum $\mathbf{L}$ has components (obtainable also from $\partial L/\partial\dot{\phi}, \partial L/\partial\dot{\theta}$)

$$L_\phi = l\dot{\phi}\sin\theta \, ml\sin\theta \perp AB, \quad L_\theta = lml\dot{\theta} \parallel AB,$$

where AB is the supporting beam in Fig. 3.15. The law of conservation of angular momentum, i.e. $d\mathbf{L}/dt = $ moment of the external forces, therefore implies

$$\frac{dL_\phi}{dt} = mg \times 1, \quad \frac{dL_\theta}{dt} \neq 0.$$

these equations are seen to agree with Eqs. (3.122), (3.123), if we identify λ_1 with $mg \times a_{11} = mg \times 1$. The angular momentum thus is not a constant of the motion (the Lagrange function contains explicitly the angle θ, so that this is not a cyclic variable).

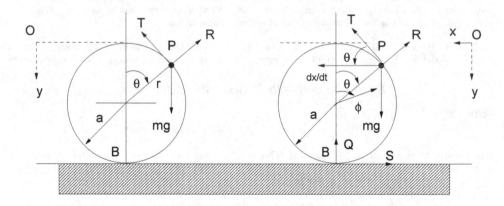

Fig. 3.16 The particle on the rolling tyre.

Example 3.19: A particle balanced on a rolling tyre

A particle of mass m is placed at the upper point of a vertically erected tyre of external radius a. Use the method of Lagrange multipliers and Euler–Lagrange equations to determine the reaction of the tyre on the particle. Also determine the height from which the particle falls down.[‡]

Solution: We distinguish between two cases: (a) The tyre fixed, (b) the tyre rolling.
(a) We assume the tyre is held fixed at a point B as in Fig. 3.16. Let T be the frictional force which the particle experiences, and R the reaction at point P. The Lagrangian is

$$L(\theta, r, \dot{\theta}, \dot{r}) = \frac{1}{2}mr^2\dot{\theta}^2 + mg(a - r\cos\theta). \tag{3.125}$$

[‡]This is unsolved problem 7 of chapter II, H. Goldstein [19].

The constraint is, if the particle leaves the tyre at $y = y_0$,

$$r - a = 0 \quad \text{up to} \quad y = y_0; \quad \text{thereafter}: \quad r > a \text{ for } y > y_0.$$

The Euler–Lagrange equations in θ and r are for $y \leq y_0$:

$$mgr \sin\theta - \frac{d}{dt}(mr^2\dot\theta) - rT = 0, \quad -mg\cos\theta + mr\dot\theta^2 + R = 0. \tag{3.126}$$

Here the specific forces (friction T and reaction R) have already been inserted — as in general considerations of elementary mechanics; the object of this example is to demonstrate that these equations can be obtained with the general Lagrangian method supplemented by constraints, and with the forces obtained from the generalized forces of the method, as outlined in earlier sections of this chapter. For $y > y_0$ in these equations T and R are zero. If we had set $r = a$ from the beginning, we would not have obtained the last equation because then we would have had only the one generalized coordinate θ. In order to determine at what angle the particle leaves the tyre, we set in the last equations: $T = 0, R = 0, r = a$, and obtain

$$mga \sin\theta = ma^2\ddot\theta = ma^2 \frac{d}{d\theta}\left(\frac{1}{2}\dot\theta^2\right), \quad -mg\cos\theta + ma\dot\theta^2 = 0. \tag{3.127}$$

From the second of these equations we obtain

$$\cos\theta = \frac{a}{g}\dot\theta^2, \tag{3.128}$$

and from the first

$$\int d\left(\frac{1}{2}\dot\theta^2\right) = \int \frac{g}{a}\sin\theta\, d\theta, \quad \frac{1}{2}[\dot\theta^2]_0^{\dot\theta} = \frac{g}{a}[-\cos\theta]_0^\theta, \quad i.e. \quad \frac{1}{2}\dot\theta^2 = \frac{g}{a}(1 - \cos\theta). \tag{3.129}$$

Inserting the last expression into Eq. (3.128), we obtain

$$\cos\theta = 2\frac{a}{g}\frac{g}{a}(1 - \cos\theta) = 2(1 - \cos\theta), \quad i.e. \quad \cos\theta = \frac{2}{3}.$$

We note that R and T are external forces independently of constraining forces, which lead the particle along the circular path. Thus R and T cannot be expressed in terms of constraints.

(b) Now we allow for a *rolling* of the tyre of radius a and mass M. We choose the origin of coordinates as in the second figure of Fig. 3.16. We let T be the frictional force and R the reaction of the tyre on the particle of mass m as shown. At time $t = 0$, *i.e.* at the instant before the particle begins to fall, the lowest point of the tyre is at point $x = 0$. We consider the falling of the particle from the tyre and the simultaneous rolling of the tyre on the plane $y = 2a$. Let ϕ be the angular velocity of the tyre, $\dot x$ its linear velocity, and θ the angular coordinate of the particle at time t.

The kinetic energy T consists of the kinetic energies of the tyre and that of the particle. We obtain the kinetic energy of the tyre from that of its centre of mass together with that of its motion around the centre of mass. The kinetic energy of the particle is obtained from computation of the components of its absolute velocity. Thus we obtain

$$T = \frac{1}{2}M\dot x^2 + \frac{1}{2}Mr^2\dot\phi^2 + \frac{1}{2}m[(\dot x - r\dot\theta\cos\theta)^2 + (r\dot\theta\sin\theta)^2]. \tag{3.130}$$

The potential energy V is simply that of the particle, since that of the tyre does not change; hence

$$V = -mg(a - r\cos\theta).$$

We obtain the reaction R and the frictional force T acting on the particle, by considering all the forces acting on it in the directions of R and T. Thus

$$ma\dot\theta^2 + m\ddot x \sin\theta = mg\cos\theta - R, \quad ma\ddot\theta - m\ddot x \cos\theta = mg\sin\theta - T. \tag{3.131}$$

Next we consider the constraints involved. First of all we have $r - a = 0$. At two points we have motion in contact: At the point of contact between the particle and the tyre, and at the point of contact between the tyre and the plane. The word *"rolling"* implies that the relative tangential velocity of the points of contact is zero. The point-like particle has no rotational energy (its radius is zero) and hence cannot roll. Instead it falls under the action of constraints responsible for the path $r - a \geq 0$ and external forces. The tyre, on the other hand, rolls on the plane, which means that $\dot{x} - a\dot{\phi} = 0$.

Euler–Lagrange equations

The constraint $\dot{x} - a\dot{\phi} = 0$ is in differential form

$$dx - ad\phi = 0. \tag{3.132}$$

This equation corresponds to the constraints which in the general theory are written $\sum_k a_{lk} dq_k = 0$ (*cf.* Eq. (3.56)) for $l = 1$, so that with $q_1 = x, q_2 = \theta, q_3 = \phi, q_4 = r$, we have $a_{11} = 1, a_{13} = -a$. With *Lagrange multipliers* λ_l the Euler–Lagrange equations are (*cf.* Eqs. (3.27) and (3.58))

$$\frac{\partial L}{\partial q_k} - \frac{d}{dt}\left(\frac{\partial L}{\partial \dot{q}_k}\right) + \sum_l \lambda_l a_{lk} + Q_k = 0. \tag{3.133}$$

Here Q_k are generalized forces. We set

$$q_1 = x, \quad q_2 = \theta, \quad q_3 = \phi, \quad q_4 = r.$$

Then $a_{11} = 1, a_{13} = -a$, all other $a_{lk} = 0$, and we write $\lambda_1 \equiv \lambda$. The Lagrangian is

$$L = T - V = \frac{1}{2}M\dot{x}^2 + \frac{1}{2}Mr^2\dot{\phi}^2 + \frac{1}{2}m[\dot{x}^2 + r^2\dot{\theta}^2 - 2r\dot{x}\dot{\theta}\cos\theta] + mg(a - r\cos\theta), \tag{3.134}$$

and the Euler–Lagrange equations for $q_k = x, \theta, \phi, r$ are respectively:

$$-\frac{d}{dt}(M\dot{x} + m\dot{x} - rm\dot{\theta}\cos\theta) + \lambda + Q_x = 0, \tag{3.135}$$

$$mr\dot{x}\dot{\theta}\sin\theta + mgr\sin\theta - \frac{d}{dt}(mr^2\dot{\theta} - mr\dot{x}\cos\theta) + Q_\theta = 0, \tag{3.136}$$

$$-\frac{d}{dt}(Mr^2\dot{\phi}) - a\lambda + Q_\phi = 0, \tag{3.137}$$

$$Mr\dot{\phi}^2 + mr\dot{\theta}^2 - m\dot{x}\dot{\theta}\cos\theta - mg\cos\theta + Q_r = 0. \tag{3.138}$$

Thus, together with the velocity or differential constraint

$$\dot{x} - a\dot{\phi} = 0, \tag{3.139}$$

we have 5 equations for 5 unknowns $(x, \theta, \phi, r, \lambda)$, provided the forces Q_k are known. With $r = a$ we can rewrite the equations as

$$M\ddot{x} + m\ddot{x} - ma\ddot{\theta}\cos\theta + ma\sin\theta\dot{\theta}^2 - \lambda - Q_x = 0, \tag{3.140}$$

$$ma^2\ddot{\theta} - ma\ddot{x}\cos\theta - mga\sin\theta - Q_\theta = 0, \tag{3.141}$$

$$Ma^2\ddot{\phi} + a\lambda - Q_\phi = 0, \tag{3.142}$$

$$Ma\dot{\phi}^2 + ma\dot{\theta}^2 - m\dot{x}\dot{\theta}\cos\theta - mg\cos\theta + Q_r = 0, \tag{3.143}$$

$$\dot{x} = a\dot{\phi}, \quad \ddot{x} = a\ddot{\phi}. \tag{3.144}$$

From Eqs. (3.140) and (3.142) we obtain by eliminating λ:

$$M\ddot{x} + m\ddot{x} - ma\ddot{\theta}\cos\theta + ma\sin\theta\dot{\theta}^2 - Q_x + Ma\ddot{\phi} - \frac{1}{a}Q_\phi = 0,$$

or with Eq. (3.144):

$$2Ma\ddot{\phi} + m\ddot{x} - ma\ddot{\theta}\cos\theta + ma\sin\theta\dot{\theta}^2 - \left(Q_x + \frac{1}{a}Q_\phi\right) = 0. \tag{3.145a}$$

Differentiating Eq. (3.143) with respect to ϕ we obtain:

$$\frac{d}{d\phi}(Ma\dot{\phi}^2) + \frac{d}{d\phi}Q_r = 0, \quad \text{or} \quad 2Ma\ddot{\phi} + \frac{d}{d\phi}Q_r = 0, \quad \text{since} \quad \ddot{\phi} = \frac{d}{d\phi}\left(\frac{1}{2}\dot{\phi}^2\right), \tag{3.145b}$$

From Eqs. (3.145a) and (3.145b) we now obtain

$$m\ddot{x} - ma\ddot{\theta}\cos\theta + ma\sin\theta\dot{\theta}^2 - \left(\frac{d}{d\phi}Q_r + Q_x + \frac{1}{a}Q_\phi\right) = 0. \tag{3.146}$$

Next we multiply Eq. (3.141) by $\cos\theta/a$, so that

$$ma\ddot{\theta}\cos\theta - m\ddot{x}\cos^2\theta - mg\cos\theta\sin\theta - \frac{\cos\theta}{a}Q_\theta = 0. \tag{3.147a}$$

Adding Eqs. (3.146) and (3.147a), we obtain

$$m\ddot{x}(1 - \cos^2\theta) + ma\dot{\theta}^2\sin\theta - mg\cos\theta\sin\theta - \left(\frac{d}{d\phi}Q_r + Q_x + \frac{1}{a}Q_\phi + \frac{1}{a}\cos\theta Q_\theta\right) = 0. \tag{3.147b}$$

The generalized forces Q_{q_k} are obtained from the relation (3.14), *i.e.*

$$Q_{q_k} = \mathbf{F}\cdot\frac{\partial\mathbf{r}}{\partial q_k}, \tag{3.148}$$

where $\mathbf{F}$ is the vectorial sum of the external forces acting at the point $\mathbf{r}$.

The (x, y)-coordinates of the particle are $\mathbf{r} = (x - a\sin\theta, a - a\cos\theta)$. Hence (note the directions of x and y indicated in Fig. 3.16)

$$
\begin{aligned}
Q_x &= \mathbf{F}\cdot(1, 0) = F_x = T\cos\theta - R\sin\theta, \quad Q_\phi = 0, \\
Q_\theta &= \mathbf{F}\cdot(-a\cos\theta, a\sin\theta) = (T\cos\theta - R\sin\theta)(-a\cos\theta) \\
&\quad + (-T\sin\theta - R\cos\theta)(a\sin\theta) = -Ta.
\end{aligned}
$$
$$\tag{3.149a}$$

We solve the equations $Q_x = T\cos\theta - R\sin\theta, Q_\theta = -Ta$ for R and T:

$$T = -\frac{1}{a}Q_\theta, \quad R = \frac{1}{\sin\theta}[T\cos\theta - Q_x] = \frac{1}{\sin\theta}\left[-\frac{1}{a}Q_\theta\cos\theta - Q_x\right]. \tag{3.149b}$$

Moreover, since $\mathbf{r} = (-r\sin\theta, r - r\cos\theta)$,

$$
\begin{aligned}
Q_r &= \mathbf{F}\cdot\frac{\partial\mathbf{r}}{\partial r} = \mathbf{F}\cdot(-\sin\theta, 1 - \cos\theta) \\
&= (T\cos\theta - R\sin\theta)(-\sin\theta) + (-T\sin\theta - R\cos\theta)(1 - \cos\theta) \\
&= R - (T\sin\theta + R\cos\theta).
\end{aligned}
$$
$$\tag{3.149c}$$

With Eqs. (3.149a) to (3.149c) we obtain:

$$\frac{d}{d\phi}Q_r + Q_x + \frac{1}{a}Q_\phi + \frac{1}{a}\cos\theta\, Q_\theta = T\cos\theta - R\sin\theta - \frac{1}{a}Ta\cos\theta = -R\sin\theta. \qquad (3.150)$$

We insert this expression into Eq. (3.147b) and divide by $\sin\theta \neq 0$. Then

$$m\ddot{x}\sin\theta + ma\dot{\theta}^2 - mg\cos\theta = -R.$$

We have thus regained via the Euler–Lagrange equation together with *Lagrange multipliers* and *generalized forces* the same expression for the reaction R, as that we obtained at the beginning from a consideration of the forces involved, *i.e.* the first of Eqs. (3.131). The second equation of Eqs. (3.131) follows from Eq. (3.147a) and $Q_\theta = -Ta$ of Eq. (3.149a).

Example 3.20: A cylinder oscillating in a cylindrical shell

A thin cylindrical shell of mass M and radius a oscillates on the inside of another but fixed cylindrical shell of radius $b > a$ as indicated in Fig. 3.17. The axes of the cylinders are parallel and horizontal. The surface of the inside of the fixed cylinder is rough, so that the other cylindrical shell oscillating on it rolls without sliding. Establish the Lagrangian L of the oscillating cylindrical shell, determine the equations of motion and solve these with the help of the method of Lagrange multpliers. What is the period of oscillation in the case of small deflections? How can the equation of motion be obtained in an elementary way from a consideration of forces? How do the considerations change when the small cylinder is replaced by a solid sphere? (Comment: Lagrange multipliers are not essential in this problem, but their use here is instructive).

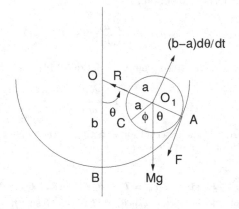

Fig. 3.17 A cylinder oscillating in a cylindrical shell.

Solution: We consider the geometry as in Fig. 3.17. Let OB be the vertical radius of the cylinder held fixed, and O_1C that particular radius of the oscillating shell which is vertical at B. Time t seconds after point C of the oscillating shell had been at B, the angle between line AO_1O and the vertical is the angle we call θ. The oscillating cylindrical shell then has rotated about an angle ϕ about its own axis, hence a total of $\theta + \phi$. The angles θ and ϕ are generalized coordinates. Since we have no sliding, but *rolling*, we have

$$a(\phi + \theta) = b\theta, \quad a\phi = (b-a)\theta, \quad a\,d\phi - (b-a)d\theta = 0. \qquad (3.151)$$

These equations have the form of

$$f(q_1, \ldots, q_n, t) = 0 \quad \text{and} \quad \sum_k \frac{\partial f}{\partial q_k}dq_k + \frac{\partial f}{\partial t}dt = 0,$$

and hence the form of *holonomic constraints* (recall that nonholonomic constraints have $\partial f/\partial t$ unequal to zero, so that the constraints are nonintegrable, *i.e.* do not yield $f(q_1, \ldots, q_n, t)$). We consider the cylindrical shell as a system of particles all of which are the same distance a away from the axis of the cylinder. We calculate the kinetic energy T as the sum of that of the motion of the centre of mass and that of the relative motion. The former is at time t

$$\frac{1}{2}M(b-a)^2\dot{\theta}^2.$$

The kinetic energy of the relative motion, *i.e.* that of the cylinder about its own axis, is, using Eq. (3.151),

$$\frac{1}{2}Ma^2\dot{\phi}^2 = \frac{1}{2}M(b-a)^2\dot{\theta}^2.$$

Hence

$$T = \frac{1}{2}M(b-a)^2\dot{\theta}^2 + \frac{1}{2}Ma^2\dot{\phi}^2 = M(b-a)^2\dot{\theta}^2.$$

The potential energy V of the smaller cylinder is

$$V = -Mg(b-a)\cos\theta \quad \text{or} \quad V = Mg(b-a)(1-\cos\theta),$$

if we choose its zero point at $\theta = 0$. Hence the Lagrangian is

$$
\begin{aligned}
L = T - V \; &= \; \frac{1}{2}M(b-a)^2\dot{\theta}^2 + \frac{1}{2}Ma^2\dot{\phi}^2 - Mg(b-a)(1-\cos\theta) \\
&\stackrel{\text{or}}{=} \; [M(b-a)^2\dot{\theta}^2 - Mg(b-a)(1-\cos\theta)].
\end{aligned}
\tag{3.152}
$$

From this we obtain

$$\frac{\partial L}{\partial\phi} = 0, \quad \frac{\partial L}{\partial\dot{\phi}} = Ma^2\dot{\phi}, \quad \frac{\partial}{\partial\theta} = -Mg(b-a)\sin\theta, \quad \frac{\partial L}{\partial\dot{\theta}} = M(b-a)^2\dot{\theta} \text{ or } [2M(b-a)^2\dot{\theta}].$$

From the Euler–Lagrange equation

$$\frac{\partial L}{\partial q_k} - \frac{d}{dt}\left(\frac{\partial L}{\partial\dot{q}_k}\right) + \sum_l \lambda_l a_{lk} = 0, \quad \text{together with} \quad \sum_k a_{lk}dq_k + a_{lt}dt = 0$$

for $q_k = \theta, \phi$ and $ad\phi - (b-a)d\theta = 0$ we obtain the equations

$$-Mg(b-a)\sin\theta - M(b-a)^2\ddot{\theta} - \lambda(b-a) = 0, \quad -Ma^2\ddot{\phi} + \lambda a = 0. \tag{3.153}$$

Together with Eq. (3.151), *i.e.* $a\phi = (b-a)\theta$, we thus have three equations for three unknowns, *i.e.* θ, ϕ, λ. From Eqs. (3.153) we obtain

$$-Mg(b-a)\sin\theta - M(b-a)^2\ddot{\theta} - Ma\ddot{\phi}(b-a) = 0, \quad \textit{i.e.} \quad g\sin\theta + (b-a)\ddot{\theta} + a\ddot{\phi} = 0.$$

Together with Eq. (3.151), *i.e.* $a\ddot{\phi} = (b-a)\ddot{\theta}$, this becomes

$$g\sin\theta + 2(b-a)\ddot{\theta} = 0. \tag{3.154}$$

We obtain the same equation if we replace right at the beginning $\dot{\phi}$ by $\dot{\theta}$ with the help of Eq. (3.151). Equation (3.154) is the desired equation. For small deflections θ we approximate $\sin\theta \approx \theta$, and the equation becomes

$$\ddot{\theta} + \frac{g}{2(b-a)}\theta = 0. \tag{3.155}$$

The period T_0 is therefore

$$T_0 = 2\pi\sqrt{\frac{2(b-a)}{g}}. \tag{3.156}$$

Finally we consider the forces F and R (*cf.* Fig. 3.17) involved. For the motion of the small cylinder we obtain*

$$Mg\sin\theta + F = -M(b-a)\ddot{\theta}, \quad \text{and} \quad Mg\cos\theta - R = -M(b-a)\dot{\theta}^2. \tag{3.157}$$

Taking the moment of forces **F** around O_1, we obtain (recall the relation between angular momentum **L** and torque **N**, $d\mathbf{L}/dt = \mathbf{r} \times \mathbf{F} \equiv \mathbf{N}$ of Eq. (2.11))

$$Fa = \frac{d}{dt}[Ma^2\dot{\phi}] = Ma^2\ddot{\phi} = Ma(b-a)\ddot{\theta}. \tag{3.158}$$

From Eqs. (3.157) and (3.158) we eliminate F and obtain

$$Mg\sin\theta + M(b-a)\ddot{\theta} = -M(b-a)\ddot{\theta}, \quad \textit{i.e.} \quad g\sin\theta + 2(b-a)\ddot{\theta} = 0.$$

This equation is seen to agree with Eq. (3.154) above obtained by the method of *Lagrange multipliers*. If the oscillating cylinder is replaced by a solid sphere, we have to replace Ma^2 by the *moment of inertia* I of the sphere, $I = 2Ma^2/5$, about one of its diameters.

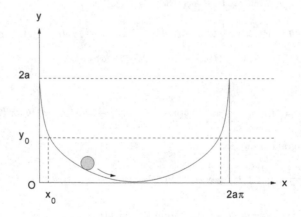

Fig. 3.18 The cycloid of the cycloidal pendulum.

Example 3.21: The cycloidal pendulum
Show that the period of oscillation of a particle on a cycloidal curve is independent of the amplitude.

Solution: For the cycloidal pendulum we require a downward cycloid as shown in Fig. 3.18, so that a particle can oscillate on it. The parameter representation of the cycloid was obtained in Example 3.15 with Eq. (3.108). We therefore select for our present purposes the equations in the following form in accordance with the curve in Fig. 3.18:

$$x = a(\theta - \sin\theta), \quad y = a(1 + \cos\theta).$$

Let y_0 be an amplitude or height such that in terms of a parameter α:

$$y_0 = a(1 + \cos\alpha), \quad 0 < \alpha < \pi.$$

---

*This assumes familiarity with radial and transversal components of acceleration — a derivation of which from first principles is given in Example 5.1. For a constant radius R to the centre of mass the components are respectively $-R\dot{\theta}^2$ and $R\ddot{\theta}$.

Hence by varying the angle α we can vary the magnitude of the amplitude. In Example 3.15 we derived the following formula (now with $v_0 = 0$ and $y_0 > y$)

$$dt = \frac{ds}{v} = \frac{ds}{\sqrt{2g(y_0 - y)}} = \frac{\sqrt{dx^2 + dy^2}}{\sqrt{2g(y_0 - y)}}.$$

Introducing the angle θ now as the independent variable, this relation yields

$$dt = \frac{\sqrt{(dx/d\theta)^2 + (dy/d\theta)^2}}{\sqrt{2g(y_0 - y)}} d\theta.$$

For a quarter of the period T we thus obtain (a prime now denoting differentiation with respect to angle θ)

$$\frac{T}{4} = \frac{1}{\sqrt{2g}} \int_\alpha^\pi \sqrt{\frac{x'^2 + y'^2}{y_0 - y}} d\theta = \sqrt{\frac{a}{g}} \int_\alpha^\pi \sqrt{\frac{1 - \cos\theta}{\cos\alpha - \cos\theta}} d\theta.$$

For $\alpha = 0$ we obtain immediately $T = 4\pi\sqrt{a/g}$. We now show that the same result follows for any value of α in $0 < \alpha < \pi$, thus proving the independence of the period of the size of the amplitude. With the relations

$$\cos\alpha - \cos\theta = 2\left[\cos^2\frac{\alpha}{2} - \cos^2\frac{\theta}{2}\right], \quad 1 - \cos\theta = 2\sin^2\frac{\theta}{2},$$

we obtain

$$\frac{T}{4} = \sqrt{\frac{a}{g}} \int_\alpha^\pi \frac{\sin(\theta/2)d\theta}{\sqrt{\cos^2(\alpha/2) - \cos^2(\theta/2)}}.$$

Setting

$$\cos\frac{\theta}{2} = u\cos\frac{\alpha}{2}, \quad \sin\frac{\theta}{2}d\theta = -2\cos\frac{\alpha}{2}du,$$

we obtain

$$\frac{T}{4} = \sqrt{\frac{a}{g}} \int_{\theta=\alpha}^\pi \frac{-2\cos(\alpha/2)du}{\cos(\alpha/2)\sqrt{1 - u^2}} = \sqrt{\frac{a}{g}} \int_{u=1}^0 \frac{-2du}{\sqrt{1 - u^2}} = -2\sqrt{\frac{a}{g}}[\sin^{-1} u]_1^0.$$

Hence

$$\frac{T}{4} = -2\sqrt{\frac{a}{g}}\left[0 - \frac{\pi}{2}\right], \quad \therefore \ T = 4\pi\sqrt{\frac{a}{g}}.$$

Thus the period T is independent of α and hence of the amplitude y_0. This result was discovered by Huygens.[†]

Example 3.22: A charged particle in an electromagnetic field

The Lagrangian of a particle of charge e and mass m moving in an electromagnetic vector potential $\mathbf{A}(q_i, t)$ and a scalar potential $\phi(q_i)$, $i = 1, 2, 3$, is defined by

$$L(q_i, \dot{q}_i) = \frac{1}{2}m\sum_i \dot{q}_i^2 + e\sum_i \dot{q}_i A_i(q_i, t) - \phi(q_i), \tag{3.159}$$

with the electric field $\mathbf{E}$ and the magnetic induction $\mathbf{B}$ given by (velocity of light $c = 1$):

$$\mathbf{E} = -\frac{\partial\mathbf{A}}{\partial t} - \nabla\phi, \quad \mathbf{B} = \nabla \times \mathbf{A}. \tag{3.160}$$

[†]R. Courant [12], p. 302.

Obtain the equation of motion of the particle in the electromagnetic field usually given as

$$\frac{d}{dt}(m\ddot{\mathbf{r}}) = e[\mathbf{E} + \dot{\mathbf{r}} \times \mathbf{B}], \quad |\mathbf{v}| \equiv |\dot{\mathbf{r}}| \ll c.$$

Solution: We obtain from the Lagrangian:

$$\frac{\partial L}{\partial \dot{q}_i} = m\dot{q}_i + eA_i, \quad \frac{\partial L}{\partial q_j} = e\dot{q}_i \frac{\partial A_i}{\partial q_j} - e\frac{\partial \phi}{\partial q_j},$$

$$\therefore \quad \frac{d}{dt}\left(\frac{\partial L}{\partial \dot{q}_j}\right) - \frac{\partial L}{\partial q_j} = 0 \quad \Rightarrow \quad m\ddot{q}_j + \frac{d}{dt}(eA_j) = e\dot{q}_i \frac{\partial A_i}{\partial q_j} - e\frac{\partial \phi}{\partial q_j}.$$

Here d/dt is the total derivative with respect to time t, *i.e.*

$$\frac{d}{dt} = \frac{\partial}{\partial t} + \sum_i \dot{q}_i \frac{\partial}{\partial q_i},$$

so that

$$m\ddot{q}_j + e\frac{\partial A_j}{\partial t} + e\frac{\partial A_j}{\partial q_i}\dot{q}_i = e\dot{q}_i \frac{\partial A_i}{\partial q_j} - e\frac{\partial \phi}{\partial q_j}, \quad \text{or} \quad m\ddot{q}_j = -e\left(\frac{\partial A_j}{\partial t} + \frac{\partial \phi}{\partial q_j}\right) + e\left(\frac{\partial A_i}{\partial q_j}\dot{q}_i - \frac{\partial A_j}{\partial q_i}\dot{q}_i\right).$$

But with Eq. (3.160),

$$\mathbf{v} \times \mathbf{B} = \mathbf{v} \times (\nabla \times \mathbf{A}) = \nabla(\mathbf{v} \cdot \mathbf{A}) - (\mathbf{v} \cdot \nabla)\mathbf{A} = \frac{\partial}{\partial \mathbf{r}}(\mathbf{v} \cdot \mathbf{A}) - \left(\mathbf{v} \cdot \frac{\partial}{\partial \mathbf{r}}\right)\mathbf{A},$$

$$i.e. \quad (\mathbf{v} \times \mathbf{B})_j = \frac{\partial}{\partial q_j}\sum_i v_i A_i - \sum_i v_i \frac{\partial}{\partial q_i}A_j,$$

so that the equation becomes

$$m\ddot{q}_j = eE_j + e(\mathbf{v} \times \mathbf{B})_j.$$

An important aspect to be noted here is that $p_i = \partial L/\partial \dot{q}_i$ is in the Lagrangian method the *conjugate momentum* and in the Hamiltonian method the *canonical momentum*, *i.e.* the expression $\mathbf{p} = m\mathbf{v} + e\mathbf{A}$. The *mechanical* or *kinematical* momentum is $m\mathbf{v}$, *i.e.* $m\mathbf{v} = \mathbf{p} - e\mathbf{A} \equiv m\dot{\mathbf{q}}$. It is the canonical momentum which in the process of quantization becomes the *covariant derivative* in the quantized theory.

Chapter 4

The Canonical or Hamilton Formalism

4.1 Introductory Remarks

We have seen in the previous chapter that the Lagrangian formalism with the Lagrange equation, which is more generally called the *Euler–Lagrange equation*, the method regains Newton's mechanics, admittedly with advantages, one being for instance the ease of incorporating constraints. Thus a lot has been achieved by the new axiomatic basis provided by Hamilton's principle of extremization of the action. The next considerable step forward in eliciting the basis of mechanics was Hamilton's recognition of the significance of linear momentum p as an individual entity of its own, and disjunct from its naive definition as the product of mass and velocity. Realizing this significance of momentum it became necessary to introduce momentum in a new basis of variables, namely $\{q, p\}$, known as *canonical variables*. Today we know that this transition to the Hamilton formalism based on this set of canonical variables was also a further and significant step in the direction of quantum mechanics. The transition from the set of n generalized coordinates $q_i, i = 1, 2, \ldots, n$, to the set of $2n$ canonical (phase space) variables $\{q_i, p_i\}$ is achieved with a *Legendre transform*. This is the step prior to that into a generalized form of mechanics which allows the transition to quantum mechanics (on a new axiomatic basis). It is evident that the Hamiltonian formalism is therefore of basic significance which reveals more than any other method the basic structure of mechanics. This chapter therefore deals with the derivation and properties of *Hamilton's equations of motion*, and basic properties of classical mechanics related to these. The most important of these properties have operator-counterparts in quantum mechanics.

4.2   Hamilton's Equations of Motion

We recall that the dynamics of a system of n degrees of freedom, *i.e.* of n generalized coordinates with no constraints, is described by n *Euler–Lagrange equations of motion, i.e. differential equations of the second order* given by

$$\frac{d}{dt}\left(\frac{\partial L}{\partial \dot{q}_i}\right) - \frac{\partial L}{\partial q_i} = 0, \quad i = 1, 2, \ldots, n. \tag{4.1}$$

The integration of these equations requires therefore the determination of $2n$ integration constants from *initial conditions, i.e.* the intial $(t = 0)$ values of all $q_i, \dot{q}_i, i = 1, 2, \ldots, n$. In Hamilton's formulation that we consider now, the independent variables are the n generalized coordinates q_i together with the n conjugate or generalized momenta p_i given by

$$p_i = \frac{\partial L(q_j, \dot{q}_j, t)}{\partial \dot{q}_i}. \tag{4.2}$$

Thus we require a change of basis variables from the set of $(q_i, \dot{q}_i, t)$ to (q_i, p_i, t). The space spanned by the dynamical (*i.e.* time-dependent) variables $q_i, p_i, i = 1, 2, \ldots, n$ is described as the "*phase space*". In the phase space formulation these variables are the basic variables, and thus coordinates and momenta are treated on the same level. We shall see that instead of the n Euler–Lagrangian differential equations of the second order, the dynamics in the phase space formulation is described by twice this number of equations, but of the first order.

We first of all define the *Hamilton function* by the equation

$$H(q, p, t) \equiv \sum_i \dot{q}_i p_i - L(q, \dot{q}, t). \tag{4.3}$$

The arguments of the Hamilton function H are q_i, p_i and t, but not $\dot{q}_i$. The function H is independent of $\dot{q}_i$, because

$$\frac{\partial H}{\partial \dot{q}_i} = p_i - \frac{\partial L}{\partial \dot{q}_i} = 0. \tag{4.4}$$

The relation (4.3), which effects the transition from the variables or generalized coordinates $q_i, \dot{q}_i$ to the variables or *canonical coordinates* q_i, p_i is known as *Legendre transform*. Since the right hand side of Eq. (4.3) involves $q_i, p_i, \dot{q}_i, t$, we obtain with Eq. (4.4):

$$\begin{aligned}
dH &= \sum_i \dot{q}_i dp_i + \sum_i p_i d\dot{q}_i - \sum_i \frac{\partial L}{\partial \dot{q}_i} d\dot{q}_i - \sum_i \frac{\partial L}{\partial q_i} dq_i - \frac{\partial L}{\partial t} dt \\
&= \sum_i \dot{q}_i dp_i - \sum_i \frac{\partial L}{\partial q_i} dq_i - \frac{\partial L}{\partial t} dt. \tag{4.5}
\end{aligned}$$

From Eq. (4.1) we obtain

$$\frac{d}{dt}p_i - \frac{\partial L}{\partial q_i} = 0.$$

Hence

$$dH = \sum_i \dot{q}_i dp_i - \sum_i \dot{p}_i dq_i - \frac{\partial L}{\partial t} dt = \sum_i \frac{\partial H}{\partial q_i} dq_i + \sum_i \frac{\partial H}{\partial p_i} dp_i + \frac{\partial H}{\partial t} dt. \quad (4.6)$$

Comparing coefficients of dp_i, dq_i, dt, we obtain

$$\dot{q}_i = \frac{\partial H}{\partial p_i}, \quad -\dot{p}_i = \frac{\partial H}{\partial q_i}, \quad -\frac{\partial L}{\partial t} = \frac{\partial H}{\partial t}. \quad (4.7)$$

These equations are known as *Hamilton's equations* or as the *canonical Hamilton equations*. They constitute a set of $2n$ differential equations of the first order in t.

Next we can convince ourselves that just like cyclic coordinates do not appear in the Lagrangian L, so they also do not appear in the Hamiltonian H. If q_n is a cyclic coordinate, then $L = L(q_1, q_2, \ldots, q_{n-1}, \dot{q}_1, \dot{q}_2, \ldots, \dot{q}_n, t)$ and

$$H = H(q_1, q_2 \ldots, q_{n-1}, p_1, p_2, \ldots, p_{n-1}, \alpha, t), \quad \text{where} \quad p_n = \alpha. \quad (4.8)$$

Whereas the formulation in terms of L is still a problem in n degrees of freedom, the formulation in terms of H, like H itself is one in $2n - 1$ degrees of freedom, the velocity $\dot{q}_n$ following from

$$\dot{q}_n = \frac{\partial H}{\partial \alpha}.$$

Evidently the Hamiltonian formalism offers some advantages when cyclic coordinates are present. To exploit this advantage, one can combine both procedures — the Hamilton method and the Lagrange method — by choosing the canonical q, p basis only for cyclic coordinates, but for the others the $q, \dot{q}$ basis. Then a *Routh function* R can be defined which acts in one case as a Hamilton function, and in the other as a Lagrange function. Since we do not make use of this method here, we refer for details to Goldstein [19].

The Legendre transform exemplified by the relation (4.3) can be considered in a more general form, and applications can be found in other contexts in theoretical physics, as illustrated later; a particularly important case is its application in thermodynamics. Following Goldstein [19]* we consider a function $f(x, y)$ of two variables x, y. Then

$$df(x, y) = \frac{\partial f}{\partial x} dx + \frac{\partial f}{\partial y} dy \equiv u\,dx + v\,dy, \quad u = \frac{\partial f}{\partial x}, \quad v = \frac{\partial f}{\partial y}. \quad (4.9)$$

---

*H. Goldstein [19], Sec. 7.1.

In analogy to the transition from L to H we now want to pass from the basis variables x, y to independent variables u, y, and — in fact — in such a way that differentials can be expressed in terms of du, dy. Then the transform of $f(x, y)$ is a new function $g(u, y)$ which is defined by the relation

$$g(u, y) = f(x, y) - ux, \tag{4.10}$$

i.e. so that $\partial g / \partial x = 0$. Then

$$dg = df - udx - xdu \overset{(4.9)}{=} vdy - xdu \quad \text{with} \quad x = -\frac{\partial g}{\partial u}, \; v = \frac{\partial g}{\partial y}, \tag{4.11}$$

so that $x = x(u, y), v = v(u, y)$. In this way x and v here are the reversals of u and y above. We illustrate these transformations in the following examples.

Example 4.1: Legendre transforms of simple cases
Determine the Legendre transforms (a) $g(u)$ of $f(x) = ax^2$ and (b) $g(x, v)$ of $f(x, y) = ax^2y^3$, and verify the results by transforming from g back to f.

Solution: (a) We have

$$f(x) = ax^2, \quad f' = 2ax = u, \quad x = u/2a, \quad g = f - ux = ax^2 - ux;$$

but $x = -\partial g / \partial u = -g'(u)$. Therefore

$$g = a(-g')^2 + ug', \quad a(g')^2 + ug' - g = 0.$$

We set $g' = \beta u, g = \beta u^2/2$, so that $a\beta^2 u^2 + \beta u^2 - \beta u^2/2 = 0, a\beta^2 + \beta/2 = 0$. It follows that

$$\beta = -\frac{1}{2a}, \quad g = -\frac{u^2}{4a}.$$

The reverse direction: Given $g = -u^2/4a$, obtain $f = g + ux = -u^2/4a + ux$. Now $u = \partial f / \partial x = f'$, so that

$$f = -\frac{(f')^2}{4a} + xf', \quad (f')^2 + 4af - 4axf' = 0.$$

We set $f' = bx, f = bx^2/2$, so that

$$b^2x^2 + \frac{4abx^2}{2} - 4abx^2 = 0, \quad b^2 - 2ab = 0, \; b = 2a, \; f = \frac{2ax^2}{2} = ax^2.$$

(b) In this case

$$f(x, y) = ax^2y^3, \quad g = f - ux, \quad x = x(u, y) = -\frac{\partial g}{\partial u}, \quad v = v(u, y),$$

$$g(u, y) = ax^2y^3 - ux = a\left(\frac{\partial g}{\partial u}\right)^2 y^3 + u\left(\frac{\partial g}{\partial u}\right).$$

We set $g(u, y) = u^2\alpha(y)$. Then $\partial g / \partial u = 2u\alpha(y)$ and without an overall multiplicative factor of u^2 we have:

$$\alpha(y) = 4a\alpha^2(y)y^3 + 2\alpha(y), \quad 4a\alpha(y)y^3 = -1, \quad \alpha(y) = -\frac{1}{4ay^3}, \; \therefore \; g(u, y) = -\frac{u^2}{4ay^3}.$$

The reverse direction: Given $g = -u^2/4ay^3$, to obtain $f(x,y) = g + ux = -u^2/4ay^3 + ux$. Now $u = \partial f/\partial x \equiv f'$.

$$\therefore \; f = -\frac{(f')^2}{4ay^3} + xf', \quad (f')^2 - 4ay^3xf' + 4ay^3f = 0.$$

We set $f = x^2\beta(y)/2$. Then

$$x^2\beta^2 - 4ay^3x^2\beta + 2ay^3x^2\beta = 0, \quad \beta = 2ay^3, \quad \therefore \; f = ax^2y^3.$$

Example 4.2: Legendre transforms and thermodynamical potentials

Starting from the analogy of the second law of thermodynamics, $dE(S,V) = TdS - pdV$ (E internal energy, S entropy, V volume, T temperature, p pressure) with $dL(q,\dot{q})$ of the Lagrangian $L(q,\dot{q})$, perform the Legendre transform to the enthalpy or total heat $H(S,p)$ in analogy to the transition from L to the Hamiltonian $H(q,p) = \dot{q}p - L(q,\dot{q})$, *i.e.* the transformation from variables S, V to variables S, p, and determine $H(S,p)$. What are the equations corresponding to Hamilton's equations in mechanics?

Solution: Comparing $dE(S,V) = TdS - pdV$ with $df(x,y) = udx + vdy$ of Eq. (4.9), we have

$$T - \left(\frac{\partial E}{\partial S}\right)_V, \quad p = -\left(\frac{\partial E}{\partial V}\right)_S.$$

We now define $H(S,p)$ so that this is independent of V, *i.e.* so that $\partial H/\partial V = 0$. Then — in analogy to $g(u,y)$ of Eq. (4.10) — we obtain

$$H(S,p) = Vp + E(S,V), \quad \text{so that} \quad \frac{\partial H}{\partial V} = p + \frac{\partial E}{\partial V} = 0$$

$$\left(\text{corresponding to} \quad \frac{\partial H(q,p)}{\partial \dot{q}} \overset{(4.3)}{=} p - \frac{\partial L}{\partial \dot{q}} = 0\right).$$

Then the analogues of Hamilton's equations are

$$\left(\frac{\partial H}{\partial S}\right)_p = \left(\frac{\partial E}{\partial S}\right)_p (= T), \quad \left(\frac{\partial H}{\partial p}\right)_S = V. \tag{4.12}$$

We add parenthetically that the formula known as the "*first Maxwell relation*" in literature on thermodynamics (and useful, for instance, for the derivation of relations between specific heats)[†] follows from these analogues of Hamilton's equations by one further partial differentiation, *i.e.* as the relation

$$\left(\frac{\partial T}{\partial V}\right)_S = -\left(\frac{\partial p}{\partial S}\right)_V \quad \text{from} \quad \left(\frac{\partial}{\partial V}\right)_S\left(\frac{\partial E}{\partial S}\right)_V = \left(\frac{\partial}{\partial S}\right)_V\left(\frac{\partial E}{\partial V}\right)_S = -\left(\frac{\partial}{\partial S}\right)_V p.$$

Without going into further details we add that in complete analogy to above , and going to different sets of variables, one obtains the "*free energy*" $F(T,V)$ and the "*free enthalpy*" or "*Gibbs function*" $G(T,p)$ as

$$F(T,V) = E(S,V) - TS, \quad G(T,p) = E(S,V) - TS + pV = F(T,V) + pV,$$

and correspondingly further "*Maxwell relations*". These can, in fact, be read-off from

$$dH = TdS + Vdp, \quad dF = -SdT - pdV, \quad dG = -SdT + Vdp.$$

Related discussion on equilibrium (minimum energy, maximum entropy conditions) can be found in Example 9.6.

[†]See F. Reif [43], Sec. 5.6.

4.3   Physical Significance of the Hamilton Function

We have just seen that if L is cyclic in a certain coordinate, then H is also cyclic in this coordinate. This means that if, for instance, a system is symmetric about a given axis, so that H is invariant under rotations about this axis, then H cannot contain the appropriate angle of rotation. This angle is then a cyclic coordinate and the corresponding angular momentum is conserved. The physical significance of the Hamilton function H can now be seen as follows. We consider first the total time derivative of H, $i.e.$

$$
\begin{aligned}
\frac{dH}{dt} &= \frac{\partial H}{\partial t} + \sum_i \left(\frac{\partial H}{\partial q_i}\dot{q}_i + \frac{\partial H}{\partial p_i}\dot{p}_i \right) \\
&= \frac{\partial H}{\partial t} + \sum_i \left(\frac{\partial H}{\partial q_i}\frac{\partial H}{\partial p_i} - \frac{\partial H}{\partial p_i}\frac{\partial H}{\partial q_i} \right) \\
&= \frac{\partial H}{\partial t} \overset{(4.7)}{=} -\frac{\partial L}{\partial t}.
\end{aligned}
\tag{4.13}
$$

Now consider the case of two restrictions: (a) The equations expressing the transformation to generalized coordinates, $i.e.$

$$
\mathbf{r}_m = \mathbf{r}_m(q_1, q_2, \ldots, q_n, t),
\tag{4.14}
$$

do not depend explicitly on t, and (b) the potential V is independent of velocities. Then, we can see this as follows, the *total energy* is given by the Hamilton function. Since under the stated conditions $\partial L/\partial t = 0$ ($i.e.$ L not an explicit function of t),

$$
\frac{dL}{dt} = \sum_j \left(\frac{\partial L}{\partial q_j}\frac{dq_j}{dt} + \frac{\partial L}{\partial \dot{q}_j}\frac{d\dot{q}_j}{dt} \right) \overset{(4.1)}{=} \sum_j \left[\frac{d}{dt}\left(\frac{\partial L}{\partial \dot{q}_j}\right)\dot{q}_j + \frac{\partial L}{\partial \dot{q}_j}\frac{d\dot{q}_j}{dt} \right].
$$

$$
= \sum_j \frac{d}{dt}\left(\dot{q}_j \frac{\partial L}{\partial \dot{q}_j} \right).
\tag{4.15}
$$

It follows that

$$
\frac{d}{dt}\left(L - \sum_j \dot{q}_j \frac{\partial L}{\partial \dot{q}_j} \right) = 0, \quad L - \sum_j \dot{q}_j \frac{\partial L}{\partial \dot{q}_j} = \text{const.},
\tag{4.16}
$$

or

$$
L - \sum_j \dot{q}_j p_j = \text{const.} \quad \text{and} \quad \overset{(4.3)}{=} -H.
\tag{4.17}
$$

It follows that

$$
H = \text{const.}
\tag{4.18}
$$

Now we take a closer look at the Hamilton function

$$H = \sum_j \dot{q}_j p_j - L. \tag{4.19}$$

We want to demonstrate that the constant in Eq. (4.18) is the *total energy*. We have

$$p_j = \frac{\partial L}{\partial \dot{q}_j} = \frac{\partial T}{\partial \dot{q}_j}, \qquad \frac{\partial V}{\partial \dot{q}_j} = 0, \tag{4.20}$$

i.e. provided the potential V is completely independent of velocities, and $L = T - V$. Hence

$$H = \sum_j \dot{q}_j \frac{\partial T}{\partial \dot{q}_j} - L. \tag{4.21}$$

Generally, with $\mathbf{v}_i = d\mathbf{r}_i(q,t)/dt$, the kinetic energy T is given by

$$T = \sum_i \frac{1}{2} m_i \mathbf{v}_i^2 = \sum \frac{1}{2} m_i \left(\sum_j \frac{\partial \mathbf{r}_i}{\partial q_j} \dot{q}_j + \frac{\partial \mathbf{r}_i}{\partial t} \right)^2. \tag{4.22}$$

When $\partial \mathbf{r}_i / \partial t = 0$, we obtain

$$T = \frac{1}{2} \sum_i m_i \left(\sum_j \frac{\partial \mathbf{r}_i}{\partial q_j} \dot{q}_j \right)^2 \equiv \sum_{j,k} a_{jk} \dot{q}_j \dot{q}_k = \frac{1}{2} \sum_j \dot{q}_j \frac{\partial T}{\partial \dot{q}_j},$$

$$\text{where } a_{jk} = \frac{1}{2} \sum_i m_i \frac{\partial \mathbf{r}_i}{\partial q_j} \cdot \frac{\partial \mathbf{r}_i}{\partial q_k}, \tag{4.23}$$

since $(1/2)\partial T / \partial \dot{q}_k = \sum_k a_{jk} \dot{q}_k$. It now follows that

$$H = \sum_j \dot{q}_j \frac{\partial T}{\partial \dot{q}_j} - L = 2T - (T - V) = T + V. \tag{4.24}$$

Thus H is the total energy, *i.e.* a conserved quantity and constant of the dynamics. One should note that if Eqs. (4.14) contain an explicit t-dependence and $\partial L/\partial t \neq 0$, then H is not a conserved quantity (*cf.* Eq. (4.24)). We illustrate this important point in the examples below.

Example 4.3: Motion of a particle in the field of a central force
Establish Hamilton's equations for the case of the 2-dimensional motion of a particle of mass m in a potential $V = V(r)$ in polar coordinates (pole at $r = 0$).

Solution: Since $V = V(r)$ the polar angle θ is a cyclic coordinate and we have

$$H = T + V, \quad T = \frac{1}{2} m v^2 = \frac{1}{2} m (\dot{r}^2 + r^2 \dot{\theta}^2). \tag{4.25}$$

Here[‡] $H = H(r,\theta,\dot{r},\dot{\theta})$. But in the Hamilton formalism we require H expressed as $H(r,\theta,p_r,p_\theta)$. We have

$$L = T - V = \frac{1}{2}m(\dot{r}^2 + r^2\dot{\theta}^2) - V(r), \quad p_r = \frac{\partial L}{\partial \dot{r}}, \quad p_\theta = \frac{\partial L}{\partial \dot{\theta}}, \tag{4.26a}$$

$$p_r = m\dot{r}, \quad p_\theta = mr^2\dot{\theta}, \quad \dot{r} = \frac{p_r}{m}, \quad \dot{\theta} = \frac{p_\theta}{mr^2}. \tag{4.26b}$$

Hence the correctly expressed Hamilton function is

$$H(r,\theta,p_r,p_\theta) = \frac{p_r^2}{2m} + \frac{p_\theta^2}{2mr^2} + V(r). \tag{4.27}$$

The momentum p_r follows as the linear momentum along $\mathbf{r}$, *i.e.* $m\dot{r}$, and p_θ as angular momentum in θ, *i.e.* $r.mr\dot{\theta} = mr^2\dot{\theta}$. Hamilton's equations are now given by (two equations for $\dot{q}$, two for $\dot{p}$)

$$\dot{r} = \frac{\partial H}{\partial p_r} = \frac{p_r}{m}, \quad \dot{\theta} = \frac{\partial H}{\partial p_\theta} = \frac{p_\theta}{mr^2}, \quad \dot{p}_r = -\frac{\partial H}{\partial r} = \frac{p_\theta^2}{mr^3} - \frac{\partial V}{\partial r}, \quad \dot{p}_\theta = -\frac{\partial H}{\partial \theta} = 0. \tag{4.28}$$

Example 4.4: Hamiltonian H not always total energy E
Let $L = T - V$ be the Lagrangian of a system of particles with kinetic energy T and velocity-independent potential energy V. The equations relating the position coordinates $\mathbf{r}_i$ of the particles to the n generalized coordinates q_j are explicitly time-dependent. Show that

$$H = \sum_j \dot{q}_j p_j - L = T_1 - T_3 + V, \; T_1 = \sum_{k,l} A_{kl}\dot{q}_k\dot{q}_l, \; A_{kl} = \sum_i \frac{1}{2}m_i \frac{\partial \mathbf{r}_i}{\partial q_k}\cdot\frac{\partial \mathbf{r}_i}{\partial q_l}, \; T_3 = \sum_i \frac{1}{2}m_i\left(\frac{\partial \mathbf{r}_i}{\partial t}\right)^2. \tag{4.29}$$

What follows for $\partial \mathbf{r}_i/\partial t = 0$?

Solution: Let the equations of the transformation be, for $i = 1,\dots,m$, $\mathbf{r}_i = \mathbf{r}_i(q_1,q_2,\dots,q_n,t)$. The total derivative with respect to t gives the velocity of the i-th mass-point,

$$\mathbf{v}_i = \frac{d\mathbf{r}_i}{dt} = \sum_{j=1}^n \frac{\partial \mathbf{r}_i}{\partial q_j}\dot{q}_j + \frac{\partial \mathbf{r}_i}{\partial t}.$$

Hence the kinetic energy T is

$$T = \sum_i \frac{1}{2}m_i \frac{d\mathbf{r}_i}{dt}\cdot\frac{d\mathbf{r}_i}{dt} = \sum_i \frac{1}{2}m_i\left(\sum_{j=1}^n \frac{\partial \mathbf{r}_i}{\partial q_j}\dot{q}_j + \frac{\partial \mathbf{r}_i}{\partial t}\right)^2 \equiv \sum_{k,l} A_{kl}\dot{q}_k\dot{q}_l + \sum_{k=1}^n B_k\dot{q}_k + C, \tag{4.30}$$

where

$$A_{kl} = \sum_i \frac{1}{2}m_i \frac{\partial \mathbf{r}_i}{\partial q_k}\cdot\frac{\partial \mathbf{r}_i}{\partial q_l} = A_{lk}, \quad B_k = \sum_i m_i \frac{\partial \mathbf{r}_i}{\partial q_k}\cdot\frac{\partial \mathbf{r}_i}{\partial t}, \quad C = \sum_i \frac{1}{2}m_i\left(\frac{\partial \mathbf{r}_i}{\partial t}\right)^2.$$

The Hamiltonian H is defined by

$$H = \sum_{j=1}^n \dot{q}_j p_j - L, \quad \text{where} \quad p_j = \frac{\partial L}{\partial \dot{q}_j} = \frac{\partial T}{\partial \dot{q}_j},$$

[‡]In order to avoid this type of expression it would be better to write E instead of H, E = total energy.

provided V is independent of any $\dot{q}_j$, $L = T - V$. Hence with Eq. (4.30)

$$H = \sum_j \dot{q}_j \frac{\partial T}{\partial \dot{q}_j} - L = 2\sum_{j,l} A_{jl}\dot{q}_j\dot{q}_l + \sum_j B_j\dot{q}_j - L.$$

If we set

$$T = T_1 + T_2 + T_3, \quad T_1 = \sum_{k,l} A_{kl}\dot{q}_k\dot{q}_l, \quad T_2 = \sum_k B_k\dot{q}_k, \quad T_3 - C,$$

then

$$H = 2T_1 + T_2 - L, \quad \text{or} \quad H = 2T_1 + T_2 - (T_1 + T_2 + T_3 - V) = T_1 - T_3 + V.$$

If Eqs. (4.29) are independent of t, we have $T_3 = 0, T_2 = 0$ and $T = T_1$, so that $H = T_1 + V = T + V = $ total energy. This example emphasizes the important point that the Hamiltonian H is the total energy (in the case of velocity-independent potentials) only if the transformation equations (4.29) do not involve t explicitly, *i.e.* when $\partial \mathbf{r}_i/\partial t = 0$ for all i.

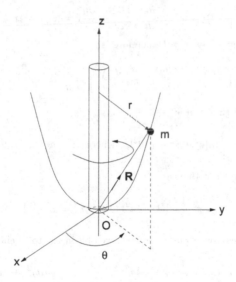

Fig. 4.1 The bead of mass m on the rotating parabolic wire.

Example 4.5: The rotating parabolic wire

A rigid parabolic wire is attached to a thin vertical tube, as depicted in Fig. 4.1. The tube (with wire attached) rotates with angular acceleration $\ddot{\theta} = \alpha = $ const. about the axis of the tube along the z-axis. A bead of mass m is placed on the frictionless wire. The shape of the wire is given by the equation $z = ar^2$.
(a) What are the equations of the transformation from the Cartesian coordinates x, y, z to cylindrical coordinates r, θ, z of the bead? Transform the kinetic energy T of the bead to cylindrical coordinates.
(b) What are the Lagrangian L and the Hamiltonian H for the motion of the bead? Is H equal to the total energy?
(c) What are Hamilton's equations?

Solution:
(a) The equations transforming to cylinder coordinates are

$$x = r\cos\theta, \quad y = r\sin\theta, \quad z = z. \tag{4.31}$$

We choose r, θ, z as generalized coordinates q_i. Since the bead moves with constant angular acceleration $\ddot{\theta} = \alpha$, we have, with constants θ_0, ω, α,

$$\theta = \theta_0 + \omega t + \frac{1}{2} \alpha t^2. \tag{4.32}$$

Thus the transformation (4.31) is explicitly time-dependent. The constraint $z = ar^2$ reduces the number of generalized coordinates 3 to 2, or 2 to 1. With Eq. (4.32) the equations of (4.31) become: $x = x(r, t), y = y(r, t), z = z(r, t)$. Since these equations contain t explicitly, the Hamilton function $H(q_i, p_i)$ is not equal to the total energy, as explained in Example 4.4. We have explicitly:

$$x = r \cos\left(\theta_0 + \omega t + \frac{1}{2}\alpha t^2\right), \quad y = r \sin\left(\theta_0 + \omega t + \frac{1}{2}\alpha t^2\right), \quad z = ar^2. \tag{4.33}$$

We see that

$$\frac{\partial x}{\partial t} = -(\omega + \alpha t)y, \quad \frac{\partial y}{\partial t} = (\omega + \alpha t)x, \quad \frac{\partial z}{\partial t} = 0,$$

$$\frac{dx}{dt} = \frac{\partial x}{\partial t} + \frac{\partial x}{\partial r}\dot{r} = -(\omega + \alpha t)y + x\frac{\dot{r}}{r}, \quad \frac{dy}{dt} = \frac{\partial y}{\partial t} + \frac{\partial y}{\partial r}\dot{r} = (\omega + \alpha t)x + y\frac{\dot{r}}{r}, \quad \frac{dz}{dt} = \frac{\partial z}{\partial r}\dot{r} = 2ar\dot{r}.$$

With $r^2 = x^2 + y^2$ the kinetic energy T therefore is

$$T = \frac{1}{2}m\left(\frac{d\mathbf{r}}{dt}\right)^2 = \frac{1}{2}m\left[\left(\frac{dx}{dt}\right)^2 + \left(\frac{dy}{dt}\right)^2 + \left(\frac{dz}{dt}\right)^2\right] = \frac{1}{2}m[\dot{r}^2 + r^2(\omega + \alpha t)^2 + 4a^2r^2\dot{r}^2]. \tag{4.34}$$

(b) The kinetic energy of the bead is

$$T = \frac{1}{2}m[\dot{r}^2 + \dot{z}^2 + r^2\dot{\theta}^2], \quad \text{where} \quad z = ar^2, \quad \dot{z} = 2ar\dot{r}. \tag{4.35}$$

With Eq. (4.34) we can rewrite this as

$$T = \frac{1}{2}m[(1 + 4a^2r^2)\dot{r}^2 + r^2(\omega + \alpha t)^2]. \tag{4.36}$$

Adding the potential energy $V = mgz = mgar^2$, we obtain the total energy E of the bead,

$$E = T + V = \frac{1}{2}m[(1 + 4a^2r^2)\dot{r}^2 + r^2(\omega + \alpha t)^2] + mgar^2, \tag{4.37}$$

and as the difference the Lagrangian

$$L(r, \dot{r}, t) = T - V = \frac{1}{2}m[(1 + 4a^2r^2)\dot{r}^2 + r^2(\omega + \alpha t)^2] - mgar^2. \tag{4.38}$$

The Hamiltonian H is defined by the Legendre transform

$$H(q_i, p_i, t) = \sum_{i=1}^{n} p_i\dot{q}_i - L(q_i, \dot{q}_i, t), \quad \text{where} \quad \dot{q}_i = \frac{\partial H}{\partial p_i}, \quad \dot{p}_i = -\frac{\partial H}{\partial q_i}.$$

In the present case

$$H(r, p_r, t) = p_r\dot{r} - L(r, \dot{r}, t), \quad \text{where} \quad p_r = \frac{\partial L}{\partial \dot{r}}. \tag{4.39}$$

Since V is independent of $\dot{r}$, we have $p_r = \partial T/\partial \dot{r} = m(1 + 4a^2r^2)\dot{r}$. The Hamiltonian is therefore:

$$H = m(1 + 4a^2r^2)\dot{r}^2 - L = \frac{1}{2}m[(1 + 4a^2r^2)\dot{r}^2 - r^2(\omega + \alpha t)^2] + mgar^2. \tag{4.40}$$

Comparing with Eq. (4.37) we see that $H \neq E$. The term responsible for this inequality is the contribution $r^2(\omega + \alpha t)^2$, which, as we observed at the beginning, has its origin in the time dependence of the transformation (4.33).

(c) With $p_r = m(1 + 4a^2r^2)\dot{r}$ we eliminate $\dot{r}$ from Eq. (4.40) and obtain

$$H(r, p_r, t) = \frac{p_r^2}{2m(1 + 4a^2r^2)} - \frac{1}{2}mr^2(\omega + \alpha t)^2 + mgar^2. \qquad (4.41)$$

Hamilton's equations now yield:

$$\dot{r} = \frac{p_r}{m(1 + 4a^2r^2)}, \quad \dot{p}_r = \frac{4a^2p_r^2 r}{m(1 + 4a^2r^2)^2} + mr(\omega + \alpha t)^2 - 2mgar.$$

4.4 Variational Principle for Hamilton's Equations

Earlier we derived the Euler–Lagrange equations from a Hamilton principle. Similarly we can obtain Hamilton's canonical equations of motion also from a variational principle. The main difference between the two cases is that here the variations of q_i and p_i are considered as being independent. From Eq. (4.3) we obtain

$$L = \sum_i \dot{q}_i p_i - H(q, p, t), \qquad (4.42)$$

and the variational principle is given by

$$\delta I = \delta \int_{t_1}^{t_2} L \, dt = 0, \quad i.e. \quad \delta \int_{t_1}^{t_2} [\sum_i p_i \dot{q}_i - H(q, p, t)] dt = 0. \qquad (4.43)$$

We assume that the significance of the parameter α as used in the previous application of the variational principle is clear, and can therefore be suppressed in the following or left understood. Thus we write δI instead of $\delta\alpha(\partial I / \partial \alpha)$. The coordinates $q_i = q_i(\alpha, t)$ depend in general explicitly on t. However (*cf.* our earlier detailed discussion in Sec. 3.5.2), for the *virtual displacements*, that we consider: (a) $\delta t = 0$, and (b) at the *fixed endpoints* (with $t = t_i, t_f$) both $\delta q_i = 0$ and $\delta p_i = 0$. Then

$$0 = \int_{t_1}^{t_2} dt \sum_i \left(\delta p_i \dot{q}_i + p_i \delta \dot{q}_i - \frac{\partial H}{\partial q_i} \delta q_i - \frac{\partial H}{\partial p_i} \delta p_i \right). \qquad (4.44)$$

But (in the second step variation at constant t)

$$\int_{t_1}^{t_2} dt p_i \delta \dot{q}_i = \int_{t_1}^{t_2} dt p_i \delta \left(\frac{dq_i}{dt} \right) = \int_{t_1}^{t_2} dt p_i \frac{d(\delta q_i)}{dt}$$

$$= p_i \delta q_i \big|_{t_1}^{t_2} - \int_{t_1}^{t_2} dt \dot{p}_i \delta q_i = - \int_{t_1}^{t_2} dt \dot{p}_i \delta q_i, \qquad (4.45)$$

where we used partial integration. Inserting this result into Eq. (4.44), it follows that

$$\int_{t_1}^{t_2} dt \sum_i \left\{ \delta p_i \left(\dot{q}_i - \frac{\partial H}{\partial p_i} \right) - \delta q_i \left(\dot{p}_i + \frac{\partial H}{\partial q_i} \right) \right\} = 0. \qquad (4.46)$$

Since the variations $\delta q_i, \delta p_i$ are independent, we obtain *Hamilton's equations,*

$$\dot{q}_i = \frac{\partial H}{\partial p_i}, \quad \dot{p}_i = -\frac{\partial H}{\partial q_i}, \qquad (4.47)$$

in agreement with Eq. (4.7). We emphasize: In the Lagrange formalism only the q_i are independent; the generalized velocities $\dot{q}_i$ are there considered to be dependent variables. Here however, in Hamilton's formalism, q_i and p_i are independent.

As a side-remark, we also point out here the so-called *"symplectic structure"* of Hamilton's equations. By this one means that their compact matrix form involves the skew-diagonal *symplectic matrix, i.e.* we can combine Hamilton's equations in the form:

$$\begin{pmatrix} \dot{q}_i \\ \dot{p}_i \end{pmatrix} = \begin{pmatrix} \mathbb{0} & \mathbb{1}_{i \times i} \\ -\mathbb{1}_{i \times i} & \mathbb{0} \end{pmatrix} \begin{pmatrix} \partial H/\partial q_i \\ \partial H/\partial p_i \end{pmatrix}. \qquad (4.48)$$

The *symplectic matrix* has unit determinant, *i.e.*

$$\begin{vmatrix} \mathbb{0} & \mathbb{1}_{i \times i} \\ -\mathbb{1}_{i \times i} & \mathbb{0} \end{vmatrix} = 1.$$

4.5   Transformation of Canonical Coordinates

The transformation from one system of canonical coordinates to another, or from one system of curvilinear coordinates to another, is an important topic that we shall deal with in detail later. However, it is convenient at this stage to consider one vital aspect of this, namely the transformation of a *volume element* $\prod_i dq_i dp_i$ to another $\prod_i du_i dv_i$ for a given set of curvilinear transformations

$$q_i = q_i(u, v), \quad p_i = p_i(u, v). \qquad (4.49)$$

Maybe the quickest way to obtain the result of the transformation is to use geometry and consider the lowest dimensional case, *i.e.* that of an area.[§]

[§]This is required, for instance, in the evaluation of multiple integrals. A rigorous treatment of the appropriate transformations together with the derivation of the Jacobian can be found in the treatise of R. Courant [12], there in Vol. II, Chapter IV, pp. 247 – 254, 374 – 381.

Thus in Fig. 4.2 we sketch coordinates q and p with origin O at $q = 0, p = 0$ and[¶] $OO' = dq, OR' = dp$. Hence

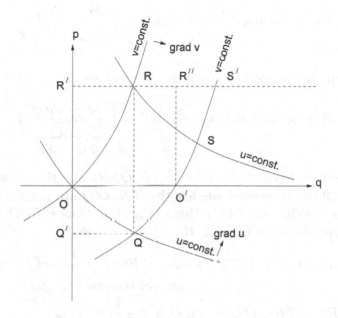

Fig. 4.2 Relating area elements $dqdp = OO'R''R'$ and $dudv = OQSR$.

$$\text{area }(OO'R''R') = dqdp. \tag{4.50}$$

We want to relate this to the area element $OQSR$ in terms of the new coordinates u and v. All the infinitesimal elements we are concerned with can be considered as straight. With this in mind, we see that

$$
\begin{aligned}
\text{area of parallelogram } (OQSR) \;&=\; \text{area of parallelogram } (OO'S'R) \\
&=\; \text{area of parallelogram } (OO'R''R') \\
&=\; dqdp. \tag{4.51}
\end{aligned}
$$

Now, as stated, the origin O has coordinates $q(u,v) = 0, p(u,v) = 0$. The point S is displaced infinitesimally in the coordinates u, v, *i.e.* it has coordinates

$$
\begin{aligned}
S(q,p) \;&=\; (q(u+du, v+dv), p(u+du, v+dv)) \\
&=\; (0,0) + \left(\frac{\partial q}{\partial u}du + \frac{\partial q}{\partial v}dv, \frac{\partial p}{\partial u}du + \frac{\partial p}{\partial v}dv \right). \tag{4.52a}
\end{aligned}
$$

[¶]In the following we write immediately dq, dp instead of $\delta q, \delta p$.

The corresponding coordinates of the point R are (since there v has the same value as at the origin):

$$R(q,p) = (q(u+du,v), p(u+du,v)) = \left(\underbrace{\frac{\partial q}{\partial u}du}_{R'R}, \underbrace{\frac{\partial p}{\partial u}du}_{OR'}\right). \qquad (4.52b)$$

Correspondingly the coordinates of the point Q are (u as at O):

$$Q(q,p) = (q(u,v+dv), p(u,v+dv)) = \left(\underbrace{\frac{\partial q}{\partial v}dv}_{Q'Q}, \underbrace{\frac{\partial p}{\partial v}dv}_{OQ'}\right). \qquad (4.52c)$$

The area we wish to compute is the area $(OQSR)$, and this is twice the area of triangle OQR. In general the lengths $R'R$, $Q'Q$ are not equal, although they may seem so in a sketch like that of Fig. 4.2. Thus $R'RQQ'$ is a polygon with two opposite edges parallel. Hence

$$\text{area of triangle } ROQ = \text{area of } R'RQQ' - \text{ area of triangle } R'RO$$
$$-\text{area of triangle } OQ'Q$$
$$= \frac{1}{2}\underbrace{R'Q'}_{R'O+OQ'}(R'R+Q'Q) - \underbrace{\frac{1}{2}R'O.R'R}_{\text{positive, }p>0} - \underbrace{\frac{1}{2}OQ'.Q'Q}_{\text{negative, }p<0}. \qquad (4.53)$$

Inserting the explicit expressions for the various lengths, we obtain

$$\text{area of triangle } ROQ$$
$$= \frac{1}{2}\left(\frac{\partial p}{\partial u}du - \frac{\partial p}{\partial v}dv\right)\left(\frac{\partial q}{\partial u}du + \frac{\partial q}{\partial v}dv\right) - \frac{1}{2}\frac{\partial p}{\partial u}du\frac{\partial q}{\partial u}du + \frac{1}{2}\frac{\partial p}{\partial v}dv\frac{\partial q}{\partial v}dv$$
$$= \frac{1}{2}\frac{\partial p}{\partial u}du\frac{\partial q}{\partial v}dv - \frac{1}{2}\frac{\partial p}{\partial v}dv\frac{\partial q}{\partial u}du = \frac{1}{2}\begin{vmatrix} \partial p/\partial u & \partial p/\partial v \\ \partial q/\partial u & \partial q/\partial v \end{vmatrix}dudv. \qquad (4.54)$$

It follows that the area of the entire element, *i.e.* the area of the parallelogram $OQSR = dqdp$ (*cf.* Eq. (4.51)) is twice that of this triangle, and we obtain

$$dqdp = \begin{vmatrix} \partial p/\partial u & \partial p/\partial v \\ \partial q/\partial u & \partial q/\partial v \end{vmatrix}dudv. \qquad (4.55)$$

We do not enter into a more general and more sophisticated derivation here, and therefore resort to the "sweeping statement" that the generalization to higher dimensions is selfevident or can be established in a similar way.[‖] The determinant appearing in the result (4.37) is known as *Jacobi determinant*;

[‖]See in particular R. Courant [12], Vol. II, Chapter IV.

this is a very important quantity, as one can realize from the importance of transformations from one coordinate system to another. Our Hamiltonian (q, p) notation above should not suggest that the result applies only there. Thus below we give a trivial example for the change from Cartesian to polar coordinates. We have introduced the Jacobi determinant here, because a quantity known as *Lagrange bracket* will be introduced below, which will be seen to be very similar to the Jacobi determinant, and the proof of the invariance of the associated *Poisson bracket* under canonical transformations utilizes this.

Example 4.6: Transforming from Cartesian to polar coordinates
In the case of 2-dimensional polar coordinates (r, θ) the element of area is seen geometrically to be $rdrd\theta$. Verify this result using the Jacobi determinant.

Solution: The transformation from Cartesian to polar coordinates in two dimensions is given by the relations $x = r\cos\theta, y = r\sin\theta$. It follows that Eq. (4.55) implies

$$dxdy = \begin{vmatrix} \partial x/\partial r & \partial x/\partial\theta \\ \partial y/\partial r & \partial y/\partial\theta \end{vmatrix} drd\theta = \begin{vmatrix} \cos\theta & -r\sin\theta \\ \sin\theta & r\cos\theta \end{vmatrix} drd\theta = rdrd\theta.$$

4.6 Lagrange and Poisson Brackets

4.6.1 The fundamental Lagrange and Poisson brackets

We now introduce the concepts of *Lagrange brackets* and *Poisson brackets*, the latter of which serves as a vital quantity to understand the step from classical mechanics to the postulates of quantum mechanics. We shall see that in a certain sense one bracket is the inverse of the other. Following our earlier arguments, we begin with the (less popular) Lagrange brackets.

Let $u_i, v_j, i = 1, 2, \ldots, n$, be $2n$ independent functions of the n coordinates q_i and the n momenta p_j, which can be inverted to yield the latter. The so-called *Lagrange bracket* of u_i and v_j is defined as the following relation with q_k, p_k appearing in numerators:

$$\{u_i, v_j\}_{q,p} = \sum_{k=1}^{n} \left(\frac{\partial q_k}{\partial u_i} \frac{\partial p_k}{\partial v_j} - \frac{\partial p_k}{\partial u_i} \frac{\partial q_k}{\partial v_j} \right). \tag{4.56}$$

Later we shall see that this expression is invariant under canonical transformations and therefore one generally drops the subscript q, p, *i.e.* one writes simply

$$\{u_i, v_j\}_{q,p} \to \{u_i, v_j\}.$$

We deduce immediately from Eq. (4.56) that

$$\{u_i, v_j\} = -\{v_j, u_i\}. \tag{4.57}$$

In these expressions u_i, v_j are coordinates of a two-dimensional domain in phase space. In particular we can choose: $u_i = q_i, v_j = q_j$. Then

$$\{u_i, v_j\} = \{q_i, q_j\} = \sum_{k=1}^{n} \left(\frac{\partial q_k}{\partial q_i} \frac{\partial p_k}{\partial q_j} - \frac{\partial p_k}{\partial q_i} \frac{\partial q_k}{\partial q_j} \right). \tag{4.58}$$

Since (in our consideration of canonical coordinates) q_i and p_j are independent coordinates, we have

$$\frac{\partial p_k}{\partial q_j} = 0 = \frac{\partial p_k}{\partial q_i}.$$

Hence

$$(a) \quad \{q_i, q_j\} = 0, \quad \text{and similarly} \quad (b) \quad \{p_i, p_j\} = 0, \tag{4.59}$$

in the latter case choosing $u = p_i, v = p_j$. However, choosing $u = q_i, v = p_j$, the expression $\{u, v\}$ becomes

$$(c) \quad \{q_i, p_j\} = \sum_{k} \left(\frac{\partial q_k}{\partial q_i} \frac{\partial p_k}{\partial p_j} - \frac{\partial p_k}{\partial q_i} \frac{\partial q_k}{\partial p_j} \right) = \sum_{k} \delta_{ki} \delta_{kj} = \delta_{ij}. \tag{4.60}$$

The relations (4.59) and (4.60) are known as *fundamental Lagrange brackets*. They are valid for any set of canonical variables.

It turns out that another set of brackets is more useful. These are the so-called *Poisson brackets* defined by the following relation with q_k, p_k now appearing in denominators:

$$[u, v]_{q,p} \equiv \sum_{k=1}^{n} \left(\frac{\partial u}{\partial q_k} \frac{\partial v}{\partial p_k} - \frac{\partial u}{\partial p_k} \frac{\partial v}{\partial q_k} \right) = -[v, u]_{q,p}, \tag{4.61}$$

with the property

$$[u, v] = -[v, u]. \tag{4.62}$$

4.6.2   Connection between Lagrange and Poisson brackets

We now proceed to establish the connection between Lagrange and Poisson brackets. We let $\{u_l\}, l = 1, 2, \dots, 2n$, be a set of $2n$ *independent functions* with the property that every u_l is a function of the n canonical q_i and the n canonical p_j, *i.e.* (observe that u_l now stands for both u_i, v_i used previously)

$$u_l = u_l(q_1, q_2, \dots, q_n, p_1, p_2, \dots, p_n), \quad \text{as well as} \quad q_i = q_i(u_l), p_i = p_i(u_l).$$

Then — this will be shown next —

$$\sum_{l=1}^{2n} \{u_l, u_i\}[u_l, u_j] = \delta_{ij}. \tag{4.63}$$

We prove this relation as follows. Inserting the explicit expressions for the brackets on the left hand side, we obtain:

$$\sum_{l=1}^{2n}\{u_l, u_i\}[u_l, u_j]$$

$$= \sum_{l=1}^{2n}\sum_{k,m=1}^{n}\left(\frac{\partial q_k}{\partial u_l}\frac{\partial p_k}{\partial u_i} - \frac{\partial p_k}{\partial u_l}\frac{\partial q_k}{\partial u_i}\right)\left(\frac{\partial u_l}{\partial q_m}\frac{\partial u_j}{\partial p_m} - \frac{\partial u_l}{\partial p_m}\frac{\partial u_j}{\partial q_m}\right). \quad (4.64)$$

We consider now the first factor on the right. Since previously

$$p_k = p_k(u, v), \quad q_k = q_k(u, v), \quad u_i = u_i(q, p), \quad v_i = v_i(q, p), \quad (4.65)$$

we have

$$\delta_{km} = \frac{dq_k}{dq_m} = \sum_{i=1}^{n}\left(\frac{\partial q_k}{\partial u_i}\frac{\partial u_i}{\partial q_m} + \frac{\partial q_k}{\partial v_i}\frac{\partial v_i}{\partial q_m}\right). \quad (4.66)$$

Renaming v_i as follows:

$$v_i := u_{n+i}, \quad i = 1, 2, \ldots, n, \quad (4.67)$$

we have

$$\frac{dq_k}{dq_m} = \sum_{i=1}^{2n}\frac{\partial q_k}{\partial u_i}\frac{\partial u_i}{\partial q_m}. \quad (4.68)$$

We use this relation in order to rewrite the first term of Eq. (4.64) (when multiplied out) in a different form:

$$\sum_{k,m=1}^{n}\left(\frac{\partial q_k}{\partial u_l}\frac{\partial p_k}{\partial u_i}\right)\left(\frac{\partial u_l}{\partial q_m}\frac{\partial u_j}{\partial p_m}\right) = \sum_{k,m=1}^{n}\frac{\partial p_k}{\partial u_i}\frac{\partial u_j}{\partial p_m}\sum_{l=1}^{2n}\left(\frac{\partial q_k}{\partial u_l}\frac{\partial u_l}{\partial q_m}\right)$$

$$= \sum_{k,m=1}^{n}\frac{\partial p_k}{\partial u_i}\frac{\partial u_j}{\partial p_m}\frac{dq_k}{dq_m} = \sum_{k,m=1}^{n}\frac{\partial p_k}{\partial u_i}\frac{\partial u_j}{\partial p_m}\delta_{km}$$

$$= \sum_{k=1}^{n}\frac{\partial p_k}{\partial u_i}\frac{\partial u_j}{\partial p_k} = \sum_{k=1}^{n}\frac{\partial u_j}{\partial p_k}\frac{\partial p_k}{\partial u_i}. \quad (4.69)$$

The last term in Eq. (4.64) has the same form, except that p and q are exchanged. Together these terms give:

$$\sum_{k=1}^{n}\left(\frac{\partial u_j}{\partial p_k}\frac{\partial p_k}{\partial u_i} + \frac{\partial u_j}{\partial q_k}\frac{\partial q_k}{\partial u_i}\right) = \frac{\partial u_j}{\partial u_i} = \delta_{ij}. \quad (4.70)$$

The remaining terms in Eq. (4.64) are first

$$-\sum_{k,m=1}^{n} \frac{\partial p_k}{\partial u_i} \frac{\partial u_j}{\partial q_m} \left(\sum_{l=1}^{2n} \frac{\partial q_k}{\partial u_l} \frac{\partial u_l}{\partial p_m} \right) = -\sum_{k,m=1}^{n} \frac{\partial p_k}{\partial u_i} \frac{\partial u_j}{\partial q_m} \left(\frac{\partial q_k}{\partial p_m} \right) = 0, \quad (4.71)$$

since q_k and p_k are independent. As the second or complementary expression we have the the expression as in Eq. (4.71) but with p and q exchanged; hence we obtain also zero. This establishes the claimed result of Eq. (4.63).

Instead of a particular set of coordinates (q_k, p_k), we could have employed any other set of canonical transformations. The result Eq. (4.63) is therefore invariant under all coordinate transformations.

What is the use of Eq. (4.63)? We shall see, that with the help of this result we can evaluate the Poisson brackets without recourse to any particular system of coordinates. We can choose, for instance,

$$(u_1, u_2, \ldots, u_{2n}) = (q_1, q_2, \ldots, q_n, p_1, p_2, \ldots, p_n). \quad (4.72)$$

We proceed from Eq. (4.63) rewritten as:

$$\sum_{\lambda=1}^{2n} \{u_\lambda, u_\rho\}[u_\lambda, u_\mu] = \delta_{\rho\mu}, \quad \lambda, \rho, \mu = 1, 2, \ldots, 2n. \quad (4.73)$$

This expression is — split up into two parts — the following, in which we then introduce indices $i, j, k = 1, 2, \ldots, n$:

$$\sum_{\lambda=1}^{n} \{u_\lambda, u_\rho\}[u_\lambda, u_\mu] + \sum_{\lambda=n+1}^{2n} \{u_\lambda, u_\rho\}[u_\lambda, u_\mu] = \delta_{\rho\mu},$$

and therefore

$$\sum_{i=1}^{n} \{u_i, u_\rho\}[u_i, u_\mu] + \sum_{\lambda=1}^{n} \{u_{\lambda+n}, u_\rho\}[u_{\lambda+n}, u_\mu] = \delta_{\rho\mu},$$

or

$$\sum_{i=1}^{n} \{u_i, u_\rho\}[u_i, u_\mu] + \sum_{i=1}^{n} \{u_{n+i}, u_\rho\}[u_{n+i}, u_\mu] = \delta_{\rho\mu}.$$

Now setting

$$u_i = q_i, \quad u_{n+i} = p_i, \quad \text{with} \quad i = 1, 2, \ldots, n,$$

we obtain

$$\sum_{i=1}^{n} \{q_i, u_\rho\}[q_i, u_\mu] + \sum_{i=1}^{n} \{p_i, u_\rho\}[p_i, u_\mu] = \delta_{\rho\mu}. \quad (4.74)$$

We consider the following cases separately:

$$
\begin{array}{lll}
(1) & \rho = k, & \mu = j, \\
(2) & \rho = k, & \mu = n + j, \\
(3) & \rho = n + k, & \mu = j, \\
(4) & \rho = n + k, & \mu = n + j.
\end{array}
\tag{4.75}
$$

- *Case (1)*: In this case Eq. (4.74) implies (using Eq. (4.60))

$$
\sum_{i=1}^{n} \underbrace{\{q_i, q_k\}}_{0}[q_i, q_j] + \sum_{i=1}^{n} \underbrace{\{p_i, q_k\}}_{-\delta_{ik}}[p_i, q_j] = \delta_{kj},
$$

  i.e.
$$
-\sum_{i=1}^{n} \delta_{ik}[p_i, q_j] = \delta_{kj},
$$

  and hence
$$
[p_k, q_j] = -\delta_{kj}, \quad \text{or} \quad [q_j, p_k] = \delta_{jk}.
\tag{4.76a}
$$

- *Case (2)*: In this case Eq. (4.74) implies

$$
\sum_{i=1}^{n} \underbrace{\{q_i, q_k\}}_{0}[q_i, p_j] + \sum_{i=1}^{n} \underbrace{\{p_i, q_k\}}_{-\delta_{ik}}[p_i, p_j] = \delta_{k,n+j} = 0,
$$

  and hence
$$
[p_k, p_j] = 0.
\tag{4.76b}
$$

- *Case (3)*: In this case Eq. (4.74) implies

$$
\sum_{i=1}^{n} \underbrace{\{q_i, p_k\}}_{\delta_{ik}}[q_i, q_j] + \sum_{i=1}^{n} \underbrace{\{p_i, p_k\}}_{0}[p_i, q_j] = \delta_{n+k,j} = 0,
$$

  and hence
$$
[q_k, q_j] = 0.
\tag{4.76c}
$$

- *Case (4)*: This case yields the same result as case (1).

Equations (4.76a), (4.76b), (4.76c) are the *fundamental Poisson brackets*. One should note that (a) we obtained these equations with no reference to a particular set of canonical variables, and (b) the invariance of these equations under canonical transformations is self-evident, as in the case of the Lagrange brackets.

The fundamental Poisson brackets are of fundamental importance. In quantum mechanics, in which q_i and p_i are operators $\hat{q}_i$ and $\hat{p}_i$ in the space of state vectors (there called Hilbert space) represented by wave functions, the corresponding relations are the postulated canonical commutation relations (with $[A, B] = AB - BA$)

$$[\hat{q}_i, \hat{q}_j] = 0, \quad [\hat{p}_i, \hat{p}_j] = 0, \quad [\hat{q}_i, \hat{p}_j] = i\hbar\delta_{ij},$$

whose validity for the representation $\hat{p} = -i\hbar d/dq$ is readily verified, $\hbar$ being Planck's constant h divided by 2π.

We emphasize the following: Originally we defined a momentum p_i in the Lagrangian formalism as the conjugate momentum associated with the generalized coordinate q_i, and we observed that also an angular momentum can be such a momentum. The generalized coordinates q_i were considered as independent variables, the momenta as dependent variables. In the Hamilton formalism we proceeded again from the generalized coordinates q_i, but then considered these together with the momenta p_i as independent variables, these together being called *canonical variables*. In proceeding like this, we did not ask ourselves originally, whether we could handle all generalized coordinates and their momenta like this. Now, having derived the *fundamental Poisson brackets*, we can answer this question: Only those variables, *i.e.* coordinates and also momenta, are canonical which satisfy the fundamental Poisson brackets. Consider a three-dimensional *angular momentum* $\mathbf{L} = (L_1, L_2, L_3)$ or (L_x, L_y, L_z). In this case a Poisson bracket of the angular momenta like

$$\text{wrong}: \quad [L_i, L_j] = 0, \quad i, j = 1, 2, 3,$$

is *not* valid, as we shall see with the help of Eqs. (4.76a), (4.76b), (4.76c) below. Instead with the help of these fundamental Poisson brackets, *i.e.*

$$\begin{aligned}
[x, y] &= 0, \quad [y, z] = 0, \quad [z, x] = 0, \\
[p_x, p_y] &= 0, \quad [p_y, p_z] = 0, \quad [p_z, p_x] = 0, \\
[x, p_x] &= 1, \quad [y, p_y] = 1, \quad [z, p_z] = 1,
\end{aligned} \qquad (4.77)$$

one obtains the relation

$$[L_i, L_j] = \epsilon_{ijk}L_k. \qquad (4.78)$$

Thus, L_x, L_y, L_z are *derived quantities*, for which fundamental Poisson brackets do not hold, and which therefore cannot be considered as canonical momenta. As an example we consider $[L_x, L_y]$:[**]

$$
\begin{aligned}
[L_x, L_y] &= [yp_z - zp_y, zp_x - xp_z] \\
&= [yp_z, zp_x] - [yp_z, xp_z] - [zp_y, zp_x] + [zp_y, xp_z] \\
&= y[p_z, z]p_x - y[p_z, p_z]x - z[p_y, p_x]z + p_y[z, p_z]x \\
&= -yp_x - 0 - 0 + xp_y \\
&= L_z
\end{aligned}
$$

in agreement with Eq. (4.78).

4.7   The Poisson Algebra and its Significance

We have seen that *Hamilton's equations* can be obtained in a number of different ways, for instance, with a Legendre transform from the Lagrange function, or from a variational principle, *i.e.*

$$
\delta \int_{t_1}^{t_2} \left[\sum_i p_i \dot{q}_i - H(q, p) \right] dt = 0,
\tag{4.79}
$$

where $H(q, p)$ is the Hamilton function, and where now, different from the derivation of the Lagrange equation of motion, both q_i and p_i are considered as independent variables. One obtains, as we know, Hamilton's equations

$$
\dot{q}_i = \frac{\partial H}{\partial p_i}, \qquad \dot{p}_i = -\frac{\partial H}{\partial q_i}.
\tag{4.80}
$$

In this Hamilton formalism it is incorrect to consider a linear momentum p as $m\dot{q}$, where m is a mass, *i.e.* it is incorrect to consider the momentum as mass times velocity. Instead the momentum is looked at as a quantity of its own, that can be observed directly as such, and not like a velocity which requires the measurement of the spatial position of the particle at two different instants of time, *i.e.* as

$$
\dot{q} = \lim_{\delta t \to 0} \frac{q(t + \delta t) - q(t)}{\delta t}.
\tag{4.81}
$$

Real quantities like q and p which are directly observable are called *observables*. A system of mass-points is therefore described by a multitude of such variables, which together describe the *state* of this system. Hence any function u of q, p, *i.e.* $u(q, p)$, is again an observable. The difference between this

---

[**]In Sec. 4.7 we derive properties of the Poisson brackets and in particular their algebra, Eqs. (4.85a) to (4.85e). The following derivation is easiest with the help of this algebra.

and an arbitrary function $f(q, p, t)$ is that the entire time-dependence of the observable $u(q, p)$ is implicitly contained in q and p. Thus we can write the total time derivative of $u(q, p)$ on using Eq. (4.80):

$$\frac{d}{dt} u(q_i, p_i) = \sum_i \left(\frac{\partial u}{\partial q_i} \dot{q}_i + \frac{\partial u}{\partial p_i} \dot{p}_i \right) = \sum_i \left(\frac{\partial u}{\partial q_i} \frac{\partial H}{\partial p_i} - \frac{\partial u}{\partial p_i} \frac{\partial H}{\partial q_i} \right). \quad (4.82)$$

If we have only one degree of freedom ($i = 1$), the expression (4.82) is a *functional determinant* or *Jacobi determinant*. We defined a *Poisson bracket* previously in Eq. (4.61), *i.e.* as

$$[A, B] := \sum_i \left(\frac{\partial A}{\partial q_i} \frac{\partial B}{\partial p_i} - \frac{\partial A}{\partial p_i} \frac{\partial B}{\partial q_i} \right). \quad (4.83)$$

With this definition we can rewrite Eq. (4.82) as[††]

$$\frac{du(q, p)}{dt} = [u(q, p), H]. \quad (4.84)$$

This equation is — in analogy with Hamilton's equations for q and p — the equation of motion of the observable $u(q, p)$. We observe that this equation contains as special cases the Hamilton equations. We can therefore look at Eq. (4.84) as a generalization of Hamilton's equations. This suggests, that it is sensible to take a closer look at the above "symbol" called Poisson bracket.

   The following properties of the Poisson bracket, called *Poisson algebra*, can be verified:

$$\text{antisymmetry} : [A, B] = -[B, A], \quad (4.85a)$$

$$\text{linearity} : \quad [A, \alpha_1 B_1 + \alpha_2 B_2] = \alpha_1 [A, B_1] + \alpha_2 [A, B_2], \quad (4.85b)$$

$$\text{complex conjugation} * : \quad [A, B]^* = [A^*, B^*], \quad (4.85c)$$

(observables are real, but could be multiplied by a complex number),

$$[A, BC] = [A, B]C + B[A, C], \quad (4.85d)$$

$$\text{Jacobi identity} : \quad [A, [B, C]] + [B, [C, A]] + [C, [A, B]] = 0. \quad (4.85e)$$

The first three properties are readily seen to hold. The fourth property (4.85d) is useful in calculations. The last property, the *Jacobi identity*, is

---

[††]But note that if u contains an explicit t-dependence,

$$\frac{du(q, p, t)}{dt} = \frac{\partial u(q, p, t)}{\partial t} + [u, H].$$

verified later in Example 5.9. A further useful property of Poisson brackets is given in Example 4.7. As long as (like here) A, B, C are commuting quantities, it is irrelevant whether we write $[A, B]C$ or $C[A, B]$. In the case of noncommuting quantities as in quantum mechanics the ordering is as in Eq. (4.85d).

Evaluating the Poisson bracket for q_i, p_i, we obtain the *fundamental Poisson brackets*, as we saw earlier, *i.e.*

$$[q_i, q_j] = 0, \quad [p_i, p_j] = 0, \quad [q_i, p_k] = \delta_{ik}. \tag{4.86}$$

We shall now make an important observation. We can convince ourselves — also by considering an explicit example — that we can solve Eq. (4.84) which combines the Hamilton equations — solely by using the properties (4.85a) to (4.85e) together with Eq. (4.86), and this means without any recourse to the original definition of the Poisson bracket. For instance, if we want to evaluate $[A, B]$, where A and B are arbitrary observables, we expand A and B in powers of q and p, and apply the rules (4.85a) to (4.85e) until only expressions in terms of the fundamental Poisson brackets are left. Since Eq. (4.86) provides the values of these brackets, the expression $[A, B]$ is completely evaluated. We consider as an example the case of the linear harmonic oscillator, and we shall see that the original definition of the Poisson bracket is never used. One can also observe that the fact that q and p are ordinary numbers and $H(q, p)$ an ordinary function is of no significance. Since multiplicative constants are irrelevant here, we choose the Hamilton function in the conveniently simplified form (*i.e.* avoiding constants)

$$H(q, p) = \frac{1}{2}(p^2 + q^2). \tag{4.87}$$

From Eq. (4.84) we obtain

$$\dot{q} = -[H, q], \quad \dot{p} = -[H, p]. \tag{4.88}$$

Inserting $H(q, p)$ from Eq. (4.87) into the first of Eqs. (4.88), we obtain

$$
\begin{aligned}
\dot{q} &= -\frac{1}{2}[(p^2 + q^2), q] \overset{(4.85b)}{=} -\frac{1}{2}([p^2, q] + [q^2, q]) \\
&\overset{(4.85a)}{=} -\frac{1}{2}(-[q, p^2] - [q, q^2]) \\
&\overset{(4.85d)}{=} -\frac{1}{2}(-[q, p]p - p[q, p] - [q, q]q - q[q, q]) = \frac{1}{2} \cdot 2p \\
&= p.
\end{aligned}
\tag{4.89}
$$

Correspondingly we derive from the second of Eqs. (4.88) the result

$$\dot{p} = -q. \tag{4.90}$$

From Eqs. (4.89) and (4.90) we deduce

$$\ddot{q} = \dot{p} = -q, \quad \ddot{q} + q = 0,$$

from which we conclude that

$$q(t) = q_0 \cos t + p_0 \sin t, \quad \text{or} \quad q(t) = q_0 + p_0 t - \frac{1}{2} q_0 t^2 - \frac{1}{3!} p_0 t^3 + \cdots . \quad (4.91)$$

4.8   Miscellaneous Examples

Example 4.7: Derivatives of Poisson brackets

Show that for $u = u(q,p), v = v(q,p)$:

$$\frac{\partial}{\partial q_i}[u,v] = \left[\frac{\partial u}{\partial q_i}, v\right] + \left[u, \frac{\partial v}{\partial q_i}\right], \quad \frac{\partial}{\partial p_i}[u,v] = \left[\frac{\partial u}{\partial p_i}, v\right] + \left[u, \frac{\partial v}{\partial p_i}\right]. \quad (4.92)$$

Solution: We have, with the definition of the Poisson bracket,

$$
\begin{aligned}
\frac{\partial}{\partial q_i}[u,v] &= \frac{\partial}{\partial q_i} \sum_k \left(\frac{\partial u}{\partial q_k} \frac{\partial v}{\partial p_k} - \frac{\partial u}{\partial p_k} \frac{\partial v}{\partial q_k} \right) \\
&= \sum_k \left(\frac{\partial^2 u}{\partial q_i \partial q_k} \frac{\partial v}{\partial p_k} + \frac{\partial u}{\partial q_k} \frac{\partial^2 v}{\partial q_i \partial p_k} - \frac{\partial^2 u}{\partial q_i \partial p_k} \frac{\partial v}{\partial q_k} - \frac{\partial u}{\partial p_k} \frac{\partial^2 v}{\partial q_i \partial q_k} \right) \\
&= \sum_k \left(\frac{\partial u}{\partial q_k} \frac{\partial}{\partial p_k} \left(\frac{\partial v}{\partial q_i} \right) - \frac{\partial u}{\partial p_k} \frac{\partial}{\partial q_k} \left(\frac{\partial v}{\partial q_i} \right) \right. \\
&\qquad\quad \left. - \frac{\partial v}{\partial q_k} \frac{\partial}{\partial p_k} \left(\frac{\partial u}{\partial q_i} \right) + \frac{\partial v}{\partial p_k} \frac{\partial}{\partial q_k} \left(\frac{\partial u}{\partial q_i} \right) \right) \\
&= \left[\frac{\partial u}{\partial q_i}, v \right] + \left[u, \frac{\partial v}{\partial q_i} \right]. \quad (4.93)
\end{aligned}
$$

The second relation is obtained correspondingly.

Example 4.8: The electron in a cylindrical magneton

An electron of mass m and charge $-e$ moves between a wire of radius a with electric potential $\phi = -\phi_0$, and earthed ($\phi = 0$) concentric cylindrical conductor of radius R. There is a constant magnetic field B parallel to the axis of the arrangement. Derive the Hamilton function and show that there are three constants of the motion.*

Solution: The scalar electric potential ϕ and the magnetic vector potential $\mathbf{A}$ are respectively in terms of cylindrical coordinates r, θ, z:

$$\phi = -\phi_0 \frac{\ln(r/R)}{\ln(a/R)}, \quad \mathbf{A} = \frac{1}{2} Br \mathbf{e}_\theta = \left(0, \frac{1}{2} Br, 0 \right). \quad (4.94)$$

The Lagrangian L is therefore:

$$
\begin{aligned}
L &= T - V = \frac{1}{2} m \dot{\mathbf{r}}^2 + e\phi - e\dot{\mathbf{r}} \cdot \mathbf{A} \\
&= \frac{1}{2} m (\dot{r}^2 + r^2 \dot{\theta}^2 + \dot{z}^2) + e\phi - \frac{1}{2} eBr^2 \dot{\theta}. \quad (4.95)
\end{aligned}
$$

---

*For additional discussion see *The Physics Coaching Class* [48], problem 2075, p. 634.

The momenta are

$$p_r = \frac{\partial L}{\partial \dot{r}} = m\dot{r}, \quad p_\theta = \frac{\partial L}{\partial \dot{\theta}} = mr^2\dot{\theta} - \frac{1}{2}eBr^2, \quad p_z = \frac{\partial L}{\partial \dot{z}} = m\dot{z}, \tag{4.96}$$

and the Hamilton function H is therefore

$$
\begin{aligned}
H & = p_r\dot{r} + p_\theta\dot{\theta} + p_z\dot{z} - L = \frac{1}{2}m(\dot{r}^2 + r^2\dot{\theta}^2 + \dot{z}^2) - c\phi \\
& = \frac{p_r^2}{2m} + \frac{1}{2mr^2}\left(p_\theta + \frac{1}{2}eBr^2\right)^2 + \frac{p_z^2}{2m} - e\phi \\
& = \frac{1}{2m}\left[p_r^2 + \left(\frac{p_\theta}{r} + \frac{1}{2}eBr\right)^2 + p_z^2\right] - e\phi. \tag{4.97}
\end{aligned}
$$

Since $\partial H/\partial t = 0$, we have $H = \text{const}$. Since $\partial H/\partial\theta = 0$, also $\dot{p}_\theta = -\partial H/\partial\theta = 0, p_\theta = \text{const}$. Similarly $\partial H/\partial z = 0$ implies $p_z = \text{const}$. Thus there are three constants of the motion. The complete integration of the equations is too involved to be pursued here.

Example 4.9: An example with a problematic quantization
Try to solve as far as possible the problem defined by the following Lagrangian and obtain the Hamiltonian (which presents a problem in quantum mechanics):

$$L(a, \phi, \dot{a}, \dot{\phi}) = \frac{1}{2}[-a\dot{a}^2 + a + a^3\dot{\phi}^2].$$

Solution: In self-evident notation we have

$$p_a = \frac{\partial L}{\partial \dot{a}} = -a\dot{a}, \quad p_\phi = \frac{\partial L}{\partial \dot{\phi}} = a^3\dot{\phi} = \text{const.}, \quad \therefore \ \dot{\phi} = \frac{p_\phi}{a^3},$$

where the constancy of p_ϕ follows from the vanishing of $\partial L/\partial\phi$ in the Euler–Lagrange equation for ϕ. The Euler–Lagrange equation in a is:

$$\frac{d}{dt}\left(\frac{\partial L}{\partial \dot{a}}\right) - \frac{\partial L}{\partial a} = \frac{d}{dt}(-a\dot{a}) - \left[-\frac{1}{2}\dot{a}^2 + \frac{1}{2} + \frac{3}{2}a^2\dot{\phi}^2\right] = 0, \quad \frac{1}{2}\dot{a}^2 + a\ddot{a} + \frac{1}{2} + \frac{3}{2a^4}p_\phi^2 = 0,$$

This can be rewritten as (using $d(\dot{a}^2/2)/da = \ddot{a}$)

$$-\left\{\frac{d}{da}\left[a\frac{\dot{a}^2}{2}\right] + \frac{1}{2} + \frac{3}{2}\frac{p_\phi^2}{a^4}\right\} = 0, \quad a\frac{\dot{a}^2}{2} + \frac{a}{2} - \frac{p_\phi^2}{2a^3} = \text{const.}$$

Hence

$$\dot{a}^2 = \frac{p_\phi^2}{a^4} + \frac{2\text{const.}}{a} - 1, \quad \int^t dt = \pm\int^a \frac{a^2\,da}{\sqrt{p_\phi^2 + 2\text{const.}a^3 - a^4}}.$$

The integral on the right is an elliptic integral and can be evaluated with the help of Tables of elliptic integrals.[†] The Hamiltonian H is given by the Legendre transform as

$$H = p_\phi\dot{\phi} + p_a\dot{a} - L = \frac{p_\phi^2}{a^3} + p_a\left(-\frac{p_a}{a}\right) - \frac{1}{2}\left[-a\frac{p_a^2}{a^2} + a + a^3\frac{p_\phi^2}{a^6}\right] = -\frac{1}{2a}p_a^2 - \frac{1}{2}a + \frac{1}{2a^3}p_\phi^2.$$

This type of Hamiltonian presents a problem when it is to be quantized. The reason is that here in the Hamiltonian of classical mechanics the canonical quantities a, p_a commute, *i.e.* can be

---

[†]*E.g.* Tables of W. Gröbner and N. Hofreiter [21] or of I. S. Gradshteyn and I. Ryzhik [20].

written in any order. Thus with this classical Hamiltonian as the jumping bord into its quantum
mechanics, one faces the problem of *ordering*, *i.e.* of how to write p_a^2/a, *i.e.* as

$$\frac{1}{a}p_a^2 \quad \text{or as} \quad p_a\frac{1}{a}p_a, \quad \text{or as} \quad p_a^2\frac{1}{a},$$

since in quantum mechanics a and p_a do not commute. Fortunately such cases are rare, and in
the few cases that arise, one can help oneself with taking *e.g.* half of one possibility and half the
other, or similarly.

Chapter 5

Symmetries and Transformations

5.1 Introductory Remarks

Symmetries are of paramount importance in physics, and their recognition is frequently a key to the solution of a query. Symmetries are related to invariances of a theory under some kinds of transformations, and these again lead to conservation laws. It is evident that such transformations play a vital role in any theory. Symmetries and transformations are therefore the topic of this chapter, but their significance will be present everywhere and permeates through the entire subject.

5.2 Symmetries

We remarked already in the preceding chapters that the Lagrange function $L(q, \dot{q}, t)$ contained in Hamilton's variational principle is part of the postulate, *i.e.* this Lagrange function is not derived. At first sight one might think that the number of possible Lagrange functions, that could be written down with reference to applications is unlimited. But this is not the case. We know that the Lagrange function has the dimension of energy, and that multiplied by (or integrated over by) time has the dimension of *action*. Thus if we want to formulate the Lagrange function L for the case of free motion of a particle of mass m in a three-dimensional Euclidean space, we do not have many possibilities. Even, for instance, an expression like

$$\frac{1}{2}m\left[\dot{x}^2 + \frac{1}{3}\dot{y}^2 + \frac{1}{5}\dot{z}^2\right]$$

is excluded, since free motion implies that the motion in any of the x, y, z-directions is possible with equal right, and hence the coefficients of the three terms must be equal. This equality is described as a *symmetry*. Symmetry is expressed in terms of "*invariance*", *i.e.* that the form of the expression is retained under transformation of the coordinates x, y, z to another set x', y', z', so that*

$$\dot{x}^2 + \dot{y}^2 + \dot{z}^2 = \dot{x}'^2 + \dot{y}'^2 + \dot{z}'^2.$$

Which transformations are relevant (initially one might think very many) is determined by the space and its geometry. This means the expression we postulate for the Lagrange function of a problem is restricted by the condition that it has to exhibit specific invariances or symmetries. This implies an enormous restriction, and applies in particular to the force-free kinetic part of the Lagrange function. If the particle is in addition subjected to some forces like the gravitational force which acts in a particular direction, we cannot expect the corresponding potential to possess the invariances of the kinetic part. Thus *e.g.*, in the expression

$$L = T - V, \quad T = \frac{1}{2}m(\dot{x}^2 + \dot{y}^2 + \dot{z}^2), \quad V = mgz,$$

the potential V violates the rotational invariance of T.

An important aspect which has to be kept in mind in the investigation of the invariance of a theory defined by Hamilton's principle is the question of or requirement of validity of a law of nature (*e.g.* Newton's second law) when the observation is made from a different reference frame, or if a transition (*i.e.* transformation) is made to a different frame. The principle of validity of one and the same law of nature in different reference frames is described as the *principle of relativity*. In the following we consider such transformations, in particular the Galilei transformation, translations and rotations (and briefly also Lorentz transformations). Thereafter we consider canonical transformations and briefly generalized coordinate transformations. These transformations then permit us to present precise definitions of scalar, vector and tensor quantities.

5.3   The Galilei Transformation

In the literature one reads sometimes that coordinate or reference frames which are at rest with respect to the sun or a fixed star (or move with constant linear velocity) are described as "*inertial frames*". Newton's laws

---

*In the case of rotation through an angle θ around the z-axis, *i.e.* in the (x, y)-plane: $\dot{x}' = \dot{x}\cos\theta + \dot{y}\sin\theta, \dot{y}' = \dot{x}\sin\theta - \dot{y}\cos\theta$.

are postulated with respect to such a frame. Put quantitatively, this says that a freely moving particle in space (*i.e.* one not subjected to some force, so that the force on the right hand side of Newton's equation of motion is zero) which naturally has a nonvanishing mass m (and hence "*inertia*") moves such that

$$m\ddot{x} = 0.$$

Thus $m\dot{x} = \text{const.}$ Any such frame with $\dot{x} = \text{const.}/m$ is then called an *inertial frame.*[†] Newton's laws were postulated with respect to such frames.

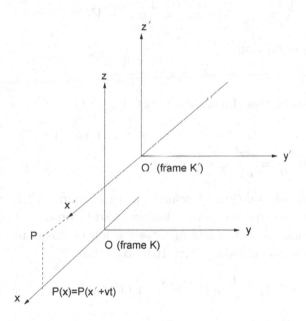

Fig. 5.1 Frame K moving in x-direction away from frame K'.

In transforming from one such system to another we have to distinguish whether we consider the transformation of the (scalar) Lagrange function (in this case one refers to the latter's *invariance*), or the vectorial equation of motion (in this case one refers to its *covariance* or form–invariance). The following transformation is referred to as the *special Galilei transformation*:

$$x \to x' = x - vt, \quad y \to y' = y, \quad z \to z' = z, \quad t' = t \tag{5.1}$$

where v is a constant (velocity) (in Einstein's Special Theory of Relativity the time given by the clock depends also on the motion; here in the strictly

---

[†]Thus for a massless particle (*e.g.* the photon or the graviton) there is no rest frame. Hence the name, since massive particles are those with inertia. See also Sec. 2.2.

nonrelativistic case not). These equations have their motivation in Newton's first law, which says that in every inertial system of coordinates the
force-free motion of a mass point m is given by $m\ddot{\mathbf{x}} = 0$. This means, the
transformation from one inertial frame to another is given by

$$\ddot{\mathbf{x}} = 0, \quad \ddot{\mathbf{x}}' = 0.$$

These equations imply for $\mathbf{x}$ parallel to $\mathbf{e}_x$ Eqs. (5.1).
    Consider the one-dimensional Lagrange function

$$L(x, \dot{x}, t) = \frac{1}{2}m\dot{x}^2 - V(x), \tag{5.2}$$

with equation of motion

$$m\ddot{x} = -\frac{\partial V}{\partial x}. \tag{5.3}$$

The simple Galilei transformation is (*cf.* Fig. 5.1)

$$x' = x - vt, \quad t' = t, \quad i.e. \quad x = x' + vt, \quad t = t', \quad dt = dt',$$

$$\dot{x} \equiv \frac{dx}{dt} = \frac{dx'}{dt'} + v = \dot{x}' + v. \tag{5.4}$$

This is the *law of addition of velocities*. In Example 12.1 we shall see the
relativistic form of this relation. The law of addition of velocities appears in
Newton's mechanics as the law of conservation of linear momentum, where
momentum is mass times velocity. It follows that

$$L(x, \dot{x}, t) = \frac{1}{2}m(\dot{x}' + v)^2 - V(x' + vt') \equiv \tilde{L}(x', \dot{x}', t'). \tag{5.5}$$

Then the Euler–Lagrange equation

$$\frac{d}{dt'}\left(\frac{\partial \tilde{L}}{\partial \dot{x}'}\right) - \frac{\partial \tilde{L}}{\partial x'} = 0$$

implies the equation

$$\frac{d}{dt'}[m(\dot{x}' + v)] + \frac{\partial}{\partial x'}V(x' + vt') = 0,$$

i.e.

$$m\frac{d^2x'}{dt'^2} = -\frac{\partial}{\partial x'}V(x' + vt'). \tag{5.6}$$

This Newton equation has the same form as that in the (x, t) coordinates,
i.e. Eq. (5.3). One says therefore, the equation is *form-invariant* or *covariant*
under Galilei transformations (*i.e.* both sides transform in the same way) in

distinction from the *invariance* of a scalar. One may also note that in view of the relation $\dot{x}' = \dot{x} - v$, also the velocity of light c would have a different value c' in the primed coordinates (this problem is solved in Einstein's Special Theory of Relativity by one of the two postulates which says that the velocity of light is the same irrespective of the moving coordinate frame).[‡]

One can extend the considerations to the so-called "*general Galilei trans-formation*"[§]

$$x_i' = \sum_j A_{ij} x_j - v_i t - c_i, \qquad t' = t - t_0, \qquad (5.7)$$

(A_{ij}, v_i, c_i, t_0 constants) which then depend on 10 constant parameters (A_{ij} implies 3 possible spatial rotations, as we shall see later). But one has to be careful here: A rotating reference frame or coordinate system (different from one which has been given a different orientation by rotation through a constant angle) is *not* an inertial frame, since the transformed acceleration contains additional terms which are described as *virtual forces*.[¶] Consider, for instance, rotation in a plane with constant angular velocity ω. In this case the change in the polar coordinates is

$$r \to r', \qquad \theta \to \theta' + \omega t,$$

so that the components of the force F, *i.e.* — these components are derived in Example 5.1 —

$$F_r = m(\ddot{r} - r\dot{\theta}^2), \qquad F_\theta = m(r\ddot{\theta} + 2\dot{r}\dot{\theta})$$

become

$$F_r' = F_r + 2mr'\omega\dot{\theta}' + m\omega^2 r', \qquad F_\theta' = F_\theta - 2m\omega\dot{r}'.$$

As one can see, ω has to be zero so that the new reference system is again an inertial system.

Consider finally the behaviour of the Lagrange function. We have (adding and subtracting a contribution $V(x')$)

$$L(x, \dot{x}, t) = L(x', \dot{x}', t') + \frac{1}{2}mv^2 + m\dot{x}'v + V(x') - V(x' + vt'). \qquad (5.8)$$

If the system is "*closed*", *i.e.* not exposed to forces from the environment, which means here $V = 0$, both L and the equation of motion are even invariant, because a constant contribution $mv^2/2$ can always be added to

---

[‡]See Sec. 12.2.
[§]This is actually a misleading or incorrect description.
[¶]This is discussed in more detail in Sec. 10.3.

the Lagrangian L, as also the total time derivative $d(mvx')/dt'$, because the variation of the corresponding part of the action vanishes, *i.e.*

$$\delta \int_{t_1}^{t_2} dt' \frac{d}{dt'}(mvx') = \delta[mvx']_{t_1}^{t_2} = 0.$$

In this case the real motion is the same in either system, whereas in the other case only in the same way, *i.e. same form*, which means with different external forces.

Example 5.1: Radial and transversal components of acceleration

Derive in an elementary way the radial and transversal accelerations of a particle on a two-dimensional trajectory, transversal meaning perpendicular to the radial direction.[||]

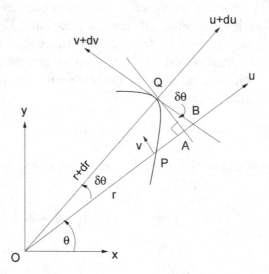

Fig. 5.2 Neighbouring points P and Q on an orbit.

Solution: Let r and θ be the polar coordinates of a particle at a point P and time t, and let Q be the point on the trajectory with coordinates $r + \delta r, \theta + \delta\theta$ at time $t + \delta t$. Also, let u be the radial velocity of the particle at point P and v its transversal velocity (so that these are the components of the velocity $\delta s/\delta t$, where δs is the arc PQ). At Q these velocities are $u + \delta u, v + \delta v$. Then, *cf.* Fig. 5.2,

$$u = \lim_{\delta t \to 0} \frac{OA - OP}{\delta t} = \lim_{\delta t \to 0} \frac{(r + \delta r)\cos\delta\theta - r}{\delta t} = \frac{dr}{dt},$$

$$v = \lim_{\delta t \to 0} \frac{AQ}{\delta t} = \lim_{\delta t \to 0} \frac{(r + \delta r)\sin\delta\theta}{\delta t} = r\frac{d\theta}{dt}.$$

Moreover:

$$\text{velocity at } Q \text{ parallel to } OP \quad = \quad (u + \delta u)\cos\delta\theta - (v + \delta v)\sin\delta\theta,$$

$$\text{velocity at } Q \text{ perpendicular to } OP \quad = \quad (u + \delta u)\sin\delta\theta + (v + \delta v)\cos\delta\theta.$$

[||]Transversal components should not be confused with tangential components of velocity or acceleration; for the latter see Sec. 7.4.

It follows therefore that the *radial acceleration* at P is

$$= \lim_{\delta t \to 0} \frac{(u + \delta u)\cos \delta\theta - (v + \delta v)\sin \delta\theta - u}{\delta t} = \frac{du}{dt} - v\frac{d\theta}{dt} = \frac{d^2 r}{dt^2} - r\left(\frac{d\theta}{dt}\right)^2,$$

and the *transversal acceleration* at P

$$= \lim_{\delta t \to 0} \frac{(u + \delta u)\sin \delta\theta + (v + \delta v)\cos \delta\theta - v}{\delta t} = u\frac{d\theta}{dt} + \frac{dv}{dt}$$

$$= \frac{dr}{dt}\frac{d\theta}{dt} + \left(\frac{dr}{dt}\frac{d\theta}{dt} + r\frac{d^2\theta}{dt^2}\right) = r\frac{d^2\theta}{dt^2} + 2\frac{dr}{dt}\frac{d\theta}{dt} = \frac{1}{r}\frac{d}{dt}(r^2\dot\theta).$$

These expressions will be referred to in Sec. 7.2 after derivation of Eqs. (7.8) and (7.9); see also Example 3.20.

5.4   Rotation and Rotation Group

5.4.1   Group property of coordinate transformations

It is fairly evident that two consecutively performed Galilei transformations or some other transformations imply another transformation from the original system to the second, so that the result of two transformations defines again a transformation. It is also immediately apparent that with two transformations one can return to the original system. In this way we recognize the *group characteristics* or *group property* of this and analogous transformations. We therefore take a closer look at these.

In order to fix or determine the location of a point in a *rigid body*, it is necessary (in a three-dimensional Euclidean space) to determine its distance to three arbitrarily chosen points $1, 2, 3$ in space. Once the positions of the points $1, 2, 3$ have been fixed, the constraints fix the positions of all other particles of the rigid body. Thus the location of the points $1, 2, 3$ determines the location of a point i, since in the case of a *rigid body* (this is what "rigid" means)

$$r_{1i} = \text{fixed}, \quad r_{2i} = \text{fixed}, \quad r_{3i} = \text{fixed}.$$

Thus the number of *degrees of freedom* of a point i in the rigid body is

$$\underbrace{3 \times 3}_{\mathbf{r}_{1,2,3}} \quad - \quad \underbrace{3}_{\text{3 constraints}} \quad = 6,$$

where the constraints are $r_{13} = c_{13}, r_{23} = c_{23}, r_{12} = c_{12}$ with c_{ij} constant. We can choose the 6 coordinates in the following way. We require three coordinates in order to fix the origin of the frame attached to or fixed in the rigid body, and we require three further coordinates in order to determine the position of an arbitrary point of the body with respect to this rigid body frame, or to determine the directions of an arbitrarily oriented Cartesian reference frame with origin at that of the rigid body frame, or — phrased differently — for the rotational orientation of the rigid body, *cf.* Fig. 5.3. The configuration space of the rigid body is therefore a 6-dimensional manifold C, which — in fact — is non-Euclidean.** Thus our first interest concerns the question, how one Cartesian reference frame can be related to another such frame with the same origin but with different relative orientation. There are several possibilities to approach this question, *e.g.* one can consider the orthogonal transformation, which corresponds to a rotation of one reference frame into the position of the other. For a better understanding of these transformations we recapitulate some basic definitions of the theory of continuous transformation groups.††

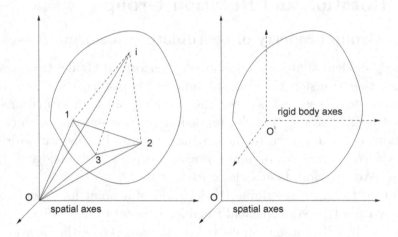

Fig. 5.3 Spatial axes and rigid body axes.

5.4.2   The group concept

We begin with some definitions.

Definition: the set $\{g\}$ of elements g is called a *group* G, if the set allows a relation called *multiplication* which associates uniquely with any two elements

---

**The non-Euclidean nature of the 6-manifold is beautifully demonstrated in the book of R. Penrose [38], pp. 217 – 220. Restricting himself to the rotational 3-manifold, Penrose considers a book with one end of a belt attached to it, the other end being held fixed. Rotating the book through 2π the belt develops a kink which cannot be shrunk away to no rotation, and thus the space is multiply-connected like a torus, on which some loops cannot be contracted continuously to a point.

††We consider only those aspects which are absolutely required for our purposes here. Anything further, also with aims at mathematical rigor, would only distract from the topic here. For this reason we do not refer to any of the many books on "Group Theory in Physics" here, since these usually contain much more than what is normally required, at least in a course like this.

of G again an element of G such that the following set of *axioms* holds:

- (a) The relation is *associative, i.e.*

$$(g_1 g_2)g_3 = g_1(g_2 g_3), \quad g_i \in G.$$

- (b) There exists a *unit element* $e \in G$, *i.e.* with every $g \in G$:

$$\exists \ e \in G \ \text{ with } \ eg = g = ge, \quad g \in G.$$

- (c) An element called *inverse element* g^{-1} exists, *i.e.*

$$\text{if} \ \ g \in G, \ \exists \ g^{-1} \in G \ \ \text{with} \ \ g^{-1}g = e = gg^{-1} \ \forall \ g \in G.$$

Definition: A *mapping* $f : G \to G'$ of a group G into a group G' is said to be *homomorphic*, if the relation between the elements is preserved, *i.e.* if for all $g_i \in G$ and $g_i \to f(g_i)$ it follows that $g_1 g_2 \to f(g_1)f(g_2)$. If every $g' \in G'$ appears as image $f(g)$ of an element $g \in G$, then f is called a *homomorphism* of G onto G'. The set of elements of G, which f maps onto the unit element $e' \in G'$ is called *kernel* K_f of the homomorphism f, put briefly: $K_f = f^{-1}(e')$. If f is a homomorphism from G onto G' with $K_f = e$, then f is unique. The mapping f is then called *isomorphic*, and the inverse map f^{-1} of f exists, which is obviously an *isomorphism* of G' onto G.

In physics the concept of groups appears predominantly in the form of *transformation groups*. Before we consider these in detail, we begin with some preliminary remarks. We shall consider rotations of a vector like $\mathbf{r} = (x, y, z)$. Writing the vector as a column matrix, the rotation is described by a matrix A as in

$$\begin{pmatrix} x' \\ y' \\ z' \end{pmatrix} = (A) \begin{pmatrix} x \\ y \\ z \end{pmatrix}.$$

In a rotation the length of the vector remains unchanged which implies

$$x'^2 + y'^2 + z'^2 = x^2 + y^2 + z^2,$$

so that the relation

$$(x', y', z') \begin{pmatrix} x' \\ y' \\ z' \end{pmatrix} = (x, y, z)A^T A \begin{pmatrix} x \\ y \\ z \end{pmatrix}$$

implies the *orthogonality relation* $A^T A = \mathbb{1}_{3\times 3}$. However — and this brings us to a more complicated but very important aspect — we can also represent a 3-component vector by a *complex* (2×2)-matrix with unit determinant, since

$$M = \begin{pmatrix} z & x - iy \\ x + iy & -z \end{pmatrix}, \ \det (M) = -z^2 - (x - iy)(x + iy) = -x^2 - y^2 - z^2.$$

The rotation is now given by the two-dimensional *similarity transformation*

$$M \to M' = UMU^{-1},$$

where U is a complex 2×2 matrix, since

$$
\begin{aligned}
\det M' &= \det(UMU^{-1}) = \det U.\det M.\det U^{-1} \\
&= \det M.\det(UU^{-1}) = \det M.
\end{aligned}
$$

The complex 2×2-matrices U introduced here describe rotations in a complex two-dimensional hyperspace. In the following we investigate both versions of description of rotations, as well as their connection (*i.e.* the isomorphism of the mapping). Thus in the following we consider examples of finite, continuous transformation groups. The transformations are uniquely characterized by a finite number of continuously variable parameters. We consider the transformations in the first place in an "active" sense, *i.e.* as transformations of the coordinates of vectors which leave the reference frame (*i.e.* the basis of the respective vector space) unchanged.

5.4.3   The orthogonal group $O(n)$

The so-called *orthogonal group* $O(n)$ consists of the real $n \times n$-matrices M, that satisfy the *orthogonality relation* (T meaning transposition)

$$M^T M = I_n, \quad I_n : \text{ unit matrix of dimension } n, \ \mathbb{1}_{n \times n}. \tag{5.9}$$

With the usual matrix multiplication as connecting relation, these matrices define a group with the unit element I_n. With $M_1^T M_1 = I_n, M_2^T M_2 = I_n$ one obtains

$$(M_1 M_2)^T (M_1 M_2) = M_2^T M_1^T M_1 M_2 = I_n.$$

Thus with M_1, M_2 belonging to $O(n)$, also $M_1 M_2$ belongs to $O(n)$. The associative law is trivially satisfied. Finally, in view of

$$(M^T)^T = M, \quad (M^T)^T M^T = I_n,$$

it follows that with every $M \in O(n)$, also the inverse $M^{-1} = M^T$ belongs to $O(n)$.

Example 5.2: Invariance of the length of a vector

Show that the invariance of the length of a three-dimensional vector **x** under rotations, *i.e.* orthogonal transformations A, implies the orthogonality condition (5.9) for the matrices A.

Solution: Invariance of the length of the vector implies that its square remains unchanged under the transformation, *i.e.* the real relation

$$\mathbf{x'}^2 = \mathbf{x}^2, \quad \text{or} \quad \sum_{i=1,2,3} x_i'^2 = \sum_{i=1,2,3} x_i^2,$$

where

$$x' = \begin{pmatrix} x_1' \\ x_2' \\ x_3' \end{pmatrix}, \quad x = \begin{pmatrix} x_1 \\ x_2 \\ x_3 \end{pmatrix}, \quad x' = Ax,$$

i.e.

$$x'^T x' = x^T A^T A x = x^T x, \quad \text{if} \quad A^T A = I_3.$$

Next we consider *some properties of the group* $O(n)$. First we observe that Eq. (5.9) implies the *boundedness of the matrix elements, i.e.*

$$\sum_m (M^T)_{km} M_{ml} = \sum_m M_{mk} M_{ml} = \delta_{kl},$$

implying

$$\sum_m (M_{mk})^2 = 1, \quad \text{and hence} \quad |M_{mk}| \le 1. \tag{5.10}$$

Thus the modulus of every matrix element is less than or equal to 1, a property we are already familiar with from the simple rotation in 2 or 3 dimensions, where the elements are given by trigonometrical expressions.

Since the matrices are real, every matrix $M \in O(n)$ has n^2 elements for which the orthogonality relation (5.9) yields

$$n \quad \text{equations } (M^T M)_{jj} = 1 \text{ from the diagonal, and}$$
$$\tfrac{1}{2}(n^2 - n) \quad \text{equations } (M^T M)_{ji} = (M^T M)_{ij}^T = (M^T M)_{ij} = 0. \tag{5.11}$$

The n equations originate from the diagonal elements of the unit matrix on the right of Eq. (5.9), and the others from the non-diagonal elements, of which there are $n^2 - n$, but in view of Eq. (5.9) and hence the second line of Eq. (5.11), only half are different. Thus altogether there are

$$n + \frac{1}{2}(n^2 - n) = \frac{1}{2}n(n + 1)$$

conditional equations. Hence every matrix $M \in O(n)$ has n^2 elements for which the orthogonality relation implies $n(n+1)/2$ equations. Thus the following number of *continuously variable independent parameters* remains:

$$n^2 - \frac{1}{2}n(n + 1) = \frac{1}{2}n(n - 1). \tag{5.12}$$

114

CHAPTER 5. Symmetries and Transformations

Example 5.3: Only one independent parameter in the case of $O(2)$
Show that there is only one independent parameter in the case of $O(2)$.

Solution: If a, b, c, d are the real elements of a typical matrix $M \in O(2)$, we have

$$MM^T = \begin{pmatrix} a & b \\ c & d \end{pmatrix} \begin{pmatrix} a & c \\ b & d \end{pmatrix} = \begin{pmatrix} a^2 + b^2 & ac + bd \\ ac + bd & c^2 + d^2 \end{pmatrix}.$$

Equating this to the two-by-two unit matrix, we obtain three equations, so that only one parameter of four is independent, *i.e.* we obtain

$$a^2 + b^2 = 1, \quad c^2 + d^2 = 1, \quad ac + bd = 0. \tag{5.13}$$

The group also allows so-called *discrete parameters*, as we now show. In view of Eq. (5.9) we have — using $\det(AB) = \det A.\det B$ — $\det(M^T M) = 1$, *i.e.* $\det M = d, d = \pm 1$. The parameter d which can assume the discrete values $+1$ or -1 is referred to as a discrete parameter. One should note that this is no restriction on a continuous parameter, as in the case of $SU(2)$ to be discussed below. The condition $\det M = \pm 1$ does not supply any additional condition on the continuous parameter, because these have to satisfy not only $M^T M = \mathbb{1}_{2 \times 2}$, but also the condition $\det M = +1$, since they have to evolve continuously from the *identity transformation*. The case $\det M = -1$ describes a *reflection* (*i.e.* so to speak a mirror image), in which case one coordinate is converted into its negative; however, this case does not evolve continuously from the identity.

5.4.4 The groups $O(2)$ and $SO(2)$

We now define a special group $SO(2)$ by demanding: The matrices

$$M = \begin{pmatrix} a & b \\ c & d \end{pmatrix} \in SO(2)$$

have to satisfy the conditions

$$M^T M = \mathbb{1}_{2 \times 2} \quad \text{and} \quad \det M = +1.$$

Here "S" of $SO(2)$ stands for "*Special*", meaning its determinant is $+1$. We can replace one of the three conditions (5.13) by the equation

$$\det M = +1, \quad \text{i.e. by} \quad ad - bc = +1. \tag{5.14}$$

Then, since (*cf.* Eq. (5.13)) $ac + bd = 0$,

$$a = -\frac{bd}{c} = \frac{ad}{d} = \frac{1 + bc}{d},$$

and hence

$$-bd^2 = c + bc^2, \quad i.e. \quad 0 = c + b(d^2 + c^2) \stackrel{(5.13)}{=} c + b,$$

i.e. $0 = c + b$, or $b = -c$. Thus there are three relations, $b = -c, a = d, a^2 + b^2 = 1$, and there is only one arbitrarily variable parameter. We make the ansatz:

$$a = \cos\varphi, \quad b = -\sin\varphi.$$

Then $\varphi = 0$ corresponds to the matrix $M = \mathbb{1}_{2\times 2}$,

$$I_2 \equiv \mathbb{1} = \begin{pmatrix} 1 & 0 \\ 0 & 1 \end{pmatrix},$$

and increasing φ describes a rotation in the positive (anti-clockwise) sense of rotation. In order to have a unique (single-valued) correspondence between the matrices $M \in SO(2)$ and the parameter values φ, we restrict the parameter φ to the domain

$$-\pi \leq \varphi \leq \pi.$$

This is necessary because the periodicity of the trigonometric functions would allow to associate with every matrix M a countably infinite number of parameter values $\varphi \pm 2n\pi$, where n is an integer. Then only $-I_2 \in SO(2)$ does not have a unique correspondence to φ in this interval, since $-I_2$ corresponds to $\varphi = \pm\pi$. Thus we have

$$M \equiv M(\varphi) = \begin{pmatrix} \cos\varphi & -\sin\varphi \\ \sin\varphi & \cos\varphi \end{pmatrix} \quad \text{with} \quad -\pi \leq \varphi \leq \pi. \tag{5.15}$$

Since $n(n-1)/2$ is 1 for $n = 2$, the angle φ is the only continuously variable parameter also of the group $O(2)$.

By an "*inversion*" one means the *reversion of all coordinates into their negatives, i.e.*

$$x \rightarrow -x, \quad y \rightarrow -y, \quad z \rightarrow -z$$

in the case of the three-dimensional Euclidean space. On the other hand, by "*reflection*" in a hyperplane, e.g. the (x, y)-plane, one means the reversion of the signs of the coordinates perpendicular to this plane (of reflection), *i.e.* in the given case the reversion $z \rightarrow -z$. In this way we can let the following matrix

$$Re_n := \begin{pmatrix} -1 & 0 & 0 & 0 & 0 \\ 0 & +1 & 0 & 0 & 0 \\ 0 & 0 & +1 & 0 & 0 \\ \cdot & \cdot & \cdot & \cdot & \cdot \\ 0 & 0 & 0 & 0 & +1 \end{pmatrix}, \quad (n \times n), \tag{5.16}$$

represent the reflection at an $(n-1)$-dimensional hyperplane. For the determinant we have $\det Re_n = -1$, and we can write

$$O(n) = SO(n) \cup Re_n SO(n). \tag{5.17}$$

In the case of $n = 2$ the set $Re_n SO(2)$ is given by

$$Re_2 M(\varphi) = \begin{pmatrix} -1 & 0 \\ 0 & 1 \end{pmatrix} \begin{pmatrix} \cos\varphi & -\sin\varphi \\ \sin\varphi & \cos\varphi \end{pmatrix} = \begin{pmatrix} -\cos\varphi & \sin\varphi \\ \sin\varphi & \cos\varphi \end{pmatrix}. \tag{5.18}$$

Thus every element of the group $O(2)$ can be written, with d denoting the discrete parameter,

$$M(d,\varphi) := \begin{pmatrix} d\cos\varphi & -d\sin\varphi \\ \sin\varphi & \cos\varphi \end{pmatrix}, \tag{5.19}$$

and

$$\det M(d,\varphi) = d = \pm 1, \quad \text{and} \quad M(1,\varphi) \in SO(2), \quad -\pi \le \varphi \le \pi.$$

We see that $O(2)$ is a mixed continuous transformation group, which is characterized by the continuous parameter φ and the discrete parameter d. From Eq. (5.19) we deduce the relation[‡‡]

$$M(d_1,\varphi_1)M(d_2,\varphi_2) = M(d_1 d_2, d_2\varphi_1 + \varphi_2).$$

5.4.5   The groups $O(3)$ and $SO(3) =: \{R\}$

We now consider three spatial Euclidean dimensions, and therein the group of transformations describing rotations, R. Since the real, *i.e.* proper, rotations must be those which evolve continuously from the identity transformation, the determinant of the matrices must be $+1$. These rotations are therefore described as *"proper rotations"*. The matrices R have three continuously variable, independent, real elements (from $n(n-1)/2$ for $n = 3$). We re-express these elements in terms of three parameters. In the problem of fixing the orientation of a rigid body, the transformation matrix, as a rotation matrix, has to be real. Then there is no difference between the property of *orthogonality*,

$$AA^T = \mathbf{1}, \quad (\text{also } A\tilde{A} = \mathbf{1}),$$

and the property of *unitarity*,

$$A^\dagger A = \mathbf{1}, \quad A^\dagger = A^{T*},$$

[‡‡] For $d_1 = d_2 = 1$, and φ_1, φ_2 arbitrary, two elements of the group $O(2)$ can be interchanged, *i.e.* they commute. These elements are precisely those of the group $SO(2)$. Thus the group $SO(2)$ is a *"commutative"* or *"Abelian"* subgroup of $O(2)$.

or between transposed and adjugated matrices; this means a real orthogonal matrix is unitary.

We add as a reminder, that every matrix transformation of the form

$$A' = BAB^{-1}$$

is called a *similarity transformation*.

In order to be able to describe the orientation of a rigid body in space, we require three independent parameters. Once we have these *generalized coordinates*, we can formulate the Lagrange function for the system and derive the equations of motion. One useful set of such parameters is the set of *three Euler angles*, which we now introduce. These angles are defined by the following three successive rotations as indicated in Fig. 5.4:

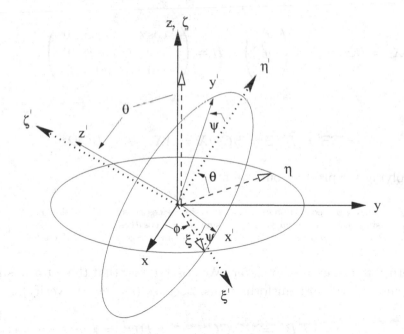

Fig. 5.4 The Euler angles ϕ, θ, ψ.

1. Rotation through angle ϕ anticlockwise around the z-axis; the new coordinate frame has coordinates ξ, η, ζ.

2. Rotation through angle θ anticlockwise around the ξ-axis to the new reference system ξ', η', ζ'.

3. Rotation through angle ψ anticlockwise around the ζ'-axis to the new reference system x', y', z'.

The angles ϕ, θ, ψ defined in this way and indicated in Fig. 5.4 determine completely the orientation of the (x', y', z') coordinate frame with respect to that of the (x, y, z)-frame, both having the same origin. The three angles, called *Euler angles*, can therefore serve as the required generalized coordinates. We have therefore:

$$
\Xi = DX, \quad \Xi = \begin{pmatrix} \xi \\ \eta \\ \zeta \end{pmatrix}, \quad X = \begin{pmatrix} x \\ y \\ z \end{pmatrix}, \quad D = \begin{pmatrix} \cos\phi & \sin\phi & 0 \\ -\sin\phi & \cos\phi & 0 \\ 0 & 0 & 1 \end{pmatrix},
$$

$$\text{(5.20a)}$$

$$
\Xi' = C\Xi, \quad \Xi' = \begin{pmatrix} \xi' \\ \eta' \\ \zeta' \end{pmatrix}, \quad C = \begin{pmatrix} 1 & 0 & 0 \\ 0 & \cos\theta & \sin\theta \\ 0 & -\sin\theta & \cos\theta \end{pmatrix}, \quad \text{(5.20b)}
$$

$$
X' = B\Xi', \quad X' = \begin{pmatrix} x' \\ y' \\ z' \end{pmatrix}, \quad B = \begin{pmatrix} \cos\psi & \sin\psi & 0 \\ -\sin\psi & \cos\psi & 0 \\ 0 & 0 & 1 \end{pmatrix}, \quad \text{(5.20c)}
$$

so that

$$
X' = B\Xi' = BC\Xi = BCDX = AX, \quad A = BCD. \quad \text{(5.21a)}
$$

Multiplying the product out, we have

$$
A = \begin{pmatrix} \cos\psi\cos\phi - \cos\theta\sin\phi\sin\psi & \cos\psi\sin\phi + \cos\theta\cos\phi\sin\psi & \sin\psi\sin\theta \\ -\sin\psi\cos\phi - \cos\theta\sin\phi\cos\psi & -\sin\psi\sin\phi + \cos\theta\cos\phi\cos\psi & \cos\psi\sin\theta \\ \sin\theta\sin\phi & -\sin\theta\cos\phi & \cos\theta \end{pmatrix}. \quad \text{(5.21b)}
$$

We return to this expression later. We observe now that the rotations B, C, D represent orthogonal transformations, because (as one can verify)

$$
BB^T = \mathbb{1}, \quad CC^T = \mathbb{1}, \quad DD^T = \mathbb{1}.
$$

It follows that the matrix A also represents an orthogonal transformation:

$$
AA^T = (BCD)(BCD)^T = \mathbb{1}.
$$

Thus $A^{-1} = A^T$. The transformations A, B, C, D represent *proper rotations, i.e.* those with unit determinant. Orthogonal transformations with determinant -1 are described as *improper* transformations. Since alternative parametrizations are very important, we now consider these.

5.4.6   The unitary groups $U(n)$ and $SU(n)$

The group $U(n)$ consists of the complex $n \times n$ matrices M which satisfy the *unitarity condition, i.e.*

$$M^{\dagger}M = I_n, \quad I_n \equiv \text{unit matrix of degree } n. \tag{5.22}$$

One can verify that these matrices define a group. Since $M^{\dagger}M = I_n$, it follows that $M^{\dagger}M$ is *Hermitian*:

$$(M^{\dagger}M)^{\dagger} = M^{\dagger}M = (\Re + i\Im), \quad i.e. \quad (\Re)_{kl} = (\Re)_{lk}, \; (\Im)_{kl} = -(\Im)_{lk}.$$

Thus the real part of the unitarity condition implies the following number of (diagonal + off-diagonal) conditions

$$n + \frac{1}{2}n(n-1) = \frac{1}{2}n(n+1),$$

and the imaginary part the number

$$\frac{1}{2}n(n-1)$$

of off-diagonal conditions. Thus altogether there are n^2 conditions. Every $M \in U(n)$ therefore possesses $2n^2 - n^2 = n^2$ real parameters. The real matrices of $U(n)$ obviously define the subgroup $O(n)$. A further subgroup of the group $U(n)$ is the *Special unitary group* $SU(n)$:

$$SU(n) := \{M | M \in U(n), \det M = +1\}. \tag{5.23}$$

We have

$$1 = \det(M^{\dagger}M) = (\det M)^* \det M = |\det M|^2.$$

It follows that $\det(M) = \exp(i\mu)$, so that the condition $\det(M) = +1$ reduces the number of continuously variable parameters by one, *i.e.* every element $M \in SU(n)$ has $n^2 - 1$ independent real parameters. We observe in particular, that the group $SU(2)$ has precisely 3 real independent parameters — exactly like the group $O(3)$. We therefore consider the group $SU(2)$ now in more detail.

The group $SU(2)$

   The group $SU(2)$ is the group of complex, unitary, unimodular 2×2 matrices (transformation matrices in a two-dimensional hyperspace with complex axes u, v, a vector in this space being called a "*spinor*")

$$U = \begin{pmatrix} \alpha & \beta \\ \gamma & \delta \end{pmatrix}, \quad U^{\dagger} = \begin{pmatrix} \alpha^* & \gamma^* \\ \beta^* & \delta^* \end{pmatrix}. \tag{5.24}$$

From $U^\dagger U = I_2$ we obtain $\beta^*\alpha + \delta^*\gamma = 0, \alpha\alpha^* + \gamma\gamma^* = 1$. Together with $\det U = \alpha\delta - \beta\gamma = +1$, these equations imply that

$$\gamma = -\beta^*, \quad \delta = \alpha^*, \quad |\alpha|^2 + |\beta|^2 = 1, \tag{5.25}$$

so that

$$U = \begin{pmatrix} \alpha & \beta \\ -\beta^* & \alpha^* \end{pmatrix} \quad \text{with} \quad |\alpha|^2 + |\beta|^2 = 1, \tag{5.26}$$

and

$$U^{-1} = U^\dagger = \begin{pmatrix} \alpha^* & -\beta \\ \beta^* & \alpha \end{pmatrix}.$$

We thus have two complex parameters together with one subsidiary condition — in complete agreement with the above counting of the number of parameters. The complex numbers α and β are named after their inventors and are therefore known as *Cayley–Klein parameters*.

Next we want to consider transformations. Let $\mathbf{s}, \mathbf{t}$ be elements of the space S_U in which an $SU(2)$ matrix P is defined as a linear operator. *e.g.*

$$\mathbf{s} = \begin{pmatrix} a + ib \\ c + id \end{pmatrix}$$

is a 2-component vector with complex components. We then demand that in a transformation (rotation) of the elements $\mathbf{s}, \mathbf{t}$, also the matrix P must be transformed, and, in fact, in such a way that the scalar product

$$(\mathbf{s}, M\mathbf{t}) = \mathbf{s}^\dagger M \mathbf{t}$$

remains invariant (*i.e.* the length):

$$(\mathbf{s}'^\dagger M' \mathbf{t}') = (\mathbf{s}^\dagger M \mathbf{t}).$$

The transformation

$$\mathbf{s}' = U\mathbf{s}, \quad \mathbf{t}' = U\mathbf{t}$$

therefore requires that

$$M' = UMU^{-1} \quad \text{and} \quad U^\dagger U = \mathbb{1}. \tag{5.27}$$

Now let P be a matrix operator in the space S_U as described above but with the special form

$$P = \begin{pmatrix} z & x - iy \\ x + iy & -z \end{pmatrix} = x\sigma_x + y\sigma_y + z\sigma_z, \tag{5.28}$$

where

$$\sigma_x = \begin{pmatrix} 0 & 1 \\ 1 & 0 \end{pmatrix}, \quad \sigma_y = \begin{pmatrix} 0 & -i \\ i & 0 \end{pmatrix}, \quad \sigma_z = \begin{pmatrix} 1 & 0 \\ 0 & 1 \end{pmatrix}. \tag{5.29}$$

Mathematically x, y, z of Eq. (5.28) can be looked at as three real parameters; physically, however, they are interpreted as the coordinates of a point in a real three-dimensional Euclidean space. The matrices $\sigma_x, \sigma_y, \sigma_z$ are known as *Pauli matrices*. We see, therefore, that the elements of the transformation group $O(3)$ and those of the group $SU(2)$ depend on three real continuously variable parameters, *i.e.* for instance x, y, z, which are the coordinates of a point in a three-dimensional Euclidean space. If we describe a rotation in this space by the group $O(3)$, the vector is represented by the column matrix or column vector

$$\begin{pmatrix} x \\ y \\ z \end{pmatrix}.$$

If we want to describe a rotation in this space with the help of the group $SU(2)$, the vector has to be represented by a matrix. Then this matrix must depend on three real continuous parameters. This is what the above matrix P does. It is important therefore to remember: In the context of $O(3)$ the vector $\mathbf{R} = (x, y, z)$ is a column matrix, whereas in the context of $SU(2)$ the vector $\mathbf{R}$ is represented by the 2×2 matrix P (an Hermitian matrix with trace zero has precisely three independent real parameters). Correspondingly the behaviour under transformations is:

$$O(3): \quad R' = UR, \ U \in O(3), \quad SU(2): \quad P' = UPU^{-1}, \ U \in SU(2). \tag{5.30}$$

Since U is unitary, we have $P' = UPU^\dagger = UPU^{-1}$. It follows that

$$\det(P') = \det(U)\det(P)\det(U^{-1}) = \det(P)\det(UU^{-1}) = \det(P).$$

Thus the transformation is simply a similarity transformation. It follows that P' is also Hermitian or selfadjoint and has trace zero, like P. Thus P' has the form

$$P' = \begin{pmatrix} z' & x' - iy' \\ x' + iy' & -z' \end{pmatrix}. \tag{5.31}$$

We also have (since P is invariant under similarity transformations)

$$\det(P) = -(x^2 + y^2 + z^2) = -(x'^2 + y'^2 + z'^2) = \det(P'). \tag{5.32}$$

We see that the length of the vector $\mathbf{r} = (x, y, z)$ remains unchanged under the transformation, *i.e.* the transformation satisfies the orthogonality condition. We thus have the following result: Corresponding to every unitary matrix U in the complex two-dimensional space there is a real orthogonal transformation in the usual three-dimensional space. The correspondence between the complex, unitary 2×2 matrices and the real, orthogonal 3×3 matrices is such, that any relation between the matrices in the one case is also satisfied by the set of matrices in the other case. The group $O(3)$ is therefore *isomorphic* to the group $SU(2)$.

The elements of the matrix M of the group $O(3)$ can be re-expressed in terms of the elements of the matrix U of $SU(2)$. We set

$$x_{\pm} = x \pm iy.$$

Then

$$
P' = \begin{pmatrix} z' & x'_- \\ x'_+ & -z' \end{pmatrix} = UPU^{\dagger} = \begin{pmatrix} \alpha & \beta \\ \gamma & \delta \end{pmatrix} \begin{pmatrix} z & x_- \\ x_+ & -z \end{pmatrix} \begin{pmatrix} \delta & -\beta \\ -\gamma & \alpha \end{pmatrix}
$$

$$
= \begin{pmatrix} (\alpha\delta + \beta\gamma)z - \alpha\gamma x_- + \beta\delta x_+ & -2\alpha\beta z + \alpha^2 x_- - \beta^2 x_+ \\ 2\gamma\delta z - \gamma^2 x_- + \delta^2 x_+ & -(\alpha\delta + \beta\gamma)z + \alpha\gamma x_- - \beta\delta x_+ \end{pmatrix}.
$$

$$(5.33)$$

From this relation we obtain the transformation equations:

$$
\begin{aligned}
x'_+ &= 2\gamma\delta z - \gamma^2 x_- + \delta^2 x_+, \\
x'_- &= -2\alpha\beta z + \alpha^2 x_- - \beta^2 x_+, \\
z' &= (\alpha\delta + \beta\gamma)z - \alpha\gamma x_- + \beta\delta x_+.
\end{aligned}
$$

$$(5.34)$$

From these equations we can deduce the matrix A of the $O(3)$ transformation

$$
\begin{pmatrix} x' \\ y' \\ z' \end{pmatrix} = (A) \begin{pmatrix} x \\ y \\ z \end{pmatrix}.
$$

$$(5.35)$$

One finds

$$
(A) = \begin{pmatrix} \frac{1}{2}(\alpha^2 - \gamma^2 + \delta^2 - \beta^2) & \frac{i}{2}(\gamma^2 - \alpha^2 + \delta^2 - \beta^2) & \gamma\delta - \alpha\beta \\ \frac{i}{2}(\alpha^2 + \gamma^2 - \beta^2 - \delta^2) & \frac{1}{2}(\alpha^2 + \gamma^2 + \beta^2 + \delta^2) & -i(\alpha\beta + \gamma\delta) \\ \beta\delta - \alpha\gamma & i(\alpha\gamma + \beta\delta) & \alpha\delta + \beta\gamma \end{pmatrix}.
$$

$$(5.36)$$

Thus we have a matrix which (together with $\alpha\delta - \beta\gamma = 1$) describes completely the orientation of a rigid body in terms of the Cayley–Klein parameters $\alpha, \beta, \gamma, \delta$.

The Cayley–Klein parameters can be re-expressed in terms of the Euler angles. This can be achieved in a variety of ways; the following procedure is particularly instructive. The angle ϕ was defined as the angle of a rotation around the z-axis, *i.e.* by the relation

$$\begin{pmatrix} \xi \\ \eta \\ \zeta \end{pmatrix} = \begin{pmatrix} \cos\phi & \sin\phi & 0 \\ -\sin\phi & \cos\phi & 0 \\ 0 & 0 & 1 \end{pmatrix} \begin{pmatrix} x \\ y \\ z \end{pmatrix}, \tag{5.37}$$

or with $\xi_\pm = \xi \pm i\eta$ by

$$\begin{pmatrix} \xi_+ \\ \xi_- \\ \zeta \end{pmatrix} = \begin{pmatrix} e^{-i\phi} & 0 & 0 \\ 0 & e^{i\phi} & 0 \\ 0 & 0 & 1 \end{pmatrix} \begin{pmatrix} x_+ \\ x_- \\ z \end{pmatrix}. \tag{5.38}$$

Comparison with Eq. (5.34) shows that this one rotation is given by the following matrix U, since $\gamma = 0$, $\delta^2 = \exp(-i\phi)$, $\beta = 0$, $\alpha^2 = \exp(i\phi)$,

$$U_\phi = \begin{pmatrix} e^{i\phi/2} & 0 \\ 0 & e^{-i\phi/2} \end{pmatrix} \overset{(5.41b)}{=} e^{i\frac{\phi}{2}\sigma_z}. \tag{5.39}$$

One finds analogously for the other two rotations:

$$U_\theta = \begin{pmatrix} \cos\frac{\theta}{2} & i\sin\frac{\theta}{2} \\ i\sin\frac{\theta}{2} & \cos\frac{\theta}{2} \end{pmatrix} = e^{i\frac{\theta}{2}\sigma_x}, \tag{5.40}$$

and (see remarks below)

$$U_\psi = \begin{pmatrix} e^{i\psi/2} & 0 \\ 0 & e^{-i\psi/2} \end{pmatrix} = \mathbb{1}\cos\frac{\psi}{2} + i\sigma_z\sin\frac{\psi}{2} = e^{i\sigma_z\psi/2}. \tag{5.41a}$$

Here in the final step we used the calculation (correspondingly in Eq. (5.39)):

$$e^{i\sigma_z\psi/2} = 1 + i\sigma_z\frac{\psi}{2} + \frac{1}{2!}\left(\frac{i\sigma_z\psi}{2}\right)^2 + \cdots = \left[1 - \frac{1}{2!}\left(\frac{\psi}{2}\right)^2\right.$$

$$\left. + \frac{1}{4!}\left(\frac{\psi}{2}\right)^4 + \cdots\right] + i\sigma_z\left[\frac{1}{1!}\left(\frac{\psi}{2}\right) - \frac{1}{3!}\left(\frac{\psi}{2}\right)^3 + \cdots\right]$$

$$= \mathbb{1}\cos\left(\frac{\psi}{2}\right) + i\sigma_z\sin\left(\frac{\psi}{2}\right). \tag{5.41b}$$

For the complete rotation we obtain therefore:

$$\begin{aligned} U &= U_\psi U_\theta U_\phi \\ &= \begin{pmatrix} e^{i(\psi+\phi)/2}\cos\frac{\theta}{2} & ie^{i(\psi-\phi)/2}\sin\frac{\theta}{2} \\ ie^{-i(\psi-\phi)/2}\sin\frac{\theta}{2} & e^{-i(\psi+\phi)/2}\cos\frac{\theta}{2} \end{pmatrix} \\ &\equiv \begin{pmatrix} \alpha & \beta \\ \gamma & \delta \end{pmatrix}. \end{aligned} \tag{5.42a}$$

Thus by comparison we obtain the Cayley–Klein parameters $\alpha, \beta, \gamma, \delta$ expressed in terms of the three Euler angles ϕ, θ, ψ:

$$
\begin{aligned}
\alpha &= e^{i(\psi+\phi)/2}\cos\frac{\theta}{2}, & \beta &= ie^{i(\psi-\phi)/2}\sin\frac{\theta}{2}, \\
\gamma &= ie^{-i(\psi-\phi)/2}\sin\frac{\theta}{2}, & \delta &= e^{-i(\psi+\phi)/2}\cos\frac{\theta}{2}.
\end{aligned}
\tag{5.42b}
$$

Together with the unit matrix

$$
\begin{pmatrix} 1 & 0 \\ 0 & 1 \end{pmatrix}
$$

the three Pauli matrices define a set of four independent matrices, so that every 2×2 matrix with 4 independent parameters may be expressed as a linear combination of these four matrices, *e.g.*

$$
\begin{aligned}
U_\theta &= \mathbb{1}\cos\left(\frac{\theta}{2}\right) + i\sigma_x\sin\left(\frac{\theta}{2}\right) = e^{i\sigma_x\theta/2} \quad (\text{rotation around the } x-\text{axis}), \\
U_\phi &= \mathbb{1}\cos\left(\frac{\phi}{2}\right) + i\sigma_z\sin\left(\frac{\phi}{2}\right) = e^{i\sigma_z\phi/2} \quad (\text{rotation around the } z-\text{axis}),
\end{aligned}
$$

$$
\tag{5.42c}
$$

and so on. Each of the three Pauli matrices therefore belongs to a rotation about a specific axis (rotation in a 2-dimensional complex hyperspace). Finally we note without further discussion that in the case of a non-trivial, physical transformation (*i.e.* rotation) there is only one vector which remains unchanged: This is the vector parallel to the axis of rotation. In Example 5.4 we deal with a number of vital properties of the very important Pauli matrices and demonstrate explicitly the isomorphism between the groups $SU(2)$ and $SO(3)$.

Example 5.4: Properties of Pauli matrices

Verify the following properties of Pauli matrices $\sigma_i, i = 1, 2, 3$, $\mathbf{n}$ being a unit vector:

(a) $U = \exp(i\theta\mathbf{n}\cdot\boldsymbol{\sigma}/2) = \mathbb{1}\cos(\theta/2) + i\mathbf{n}\cdot\boldsymbol{\sigma}\sin(\theta/2)$.

(b) $(\boldsymbol{\sigma}\cdot\mathbf{A})(\boldsymbol{\sigma}\cdot\mathbf{B}) = \mathbf{A}\cdot\mathbf{B} + i\boldsymbol{\sigma}\cdot(\mathbf{A}\times\mathbf{B})$.

(c) Let $R := \mathbf{r}\cdot\boldsymbol{\sigma}$. With the help of (a) and (b) show that the rotation $R' \equiv \mathbf{r}'\cdot\boldsymbol{\sigma} = URU^{-1}$ leads to the following expression for $\mathbf{r}'$:

$$
\mathbf{r}' = \mathbf{r}\cos\theta + \mathbf{r}\times\mathbf{n}\sin\theta + (1 - \cos\theta)\mathbf{n}(\mathbf{n}\cdot\mathbf{r}).
$$

(d) Verify that the components $J_i = \sigma_i/2$ satisfy the relation

$$
[J_i, J_j] = i\sum_k \epsilon_{ijk} J_k.
$$

(e) Matrices t_i, $i = 1, 2, 3$, with $(t_k)_{ij} = -i\epsilon_{ijk}$, satisfy the commutation relation given in (d), and $R(\mathbf{n}, \theta) = \exp(i\theta \mathbf{n} \cdot \mathbf{t})$ describes the rotation (through angle θ) around an axis along direction $\mathbf{n}$. Show that

$$(\mathbf{n} \cdot \mathbf{t})^3 = (\mathbf{n} \cdot \mathbf{t}), \quad \text{and hence} \quad R(\mathbf{n}, \theta) = 1 + i(\mathbf{n} \cdot \mathbf{t})\sin\theta + (\mathbf{n} \cdot \mathbf{t})^2(\cos\theta - 1).$$

(f) Finally verify that $R(\mathbf{n}, \theta)\mathbf{r} = \mathbf{r}'$.

Solution: The relation (a) follows from expansion of the exponential as in Eq. (5.41b) together with the use of the following properties of Pauli matrices:

$$\sigma_i \sigma_k + \sigma_k \sigma_i = 2\delta_{ik}, \quad \sigma_i \sigma_k = \delta_{ik} + i\sum_l \epsilon_{ikl}\sigma_l, \quad i, k, l = 1, 2, 3,$$

where $\epsilon_{ikl} = +1$ for i, k, l an even permutation of $1, 2, 3$, and -1 in the case of an odd permutation, and zero if two indices are equal. From these relations we also obtain the result (b):

$$(\boldsymbol{\sigma} \cdot \mathbf{A})(\boldsymbol{\sigma} \cdot \mathbf{B}) = \sum_{i,j} \sigma_i \sigma_j A_i B_j = \sum_{i,j,l} (\delta_{ij} + i\epsilon_{ijl}\sigma_l) A_i B_j = \mathbf{A} \cdot \mathbf{B} + i\boldsymbol{\sigma} \cdot (\mathbf{A} \times \mathbf{B}).$$

We use this result to evaluate next the expression $[U, R] = UR - RU$. We have

$$
\begin{aligned}
UR - RU &= \left[1\cos\frac{\theta}{2} + i\mathbf{n} \cdot \boldsymbol{\sigma}\sin\frac{\theta}{2}\right](\mathbf{r} \cdot \boldsymbol{\sigma}) - (\mathbf{r} \cdot \boldsymbol{\sigma})\left[1\cos\frac{\theta}{2} + i\mathbf{n} \cdot \boldsymbol{\sigma}\sin\frac{\theta}{2}\right] \\
&= i\sin\frac{\theta}{2}[(\mathbf{n} \cdot \boldsymbol{\sigma})(\mathbf{r} \cdot \boldsymbol{\sigma}) - (\mathbf{r} \cdot \boldsymbol{\sigma})(\mathbf{n} \cdot \boldsymbol{\sigma})] \\
&= i\sin\frac{\theta}{2}[(\mathbf{n} \cdot \mathbf{r}) + i\boldsymbol{\sigma} \cdot (\mathbf{n} \times \mathbf{r}) - (\mathbf{n} \cdot \mathbf{r}) - i\boldsymbol{\sigma} \cdot (\mathbf{r} \times \mathbf{n})] = -2\sin\frac{\theta}{2}\boldsymbol{\sigma} \cdot (\mathbf{n} \times \mathbf{r}).
\end{aligned}
$$

With

$$U^{-1} = \exp\left[i\frac{(-\theta)}{2}\mathbf{n} \cdot \boldsymbol{\sigma}\right] = 1\cos\frac{\theta}{2} - i\mathbf{n} \cdot \boldsymbol{\sigma}\sin\frac{\theta}{2}$$

(corresponding to a rotation in the reverse sense to that of U, or simply using the unitarity of U, $U^{-1} = U^\dagger$), it follows that

$$URU^{-1} = R + [U, R]U^{-1} = (\mathbf{r} \cdot \boldsymbol{\sigma}) + \left[-2\sin\frac{\theta}{2}\boldsymbol{\sigma} \cdot (\mathbf{n} \times \mathbf{r})\right] \times \left[1\cos\frac{\theta}{2} - i\mathbf{n} \cdot \boldsymbol{\sigma}\sin\frac{\theta}{2}\right].$$

Using the result (b) this can be simplified to

$$
\begin{aligned}
URU^{-1} &= (\mathbf{r} \cdot \boldsymbol{\sigma}) - \sin\theta\,\boldsymbol{\sigma} \cdot (\mathbf{n} \times \mathbf{r}) + 2i\sin^2\frac{\theta}{2}\boldsymbol{\sigma} \cdot (\mathbf{n} \times \mathbf{r})(\mathbf{n} \cdot \boldsymbol{\sigma}) \\
&= (\mathbf{r} \cdot \boldsymbol{\sigma}) - \sin\theta\,\boldsymbol{\sigma} \cdot (\mathbf{n} \times \mathbf{r}) + 2i\sin^2\frac{\theta}{2}[(\mathbf{n} \times \mathbf{r}) \cdot \mathbf{n} + i\boldsymbol{\sigma} \cdot \{(\mathbf{n} \times \mathbf{r}) \times \mathbf{n}\}] \\
&= (\mathbf{r} \cdot \boldsymbol{\sigma}) - \sin\theta\,\boldsymbol{\sigma} \cdot (\mathbf{n} \times \mathbf{r}) - 2\sin^2\frac{\theta}{2}\boldsymbol{\sigma} \cdot \{\mathbf{n}^2\mathbf{r} - (\mathbf{n} \cdot \mathbf{r})\mathbf{n}\} \\
&= (\mathbf{r} \cdot \boldsymbol{\sigma}) + (\mathbf{r} \times \mathbf{n}) \cdot \boldsymbol{\sigma}\sin\theta + (\cos\theta - 1)[\boldsymbol{\sigma} \cdot \mathbf{r} - (\mathbf{n} \cdot \mathbf{r})(\mathbf{n} \cdot \boldsymbol{\sigma})] \\
&= [\mathbf{r}\cos\theta + \mathbf{r} \times \mathbf{n}\sin\theta + (1 - \cos\theta)\mathbf{n}(\mathbf{n} \cdot \mathbf{r})] \cdot \boldsymbol{\sigma}.
\end{aligned}
$$

Since

$$R \equiv \mathbf{r} \cdot \boldsymbol{\sigma} = \begin{pmatrix} z & x - iy \\ x + iy & -z \end{pmatrix}, \quad \text{and} \quad R' = \mathbf{r}' \cdot \boldsymbol{\sigma} = URU^{-1},$$

the expression for $\mathbf{r}'$ is:

$$\mathbf{r}' = \mathbf{r}\cos\theta + \mathbf{r} \times \mathbf{n}\sin\theta + (1 - \cos\theta)\mathbf{n}(\mathbf{n} \cdot \mathbf{r}). \tag{5.43}$$

We see therefore: Constructing from the vector $\mathbf{r}$ the 2×2 matrix R, the unitary transformation U, i.e. $R' = URU^{-1}$, corresponds to a rotation in three dimensions. That $\boldsymbol{\sigma}$ transforms like a vector can be deduced from the last result, namely that

$$\boldsymbol{\sigma}' = U\boldsymbol{\sigma}U^{-1} = \boldsymbol{\sigma}\cos\theta + \boldsymbol{\sigma} \times \mathbf{n}\sin\theta + (1 - \cos\theta)\mathbf{n}(\mathbf{n}\cdot\boldsymbol{\sigma}).$$

Next, to verify that $J_i = \sigma_i/2$ satisfies the commutation relation (d), we use the relation $\sigma_i\sigma_k = \delta_{ik} + i\sum_l \epsilon_{ikl}\sigma_l, i, k, l = 1, 2, 3.$ Thus

$$\left[\frac{1}{2}\sigma_i, \frac{1}{2}\sigma_j\right] = \frac{1}{4}(\sigma_i\sigma_j - \sigma_j\sigma_i) = \frac{i}{4}\sum_l(\epsilon_{ijl} - \epsilon_{jil})\sigma_l = i\sum_l \epsilon_{ijl}\left(\frac{1}{2}\sigma_l\right).$$

Next we consider the rotation $R(\mathbf{n}, \theta) = \exp(i\theta\mathbf{n}\cdot\mathbf{t})$ of an ordinary three-component vector $\mathbf{r} = (x, y, z)$. Expanding R we have

$$R(\mathbf{n}, \theta) = 1 + \frac{i\theta(\mathbf{n}\cdot\mathbf{t})}{1!} + \frac{(i\theta)^2(\mathbf{n}\cdot\mathbf{t})^2}{2!} + \cdots.$$

Consider now $(\mathbf{n}\cdot\mathbf{t})^3 = \sum_{i,j,k} n_i n_j n_k t_i t_j t_k$. Here (see (e))

$$(t_i t_j t_k)_{\alpha\delta} = \sum_{\beta,\gamma}(t_i)_{\alpha\beta}(t_j)_{\beta\gamma}(t_k)_{\gamma\delta} = \sum_{\beta,\gamma}(-i)^2\epsilon_{\alpha\beta i}\epsilon_{\beta\gamma j}(t_k)_{\gamma\delta} = -\sum_{\beta\gamma}\epsilon_{i\alpha\beta}\epsilon_{\gamma j\beta}(t_k)_{\gamma\delta}.$$

Now using

$$\sum_k \epsilon_{ijk}\epsilon_{lmk} = \delta_{il}\delta_{jm} - \delta_{im}\delta_{jl}, \quad\text{or}\quad \sum_\beta \epsilon_{i\alpha\beta}\epsilon_{\gamma j\beta} = \delta_{i\gamma}\delta_{\alpha j} - \delta_{ij}\delta_{\alpha\gamma},$$

$$(t_i t_j t_k)_{\alpha\delta} = -\sum_\gamma(\delta_{i\gamma}\delta_{\alpha j} - \delta_{ij}\delta_{\alpha\gamma})(t_k)_{\gamma\delta} = -(t_k)_{i\delta}\delta_{\alpha j} + \delta_{ij}(t_k)_{\alpha\delta},$$

we obtain

$$\begin{aligned}\{(\mathbf{n}\cdot\mathbf{t})^3\}_{\alpha\delta} &= \sum_{i,j,k} n_i n_j n_k[-\delta_{\alpha j}(t_k)_{i\delta} + (t_k)_{\alpha\delta}\delta_{ij}] = \sum_{i,k}[-n_i n_\alpha n_k(t_k)_{i\delta} + n_i n_i n_k(t_k)_{\alpha\delta}] \\ &= \sum_{i,k}[-n_i n_\alpha n_k(t_k)_{i\delta} + (\mathbf{n}\cdot\mathbf{t})_{\alpha\delta}].\end{aligned}$$

Consider the first contribution here:

$$\Gamma \equiv \sum_{i,k}[-n_i n_\alpha n_k(t_k)_{i\delta}] = i\sum_i n_i n_\alpha n_k\epsilon_{i\delta k} = -i\sum_{i,k} n_i n_\alpha n_k\epsilon_{k\delta i} = -i\sum_{i,k} n_i n_\alpha n_k\epsilon_{ik\delta} = -\Gamma.$$

Thus $\Gamma = 0$ and hence $(\mathbf{n}\cdot\mathbf{t})^3_{\alpha\delta} = (\mathbf{n}\cdot\mathbf{t})_{\alpha\delta}$, i.e. $(\mathbf{n}\cdot\mathbf{t})^3 = (\mathbf{n}\cdot\mathbf{t})$. With this result we can simplify the above expansion of R:

$$\begin{aligned}R(\mathbf{n}, \theta) &= 1 + \frac{i\theta(\mathbf{n}\cdot\mathbf{t})}{1!} + \frac{(i\theta)^2}{2!}(\mathbf{n}\cdot\mathbf{t})^2 + \frac{(i\theta)^3}{3!}(\mathbf{n}\cdot\mathbf{t}) + \frac{(i\theta)^4}{4!}(\mathbf{n}\cdot\mathbf{t})^2 + \cdots \\ &= 1 + i(\mathbf{n}\cdot\mathbf{t})\sin\theta + (\mathbf{n}\cdot\mathbf{t})^2(\cos\theta - 1).\end{aligned}$$

We now use the explicit expressions following from $(t_k)_{ij} = -i\epsilon_{ijk}$:

$$t_1 = \begin{pmatrix} 0 & 0 & 0 \\ 0 & 0 & -i \\ 0 & i & 0 \end{pmatrix}, \quad t_2 = \begin{pmatrix} 0 & 0 & i \\ 0 & 0 & 0 \\ -i & 0 & 0 \end{pmatrix}, \quad t_3 = \begin{pmatrix} 0 & -i & 0 \\ i & 0 & 0 \\ 0 & 0 & 0 \end{pmatrix}. \tag{5.44}$$

Then for $\mathbf{n}\cdot\mathbf{t}=n_1t_1+n_2t_2+n_3t_3$ we obtain:

$$\mathbf{n}\cdot\mathbf{t}=\begin{pmatrix} 0 & -in_3 & in_2 \\ in_3 & 0 & -in_1 \\ -in_2 & in_1 & 0 \end{pmatrix},\quad (\mathbf{n}\cdot\mathbf{t})^2=\begin{pmatrix} n_2^2+n_3^2 & -n_1n_2 & -n_1n_3 \\ -n_1n_2 & n_1^2+n_3^2 & -n_2n_3 \\ -n_1n_3 & -n_2n_3 & n_1^2+n_2^2 \end{pmatrix}.$$

Applying $R(\mathbf{n},\theta)$ to the vector $\mathbf{r}=(x,y,z)$, we obtain:

$$
\begin{aligned}
R(\mathbf{n},\theta)\mathbf{r} &= [\mathbf{1}+i(\mathbf{n}\cdot\mathbf{t})\sin\theta+(\mathbf{n}\cdot\mathbf{t})^2(\cos\theta-1)]\begin{pmatrix} x \\ y \\ z \end{pmatrix} \\
&= \begin{pmatrix} x+i\sin\theta(-in_3y+in_2z)+(\cos\theta-1)[x(n_2^2+n_3^2)-yn_1n_2-zn_1n_3] \\ y+i\sin\theta(in_3x-in_1z)+(\cos\theta-1)[-xn_1n_2+y(n_1^2+n_3^2)-zn_2n_3] \\ z+i\sin\theta(-in_2x+in_1y)+(\cos\theta-1)[-xn_1n_3-yn_2n_3+z(n_1^2+n_2^2)] \end{pmatrix} \\
&= \mathbf{r}+\sin\theta(\mathbf{r}\times\mathbf{n})+(\cos\theta-1)[\mathbf{r}-\mathbf{n}(\mathbf{r}\cdot\mathbf{n})] \\
&= \mathbf{r}\cos\theta+(\mathbf{r}\times\mathbf{n})\sin\theta+\mathbf{n}(\mathbf{n}\cdot\mathbf{r})(1-\cos\theta)=\mathbf{r}'.
\end{aligned}
$$

This result is seen to be in agreement with Eq. (5.43). This exercise thus demonstrates the isomorphism between the group $SU(2)$ (the group of complex, unitary, unimodular 2×2 matrices) and the group $SO(3)$ (the group of real, orthogonal 3×3 matrices with determinant $+1$). Thus both, the 2×2 matrices $\sigma_i/2$ and the 3×3 matrices t_i satisfy the fundamental commutation relation (d). In the former case the rotation is described by the unitary similarity transformation URU^{-1} of the matrix R, in the latter case by the orthogonal transformation of the vector $\mathbf{r}$.

5.4.7 The infinitesimal rotation of a vector

Our aim is to investigate the motion of a rigid body as a function of time. To this end it is useful to consider infinitesimal quantities, *i.e.* infinitesimal rotations. Writing x_1,x_2,x_3 instead of x,y,z, the infinitesimal orthogonal transformation of the coordinates x_1,x_2,x_3 of the vector $\mathbf{r}=(x_1,x_2,x_3)$ with respect to a fixed reference frame with axes $(x,0,0),(0,y,0),(0,0,z)$ can be written:

$$x_i'=x_i+\sum_j\epsilon_{ij}x_j=\sum_j(\delta_{ij}+\epsilon_{ij})x_j,\quad\epsilon_{ij}\text{ small,}\qquad(5.45a)$$

or in matrix form:

$$x'=(\mathbf{1}+\epsilon)x,\qquad dx=x'-x=\epsilon x\qquad(5.45b)$$

(rotation of vector $\mathbf{x}$ with respect to a fixed reference frame). We observe first that two consecutively performed *finite* transformations, *i.e.* rotations, do not commute in general,[*] since

$$(\mathbf{1}+\epsilon_1)(\mathbf{1}+\epsilon_2)=\mathbf{1}+\epsilon_1+\epsilon_2+\epsilon_1\epsilon_2,\quad\text{where }\epsilon_1,\epsilon_2\text{ are matrices.}$$

---

[*]This is nicely illustrated in the book of H. Goldstein [19], Sec. 4.7, by a series of rotations of an elongated box in various directions, and in the book of H. Aratyn and C. Rasinariu [2], p. 169, by a series of rotations of a six-sided die, as the reader can also verify himself with a die.

Here in general $\epsilon_1\epsilon_2 \neq \epsilon_2\epsilon_1$. However, in the case of *infinitesimal* rotations we do have *commutativity*, since there

$$(\mathbb{1} + \epsilon_1)(\mathbb{1} + \epsilon_2) \approx \mathbb{1} + \epsilon_1 + \epsilon_2.$$

Also, the *inverse* of the matrix $\mathbb{1} + \epsilon$ is simply $(\mathbb{1} + \epsilon)^{-1} \simeq \mathbb{1} - \epsilon$. Consider as an example a rotation around the z-axis:

$$A = \begin{pmatrix} \cos\phi & \sin\phi & 0 \\ -\sin\phi & \cos\phi & 0 \\ 0 & 0 & 1 \end{pmatrix}. \tag{5.46}$$

As an infinitesimal rotation $(\phi \rightarrow d\phi)$ this is

$$\mathbb{1} + \epsilon = \begin{pmatrix} 1 & d\phi & 0 \\ -d\phi & 1 & 0 \\ 0 & 0 & 1 \end{pmatrix}, \quad i.e. \quad \epsilon = \begin{pmatrix} 0 & d\phi & 0 \\ -d\phi & 0 & 0 \\ 0 & 0 & 0 \end{pmatrix}.$$

Since the infinitesimal transformation is still an orthogonal transformation, the same must hold for $\mathbb{1} + \epsilon$ as for A:

$$A^{-1} = A^T, \quad i.e. \quad (\mathbb{1} - \epsilon) = (\mathbb{1} + \epsilon)^T = \mathbb{1} + \epsilon^T,$$

or

$$-\epsilon = \epsilon^T, \quad \text{or} \quad \epsilon_{ij} = -\epsilon_{ji}. \tag{5.47}$$

It follows that $\epsilon_{ii} = 0, i = 1, 2, 3$. Thus ϵ has the form

$$\epsilon = \begin{pmatrix} 0 & d\Omega_3 & -d\Omega_2 \\ -d\Omega_3 & 0 & d\Omega_1 \\ d\Omega_2 & -d\Omega_1 & 0 \end{pmatrix}, \quad \epsilon_{12} = d\Omega_3, \quad \text{etc.} \tag{5.48}$$

Inserting this expression into $dx = \epsilon x$, *i.e.* into Eq. (5.45b), we obtain

$$\begin{aligned} dx_1 &= x_2 d\Omega_3 - x_3 d\Omega_2, \\ dx_2 &= x_3 d\Omega_1 - x_1 d\Omega_3, \\ dx_3 &= x_1 d\Omega_2 - x_2 d\Omega_1, \end{aligned} \tag{5.49}$$

i.e. with $\mathbf{r} = (x_1, x_2, x_3)$ and $d\mathbf{\Omega} = (d\Omega_1, d\Omega_2, d\Omega_3)$:

$$d\mathbf{r} = \mathbf{r} \times d\mathbf{\Omega}. \tag{5.50}$$

Here the vector $\mathbf{r}$ is rotated through the infinitesimal angle $d\mathbf{\Omega}$ which changes $\mathbf{r}$ by $d\mathbf{r}$. One should observe that $d\mathbf{\Omega}$ is an *axial vector* (like angular momentum).[†]

[†]Whereas a vector changes sign under $\mathbf{r} \rightarrow -\mathbf{r}$, the axial vector does not.

As an example we consider the rotation of the reference frame, not of the vector, through an angle $-d\Omega$ and around the z-axis as indicated in Fig. 5.5 (perpendicular to the paper and out of this). Here $\mathbf{r}$ is the position vector of a point P in a frame R, and $\mathbf{r}'$ is the position vector of the same point with respect to the same reference frame but after rotation of the vector $\mathbf{r}$ around the origin to its new position $\mathbf{r}'$. The same configuration is achieved with a rotation in the opposite sense of the frame R to a frame R', as indicated in Fig. 5.5. We call the new reference frame R'. Then $d\mathbf{r}$ is the change of the vector $\mathbf{r}$ referred to the original frame R (or with respect to a spatial frame which is identical with R). We deduce from Eq. (5.50) that

$$d\mathbf{r} = 0, \quad \text{if} \quad \mathbf{r} \parallel d\Omega.$$

The vector $d\mathbf{r}$ does not have the same direction as $\mathbf{r}$!

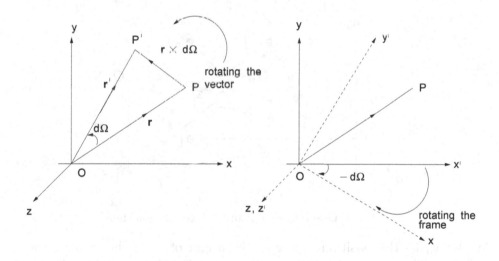

Fig. 5.5 Rotation of a vector and rotation of the frame.

We summarize: Only infinitesimal rotations commute, not finite rotations. In the case of infinitesimal rotations one obtains automatically a vector, *i.e.* $d\Omega$, which determines the rotation. From Eq. (5.50) we deduce that $d\mathbf{r} = 0$ when the vector $\mathbf{r}$ is parallel to the vector $d\Omega$. Thus a vector parallel to $d\Omega$ remains unchanged in a rotation. Such a vector is only the vector parallel to the axis of rotation. Its modulus is obviously the magnitude of the angle of rotation.[‡]

[‡]These aspects can be studied in more detail. See, for instance, the treatment of Euler's theorem in the book of H. Goldstein [19].

5.5   Rotating Reference Frames

In the following we attempt first a *geometrical explanation* of the transformation of a vector. To help a better understanding we mention immediately a crucial point, which when overlooked can make the arguments incomprehensible. Compared with a scalar, a vector is a quantity with a direction — and a direction results from two separated points. Thus, whereas a scalar may be visualized as a fly which is stationary at a point on a rigid body, the vector would have to be seen as the fly taking off to somewhere, thus tracing a path with a direction. We now consider rotating reference frames which are accelerated, non-inertial frames. Let S for "spatial" be an *inertial or space-fixed system* with origin O_S and point vectors $\mathbf{r} = x_1\mathbf{e}_1 + x_2\mathbf{e}_2 + x_3\mathbf{e}_3$, and R for *"rigid body"* with point vectors $\mathbf{r}' = x_1'\mathbf{e}_1' + x_2'\mathbf{e}_2' + x_3'\mathbf{e}_3'$ a frame (for instance fixed in a rigid body) which is moving with respect to the frame S. This correspondence is illustrated in Fig. 5.6.

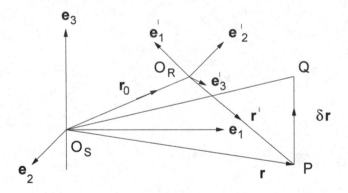

Fig. 5.6 One reference frame relative to another.

We let $\mathbf{r}_0$ be the position vector of the origin of R in the frame S, as in Fig. 5.6. Then the position vector of a point P (with now $\mathbf{e}_i' = \mathbf{e}_i'(t)$) is in general (*i.e.* allowing also a point to wander in R, like the position of an ant on a rigid body) given by

$$\mathbf{r}(t) = \mathbf{r}_0(t) + \mathbf{r}'(t) = \mathbf{r}_0(t) + \sum_i x_i'(t)\mathbf{e}_i'(t). \qquad (5.51)$$

Therefore we have for the velocities

$$\dot{\mathbf{r}}(t) = \dot{\mathbf{r}}_0(t) + \sum_i \{\dot{x}_i'(t)\mathbf{e}_i'(t) + x_i'(t)\dot{\mathbf{e}}_i'(t)\}. \qquad (5.52)$$

Seen from the spatially fixed frame S, the expression $\sum_i x_i'\dot{\mathbf{e}}_i'(t), \dot{x}_i' = 0$, is the velocity of a point fixed in the rotating frame R, like the point Q in

Fig. 5.7, so that

$$\overrightarrow{PQ} = \delta t \sum_i x_i' \dot{\mathbf{e}}_i'(t) = \delta t \left(\frac{d\mathbf{\Omega}}{dt} \times \mathbf{r} \right). \tag{5.53}$$

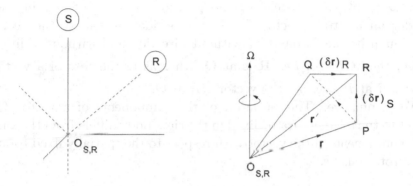

Fig. 5.7 Frame R rotating with respect to frame S.

This expression should be compared with that of the infinitesimal rotation in a time interval dt given by Eq. (5.50). There the vector was rotated, here the reference frame. The other contribution in $\dot{\mathbf{r}}$, *i.e.* $\sum_i \dot{x}_i' \mathbf{e}_i'(t)$, is the velocity contribution of the wandering point for a fixed reference frame, *i.e.* in Fig. 5.7 the expression $\overrightarrow{QR} / \delta t$. Now let $O_{S,R}$ be the joint origin of a body-fixed reference frame R and of a space-fixed reference frame S (here the so-called inertial system). We consider S as fixed, and R as rotating with respect to S. Let $d\mathbf{\Omega}/dt$ be the angular velocity of the rotation of R about $O_{S,R}$. Let $\mathbf{r}(t)$ be the time-varying position vector of a point P. In the time interval δt the point P (vector $\mathbf{r}$) moves (like a fly) to R (position $\mathbf{r}'$, *cf.* Fig. 5.7), so that $\overrightarrow{OR}$ is its new position with respect to the fixed reference frame S; *i.e.* $\overrightarrow{PR} = (\delta \mathbf{r})_S$ is the change in the position with respect to S. Looked at from the reference frame R, the initial point P (looked at as fixed in frame S) has moved to the point Q in the time interval δt with

$$\overrightarrow{PQ} = \delta t \left(\frac{d\mathbf{\Omega}}{dt} \times \mathbf{r} \right) = \delta \mathbf{\Omega} \times \mathbf{r}. \tag{5.54}$$

The change of the position vector of P with respect to R is then $(\delta \mathbf{r})_R = \overrightarrow{QR}$. Since

$$\overrightarrow{PR} = \overrightarrow{PQ} + \overrightarrow{QR},$$

we obtain

$$(\delta\mathbf{r})_S = (\delta\mathbf{r})_R + \delta t\left(\frac{d\boldsymbol{\Omega}}{dt} \times \mathbf{r}\right),$$

i.e.

$$\left(\frac{d\mathbf{r}}{dt}\right)_S = \left(\frac{d\mathbf{r}}{dt}\right)_R + \boldsymbol{\omega} \times \mathbf{r}, \tag{5.55}$$

where $\boldsymbol{\omega} = d\boldsymbol{\Omega}/dt$. Instead of the position vector $\mathbf{r}$ of a point, we can consider an arbitrary vector $\mathbf{G}$. If we consider instead of the vector $\mathbf{r}$ a scalar quantity, *i.e.* a quantity without direction and without a directional change, then $\overrightarrow{QR} = 0$, *i.e.* $\mathbf{R}$ is at $\mathbf{Q}$, whereas in the case of a vector, the vector $\overrightarrow{PR}$ at P becomes the vector $\overrightarrow{QR}$ at Q.

We summarize: The variation of the components of a vector $\mathbf{G}$ with respect to the reference frame fixed in the rigid body differs from the variation of the components of the vector with respect to the spatially fixed frame only by the rotation, *i.e.*

$$(d\mathbf{G})_{\text{rigid body, R}} = (d\mathbf{G})_{\text{space, S}} + (d\mathbf{G})_{\text{rotation}}. \tag{5.56}$$

The variation resulting solely from an infinitesimal rotation of the frame R (note: frame, not vector) follows as the negative of that of Eq. (5.50):

$$(d\mathbf{G})_{\text{rotation}} = -\mathbf{G} \times d\boldsymbol{\Omega}.$$

This result, here obtained geometrically, will be rederived in Sec. 5.9.2 (*cf.* Eq. (5.153a)) with the help of the generator of rotations. We thus arrive at the important relation[§]

$$\left(\frac{d\mathbf{G}}{dt}\right)_{\text{space}} = \left(\frac{d\mathbf{G}}{dt}\right)_{\text{rigid body}} \pm \boldsymbol{\omega} \times \mathbf{G}, \quad \text{where} \quad \boldsymbol{\omega} = \frac{d\boldsymbol{\Omega}}{dt}, \tag{5.57}$$

where $\boldsymbol{\omega}$ is the angular velocity of the rigid body (and hence its frame, with $+$ for rotation of the vector with the frame left fixed, and $-$ for rotation of the reference frame with the vector left fixed). The two reference frames have a joint origin $O_{S,R}$.

The frame fixed on the rigid body is useful for establishing the equations of motion of a rigid body. The angular frequency $\boldsymbol{\omega}_\phi$ is a vector of modulus $\dot{\phi}$; the direction of this vector is that of the axis of its rotation. According to our earlier definition of the angle ϕ (see Sec. 5.4.5, Fig. 5.4), the vector $\boldsymbol{\omega}_\phi$ is a vector parallel to the spatially fixed z-axis. We want to see now, how we obtain the components of this vector along the axes of the rigid body frame.

[§]For a similar discussion see *e.g.* H. Aratyn and C. Rasinariu [2], p. 154.

With this aim we look back at Eqs. (5.20a) to (5.20c) and Eqs. (5.21a) and (5.21b). Equation (5.21a) provides the matrix (A) of the transformation

$$(x_{\text{rigid body}}) = (A)(x_{\text{spatial frame}}) = (A)(x_{\text{rigid body}})_{\text{before rotation}} = \begin{pmatrix} x \\ y \\ z \end{pmatrix}.$$
$$(5.58)$$

Let (x', y', z') be the coordinates of the rigid body-fixed reference frame. Then — looking back at the expression of the matrix (A) given in Eq. (5.21b) — we have (recall again that $\boldsymbol{\omega}_\phi = \dot{\phi}\mathbf{e}_z$):

$$(\boldsymbol{\omega}_{\phi,\text{body}}) = A(\boldsymbol{\omega}_{\phi,\text{space}}) = A \begin{pmatrix} 0 \\ 0 \\ \dot{\phi} \end{pmatrix} \overset{(5.21b)}{=} \begin{pmatrix} \dot{\phi}\sin\theta\sin\psi \\ \dot{\phi}\sin\theta\cos\psi \\ \dot{\phi}\cos\theta \end{pmatrix} \leftrightarrow \begin{pmatrix} \omega_{x'} \\ \omega_{y'} \\ \omega_{z'} \end{pmatrix}.$$
$$(5.59)$$

The column on the right indicates the frequencies to which the components contribute. In a similar way we can deal with the other rotations. The direction of $\boldsymbol{\omega}_\theta$ coincides with that of the ξ'-axis. This means that in Eq. (5.20b) we make the replacement:

$$\Xi' = (\xi', \eta', \zeta') \longrightarrow (\dot{\theta}, 0, 0).$$

Hence, with Eq. (5.20c), we obtain

$$(\boldsymbol{\omega}_\theta)_{\text{body}} = (B)(\boldsymbol{\omega}_\theta)_{\text{space}} = (B) \begin{pmatrix} \dot{\theta} \\ 0 \\ 0 \end{pmatrix} = \begin{pmatrix} \dot{\theta}\cos\psi \\ -\dot{\theta}\sin\psi \\ 0 \end{pmatrix} \leftrightarrow \begin{pmatrix} \omega_{x'} \\ \omega_{y'} \\ \omega_{z'} \end{pmatrix}.$$
$$(5.60)$$

Finally: Since $\boldsymbol{\omega}_\psi$ is directed along the axis of z', this case requires no transformation. We simply add $\dot{\psi}$ to the other contributions of the z' component. Thus adding these components of $\boldsymbol{\omega}$ with respect to the rigid body frame, we obtain the following important equations which we shall refer back to in later contexts:

$$\begin{aligned} \omega_{x'} &= \dot{\phi}\sin\theta\sin\psi + \dot{\theta}\cos\psi, \\ \omega_{y'} &= \dot{\phi}\sin\theta\cos\psi - \dot{\theta}\sin\psi, \\ \omega_{z'} &= \dot{\phi}\cos\theta + \dot{\psi}. \end{aligned}$$
$$(5.61)$$

These equations are probably most easily derived directly from Fig. 5.4 which defines the Euler angles — this requires, of course, a familiarity with the diagram and one has to remember that $\boldsymbol{\omega}_\phi, \boldsymbol{\omega}_\theta, \boldsymbol{\omega}_\psi$ are vectors along the axes $\mathbf{z}, \boldsymbol{\xi}', \mathbf{z}'$.

5.6   Definition of Scalars, Vectors, Tensors

So far we used the concepts of scalars, vectors and tensors without a precise definition. Now, after the investigation of orthogonal transformations, we can do this. We consider here, at least initially, only such quantities which are defined by orthogonal transformations, but it will be apparent that for instance a Lorentz scalar is defined correspondingly with respect to Lorentz transformations. It suffices in these cases to consider again only infinitesimal transformations since every finite transformation results from a series of infinitesimal transformations. We consider the infinitesimal rotation of a vector $\mathbf{x} = (x_1, x_2, x_3)$ again in the form[¶]

$$x_i' = x_i + \sum_j \epsilon_{ij} x_j, \quad \epsilon_{ij} = -\epsilon_{ji} \text{ and infinitesimal.} \qquad (5.62)$$

Then

$$dx_i \equiv x_i' - x_i = \sum_j \epsilon_{ij} x_j,$$

with

$$\sum_j \frac{\partial x_i'}{\partial x_j} x_j = \sum_j (\delta_{ij} + \epsilon_{ij}) x_j = x_i + \sum_j \epsilon_{ij} x_j = x_i',$$

$$\therefore \ x_i' = \sum_j \frac{\partial x_i'}{\partial x_j} x_j. \qquad (5.63)$$

A *scalar function* $S(x)$ is now *defined* by the following property with respect to orthogonal transformations (where appropriate summation over indices appearing twice is understood):

$$S'(x_i') = S(x_i') = S(x_i + \epsilon_{ij} x_j) = S(x_i) + \delta S, \qquad (5.64)$$

where

$$\delta S = \sum_{i,j} \epsilon_{ij} x_j \left(\frac{\partial S}{\partial x_i} \right).$$

Thus, if S is, for instance, $S = x^2$:

$$S = x^2 = \sum_{i=1}^{3} x_i x_i, \qquad \frac{\partial S}{\partial x_i} = 2x_i,$$

[¶]Browsing through pages and chapters, a reader may feel uneasy about the distinction between quantities with lower indices and those with higher indices. As long as we are only concerned with $\mathbf{x} \in \mathbb{R}^3$, we choose whichever is convenient in the context. More generally and in tensor calculus we have to distinguish very clearly between covariant (lower) and contravariant (upper) indices, and the corresponding transformations. The mixed tensor of Eq. (14.1) shows the difference.

and hence

$$\delta S = 2 \sum_{i,j} \epsilon_{ij} x_j x_i = \sum_{i,j} (\epsilon_{ij} + \epsilon_{ji}) x_j x_i = 0,$$

since $\epsilon_{ij} = -\epsilon_{ji}$. Hence $S'(x') = S(x)$.

The *definition* of an $SO(3)$ *vector function* $G_k(x), k = 1, 2, 3$, on the other hand, is given by its transformation under the orthogonal transformations of the group $SO(3)$:

$$
\begin{aligned}
G_k(x_i) \to G'_k(x_i) &= \sum_l \frac{\partial x'_k}{\partial x_l} G_l(x_i) \simeq \sum_l (\delta_{kl} + \epsilon_{kl}) G_l(x_i) \\
&= G_k(x_i) + \sum_l \epsilon_{kl} G_l(x_i). \quad (5.65)
\end{aligned}
$$

This rotation describes the *"active"* or *"objective"* rotation of the vector, and hence of the physical system, whereas the *"passive"* or *"subjective"* rotation of the variables x_i is described by the following transformation:

$$
\begin{aligned}
G_k(x_i) \to G'_k(x_i) &= \bar{G}_k(x'_i) \simeq G_k \left(x_i + \sum_j \epsilon_{ij} x_j \right) \\
&\simeq G_k(x_i) + \sum_{j,i} \epsilon_{ij} x_j \left(\frac{\partial G_k}{\partial x_i} \right)_{x_i}. \quad (5.66)
\end{aligned}
$$

We now consider two examples in order to convince ourselves of the equivalence of these rotations.

Example 1: Let $G_k(x) := \sum_l x_k x_l^2$. Then

$$
\begin{aligned}
\sum_{i,j} \epsilon_{ij} x_j \left(\frac{\partial G_k}{\partial x_i} \right) &= \sum_{i,j,l} \epsilon_{ij} x_j [\delta_{ki} x_l^2 + 2 x_k x_i] \\
&= \sum_{j,l} \epsilon_{kj} x_j x_l^2 + 0 \quad \left(\text{since } \sum_{i,j} \epsilon_{ij} x_i x_j = 0 \right) \\
&= \sum_j \epsilon_{kj} G_j(x).
\end{aligned}
$$

Hence

$$G'_k(x) \overset{(5.66)}{=} G_k(x) + \sum_j \epsilon_{kj} G_j(x)$$

in agreement with Eq. (5.65).

Example 2: Let the angular momentum component k be

$$G_k := L_k = \sum_{l,m} \epsilon_{klm} x_l p_m = \frac{1}{2} \sum_{l,m} \epsilon_{klm} (x_l p_m - x_m p_l).$$

Note that the double sum necessitates the factor $1/2$. Here we have to remember that $L_k = L_k(x_i, p_i)$, *i.e.*

$$
L'_k(x_i, p_i) \overset{(5.66)}{=} L_k(x'_i, p'_i)
$$

$$
\overset{(5.66)}{=} L_k(x_i, p_i) + \sum_{i,j} \epsilon_{ij} x_j \left(\frac{\partial L_k}{\partial x_i} \right) + \sum_{i,j} \epsilon_{ij} p_j \left(\frac{\partial L_k}{\partial p_i} \right)
$$

$$
= L_k(x_i, p_i) + \sum_{i,j,m} \epsilon_{ij} (x_j \epsilon_{kim} p_m + p_j \epsilon_{kli} x_l)
$$

$$
= L_k(x_i, p_i) + \sum_{i,j,l} \epsilon_{ij} \epsilon_{ikl} (x_l p_j - x_j p_l). \tag{5.67}
$$

We emphasize: Here ϵ_{ij} is the infinitesimal element of an $SO(2)$ rotation matrix (*cf.* Eq. (5.15) for φ small), whereas ϵ_{ikl} is the Levi–Civita symbol or tensor). Later (*cf.* Eq. (5.75)) we obtain the relation between the rotation matrix element and the infinitesimal angle. Since $\epsilon_{ikl} \neq 0$ only for i, k, l different, and $\epsilon_{ij} \neq 0$ only for i, j different, it follows that $j \neq i$, *i.e.* j can only be k or l. For $j = l$ the bracket on the right vanishes. Hence $j = k$ and therefore with no summation over k:

$$
L'_k(x_i, p_i) = L_k(x_i, p_i) + \sum_{i,l} \epsilon_{ik} \epsilon_{ikl} (x_l p_k - x_k p_l)
$$

$$
= L_k(x_i, p_i) + \sum_{i,l} \epsilon_{ki} \epsilon_{ilk} (x_l p_k - x_k p_l)
$$

$$
= L_k(x_i, p_i) + \sum_i \epsilon_{ki} L_i(x_i, p_i) \tag{5.68}
$$

as expected on comparison with Eq. (5.65).

The relation describing the transformation (*i.e.* Eq. (5.66)) also results from a *unitary transformation* $U(x)$ as we now demonstrate. Let $M_{ab}(x), a, b = 1, 2, 3$, be an (Hermitian) differential operator, the *generator of the transformation*, and

$$
U(x) = \exp \left\{ \frac{i}{2} \epsilon_{ab} M_{ab}(x) \right\}, \quad \epsilon_{ab} = -\epsilon_{ba} \text{ infinitesimal.} \tag{5.69}
$$

Then (summations understood)

$$
G'_k(x) = e^{-i\epsilon_{ab} M_{ab}(x)/2} G_k(x) e^{i\epsilon_{ab} M_{ab}(x)/2}
$$

$$
\simeq \left[1 - \frac{1}{2} i\epsilon_{ab} M_{ab}(x) \right] G_k(x) \left[1 + \frac{1}{2} i\epsilon_{ab} M_{ab}(x) \right]
$$

$$
= G_k(x) - \frac{1}{2} i\epsilon_{ab} [M_{ab}(x), G_k(x)], \tag{5.70}
$$

where $[A, B] = AB - BA$. Setting

$$M_{ab}(x) = -i\left(x_a\frac{\partial}{\partial x_b} - x_b\frac{\partial}{\partial x_a}\right) = -M_{ba}(x), \tag{5.71}$$

so that applied to an arbitrary function $F(x)$ (remember $[\ldots,\ldots]$ is an operator):

$$
\begin{aligned}
[M_{ab}(x), G_k(x)]F(x) &= -i\left(x_a\frac{\partial G_k(x)}{\partial x_b} - x_b\frac{\partial G_k(x)}{\partial x_a}\right)F(x) - iG_k(x) \times \\
&\quad \left(x_a\frac{\partial F(x)}{\partial x_b} - x_b\frac{\partial F(x)}{\partial x_a}\right) - G_k(x)M_{ab}(x)F(x) \\
&= -i\left(x_a\frac{\partial G_k(x)}{\partial x_b} - x_b\frac{\partial G_k(x)}{\partial x_a}\right)F(x) \\
&= \{M_{ab}(x)G_k(x)\}F(x),
\end{aligned}
$$

and we have formally (by dropping $F(x)$ on both sides) the relation:

$$[M_{ab}(x), G_k(x)] = M_{ab}(x)G_k(x).$$

Then, since, with $\epsilon_{ab} = -\epsilon_{ba}$ and summations understood,

$$
\begin{aligned}
\epsilon_{ab}x_a\partial_b G_k(x) &= \frac{1}{2}(\epsilon_{ab}x_a\partial_b + \epsilon_{ba}x_b\partial_a)G_k(x) \\
&= \frac{1}{2}\epsilon_{ab}\left(x_a\frac{\partial}{\partial x_b} - x_b\frac{\partial}{\partial x_a}\right)G_k(x) = \frac{1}{2}i\epsilon_{ab}M_{ab}(x)G_k(x),
\end{aligned}
$$

we obtain

$$G'_k(x) \overset{(5.70)}{=} G_k(x) - \sum_{a,b}\epsilon_{ab}x_a\left(\frac{\partial G_k(x)}{\partial x_b}\right) = G_k(x) + \sum_{a,b}\epsilon_{ba}x_a\left(\frac{\partial G_k(x)}{\partial x_b}\right), \tag{5.72}$$

i.e. we obtain the same expression as in Eq. (5.66).

One may also want to see the connection with the expression we obtained previously by geometric arguments, *i.e.* the result Eq. (5.54), *i.e.* with

$$\mathbf{r}' - \mathbf{r} = \delta\mathbf{\Omega} \times \mathbf{r}, \qquad x'_j - x_j = \sum_{k,l}\epsilon_{jkl}d\Omega_k x_l. \tag{5.73}$$

Setting $G'_k(x) = x'_k, G_k(x) = x_k$, we obtain from Eq. (5.65)

$$x'_j - x_j = \sum_l \epsilon_{jl}x_l, \tag{5.74}$$

so that we obtain by comparison with Eq. (5.73)

$$\epsilon_{jl} = \sum_k \epsilon_{jkl} d\Omega_k = -\epsilon_{lj}. \tag{5.75}$$

This is the expected relation between the infinitesimal angle $d\Omega_k$ of the rotation about the x_k-axis and the corresponding element ϵ_{jl} of the $SO(2)$ rotation matrix in the plane of the other two coordinates.

Finally we come to tensors. A *tensor of rank 2*, T_{ik}, is defined in close analogy to the vector G_k (*cf.* Eq. (5.65)): Under orthogonal transformations the quantity defined as this tensor transforms according to the relation:

$$T_{ik}(x_j) \rightarrow T'_{ik}(x_j) = \sum_{l,m} \frac{\partial x'_i}{\partial x_l} \frac{\partial x'_k}{\partial x_m} T_{lm}(x_j). \tag{5.76}$$

The generalization to higher rank tensors proceeds correspondingly.

5.7   The Theorem of E. Noether

A theorem discovered by E. Noether says in considerable generality, that a consequence of the invariance of a theory — this means the Lagrange function or action integral, since this is what defines a specific theory — under a continuous transformation group (*i.e.* one which depends on continuous parameters like translational shifts or angular rotations) is the existence of *conservations laws*, *i.e.* constants of the motion, in fact one such conservation law in association with every continuous parameter of the group. We already encountered the simplest cases. If the Lagrange function $L(x_i, \dot{x}_i)$ is independent of $x_i, i = 1, 2, 3$, then L is invariant under the translations:

$$x_i \rightarrow x'_i = x_i + a_i, \quad a_i = \text{const.}, \tag{5.77}$$

and

$$\frac{\partial L}{\partial x_i} = 0,$$

so that the Euler–Lagrange equation becomes

$$\frac{d}{dt}\left(\frac{\partial L}{\partial \dot{x}_i}\right) = 0, \quad i.e. \quad p_i = \frac{\partial L}{\partial \dot{x}_i} = \text{const.}$$

Analogous considerations apply, of course, in the case of the conservation of angular momentum L for invariance of the Lagrange function $L(\theta, \dot{\theta})$ under transformations $\theta \rightarrow \theta' = \theta + \alpha, \alpha = \text{const.}$ In electrodynamics the conservation of charge follows similarly from the invariance of the Lagrangian analogue there under gauge transformations $\mathbf{A} \to \mathbf{A}' = \mathbf{A} + \nabla\chi$, where $\mathbf{A}$ is the electromagnetic vector potential. There are, of course, also other cases.

Here we do not want to establish the Noether theorem in all generality and with the utmost of mathematical rigour, but in a simplified form. As we observed above, in most cases the argument is based on the equation of motion. The cases of linear momentum and angular momentum are fairly clear. How does the law of conservation of energy follow? The answer is: From the invariance under translations in time t, $i.e.$ $t \to t' = t + \delta t$. We set $\delta L = L' - L$. Then

$$
\begin{aligned}
L' &= L(q(t'), \dot{q}(t'), t') \\
&= L(q(t) + \delta t \dot{q}(t), \dot{q}(t) + \delta t \ddot{q}(t), t + \delta t) \\
&= L(q(t), \dot{q}(t), t) + \delta t \left[\frac{\partial L}{\partial t} + \dot{q}(t) \frac{\partial L}{\partial q(t)} + \ddot{q}(t) \frac{\partial L}{\partial \dot{q}(t)} \right]. \quad (5.78)
\end{aligned}
$$

Hence

$$
\delta L = L' - L = \delta t \left[\frac{\partial L}{\partial t} + \dot{q}(t) \frac{\partial L}{\partial q(t)} + \ddot{q}(t) \frac{\partial L}{\partial \dot{q}(t)} \right] = \delta t \frac{dL}{dt}. \quad (5.79)
$$

In case the Lagrange function has no explicit dependence on t (instead only implicit dependence on t through $q, \dot{q}$), we have $\partial L / \partial t = 0$. Then, with the help of the equation of motion in the second step,

$$
\delta t \frac{dL}{dt} = \delta t \left[\dot{q} \frac{\partial L}{\partial q} + \ddot{q} \frac{\partial L}{\partial \dot{q}} \right] = \delta t \left[\dot{q} \frac{d}{dt} \left(\frac{\partial L}{\partial \dot{q}} \right) + \ddot{q} \frac{\partial L}{\partial \dot{q}} \right]. \quad (5.80)
$$

Thus, since $p = \partial L / \partial \dot{q}$,

$$
\delta t \left[\frac{dL}{dt} - \dot{q}\dot{p} - \ddot{q}p \right] = \delta t \frac{d}{dt} [L - \dot{q}p] = 0,
$$

and hence, H being the Hamilton function ($cf.$ Eq. (4.42)),

$$
\frac{d}{dt}[-H] = 0. \quad (5.81)
$$

Hence H is a constant of the motion, $i.e.$ a conserved quantity, in the present case the total energy.

Truly nontrivial applications of the Noether theorem arise in the consideration of fields in space (like the electromagnetic field) where the Lagrange function L is the spatial integral over the Lagrangian density $\mathcal{L}$, like for instance in the following, where $\phi(\mathbf{x}, t)$ is a $scalar\ field$:

$$
L = \int_{\mathbb{R}^3} d\mathbf{x} \mathcal{L}(\phi, \partial_i \phi, \dot{\phi}), \quad S = \int L dt.
$$

Here we have to vary with respect to spacetime which we have not yet introduced (but see Sec. 6.4), and hence we do not go deeper into this at present.

Example 5.5: Phase independence
Show that The Lagrange function $L = \lambda \phi \phi^*$ possesses an invariance.

Solution: Since $L = \lambda |\phi|^2$, we see that L is independent of a phase of ϕ, *i.e.*

$$\phi \to \phi' = \phi e^{i\alpha}, \quad \delta\phi = \phi' - \phi = \phi(1 + i\alpha) - \phi = i\alpha\phi,$$

and

$$\delta L = \lambda(\delta\phi^*)\phi + \lambda\phi^*\delta\phi = \lambda(-i\alpha\phi^*)\phi + \lambda\phi^*(i\alpha\phi) = 0.$$

But

$$L = L(\phi, \phi^*, \dot{\phi}, \dot{\phi}^*) = L(|\phi|, \alpha; |\dot{\phi}|, \dot{\alpha}).$$

Here

$$\frac{\partial L}{\partial \alpha} = 0 \quad \text{implies} \quad \frac{d}{dt}\left(\frac{\partial L}{\partial \dot{\alpha}}\right) = 0, \quad i.e. \quad \frac{\partial L}{\partial \dot{\alpha}} = \text{const.}$$

Thus $p_\alpha = \partial L/\partial \dot{\alpha} = \text{const.}$ is the conservation law.

5.8   Canonical Transformations

It is evident that the transition from configuration space coordinates or position space coordinates to phase space coordinates suggests that we now investigate in some detail the significance of transformations in the latter space, in which — as we saw earlier — q_i and p_i are treated as on the same level.

The advantage of Hamilton's formulation is not its use as a calculational method, but rather the deeper insight it provides into the formal structure of mechanics. Attributing the same status to generalized coordinates and generalized, canonical momenta as independent variables provides a greater freedom in the choice of physical quantities, which is particularly useful with a view towards quantum mechanics and statistical mechanics.

Consider the case of $H = E = \text{const.}$ and all q_i cyclic, *i.e.* $\partial L/\partial q_i = 0, i = 1, 2, \ldots, n$, and so

$$\frac{d}{dt}\left(\frac{\partial L}{\partial \dot{q}_i}\right) = 0, \qquad \frac{\partial L}{\partial \dot{q}_i} \equiv p_i = \text{const.} = \alpha_i. \tag{5.82}$$

Then

$$H = H(\alpha_1, \alpha_2, \alpha_3, \ldots, \alpha_n),$$

since cyclic q_i do not appear in H. We also have (since $H=$ const.)

$$\dot{q}_i = \frac{\partial H}{\partial p_i} = \frac{\partial H}{\partial \alpha_i} \equiv \omega_i = \text{const.} \tag{5.83}$$

It follows that

$$q_i = \omega_i t + \beta_i, \qquad (5.84)$$

with constants β_i which can be determined from the *initial conditions*. Is this case only of academic interest? Not at all! A given mechanical system can be described by several sets of generalized coordinates, *e.g.* to describe the motion of a particle in a plane, we can choose

$$q_1 = x, \quad q_2 = y, \qquad \text{or} \qquad q_1 = r, \quad q_2 = \theta.$$

In the case of *central forces* (Chapter 7) neither x nor y is cyclic, however the angle θ. Thus *the number of cyclic coordinates depends on the choice of the generalized coordinates*. Thus the question arises, how one can find a set of generalized coordinates with the maximum number of cyclic coordinates, and hence how one transforms from a given set to a new set, which is such that all coordinates are cyclic. Transformations of the form

$$q_i \longrightarrow Q_i = Q_i(q_j, t), \qquad (5.85)$$

which transform coordinates q_i into coordinates Q_i, like for instance orthogonal transformations, are called *point transformations*. In Hamilton's formalism, however, the momenta are variables of equivalent standing, and hence the transformations of one set of *canonical variables* into another set of canonical variables, *i.e.*

$$Q_i = Q_i(q, p, t), \quad P_i = P_i(q, p, t) \qquad (5.86)$$

are called *canonical transformations*. In fact, the new coordinates are *again canonical* if a function $K(Q, P, t)$ exists, such that the corresponding Hamilton equations are valid, *i.e.*

$$\dot{Q}_i = \frac{\partial K}{\partial P_i}, \quad \dot{P}_i = -\frac{\partial K}{\partial Q_i}. \qquad (5.87)$$

Thus in these cases the canonical transformations leave the Hamilton equations invariant, *i.e. form-invariant*.

5.8.1 Generators of canonical transformations

We recall once again *Hamilton's principle* from which Hamilton's equations can be derived:

$$\delta \int_{t_1}^{t_2} L \, dt = 0, \quad i.e. \quad \delta \int_{t_1}^{t_2} \left[\sum_i \dot{q}_i p_i - H(q, p, t) \right] dt = 0. \qquad (5.88)$$

Thus for the equations (5.87) to result, we must have in a corresponding way
the validity of the principle

$$\delta \int_{t_1}^{t_2} \left[\sum_i \dot{Q}_i P_i - K(Q,P,t) \right] dt = 0. \tag{5.89}$$

Equations (5.88) and (5.89) have to hold simultaneously. This is the case
when the integrands differ at most only by a total time derivative, *i.e.* when

$$\left[\sum_i \dot{q}_i p_i - H(q,p,t) \right] - \left[\sum_i \dot{Q}_i P_i - K(Q,P,t) \right] = \frac{dF}{dt} \tag{5.90}$$

(instead of the minus sign for the difference on the left we could also have a
plus sign, or -538; then F would simply be defined in a different way). We
can see this as follows. With Eq. (5.88) we obtain from Eq. (5.90):

$$\delta \int_{t_1}^{t_2} \left[\sum_i P_i \dot{Q}_i - K(Q,P,t) + \frac{dF}{dt} \right] dt = 0,$$

i.e.

$$\delta \int_{t_1}^{t_2} \left[\sum_i P_i \dot{Q}_i - K(Q,P,t) \right] dt = 0,$$

since

$$\delta[F(1) - F(2)] = 0,$$

because the delta variation vanishes at the endpoints of the integration, in
fact quite generally:

$$\delta F = \sum_i \left[\frac{\partial F}{\partial q_i} \delta q_i + \frac{\partial F}{\partial p_i} \delta p_i + \frac{\partial F}{\partial Q_i} \delta Q_i + \frac{\partial F}{\partial P_i} \delta P_i \right] + \frac{\partial F}{\partial t} \delta t,$$

which vanishes at the limits with

$$(\delta q_i)_{1,2} = 0, \quad (\delta p_i)_{1,2} = 0, \quad (\delta Q_i)_{1,2} = 0, \quad (\delta P_i)_{1,2} = 0.$$

The arbitrary function F is called the *generating function of the transformation*, because — as we shall see — the transformation equations are com-
pletely determined, once F is known.[||] If this is so, the function F must
basically be considered to be a function of the old as well as the new variables, and this means of $4n$ variables as well as t. Since, however, the two
sets of coordinates are connected by $2n$ transformation equations, only $2n$

[||] In Example 5.16 necessary and sufficient conditions for the integrability of dF are derived.

variables are independent and F can therefore be considered to be a function of any of the following four forms:

$$F_1(q, Q, t), \quad F_2(q, P, t), \quad F_3(p, Q, t), \quad F_4(p, P, t).$$

One should note: In the version F_1 the variables $q_i, Q_i, i = 1, 2, \ldots, n$, are independent, which means that in the case of point transformations $Q_i = Q_i(q, t)$, the generating functions of type F_1 are excluded, since in the point transformation the variables q_i are connected with the variables Q_i.

$F_1(q, Q, t)$:

We assume first that the version $F_1(q, Q, t)$ is for some reason suitable. Then we have

$$\sum_i \dot{q}_i p_i - H(q, p, t) = \sum_i \dot{Q}_i P_i - K(Q, P, t) + \frac{d}{dt} F_1(q, Q, t). \qquad (5.91)$$

Since $F_1 \equiv F_1(q, Q, t)$, we have

$$\frac{dF_1}{dt} = \frac{\partial F_1}{\partial t} + \sum_i \frac{\partial F_1}{\partial q_i} \dot{q}_i + \sum_i \frac{\partial F_1}{\partial Q_i} \dot{Q}_i,$$

so that Eq. (5.91) becomes

$$\sum_i \dot{q}_i p_i - H(q, p, t) = \sum_i \dot{Q}_i P_i - K(Q, P, t) + \frac{\partial F_1}{\partial t}$$
$$+ \sum_i \frac{\partial F_1}{\partial q_i} \dot{q}_i + \sum_i \frac{\partial F_1}{\partial Q_i} \dot{Q}_i. \qquad (5.92)$$

Here q_i, Q_i are independent, *i.e.* $\delta q_i, \delta Q_i$, which result from $\dot{q}_i, \dot{Q}_i = dQ_i/dt$. Hence, multiplying Eq. (5.92) by dt, and remembering that dq_i, dQ_i, dt are independent, we obtain

$$(a) \quad p_i = \frac{\partial F_1(q, Q, t)}{\partial q_i}, \qquad (b) \quad P_i = -\frac{\partial F_1(q.Q, t)}{\partial Q_i},$$

$$(c) \quad K = H + \frac{\partial F_1(q, Q, t)}{\partial t}. \qquad (5.93)$$

Each of cases (a) and (b) imply n equations. Since $F_1 = F_1(q, Q, t)$, the n equations of case (a) are equations in q_i, Q_i, p_i and t, which can be solved for Q_i, *i.e.*

$$Q_i = Q_i(p_i, q_i, t).$$

This gives the first set of n equations. The other set of n equations follows from solving the equations of case (b) in Eq. (5.93) and inserting $Q_i = Q_i(p, q, t)$:

$$P_i = P_i(q_i, Q_i, t) = P_i(q_i, p_i, t).$$

Finally the case (c) of Eq. (5.93) provides the connection between the Hamilton function $H(q, p, t)$ and the new Hamilton function $K(Q, P, t)$.

$F_2(q,P,t)$:

We now assume that the version F_2 is what we want, *i.e.* $F_2 = F_2(q, P, t)$. The transformation from the set of variables (q_i, Q_i, t) to the set (q_i, P_i, t) can be performed with the help of a Legendre transform in analogy to the transformation from $(q, \dot{q}, t)$ to (q, p, t) with

$$H(q, p, t) = \sum_i \dot{q}_i p_i - L(q, \dot{q}, t).$$

This analogy suggests to define F_2 by the relation

$$F_2(q, P, t) = \sum_i Q_i P_i + F_1(q, Q, t), \quad \text{with} \quad \frac{\partial F_2}{\partial P_i} = Q_i. \qquad (5.94)$$

Since

$$0 = \frac{\partial F_2}{\partial Q_i} = P_i + \frac{\partial F_1}{\partial Q_i} \overset{(5.93)}{=} P_i - P_i = 0, \qquad \frac{\partial F_2}{\partial q_i} = \frac{\partial F_1}{\partial q_i} \overset{(5.93)}{=} p_i,$$

it follows that, since from Eq. (5.94)

$$F_1(q, Q, t) = F_2(q, P, t) - \sum_i Q_i P_i, \qquad (5.95)$$

this expression substituted into Eq. (5.91) yields

$$\begin{aligned}
\sum_i \dot{q}_i p_i - H(q, p, t) &= \sum_i \dot{Q}_i P_i - K(Q, P, t) + \frac{d}{dt}[F_2(q, P, t) - \sum_i Q_i P_i] \\
&= -\sum_i \dot{P}_i Q_i - K + \frac{d}{dt} F_2(q, P, t) \\
&= -\sum_i \dot{P}_i Q_i - K + \frac{\partial F_2}{\partial t} + \sum_i \frac{\partial F_2}{\partial q_i} \dot{q}_i + \sum_i \frac{\partial F_2}{\partial P_i} \dot{P}_i.
\end{aligned}$$

$$(5.96)$$

Since q_i, P_i are here independent variables, we obtain (again by considering the equation as multiplied by dt):

$$(a) \;\; p_i = \frac{\partial F_2(q, P, t)}{\partial q_i}, \quad (b) \;\; Q_i = \frac{\partial F_2(q, P, t)}{\partial P_i}, \quad K = H + \frac{\partial F_2(q, P, t)}{\partial t}.$$

$$(5.97)$$

These equations can be solved for $P_i = P_i(q, p, t)$. These inserted into (b) give $Q_i = Q_i(q, p, t)$.

In the special case when $K = 0$, and therefore $\dot{Q}_i = \partial K / \partial P_i = 0, \dot{P}_i = -\partial K / \partial Q_i = 0$, the new coordinates $Q_i \equiv \beta_i, P_i \equiv \alpha_i$ are all constant, and

$$H\left(q_i, \frac{\partial F_2}{\partial q_i}, t\right) + \frac{\partial F_2(q, \alpha, t)}{\partial t} = 0, \qquad p_i - \frac{\partial F_2}{\partial q_i}.$$

This equation with $F_2 \equiv S(q_i, \alpha_i, t)$ is known as the *Hamilton–Jacobi equation*, and the *generating function* S is known as *Hamilton's action*. The Hamilton–Jacobi equation is a partial differential equation in the variables q_i, t. One is then interested in the solution q_i of the equation as $q_i = q_i(\alpha_i, \beta_i, t)$, where α_i and β_i are respectively the constant (initial) values of P_i and Q_i, the latter (*cf.* Eq. (5.97)) as $Q_i = \beta_i = \partial S / \partial \alpha_i$. For the application of this equation to the harmonic oscillator see Example 5.17.

$F_3(p, Q, t)$:

We now come to the case of $F_3 = F_3(p, Q, t)$. This type of generating function of canonical transformations can also be linked by a Legendre transform to F_1:

$$F_3(p, Q, t) = F_1(q, Q, t) - \sum_i q_i p_i. \qquad (5.98)$$

We note that this definition is such that partial differentiation with respect to q_i yields as required:

$$0 = \frac{\partial F_3(p, Q, t)}{\partial q_i} = \frac{\partial F_1(q, Q, t)}{\partial q_i} - p_i \stackrel{(5.93)(a)}{=} p_i - p_i = 0.$$

Inserting the relation (5.98) for $F_1(q, Q, t)$ into Eq. (5.91), we obtain

$$\sum_i \dot{q}_i p_i - H(q, p, t) = \sum_i \dot{Q}_i P_i - K(Q, P, t) + \frac{d}{dt}\left[\sum_i q_i p_i + F_3(p, Q, t)\right]$$

$$= \sum_i \dot{Q}_i P_i - K + \sum_i \dot{p}_i q_i + \sum_i \dot{q}_i p_i + \frac{\partial F_3}{\partial t}$$

$$+ \sum_i \frac{\partial F_3}{\partial Q_i}\dot{Q}_i + \sum_i \frac{\partial F_3}{\partial p_i}\dot{p}_i. \qquad (5.99)$$

Since Q_i, p_i are here independent variables, we obtain in a way analogous to that above:

$$(a) \quad q_i = -\frac{\partial F_3(p, Q, t)}{\partial p_i}, \quad (b) \quad P_i = -\frac{\partial F_3(p, Q, t)}{\partial Q_i},$$

$$(c) \quad K = H + \frac{\partial F_3(p, Q, t)}{\partial t}. \qquad (5.100)$$

Equations (5.100) (a) give $Q_i = Q_i(q, p, t)$, and Eqs. (5.100) (b) with this $P_i = P_i(q, p, t)$.

$F_4(p,P,t)$:
    Finally we come to the case of F_4. This time we use a double Legendre transform in order to link this with F_1. We set first

$$
\begin{aligned}
F_4(p, P, t) &:= F_2(q, P, t) - \sum_i q_i p_i \\
&\overset{(5.94)}{=} F_1(q, Q, t) + \sum_i Q_i P_i - \sum_i q_i p_i.
\end{aligned}
\tag{5.101}
$$

Equation (5.91) then implies

$$
\begin{aligned}
\sum_i \dot{q}_i p_i - H &= \sum_i \dot{Q}_i P_i - K + \frac{d}{dt} F_1(q, Q, t) \\
&= \sum_i \dot{Q}_i P_i - K + \frac{d}{dt} [F_4(p, P, t) - \sum_i Q_i P_i + \sum_i q_i p_i] \\
&= \sum_i \dot{Q}_i P_i - K - \sum_i Q_i \dot{P}_i - \sum_i \dot{Q}_i P_i + \sum_i \dot{q}_i p_i + \sum_i \dot{p}_i q_i \\
&\quad + \frac{\partial F_4}{\partial t} + \sum_i \frac{\partial F_4}{\partial p_i} \dot{p}_i + \sum_i \frac{\partial F_4}{\partial P_i} \dot{P}_i.
\end{aligned}
\tag{5.102}
$$

Proceeding as before and identifying coefficients, we now obtain:

$$
(a)\ \ q_i = -\frac{\partial F_4}{\partial p_i}, \quad (b)\ \ Q_i = \frac{\partial F_4}{\partial P_i}, \quad (c)\ \ K = H + \frac{\partial F_4}{\partial t}.
\tag{5.103}
$$

Example 5.6: F_2 for orthogonal transformations
Determine the generating function F_2 for the case of orthogonal transformations.

Solution: The orthogonal transformations we considered earlier are special cases of point transformations (*cf.* Eq. (5.85)) which we can write:

$$
Q_i = \sum_k a_{ik} q_k.
$$

Since in terms of F_2 the new coordinates Q_i are given by $Q_i = \partial F_2 / \partial P_i$ (*cf.* (5.97)(b)), we obtain by integration

$$
F_2 = \sum_{i,k} a_{ik} q_k P_i + R(q, t).
$$

Then

$$
p_k \overset{(5.97)}{=} \frac{\partial F_2}{\partial q_k} = \sum_i a_{ik} P_i + \frac{\partial R(q, t)}{\partial q_k}.
$$

Solving this equation we obtain P_i. Actually we should argue: Given F_2, then Q_i and p_i follow.

5.8.2 Invariance of Poisson brackets

The fundamental significance of Poisson brackets can be appreciated by the importance of the following theorem, in the proof of which they are required.

Let F and G be two arbitrary functions. The *Poisson bracket* of these functions with respect to the set of coordinates or canonical variables q_i, p_j, where $i, j = 1, 2, \ldots, n$, is defined as the quantity (*cf.* Eq. (4.61))

$$\{F, G\}_{q,p} = \sum_j \left[\frac{\partial F}{\partial q_j} \frac{\partial G}{\partial p_j} - \frac{\partial F}{\partial p_j} \frac{\partial G}{\partial q_j} \right]. \tag{5.104}$$

With the help of the fundamental Poisson brackets (Eqs. (4.86)) it can be shown — this is the theorem — that

$$\{F, G\}_{q,p} = \{F, G\}_{Q,P}, \tag{5.105}$$

i.e. that $\{F, G\}$ is independent of any particular set of canonical coordinates. In proving this relation, we proceed as follows. The canonical transformation is given by the relations

$$q_j = q_j(Q, P), \quad p_j = p_j(Q, P). \tag{5.106}$$

Hence

$$\{F, G, \}_{q,p} = \sum_j \left(\frac{\partial F}{\partial q_j} \frac{\partial G}{\partial p_j} - \frac{\partial F}{\partial p_j} \frac{\partial G}{\partial q_j} \right)$$

$$= \sum_{j,k} \left[\frac{\partial F}{\partial q_j} \left(\frac{\partial G}{\partial Q_k} \frac{\partial Q_k}{\partial p_j} + \frac{\partial G}{\partial P_k} \frac{\partial P_k}{\partial p_j} \right) - \frac{\partial F}{\partial p_j} \left(\frac{\partial G}{\partial Q_k} \frac{\partial Q_k}{\partial q_j} + \frac{\partial G}{\partial P_k} \frac{\partial P_k}{\partial q_j} \right) \right]$$

$$= \sum_k \left[\frac{\partial G}{\partial Q_k} \{F, Q_k\}_{q,p} + \frac{\partial G}{\partial P_k} \{F, P_k\}_{q,p} \right]. \tag{5.107}$$

Replacing here F by Q_k, and G by F, we obtain

$$\{Q_k, F\}_{q,p} = \sum_j \left[\frac{\partial F}{\partial Q_j} \{Q_k, Q_j\}_{q,p} + \frac{\partial F}{\partial P_j} \{Q_k, P_j\}_{q,p} \right]. \tag{5.108}$$

We had seen previously (*cf.* Eq. (4.84)) that for a function $u(q, p)$

$$\frac{du(q, p)}{dt} = \{u(q, p), H(q, p)\}, \tag{5.109}$$

and therefore**

$$\dot{q} = \{q, H\}, \quad \dot{p} = \{p, H\}, \tag{5.110}$$

**In Chapter 4 we used brackets [,] for Poisson brackets and {,} for Lagrange brackets. Since Lagrange brackets are not used much, one generally uses the brackets {,} for Poisson brackets and retains the brackets [,] for the corresponding operator cases in quantum mechanics.

or with Hamilton's equations

$$\{q, H\} = \frac{\partial H}{\partial p}, \quad \{p, H\} = -\frac{\partial H}{\partial q}. \tag{5.111}$$

The form-invariance of Hamilton's equations under canonical transformations (which defines the latter) implies that in the transformed system with Hamilton function $K(Q, P)$:

$$\{Q, K\} = \frac{\partial K}{\partial P}, \quad \{P, K\} = -\frac{\partial K}{\partial Q}. \tag{5.112}$$

Replacing in Eq. (5.108) the completely arbitrary function F by K, we obtain

$$\{Q_k, K\}_{q,p} = \sum_j \left[\frac{\partial K}{\partial Q_j} \{Q_k, Q_j\}_{q,p} + \frac{\partial K}{\partial P_j} \{Q_k, P_j\}_{q,p} \right], \tag{5.113}$$

and we see from Eq. (5.112) that this has to be equal to $\partial K / \partial P_k$. From the right hand side of Eq. (5.113) we therefore deduce that

$$\{Q_k, Q_j\}_{q,p} = 0, \quad \{Q_k, P_j\}_{q,p} = \delta_{kj}. \tag{5.114}$$

We obtain by a similar procedure

$$\{P_k, P_j\}_{q,p} = 0, \quad \{P_k, Q_j\}_{q,p} = -\delta_{kj}. \tag{5.115}$$

Inserting the expressions of Eq. (5.114) into the right hand side of Eq. (5.108), we obtain

$$\{F, Q_k\}_{q,p} = -\frac{\partial F}{\partial P_k}. \tag{5.116}$$

Replacing in Eq. (5.107) F by P_j and G by F, we obtain

$$\{P_j, F\}_{q,p} = \sum_k \left[\frac{\partial F}{\partial Q_k} \{P_j, Q_k\}_{q,p} + \frac{\partial F}{\partial P_k} \{P_j, P_k\}_{q,p} \right]. \tag{5.117}$$

With Eqs. (5.114) and (5.115) we obtain

$$\{F, P_j\}_{q,p} = \frac{\partial F}{\partial Q_j}. \tag{5.118}$$

Equations (5.116) and (5.118) are of considerable importance by themselves. However, now we insert their results into Eq. (5.107) and obtain

$$\{F, G\}_{q,p} = \sum_k \left[\frac{\partial G}{\partial Q_k} \left(-\frac{\partial F}{\partial P_k} \right) + \frac{\partial G}{\partial P_k} \left(\frac{\partial F}{\partial Q_k} \right) \right] = \{F, G\}_{Q,P}. \tag{5.119}$$

This shows that the Poisson bracket of any two arbitrary functions $F(q, p)$ and $G(q, p)$ is equal to the Poisson bracket of these functions in terms of any other set of canonical variables. This is an extremely important result of classical mechanics and establishes the above claim. We can therefore drop the subscripts q, p or Q, P. These considerations demonstrate also that the form-invariance of Hamilton's equations under canonical transformations implies the corresponding invariance of the fundamental Poisson brackets and *vice versa*. For a test of whether a given transformation is a canonical one, it suffices therefore to check the invariance of the Poisson-brackets, which is easier in general. The following examples illustrate various aspects and uses of canonical transformations.

Example 5.7: Transformation to initial values
The Hamilton function of the harmonic oscillator is given by $H(q, p) = p^2/2 + \omega^2 q^2/2$. Determine the canonical transformation to the initial values of the phase space coordinates, *i.e.* $q(t), p(t) \to q(0), p(0)$.

Solution: Applying Hamilton's equations to the given $H(q, p)$, we obtain

$$\dot{q} - \frac{\partial H}{\partial p} = p, \quad \dot{p} = -\frac{\partial H}{\partial q} = -\omega^2 q, \quad \therefore \ \ddot{q}(t) + \omega^2 q(t) = 0.$$

We obtain therefore with constants a, a^* the real solutions

$$q(t) = \frac{1}{\sqrt{2\omega}} \left[ae^{-i\omega t} + a^* e^{i\omega t} \right], \quad p(t) = \dot{q}(t) = \frac{-i\omega}{\sqrt{2\omega}} \left[ae^{-i\omega t} - a^* e^{i\omega t} \right].$$

Expressing a, a^* in terms of q and p, we obtain

$$a = \sqrt{\frac{\omega}{2}} \left[q(t) + \frac{1}{-i\omega} p(t) \right] e^{i\omega t}, \quad a^* = \sqrt{\frac{\omega}{2}} \left[q(t) - \frac{1}{-i\omega} p(t) \right] e^{-i\omega t}.$$

Thus

$$q(0) = \frac{1}{\sqrt{2\omega}} (a + a^*) = \left[q(t) \cos \omega t - \frac{p(t)}{\omega} \sin \omega t \right],$$

$$p(0) = \frac{-i\omega}{\sqrt{2\omega}} (a - a^*) = \omega \left[q(t) \sin \omega t + \frac{p(t)}{\omega} \cos \omega t \right].$$

This is the transformation $\{q(0), p(0)\} \leftrightarrow \{q(t), p(t)\}$.

Example 5.8: An explicit canonical transformation
Show that the following transformation is a canonical transformation:[††]

$$Q = \ln\left(\frac{\sin p}{q}\right), \quad P = q \cot p; \quad \frac{\partial Q}{\partial q} = -\frac{1}{q}, \quad \frac{\partial Q}{\partial p} = \cot p, \quad \frac{\partial P}{\partial q} = \cot p, \quad \frac{\partial P}{\partial p} = -\frac{q}{\sin^2 p}.$$

Solution: We have (recall $d\cot x/dx = -1/\sin^2 x$):

$$\dot{Q} = \frac{-(\sin p)\dot{q}/q^2 + (\cos p)\dot{p}/q}{(\sin p)/q} = -\frac{\dot{q}}{q} + (\cot p)\dot{p}, \quad \dot{P} = \dot{q}(\cot p) - \frac{q\dot{p}}{\sin^2 p}.$$

[††]This is problem 1 of H. Goldstein [19], Chapter VIII.

We assume

$$\dot{q} = \frac{\partial H}{\partial p}, \quad \dot{p} = -\frac{\partial H}{\partial q}, \quad H(q,p) = K(Q,P),$$

and verify that

$$\dot{Q} = \frac{\partial K}{\partial P}, \quad \dot{P} = -\frac{\partial K}{\partial Q}.$$

We have, since $Q = Q(q,p), P = P(q,p)$,

$$\dot{Q} = \frac{\partial Q}{\partial q}\dot{q} + \frac{\partial Q}{\partial p}\dot{p} = -\frac{1}{q}\frac{\partial H}{\partial p} - \cot p\frac{\partial H}{\partial q},$$

$$\dot{P} = \frac{\partial P}{\partial q}\dot{q} + \frac{\partial P}{\partial p}\dot{p} = \cot p\frac{\partial H}{\partial p} + \frac{q}{\sin^2 p}\frac{\partial H}{\partial q}.$$

Since $H(q,p) = K(Q,P)$, we obtain

$$\dot{Q} = -\frac{1}{q}\left(\frac{\partial K}{\partial Q}\frac{\partial Q}{\partial p} + \frac{\partial K}{\partial P}\frac{\partial P}{\partial p}\right) - \cot p\left(\frac{\partial K}{\partial Q}\frac{\partial Q}{\partial q} + \frac{\partial K}{\partial P}\frac{\partial P}{\partial q}\right)$$

$$= -\frac{1}{q}\left(\frac{\partial K}{\partial Q}\cot p - \frac{\partial K}{\partial P}\frac{q}{\sin^2 p}\right) - \cot p\left(-\frac{1}{q}\frac{\partial K}{\partial Q} + \frac{\partial K}{\partial P}\cot p\right)$$

$$= \frac{\partial K}{\partial P}\left(\frac{1}{\sin^2 p} - \cot^2 p\right) = \frac{\partial K}{\partial P},$$

$$\dot{P} = \cot p\left(\frac{\partial K}{\partial Q}\frac{\partial Q}{\partial p} + \frac{\partial K}{\partial P}\frac{\partial P}{\partial p}\right) + \frac{q}{\sin^2 p}\left(\frac{\partial K}{\partial Q}\frac{\partial Q}{\partial q} + \frac{\partial K}{\partial P}\frac{\partial P}{\partial q}\right)$$

$$= \cot p\left(\frac{\partial K}{\partial Q}\cot p - \frac{\partial K}{\partial P}\frac{q}{\sin^2 p}\right) + \frac{q}{\sin^2 p}\left(-\frac{\partial K}{\partial Q}\frac{1}{q} + \frac{\partial K}{\partial P}\cot p\right) = -\frac{\partial K}{\partial Q}.$$

Example 5.9: Explicit canonical transformations

Are the following transformations (the first has $q \leftrightarrow p$ in Example 5.8) canonical transformations?[‡‡]

$$(a) \quad Q = \ln\left(\frac{\sin q}{p}\right), \quad P = p\cot q, \quad (b) \quad Q = q^\alpha \cos\beta p, \quad P = q^\alpha \sin\beta p.$$

For what values of α and β is the second transformation canonical?

Solution: Proceeding in the case of the first transformation as in Example 5.8, one obtains $\dot{Q} = -\partial K/\partial P$. The minus sign indicates that the transformation is not canonical. This result can also be seen from the following relation, which in the case of a canonical transformation would have to be $+1$. We have

$$\frac{\partial Q}{\partial q}\frac{\partial P}{\partial p} - \frac{\partial Q}{\partial p}\frac{\partial P}{\partial q} = \frac{\cos q}{\sin q}\frac{\cos q}{\sin q} - \frac{1}{p}\frac{p}{\sin^2 q} = \frac{\cos^2 q - 1}{\sin^2 q} = -1.$$

In the case of the second transformation we start again with Hamilton's equations as in Example 5.8, and with $H(q,p) = K(Q,P)$. Then

$$\dot{Q} = \alpha q^{\alpha-1}\dot{q}\cos\beta p - \beta q^\alpha \dot{p}\sin\beta p = \alpha q^{\alpha-1}\cos\beta p\frac{\partial H}{\partial p} + \beta q^\alpha \sin\beta p\frac{\partial H}{\partial q}$$

$$= \alpha q^{\alpha-1}\cos\beta p\left(\frac{\partial K}{\partial Q}\frac{\partial Q}{\partial p} + \frac{\partial K}{\partial P}\frac{\partial P}{\partial p}\right) + \beta q^\alpha \sin\beta p\left(\frac{\partial K}{\partial Q}\frac{\partial Q}{\partial q} + \frac{\partial K}{\partial P}\frac{\partial P}{\partial q}\right).$$

[‡‡]The second part is problem 4 of H. Goldstein [19], Chapter VIII.

However,

$$\frac{\partial P}{\partial q} = \alpha q^{\alpha-1} \sin \beta p, \quad \frac{\partial P}{\partial p} = \beta q^{\alpha} \cos \beta p, \quad \frac{\partial Q}{\partial q} = \alpha q^{\alpha-1} \cos \beta p, \quad \frac{\partial Q}{\partial p} = -q^{\alpha} \beta \sin \beta p.$$

Hence

$$
\begin{aligned}
\dot{Q} &= \alpha q^{\alpha-1} \cos \beta p \left(-q^{\alpha} \beta \sin \beta p \frac{\partial K}{\partial Q} + \beta q^{\alpha} \cos \beta p \frac{\partial K}{\partial P} \right) \\
&\quad + \beta q^{\alpha} \sin \beta p \left(\alpha q^{\alpha-1} \cos \beta p \frac{\partial K}{\partial Q} + \alpha q^{\alpha-1} \sin \beta p \frac{\partial K}{\partial P} \right) \\
&= \alpha \beta q^{2\alpha-1} (\cos^2 \beta p + \sin^2 \beta p) \frac{\partial K}{\partial P} = \alpha \beta q^{2\alpha-1} \frac{\partial K}{\partial P} \\
&=: \frac{\partial K}{\partial P} \quad \Longleftrightarrow \quad \alpha = \frac{1}{2}, \ \beta = 2.
\end{aligned}
$$

Analogously we can proceed with $\dot{P} = \alpha q^{\alpha-1} \dot{q} \sin \beta p + \beta q^{\alpha} \dot{p} \cos \beta p, \dots$ Also, the following Poisson bracket has to be -1:

$$
\begin{aligned}
\frac{\partial P}{\partial q} \frac{\partial Q}{\partial p} - \frac{\partial P}{\partial p} \frac{\partial Q}{\partial q} &= (\alpha q^{\alpha-1} \sin \beta p)(-q^{\alpha} \beta \sin \beta p) - (\beta q^{\alpha} \cos \beta p)(\alpha q^{\alpha-1} \cos \beta p) \\
&= -\alpha \beta q^{2\alpha-1} (\sin^2 \beta p + \cos^2 \beta p) = -\alpha \beta q^{2\alpha-1} \overset{!}{=} -1 \quad \Longleftrightarrow \quad \alpha = \frac{1}{2}, \ \beta = 2.
\end{aligned}
$$

One could also proceed as follows. From the given Q and P we obtain

$$\tan \beta p = \frac{P}{Q}, \quad q^{2\alpha} = P^2 + Q^2, \quad \therefore \ q = (P^2 + Q^2)^{1/2\alpha}, \quad p = \frac{1}{\beta} \tan^{-1}\left(\frac{P}{Q}\right).$$

From these equations we obtain by differentiation (we skip details)

$$\frac{\partial q}{\partial P} = \frac{\sin \beta p}{\alpha q^{\alpha-1}}, \quad \frac{\partial q}{\partial Q} = \frac{\cos \beta p}{\alpha q^{\alpha-1}}, \quad \frac{\partial p}{\partial P} = \frac{\cos \beta p}{\beta q^{\alpha}}, \quad \frac{\partial p}{\partial Q} = -\frac{\sin \beta p}{\beta q^{\alpha}}.$$

It follows that

$$\frac{\partial q}{\partial Q} \frac{\partial p}{\partial P} - \frac{\partial q}{\partial P} \frac{\partial p}{\partial Q} = \frac{(\cos \beta p)^2}{\alpha q^{\alpha-1} \beta q^{\alpha}} + \frac{(\sin \beta p)^2}{\alpha q^{\alpha-1} \beta q^{\alpha}} = \frac{1}{\alpha \beta q^{2\alpha-1}} \overset{!}{=} 1 \quad \Longleftrightarrow \quad \alpha = \frac{1}{2}, \ \beta = 2.$$

In the present case it is not too difficult to find also the generator $F_3(p, Q, t)$. For F_3 we have (see Eq. (5.100))

$$q_i = -\frac{\partial F_3}{\partial p_i}, \quad P_i = -\frac{\partial F_3}{\partial Q_i}.$$

In the present case we have (expressed in terms of p and Q, and with $\alpha = 1/2, \beta = 2$):

$$q = \frac{Q^2}{\cos^2 2p}, \quad P = q^{1/2} \sin 2p = Q \tan 2p, \quad i.e. \quad \frac{Q^2}{\cos^2 2p} = -\frac{\partial F_3}{\partial p}, \quad Q \tan 2p = -\frac{\partial F_3}{\partial Q}.$$

From these relations we deduce that $F_3 = -[Q^2 \tan 2p]/2$.

5.9   Conserved Quantities

We had the equation (*cf.* Eqs. (4.84), (5.109))

$$\frac{du}{dt} = \{u, H\} + \frac{\partial u}{\partial t}. \tag{5.120}$$

Setting here $u = q_i$ and p_i respectively, we obtain Hamilton's equations with

$$\frac{\partial q_i}{\partial t} = 0, \quad \frac{\partial p_i}{\partial t} = 0.$$

(Recall: Originally we introduced u_i, v_i as the parameters of a point in the two-dimensional hyperplane, and then as examples we replaced u_i, v_i by q_i, p_i. The explicit time-dependence is contained in the original coordinates $\mathbf{r}_i$, *i.e.* in $\mathbf{r}_i = \mathbf{r}_i(q_1, q_2, \ldots, q_n, t)$, and therefore $\partial q_i/\partial t = 0$). For $u = H$ we obtain

$$\frac{dH}{dt} = \frac{\partial H}{\partial t}.$$

We now restrict ourselves to systems for which the functions u (*i.e.* arbitrary dynamical quantities that are of interest to us) do not depend explicitly on t (*i.e.* $\partial u/\partial t = 0$). What kind of systems are these? Since for these

$$\frac{du}{dt} = \{u, H\},$$

this implies, that for such systems (since $\{H, H\}=0$)

$$\frac{dH}{dt} = 0, \quad H = \text{constant in time.}$$

Thus any function u with $\{u, H\} = 0$ is a constant of motion (*i.e.* in time), and therefore a conserved quantity. Conversely, if a quantity u is constant, then we must have $\{u, H\} = 0$. Conserved quantities can therefore be found, by searching for those quantities which *"commute"* with H, *i.e.* for which $\{u, H\} = 0$. If two conserved quantities u, v are known, one can obtain others w from the *Jacobi identity*:

$$\{u, \{v, w\}\} + \{v, \{w, u\}\} + \{w, \{u, v\}\} = 0, \tag{5.121}$$

where u, v, w are functions of q, p. We verify this relation in Example 5.10. If u and v are conserved quantities, so that

$$\{H, u\} = 0, \quad \{H, v\} = 0,$$

it follows that for $w = H$:

$$\{H, \{u, v\}\} = 0. \tag{5.122}$$

Thus the Poisson bracket of two conserved quantities is itself a conserved quantity. This result is known as *Poisson's theorem.*

Example 5.10: The Jacobi identity
Verify the Jacobi identity (5.121).

Solution: We have

$$\{v, w\} = \sum_{i=1}^{n} \left[\frac{\partial v}{\partial q_i} \frac{\partial w}{\partial p_i} - \frac{\partial w}{\partial q_i} \frac{\partial v}{\partial p_i} \right] \equiv D_v w, \quad D_v \equiv \sum_{i=1}^{n} \left[\frac{\partial v}{\partial q_i} \frac{\partial}{\partial p_i} - \frac{\partial v}{\partial p_i} \frac{\partial}{\partial q_i} \right] \equiv \sum_{i=1}^{2n} \alpha_i \frac{\partial}{\partial \xi_i}, \tag{5.123}$$

with, for $i = 1, \dots, n$,

$$\alpha_i = \frac{\partial v}{\partial q_i}, \quad \frac{\partial}{\partial \xi_i} = \frac{\partial}{\partial p_i}, \quad \text{and} \quad \alpha_{n+i} = -\frac{\partial v}{\partial p_i}, \quad \frac{\partial}{\partial \xi_{n+i}} = \frac{\partial}{\partial q_i} \quad \text{for} \quad i = 1, 2, \dots, n. \tag{5.124}$$

Then

$$\{u, \{v, w\}\} + \{v, \{w, u\}\} = \{u, \{v, w\}\} - \{v, \{u, w\}\}$$

$$= \{u, D_v w\} - \{v, D_u w\} \overset{(5.123)}{=} D_u D_v w - D_v D_u w$$

$$\overset{(5.123)}{=} D_u \left(\sum_i \alpha_i \frac{\partial w}{\partial \xi_i} \right) - D_v \left(\sum_i \beta_i \frac{\partial w}{\partial \eta_i} \right) \quad \left(\text{where } D_u = \sum_{i=1}^{2n} \beta_i \frac{\partial}{\partial \eta_i} \right)$$

$$= \sum_{i,k} \left[\beta_i \frac{\partial}{\partial \eta_i} \left(\alpha_k \frac{\partial w}{\partial \xi_k} \right) - \alpha_k \frac{\partial}{\partial \xi_k} \left(\beta_i \frac{\partial w}{\partial \eta_i} \right) \right]$$

$$= \sum_{i,k} \left[\beta_i \frac{\partial \alpha_k}{\partial \eta_i} \frac{\partial w}{\partial \xi_k} + \beta_i \alpha_k \frac{\partial^2 w}{\partial \eta_i \partial \xi_k} - \alpha_k \beta_i \frac{\partial^2 w}{\partial \xi_k \partial \eta_i} - \alpha_k \frac{\partial \beta_i}{\partial \xi_k} \frac{\partial w}{\partial \eta_i} \right]$$

$$= \sum_{i,k} \left[\beta_i \frac{\partial \alpha_k}{\partial \eta_i} \frac{\partial w}{\partial \xi_k} - \alpha_k \frac{\partial \beta_i}{\partial \xi_k} \frac{\partial w}{\partial \eta_i} \right] \equiv \sum_k \left[A_k \frac{\partial w}{\partial p_k} + B_k \frac{\partial w}{\partial q_k} \right], \tag{5.125}$$

where $\partial w / \partial p_k$ and $\partial w / \partial q_k$ contain contributions $\partial w / \partial \xi, \partial w / \partial \eta$. It remains to determine the quantities A_k and B_k, which are functions of u, v, but not of w.

Consider the special case of $w = p_i$. In this case Eq. (5.125) implies (since evaluating $\{v, p_i\}$ yields $\partial v / \partial q_i$ etc.)

$$\{u, \{v, p_i\}\} - \{v, \{u, p_i\}\} = A_i = \left\{ u, \frac{\partial v}{\partial q_i} \right\} + \left\{ \frac{\partial u}{\partial q_i}, v \right\} = \frac{\partial}{\partial q_i} \{u, v\}. \tag{5.126}$$

In the last step we used here the result of Example 4.7. Similarly one finds for $w = q_i$ that

$$B_i = -\frac{\partial}{\partial p_i} \{u, v\}. \tag{5.127}$$

Inserting these expressions into Eq. (5.125), we obtain

$$\{u, \{v, w\}\} + \{v, \{w, u\}\} = \sum_k \left[\frac{\partial}{\partial q_k} \{u, v\} \frac{\partial w}{\partial p_k} - \frac{\partial}{\partial p_k} \{u, v\} \frac{\partial w}{\partial q_k} \right] = \{\{u, v\}, w\}. \tag{5.128}$$

5.9.1   Infinitesimal canonical transformations

We now consider canonical transformations in their infinitesimal form and examine their significance, particularly with a view to conserved quantities. The infinitesimal transformation can be written as

$$q_i \to Q_i = q_i + \delta q_i, \qquad p_i \to P_i = p_i + \delta p_i. \tag{5.129}$$

Here both (q_i, p_i) and (Q_i, P_i) are sets of canonical variables. Since $\delta q_i, \delta p_i$ are only infinitesimal deviations, also the generating function of the infinitesimal transformation can differ from that of the identity transformation only by an infinitesimal amount. Thus we ask first: *What is the generating function of the identity transformation?* Consider the expression

$$F = \sum_i q_i P_i. \tag{5.130}$$

Since in the case of this function, $F = F(q, P)$, we know from our preceding considerations that (*cf.* the case of the generating function F_2, Eq. (5.97))

$$p_i = \frac{\partial F}{\partial q_i} = P_i, \qquad Q_i = \frac{\partial F}{\partial P_i} = q_i. \tag{5.131}$$

We see that F is the generating function of the identity transformation. We expect therefore that in the case of its infinitesimal transformation, we can write the generating function

$$F = \sum_i q_i P_i + \epsilon G(q, P), \tag{5.132}$$

where ϵ is an infinitesimal parameter. It follows that (to first order in ϵ)

$$p_i = \frac{\partial F}{\partial q_i} = P_i + \epsilon \frac{\partial G}{\partial q_i} \implies \delta p_i = P_i - p_i = -\epsilon \frac{\partial G}{\partial q_i}, \tag{5.133a}$$

$$Q_i = \frac{\partial F}{\partial P_i} = q_i + \epsilon \frac{\partial G}{\partial P_i} \implies \delta q_i = Q_i - q_i = \epsilon \frac{\partial G}{\partial P_i} = \epsilon \frac{\partial G}{\partial p_i}. \tag{5.133b}$$

We consider therefore G as $G(q, P)$ instead of $G(q, p)$. The function G is also described as *generator* or *generating function*.

As an example, we consider the case of

$$G := H(q, p), \qquad \epsilon = dt. \tag{5.134}$$

Then, using Hamilton's equations,

$$\delta q_i \;\; \Longrightarrow \;\; dq_i = \;\; dt\frac{\partial H}{\partial p_i} = \dot{q}_i dt \;\; (= dq_i),$$

$$\delta p_i \;\; \Longrightarrow \;\; dp_i = -dt\frac{\partial H}{\partial q_i} = \dot{p}_i dt. \tag{5.135}$$

This means, the transformation $q_i, p_i \longrightarrow Q_i, P_i$ transforms the time-t canonical coordinates q_i, p_i into time-$t + dt$ canonical coordinates $q_i + dq_i, p_i + dp_i$. This means, the motion of the system in the time interval dt is given by that particular transformation whose generating function $G = H$ is the Hamilton function. Thus *the motion of a mechanical system corresponds exactly to the continuous generation of a canonical transformation, and the Hamilton function is the generator of this motion* in the course of time (*i.e.* H is an evolution quantity). It is obvious that this connection is of fundamental importance, and re-appears in operator form in quantum mechanics.

Conversely one can enquire about a canonical function, which transforms the coordinates of the system at time t back into its initial configuration, *i.e.* back to the constant initial conditions of the generalized coordinates and momenta as, for instance, in Example 5.7. The search of such a transformation is equivalent to the complete solution of the problem and will be used later in Sec. 6.2.3.

5.9.2 Infinitesimal transformations and Poisson brackets

Our next objective is to see the connection between infinitesimal canonical transformations and Poisson brackets. We assume $u = u(q, p)$ to be a function of q and p. Then we have for the variation of the function u as a result of an infinitesimal canonical transformation

$$\delta u = u(q_i + \delta q_i, p_i + \delta p_i) - u(q_i, p_i) = \sum_i \left(\frac{\partial u}{\partial q_i}\delta q_i + \frac{\partial u}{\partial p_i}\delta p_i \right) \tag{5.136}$$

after Taylor expansion. Inserting here Eqs. (5.133a) and (5.133b), we obtain

$$\delta u = \epsilon \sum_i \left(\frac{\partial u}{\partial q_i}\frac{\partial G}{\partial p_i} - \frac{\partial u}{\partial p_i}\frac{\partial G}{\partial q_i} \right) = \epsilon\{u, G\}. \tag{5.137}$$

For $u = H$ this becomes

$$\delta H = \epsilon\{H, G\}. \tag{5.138}$$

Thus this is the change which the Hamilton function undergoes under a canonical transformation. We know, however (see Eq. (5.120) and thereafter), that if G is a conserved quantity, then

$$\{H, G\} = 0, \tag{5.139}$$

and hence according to Eq. (5.132): The constant G generates such a canonical transformation for which $\delta H = 0$, *i.e.* under which the Hamilton function remains unchanged. The conserved quantities or constants of motion are therefore the generators of those infinitesimal canonical transformations which leave H invariant. Hence one can determine the conserved quantities of the motion by examining the symmetry properties of the Hamilton function (conserved quantities are therefore related to symmetry properties). The following Examples illustrate some special cases.

Example 5.11: Linear momentum conserved and generator

Consider a Hamilton function $H(q, p)$ which is invariant under translations in (say) direction i, and show that the generator G is the momentum p_i.

Solution: Since H is invariant under translational shifts $q_i \rightarrow q_i + \delta q_i$, this implies q_i is a cyclic variable. The canonical transformation we have to consider is therefore

$$q_i \rightarrow q_i + \delta q_i, \quad \delta q_i \equiv \epsilon_i, \quad p_i \rightarrow p_i,$$

or

$$\delta q_i = \sum_j \epsilon_j \delta_{ji} = \epsilon_i \equiv \epsilon, \quad \delta p_i = 0.$$

From Eqs. (5.133a) and (5.133b) we obtain therefore

$$\delta q_i = \epsilon \frac{\partial G}{\partial p_i}.$$

Thus this implies $\delta q_i = \epsilon_i$ for $G = p_i$, so that $\delta q_i = \sum_j \epsilon_j \delta_{ji}, \epsilon \equiv \epsilon_i$. Thus the generating function G is the momentum conjugate to q_i. We thus obtain the law of conservation of linear momentum.

Example 5.12: Angular momentum as generator of rotations

Consider infinitesimal rotations in q_i and $p_i, i \rightarrow x, y, z$, and determine their generator G.

Solution: Without loss of generality we consider an infinitesimal rotation through angle $-\delta\theta$ around the z-axis of a Cartesian system of coordinates. From the following formula for the finite rotation through angle θ

$$\begin{pmatrix} X \\ Y \end{pmatrix} = \begin{pmatrix} \cos\theta & \sin\theta \\ -\sin\theta & \cos\theta \end{pmatrix} \begin{pmatrix} x \\ y \end{pmatrix}$$

we obtain with substitution $\theta \rightarrow -\delta\theta$:

$$\begin{pmatrix} X \\ Y \end{pmatrix} = \begin{pmatrix} 1 & -\delta\theta \\ \delta\theta & 1 \end{pmatrix} \begin{pmatrix} x \\ y \end{pmatrix}, \quad Z = z. \tag{5.140}$$

We therefore obtain

$$X - x = \delta x = -y\delta\theta, \quad Y - y = \delta y = x\delta\theta, \quad \delta z = 0, \tag{5.141}$$

and similarly

$$\delta p_x = -p_y \delta\theta, \quad \delta p_y = p_x \delta\theta, \quad \delta p_z = 0. \tag{5.142}$$

We expect — and verify below — that with reference to Eqs. (5.133a) and (5.133b) the generating function G has the following form :

$$G = (xp_y - yp_x), \quad \text{with } \epsilon = d\theta. \tag{5.143}$$

In verifying this relation we construct in conformity with Eqs. (5.133a) and (5.133b) the infinitesimal relations $(i \to x, y, z)$

$$\delta q_i = \epsilon \frac{\partial G}{\partial p_i}, \quad \delta p_i = -\epsilon \frac{\partial G}{\partial q_i}. \tag{5.144}$$

Applying here the expression (5.143) we obtain the relations:

$$
\left.
\begin{aligned}
\delta x &\stackrel{(5.133b)}{=} \delta\theta \frac{\partial G}{\partial p_x} \stackrel{(5.143)}{=} -y\delta\theta \\
\delta y &= \delta\theta \frac{\partial G}{\partial p_y} = x\delta\theta
\end{aligned}
\right\},
\quad
\left.
\begin{aligned}
\delta p_x &= -\delta\theta \frac{\partial G}{\partial x} \stackrel{(5.143)}{=} -p_y\delta\theta \\
\delta p_y &= -\delta\theta \frac{\partial G}{\partial y} = p_x\delta\theta
\end{aligned}
\right\}.
\tag{5.145}
$$

The results of the partial differentiations involved here and applied to G of Eq. (5.143) are seen to be in agreement with the relations contained in Eq. (5.141). The physical significance of G is obviously that of the z-component of angular momentum $\mathbf{L}$,

$$G = L_z \equiv \mathbf{L} \cdot \mathbf{e}_z. \tag{5.146}$$

In Example 5.12 the z-axis was arbitrarily chosen as the axis of rotation. In general we have

$$G = \mathbf{L} \cdot \mathbf{n}, \tag{5.147}$$

where $\mathbf{n}$ is a unit vector pointing along the direction of the axis of rotation. Hence the generating function of an infinitesimal rotation is the component of the angular momentum along the axis of rotation. Above we had the relation (5.137), *i.e.*

$$\delta u(q, p) = \epsilon\{u, G\}, \tag{5.148}$$

where $u(q, p)$ is an arbitrary function. Replacing now u by the vector function $\mathbf{F}(q, p) \equiv (A(q, p), B(q, p), C(q, p))$ and G by $\mathbf{L} \cdot \mathbf{n}$, and ϵ by $\delta\theta$, we obtain

$$\delta\mathbf{F} = \delta\theta\{\mathbf{F}, \mathbf{L} \cdot \mathbf{n}\}. \tag{5.149}$$

Then

$$
\begin{aligned}
\delta A &= A(Q, P) - A(q, p) = \delta\theta\{A, \mathbf{L} \cdot \mathbf{n}\}, \\
\delta B &= B(Q, P) - B(q, p) = \delta\theta\{B, \mathbf{L} \cdot \mathbf{n}\}, \\
\delta C &= C(Q, P) - C(q, p) = \delta\theta\{C, \mathbf{L} \cdot \mathbf{n}\}.
\end{aligned}
\tag{5.150}
$$

Under an infinitesimal rotation through angle $-\delta\theta$ around the z-axis, the components A, B, C of the vector $\mathbf{F}$ transform in conformity with the transformation of a vector, *i.e.* (*cf.* Eq. (5.46), there with $\phi \to -\delta\theta$)

$$
\begin{aligned}
A(Q, P) &= A(q, p) - B(q, p)\delta\theta, \\
B(Q, P) &= B(q, p) + A(q, p)\delta\theta, \\
C(Q, P) &= C(q, p)
\end{aligned}
\tag{5.151}
$$

$(-\delta\theta$ for the transformation from $(q,p) \to (Q,P)$, $+\delta\theta$ for the transformation from $(Q,P) \to (q,p)$). Hence

$$
\begin{aligned}
\delta A &= A(Q,P) - A(q,p) = -B(q,p)\delta\theta, \\
\delta B &= B(Q,P) - B(q,p) = A(q,p)\delta\theta, \\
\delta C &= C(Q,P) - C(q,p) = 0.
\end{aligned}
\tag{5.152}
$$

Summarizing we have (*cf.* Eqs. (5.50), (5.57)):

$$
\delta\mathbf{F} = \overbrace{\mathbf{n}\delta\theta}^{d\Omega} \times\mathbf{F},
\tag{5.153a}
$$

because

$$
\delta F_x = \delta A = -\delta\theta F_y = -\delta\theta B
\tag{5.153b}
$$

(recall: $\mathbf{n}$ is parallel to the z-axis, $\mathbf{n} \times \mathbf{F}$ is parallel to the x-axis, hence $\mathbf{F}$ is parallel to the y-axis). This is the – in applications important – result we obtained earlier geometrically in sections 5.4.7 and 5.5 for a position vector — and we postulated that this applies to any vector in the 3-dimensional Euclidean space under consideration here.

5.9.3   Angular momenta and Poisson brackets

We continue a little further the study of angular momentum. The Poisson bracket relations we shall obtain below are those which correspond exactly to the very important commutator relations in *quantum mechanics*, and one can say, an understanding here immediately allows an understanding there, although we are here concerned with quantities which are c-numbers and in quantum mechanics with quantities which are operators in the space of state vectors (called Hilbert space).

From Eqs. (5.149) and (5.153a) we obtain:

$$
\mathbf{n} \times \mathbf{F} = \{\mathbf{F}, \mathbf{L} \cdot \mathbf{n}\}.
\tag{5.154}
$$

Here the vector $\mathbf{F}$ is arbitrary. Setting, for example, $\mathbf{F} = \mathbf{L}$, we obtain

$$
\mathbf{n} \times \mathbf{L} = \{\mathbf{L}, \mathbf{L} \cdot \mathbf{n}\},
\tag{5.155}
$$

and hence

$$
(\mathbf{n} \times \mathbf{L})_x = \{\mathbf{L}, \mathbf{L} \cdot \mathbf{n}\}_x = \{L_x, \mathbf{L} \cdot \mathbf{n}\}.
\tag{5.156}
$$

If the unit vector $\mathbf{n}$ points in the direction of the y-axis, we obtain

$$
L_z = \left(\frac{\mathbf{y}}{|\mathbf{y}|} \times \mathbf{L}\right)_x = \begin{vmatrix} \mathbf{e}_x & \mathbf{e}_y & \mathbf{e}_z \\ 0 & 1 & 0 \\ L_x & L_y & L_z \end{vmatrix}_x = \{L_x, L_y\}.
\tag{5.157}
$$

With the cyclicity of the indices we may conclude that

$$\{L_i, L_j\} = L_k, \quad i, j, k \text{ in cyclic order} \tag{5.158}$$

(hence L_i, L_j are not canonical momenta, since for the latter $\{p_i, p_j\} = 0$).

If L_x, L_y are conserved quantities, then Poisson's theorem (5.122) tells us that $\{L_x, L_y\} = L_z$ is also a conserved quantity. Thus if two components are constants of motion, the entire angular momentum is a constant.

Finally, we consider the following scalar triple product which is zero for two identical vectors (since in this product scalar and vector multiplications may be interchanged):

$$
\begin{aligned}
0 &= 2\mathbf{L} \cdot (\mathbf{n} \times \mathbf{L}) \\
&\overset{(5.155)}{=} 2\mathbf{L} \cdot \{\mathbf{L}, \mathbf{L} \cdot \mathbf{n}\} \\
&= 2\mathbf{L} \cdot \sum_i \left(\frac{\partial \mathbf{L}}{\partial q_i} \frac{\partial (\mathbf{L} \cdot \mathbf{n})}{\partial p_i} - \frac{\partial \mathbf{L}}{\partial p_i} \frac{\partial (\mathbf{L} \cdot \mathbf{n})}{\partial q_i} \right) \\
&= \sum_i \left(\frac{\partial (\mathbf{L} \cdot \mathbf{L})}{\partial q_i} \frac{\partial (\mathbf{L} \cdot \mathbf{n})}{\partial p_i} - \frac{\partial (\mathbf{L} \cdot \mathbf{L})}{\partial p_i} \frac{\partial (\mathbf{L} \cdot \mathbf{n})}{\partial q_i} \right) \\
&= \{\mathbf{L} \cdot \mathbf{L}, \mathbf{L} \cdot \mathbf{n}\} = \{L^2, \mathbf{L} \cdot \mathbf{n}\}.
\end{aligned}
$$

Thus we have

$$\{L^2, \mathbf{L} \cdot \mathbf{n}\} = 0. \tag{5.159}$$

Since for *canonical momenta* $\{p_i, p_j\} = 0$, Eq. (5.158) tells us that the angular momentum components L_x, L_y cannot be chosen as canonical momenta, however, this is possible with L^2 and $\mathbf{L} \cdot \mathbf{n}$ (*e.g.* L_z). In quantum mechanics $\mathbf{L}, L^2, \mathbf{L} \cdot \mathbf{n}$ *etc.* all are operators in a Hilbert space; also the commutator $[\cdots, \cdots]$ which replaces the Poisson bracket is an operator. Equation (5.159) then says: Only L^2 and $\mathbf{L} \cdot \mathbf{n}$ are simultaneously diagonalizable (commuting operators have a common system of eigenvectors), *i.e.* for instance L^2 and L_z, but not L_x, L_y and L_z. This means only the eigenvalues of L^2 and L_z are then so-called "*good*" quantum numbers (an old terminology in quantum mechanics). Equation (5.159) is therefore one of the most important results in our classical considerations here and is of the utmost significance in quantum mechanics (there in particular in the case of hydrogen-like atoms), *i.e.* there in the corresponding context with the Poisson bracket replaced by the corresponding operator. One should also note, that we are here considering a three-dimensional angular momentum in a space with three spatial dimensions.

5.10   Miscellaneous Examples

Example 5.13: $SO(2)$ invariant tensors
Show that the quantities $\epsilon_{ij}, \delta_{ij}$ are $SO(2)$ invariant tensors.

Solution: An $SO(2)$ second rank tensor T_{ij} transforms according to the equation

$$T'_{ij} = \sum_{k,l} \frac{\partial x'_i}{\partial x_k} \frac{\partial x'_j}{\partial x_l} T_{kl},$$

under $SO(2)$ transformations (θ assumed to be infinitesimal)

$$x'_i = x_i + \theta \epsilon_{im} x_m, \quad \epsilon_{im} = -\epsilon_{mi}.$$

Hence

$$T'_{ij} = \sum_{k,l} (\delta_{ik} + \theta \epsilon_{ik})(\delta_{jl} + \theta \epsilon_{jl}) T_{kl} = \sum_{k,l} [\delta_{ik}\delta_{jl} + \theta(\epsilon_{ik}\delta_{jl} + \epsilon_{jl}\delta_{ik})] T_{kl}.$$

Hence the invariance of the above quantities:

$$
\begin{aligned}
\epsilon'_{ij} &= \sum_{k,l} [\delta_{ik}\delta_{jl} + \theta(\epsilon_{ik}\delta_{jl} + \epsilon_{jl}\delta_{ik})]\epsilon_{kl} = \epsilon_{ij} + \theta(\epsilon_{ik}\epsilon_{kj} + \epsilon_{jl}\epsilon_{il}) \\
&= \epsilon_{ij} + \theta(\epsilon_{ik}\epsilon_{kj} - \epsilon_{il}\epsilon_{lj}) = \epsilon_{ij},
\end{aligned}
$$

$$\delta'_{ij} = \sum_{k,l} [\delta_{ik}\delta_{jl} + \theta(\epsilon_{ik}\delta_{jl} + \epsilon_{jl}\delta_{ik})]\delta_{kl} = \delta_{ij} + \theta(\epsilon_{ij} + \epsilon_{ji}) = \delta_{ij}.$$

Example 5.14: Invariance of an action integral
Show that the following action S is invariant under the infinitesimal transformation $e(t) \to e(t) + \delta e(t)$ with $\delta e(t) = e\dot{\epsilon} + \dot{e}\epsilon = d(e\epsilon)/dt$, and $x_i \to x_i + \delta x_i$ with $\delta x_i = \epsilon(t)\dot{x}_i(t)$, where dots denote differentiation with respect to t, and $\epsilon(t_0) = \epsilon(t_1) = 0$,

$$S = \frac{1}{2} \int_{t_0}^{t_1} dt \left[\frac{1}{e(t)} \dot{\mathbf{x}}^2(t) - e(t) m^2 \right], \quad i = 1,2,3.$$

Solution: We have, with $\delta\dot{\mathbf{x}} = d(\delta\mathbf{x})/dt = d(\epsilon(t)\dot{\mathbf{x}})/dt$ as given,

$$
\begin{aligned}
\delta S &= \frac{1}{2} \int_{t_0}^{t_1} dt \left[-\frac{\delta e}{e^2} \dot{\mathbf{x}}^2(t) - m^2 \delta e + \frac{2}{e}\dot{\mathbf{x}} \cdot \frac{d}{dt}\{\epsilon(t)\dot{\mathbf{x}}\} \right] \\
&= \frac{1}{2} \int_{t_0}^{t_1} dt \left[-m^2 \delta e + \dot{\mathbf{x}}^2 \left\{ -\frac{\delta e}{e^2} + \frac{2\dot{\epsilon}}{e} \right\} + \frac{2}{e}\epsilon\dot{\mathbf{x}} \cdot \ddot{\mathbf{x}} \right] \\
&= \frac{1}{2} \int_{t_0}^{t_1} dt \left[-m^2 \delta e + \dot{\mathbf{x}}^2 \left\{ -\frac{\delta e}{e^2} + \frac{2\dot{\epsilon}}{e} \right\} + \frac{\epsilon}{e}\frac{d}{dt}\dot{\mathbf{x}}^2 \right],
\end{aligned}
$$

and with partial integration of the last term,

$$
\begin{aligned}
\delta S &= \frac{1}{2} \int_{t_0}^{t_1} dt \left[-m^2 \delta e + \dot{\mathbf{x}}^2 \left\{ -\frac{e\dot{\epsilon} + \epsilon\dot{e}}{e^2} + \frac{2\dot{\epsilon}}{e} - \frac{\dot{\epsilon}}{e} + \frac{\epsilon\dot{e}}{e^2} \right\} \right] \\
&= -\frac{1}{2} \int_{t_0}^{t_1} dt\, m^2 \delta e = -\frac{m^2}{2} \int_{t_0}^{t_1} dt \frac{d(e\epsilon)}{dt} = -\frac{m^2}{2} (e\epsilon) \Big|_{t_0}^{t_1} = 0.
\end{aligned}
$$

Example 5.15: An example where $H = L$

Show that in the case of the following Lagrangian L the Hamiltonian H is equal to L:

$$L(r, \theta, \phi, \psi, \dot{r}, \dot{\theta}, \dot{\phi}, \dot{\psi}) = \pi\left(1 - \frac{2}{r}\right)(\dot{r}^2 + r^2\dot{\theta}^2 + r^2\sin^2\theta\dot{\phi}^2) + \frac{4\pi}{(1 - 2/r)}(\dot{\psi} + \cos\theta\dot{\phi})^2.$$

Solution: We derive in the usual procedure the Euler–Lagrange equations in ψ, θ, ϕ, r. In these cases we have respectively

$$\frac{\partial L}{\partial \psi} = 0 \;\Rightarrow\; \frac{d}{dt}\left(\frac{\partial L}{\partial \dot{\psi}}\right) = 0, \quad p_\psi = \frac{\partial L}{\partial \dot{\psi}} = \frac{8\pi}{(1 - 2/r)}(\dot{\psi} + \dot{\phi}\cos\theta) = q = \text{const.}, \tag{5.160}$$

and analogously (inserting the constant q)

$$\frac{\partial L}{\partial \theta} = \pi\left(1 - \frac{2}{r}\right)r^2\dot{\phi}^2 2\sin\theta\cos\theta - q\dot{\phi}\sin\theta, \quad p_\theta = \frac{\partial L}{\partial \dot{\theta}} = \pi\left(1 - \frac{2}{r}\right)r^2 2\dot{\theta}, \tag{5.161}$$

which implies

$$\frac{d}{dt}\left[\pi\left(1 - \frac{2}{r}\right)r^2 2\dot{\theta}\right] = \pi\left(1 - \frac{2}{r}\right)r^2\dot{\phi}^2 2\sin\theta\cos\theta - q\dot{\phi}\sin\theta, \tag{5.162}$$

$$\frac{\partial L}{\partial \phi} = 0 \;\Rightarrow\; p_\phi = \frac{\partial L}{\partial \dot{\phi}} = p_z = \text{const.} = 2\pi\left(1 - \frac{2}{r}\right)r^2\dot{\phi}\sin^2\theta + q\cos\theta, \tag{5.163}$$

$$p_r = \frac{\partial L}{\partial \dot{r}} = 2\pi\left(1 - \frac{2}{r}\right)\dot{r}. \tag{5.164}$$

We skip the lengthy expression for $\partial L/\partial r$, which is not required. We now have all expressions required to enable us to re-express all momenta in terms of velocities. Then

$$
\begin{aligned}
H &= p_r\dot{r} + p_\theta\dot{\theta} + p_\phi\dot{\phi} + p_\psi\dot{\psi} - L \\
&= 2\pi\left(1 - \frac{2}{r}\right)\dot{r}^2 + \pi\left(1 - \frac{2}{r}\right)r^2 2\dot{\theta}^2 + 2\pi\left(1 - \frac{2}{r}\right)r^2\sin^2\theta\dot{\phi}^2 + \dot{\phi}q\cos\theta + q\dot{\psi} \\
&\quad -\pi\left(1 - \frac{2}{r}\right)(\dot{r}^2 + r^2\dot{\theta}^2 + r^2\sin^2\theta\dot{\phi}^2) - \frac{4\pi}{(1 - 2/r)}(\dot{\psi} + \cos\theta\dot{\phi})^2 \\
&= \pi\left(1 - \frac{2}{r}\right)(\dot{r}^2 + r^2\dot{\theta}^2 + r^2\sin^2\theta\dot{\phi}^2) - \pi\left(1 - \frac{2}{r}\right)\frac{q^2}{16\pi^2} + q(\dot{\psi} + \dot{\phi}\cos\theta) \\
&= \pi\left(1 - \frac{2}{r}\right)(\dot{r}^2 + r^2\dot{\theta}^2 + r^2\dot{\phi}^2\sin^2\theta) - \pi\left(1 - \frac{2}{r}\right)\frac{q^2}{16\pi^2} + \frac{q^2}{8\pi}\left(1 - \frac{2}{r}\right) \\
&= \pi\left(1 - \frac{2}{r}\right)(\dot{r}^2 + r^2\dot{\theta}^2 + r^2\sin^2\theta\dot{\phi}^2) + \frac{q^2}{16\pi}\left(1 - \frac{2}{r}\right) = L.
\end{aligned}
$$

This is an example motivated by literature* which may illustrate the wide occurrence of Lagrangian theory in fundamental physics, as mentioned in Sec. 3.1.

Example 5.16: Canonical transformations, integrability, generators

Let $q_k = q_k(Q, P, t), p_k = p_k(Q, P, t)$ be a canonical transformation in the restricted sense that $H(q_k, p_k, t) = K(Q_k, P_k, t)$ with generator F for which $dF(Q, P) = \sum_k(p_k dq_k - P_k dQ_k)$. (a) In the latter relation express all quantities in terms of Q_k, P_k and derive expressions for $\partial F/\partial Q_k, \partial F/\partial P_k$. (b) By considering second order derivatives show that necessary and sufficient conditions for the integrability of dF are three relations, of which two are

$$\frac{\partial q_k}{\partial Q_r}\frac{\partial p_k}{\partial Q_l} - \frac{\partial p_k}{\partial Q_r}\frac{\partial q_k}{\partial Q_l} = 0, \quad \frac{\partial q_k}{\partial P_r}\frac{\partial p_k}{\partial P_l} - \frac{\partial p_k}{\partial P_r}\frac{\partial q_k}{\partial P_l} = 0. \tag{5.165}$$

---

*G. W. Gibbons and N. S. Manton [17].

What is the third relation? (c) Is the following transformation a canonical transformation:

$$q_{1,2} = (+,-)\sqrt{\frac{Q_1}{k_1}} \cos P_1 + \sqrt{\frac{Q_2}{k_2}} \cos \Gamma_2, \qquad p_{1,2} = (+,-)\sqrt{k_1 Q_1} \sin P_1 + \sqrt{k_2 Q_2} \sin P_2. \quad (5.166)$$

(d) What is the generator of this transformation? (e) Transform the following Hamilton function to Q_k, P_k, find in terms of the latter the equations of motion, integrate these and transform back:

$$H = p_1^2 + p_2^2 + \frac{1}{2}k_1^2(q_1 - q_2)^2 + \frac{1}{2}k_2^2(q_1 + q_2)^2. \quad (5.167)$$

This is the Hamilton function of two coupled harmonic oscillators.

Solution: (a) Since generally $H(q_k, p_k, t) = K(Q_k, P_k, t) + \partial F/\partial t$, the generating functional F is not explicitly time-dependent, *i.e.* $\partial F/\partial t = 0$ or $F = F(Q, P)$. Thus we have

$$dF(Q, P) = \sum_k \left(\frac{\partial F}{\partial Q_k} dQ_k + \frac{\partial F}{\partial P_k} dP_k \right).$$

But $q_k = q_k(Q, P, t)$ so that we also have

$$
\begin{aligned}
dF(Q, P) &= \sum_k (p_k dq_k - P_k dQ_k) = \sum_{k,i} p_k \left(\frac{\partial q_k}{\partial P_i} dP_i + \frac{\partial q_k}{\partial Q_i} dQ_i + \frac{\partial q_k}{\partial t} dt \right) - \sum_k P_k dQ_k \\
&= \sum_{k,i} p_i \frac{\partial q_i}{\partial P_k} dP_k + \sum_k \left(\sum_i p_i \frac{\partial q_i}{\partial Q_k} - P_k \right) dQ_k + \sum_k p_k \dot{q}_k dt.
\end{aligned}
$$

By comparison of this relation with the preceding one we obtain

$$\frac{\partial F}{\partial P_k} = \sum_i p_i \frac{\partial q_i}{\partial P_k}, \qquad \frac{\partial F}{\partial Q_k} = \sum_i p_i \frac{\partial q_i}{\partial Q_k} - P_k, \qquad \frac{\partial F}{\partial t} = 0 = \sum_k p_k \dot{q}_k. \quad (5.168)$$

(b) The commutability of two successive derivatives implies necessarily (and therefore provides necessary conditions for the compatibility of Eqs. (5.168) with the total differential dF):

$$\frac{\partial^2 F}{\partial Q_l \partial Q_r} = \frac{\partial}{\partial Q_r} \left(\sum_k p_k \frac{\partial q_k}{\partial Q_l} - P_l \right) = \frac{\partial}{\partial Q_l} \left(\sum_k p_k \frac{\partial q_k}{\partial Q_r} - P_r \right) = \frac{\partial^2 F}{\partial Q_r \partial Q_l},$$

$$\frac{\partial^2 F}{\partial P_l \partial P_r} = \frac{\partial}{\partial P_r} \left(\sum_k p_k \frac{\partial q_k}{\partial P_l} \right) = \frac{\partial}{\partial P_l} \left(\sum_k p_k \frac{\partial q_k}{\partial P_r} \right) = \frac{\partial^2 F}{\partial P_r \partial P_l},$$

$$\frac{\partial^2 F}{\partial Q_l \partial P_r} = \frac{\partial}{\partial P_r} \left(\sum_k p_k \frac{\partial q_k}{\partial Q_l} - P_l \right) = \frac{\partial}{\partial Q_l} \left(\sum_k p_k \frac{\partial q_k}{\partial P_r} \right) = \frac{\partial^2 F}{\partial P_r \partial Q_l}. \quad (5.169)$$

From the first of this set of relations, and there from the equality in the middle, we obtain:

$$\sum_k \frac{\partial p_k}{\partial Q_r} \frac{\partial q_k}{\partial Q_l} - \frac{\partial P_l}{\partial Q_r} + \sum_k p_k \frac{\partial^2 q_k}{\partial Q_l \partial Q_r} - \sum_k \frac{\partial p_k}{\partial Q_l} \frac{\partial q_k}{\partial Q_r} + \frac{\partial P_r}{\partial Q_l} - \sum_k p_k \frac{\partial^2 q_k}{\partial Q_r \partial Q_l} = 0.$$

The second order derivatives cancel. Since Q, P are independent variables, it then follows that

$$\sum_k \left(\frac{\partial q_k}{\partial Q_r} \frac{\partial p_k}{\partial Q_l} - \frac{\partial p_k}{\partial Q_r} \frac{\partial q_k}{\partial Q_l} \right) = 0.$$

Corresponding equations follow from the second and third of the relations (5.169). Thus one obtains as necessary and sufficient integrability conditions for dF the three relations:

$$\sum_k \left(\frac{\partial q_k}{\partial Q_r} \frac{\partial p_k}{\partial Q_l} - \frac{\partial p_k}{\partial Q_r} \frac{\partial q_k}{\partial Q_l} \right) = 0 = \sum_k \left(\frac{\partial q_k}{\partial P_r} \frac{\partial p_k}{\partial P_l} - \frac{\partial p_k}{\partial P_r} \frac{\partial q_k}{\partial P_l} \right), \quad \sum_k \left(\frac{\partial q_k}{\partial Q_r} \frac{\partial p_k}{\partial P_l} - \frac{\partial p_k}{\partial Q_r} \frac{\partial q_k}{\partial P_l} \right) = \delta_{rl}.$$

(5.170)

These three conditions follow necessarily if dF is to be a total differential and hence integrable. The proof that these three conditions are also sufficient results as follows: Assuming the three conditions are satisfied, the coefficients of $d(dF)$ — all written out — are all zero, i.e. from $d(dF) = 0$ it follows that dF is integrable.

(c) It is to be shown that the given transformation is canonical, thus it is to be shown that q_i, p_i are canonical variables if Q_i, P_i are canonical variables. It is given that

$$H(Q,P,t) = K(q,p,t), \quad Q_i, P_i \quad \text{are canonical} \quad \Longrightarrow \quad \dot{Q}_i = \frac{\partial H}{\partial P_i}, \ \dot{P}_i = -\frac{\partial H}{\partial Q_i},$$

and we have to show that

$$\dot{q}_i = \frac{\partial K}{\partial p_i}, \quad \dot{p}_i = -\frac{\partial K}{\partial q_i}.$$

From d/dt of Eq. (5.166) we obtain $\dot{q}_{1,2}$

$$= \pm \left(\frac{1}{2\sqrt{Q_1/k_1}} \frac{1}{k_1} \dot{Q}_1 \cos P_1 - \sqrt{\frac{Q_1}{k_1}} \dot{P}_1 \sin P_1 \right) + \left(\frac{1}{2\sqrt{Q_2/k_2}} \frac{1}{k_2} \dot{Q}_2 \cos P_2 - \sqrt{\frac{Q_2}{k_2}} \dot{P}_2 \sin P_2 \right)$$

$$= \pm \left(\frac{1}{2\sqrt{Q_1 k_1}} \frac{\partial H}{\partial P_1} \cos P_1 + \sqrt{\frac{Q_1}{k_1}} \frac{\partial H}{\partial Q_1} \sin P_1 \right) + \left(\frac{1}{2\sqrt{Q_2 k_2}} \frac{\partial H}{\partial P_2} \cos P_2 + \sqrt{\frac{Q_2}{k_2}} \frac{\partial H}{\partial Q_2} \sin P_2 \right).$$

For reasons of brevity we write out in the following only those contributions which add up to a nonvanishing quantity; thus in the expressions following immediately we do not give e.g. the terms in $\partial K/\partial q_{1,2}$, since their contributions add up to zero. Thus with this proviso we now obtain therefore for $\dot{q}_{1,2}$ (replacing $H(Q,P)$ by $K(q,p)$):

$$\dot{q}_{1,2} = \pm \frac{1}{2\sqrt{Q_1 k_1}} \frac{\partial K}{\partial p_{1,2}} \frac{\partial p_{1,2}}{\partial P_1} \cos P_1 \pm \sqrt{\frac{Q_1}{k_1}} \frac{\partial K}{\partial p_{1,2}} \frac{\partial p_{1,2}}{\partial Q_1} \sin P_1 + \frac{1}{2\sqrt{Q_2 k_2}} \frac{\partial K}{\partial p_{1,2}} \frac{\partial p_{1,2}}{\partial P_2} \cos P_2$$

$$+ \sqrt{\frac{Q_2}{k_2}} \frac{\partial K}{\partial p_{1,2}} \frac{\partial p_{1,2}}{\partial Q_2} \sin P_2 = \frac{\partial K}{\partial p_{1,2}} \left[\frac{1}{2\sqrt{Q_1 k_1}} \cos^2 P_1 \sqrt{k_1 Q_1} + \sqrt{\frac{Q_1}{k_1}} \sin^2 P_1 \frac{k_1}{2\sqrt{Q_1 k_1}} \right.$$

$$\left. + \frac{1}{2\sqrt{Q_2 k_2}} \cos^2 P_2 \sqrt{k_2 Q_2} + \sqrt{\frac{Q_2}{k_2}} \sin^2 P_2 \frac{k_2}{2\sqrt{Q_2 k_2}} \right]$$

$$= \frac{\partial K}{\partial p_{1,2}} \left[\frac{1}{2} (\cos^2 P_1 + \sin^2 P_1) + \frac{1}{2} (\sin^2 P_2 + \cos^2 P_2) \right] = \frac{\partial K}{\partial p_{1,2}}.$$

Similarly we obtain from d/dt of Eq. (5.166) (again as above omitting contributions which add up to zero) $\dot{p}_{1,2}$:

$$= \pm \left(\frac{k_1}{2\sqrt{k_1 Q_1}} \dot{Q}_1 \sin P_1 + \sqrt{k_1 Q_1} \dot{P}_1 \cos P_1 \right) + \left(\frac{k_2}{2\sqrt{k_2 Q_2}} \dot{Q}_2 \sin P_2 + \sqrt{k_2 Q_2} \dot{P}_2 \cos P_2 \right)$$

$$= \pm \left(\frac{k_1}{2\sqrt{k_1 Q_1}} \frac{\partial H}{\partial P_1} \sin P_1 - \sqrt{k_1 Q_1} \cos P_1 \frac{\partial H}{\partial Q_1} \right) + \left(\frac{k_2}{2\sqrt{k_2 Q_2}} \frac{\partial H}{\partial P_2} \sin P_2 - \sqrt{k_2 Q_2} \cos P_2 \frac{\partial H}{\partial Q_2} \right),$$

$$\dot{p}_{1,2} = \mp \frac{\partial K}{\partial q_{1,2}} \left(- \frac{k_1}{2\sqrt{k_1 Q_1}} \frac{\partial q_{1,2}}{\partial P_1} \sin P_1 + \sqrt{k_1 Q_1} \cos P_1 \frac{\partial q_{1,2}}{\partial Q_1} \right)$$

$$+ \frac{\partial K}{\partial q_{1,2}} \left(\frac{k_2}{2\sqrt{k_2 Q_2}} \frac{\partial q_{1,2}}{\partial P_2} \sin P_2 - \sqrt{k_2 Q_2} \cos P_2 \frac{\partial q_{1,2}}{\partial Q_2} \right)$$

$$= \mp \frac{\partial K}{\partial q_{1,2}} \left[\left(\pm \frac{k_1}{2\sqrt{k_1 Q_1}} \sqrt{\frac{Q_1}{k_1}} \sin^2 P_1 \pm \frac{\sqrt{k_1 Q_1}}{2} \cos^2 P_1 \frac{1}{k_1 \sqrt{Q_1/k_1}} \right) \right.$$

$$\left. \pm \left(\frac{k_2}{2\sqrt{k_2 Q_2}} \sqrt{\frac{Q_2}{k_2}} \sin^2 P_2 + \frac{\sqrt{k_2 Q_2}}{2} \cos^2 P_2 \frac{1}{k_2 \sqrt{Q_2/k_2}} \right) \right]$$

$$= - \frac{\partial K}{\partial q_{1,2}} \left[\frac{1}{2} (\cos^2 P_1 + \sin^2 P_1) + \frac{1}{2} (\cos^2 P_2 + \sin^2 P_2) \right] = - \frac{\partial K}{\partial q_{1,2}}.$$

(d) Our next task is to determine the generating function of the canonical transformation (5.166). We have

$$dF(Q,P) = \sum_l \left[\frac{\partial F}{\partial Q_l} dQ_l + \frac{\partial F}{\partial P_l} dP_l \right] \overset{(5.168)}{=} \sum_{k,l} \left[\left(p_k \frac{\partial q_k}{\partial Q_l} - P_l \right) dQ_l + p_k \frac{\partial q_k}{\partial P_l} dP_l \right]$$

$$= \left(p_1 \frac{\partial q_1}{\partial Q_1} + p_2 \frac{\partial q_2}{\partial Q_1} \right) dQ_1 + \left(p_1 \frac{\partial q_1}{\partial Q_2} + p_2 \frac{\partial q_2}{\partial Q_2} \right) dQ_2$$

$$+ \left(p_1 \frac{\partial q_1}{\partial P_1} + p_2 \frac{\partial q_2}{\partial P_1} \right) dP_1 + \left(p_1 \frac{\partial q_1}{\partial P_2} + p_2 \frac{\partial q_2}{\partial P_2} \right) dP_2 - P_1 dQ_1 - P_2 dQ_2. \quad (5.171)$$

However, with the help of the given transformations (5.166):

$$p_1 \frac{\partial q_1}{\partial Q_{1,2}} + p_2 \frac{\partial q_2}{\partial Q_{1,2}} = [\sqrt{k_1 Q_1} \sin P_1 + \sqrt{k_2 Q_2} \sin P_2] \frac{\cos P_{1,2}}{2\sqrt{Q_{1,2} k_{1,2}}}$$

$$\mp [-\sqrt{k_1 Q_1} \sin P_1 + \sqrt{k_2 Q_2} \sin P_2] \frac{\cos P_{1,2}}{2\sqrt{Q_{1,2} k_{1,2}}} = \cos P_{1,2} \sin P_{1,2},$$

$$p_1 \frac{\partial q_1}{\partial P_{1,2}} + p_2 \frac{\partial q_2}{\partial P_{1,2}} = [\sqrt{k_1 Q_1} \sin P_1 + \sqrt{k_2 Q_2} \sin P_2](-)\sqrt{\frac{Q_{1,2}}{k_{1,2}}} \sin P_{1,2}$$

$$+ [-\sqrt{k_1 Q_1} \sin P_1 + \sqrt{k_2 Q_2} \sin P_2](\pm)\sqrt{\frac{Q_{1,2}}{k_{1,2}}} \sin P_{1,2}$$

$$= -2 Q_{1,2} \sin^2 P_{1,2}.$$

From these relations we obtain for Eq. (5.171):

$$dF(Q,P) = \cos P_1 \sin P_1 dQ_1 + \sin P_2 \cos P_2 dQ_2 - 2 Q_1 \sin^2 P_1 dP_1$$

$$- 2 Q_2 \sin^2 P_2 dP_2 - P_1 dQ_1 - P_2 dQ_2. \quad (5.172)$$

Since

$$d(Q_1 \cos P_1 \sin P_1) = \cos P_1 \sin P_1 dQ_1 + Q_1 dP_1 - 2 Q_1 \sin^2 P_1 dP_1,$$

we obtain

$$dF = d(Q_1 \cos P_1 \sin P_1) - Q_1 dP_1 + d(Q_2 \cos P_2 \sin P_2) - Q_2 dP_2 - P_1 dQ_1 - P_2 dQ_2$$

$$= d(Q_1 \cos P_1 \sin P_1 + Q_2 \cos P_2 \sin P_2 - Q_1 P_1 - Q_2 P_2)$$

and therefore

$$F(Q,P) = Q_1 \cos P_1 \sin P_1 + Q_2 \sin P_2 \cos P_2 - P_1 Q_1 - P_2 Q_2 + \text{const.} \quad (5.173)$$

The constant is independent of $Q_i, P_i, t, i = 1, 2$. The result (5.173) for F is the requested generating function.

(e) We obtain from the given transformation (5.166):

$$
\begin{aligned}
p_1^2 &= k_1 Q_1 \sin^2 P_1 + k_2 Q_2 \sin^2 P_2 + 2\sqrt{k_1 k_2 Q_1 Q_2} \sin P_1 \sin P_2, \\
p_2^2 &= k_1 Q_1 \sin^2 P_1 + k_2 Q_2 \sin^2 P_2 - 2\sqrt{k_1 k_2 Q_1 Q_2} \sin P_1 \sin P_2,
\end{aligned}
$$

$$
\frac{1}{2} k_1^2 (q_1 - q_2)^2 = 2 Q_1 k_1 \cos^2 P_1, \qquad \frac{1}{2} k_2^2 (q_1 + q_2)^2 = 2 Q_2 k_2 \cos^2 P_2,
$$

so that $H(q, p)$ given by Eq. (5.167) becomes

$$
\begin{aligned}
H(q, p) &= 2 k_1 Q_1 \sin^2 P_1 + 2 k_2 Q_2 \sin^2 P_2 + 2 Q_1 k_1 \cos^2 P_1 + 2 Q_2 k_2 \cos^2 P_2 \\
&= 2 k_1 Q_1 + 2 k_2 Q_2 = K(Q, P).
\end{aligned} \tag{5.174}
$$

We obtain the equations of motion from the Hamilton equations

$$
\dot{Q}_i = \frac{\partial K}{\partial P_i}, \qquad \dot{P}_i = -\frac{\partial K}{\partial Q_i}, \qquad i = 1, 2,
$$

i.e.

$$
\dot{Q}_1 = 0, \quad \dot{Q}_2 = 0, \quad \dot{P}_1 = -2k_1, \quad \dot{P}_2 = -2k_2.
$$

Integrating the resulting equations we obtain with constants c_1, c_2, d_1, d_2 (*e.g.* initial values):

$$
Q_1 = c_1, \qquad Q_2 = c_2, \qquad P_1 = -2k_1 t + d_1, \qquad P_2 = -2k_2 t + d_2. \tag{5.175}
$$

Transforming back to the variables q_i, p_i, *i.e.* inserting the results (5.175) into the given transformation (5.166), we obtain the superposition of two harmonic functions:

$$
\begin{aligned}
q_{1,2} &= \pm\sqrt{\frac{c_1}{k_1}} \cos(-2k_1 t + d_1) + \sqrt{\frac{c_2}{k_2}} \cos(-2k_2 t + d_2), \\
p_{1,2} &= \pm\sqrt{c_1 k_1} \sin(-2k_1 t + d_1) + \sqrt{c_2 k_2} \sin(-2k_2 t + d_2).
\end{aligned}
$$

Example 5.17: Hamilton–Jacobi equation for the harmonic oscillator

Solve the Hamilton–Jacobi equation for the Hamilton function

$$
H(q, p) = \frac{p^2}{2m} + \frac{1}{2} k q^2.
$$

Solution: The Hamilton–Jacobi equation was defined in Sec. 5.8.1, and the generating function F_2 was written $S(q, \alpha, t)$ with $p = \partial S / \partial q$ (recall that the transformed momentum P_i has the constant value α_i). The equation is therefore in the present case

$$
\frac{1}{2m} \left(\frac{\partial S}{\partial q} \right)^2 + \frac{1}{2} k q^2 + \frac{\partial S}{\partial t} = 0, \qquad S = S(q, \alpha, t), \qquad \alpha = \text{const.}
$$

The t-dependence can be separated off by setting

$$
S(q, \alpha, t) = W(q, \alpha) - \alpha t.
$$

Then (note that choosing the constant multiplying t as α, fixes this constant as the total energy — this follows from the third relation of Eqs. (5.97), which for $K = 0$ implies $H = -\partial S(q, \alpha, t)/\partial t$)

$$
\frac{1}{2m} \left(\frac{\partial W}{\partial q} \right)^2 + \frac{1}{2} k q^2 = \alpha, \qquad \frac{\partial W}{\partial q} = \sqrt{2m \left(\alpha - \frac{k q^2}{2} \right)}.
$$

This equation can be rewritten as

$$W(q,\alpha) = \sqrt{mk} \int dq \sqrt{\frac{2\alpha}{k} - q^2}, \quad S(q,\alpha,t) = \sqrt{mk} \int dq \sqrt{\frac{2\alpha}{k} - q^2} - \alpha t.$$

Since α is the constant transformed momentum P (cf. the discussion after Eq. (5.97)), we have for the constant value β of the transformed coordinate Q:[†]

$$\beta = \frac{\partial S(q,\alpha,t)}{\partial \alpha} = \sqrt{\frac{m}{k}} \int^q \frac{dq}{\sqrt{2\alpha/k - q^2}} - t = \sqrt{\frac{m}{k}} \sin^{-1}\left(q\sqrt{\frac{k}{2\alpha}}\right) - t.$$

It follows that

$$q = \sqrt{\frac{2\alpha}{k}} \sin\left[\sqrt{\frac{k}{m}}(\beta + t)\right] = \sqrt{\frac{2\alpha}{k}} \sin[\omega(t+\beta)], \quad \omega = \sqrt{\frac{k}{m}}.$$

Since $p = \partial S(q,\alpha,t)/\partial q = \partial W(q,\alpha)/\partial q$, we obtain — by inserting this into H — the total energy E:

$$H = E = \left(\alpha - \frac{1}{2}kq^2\right) + \frac{1}{2}kq^2 = \alpha.$$

Thus α has been identified as the total energy (thus if we had left this undetermined earlier, we would have found this now). Further, if at time $t = 0$ we have $q(t) = 0$, the constant $\beta = 0$. Thus we obtain the expected and typical harmonic oscillator solution

$$q(t) = q(0) \sin \omega t.$$

[†]Using formula 320.01 of H. B. Dwight [13], p. 62.

Chapter 6

Looking Beyond Classical Mechanics

6.1 Introductory Remarks

In order to help an understanding of some basic aspects of other branches of physics — particularly statistical mechanics, relativity and field theory (the latter including electrodynamics) — whose roots can be seen in classical mechanics but are not always given substantial emphasis in these, we digress here somewhat into an extension of the previous considerations of classical mechanics. In particular we consider (a) the significance of Hamiltonian mechanics in the formulation of classical probabilities, (b) the extension of spatial rotations and translations to spacetime Lorentz and Poincaré transformations, and (c) the introduction of the concept of fields and the construction of Lagrangians of field theories such as electrodynamics. The following considerations are only meant to be brief introductions to these topics, which — in addition — are restricted to some selected aspects. Topic (a) — although really belonging to statistical mechanics — is usually touched in texts on classical mechanics (as in those of Goldstein [19] and Kibble and Berkshire [25]) in view of its connection with Hamilton's formalism, but is also of considerable relevance in relation to quantum mechanics. The Liouville equation, as derived in this context — and distinct from Liouville's theorem (which establishes the time-independence of a phase space element) — is the equation of motion of an ensemble and appears in one form or another in diverse branches of physics, as in electrodynamics and in quantum statistics. Topics (b) and (c), particularly the former, will be considerably elaborated on in later chapters. Electrodynamics, specifically Maxwell's equations, will frequently be referred to for purposes of illustration.

6.2   Aspects of Classical Statistics

6.2.1   Classical probabilities

We recall from our consideration of Hamilton's formulation of mechanics in Chapter 4 the equation (*cf.* Eq. (4.84))

$$\frac{du(q,p)}{dt} = \{u(q,p), H(q,p)\}, \tag{6.1}$$

where $u(q,p)$ is an observable, which specialized to q and p yields Hamilton's equations (4.47). The bracket $\{,\}$ is the *Poisson bracket* of Chapter 4, there written $[,]$. We recall that the motion of a mass-point in classical mechanics is "*deterministic*", meaning that if the initial conditions of the particle's motion are known, *i.e.* its position and velocity or momentum at a certain instant of time t_i, then the equations of motion (Newton, Lagrange or Hamilton) determine the exact position of the particle at any later time t_f. However, one can imagine a situation in the framework of classical mechanics, that the initial conditions of the particle are not known precisely, and that all one can say is that the particle's initial location in phase space is only known approximately, *i.e.* as that in the neighbourhood of a certain point in phase space. Phrased differently this means, all one can say is that there is a probability less than 1 that the particle is located at a certain point in phase space, and consequently in reality it is somewhere in a small domain around this point, say in a domain $\triangle q \triangle p$ around a point (q,p) in phase space. Then (assuming the total phase space volume V_{ps} normalized to (and therefore equal to) 1) $\int \triangle q \triangle p$ over the entire phase space is 1, meaning that the particle is localized with 100% probability somewhere in the entire space. If we integrated only over one half of the space, the probability of the particle to be therein would be 50%, and so on. The product $\triangle q \triangle p$ is therefore described as the "*classical a priori probability*"* or "*classical a priori weighting*", *i.e.* as that assumed from the very beginning for the particle to be in a certain part of phase space, *i.e.* in that with q-value between q and $q + \triangle q$ and p-value between p and $p + \triangle p$. This is still a static consideration — we shall show below that the quantity $\triangle q \triangle p$ is independent of time t, a result known as *Liouville's theorem*. The expression $\triangle q \triangle p$ as a phase space *a priori* probability must be independent of time, since insertion of a time dependence would be insertion of known information in contradiction to its *a priori* understanding, '*a priori*' meaning 'from what is before', *i.e.* without observation, and so without prior knowledge (*e.g.* if it were known that the

---

*The interpretation here seems to agree with that of A. Ben–Naim [4], p. 31, who illustrates the *a priori* probability by the equal probabilities of the six faces of a die.

particle was at time t_0 at rest at $q(t_0)$, this would be an initial condition with which the equations of motion would determine the exact whereabouts of the particle at all later times, this being the "*deterministic*" property of classical physics).[†] Thus initially the particle may be at the phase space point (q_0, p_0) with *a priori* probability $(\triangle q \triangle p)_{q_0, p_0}$. This is a purely classical consideration and a classical concept of probability, and we can enquire about the particle's probability to be at a later time at some other point in phase space. Of course, particles move and hence their phase space locations according to Hamilton's equations, and hence also physical probabilities can vary with time, and thus it makes sense to introduce in addition a *probability density* $\rho(q, p, t) \triangle q \triangle p$ with phase space particle density $\rho(q, p, t)$, such that the total probability is

$$\int \rho(q, p, t) \triangle q \triangle p = 1.$$

We can look at $\rho(q, p, t)$ as providing the time-dependence not contained in the *a priori* probability, *i.e.* how this varies with time t around a specific point (q, p) in phase space, since — as stated — the product $\triangle q \triangle p$ is independent of time t, and thus is not dynamical. Thus, if there is only one particle and this is located precisely at (q_0, p_0), this phase space density would be the delta function quantity $\delta(q - q_0)\delta(p - p_0)$, and with $\triangle q \to dq, \triangle p \to dp$, the integral over the entire phase space yields 1, as above. Examples 6.1 to 6.5 illustrate the significance of this classical *a priori* probability, which is there re-expressed in terms of the classical particle energy ϵ. In classical mechanics the coordinates (q_i, p_i) specify what one may call — in analogy with quantum mechanics — the state of the system; in quantum mechanics the state is represented by a wave function. For ease of calculation it is useful to relate the two. This is achieved in the following way. Consider the phase space area (really volume, but it is more convenient to visualize it here as an area) $\triangle q \triangle p$ with q, p at its centre. This area encloses other values of q, p which are possible system coordinates and hence "states" of the system. All those contained in $\triangle q \triangle p$ can be lumped together with the centre state of energy ϵ, and their number in $\triangle q \triangle p$ is called "*degeneracy*" of the state specified by q, p. Effectively one thus discretizes classical mechanics to get closer to quantum mechanics. The quantal *a priori* probability is then the quantum mechanical degeneracy, *i.e.* the number of different states or wave functions all having the same energy ϵ.

But we can also look at this in a different way. We can assume a large number of copies of the actual system, this is called an *ensemble*, with the

---

[†]Correspondingly one has *e.g.* a two-dimensional configuration space *a priori* probability $\triangle x \triangle y$.

ensemble *density* $\rho(q, p, t)$ expressing the fraction of the total number of these at time t at the point (q, p). We thus arrive at the picture of a multitude of systems, each described by its canonical variables (not all at the same point, but most of them crowding together), which travels through phase space. We thus require in addition to the equations of motion of the individual systems (*i.e.* Hamilton's equations) also an equation for $\rho(q, p, t)$, *i.e.* the equation which describes the motion through phase space of the entire ensemble. This equation is known as the *Liouville equation*. This is therefore the equation we derive next.[‡] However, before we do this we familiarize ourselves first with *Liouville's theorem*. This says that $\triangle q \triangle p$ is independent of time t, *i.e.*[§]

$$\frac{d}{dt}(\triangle q \triangle p) = 0 \quad \text{in view of} \quad \frac{\partial \dot{q}}{\partial q} + \frac{\partial \dot{p}}{\partial p} = 0. \tag{6.2}$$

We leave the verification of this result to Example 6.6 and illustrating cases to Example 6.7.

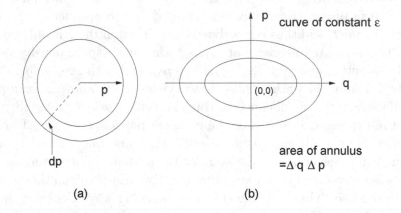

Fig. 6.1 (a) The spherical annulus, (b) ellipse as curve of constant energy.

Example 6.1: Calculation of *a priori* probability in two cases

Calculate the *a priori* probability g_i (a) for a particle of mass m in a volume V (or this probability per particle for a perfect gas confined to a volume V), and (b) for a particle of mass m subjected to a restoring force proportional to q (*i.e.* harmonic oscillator). Note: Here in case (b) the energy ϵ is quadratic in q and in p. In Example 6.4 we consider the case of ϵ linear in $|q|$ and $|p|$.

Solution: In case (a) the required classical *a priori* probability g_i is proportional to $V dp_x dp_y dp_z \equiv$

[‡]H. Goldstein [19] deals with Liouville's equation and Liouville's theorem in Sec. 8.8 of his book, and obtains one as a consequence of the other. His verbal arguments there are difficult to follow. We therefore prefer to distinguish very clearly between the two cases.

[§]Sometimes, as *e.g.* in the book of T. W. B. Kibble and F. H. Berkshire [25], p. 289, the second equation, which is effectively the incompressibility condition of a fluid, *i.e.* the vanishing of the divergence of a velocity, is referred to as Liouville's theorem.

$V4\pi p^2 dp$.[¶] Here, cf. Fig. 6.1, $p^2 = p_x^2 + p_y^2 + p_z^2$, and the volume of the annulus of width dp is $4\pi p^2 dp$ equal to the equivalent element in spherical coordinates. Since energy $\epsilon = p^2/2m$, we have

$$d\epsilon = \frac{p}{m} dp, \quad \text{or} \quad p^2 dp = \sqrt{2}m^{3/2}\epsilon^{1/2}d\epsilon, \quad \text{so that} \quad g_i \propto V\epsilon^{1/2}d\epsilon.$$

In case (b) of an harmonic oscillator of natural frequency ν, the energy is

$$\epsilon = \frac{p^2}{2m} + 2\pi^2 m\nu^2 q^2 \quad \text{or} \quad \frac{p^2}{2m\epsilon} + \frac{q^2}{\epsilon/2\pi^2 m\nu^2} = 1,$$

and thus represents an ellipse in phase space, as shown in Fig. 6.1. The a priori probability is the probability that the particle has energy in the range ϵ to $\epsilon + d\epsilon$. The area of an ellipse with semi-axes of lengths a and b is given by πab. Hence the area of the ellipse enclosed by the curve of constant energy ϵ is

$$\oint dqdp = \pi\sqrt{2m\epsilon}\sqrt{\frac{\epsilon}{2\pi^2 m\nu^2}} = \frac{\epsilon}{\nu}.$$

Hence the a priori probability g_i is $q \propto d\epsilon/\nu$ (comparing with case (a), we see that there is no extra energy factor here).

Example 6.2: A priori weightings in some simple cases

(a) Show that if the energy of particles is independent of spatial coordinates, the a priori weighting g_i per particle i is proportional to the accessible volume. (b) Show that if the potential energy of a particle depends only upon the angles θ and ϕ of inclination to a given direction, the a priori weighting g_i per particle os proportional to $\sin\theta d\theta d\phi$ in spherical polar coordinates.

Solution: (a) We have $g_i = \triangle q\triangle p$ per particle. Since energy $\epsilon = p^2/2m$ and independent of q, g_i is proportional to $dq_x dq_y dq_z$ and therefore to the volume. (b) In spherical polar coordinates r, θ, ϕ the volume element is $rd\theta.dr.r\sin\theta d\phi$. Hence $g_i \propto \sin\theta d\theta d\phi$.

Example 6.3: Perfect gas with specified movement

Obtain the a priori weighting g_i per particle for a perfect gas with movement confined to (a) one direction of length L, (b) a plane of area A, (c) a volume V.

Solution: (a) We have a priori probability $g_i = \triangle q\triangle p$. Therefore $\triangle q$ is L. The energy of a particle is $\epsilon = p^2/2m, p = \sqrt{2m\epsilon}$, so that $d\epsilon = pdp/m, dp = md\epsilon/p = md\epsilon/\sqrt{2m\epsilon}$. It follows that $g_i \propto L\epsilon^{-1/2}d\epsilon$. (b) Using the foregoing for the energy, we have $g_i = \triangle q\triangle p = (dq_x dq_y)(dp_x dp_y)$ or $(dq_x dq_y)(pdp)$ implying $g_i \propto A\sqrt{2m\epsilon}(md\epsilon/\sqrt{2m\epsilon}) \propto Ad\epsilon$, i.e. $g_i \propto Ad\epsilon$. (c) Cf. Example 6.1 (a).

Example 6.4: Rhombus boundary as curve of constant energy

If the energy of a certain system per particle is given by

$$\epsilon = \alpha|p| + \beta|q|,$$

where α and β are constants, the (q, p) curve for constant ϵ is the boundary of a rhombus of area $\oint dqdp = 2\epsilon^2/\alpha\beta$ as indicated in Fig. 6.2. Hence show that the a priori weighting g_i is given by $g_i \propto 4\epsilon d\epsilon/\alpha\beta$. Note: Here the energy ϵ is linear in $|q|$ and $|p|$, whereas in Example 6.1(b) it is quadratic in q and in p.

Solution: When $|p| = 0, \epsilon = \beta|q|, |q| = \epsilon/\beta$, i.e. $q = \pm\epsilon/\beta$. Similarly $p = \pm\epsilon/\alpha$. Any q and any p of the 4 values derived satisfy the above equation. Hence the curve is the boundary of a rhombus.

[¶]In quantum statistics this divided by h^3, h Planck's constant, is the number of degenerate states (i.e. of the same energy) with momentum between p and $p + dp$, and thus the quantal a priori probability corresponding to a certain momentum range.

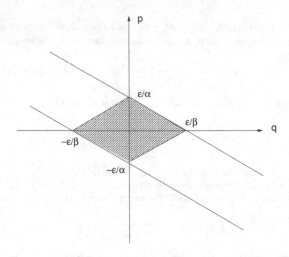

Fig. 6.2 The rhombus as curve of constant energy.

The area of this rhombus is (twice the area of the triangle in the upper half plane)

$$A = 2\left[\frac{1}{2}2\frac{\epsilon}{\beta}\frac{\epsilon}{\alpha}\right] = \frac{2\epsilon^2}{\alpha\beta}, \quad \text{so that} \quad \oint dqdp = \frac{2\epsilon^2}{\alpha\beta}.$$

It follows that g_i is the differential of A, i.e. $g_i = \triangle q\triangle p = (4\epsilon/\alpha\beta)d\epsilon$.

Example 6.5: *A priori* **weighting of a molecule**

If the rotational energy of a diatomic molecule with moment of inertia I is

$$E = \frac{1}{2I}\left(p_\theta^2 + \frac{p_\phi^2}{\sin^2\theta}\right), \quad \text{so that} \quad 1 = \frac{p_\theta^2}{2IE} + \frac{p_\phi^2}{2IE\sin^2\theta},$$

in spherical polar coordinates, the (p_θ, p_ϕ)-curve for constant E and θ is an ellipse of area[||] $\oint dp_\theta dp_\phi = \pi\sqrt{2IE}\sqrt{2IE\sin^2\theta} = 2\pi IE\sin\theta$. Show that the total volume of phase space covered for constant E is $8\pi^2 IE$, and hence the *a priori* weighting $g_i \propto 8\pi^2 IdE$.

Solution: Integrating over the angles we have

$$\int_{\phi=0}^{\phi=2\pi}\int_{\theta=0}^{\theta=\pi} 2I\pi E\sin\theta d\theta d\phi = 8\pi^2 IE.$$

Hence $g_i \propto 8\pi^2 IdE$.

Example 6.6: **Liouville's theorem**

Show that $\triangle q\triangle p$ is independent of time t, which means, this has the same value at a time t_0, as at a time $t_0' \neq t_0$.

Solution: We consider

$$\frac{d}{dt}\ln(\triangle q\triangle p) = \frac{1}{\triangle q\triangle p}\frac{d}{dt}(\triangle q\triangle p) = \frac{d(\triangle q)}{dt}\frac{1}{\triangle q} + \frac{d(\triangle p)}{dt}\frac{1}{\triangle p}.$$

[||]The area of an ellipse with semi-axes of lengths a and b is πab.

Here $d(\triangle q)/dt$ is the rate at which the q-walls of the phase space element move away from the centre of the element as illustrated in Fig. 6.3, *i.e.* the velocities of these walls to the right and the left are

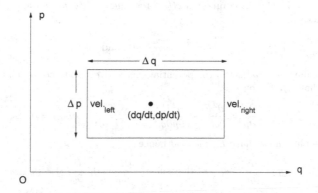

Fig. 6.3 Motion of q-walls

$$\text{vel.}_{\text{right}} \equiv \dot{q} + \frac{\partial \dot{q}}{\partial q}\frac{\triangle q}{2} \quad \text{to the right} \quad \text{and} \quad \text{vel.}_{\text{left}} \equiv \dot{q} - \frac{\partial \dot{q}}{\partial q}\frac{\triangle q}{2} \quad \text{to the left.}$$

Hence from the difference:

$$\frac{d(\triangle q)}{dt} = \frac{\partial \dot{q}}{\partial q}\triangle q, \quad \text{and similarly} \quad \frac{d(\triangle p)}{dt} = \frac{\partial \dot{p}}{\partial p}\triangle p,$$

and with with Hamilton's equations (4.47):

$$\frac{d}{dt}\ln(\triangle q\triangle p) = \frac{\partial \dot{q}}{\partial q} + \frac{\partial \dot{p}}{\partial p} = \frac{\partial^2 H}{\partial q\partial p} - \frac{\partial^2 H}{\partial p\partial q} = 0.$$

Observe that here the expression after the first equal-to sign has (in Cartesian coordinates) the form of $\boldsymbol{\nabla} \cdot \mathbf{v}$, $\mathbf{v}$ the velocity. In other contexts, like hydrodynamics, this equal to zero is known as the *condition of incompressibility*. Thus the phase space volume $\triangle q\triangle p$, or more generally, $\triangle q_1\triangle p_1 \ldots \triangle q_n\triangle p_n$, remains constant but its shape can change as the swarm of system points travels in phase space as indicated in Fig. 6.4. For the configuration space case *cf.* Example 6.11.

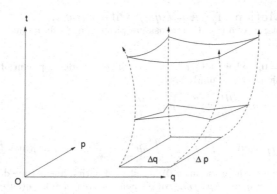

Fig. 6.4 Constant phase space volume $\triangle q\triangle p$ changes shape with time.

Example 6.7: Illustrations of Liouville's theorem

Consider (a) a particle of mass m moving in an electric field, potential energy $V(q)$, and (b) the photon travelling with velocity c, and show that Liouville's theorem is obeyed.

Solution: We write the Hamilton function $H(q,p)$ as energy $E(q,p)$ and apply Hamilton's equations. Then in case (a):

$$E(q,p) = \frac{p^2}{2m} + V(q), \quad \dot{q} = \frac{\partial E}{\partial p} = \frac{p}{m}, \quad \frac{\partial \dot{q}}{\partial q} = 0 \;\; \text{and} \;\; \dot{p} = -\frac{\partial E}{\partial q} = -\frac{\partial V}{\partial q}, \quad \frac{\partial \dot{p}}{\partial q} = 0,$$

the latter being minus the gradient of the potential energy, *i.e.* the force, which is here the electric field. It follows that (*cf.* Example 6.6):

$$\frac{\partial \dot{q}}{\partial q} + \frac{\partial \dot{p}}{\partial p} = \frac{\partial^2 E}{\partial q \partial p} - \frac{\partial^2 E}{\partial p \partial q} = 0, \quad \therefore \;\; \frac{d}{dt}(\triangle q \triangle p) = 0.$$

In case (b) of the photon, we have $E = pc$ and hence

$$\dot{q} = \frac{\partial E}{\partial p} = c, \quad \text{so that since} \quad \frac{\partial E}{\partial q} = 0, \; \frac{\partial}{\partial q}(\dot{q}) = 0,$$

the conclusions are the same as in case (a).

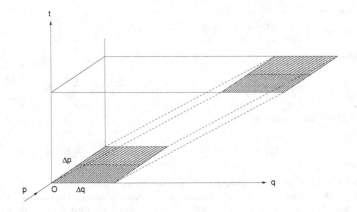

Fig. 6.5 Motion of a phase space element.

Example 6.8: Motion of $\triangle A = \triangle q \triangle p$ with time t.

Show how the boundaries of a $(1+1)$-dimensional phase space volume move with time in the case of potential $V = 0$.

Solution: Since $V = 0$, we have $H(q,p) = p^2/2m$. The time-development of q and p is determined by Hamilton's equations. These imply here:

$$\dot{q} = \frac{\partial H}{\partial p} = \frac{p}{m}, \quad \dot{p} = -\frac{\partial H}{\partial q} = 0, \quad p = \text{const.} = p_0,$$

and hence

$$q(t) - q(0) = \int_0^t dt \frac{p}{m} = \int_0^t dt \frac{p_0}{m} = \frac{p_0}{m}t, \quad q(t) = q(0) + \frac{p_0}{m}t.$$

Thus at time $t > 0$ the boundary point $q(0)$ has shifted with a linear time dependence to $q(t)$, and a point $q_1(0)$ to $q_1(t) = q_1(0) + p_0 t/m$. An element of area $[q_1(t) - q(t)]p_0 = \text{const.}$ then moves with time as indicated in Fig. 6.5.

6.2.2   The Liouville equation

We now assume a large number of identical systems — the entire collection is called an *ensemble* — all of whose initial locations are possible locations of our system in a domain G_0 around the point q_0, p_0 as illustrated in Fig. 6.6. Thus we assume a large number of identical sytems, whose positions in phase space are characterized by points. We consider the totality of these systems which is described by a density of points ρ (number dn of points per unit infinitesimal volume) in phase space, *i.e.* by

$$\frac{dn}{dqdp} = \rho(q, p, t), \quad dqdp = \prod_i dq_i dp_i. \tag{6.3}$$

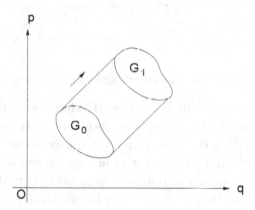

Fig. 6.6   The system moving from domain G_0 to domain G_1.

Thus dn is that number of systems which at time t are contained in the domain $q, q + dq; p, p + dp$. The total number of systems N is obtained by integrating over the whole of phase space, *i.e.*

$$\int dn = \int \rho(q, p, t) dqdp = N. \tag{6.4}$$

With a suitable normalization we can write this

$$\int W(q, p, t) dqdp = 1, \quad W = \frac{1}{N}\rho(q, p, t). \tag{6.5}$$

Thus W is the *probability to find the system at time t around* (q, p) (*i.e.* in the interval $(q, q + dq; p, p + dp)$ around (q, p)). Since W has the dimension of a reciprocal action, it is suggestive to introduce a factor $2\pi\hbar$ with every pair $dpdq$ without, however, leaving the basis of classical mechanics ($h = 2\pi\hbar$

being the Planck constant, but this is irrelevant here in classical mechanics)!
Hence we set

$$\int W(q,p,t)\frac{dqdp}{2\pi\hbar} = 1. \qquad (6.6)$$

We can consider $2\pi\hbar$ as a unit of area in (here the $(1+1)$-dimensional) phase
space.

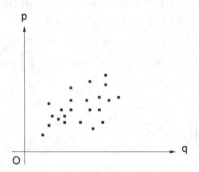

Fig. 6.7 The ensemble in phase space.

We are now interested in how n or W changes in time, *i.e.* how the
system (of points as in Fig. 6.7) moves about in phase space. The equation
of motion for n or W is the so-called *Liouville equation*. In order to derive
this equation, we consider the domain G in Fig. 6.8 and establish an equation
for the change of the number of points or systems in G in the time interval dt.
In doing this, we take into account, that in our consideration no additional
points are created or destroyed.

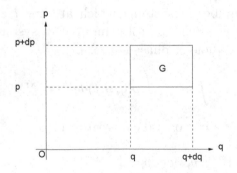

Fig. 6.8 The region G.

The number of points at time $t + dt$ in domain G, *i.e.* $\rho(q,p,t+dt)dqdp$,
is equal to the number in G at time t plus the number that went into G in
the time interval dt minus the number that left G in the time interval dt.

We set $v_q(q,p) = (dq/dt)_{p=\text{const.}}$ and $v_p(q,p) = (dp/dt)_{q=\text{const.}}$ which denote respectively the velocities in directions q and p. Then the difference between the number of systems at time $t + dt$ and the number at time t passing through the area $dqdp$ is (with the dots indicating corresponding flux in the direction of p):

$$\rho(q,p,t+dt)dqdp - \rho(q,p,t)dqdp$$

$$= \rho(q,p,t)\left(\frac{dq}{dt}\right)_{q,p} dpdt - \rho(q+dq,p,t)\left(\frac{dq}{dt}\right)_{q+dq,p} dpdt + \cdots,$$

and thus

$$\rho(q,p,t+dt)dqdp - \rho(q,p,t)dqdp$$

$$= \rho(q,p,t)v_q(q,p)dpdt - \rho(q+dq,p,t)v_q(q+dq,p)dpdt$$

$$+ \rho(q,p,t)v_p(q,p)dqdt - \rho(q,p+dp,t)v_p(q,p+dp)dqdt.$$

Dividing both sides by $dqdpdt$ this becomes

$$\frac{\rho(q,p,t+dt) - \rho(q,p,t)}{dt}$$

$$= \frac{\rho(q,p,t)v_q(q,p) - \rho(q+dq,p,t)v_q(q+dq,p)}{dq}$$

$$+ \frac{\rho(q,p,t)v_p(q,p) - \rho(q,p+dp,t)v_p(q,p+dp)}{dp},$$

or

$$\frac{\partial \rho}{\partial t} = -\frac{\partial}{\partial q}[\rho v_q(q,p)] - \frac{\partial}{\partial p}[\rho v_p(q,p)].$$

However,

$$v_q(q,p) = \dot{q} = \frac{\partial H}{\partial p}, \quad v_p(q,p) = \dot{p} = -\frac{\partial H}{\partial q},$$

so that

$$\frac{\partial \rho}{\partial t} = -\frac{\partial}{\partial q}\left(\rho \frac{\partial H}{\partial p}\right) + \frac{\partial}{\partial p}\left(\rho \frac{\partial H}{\partial q}\right)$$

$$= -\frac{\partial \rho}{\partial q}\frac{\partial H}{\partial p} + \frac{\partial \rho}{\partial p}\frac{\partial H}{\partial q} \overset{\text{also}}{=} -\frac{\partial \rho}{\partial q}\dot{q} - \frac{\partial \rho}{\partial p}\dot{p}$$

$$= \{H, \rho\}. \tag{6.7}$$

Hence

$$\frac{\partial \rho}{\partial t} = \{H, \rho\}. \tag{6.8}$$

This is the *Liouville equation* which describes the motion of the ensemble or, put differently, the probable motion of the system under consideration. Using the result of Example 6.6, *i.e.*

$$\frac{\partial \dot{q}}{\partial q} + \frac{\partial \dot{p}}{\partial p} = 0,$$

we can also rewrite Eq. (6.7) as

$$\frac{\partial \rho}{\partial t} = -\frac{\partial}{\partial q}(\rho \dot{q}) - \frac{\partial}{\partial p}(\rho \dot{p}) = -\text{div}(\text{current density}). \qquad (6.9)$$

These equations have a deep meaning, as we explain below. Comparison of Eq. (6.8) with Eq. (6.1) shows that ρ and u satisfy very similar equations. With Eqs. (6.5) and (6.6) we can also write

$$\frac{\partial W(q,p,t)}{\partial t} = \{H(q,p), W(q,p,t)\} \quad \text{with} \quad \int W(q,p,t)\frac{dqdp}{2\pi\hbar} = 1. \qquad (6.10)$$

The generalization to n degrees of freedom is evident: The volume element of phase space is

$$\prod_i \left(\frac{dq_i dp_i}{2\pi\hbar}\right) := \left(\frac{dqdp}{2\pi\hbar}\right)^n \quad \text{with} \quad \int W(q,p,t)\left(\frac{dqdp}{2\pi\hbar}\right)^n = 1, \qquad (6.11)$$

where

$$W(q,p,t)\left(\frac{dqdp}{2\pi\hbar}\right)^n$$

is the probability for the system to be at time t in the volume $q, q+dq; p, p+dp$.

We deduce from the Liouville equation the important consequence that

$$\frac{d\rho(q,p,t)}{dt} = 0, \qquad (6.12)$$

since the total derivative is made up of precisely the partial derivatives contained in Eq. (6.7). Equation (6.12) implies that ρ is a constant in time, and hence that equal phase space volumes contain the same number of systems, and this means — since these systems are contained in a finite part V of phase space — that

$$\frac{d}{dt}\int_V \rho dV = \frac{dN}{dt} = 0.$$

We have in particular, since no systems are created or destroyed, that

$$dV \propto \left(\frac{dq_0 dp_0}{2\pi\hbar}\right)^n = \left(\frac{dqdp}{2\pi\hbar}\right)^n, \qquad (6.13)$$

if q_0, p_0 are the initial values of q, p (*cf.* Example 6.6). Thus in Fig. 6.6 the area G_0 is equal to the area G_1. Thus we can say, the zero on the right hand side of Eq. (6.12) expresses the fact that no systems, *i.e.* elements, of the ensemble in a volume $\triangle q \triangle p$ are annihilated in this volume, nor are any new ones created. However, there are local (*i.e.* non-equilibrium) fluctuations of their number expressed by $\partial \rho / \partial t$. If this is zero, these fluctuations are removed, and Eq. (6.9) implies a *steady current* of systems through the volume $\triangle q \triangle p$. This is a state of *equilibrium*, and $\partial \rho / \partial t = 0$ is therefore the condition of *static equilibrium*. This is a condition which one encounters in diverse areas of physics — *e.g.* in statistical mechanics and in electrodynamics (there the condition for stationary currents).[*] If a system is interfered with in some way, *e.g.* particles are added at some time t_0, then the system undergoes a non-equilibrium and it is only after some time that the system is in a new state of equilibrium (*i.e.* until it has "relaxed" to a steady state after a certain length of time called "relaxation time"). Thus basic statistical mechanics considers equilibrium statistical mechanics (which is frequently not mentioned).

6.2.3 Probable values of observables

Let $u = u(q, p)$ be an observable. We define as *probable value* or *expectation value* of $u(q, p)$ the following expression:

$$\langle u \rangle = \int u(q, p) W(q, p, t) \left(\frac{dqdp}{2\pi\hbar} \right)^n. \tag{6.14}$$

With Eq. (6.1), *i.e.*

$$\frac{du(q, p)}{dt} = \{u, H\},$$

we described the time variation of the observable $u(q, p)$. We now enquire about the time variation of the expectation value $\langle u \rangle$ of u. We shall see that we have two possibilities for this, *i.e.* for

$$\frac{d}{dt} \langle u \rangle = \frac{d}{dt} \int u(q, p) W(q, p, t) \left(\frac{dqdp}{2\pi\hbar} \right)^n. \tag{6.15}$$

The first and most immediate possibility is — as indicated – that the density or probability $W(q, p, t)$ depends *explicitly* on time t (if determined at a fixed point in phase space), and the time variation $d\langle u \rangle / dt$ is attributed to the fact

---

[*]For the application in electrodynamics see for instance H. J. W. Müller–Kirsten [34], p. 111.

that it is this probability (that $u(q,p)$ assumes certain values) that depends explicitly on time. Then Eq. (6.15) becomes

$$
\begin{aligned}
\frac{d}{dt}\langle u \rangle &= \int u(q,p) \frac{\partial}{\partial t} W(q,p,t) \left(\frac{dqdp}{2\pi\hbar}\right)^n \\
&= \int u(q,p)\{H(q,p), W(q,p,t)\} \left(\frac{dqdp}{2\pi\hbar}\right)^n,
\end{aligned}
\tag{6.16}
$$

where we used Eq. (6.10).

However, we can also employ a more complicated consideration.[†] Solving the equations of motion for q, p, we can express these in terms of their initial values q_0, p_0, *i.e.* at $t = 0$, so that (*cf.* Example 5.7)

$$
q = g(q_0, p_0, t), \quad p = f(q_0, p_0, t),
\tag{6.17}
$$

and hence

$$
u(q,p) \equiv u(q,p,0) = u(g(q_0, p_0, t), f(q_0, p_0, t), 0) \equiv u_0(q_0, p_0, t).
\tag{6.18}
$$

The distribution of the canonical variables is given by $W(q,p,t)$. Thus we can write, since $W \propto \rho$ is constant in time according to Eq. (6.12):

$$
\begin{aligned}
W(q,p,t) &= W(g(q_0, p_0, t), f(q_0, p_0, t), t) \\
&= W(q_0, p_0, 0) \equiv W_0(q_0, p_0) \quad \text{at time } t = 0,
\end{aligned}
\tag{6.19}
$$

i.e. W is the density in the neighbourhood of a given point in phase space and has an *implicit* dependence on time t. With these expressions we obtain for the expectation value $\langle u \rangle_0$ (compare the time-dependence with that in Eq. (6.17)):

$$
\langle u \rangle_0 = \int u_0(q_0, p_0, t) W_0(q_0, p_0) \left(\frac{dq_0 dp_0}{2\pi\hbar}\right)^n.
\tag{6.20}
$$

In this expression the time t is contained explicitly in the observable $u(q,p) \equiv u_0(q_0, p_0, t)$. We expect, of course, that

$$
\langle u \rangle = \langle u \rangle_0.
\tag{6.21}
$$

We verify this claim as follows.

Reversing Eq. (6.17), we have

$$
q_0 = \tilde{g}(q,p,t), \quad p_0 = \tilde{f}(q,p,t),
\tag{6.22}
$$

[†]See also H. Goldstein [19], Sec. 8.8.

so that on the other hand with Eq. (6.17)

$$q \equiv g(\tilde{g}(q,p,t), \tilde{f}(q,p,t), t), \quad p \equiv f(\tilde{g}(q,p,t), \tilde{f}(q,p,t), t). \tag{6.23}$$

Inserting these expressions into u_0 we obtain

$$
\begin{aligned}
u_0(q_0, p_0, t) &= u_0(\tilde{g}(q,p,t), \tilde{f}(q,p,t), t) \\
&\overset{(6.18)}{=} u(g(\tilde{g}(q,p,t), \tilde{f}(q,p,t), t), f(\tilde{g}(q,p,t), \tilde{f}(q,p,t), t), 0) \\
&\overset{(6.23),(6.19)}{=} u(q,p,0) \equiv u(q,p).
\end{aligned}
\tag{6.24}
$$

Moreover, (cf. Eq. (6.19))

$$W_0(q_0, p_0) = W(q, p, t). \tag{6.25}$$

Inserting Eqs. (6.24) and (6.25) and (6.13) into Eq. (6.20), we obtain

$$\langle u \rangle_0 = \int u(q,p) W(q,p,t) \left(\frac{dq\,dp}{2\pi\hbar} \right)^n = \langle u \rangle,$$

as had to be shown.

Taking now the total time derivative of $\langle u \rangle_0$, Eq. (6.20), we obtain an expression which is different from that in Eq. (6.16), i.e.

$$
\begin{aligned}
\frac{d}{dt}\langle u \rangle_0 &= \frac{d}{dt} \int u_0(q_0, p_0, t) W_0(q_0, p_0) \left(\frac{dq_0\,dp_0}{2\pi\hbar} \right)^n \\
&= \int \frac{\partial u_0(q_0, p_0, t)}{\partial t} W_0(q_0, p_0) \left(\frac{dq_0\,dp_0}{2\pi\hbar} \right)^n.
\end{aligned}
\tag{6.26}
$$

We deal with the partial derivative with the help of Eq. (6.18):

$$
\begin{aligned}
\frac{\partial u_0(q_0, p_0, t)}{\partial t} &= \left(\frac{\partial u}{\partial q} \right)_{\substack{q=g(q_0,p_0,t), \\ p=f(q_0,p_0,t)}} \frac{dq}{dt} + \left(\frac{\partial u}{\partial p} \right)_{\substack{q=g(q_0,p_0,t), \\ p=f(q_0,p_0,t)}} \frac{dp}{dt} \\
&= \left(\frac{\partial u}{\partial q} \right)_{\substack{q=g(q_0,p_0,t), \\ p=f(q_0,p_0,t)}} \frac{\partial H}{\partial p} - \left(\frac{\partial u}{\partial p} \right)_{\substack{q=g(q_0,p_0,t), \\ p=f(q_0,p_0,t)}} \frac{\partial H}{\partial q} \\
&= -(\{H(q,p), u(q,p)\})_{\substack{q=g(q_0,p_0,t), \\ p=f(q_0,p_0,t)}}.
\end{aligned}
\tag{6.27}
$$

Substituting this into Eq. (6.26) we obtain

$$\frac{d}{dt}\langle u \rangle_0 = -\int (\{H(q,p), u(q,p)\})_{\substack{q=g(q_0,p_0,t), \\ p=f(q_0,p_0,t)}} W_0(q_0, p_0) \left(\frac{dq_0\,dp_0}{2\pi\hbar} \right)^n. \tag{6.28}$$

Here we perform the transformation (6.22) and use (6.13) and (6.19), so that

$$\frac{d}{dt}\langle u\rangle_0 = -\int \{H(q,p), u(q,p)\}W(q,p,t)\left(\frac{dqdp}{2\pi\hbar}\right)^n. \qquad (6.29)$$

This expression contains $\{H,u\}W$ instead of $u\{H,W\}$ in Eq. (6.16). However, from the properties of the Poisson bracket, we obtain

$$\{H, uW\} = \{H, u\}W + u\{H, W\}. \qquad (6.30)$$

The phase-space integral of a Poisson bracket like

$$I := \int \{H, uW\}\left(\frac{dqdp}{2\pi\hbar}\right)^n$$

vanishes under certain conditions. Consider

$$\begin{aligned}
\{H, uW\} &= \sum_i \left[\frac{\partial H}{\partial q_i}\frac{\partial(uW)}{\partial p_i} - \frac{\partial H}{\partial p_i}\frac{\partial(uW)}{\partial q_i}\right] \\
&= \sum_i \left[\frac{\partial}{\partial p_i}\left(\frac{\partial H}{\partial q_i}uW\right) - \frac{\partial}{\partial q_i}\left(\frac{\partial H}{\partial p_i}uW\right)\right]. \qquad (6.31)
\end{aligned}$$

If for all i:

$$\lim_{p_i\to\pm\infty} \frac{\partial H}{\partial q_i}uW = 0, \qquad \lim_{q_i\to\pm\infty} \frac{\partial H}{\partial p_i}uW = 0, \qquad (6.32)$$

(which is reasonable since the density vanishes at infinity), we obtain zero after partial integration of I and hence from Eqs. (6.29) and (6.30) the relation

$$\frac{d}{dt}\langle u\rangle_0 = \int u(q,p)\{H(q,p), W(q,p,t)\}\left(\frac{dqdp}{2\pi\hbar}\right)^n \qquad (6.33)$$

in agreement with Eq. (6.16). Alternatively we could deduce from Eqs. (6.16) and (6.33) the Liouville equation.

The considerations we just performed demonstrate that we have two ways of treating the time-dependence of expectation values of observables: The explicit time-dependence can either be contained in the probability W or in the (transformed) observables. In the first case, described as "*Schrödinger picture*", the observable u is treated as a function $u(q,p)$, and the time variation of $\langle u(q,p)\rangle$ is attributed to the probability $W(q,p,t)$ of u assuming certain values q and p. This time-dependence is described by the *Liouville equation*

$$\frac{\partial W(q,p,t)}{\partial t} = \{H(q,p), W(q,p,t)\}.$$

In the other case, called *"Heisenberg picture"*, the probability of the initial values $W_0(q_0, p_0)$ is assumed,[‡] and the explicit time-dependence is transferred into the correspondingly transformed observables $u_0(q_0, p_0, t)$. The equation of motion is then that of an observable, *i.e.* (*cf.* Eq. (6.1))

$$\frac{du(q,p)}{dt} = \{u(q,p), H(q,p)\},$$

the reason being that — since q_0, p_0 are constant initial values — we have

$$\frac{\partial u_0(q_0, p_0, t)}{\partial t} \stackrel{(6.24)}{=} \frac{du_0(q_0, p_0, t)}{dt} = \frac{du(q,p)}{dt}$$

$$\stackrel{(6.1)}{=} \{u(q,p), H(q,p)\}. \tag{6.34}$$

We thus also recognize the connection between the Liouville equation, as the equation of motion of an ensemble or of a probability distribution on the one hand, and the equation of motion (6.1) of an observable on the other.

With the above considerations we achieved a general formulation of classical mechanics. This formulation deals with observables u representing physical quantities, and probabilities W, which describe the state of a system. The time dependence of the expectation values can be dealt with in two different ways, as we observed, which are described as "Schrödinger picture" and "Heisenberg picture"— all this on a purely classical level but with the use of canonical transformations.[§]

These considerations point the way to a generalization which results if we permit u and W to belong to a more general class of mathematical quantities. This is then the topic of quantum mechanics.

6.3   Spacetime Formulations

Different from mechanics, electrodynamics, for instance, deals with *fields* which spread through the whole of space. In this case one therefore has also *field densities* in space, and correspondingly a *Lagrange density* $\mathcal{L}$, a *Hamilton density* $\mathcal{H}$, and so on. The electromagnetic case, with the *vector potential* $\mathbf{A}(\mathbf{x}, t)$ is already a very complicated quantity. We therefore refer here as an illustration to the much simpler, but less physical case, of a *scalar field* $\phi(\mathbf{x}, t)$ with *action integral* (with velocity of light $c = 1$)

$$S = \int dt L, \quad L = \int d\mathbf{x} \mathcal{L}(\phi, \partial_\mu \phi), \quad S = \int d^4 x \mathcal{L}(\phi, \partial_\mu \phi). \tag{6.35}$$

[‡]This means: Only in the Heisenberg picture $\partial W/\partial t = 0$.

[§]The author learned this approach from lectures of H. Koppe [26] at the university of Munich around 1964.

The derivation of the *Euler–Lagrange equation* for such a case will be dealt with in Sec. 6.4. We now demand that (in addition to some other properties) $\mathcal{L}$ be invariant under Lorentz transformations. It will be seen that we can construct expressions which are immediately seen to possess this invariance. In order to achieve this, we require the definition of *covariant* and *contravariant vectors*. To arrive at a definition of these is therefore our first objective.

6.3.1   Spacetime (Lorentz) transformations

It is not customary but very instructive to introduce here, at this stage of our treatment of classical mechanics, transformations called *"Lorentz transformations"*, which correspond in four-dimensional *spacetime* to the three-dimensional rotations or orthogonal transformations we considered earlier, although their physical interpretation goes far beyond that of the latter and is the topic of Einstein's Special Theory of Relativity. In this introduction of these Lorentz transformations we lean again on the concepts of group theory. We recall that rotations in a three-dimensional Euclidean space are elements of the group $SO(3)$, and as such are those which leave invariant (*i.e.* unchanged) the value of the square of the distance between two points in the Euclidean space $\mathbb{R}^3$, *i.e.*

$$SO(3): \qquad x^2 + y^2 + z^2 = x'^2 + y'^2 + z'^2.$$

This holds, for instance, for a rotation about the z-axis:

$$\begin{cases} x & \longrightarrow & x' = & x\cos\theta + y\sin\theta, \\ y & \longrightarrow & y' = & -x\sin\theta + y\cos\theta, \\ z & \longrightarrow & z' = & z. \end{cases}$$

In the theory of relativity the three-dimensional Euclidean configuration space is augmented by a further dimension with length given by time t multiplied by the velocity of light c. The new (flat) space is called *spacetime* or *Minkowski spacetime* M_4. The square of the distance between two points in this spacetime is given by

$$-c^2 t^2 + x^2 + y^2 + z^2 \quad \text{or} \quad c t^2 - x^2 - y^2 - z^2.$$

The transformations that leave this expression invariant are the so-called *Lorentz transformations*. Just like the rotations we considered earlier are elements of the *compact group* $SO(3)$ ("compact" because its parameters which are angles (recall three angles in the case of $O(3)$, two angles in the case of the special group $SO(3)$, where 'S' implies an extra condition) are confined to a bounded and closed domain between 0 and 2π), so here the Lorentz transformations are elements of the noncompact Lorentz group $SO(3,1)$ or $SO(1,3)$. Thus the expression left invariant by these transformations is

$$SO(1,3): \quad -\underbrace{c^2 t^2}_{x_0^2} + x^2 + y^2 + z^2 = -\underbrace{c^2 t'^2}_{x_0'^2} + x'^2 + y'^2 + z'^2.$$

In the case of $SO(3,1)$ we have one additional parameter which is not bounded, and this is indicated by the separated "1" in $SO(3,1)$, and makes this indicative of the noncompactness of the group (parameter).

The Lorentz transformation in the plane of (ct, x) is therefore (one time plus one space dimensions, hence no angles, no trigonometric factors), as we verify below:

$$\begin{cases} x_0 & \longrightarrow & x_0' = & x_0 \cosh \zeta - x \sinh \zeta, \\ x & \longrightarrow & x' = & -x_0 \sinh \zeta + x \cosh \zeta, \\ y & \longrightarrow & y' = & y, \\ z & \longrightarrow & z' = & z. \end{cases} \tag{6.36}$$

With these expressions we have

$$\begin{aligned} -{x_0'}^2 + {x'}^2 &= -(x_0 \cosh \zeta - x \sinh \zeta)^2 + (-x_0 \sinh \zeta + x \cosh \zeta)^2 \\ &= -x_0^2(\cosh^2 \zeta - \sinh^2 \zeta) + x^2(\cosh^2 \zeta - \sinh^2 \zeta) \\ &= -x_0^2 + x^2. \end{aligned}$$

Thus the square of a *spacetime distance* remains invariant. The parameter ζ is called *"boost parameter"* or *rapidity*. There are other ways to define this parameter which also explain their names. Thus one can set

$$\tanh \zeta \equiv \beta = \frac{v}{c}, \quad \gamma\beta = \sinh \zeta, \quad \cosh \zeta \equiv \gamma = \frac{1}{\sqrt{1-\beta^2}}. \tag{6.37}$$

With these substitutions, v having the meaning of velocity, one obtains the customary form of the special Lorentz transformations we are considering here in the Special Theory of Relativity, *i.e.*

$$\begin{aligned} x_0' &= \gamma(x_0 - \beta x), \\ x' &= \gamma(x - \beta x_0), \\ y' &= y, \\ z' &= z. \end{aligned} \tag{6.38}$$

We thus have a simple way to visualize the Lorentz transformation as something like a "rotation" (in quotes, since it is not a real rotation) in spacetime, here in the plane of (ct, x). We have here the noncompact cosh and sinh instead of cos and sin in view of the factor "i" in

$$(ix_0')^2 + x'^2 = -x_0'^2 + x'^2.$$

In the new parameters, the quantity v has the meaning of velocity, *i.e.* the velocity of the transformed reference frame relative to that of the rectilinear motion of the first frame. The invariance of a theory under Lorentz transformations thus expresses the independence of the theory of any particular reference frame which moves uniformly with respect to any other frame (note again: no acceleration). Thus the *mathematics* of the transformation is very easy to understand. It is their *physical interpretation* with the transformation of time (*i.e.* x_0) which was one of the great insights of Einstein. We consider this, *i.e.* the Special Theory of Relativity, in more detail in Chapters 11 and 12.

6.3.2   The Poincaré group

Adding simple translations (in 3 spatial directions and one time) to the Lorentz transformations, one obtains the *inhomogeneous Lorentz transformations* which are also known as the *transformations of the Poincaré group* $\mathcal{P}$. A vector of the *covariant form* in four-dimensional or Minkowski spacetime M_4 is charaterized by lower Greek indices, *i.e.* as $x_\mu, \mu = 0, 1, 2, 3$. With the notation: $x_0 = ct, x_1 = -x, x_2 = -y, x_3 = -z$, a path element ds in this spacetime is given by

$$(ds)^2 = (dx_0)^2 - \sum_{i=1}^{3}(dx_i)^2. \tag{6.39}$$

With this definition, we define the following *metric* or *covariant metric tensor* of the space M_4:

$$(\eta_{\mu\nu}) = \begin{pmatrix} +1 & 0 & 0 & 0 \\ 0 & -1 & 0 & 0 \\ 0 & 0 & -1 & 0 \\ 0 & 0 & 0 & -1 \end{pmatrix} \equiv (\eta^{\mu\nu}), \quad \det(\eta_{\mu\nu}) \neq 0. \tag{6.40}$$

On the other hand the unit matrix is defined by

$$(\mathbb{1}) = (\delta_\mu^\rho) \equiv (\eta_\mu{}^\rho), \quad \eta_\mu{}^\rho \eta_{\rho\nu} = \eta_{\mu\nu}, \tag{6.41}$$

so that the *contravariant* counterpart of $\eta_{\mu\nu}$, i.e. $\eta^{\mu\nu}$, is given by

$$\eta_{\mu\nu}\eta^{\nu\rho} = \delta_\mu^\rho. \tag{6.42}$$

Thus, as indicated on the right of Eq. (6.40), this must be identical with $\eta_{\mu\nu}$. The *contravariant vector* x^μ with upper indices is defined by

$$x^\mu = \eta^{\mu\nu} x_\nu, \quad \text{so that} \quad dx^\mu = \eta^{\mu\nu} dx_\nu, \qquad (6.43a)$$

and hence, with $i = 1, 2, 3$,

$$(dx^\mu) = \begin{pmatrix} +1 & 0 & 0 & 0 \\ 0 & -1 & 0 & 0 \\ 0 & 0 & -1 & 0 \\ 0 & 0 & 0 & -1 \end{pmatrix} \begin{pmatrix} dx_0 \\ dx_1 \\ dx_2 \\ dx_3 \end{pmatrix} = (dx_0, -dx_i), \qquad (6.43b)$$

and (summation over identical indices being understood)

$$dx_\mu dx^\mu = (dx_0, dx_i) \begin{pmatrix} dx_0 \\ -dx_i \end{pmatrix} = (dx_0)^2 - (dx_i)^2 = (ds)^2. \qquad (6.43c)$$

With the help of $\eta^{\mu\nu}$ we can "raise" and "lower" indices, and summation over an upper and an identical lower index is understood. We shall become familiar with such manipulations in the following. A quantity which is such that all indices are summed over, is a *Lorentz invariant*, e.g.

$$dx^\mu dx_\mu = (ds)^2$$

(the disappearance of indices on the right is referred to as their "contraction").

The Poincaré group defined above is a 10-parameter group consisting of the following transformations:
(a) *Translations*
There are 4 translations with 4 independent real parameters a_μ:

$$T_4: \quad x'_\mu = x_\mu + a_\mu.$$

The infinitesimal transformation is

$$\delta x_\mu = x'_\mu - x_\mu = \alpha_\mu = \text{infinitesimal.}$$

(b) *The proper homogeneous Lorentz transformations*
These are given by the following equations and involve 6 independent real parameters, which are the elements of the 4×4 matrix $(l_\mu{}^\nu)$:

$$L_6^+: \quad x'_\mu = l_\mu{}^\nu x_\nu, \quad \text{equivalently} \quad x'^\mu = \eta^{\mu\nu} l_\nu{}^\rho x_\rho = l^{\mu\rho} x_\rho \qquad (6.44)$$

with (see below for explanation)

$$\underbrace{l^T{}_\mu{}^\kappa l_\kappa{}^\rho = \delta_\mu^\rho, \quad \det(l_\mu{}^\nu) = +1, \quad l_\mu{}^\nu \text{ real.}}_{\therefore L^+} \qquad (6.45)$$

Here the last condition follows from the requirement of invariance of the square of a spacetime distance, *i.e.*

$$x^{\mu'} x'_{\mu} = x^{\mu} x_{\mu}. \tag{6.46}$$

The superscript T for transposition is defined by the relations

$$l^{T\,\kappa\nu} := l^{\nu\kappa}, \qquad l^{T}_{\mu}{}^{\kappa} := l^{\kappa}{}_{\mu} (\neq l_{\mu}{}^{\kappa}). \tag{6.47}$$

The six independent parameters correspond to the three angles of rotation and the velocities of the three Lorentz transformations, one each in a spatial direction. The matrices (l) define a 4×4 representation of the non-compact Lorentz group $SO(3,1)$ ("3" referring to the three angles of rotation with compact domains of validity, and "1" referring to the non-compact parameter of a Lorentz transformation in the 3+1-dimensional Minkowski spacetime M_4. Actually the vectors x^{μ} and x_{μ} belong to different spaces T, T^*. But since a metric is defined on the 4-dimensional manifold $\mathbb{R}^4$ (*i.e.* the Minkowski metric $(\eta_{\mu\nu})$, which together with $\mathbb{R}^4$ defines the Minkowski space M_4), it suffices to concentrate on only one of these two spaces. The covariant components of a vector $x^{\nu} \in T$ then follow from $x_{\mu} = \eta_{\mu\nu} x^{\nu}$.

The field tensor of electrodynamics is given by the following expression:

$$F_{\mu\nu}(x) := \partial_{\mu} A_{\nu}(x) - \partial_{\nu} A_{\mu}(x). \tag{6.48}$$

One can show that this expression is a covariant tensor of the second rank. To this end one starts from the transformation of a covariant vector, *i.e.*

$$A_{\nu}(x') = \frac{\partial x^{\mu}}{\partial x'^{\nu}} A_{\mu}(x). \tag{6.49}$$

In order to verify this relation, we have to show that with

$$x'_{\mu} = l_{\mu}{}^{\nu} x_{\nu} \tag{6.50}$$

Eq. (6.49) implies the following relation:

$$A_{\mu}(x') = l_{\mu}{}^{\nu} A_{\nu}(x). \tag{6.51}$$

Since $l^{T}{}_{\mu}{}^{\kappa} l_{\kappa}{}^{\rho} = \delta_{\mu}^{\rho}$, we deduce from Eq. (6.50):

$$x_{\mu} = l^{T}{}_{\mu}{}^{\kappa} x'_{\kappa}, \quad \text{or} \quad x^{\mu} = \eta^{\mu\rho} x_{\rho} = \eta^{\mu\rho} l^{T}{}_{\rho}{}^{\kappa} x'_{\kappa} = \eta^{\mu\rho} l^{T}{}_{\rho}{}^{\kappa} \eta_{\kappa\nu} x'^{\nu},$$

so that

$$\frac{\partial x^{\mu}}{\partial x'^{\nu}} = \eta^{\mu\rho} l^{T}{}_{\rho}{}^{\kappa} \eta_{\kappa\nu} = l^{T\mu}{}_{\nu} = l_{\nu}{}^{\mu},$$

and hence

$$\frac{\partial x^\mu}{\partial x'^\nu} A_\mu(x) = l_\nu{}^\mu A_\mu(x),$$

i.e. Eq. (6.49) implies

$$A_\nu(x') = l_\nu{}^\kappa A_\kappa(x),$$

which had to be shown.

6.3.3   Derivatives

Since it is necessary to keep track of minus signs in raising and lowering indices, we consider derivatives in detail. We define

$$\left. \begin{array}{l} (\partial^\mu) \equiv (\frac{\partial}{\partial x_\mu}) = (\frac{\partial}{\partial x_0}, -\boldsymbol{\nabla}) \\[2mm] (\partial_\mu) \equiv (\frac{\partial}{\partial x^\mu}) = (\frac{\partial}{\partial x^0}, \ \boldsymbol{\nabla}) \end{array} \right\} \ \partial^\mu \partial_\mu = \frac{\partial^2}{\partial x_0^2} - \boldsymbol{\nabla}^2 \equiv -\Box^2. \tag{6.52}$$

The *chain rule* is (using in the last step $x^{\rho\prime} = \eta^{\rho\kappa} x'_\kappa = \eta^{\rho\kappa} l_\kappa{}^\mu x_\mu$)

$$(\partial^\mu) \equiv \left(\frac{\partial}{\partial x_\mu} \right) = \frac{\partial x^{\rho\prime}}{\partial x_\mu} \frac{\partial}{\partial x^{\rho\prime}} = \eta^{\rho\kappa} l_\kappa{}^\mu \frac{\partial}{\partial x^{\rho\prime}} = l^{\rho\mu} \partial'_\rho. \tag{6.53}$$

Hence

$$\partial^\mu = l^{\rho\mu} \partial'_\rho \quad \text{and} \quad \partial_\nu = \eta_{\nu\mu} \partial^\mu = \eta_{\nu\mu} l^{\rho\mu} \partial'_\rho,$$

so that

$$\partial_\nu = \partial'_\rho l^\rho{}_\nu. \tag{6.54}$$

This relation is identically satisfied if ∂'_ρ transforms like the vector x'_ρ, because then:

$$\partial'_\rho l^\rho{}_\nu = l^\rho{}_\nu \partial'_\rho = l^\rho{}_\nu l_\rho{}^\kappa \partial_\kappa \stackrel{cf.(6.47)}{=} l_\nu^T{}^\rho l_\rho{}^\kappa \partial_\kappa \stackrel{cf.(6.45)}{=} \delta_\nu^\kappa \partial_\kappa = \partial_\nu. \tag{6.55}$$

This means, the derivatives also transform like Lorentz vectors. We can now consider the transformation of the *volume element* in Minkowski space:

$$dx'_0 dx'_1 dx'_2 dx'_3 \ = \ \begin{vmatrix} \frac{\partial x'_0}{\partial x_0} & \frac{\partial x'_0}{\partial x_1} & \frac{\partial x'_0}{\partial x_2} & \frac{\partial x'_0}{\partial x_3} \\[2mm] \frac{\partial x'_1}{\partial x_0} & \frac{\partial x'_1}{\partial x_1} & \frac{\partial x'_1}{\partial x_2} & \frac{\partial x'_1}{\partial x_3} \\[2mm] \frac{\partial x'_2}{\partial x_0} & \frac{\partial x'_2}{\partial x_1} & \frac{\partial x'_2}{\partial x_2} & \frac{\partial x'_2}{\partial x_3} \\[2mm] \frac{\partial x'_3}{\partial x_0} & \frac{\partial x'_3}{\partial x_1} & \frac{\partial x'_3}{\partial x_2} & \frac{\partial x'_3}{\partial x_3} \end{vmatrix} \ dx_0 dx_1 dx_2 dx_3$$

$$= \ dx_0 dx_1 dx_2 dx_3. \tag{6.56}$$

Hence the volume element is invariant under transformations of the proper homogeneous Lorentz group L^+ (*cf.* Eq. (6.45)).

6.4   From Particles to Fields

In courses on electrodynamics the four Maxwell equations are mostly presented as extracted from observation. However, they can also be looked at and derived as the Euler–Lagrange equations of the electromagnetic field. The electromagnetic field is already a fairly complicated quantity as we know from basic electrodynamics. Since we shall later be concerned with the gravitational field which is a tensor field (whereas the electromagnetic field is a vector field), we show here how one proceeds in such cases along lines parallel to the case in mechanics.

6.4.1   Euler–Lagrange equations

We consider the *Lagrangian density* of a scalar field $\phi(x)$ already alluded to in Sec. 6.3, *i.e.*

$$\mathcal{L}(\phi(x), \partial_\mu(x)).$$

Here the variable $x_\mu \in \mathsf{M}_4$ belongs to the Minkowski manifold, and the field $\phi(x)$ spans an infinitely dimensional configuration space for all values of x_μ, *i.e.* the field at each value of x_μ corresponds precisely to one generalized coordinate in our earlier sense. One should note that the density $\mathcal{L}$ has no explicit dependence on x_μ. With the help of the appropriately extended Hamilton principle, we obtain the equations of motion, called *Euler–Lagrange equations*. Thus the *action* (integral) S and the *Lagrangian* L are given by

$$S = \int_{t_1}^{t_2} dt L(t), \quad L(t) = \int_{\mathbb{R}^3} d\mathbf{x} \mathcal{L}(\phi(x), \partial_\mu \phi(x)), \quad x_\mu \in \mathsf{M}_4. \qquad (6.57)$$

Hence we consider the variation

$$0 = \delta S = \delta \int_{t_1}^{t_2} \int_{\mathbb{R}^3} dt d\mathbf{x} \mathcal{L}(\phi(x), \partial_\mu \phi(x)) \qquad (6.58)$$

with the boundary conditions

$$\delta\phi(x) = 0, \quad t = t_1, t_2. \qquad (6.59)$$

Parallel to our earlier considerations we obtain with a partial integration

$$0 = \int_{t_1}^{t_2} \int_{\mathbb{R}^3} dt d\mathbf{x} \left[\frac{\partial \mathcal{L}}{\partial \phi} \delta\phi + \frac{\partial \mathcal{L}}{\partial(\partial_\mu \phi)} \delta(\partial_\mu \phi) \right]$$

$$= \int_{t_1}^{t_2} \int_{\mathbb{R}^3} dt d\mathbf{x} \left[\frac{\partial \mathcal{L}}{\partial \phi} - \partial_\mu \left(\frac{\partial \mathcal{L}}{\partial(\partial_\mu \phi)} \right) \right] \delta\phi + \int d\sigma_\mu \left[\frac{\partial \mathcal{L}}{\partial(\partial_\mu \phi)} \delta\phi \right]_{t_1}^{t_2}. \quad (6.60)$$

Here σ_μ is the appropriate surface element (this surface integral results from the use of Gauss' divergence theorem applied to the spacetime integral over the total divergence). The surface integral vanishes as a result of the assumed vanishing of the field at infinity (hence one can also add an arbitrary such divergence to $\mathcal{L}$ which is therefore uniquely defined only up to a total divergence). The *Euler–Lagrange equations* of the *scalar field theory* therefore are

$$\partial_\mu\left(\frac{\partial \mathcal{L}}{\partial(\partial_\mu \phi)}\right) - \frac{\partial \mathcal{L}}{\partial \phi} = 0. \tag{6.61}$$

This equation is Lorentz covariant (in the sense of a scalar) or rather invariant provided the action S is.¶

Considerations like those above are of broad generality and can for instance immediately be applied to electrodynamics; we do this in Example 6.9.

Example 6.9: Maxwell equations from Euler–Lagrange equations
Derive Maxwell's equations of the free electromagnetic field from the following action integral (note that here S differs from that in Eq. (6.57) by a factor c in x_0)

$$S \equiv \int dx_0 \int d^3x \mathcal{L}(A_\mu, \partial_\nu A_\mu) = -\frac{1}{4}\int d^4x F_{\mu\nu}F^{\mu\nu}, \quad F_{\mu\nu} := \partial_\mu A_\nu - \partial_\nu A_\mu, \quad A_\mu = (A_0, \mathbf{A}),$$

with $\mathbf{B} = \nabla \times \mathbf{A}$ ($\therefore B_1 = \partial_2 A_3 - \partial_3 A_2$, etc. and $B_i = \epsilon_{ijk}F_{jk}$, and also $F_{23} = B_1 = \epsilon_{231}B_1$, etc. and $F_{ij} = \epsilon_{ijk}B_k$) and $F_{0j} = -E_j/c$.

Solution: With ϕ in Eq. (6.61) replaced by A_ν, and since $\mathcal{L}$ contains A_ν only in derivatives, the four ($\nu = 0,1,2,3$) Euler–Lagrange equations are

$$\partial_\mu\left[\frac{\partial \mathcal{L}}{\partial(\partial_\mu A_\nu)}\right] = 0, \quad \partial_\mu F^{\mu\nu} = 0.$$

The canonical momentum conjugate to A_μ is

$$\pi_\mu := \frac{\partial \mathcal{L}}{\partial(\partial_0 A^\mu)} = -F_{0\mu} \quad \text{with} \quad \pi_0 = 0, \ \pi_i = F_{i0} := \frac{1}{c}E_i.$$

The component A_0 plays a special role, that we do not elaborate on here, although this is well understood, also the fact that we cannot do without A_0, since otherwise one would not obtain the Gauss law in case (b) below. As Euler–Lagrange equations we obtain now the Maxwell equations (with $x_0 = ct$):
(a) Performing the differentiations we obtain for $\nu = j$:

$$\partial_0(F^{0j}) + \partial_i(F^{ij}) = 0, \quad i.e. \quad \partial_0 F_{0j} - \partial_i F_{ij} = 0.$$

Now:

$$\sum_i \partial_i F_{ij} = \sum_i \partial_i \sum_k \epsilon_{ijk}B_k = \sum_{i,k}\epsilon_{ijk}\partial_i B_k = -\sum_{i,k}\epsilon_{ikj}\partial_i B_k = -\sum_{i,k}\epsilon_{jik}\partial_i B_k = -(\nabla \times \mathbf{B})_j.$$

¶The invariance of Eq. (6.61) can be seen by observing that differentiating an invariant like $(\partial_\mu\phi)(\partial^\mu\phi)$ with respect to $(\partial_\mu\phi)$ leaves $(\partial^\mu\phi)$, *i.e.* with upper index. Thus with ∂_μ applied, the expression $\partial_\mu(\partial^\mu\phi)$ is seen to be invariant.

Hence:

$$\frac{1}{c^2}\frac{\partial \mathbf{E}}{\partial t} - \mathbf{\nabla} \times \mathbf{B} = 0.$$

In electrodynamics this equation is known as *Ampére's law.*
(b)

$$\frac{\partial}{\partial x_i}\left(\frac{\partial \mathcal{L}}{\partial(\partial_i A_0)}\right) = 0, \quad \text{or} \quad \mathbf{\nabla} \cdot \mathbf{E} = O(A_0) = 0 \quad \text{with} \quad A_0 = 0,$$

which is the *Gauss law.* The remaining two Maxwell equations follow from the *Jacobi identity*$^{\|}$ for the field tensor $F_{\mu\nu}$, *i.e.* from

$$\epsilon^{\alpha\beta\rho\sigma}\partial_\alpha F_{\rho\sigma} = -\epsilon^{\beta\alpha\rho\sigma}\partial_\alpha F_{\rho\sigma} = -\frac{1}{3}\epsilon^{\beta\alpha\rho\sigma}[\partial_\alpha F_{\rho\sigma} + \partial_\rho F_{\sigma\alpha} + \partial_\sigma F_{\alpha\rho}] = 0,$$

since $F_{\rho\sigma} = \partial_\rho A_\sigma - \partial_\sigma A_\rho$. One writes and obtains (when written out**) by matrix multiplication (observe that in the following equation the first term on the far right is a scalar)

$$0 = \partial_\alpha \epsilon^{\alpha\beta\rho\sigma} F_{\rho\sigma} \equiv \partial_\alpha {}^*F^{\alpha\beta} = \left(\mathbf{\nabla}\cdot\mathbf{B}, -\frac{\partial\mathbf{B}}{\partial x^0} - \frac{1}{c}\mathbf{\nabla}\times\mathbf{E}\right),$$

where $^*F$ is the *dual tensor.* With this we obtain the equations as before except with the interchange $\mathbf{E}/c \rightarrow \mathbf{B}, \mathbf{B} \rightarrow -\mathbf{E}/c$, *i.e.*
(c) *Faraday's law*

$$\frac{\partial\mathbf{B}}{\partial t} + \mathbf{\nabla}\times\mathbf{E} = 0,$$

and the equation expressing the nonexistence of single magnetic poles, *i.e.*
(d)

$$\mathbf{\nabla}\cdot\mathbf{B} = O.$$

6.4.2   The Noether theorem

Having introduced the four-dimensional spacetime and thereafter the concept of fields (for reasons of simplicity here restricted to the case of the scalar field, although we can keep in mind the electromagnetic vector field A_μ or the electromagnetic field tensor $F_{\mu\nu}$ with the electric and magnetic fields $\mathbf{E}$ and $\mathbf{B}$ as components), we can now consider the more general form of Noether's theorem. Let

$$S = \int d^4x \mathcal{L}(\phi, \partial_\mu\phi)$$

be the action integral defining the theory. We consider first an infinitesimal transformation of the field, *i.e.*

$$\phi(x) \rightarrow \phi'(x) = \phi(x') + \delta\phi(x). \tag{6.62}$$

$^{\|}$This is the case in the back of the mind of people who argue that one does not require necessarily a Lagrangian to obtain an equation of motion.

**See *e.g.* H. J. W. Müller–Kirsten [34], pp. 399, 412.

We assume that without the use of the equations of motion, the variation yields an expression

$$\delta\mathcal{L} = \partial_\mu \Lambda^\mu, \tag{6.63}$$

where Λ^μ is a spacetime vector. In our earlier one-dimensional (*i.e.* time) consideration we had

$$S = \int dt L(q, \dot{q}) \quad \text{and} \quad \delta S = \int dt \delta L,$$

with L uniquely defined only up to a total time derivative, say $dF(q, \dot{q})/dt$, *i.e.*

$$\delta L = \frac{d}{dt} F(q, \dot{q}) \implies \delta S = \int dF(q, \dot{q}) = [F(q, \dot{q})]_{t_i}^{t_f} = 0$$

with F vanishing at the limits. We described as symmetries of a theory or of S those transformations for which $\delta S = 0$ up to a time derivative. Similarly in the present case of four-dimensional spacetime, the Lagrangian density is defined uniquely only up to a four-dimensional divergence, and hence a transformation for which

$$\delta\mathcal{L} = \partial_\mu \Lambda^\mu \tag{6.64}$$

is said to be a *symmetry* of the theory. Then

$$\delta S = \int d^4 x \delta\mathcal{L} = \int d^4 x \partial_\mu \Lambda^\mu = \int d\sigma_\mu \Lambda^\mu.$$

With the integration surface extended to infinity, where the fields involved vanish, this implies $\delta S = 0$.

On the other hand, using the equation of motion, *i.e.*

$$\partial_\mu \left(\frac{\partial\mathcal{L}}{\partial(\partial_\mu\phi)} \right) - \frac{\partial\mathcal{L}}{\partial\phi} = 0, \tag{6.65}$$

we have

$$\delta\mathcal{L} = \frac{\partial\mathcal{L}}{\partial\phi}\delta\phi + \frac{\partial\mathcal{L}}{\partial(\partial_\mu\phi)}\underbrace{\delta(\partial_\mu\phi)}_{\partial_\mu(\delta\phi)}$$

$$\overset{(6.65)}{=} \partial_\mu\left(\frac{\partial\mathcal{L}}{\partial(\partial_\mu\phi)}\right)\delta\phi + \frac{\partial\mathcal{L}}{\partial(\partial_\mu\phi)}\partial_\mu(\delta\phi) = \partial_\mu\left[\frac{\partial\mathcal{L}}{\partial(\partial_\mu\phi)}\delta\phi\right]. \tag{6.66}$$

From Eqs. (6.63) and (6.66) we obtain the *continuity equation* with *Noether current* J^μ, *i.e.*

$$\partial_\mu J^\mu = 0, \quad J^\mu = \Lambda^\mu - \frac{\partial\mathcal{L}}{\partial(\partial_\mu\phi)}\delta\phi. \tag{6.67}$$

This is the claim of *Noether's theorem*: Every continuous symmetry of a theory (as defined here, the parameter of the symmetry group being *e.g.* the parameter ϵ^λ in the following Example 6.10) is associated with a *conserved current* (density).

Example 6.10: Noether current and translations

By considering four-dimensional translations determine the Noether current in the case of a scalar field $\phi(x)$ of mass m.

Solution: We consider translations

$$ x'^\mu \longrightarrow x^\mu + \epsilon^\mu, \quad \epsilon^\mu = \text{const.} $$

Since $\phi(x)$ is a scalar field, we have (*cf.* Sec. 5.6) $\phi'(x') = \phi(x')$. With Taylor expansion we obtain

$$ \phi'(x') = \phi(x + \epsilon) = \phi(x) + \epsilon^\lambda \partial_\lambda \phi(x), $$

so that

$$ \delta\phi(x) \equiv \phi'(x') - \phi(x) = \epsilon^\lambda \partial_\lambda \phi. \tag{6.68} $$

The Lagrangian density of the mass-m scalar field $\phi(x)$ can be written

$$ \mathcal{L} = \frac{1}{2}\partial^\mu\phi\partial_\mu\phi - \frac{1}{2}m^2\phi^2, $$

so that

$$
\begin{aligned}
\delta\mathcal{L} &= \partial^\mu\phi\partial_\mu(\delta\phi) - m^2\phi(\delta\phi) = \partial^\mu\phi\epsilon^\lambda\partial_\mu\partial_\lambda\phi - m^2\phi\epsilon^\lambda\partial_\lambda\phi \\
&= \epsilon^\lambda\partial_\lambda\left\{\frac{1}{2}\partial^\mu\phi\partial_\mu\phi - \frac{1}{2}m^2\phi^2\right\} = \partial_\lambda(\epsilon^\lambda\mathcal{L}) \equiv \partial_\lambda\Lambda^\lambda,
\end{aligned}
$$

with

$$ \Lambda^\lambda := \epsilon^\lambda\mathcal{L}. \tag{6.69} $$

Inserting this expression into Eq. (6.67) and using Eq. (6.68), we obtain

$$ J^\mu = \Lambda^\mu - \frac{\partial\mathcal{L}}{\partial(\partial_\mu\phi)}\delta\phi = \epsilon^\mu\mathcal{L} - \epsilon^\lambda\partial_\lambda\phi.\frac{\partial\mathcal{L}}{\partial(\partial_\mu\phi)} = \epsilon^\nu\left\{\delta^\mu_\nu\mathcal{L} - \frac{\partial\mathcal{L}}{\partial(\partial_\mu\phi)}\partial_\nu\phi\right\} \equiv \epsilon^\nu T_\nu{}^\mu, \tag{6.70} $$

where $T_\nu{}^\mu$ is the *canonical energy-momentum tensor*. We observe here that in the case of the one-dimensional space with only the time variable the quantity in braces is $\mathcal{L} - p\dot{q}$, which is minus the Hamiltonian of the Legendre transform (4.3) (*cf.* also Eqs. (5.78) to (5.81)).

6.4.3 Curved spacetime

We hint here briefly at some diversifications, which arise in more detail later, specifically in Chapter 13.

We saw that in Minkowski spacetime the square of the path-element ds is given by

$$ ds^2 = dx^\mu dx_\mu = (dx^0)^2 - (d\mathbf{x})^2 = c^2 dt^2 - (d\mathbf{x})^2. $$

If we introduce for instance spatial cylinder coordinates, *i.e.*

$$x = r\cos\psi, \quad y = r\sin\psi, \quad z = z,$$

with

$$dx = -r\sin\psi d\psi + dr\cos\psi, \quad dy = r\cos\psi d\psi + dr\sin\psi, \quad dz = dz,$$

we have

$$ds^2 = c^2 dt^2 - (dr^2 + r^2 d\psi^2 + dz^2).$$

With a uniform (constant ω) rotation, *i.e.*

$$\phi = \psi - \omega t, \quad d\psi = d\phi + \omega dt,$$

this becomes

$$
\begin{aligned}
ds^2 &= c^2 dt^2 - dr^2 - r^2(d\phi + \omega dt)^2 - dz^2 \\
&= (c^2 - \omega^2 r^2)dt^2 - (dr^2 + r^2 d\phi^2 + 2\omega r^2 d\phi dt + dz^2) \\
&\equiv g_{\mu\nu}(x)dx^\mu dx^\nu.
\end{aligned}
$$

The quantity $g_{\mu\nu}(x)$ is the "*metric tensor*" of the space. This is no longer an example of Minkowskian geometry. Consider the following action integral:

$$S = \int ds \left\{ g_{\mu\nu}(x)\frac{dx^\mu}{ds}\frac{dx^\nu}{ds} \right\}. \tag{6.71}$$

We show in Example 6.11 that this integral is invariant under the transformation $s \rightarrow \lambda = s + \epsilon$. This is not sufficiently general for a sensible theory, and is here chosen only as an example. With $g_{\mu\nu}(x)dx^\mu dx^\nu$ from the previous equation inserted here, we can derive the equations of motion in the variables r, z, ϕ, t from the following expression:

$$
\begin{aligned}
S = \int ds \Bigg[(c^2 - \omega^2 r^2)\left(\frac{dt}{ds}\right)^2 - \left\{ \left(\frac{dr}{ds}\right)^2 + r^2\left(\frac{d\phi}{ds}\right)^2 \right. \\
\left. + 2\omega r^2\left(\frac{d\phi}{ds}\right)\left(\frac{dt}{ds}\right) + \left(\frac{dz}{ds}\right)^2 \right\} \Bigg].
\end{aligned}
\tag{6.72}
$$

If we want to construct the action integral of a theory with respect to a curved space, we have to take into account the nonplanarity of this space. We demand again that the Lagrangian (or Lagrangian density in the case of fields) be a scalar, but not only (globally) under transformations of the Poincaré group, but also (locally) under *generalized coordinate transformations* — also called *diffeomorphisms* — in the curved spacetime. Infinitesimally these are *e.g.* (ϵ being a parameter of small value)

$$x^\mu \to x'^\mu = x^\mu - \epsilon \xi^\mu(x), \tag{6.73}$$

so that in the case of a scalar field $\phi(x)$:

$$\begin{aligned} \delta\phi(x) &= \phi'(x') - \phi(x) = \phi(x') - \phi(x) \\ &= \phi(x^\mu - \epsilon\xi^\mu(x)) - \phi(x) = -\epsilon\xi^\lambda(x)\partial_\lambda\phi. \end{aligned} \tag{6.74}$$

Now, how about $g^{\mu\nu}(x)$? Since this is a tensor, we have[††]

$$g'^{\mu\nu}(x') = \frac{\partial x'^\mu}{\partial x^\alpha}\frac{\partial x'^\nu}{\partial x^\beta}g^{\alpha\beta}(x), \qquad g'_{\mu\nu}(x') = \frac{\partial x^\alpha}{\partial x'^\mu}\frac{\partial x^\beta}{\partial x'^\nu}g_{\alpha\beta}(x). \tag{6.75}$$

Hence

$$\begin{aligned} g'^{\mu\nu}(x') &= (\delta^\mu_\alpha - \epsilon\partial_\alpha\xi^\mu(x))(\delta^\nu_\beta - \epsilon\partial_\beta\xi^\nu(x))g^{\alpha\beta}(x) \\ &\simeq g^{\mu\nu}(x) - \epsilon(\delta^\mu_\alpha\partial_\beta\xi^\nu(x) + \delta^\nu_\beta\partial_\alpha\xi^\mu(x))g^{\alpha\beta}(x). \end{aligned} \tag{6.76}$$

If we assume the metric tensor $g^{\alpha\beta}(x)$ to be close to that of Minkowski space, *i.e.* $g^{\alpha\beta}(x) \simeq \eta^{\alpha\beta} + \epsilon\gamma^{\alpha\beta}$, we have to first order in ϵ:

$$\begin{aligned} g'^{\mu\nu}(x') &\simeq g^{\mu\nu}(x) - \epsilon(\eta^{\mu\beta}\partial_\beta\xi^\nu(x) + \eta^{\alpha\nu}\partial_\alpha\xi^\mu(x)) \\ &= g^{\mu\nu}(x) - \epsilon(\partial^\mu\xi^\nu(x) + \partial^\nu\xi^\mu(x)). \end{aligned} \tag{6.77}$$

Thus

$$\delta g^{\mu\nu}(x) = g'^{\mu\nu}(x') - g^{\mu\nu}(x) \simeq -\epsilon(\partial^\mu\xi^\nu(x) + \partial^\nu\xi^\mu(x)). \tag{6.78}$$

The requirement that this vanish implies a condition on the vector field $\xi^\mu(x)$. Such vectors $\xi^\mu(x)$ are known as *Killing vectors*. Observers in reference frames which are related in this way observe the same field $g^{\mu\nu}(x)$, *i.e.* the same gravitational field. We mention this here only in passing, and we shall not need this later.

The ultimately invariant action integral is then (more generally also with a derivative of $g^{\mu\nu}$)

$$S = \int d^4x \sqrt{-g}\mathcal{L}(g^{\mu\nu}, \phi, \partial_\mu\phi), \tag{6.79}$$

[††]The definition of and distinction between contravariant tensors (those with upper indices) and covariant tensors (those with lower indices) is treated in Sec. 12.5.2. See *e.g.* Eqs. (12.23) and (14.1).

where $g = \det(g_{\mu\nu})$, and the minus sign in $\sqrt{-g}$ is introduced in order to ensure a real quantity in the limit of a flat Minkowski spacetime. The appearance of the factor $\sqrt{-g}$ here follows from the requirement of invariance of the volume element. Consider the analogy to that in the two-dimensional case. In this case the element of area is

$$d^2x\sqrt{\det g_{ab}(x)}.$$

With transformations (*cf.* the Jacobian in Eq. (4.55))

$$x_a \rightarrow x'_a = x'_a(x), \quad i.e. \quad d^2x' = \det\left(\frac{\partial x'_a}{\partial x_b}\right)d^2x,$$

$$\tilde{g}_{ab}(x') = \frac{\partial x^c}{\partial x'^a}\frac{\partial x^d}{\partial x'^b}g_{cd}(x),$$

one obtains

$$\det\tilde{g}_{ab}(x') = \det\left(\frac{\partial x}{\partial x'}\right)\det\left(\frac{\partial x}{\partial x'}\right)\det g_{cd}(x) \tag{6.80}$$

and

$$\sqrt{\det\tilde{g}_{ab}(x')}d^2x' = \sqrt{\det\left(\frac{\partial x}{\partial x'}\right)\det\left(\frac{\partial x}{\partial x'}\right)}\sqrt{\det g_{cd}(x)}\det\left(\frac{\partial x'}{\partial x}\right)d^2x$$

$$= \sqrt{\det g_{ab}(x)}d^2x. \tag{6.81}$$

With these considerations — which point the way in the direction of general relativity — we end the discussion here. Our aim here was solely to open our eyes for a class of transformations which is much larger than that we considered initially.

6.5 Miscellaneous Examples

Example 6.11: Configuration space *a priori* probability
Show that in the case of two dimensions (as an example), the configuration space *a priori* probability $\triangle x\triangle y$ is time-independent as a result of the incompressibility condition.

Solution: For simplicity we consider the case of two dimensions and there the quantity $\triangle x\triangle y$. Then, in analogy to the case of the phase space *a priori* probability of Example 6.6, we have

$$\frac{d}{dt}\ln(\triangle x\triangle y) = \frac{1}{\triangle x\triangle y}\frac{d}{dt}(\triangle x\triangle y) = \frac{1}{\triangle x}\frac{d}{dt}(\triangle x) + \frac{1}{\triangle y}\frac{d}{dt}(\triangle y).$$

Using arguments as in Example 6.6 (which we do not repeat here) we obtain

$$\frac{d}{dt}\ln(\triangle x\triangle y) = \frac{1}{\triangle x}\frac{\partial\dot{x}}{\partial x}\triangle x + \frac{1}{\triangle y}\frac{\partial\dot{y}}{\partial y}\triangle y = \frac{\partial\dot{x}}{\partial x} + \frac{\partial\dot{y}}{\partial y} \rightarrow 0.$$

Thus the configuration space *a priori* probability $\triangle x \triangle y$ is time-independent for points satisfying the incompressibility condition. This is what one expects on the basis of the lack of any information about the system, as argued earlier.

Example 6.12: Invariance of an action under parameter translation

Show that the action integral

$$S = \int ds L(x, \dot{x}), \quad L(x, \dot{x}) = g_{\mu\nu}(x)\dot{x}^\mu(s)\dot{x}^\nu(s), \quad \mu, \nu = 1, 2, \ldots, n,$$

is invariant under translations ϵ of the parameter s, *i.e.* under transformations $s \to s' = s + \epsilon$.

Solution: Consider the given action integral with s replaced by $s + \epsilon$:

$$
\begin{aligned}
\int ds [L(x, \dot{x})]_{s \to s+\epsilon} &= \int ds g_{\mu\nu}(x(s+\epsilon)) \left[\frac{d}{ds} x^\mu(s+\epsilon) \right] \left[\frac{d}{ds} x^\nu(s+\epsilon) \right] \\
&= \int ds g_{\mu\nu}(x(s) + \epsilon\dot{x}(s)) \frac{d}{ds}[x^\mu(s) + \epsilon\dot{x}^\mu(s)] \frac{d}{ds}[x^\nu(s) + \epsilon\dot{x}^\nu(s)] \\
&= \int ds \left[g_{\mu\nu}(x(s)) + \epsilon \frac{\partial g_{\mu\nu}}{\partial x^\lambda}\dot{x}^\lambda \right] \frac{d}{ds}[x^\mu(s) + \epsilon\dot{x}^\mu(s)] \frac{d}{ds}[x^\nu(s) + \epsilon\dot{x}^\nu(s)] \\
&= \int ds g_{\mu\nu}(x(s)) \frac{dx^\mu}{ds} \frac{dx^\nu}{ds} + \epsilon \int ds g_{\mu\nu}(x(s)) \frac{dx^\mu}{ds} \frac{d\dot{x}^\nu}{ds} \\
&\quad + \epsilon \int ds g_{\mu\nu}(x(s)) \frac{d\dot{x}^\mu}{ds} \frac{dx^\nu}{ds} + \epsilon \int ds \frac{\partial g_{\mu\nu}(x(s))}{\partial x^\lambda}\dot{x}^\lambda \frac{dx^\mu}{ds} \frac{dx^\nu}{ds} \\
&= \int ds g_{\mu\nu}(x(s))\dot{x}^\mu\dot{x}^\nu + \epsilon \int ds \frac{d}{ds}[g_{\mu\nu}(x)\dot{x}^\mu\dot{x}^\nu].
\end{aligned}
$$

We see that this is equal to the given action integral S provided

$$\delta S = \epsilon \int_{s_i}^{s_f} ds \frac{d\Lambda}{ds} = 0, \quad \Lambda = [g_{\mu\nu}(x)\dot{x}^\mu\dot{x}^\nu].$$

Thus the requirement is that this total derivative vanish. We recall that the Lagrangian is uniquely defined only up to a total derivative (or divergence). We see that the invariance here expresses a conservation law — the conservation of the quantity Λ (effectively the kinetic energy), with $\Lambda(s_f) = \Lambda(s_i) = $ const.

Chapter 7

Two-Body Central Forces

7.1 Introductory Remarks

The topic of central forces that we now come to is of overwhelming importance in physics and thus deserves a detailed study. The importance of the topic can immediately be appreciated by recalling that the Coulomb potential which plays a central role in electrodynamics and atomic physics, and Newton's gravitational potential which is of analogous significance in celestial physics, that these two basic potentials span the entire spectrum from microscopic to macroscopic phenomena. In the following we deal mostly with Newton's gravitational potential and thus the Kepler problem of planetary motion.* But it is wellknown that the Bohr model of the atom — roughly speaking the Kepler problem with discretized energy, which is in this form frequently called "old quantum theory" — has many analogies, particularly in a classical limit, and although the Bohr model is incorrect, numerous such inferences can also be derived (at least in a good approximation) from the correct quantum theory. Our treatment below therefore is also of considerable importance for an understanding of some crucial aspects of the distinction between the Bohr model and proper quantum mechanics of atomic physics, then of planetary motion, and of its relation to general relativity, the latter in view of the corrections it provides to the Newtonian treatment. The trajectories of classically moving particles will be seen to be predominantly the curves of conical sections, so that the latter play an important role. Most, if not all properties of these conical sections, which are of relevance here, will be derived in the context in which they appear. As a useful reference on the analytical geometry of conics we refer to the book of

---

*For an elementary derivation of Kepler's laws from conservation of energy and angular momentum and with a minimum of mathematics see E. Vogt [49].

Sommerville.[†] The predominant example we consider here is the calculation of the perihelion anomaly of the planet Mercury, the proper justification of which will be obtained in later chapters on general relativity.

7.2   Equations of Motion

We consider a conservative system of two masspoints of masses m_1, m_2 and position coordinates $\mathbf{r}_1, \mathbf{r}_2$. This system has 6 *degrees of freedom*, hence also 6 generalized coordinates, which we choose as the components of vectors $\mathbf{R}$ and $\mathbf{r}$ which are defined as follows:

$$\mathbf{R} = \frac{m_1\mathbf{r}_1 + m_2\mathbf{r}_2}{m_1 + m_2}, \quad \mathbf{r} = \mathbf{r}_2 - \mathbf{r}_1. \tag{7.1}$$

Here $\mathbf{R}$ is the position vector of the *centre of mass* and $\mathbf{r}$ is the vector from one particle to the other as shown in Fig. 7.1.

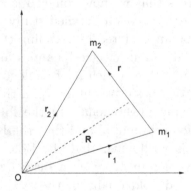

Fig. 7.1 Centre of mass vector coordinate $\mathbf{R}$ and and relative coordinate $\mathbf{r}$.

We let $\mathbf{r}_i'$ be the position vector of the point mass m_i with respect to the centre of mass, *i.e.*

$$\mathbf{r}_i' = \mathbf{r}_i - \mathbf{R}, \quad i = 1, 2.$$

Then

$$m_1\mathbf{r}_1' + m_2\mathbf{r}_2' = m_1(\mathbf{r}_1 - \mathbf{R}) + m_2(\mathbf{r}_2 - \mathbf{R}) \overset{(7.1)}{=} 0.$$

It follows that

$$\begin{aligned} \mathbf{r}_1' &= \mathbf{r}_1 - \mathbf{R} = \mathbf{r}_1 - \frac{m_1\mathbf{r}_1 + m_2\mathbf{r}_2}{m_1 + m_2} = -\frac{m_2}{m_1 + m_2}\mathbf{r}, \\ \mathbf{r}_2' &= \frac{m_1}{m_1 + m_2}\mathbf{r}. \end{aligned} \tag{7.2}$$

[†]D. M. Y. Sommerville [46].

We saw earlier (*cf.* Eq. (2.39)) that for a system of masspoints the overall kinetic energy T is equal to the kinetic energy of the centre of mass plus the kinetic energy T' of the motion around the centre of mass. Hence we have

$$T = \frac{1}{2}(m_1 + m_2)\dot{\mathbf{R}}^2 + \frac{1}{2}m_1\dot{\mathbf{r}}_1'^2 + \frac{1}{2}m_2\dot{\mathbf{r}}_2'^2 = \frac{1}{2}(m_1 + m_2)\dot{\mathbf{R}}^2 + \frac{1}{2}m\dot{\mathbf{r}}^2, \quad (7.3)$$

where

$$m = \frac{m_1 m_2}{m_1 + m_2}$$

is described as the "*reduced mass*". The entire Lagrangian is therefore

$$L(\mathbf{R}, \mathbf{r}, \dot{\mathbf{R}}, \dot{\mathbf{r}}) = \frac{1}{2}(m_1 + m_2)\dot{\mathbf{R}}^2 + \frac{1}{2}m\dot{\mathbf{r}}^2 - V(\mathbf{r}), \quad (7.4)$$

assuming that the potential V depends only on the relative separation of the masspoints. Since the system is to be conservative, the *Euler–Lagrange equations of motion* are

$$\frac{d}{dt}\left(\frac{\partial L}{\partial \dot{q}_j}\right) - \frac{\partial L}{\partial q_j} = 0, \quad \{q_j\} = \{\mathbf{R}, \mathbf{r}\}. \quad (7.5)$$

First we observe that L is independent of $\mathbf{R}$. Thus $\mathbf{R}$ is a *cyclic coordinate*. It follows that for $\{q_j\} = \{\mathbf{R}\}, j = 1, 2, 3$, we obtain

$$\frac{d}{dt}(\dot{\mathbf{R}}) = 0, \quad \dot{\mathbf{R}} = \text{const.} \quad (7.6)$$

This means the centre of mass moves uniformly or is at rest. The equations of motion in $\mathbf{r}$ therefore contain neither $\mathbf{R}$ nor $\dot{\mathbf{R}}$. The motion of the centre of mass is therefore in the following of no more interest. If we ignore in the Lagrangian (7.4) the constant term $(m_1 + m_2)\dot{\mathbf{R}}^2/2$, the remaining part represents exactly the Lagrange function, that one obtains if one considers the motion of a particle of mass m in the potential $V(\mathbf{r})$. The original two-body problem has therefore been reduced to an effective one-body problem. We observe in addition, that L is invariant under rotations about any fixed axis, since

$$V(\mathbf{r}) = V(|\mathbf{r}|) = V(r)$$

is a function depending solely on the radial distance from the force centre at $r = 0$. The invariance of $\dot{\mathbf{r}}^2$ under rotations, *e.g.* under the transformation

$$x' = x \cos\theta + y \sin\theta, \quad y' = -x \sin\theta + y \cos\theta, \quad z' = z$$

through a *fixed* angle θ around the z-axis, can readily be checked, since

$$\dot{x}' = \dot{x} \cos\theta + \dot{y} \sin\theta, \quad \dot{y}' = -\dot{x} \sin\theta + \dot{y} \cos\theta, \quad \dot{z}' = \dot{z},$$

so that

$$\dot{\mathbf{r}}'^2 = \dot{x}'^2 + \dot{y}'^2 + \dot{z}'^2 = \dot{x}^2 + \dot{y}^2 + \dot{z}^2 = \dot{\mathbf{r}}^2.$$

Evidently this result applies also for arbitrary rotations around the other two axes and hence in general (however beware, rotations through finite angles in general do not commute, i.e. depend on the order in which they are performed; for the verification of the invariance it suffices to consider infinitesimal rotations, and these always commute as we noted in Sec. 5.4.7).

Hence the Lagrange function L is invariant under rotations in a three-dimensional Euclidean space. This is equivalent to saying that L does not contain an explicit dependence on angles, and consequently

$$\text{angular momentum} \quad \mathbf{L} = \text{const.}$$

Since $\mathbf{L} = \mathbf{r} \times \mathbf{p}$ is perpendicular to $\mathbf{r}$, it follows that the vector $\mathbf{r}$ is always perpendicular to this constant vector $\mathbf{L}$. This means the vector $\mathbf{r}$ lies in a plane. We can therefore switch to polar coordinates r, θ in a plane with the *pole* at the origin. Then the Lagrange function becomes

$$L(r, \theta, \dot{r}, \dot{\theta}) = \frac{1}{2}m(\dot{r}^2 + r^2\dot{\theta}^2) - V(r). \tag{7.7}$$

We therefore obtain two Lagrange equations of motion. The equation in the angular variable θ is

$$\frac{d}{dt}\left(\frac{\partial L}{\partial \dot{\theta}}\right) = \frac{d}{dt}p_\theta = 0, \quad i.e. \quad \frac{d}{dt}(mr^2\dot{\theta}) = 0, \quad mr^2\dot{\theta} = l = \text{const.} \tag{7.8}$$

Comparing this equation with the *transversal acceleration* derived in Example 5.1 (the velocity components were also derived there from first principles), we see that the present equation says that the particle has at no time a transversal acceleration. The Lagrange equation in the polar coordinate r is

$$\frac{d}{dt}(m\dot{r}) - mr\dot{\theta}^2 + \frac{\partial V(r)}{\partial r} = 0. \tag{7.9}$$

Again referring to the results of Example 5.1, we see that the force $f = -\partial V/\partial r$ acts in the radial direction. Here we restricted ourselves to direction-independent and hence also angle-independent potentials — hence the term "*central forces*". In general, e.g. in the case of the gravitational field of the Earth, the potential V can depend on angles, e.g. in the general form (*cf.* Example 2.7)

$$V(r, \phi, \theta) = \frac{\mu}{r}\sum_{l=0}^{\infty}\sum_{m=0}^{\infty}(c_{lm}\cos m\theta + d_{lm}\sin m\theta)P_l^m(\sin \phi), \tag{7.10}$$

where $\mu = GM_{\text{Earth}} = 3986.03$ km^{3}s^{-2}, $r_0 = 6378.16$ km the largest equatorial radius of the Earth, θ the longitude angle measured from Greenwich to the East and ϕ the angular altitude as shown in Fig. 7.2 (*cf.* also Fig. 10.18 where $\theta \leftrightarrow \phi$). The functions $P_l^m(x)$ are the *associated Legendre polynomials* and $P_l(x)$ the *Legendre polynomials* given by

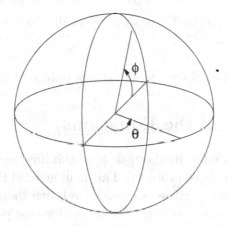

Fig. 7.2 Longitude angle θ and and latitude angle ϕ.

$$P_l^m(x) = (1 - x^2)^{m/2} \frac{d^m}{dx^m} P_l(x), \quad \text{and} \quad P_l(x) = \frac{1}{2^l l!} \frac{d^l}{dx^l}(x^2 - 1)^l. \quad (7.11)$$

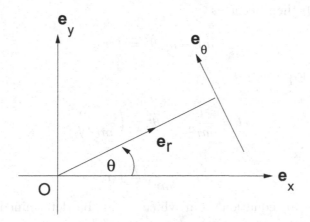

Fig. 7.3 The directions of unit vectors.

We can obtain the *transversal components* of the *velocity* (above referred to Example 5.1) also as follows. We have, *cf.* Fig. 7.3, $\mathbf{r} = (x, y)$, $x = r \cos \theta$, $y =$

$r \sin \theta$, so that

$$
\begin{aligned}
\dot{x} &= \dot{r} \cos \theta - r\dot{\theta} \sin \theta, \quad \dot{y} = \dot{r} \sin \theta + r\dot{\theta} \cos \theta, \quad \dot{\mathbf{r}} = \dot{x}\mathbf{e}_x + \dot{y}\mathbf{e}_y, \\
\dot{\mathbf{r}} \cdot \mathbf{e}_r &= \dot{x}(\mathbf{e}_x \cdot \mathbf{e}_r) + \dot{y}(\mathbf{e}_y \cdot \mathbf{e}_r) = \dot{x} \cos \theta + \dot{y} \sin \theta = \dot{r}, \\
\dot{\mathbf{r}} \cdot \mathbf{e}_\theta &= \dot{x}(\mathbf{e}_x \cdot \mathbf{e}_\theta) + \dot{y}(\mathbf{e}_y \cdot \mathbf{e}_\theta) \\
&= \dot{x} \cos(\pi/2 + \theta) + \dot{y} \cos \theta = -\dot{x} \sin \theta + \dot{y} \cos \theta \\
&= -\sin \theta(\dot{r} \cos \theta - r\dot{\theta} \sin \theta) + \cos \theta(\dot{r} \sin \theta + r\dot{\theta} \cos \theta) \\
&= r\dot{\theta}. \tag{7.12}
\end{aligned}
$$

These expressions are referred to at various points of this text.

7.3   Solution of the Equations

We are now confronted with the task of obtaining the complete solution of the two-body central force problem. This is in general the problem to obtain either the orbit of the particle (*i.e.* of its relative motion) if the force (*i.e.* potential) is given, or to determine the law of force or potential if the orbit is given. We have two equations with maximally two derivatives of the variables r and θ. The complete solution therefore requires four integrations. But we begin with some general aspects. We set

$$
-\frac{\partial V}{\partial r} \equiv f(r). \tag{7.13}
$$

Equation (7.9) then becomes

$$
m\ddot{r} - mr\dot{\theta}^2 = f(r), \tag{7.14}
$$

and with l of Eq. (7.8):

$$
\dot{\theta} = \frac{l}{mr^2}, \quad \dot{\theta}^2 = \left(\frac{l}{mr^2}\right)^2, \tag{7.15}
$$

$$
m\ddot{r} - \frac{l^2}{mr^3} = f(r). \tag{7.16}
$$

We thus have an equation from which r can be determined. One of the four integration constants is the constant *angular momentum l* contained in Eq. (7.16). A second constant is the *total energy E*, *i.e.*

$$
\text{const.} = E = \frac{1}{2}m(\dot{r}^2 + r^2\dot{\theta}^2) + V(r) = \frac{1}{2}m\dot{r}^2 + \frac{1}{2}\frac{l^2}{mr^2} + V(r) \tag{7.17}
$$

with Eq. (7.15). The other two integrations can be performed at least formally as follows. Solving Eq. (7.17) for $\dot{r}$, we obtain

$$\dot{r} = \sqrt{\frac{2}{m}\left(E - V - \frac{l^2}{2mr^2}\right)},$$

or

$$t - t_0 = \int_{r_0}^{r} \frac{dr}{\sqrt{(2/m)(E - V - l^2/2mr^2)}} \equiv t(r), \qquad (7.18)$$

where $r = r_0$ at time $t = t_0$. Thus r_0 is the third constant of integration. We obtain the fourth by integrating Eq. (7.15):

$$\theta = l \int_{t_0}^{t} \frac{dt}{m[r(t)]^2} + \theta_0, \qquad (7.19)$$

where $\theta = \theta_0$ at time $t = t_0$. Here $r = r(t)$ is the expression in terms of time obtained from Eq. (7.18). The four constants of integration of the problem are therefore the total energy E, the (total) angular momentum l, and the initial values of the two coordinates, i.e. r_0, θ_0. We could have replaced E and l by the initial values of $\dot{r}$ and $\dot{\theta}$. However, in *quantum mechanics* such constants become meaningless, since there one has only probability statements about the position and momentum or velocity of a particle, whereas E and l retain their significance as total energy and angular momentum. Thus the correspondence with quantum mechanics here becomes particularly conspicuous in the description of a system in terms of its energy and angular momentum.

We consider first the motion of the particle in a potential in a more general way. The total energy given by Eq. (7.17) can be written

$$E = V' + \frac{1}{2}m\dot{r}^2, \qquad (7.20)$$

where the fictitious potential V' is the sum of the actual potential V and the *centrifugal potential* $l^2/2mr^2$, i.e.

$$V' = V + \frac{1}{2}\frac{l^2}{mr^2}. \qquad (7.21)$$

The potential V' consisting of these two contributions can conveniently be called the "*effective potential*" — particularly so, since in the relativistic case considered in Chapter 16 we shall encounter a corresponding potential (*cf.* Eq. (16.83)), there also called effective potential. A particularly instructive example is the case of the *attractive Coulomb potential*, i.e.

$$V = -\frac{k}{r}, \quad k > 0. \qquad (7.22)$$

This potential together with the centrifugal part and their sum is shown schematically in Fig. 7.4.

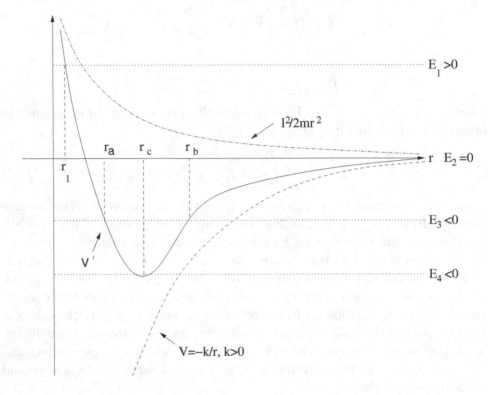

Fig. 7.4 Coulomb and centrifugal potentials and their sum.

We consider a particle with energy $E_1 > 0$. When $E_1 = V'$, i.e. at $r = r_1$ as in Fig. 7.4, its velocity $\dot{r} = 0$. When $r < r_1$, we have $V' > E_1$ and hence $m\dot{r}^2/2 < 0$, i.e. $\dot{r}$ is imaginary. Since this is physically not possible, the particle reverses its direction at the distance $r = r_1$ from the force centre and is reflected by the *centrifugal barrier* $l^2/2mr^2$. Thus the particle with energy E_1 comes from infinity, and is reflected at $r = r_1$ and escapes back to infinity. One can show that its path or orbit is a *hyperbola* (see below). At energy $E_2 = 0$, i.e. for $E_2 = \lim_{r\to\infty}(l^2/2mr^2)$ we have a similar situation, but we shall see that the particle's orbit then has the shape of a *parabola*. Consider now the case of energy $E_3 < 0$ indicated in Fig. 7.4. In the domain $r < r_a$ we have again $V' > E_3$ and $\dot{r}^2 < 0$. The same applies for the domain $r > r_b$, since here also $V' > E_3$. Thus the particle with energy E_3 is trapped in the domain $r_a < r < r_b$ and travels or oscillates therein between r_a and r_b back and forth. The path or orbit the particle executes thereby is not necessarily closed, but in the case of the Coulomb potential under consideration here we shall see that the orbit is an *ellipse*, *i.e.* a closed path. A case like this is referred to as a case of *"binding"*, since the total energy is negative and the two particles of the 2-body problem are so-to-speak bound to one another.

In the case of the energy $E_4 < 0$, as indicated in Fig. 7.4, we have $r = r_c$, *i.e.* $r = $ const., and this is the one and only allowed distance of the particle from the force centre. Its orbit is therefore a *circle*. Other potentials lead to different situations; thus, for instance, the strongly singular potential

$$V(r) = -\frac{a}{r^3}, \quad a > 0,$$

leads to a potential V' of a very different shape as indicated in Fig. 7.5.

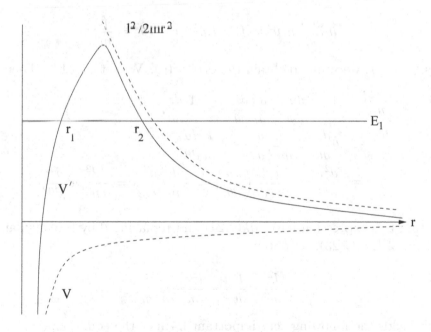

Fig. 7.5 Singular and centrifugal potentials and their sum V'.

In this case it is the domain $r_1 < r < r_2$, in which for energy $E = E_1$ we have $V' > E_1$, and hence $m\dot{r}^2/2 < 0$. This means, the motion of the particle is either restricted, *i.e.* bound, and the particle oscillates in the domain $0 < r < r_1$ (and can be sucked in if the potential is attractive), or the motion is unbound in the domain $r > r_2$, and the particle coming from $r > r_2$ is repelled and escapes to infinity.

In these considerations of the motion of the particle (replacing the relative motion of the 2-particle problem) we used only the terms of energy, potential, and angular momentum of the particle, not, however, its position coordinates and velocities. It is therefore not astonishing that these considerations retain their validity in the corresponding quantum mechanical formulation. We emphasize that one has binding when $E < 0$ and the potential is attractive, i.e. the force is directed towards its centre and $\lim_{r\to\infty} V(r) = 0$.

7.4   Differential Equation of the Orbit

We begin with Eqs. (7.8) and (7.14), i.e.

$$l = mr^2\dot{\theta}, \tag{7.23}$$

(so that $r = r[\theta(t)]$) and

$$m\ddot{r} - mr\dot{\theta}^2 = f, \quad i.e. \quad \ddot{r} - r\dot{\theta}^2 = \frac{f}{m}. \tag{7.24}$$

Thus, given f, we want to obtain the orbit $r(\theta)$. We set $u = 1/r$, Then

$$
\begin{aligned}
u &= \frac{1}{r}, \quad \frac{du}{d\theta} = \frac{du}{dr}\frac{dr}{d\theta} = -\frac{1}{r^2}\frac{dr}{d\theta}, \\
\frac{dr}{dt} &= \dot{\theta}\frac{dr}{d\theta} = \frac{l}{mr^2}\frac{dr}{d\theta} = -\frac{l}{m}\frac{du}{d\theta}, \\
\ddot{r} &\equiv \frac{d^2r}{dt^2} = -\frac{l}{m}\frac{d}{dt}\left(\frac{du}{d\theta}\right) = -\frac{l}{m}\dot{\theta}\frac{d^2u}{d\theta^2} = -\frac{l^2}{m^2}u^2\frac{d^2u}{d\theta^2}.
\end{aligned}
$$

Inserting this expression into Eq. (7.24) and replacing $\dot{\theta}$ by the angular momentum of Eq. (7.23), we obtain

$$-\frac{l^2}{m^2}u^2\frac{d^2u}{d\theta^2} - \frac{l^2u^3}{m^2} = \frac{f}{m},$$

which yields the following very important form of the equation:

$$\frac{d^2u}{d\theta^2} + u = -\frac{mf}{l^2}\frac{1}{u^2}. \tag{7.25}$$

This is the differential equation of the orbit describing the relative motion of the two particles. Here f is a function of r or, alternatively, of u. Integrating the equation we obtain the polar equation of the orbit. On the other hand, if the equation of the orbit is given, it is necessary to convert this first into its polar form, and to eliminate from this θ with the help of Eq. (7.25), in order to obtain the force f as a function of r.

A different question, which presents itself, is: How does one obtain the velocity v of the particle along its trajectory? We use the infinitesimal geometry depicted in Fig. 7.6. In Fig. 7.6 the element δs is the piece PQ of the trajectory, and p is the length of the perpendicular from the pole O to the tangent at P. The area δA of the triangle OPQ is given by

$$\delta A = \frac{1}{2}(r + \delta r)r\delta\theta = \frac{1}{2}r^2\delta\theta, \quad \text{but also} \quad \delta A = \frac{1}{2}p\delta s. \tag{7.26}$$

Hence

$$\frac{\delta A}{\delta t} = \frac{1}{2}r^2\frac{\delta\theta}{\delta t} = \frac{1}{2}p\frac{\delta s}{\delta t} = \frac{1}{2}pv, \quad p = ON, \tag{7.27}$$

so that (*cf.* Eq. (7.23))

$$\frac{1}{2}\frac{l}{m} = \frac{1}{2}pv, \quad v = \frac{ds}{dt}, \quad i.e. \quad pv = \frac{l}{m} = \text{const.} \tag{7.28}$$

Hence velocity $v \propto 1/p$. But now (*cf.* Fig. 7.6) with ϕ the angle between $\mathbf{r}$ and the tangent,

$$p = r\sin\phi, \quad \frac{1}{p^2} = \frac{u^2}{\sin^2\phi} = u^2(1 + \cot^2\phi). \tag{7.29}$$

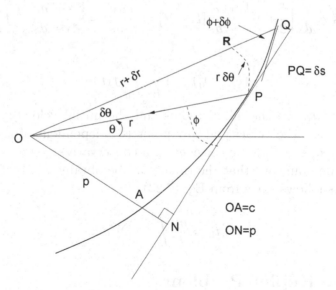

Fig. 7.6 The perpendicular p to the tangent at P.

Since the angle OQP differs only minimally from ϕ, we obtain from triangle RQP:

$$\tan\phi = \left(\frac{r\delta\theta}{\delta r}\right)_{\delta r,\delta\theta\to 0} = r\frac{d\theta}{dr} = -u\frac{d\theta}{du}, \quad u^2\cot^2\phi = \left(\frac{du}{d\theta}\right)^2, \tag{7.30}$$

because $u = 1/r$. Analogously we have

$$\sin\phi = r\frac{d\theta}{ds}, \quad \cos\phi = \frac{dr}{ds}.$$

It therefore follows from Eqs. (7.28), (7.29) that the velocity v is given by

$$v^2 = \frac{l^2}{m^2}\frac{1}{p^2} = \frac{l^2}{m^2}\left[u^2 + \left(\frac{du}{d\theta}\right)^2\right]. \tag{7.31}$$

This equation permits us to derive the velocity v from the trajectory $u(\theta)$ or $r(\theta)$. However, the velocity can also be obtained without a knowledge of the orbit from the law of force. We see this as follows. We have for the *tangential acceleration*[‡]

$$\frac{d^2 s}{dt^2} = \frac{d}{dt}\frac{ds}{dt} = \frac{d}{dt}(v) = \frac{dv}{ds}\frac{ds}{dt} = \frac{dv}{ds}v = \frac{1}{2}\frac{dv^2}{ds}. \tag{7.32}$$

It follows that (note that $v^2 = \dot{r}^2 + r^2\dot{\theta}^2$, as demonstrated later in Example 11.1)

$$m\ddot{s} = \frac{d}{ds}\left[\frac{1}{2}mv^2\right] = \frac{d}{ds}[E - V] = -\frac{dV}{ds} = -\frac{\partial V}{\partial r}\frac{dr}{ds} = f\frac{dr}{ds},$$

i.e.

$$\frac{1}{2}(v^2 - v_0^2) = +\frac{1}{m}\int_{r_0}^{r} f(r)dr, \tag{7.33}$$

where r_0 and v_0 are the initial radius and the initial velocity respectively. We observe that the velocity at any point is independent of the trajectory — as expected, since we are considering a conservative system.

Finally we point out that the time t can be obtained only if the orbit is known. This follows since from Eq. (7.23)

$$t - t_0 = \frac{m}{l}\int_{\theta_0}^{\theta}[r(\theta)]^2 d\theta. \tag{7.34}$$

7.5   The Kepler Problem

A particularly important case of a central force is that of the force proportional to the inverse square of the separation from a fixed point. This case is important in view of its historical development in connection with the

---

[‡]This should not be confused with the transversal acceleration that we considered earlier in Example 5.1; the latter is orthogonal to the radial vector **r**.

motion of planets and is there known as the *Kepler problem*, it is important
in view of its relation to the quantum mechanics of hydrogen-like atoms as
pointed out earlier, and it is important in connection with Einstein's general
relativity, there in view of the correction the latter provides to the Newtonian
calculation of the perihelion anomaly (see later).

We begin with *Newton's gravitational law* which postulates the force acting between massive objects to be attractive and here conveniently written
(*cf.* Sec. 2.6)[§]

$$f = -\frac{m\mu}{r^2} = -m\mu u^2, \quad \mu = \text{const.} \tag{7.35}$$

Equation (7.25) thus implies in this case the following equation of the orbit:

$$\frac{d^2u}{d\theta^2} + u = \frac{m^2\mu}{l^2}. \tag{7.36}$$

Since $l = mr^2\dot\theta$, the dependence on m drops out. Setting

$$U \equiv u - \frac{m^2\mu}{l^2} = \frac{1}{r} - \frac{m^2\mu}{l^2}, \tag{7.37}$$

the equation assumes the simple form

$$\frac{d^2U}{d\theta^2} + U = 0. \tag{7.38}$$

The solution of this equation is, with A and α constants:

$$\frac{1}{r} = u = \frac{m^2\mu}{l^2} + A\cos(\theta + \alpha) \equiv \frac{m^2\mu}{l^2}[1 + e\cos(\theta + \alpha)], \tag{7.39}$$

where

$$e = \frac{l^2 A}{m^2\mu}. \tag{7.40}$$

We next recapitulate some geometry of *conical sections*, which are required
for an understanding of the implications of this equation. Specifically we
deal with the *polar equation of a conical section*. We refer in this connection
to Fig. 7.7 which summarizes also various names which have a long history
from the investigation of planetary motion.

A *conical section* (not necessarily an ellipse) is defined by the following
relation, where SP and PM are distances as explained in Fig. 7.7,

$$SP = e.PM, \quad i.e. \quad r = e[SX - r\cos(\theta - \alpha')]. \tag{7.41}$$

[§]Note that m is the reduced mass defined by Eq. (7.3). Thus in the case of a single mass m
acted upon by the force f we have $\mu = MG$ (in units with $c = 1$), but actually MG/c^2, where M
is the mass of the sun and G Newton's gravitational constant. In the case of the proper 2-particle
system we set $\mu = G(m_1 + m_2)$ as in Example 7.5, so that $f = -Gm_1 m_2/r^2$.

The locus of the point P is then the ellipse (or other conical section). It follows that
$$r[1 + e\cos(\theta - \alpha')] = e.SX = \text{const.} \equiv l'.$$

The parameter e is the *eccentricity*[¶] and the distance $l' \equiv SR$ (*cf.* Fig. 7.7) is called the "*semi-latus rectum*". From the definition (7.41) numerous properties of the geometry of conical sections can be deduced.[‖] We obtain therefore the following equation

$$\frac{l'}{r} = 1 + e\cos(\theta - \alpha'). \tag{7.42}$$

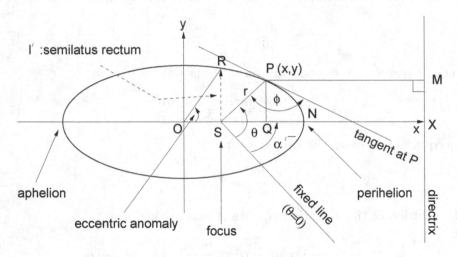

Fig. 7.7 The elliptic orbit.

[¶]In the case of the Earth $e \simeq 1/60$.

[‖]Referring to Fig. 7.7 we set $SN = k, QN = \xi, PQ = y$. Then from Fig. 7.7 and Eq. (7.41) we obtain (since with Eq. (7.41) we have also $SN = e.NX$)

$$SP/e = PM = QX = QN + NX = \xi + k/e, \quad SP^2 = SQ^2 + QP^2, \quad e^2(\xi + k/e)^2 = (k - \xi)^2 + y^2.$$

Multiplying out and completing the square the equation becomes

$$\xi^2(e^2 - 1) + 2k\xi(e - 1) = y^2, \quad [\xi - k/(1 - e)]^2 + y^2/(1 - e^2) = k^2/(1 - e)^2.$$

With $x := \xi - k/(1 - e), a^2 := k^2/(1 - e)^2, b^2 := k^2(1 + e)/(1 - e)$, and therefore $b^2/a^2 = (1 - e^2)$, the equation becomes the standard equation of an ellipse,

$$\frac{x^2}{a^2} + \frac{y^2}{b^2} = 1.$$

From these relations one deduces now that $OS = ae, l' = b^2/a, OX = a/e$. Corresponding results are obtained for the hyperbola,

$$\frac{x^2}{a^2} - \frac{y^2}{b^2} = 1.$$

by a change of sign of b^2. This hyperbola will be required in the treatment of scattering in Sec. 7.12.

Equation (7.42) is the *polar equation of a conical section*: For $e < 1$ the curve is an *ellipse*, for $e = 1$ the curve is a *parabola*, for $e > 1$ a *hyperbola*, and for $e = 0$ a *circle*. We observe now that Eq. (7.39) is the equation of such a conical section. In order to obtain its value of the eccentricity e, it is necessary to obtain the values of both l' and e, *i.e.* l' and A. To this end we return to Eq. (7.31),

$$v^2 = \frac{l^2}{m^2}\left[u^2 + \left(\frac{du}{d\theta}\right)^2\right]. \tag{7.43}$$

However, we also have (since $V = -m\mu/r$, $f = -\partial V/\partial r = -m\mu/r^2$)

$$E = \frac{1}{2}mv^2 + V(r) = \frac{1}{2}mv^2 - \frac{m\mu}{r}. \tag{7.44}$$

Extracting from this equation v^2 and equating this to v^2 of Eq. (7.43), we obtain

$$v^2 = \frac{2}{m}(E + m\mu u) = \frac{l^2}{m^2}\left[u^2 + \left(\frac{du}{d\theta}\right)^2\right],$$

so that

$$\left(\frac{du}{d\theta}\right)^2 = \frac{2Em}{l^2} + \frac{2\mu m^2}{l^2}u - u^2 = -\left(u - \frac{\mu m^2}{l^2}\right)^2 + \frac{2Em}{l^2} + \frac{\mu^2 m^4}{l^4}. \tag{7.45}$$

From this we obtain

$$\frac{du}{d\theta} \stackrel{(+)}{=} \sqrt{a^2 - \left(u - \frac{\mu m^2}{l^2}\right)^2}, \quad a^2 \equiv \frac{1}{l^4}(2Eml^2 + \mu^2 m^4). \tag{7.46}$$

With the lower sign as indicated we have

$$\int d\theta = -\int \frac{du}{\sqrt{a^2 - (u - \mu m^2/l^2)^2}},$$

and hence

$$\theta - \alpha' = \cos^{-1}\left(\frac{u - \mu m^2/l^2}{a}\right), \quad \alpha' = \text{const.} \tag{7.47}$$

It follows that

$$u - \frac{\mu m^2}{l^2} = a\cos(\theta - \alpha') \quad \text{and} \quad \frac{l^2}{\mu m^2}\frac{1}{r} = 1 + \frac{l^2}{\mu m^2}\sqrt{\frac{2Eml^2 + m^4\mu^2}{l^4}}\cos(\theta - \alpha'),$$

or

$$\frac{l^2}{\mu m^2}\frac{1}{r} = 1 + \frac{1}{\mu}\sqrt{\frac{2El^2}{m^3} + \mu^2}\cos(\theta - \alpha'). \tag{7.48}$$

Comparison with Eq. (7.42) determines the *semi-latus rectum* l' and the eccentricity e as

$$l' = \frac{l^2}{\mu m^2} \quad \text{and} \quad e = \sqrt{1 + \frac{2El^2}{\mu^2 m^3}} \overset{(7.40)}{=} \frac{l^2 A}{m^2 \mu}. \tag{7.49}$$

We thus arrive at the following conclusions:

$$
\begin{array}{llll}
\text{If} & -\mu^2 m^3/2l^2 < E < 0, & \text{then} & 0 < e < 1 & \text{(ellipse)}, \\
\text{if} & E > 0, & \text{then} & e > 1 & \text{(hyperbola)}, \\
\text{if} & E = 0, & \text{then} & e = 1 & \text{(parabola)}, \\
\text{if} & E = -\mu^2 m^3/2l^2, & \text{then} & e = 0 & \text{(circle)}, \\
\text{if} & E < -\mu^2 m^3/2l^2, & \text{then} & \text{not possible} & \text{(velocity imaginary)}.
\end{array}
\tag{7.50}
$$

These results summarize *Kepler's first law*: The orbits of the planets are conics with the sun at one focus.

It is instructive to obtain the same result in another way, which demonstrates how the shape of the orbit (ellipse, parabola, etc.) is determined by the initial conditions. We saw above that (*cf.* Eq. (7.30))

$$\tan\phi = \frac{rd\theta}{dr} = -u\frac{d\theta}{du}. \tag{7.51}$$

Differentiating Eq. (7.39) we obtain

$$\frac{du}{d\theta} = -A\sin(\theta + \alpha), \quad A = \frac{m^2\mu}{l^2}e. \tag{7.52}$$

For $\phi = \beta$ and $u = 1/c$ at $\theta = 0$, we deduce from Eq. (7.51):

$$\tan\beta = \frac{1}{cA\sin\alpha}. \tag{7.53}$$

For simplicity we choose $\beta = 90°$ which means that in Fig. 7.7 the point P is at the perihelion N. Then

$$cA\sin\alpha = 0, \quad \alpha = 0.$$

From Eq. (7.39) follows moreover (with $\theta = 0, r = c$):

$$\frac{1}{c} = \frac{m^2\mu}{l^2} + A\cos\alpha = A + \frac{m^2\mu}{l^2}, \quad \text{since } \alpha = 0,$$

so that

$$A = \frac{1}{c} - \frac{m^2\mu}{l^2}. \tag{7.54}$$

If, however, as assumed, the initial direction of motion is (at $t = 0, \theta = 0$) perpendicular to the radial vector and with velocity v_0, then it follows from Eq. (7.29), *i.e.*

$$p = r \sin \phi, \quad \text{that} \quad p = c,$$

where p is the length of the perpendicular SN in Fig. 7.7. From Eq. (7.28) we know that $pv = l/m$, so that ($v = v_0$ at time $t = 0, \theta = 0, r = c$)

$$cv_0 = \frac{l}{m}.$$

Inserting this value of l/m in Eq. (7.54), we obtain

$$A = \frac{1}{c} - \frac{\mu}{c^2 v_0^2}. \tag{7.55}$$

With Eq. (7.40) we obtain therefore

$$e = \frac{l^2 A}{m^2 \mu} = \frac{c^2 v_0^2}{\mu} A = \frac{v_0^2 c}{\mu} - 1,$$

i.e.

$$e = \frac{v_0^2 c}{\mu} - 1. \tag{7.56}$$

This relation expresses the eccentricity of the conical orbit in terms of the initial velocity. Thus if $v_0^2 c/\mu > 2$, *i.e.* $v_0^2 > 2\mu/c$, then $e > 1$ and the orbit is a hyperbola. If on the other hand, $v_0^2 = 2\mu/c$, *i.e.* $e = 1$, the orbit is a parabola. For a small initial velocity with $v_0^2 < 2\mu/c$, we have $e < 1$ and the orbit is an ellipse. For a circular orbit, *i.e.* $e = 0$, the initial velocity is given by $v_0^2 = \mu/c$. The velocity $v_0 = \sqrt{\mu/c}$ is described[**] as the *first cosmic velocity*. The first cosmic velocity of a terrestrial satellite is 7.9 km s^{-1}. This velocity is just sufficient for a satellite to encircle a spherical celestial body immediately above its surface along a great circle. In order to escape the gravitational field of the Earth, one requires at least a parabolic initial velocity. This velocity is described as the *second intitial cosmic velocity* (*cf.* Example 2.3 for the *velocity of escape* from the Earth) and is given by $\sqrt{2} \times 7.9$ km s$^{-1} = 11.2$ km s^{-1}. With the *third cosmic velocity*, 16.7 km s^{-1}, one can escape from the solar system (ignoring atmospheric friction effects).

We can explore the physical meaning of the above conditions for the shape of the orbit. For this purpose we refer to Fig. 7.6. There the piece PQ is the element δs of the orbit. The force acts in the direction towards

---

[**]See *e.g.* R. H. Giese [18] or H. Bucerius and M. Schneider [7].

the pole O. The work performed by the force $f = -\partial V/\partial r$ in moving the particle from position P to Q is (with $\cos(\phi + \delta\phi) \simeq \cos\phi$)

$$f\delta r = f\frac{\delta r}{\delta s}\delta s = f\cos\phi\,\delta s.$$

If the particle is initially at the $r = c$, its gain in potential energy is

$$-\int_c^r f\,dr = +\int_c^r \frac{m\mu}{r^2}dr = \left[-\frac{m\mu}{r}\right]_c^r = \frac{m\mu}{c} - \frac{m\mu}{r}.$$

If the initial velocity is v_0, we obtain from the law of conservation of energy,

$$\frac{1}{2}mv_0^2 - \frac{m\mu}{c} = \frac{1}{2}mv^2 - \frac{m\mu}{r},$$

the equation

$$\frac{1}{2}m(v^2 - v_0^2) = -m\mu\left(\frac{1}{c} - \frac{1}{r}\right). \tag{7.57}$$

If the particle travels from distance $r = c$ to $r = \infty$, we obtain for its velocity at infinity:

$$\frac{1}{2}v_0^2 - \frac{1}{2}v_\infty^2 = \frac{\mu}{c}, \quad i.e. \quad \frac{1}{2}v_\infty^2 = \frac{1}{2}v_0^2 - \frac{\mu}{c}.$$

But $v_\infty^2/2$ is necessarily positive, hence

$$v_0^2 > \frac{2\mu}{c}.$$

If $v_0^2 > 2\mu/c$, the particle can travel all the way to $r = \infty$, and its orbit is a hyperbola. If $v_0^2 = 2\mu/c$, the particle can just barely reach $r = \infty$. When $v_0^2 < 2\mu/c$, the particle is unable to reach infinity, its orbit is an ellipse. One should note that the latter considerations do not require anymore that $\phi = \pi/2$ at $r = c$.

7.6   Tangential Equations of Orbits

We observed above that the *pole O* of the polar coordinates does not necessarily sit at the centre of a curve, as in the case of the ellipse. In fact we can put the pole at many other positions and then obtain different types of equations. It is very instructive to realize this in the case of a few typical examples. It is customary and convenient to express these equations in terms of the polar coordinate r and the length p of the perpendicular to the orbit at radial coordinate r. The resulting equations are then known as *tangential* or (p, r)-*equations*.[††] We see from Fig. 7.8 that

$$p = r \sin \phi \quad \text{with} \quad \tan\phi \overset{(7.30)}{=} r\frac{d\theta}{dr}, \quad r = r(\theta). \tag{7.58}$$

Here the second equation is that of Eq. (7.51). Eliminating θ and ϕ from these equations, we obtain the desired equations in terms of p and r.

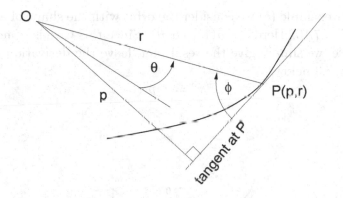

Fig. 7.8 The tangent to the orbit.

We consider a few cases. The first case (a) is that of a circular orbit of radius a with pole at the centre as shown in Fig. 7.9(a). Clearly the (p, r)-equation in this case is simply

$$r = p. \tag{7.59a}$$

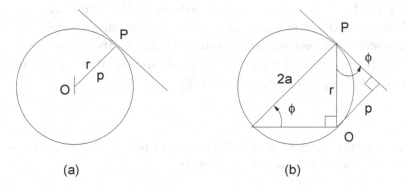

(a)   (b)

Fig. 7.9 Circular orbit with pole (a) at centre, (b) on the circle.

[††]The best reference the author knows on this topic is the book of A. S. Ramsey [41], Chapter XII.

In our second case (b) we put the pole on the circumference of the circle as shown in Fig. 7.9(b). In this case we have

$$r = 2a \sin \phi, \quad \text{or} \quad p = r \sin \phi.$$

Eliminating ϕ we obtain the required equation

$$r^2 = 2ap. \tag{7.59b}$$

As a further example (c) we consider the orbit with the shape of a *cardioid* as shown in Fig. 7.10. Here $r = a(1 + \cos \theta)$. Since this case is of no particular interest here, we simply give the result and leave the derivation of the force to Example 7.3 below:

$$r^3 = 2ap^2. \tag{7.60a}$$

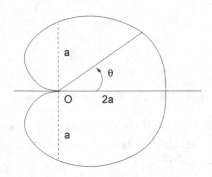

Fig. 7.10 The cardioid.

Clearly of more interest to us, in the present context, are the tangential versions of the equations of conical sections. We quote here a few such equations from the literature[‡‡] and derive below only the last case, which is of particular interest for our purposes here, *i.e.* the polar equation of an ellipse with the pole at its centre. Thus (d) the equation of a *parabola* with pole at the focus $(2a/r = 1 + \cos \theta)$ is

$$p^2 = ar. \tag{7.60b}$$

(e) The tangential equation of the *ellipse with pole at the focus* and semi-axes of lengths a and b, with $a > b$, is given by

$$\frac{b^2}{p^2} = \frac{2a}{r} - 1, \tag{7.61}$$

‡‡A. S. Ramsey [41], pp. 160 − 162.

whereas (f) the equation of the *ellipse with pole at the centre* is given by

$$r^2 + \frac{a^2 b^2}{p^2} = a^2 + b^2. \tag{7.62}$$

We explore this last case of the ellipse with pole at the centre in more detail. In Examples 7.1 we derive Eq. (7.62) and in Example 7.2 we verify (not by using the Cartesian but the tangential formulas here) that in terms of the polar coordinates r and θ *referred to the centre of the ellipse as pole* the equation is given by

$$\frac{1}{r^2} = \frac{\cos^2 \theta}{a^2} + \frac{\sin^2 \theta}{b^2}. \tag{7.63a}$$

With an angular shift of $\pi/2$ to $\phi = \theta + \pi/2$ the equation becomes

$$\frac{1}{r^2} = \frac{\sin^2 \phi}{a^2} + \frac{\cos^2 \phi}{b^2}. \tag{7.63b}$$

The result (7.63a) will be required in Example 10.3. For better clarity the ellipse (with $a > b$) and the coordinates are shown in Fig. 7.11.

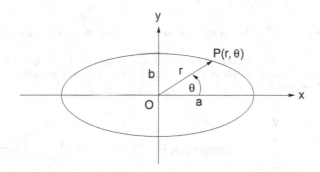

Fig. 7.11 The ellipse with polar coordinates r, θ and pole at the centre.

Example 7.1: Tangential polar equation of ellipse with pole at centre

Derive the tangential polar equation of an ellipse with pole at the centre, *i.e.* Eq. (7.62), $r^2 = a^2 + b^2 - a^2 b^2/p^2$.

Solution: In the first place we require the parametrization of an ellipse in terms of its *eccentric angle* ϑ, also described as *eccentricity*. Consider a point P on the ellipse $x^2/a^2 + y^2/b^2 = 1, a > b$ with x-coordinate parametrized in terms of ϑ as $x = a \cos \vartheta$. The y-coordinate of the point P then follows from the equation as $y = b \sin \vartheta$. Figure 7.12 shows the ellipse surrounded by the *auxiliary circle* of radius a. Since there

$$\frac{QN}{PN = y} = \frac{a \sin \vartheta}{b \sin \vartheta} = \frac{a}{b},$$

the ordinate of every point on the ellipse bears a constant ratio to the ordinate of the corresponding point on the auxiliary circle. There are several ways one can introduce the concept of *conjugate diameters* PP', DD' and *conjugate points* P, D on an ellipse. For our purposes here it suffices to

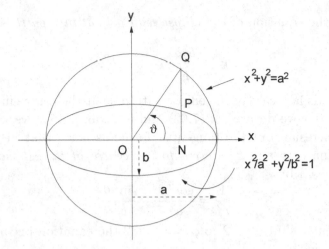

Fig. 7.12 The ellipse with auxiliary circle and eccentric angle ϑ.

define the diameter conjugate to PP' as the diameter DD' which has the eccentric angle ϑ displaced by $\pi/2$, *i.e.* DD' in Fig. 7.13 has eccentric angle $\varphi = \vartheta + \pi/2$. The only pair of perpendicular diameters are the major and minor axes. With points $P = (a\cos\vartheta, b\sin\vartheta)$, $D = (-a\sin\vartheta, b\cos\vartheta)$ as extremities of conjugate diameters as in Fig. 7.13, the squares of their distances from the centre O are given by

$$OP^2 = a^2\cos^2\vartheta + b^2\sin^2\vartheta, \quad OD^2 = a^2\sin^2\vartheta + b^2\cos^2\vartheta, \quad \therefore OP^2 + OD^2 = a^2 + b^2.$$

The equation of a tangent to an ellipse is derived in Example 8.17, Since in the present case the tangent has to pass through the point P, we replace there (*i.e.* in Example 8.17) (x_0, y_0) by the coordinates of P and obtain

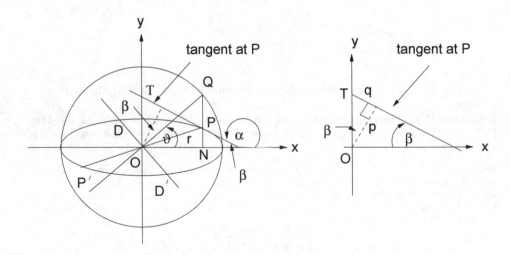

Fig. 7.13 Diameter PP' and its conjugate diameter DD'.

$$\frac{x\cos\vartheta}{a} + \frac{y\sin\vartheta}{b} = 1, \quad y-\text{intercept} = \frac{b}{\sin\vartheta}.$$

From this we obtain (with the angle β defining the slope of the tangent as in Fig. 7.13)

$$\frac{dy}{dx} = -\frac{b\cos\vartheta}{a\sin\vartheta} = -\tan\beta.$$

Then (*cf.* Fig. 7.13)

$$(\tan\beta)^2 = \frac{q^2}{p^2} = \frac{(y-\text{intercept})^2 - p^2}{p^2}, \quad\text{or}\quad \left(\frac{b\cos\vartheta}{a\sin\vartheta}\right)^2 = \frac{(b/\sin\vartheta)^2 - p^2}{p^2},$$

or

$$p^2 = \frac{(b/\sin\vartheta)^2}{1 + (b\cos\vartheta/a\sin\vartheta)^2} = \frac{a^2 b^2}{a^2 \sin^2\vartheta + b^2\cos^2\vartheta} = \frac{a^2 b^2}{OD^2}, \quad \frac{a^2 b^2}{p^2} = OD^2.$$

Hence* $p.OD = ab$. Setting $OP = r$, we obtain from this and the first equation, *i.e.* $OD^2 = a^2 + b^2 - OP^2$, the requested equation

$$\frac{a^2 b^2}{p^2} = a^2 + b^2 - r^2.$$

Example 7.2: Polar equation of an ellipse with pole at the centre

Verify, that the tangential polar equation of an ellipse with semi-axes of lengths a and b and *pole at the origin* can be rewritten as

$$\frac{1}{r^2} = \frac{\cos^2\theta}{a^2} + \frac{\sin^2\theta}{b^2},$$

also obtainable from the usual Cartesian form $1 = x^2/a^2 + y^2/b^2$ by setting $x = r\cos\theta, y = r\sin\theta$.

Solution: We recall from Eq. (7.30) that the angle of inclination ϕ of the tangent to the radial vector from the pole at O is related to the actual polar angle θ by the relation $\tan\phi = r(d\theta/dr)$, and the length p of the normal from O to the tangent is given by $p = r\sin\phi$ (*cf.* Eq. (7.29)). Hence we have the relations

$$p^2 = r^2 \sin^2\phi, \quad \tan^2\phi = r^2\left(\frac{d\theta}{dr}\right)^2 \equiv r^2\theta'^2, \quad \sin^2\phi = \frac{\tan^2\phi}{1+\tan^2\phi} = \frac{r^2\theta'^2}{1+r^2\theta'^2}.$$

Inserting p^2 into the tangential polar equation (7.62), *i.e.* $p^2(a^2 + b^2 - r^2) = a^2 b^2$, we obtain

$$p^2 = \frac{r^4\theta'^2}{(1 + r^2\theta'^2)} = \frac{a^2 b^2}{a^2 + b^2 - r^2}.$$

This equation can be rearranged as

$$\theta'^2[r^4(a^2 + b^2 - r^2) - a^2 b^2 r^2] = a^2 b^2, \quad\text{or}\quad \theta'^2(r^2 - a^2)(b^2 - r^2) = \frac{a^2 b^2}{r^2}.$$

One can now extract $\theta' = d\theta/dr$ and integrate the equation. For our verification here it suffices to differentiate the known result above and check that we obtain the equation just derived. Thus from Eq. (7.63a) we obtain

$$\frac{1}{r^2} = \frac{(1-\sin^2\theta)}{a^2} + \frac{\sin^2\theta}{b^2}, \quad \frac{1}{r^2} - \frac{1}{a^2} = \sin^2\theta\left(\frac{a^2 - b^2}{a^2 b^2}\right), \quad \sin^2\theta = \left(\frac{a^2 b^2}{a^2 - b^2}\right)\frac{a^2 - r^2}{a^2 r^2},$$

---

*This result can be seen in A. S. Ramsey [41], p. 159, D. M. Y. Sommerville [46], p. 43.

and differentiating the version in the middle (the last is substituted for $\sin^2\theta$ later) we have

$$\sin\theta\cos\theta d\theta = -\frac{a^2 b^2}{a^2 - b^2}\frac{dr}{r^3}, \qquad \frac{d\theta}{dr} = \frac{a^2 b^2}{b^2 - a^2}\frac{1}{r^3\sin\theta\cos\theta},$$

or, by squaring the expression and substituting for $\sin^2\theta$,

$$\begin{aligned}
\left(\frac{d\theta}{dr}\right)^2 &= \left(\frac{a^2 b^2}{b^2 - a^2}\right)^2 \frac{1}{r^6 \sin^2\theta(1 - \sin^2\theta)} \\
&= -\left(\frac{a^2 b^2}{b^2 - a^2}\right)\frac{a^4(a^2 - b^2)r^2}{r^4(a^2 - r^2)[(a^2 - b^2)a^2 r^2 - a^2 b^2(a^2 - r^2)]} \\
&= \frac{b^2 a^4}{r^2(a^2 - r^2)[a^2(r^2 - b^2)]} = \frac{a^2 b^2}{r^2(a^2 - r^2)(r^2 - b^2)},
\end{aligned}$$

which is seen to agree with the expression above.

Our next question is, how — given the law of the force — one obtains the (p, r)-equation of the orbit, and then *vice versa*, how, given the orbit, one can deduce the law of the force. We begin again with the total energy, *i.e.*

$$E = \frac{1}{2}mv^2 + V(r), \qquad V(r) = -\int_c^r f(r)dr. \tag{7.64}$$

With Eq. (7.28), *i.e.* $pv = l/m = $ const., it follows that

$$E = \frac{1}{2}\frac{m}{p^2}\left(\frac{l}{m}\right)^2 - \int f(r)dr. \tag{7.65}$$

Differentiation of this equation with respect to r implies

$$\frac{l^2/m}{p^3}\frac{dp}{dr} = -f(r). \tag{7.66}$$

Thus given the orbit, *i.e.* p as a function of r, we obtain the force $f(r)$.

Example 7.3: The law of force of a cardioid
Given Eq. (7.60a), *i.e.* the tangential polar equation of a cardioid, derive the law of its force.[†]

Solution: Differentiating Eq. (7.60a), *i.e.* $r^3 = 2ap^2$, we obtain

$$3r^2 = 4ap\frac{dp}{dr}.$$

Inserting this expression into Eq. (7.66), we obtain the force f, *i.e.*

$$f = -\frac{l^2/m}{p^3}\frac{3r^2}{4ap} = -\frac{(l^2/m)3r^2 4a^2}{4ar^6} = -\frac{3l^2}{m}\frac{a}{r^4}.$$

The force is therefore attractive and proportional to $1/r^4$.

[†]For a different derivation see *The Physics Coaching Class* [48], problem 1073, p. 115.

Example 7.4: Pole not at centre of circular orbit

A point-like mass m moves on a circle of radius R with constant areal velocity with respect to an eccentrically located origin or pole, as indicated in Fig. 7.14. What is its angular acceleration at the point $\mathbf{r}$?

Solution: We begin with *Kepler's area law* or law of constant areal velocity expressed by, A meaning area,

$$A = \frac{1}{2}r^2\dot{\theta} = \text{const.}, \quad i.e. \quad l - mr^2\dot{\theta} = \text{const.}$$

The expression requested is the acceleration given by

$$\ddot{\theta} = \frac{d}{d\theta}\left(\frac{1}{2}\dot{\theta}^2\right) = \frac{d}{d\theta}\left[\frac{1}{2}\left(\frac{l}{mr^2}\right)^2\right] = \frac{d}{dr}\left[\frac{1}{2}\left(\frac{l}{mr^2}\right)^2\right]\frac{dr}{d\theta}.$$

In Fig. 7.14 the point M is the centre of the circle. From the triangle OPM we obtain

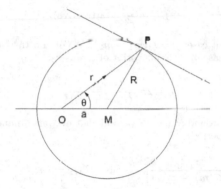

Fig. 7.14   Circular orbit with off-centric pole.

$$R^2 = a^2 + r^2 - 2ar\cos\theta, \quad \therefore \quad 0 = 2rdr - 2adr\cos\theta + 2ar\sin\theta d\theta,$$

$$\frac{dr}{d\theta} = \frac{2ar\sin\theta}{2a\cos\theta - 2r} = \frac{ar\sqrt{1-\cos^2\theta}}{a\cos\theta - r}.$$

It follows that

$$\ddot{\theta} = -2\frac{l^2}{m^2r^4}\left(\frac{a\sqrt{1-\cos^2\theta}}{a\cos\theta - r}\right), \quad \text{where} \quad \cos\theta = \frac{a^2 + r^2 - R^2}{2ar}.$$

Example 7.5: Ratio of mass of the sun to that of Mars

The planet Mars encircles the sun in 687 days, and the satellite Deimos the planet Mars in 30 hours. The ratio of the mean distances Deimos–Mars, Mars–sun is 1 to 10 000. Compute from this the ratio of the mass of the sun m_s to that of Mars, m_M.

Solution: The period $T = 2\pi/n$ of motion of a body of mass m around the pole of an attractive central force f proportional to $1/r^2$, where r is the distance of the body from the pole, is given by the expression (rederived below):

$$T = \frac{2\pi a^{3/2}}{\sqrt{\mu}} = 2\pi\left(\frac{a^3}{\mu}\right)^{1/2}, \tag{7.67}$$

where a is the length of the semi-major axis of the elliptic orbit, and μ is the gravitational constant multiplied by the mass of the body at the centre of the force, *i.e.* $f = -\mu m/r^2$. The formula (7.67) is obtained as follows. Since $r^2\dot\theta = l/m = $ const., where l is the angular momentum, we obtain for the element of area $dA = r^2 d\theta/2$, so that

$$\frac{dA}{dt} = \frac{1}{2}\frac{l}{m}, \quad \text{and} \quad \therefore A = \frac{1}{2}\frac{l}{m}T = \pi ab,$$

the latter expression being the formula for the area of an ellipse in terms of the lengths a and b of the semi-major and -minor axes respectively. With the help of the following relation $m/l = \sqrt{a/\mu}/b$ which will be obtained below, *cf.* Eq. (7.75), it follows that

$$T = \frac{2m\pi ab}{l} = 2\pi a \frac{b}{b}\sqrt{\frac{a}{\mu}} = \frac{2\pi a^{3/2}}{\sqrt{\mu}}.$$

Since we defined with Eq. (7.35) μ by the relation for the force f experienced by the reduced mass $m = m_1 m_2/(m_1 + m_2)$ at a separation r from the centre of the force, *i.e.*

$$f = -\frac{\mu m}{r^2} \quad \text{(attractive)},$$

we have for μ in the present case: $\mu = G(m_1 + m_2)$. We obtain therefore from Eq. (7.67) for the period of mass m_2 in the case of a circular orbit of radius r_c:

$$T = 2\pi\left[\frac{r_c^3}{G(m_1 + m_2)}\right]^{1/2}. \tag{7.68}$$

For the period of Deimos around Mars, T_{DM}, and that of Mars around the sun, T_{MS}, we obtain therefore respectively the periods:

$$T_{DM} = 2\pi\left[\frac{r_{DM}^3}{G(m_D + m_M)}\right]^{1/2} \quad \text{and} \quad T_{MS} = 2\pi\left[\frac{r_{MS}^3}{G(m_S + m_M)}\right]^{1/2}.$$

We now consider the ratio

$$\frac{T_{DM}^2}{T_{MS}^2} = \frac{r_{DM}^3(m_S + m_M)}{r_{MS}^3(m_D + m_M)}, \quad \therefore \frac{T_{DM}^2 r_{MS}^3}{T_{MS}^2 r_{DM}^3} \simeq \frac{m_S}{m_M}, \text{ if } m_D \ll m_M \ll m_S.$$

Inserting the given values, we obtain $m_S \simeq 3 \times 10^6 m_M$.

7.7   Maxima and Minima of Velocities

We now explore the implications of the law of conservation of energy in the form (*i.e.* kinetic energy $= -$ potential energy $+$ const.):

$$v^2 = \frac{2\mu}{r} + \text{const.} \tag{7.69}$$

We return to the force of the Kepler problem, *i.e.* $f = -m\mu/r^2$. Inserting this into Eq. (7.66), we obtain

$$\frac{l^2/m}{p^3}\frac{dp}{dr} = \frac{m\mu}{r^2}, \quad \text{or} \quad \frac{l^2}{m^2}\frac{dp}{p^3} = \mu\frac{dr}{r^2}. \tag{7.70}$$

Integration yields

$$-\frac{1}{2}\frac{l^2}{m^2 p^2} = -\frac{\mu}{r} + C, \quad C = \text{const.}, \tag{7.71}$$

where with Eq. (7.28) $pv = l/m$, *i.e.* $l/pm = v$. This equation is therefore identical with the law of conservation of energy in the form of Eq. (7.57), and can also be written as

$$\frac{l^2}{p^2 m^2} = v^2 = \frac{2\mu}{r} + C. \tag{7.72}$$

The (p,r)-equations of conical sections with pole at the focus are[‡] (*cf.* also Eq. (7.61))

$$\frac{b^2}{p^2} = \frac{2a}{r} + \begin{pmatrix} +1 \\ -1 \\ 0 \end{pmatrix} \quad \text{depending on whether} \quad \begin{cases} \text{hyperbola,} \\ \text{ellipse,} \\ \text{parabola,} \end{cases} \tag{7.73}$$

or

$$\frac{b^2}{p^2}\frac{\mu}{a} = \frac{2\mu}{r} + \frac{\mu}{a}\begin{pmatrix} +1 \\ -1 \\ 0 \end{pmatrix}.$$

It follows that the orbit in the present case is a hyperbola, parabola or ellipse depending on whether $C > 0, < 0$ or 0. Comparing Eqs. (7.69) and (7.72), we can rewrite Eq. (7.69) as

$$\frac{l^2}{p^2 m^2} = \frac{\mu b^2}{a p^2} = \frac{2\mu}{r} + \frac{\mu}{a}\begin{pmatrix} +1 \\ -1 \\ 0 \end{pmatrix}, $$

and we obtain (comparing the left hand side of this equation with that of Eq. (7.73))

$$\frac{l}{m} = b\sqrt{\frac{\mu}{a}}, \quad C = \frac{\mu}{a}\begin{pmatrix} +1 \\ -1 \\ 0 \end{pmatrix} \iff \begin{cases} \text{hyperbola,} \\ \text{ellipse,} \\ \text{parabola.} \end{cases} \tag{7.74}$$

Thus for a *conical section* we have from Eq. (7.74):

$$\frac{l}{m} = b\sqrt{\frac{\mu}{a}}, \tag{7.75}$$

[‡]A. S. Ramsey [41], pp. 161 – 163.

and from Eqs. (7.69) and (7.74) for an *ellipse*:

$$v^2 = \frac{2\mu}{r} - \frac{\mu}{a}. \tag{7.76}$$

Hence the velocity reaches its maximum value where r is smallest, *i.e.* at the *perihelion*, and is a minimum, where r is largest, *i.e.* at the *aphelion* (*cf.* Example 7.8).

7.8 Same Orbit, Different Forces

We convince ourselves now that different forces can lead to the same orbit. The idea is straight-forward: In the case of one and the same ellipse the force may be directed to the focus or, alternately, towards the centre. The forces are different.

(1) We consider first the case of a central force directed towards the centre. The question is: What is the force? The (p, r)-equation of the ellipse with *pole at the centre* is (*cf.* Eq. (7.62))

$$\frac{a^2 b^2}{p^2} = a^2 + b^2 - r^2. \tag{7.77}$$

Differentiation with respect to r implies

$$\frac{a^2 b^2}{p^3} \frac{dp}{dr} = r, \qquad \frac{dp}{dr} = \frac{rp^3}{a^2 b^2}. \tag{7.78}$$

Using Eq. (7.66), *i.e.* $f = -(l^2/mp^3)dp/dr$, we obtain the force as

$$f = -\frac{l^2}{mp^3} \frac{rp^3}{a^2 b^2} = -\frac{(l^2/m)r}{a^2 b^2} \equiv -\frac{\partial V(r)}{\partial r}, \qquad V(r) \propto r^2. \tag{7.79}$$

We see that the potential is now proportional to the square of the distance from the pole.

(2) As the second case we recall the elliptic orbit with a central force directed towards the focus, and again we wish to derive the force. The (p, r)-equation of the ellipse with *pole at a focus* is (*cf.* Eq. (7.61))

$$\frac{b^2}{p^2} = \frac{2a}{r} - 1.$$

Differentiation with respect to r yields

$$-2\frac{b^2}{p^3}\frac{dp}{dr} = -\frac{2a}{r^2}, \qquad \frac{dp}{dr} = \frac{ap^3}{b^2 r^2},$$

so that with Eq. (7.66) the force is obtained as what we expect:

$$f = -\frac{(l^2/m)}{p^3}\frac{dp}{dr} = -\frac{l^2}{m}\frac{a}{b^2 r^2}.\qquad(7.80)$$

Thus, as we know already, the force is proportional to the square of the reciprocal of the distance from the pole, the potential $V(r) \propto 1/r$.

7.9 Period

The time t of the motion of a particle on its central force trajectory is obtained from Eq. (7.15), *i.e.* from

$$r^2\dot{\theta} = \frac{l}{m} \quad\text{or}\quad \frac{dA}{dt} = \frac{1}{2}\frac{l}{m}, \qquad dt = \frac{2m}{l}dA,\qquad(7.81)$$

since the infinitesimal area dA is given by $dA = r^2 d\theta/2$ (*cf.* Eq. (7.26)). This is *Kepler's second law* which states that the radius vector sweeps out equal areas in equal times. For the *period* T, *i.e.* the time required for one complete journey around the orbit of the (closed) ellipse with lengths of its semi-axes a and b with $a > b$, we have $A = \pi ab$ and hence

$$T = \frac{2m}{l}\pi ab = \frac{m}{l}2\pi ab.\qquad(7.82)$$

However (*cf.* Eq. (7.75) and for the second relation Eq. (7.49))

$$\frac{l}{m} = b\sqrt{\frac{\mu}{a}}, \quad\text{and}\quad l' = \frac{l^2}{\mu m^2} = \frac{b^2}{a} = a(1 - e^2),\qquad(7.83)$$

the latter relation having been derived in the footnote following the introduction of l' between Eqs. (7.41) and (7.42). Hence we obtain for the period T:

$$T = 2\pi a\frac{b}{b}\sqrt{\frac{a}{\mu}} = \frac{2\pi a^{3/2}}{\sqrt{\mu}}, \quad\text{frequency}\quad n = \frac{2\pi}{T} = \left(\frac{\mu}{a^3}\right)^{1/2}.\qquad(7.84)$$

Thus, *the square of the period is proportional to the third power of the length of the semi-major axis*, a statement known as *Kepler's third law*. Applied to the motion of planets in general, the statement says: The squares of the periods of the different planets are proportional to the third power of their semi-major axes.

In the case of an ellipse, with the force directed towards a focus, we have from Eq. (7.42):

$$\frac{l'}{r} = 1 + e\cos\theta,\qquad(7.85)$$

and thus, with $r^2\dot{\theta} = l/m$, we have

$$\frac{dt}{d\theta} = \frac{r^2 m}{l} = \frac{m}{l}\frac{l'^2}{(1+e\cos\theta)^2}, \tag{7.86}$$

and thus

$$t = \frac{ml'^2}{l}\int_{\theta_1}^{\theta_2}\frac{d\theta}{(1+e\cos\theta)^2}. \tag{7.87}$$

With Eq. (7.83) we can re-express this as (in the second step using again $b^2 = a^2(1-e^2)$, see after Eq. (7.41))

$$
\begin{aligned}
t &= \frac{1}{b}\sqrt{\frac{a}{\mu}}a^2(1-e^2)^2\int_{\theta_1}^{\theta_2}\frac{d\theta}{(1+e\cos\theta)^2}\\
&= \frac{a^{3/2}(1-e^2)^{3/2}}{\sqrt{\mu}}\int_{\theta_1}^{\theta_2}\frac{d\theta}{(1+e\cos\theta)^2}\\
&= \frac{l'^{3/2}}{\sqrt{\mu}}\int_{\theta_1}^{\theta_2}\frac{d\theta}{(1+e\cos\theta)^2}.
\end{aligned} \tag{7.88}
$$

Integration with $\theta_1 = 0, \theta_2 = \theta$ yields (the integral is evaluated in Example 7.9, Eqs. (7.129a), (7.129b)):

$$t = \frac{a^{3/2}}{\sqrt{\mu}}\left[2\tan^{-1}\left\{\left(\frac{1-e}{1+e}\right)^{1/2}\tan\frac{\theta}{2}\right\} - \frac{e\sqrt{1-e^2}\sin\theta}{1+e\cos\theta}\right]. \tag{7.89}$$

Example 7.6: Period of the planet Jupiter

The planet Jupiter has an almost circular orbit like the Earth but with a radius approximately 5.2 times that of the Earth. Determine T_J, the length of a year on Jupiter, in terms of that of T_E, the length of a year on Earth.

Solution: According to the above (*i.e.* Kepler's third law, Eq. (7.84)) we have (since μ is essentially Newton's gravitational constant multiplied by the mass of the sun, *cf.* Eq. (7.35))

$$\left(\frac{T_J}{T_E}\right)^2 = \left(\frac{5.2}{1}\right)^3,$$

Hence

$$T_J = T_E(5.2)^{3/2} \simeq 11.9 T_E.$$

Thus a Jupiter year is approximately equal to 11.9 years on Earth.

7.10   Perihelion Precession of Mercury

We first explain what the *apsides* of a planetary orbit are. An *apsis* is a point on the trajectory of a particle in a central-force problem at which the orthogonal to the trajectory passes through the centre of the force; the length of the radius vector at this point is described as the length of that apsis. Thus, in our earlier terminology, at an apsis $\mathbf{v} \perp \mathbf{r}, p = r$, and the line connecting the apsides divides the orbit into symmetrical parts; this results from the fact that if the velocity is reversed, the particle must traverse its trajectory in the reverse direction. From our earlier Eq. (7.31), *i.e.*

$$v^2 = \frac{l^2}{m^2}\left[u^2 + \left(\frac{du}{d\theta}\right)^2\right] = \frac{l^2}{m^2}\frac{1}{p^2}, \qquad (7.90)$$

we see that $u(\theta)$ is a function of v^2. The apsides are found by setting (see explanation below)

$$\frac{dr}{d\theta}, \quad \frac{du}{d\theta} \quad \text{or} \quad \frac{dr}{dt} = 0, \quad \text{and} \quad p = r.$$

These conditions follow from the fact that with $p = r = 1/u$, Eq. (7.90) implies $du/d\theta = 0$. Also (see Eq. (7.51)):

$$(\tan\phi)_{\phi=\pi/2} = r\frac{d\theta}{dr}, \qquad \text{i.e.} \qquad \frac{dr}{d\theta} = 0.$$

Moreover, using the relation $l = mr^2\dot{\theta}$, we have

$$\frac{dr}{dt} = \frac{dr}{du}\frac{du}{dt} = -r^2\frac{du}{dt} = -\frac{l}{m\dot{\theta}}\frac{du}{dt} = -\frac{l}{m}\frac{du}{d\theta},$$

so that

$$\frac{dr}{dt} = -\frac{l}{m}\frac{du}{d\theta}, \quad \text{which is zero when} \quad \frac{du}{d\theta} = 0.$$

We now come to a very important application of our considerations, the precession of the *perihelion apsis* of the planet Mercury (point N on the ellipse in Fig. 7.7). In the general theory of relativity, where this provides one of the spectacular confirmations, this is known as the *Schwarzschild solution* (of Einstein's equation). Astronomical observations had shown that the motion of the planets is very well explained by Kepler's laws and Newton's gravitational law. Although the orbits of all planets show minute deviations from the elliptic shape resulting from Newton's law, this deviation (*cf.* Fig. 7.15) is reasonably noticeable only in the case of the planet Mercury. The deviations of the planetary orbits from the elliptic shape discussed up to this point are largely due to perturbations due to the presence of the other planets. Such perturbations result in a slow but continuous displacement of the axis connecting the apsides, and this is explained by Einstein's gravitational law. In the following we do not enter into the theory of relativity

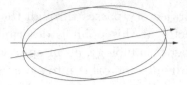

Fig. 7.15 The perihelion shift.

(which we leave to later chapters); instead we assume a certain perturbation which will be seen to be that resulting from Einstein's theory. We saw in Sec. 7.5 that in Newton's theory (*i.e.* with Newton's law of gravitation) the differential equation of the orbit is given by Eq. (7.36), *i.e.* the equation

$$\frac{d^2u}{d\theta^2} + u = \left(\frac{m^2}{l^2}\right)\mu. \tag{7.91}$$

The solution is, as derived above,

$$u = \mu\left(\frac{m^2}{l^2}\right)[1 + e\cos(\theta - \alpha)] \equiv u_0. \tag{7.92}$$

Since for $du/d\theta = 0$ we have $\theta = \alpha$, the angle α is the *angular coordinate of the perihelion apsis*. According to *Einstein's General Theory of Relativity* the equation of the orbit is actually — we cite this here, since this is all we need at present, the full derivation of this equation will be given in Sec. 16.5, there it is Eq. (16.100) — (*cf.* also Example 3.13, Eq. (3.99))

$$\frac{d^2u}{d\theta^2} + u = \left(\frac{m^2}{l^2}\right)\mu + 3\mu u^2. \tag{7.93}$$

The additional term $3\mu u^2$ — here added with the constant derived from Einstein's theory instead of using an arbitrary constant — is small.[§] Its relation to $\mu(m^2/l^2)$ is[¶]

$$3\left(\frac{l^2}{m^2}\right)u^2 \overset{(7.28)}{=} \frac{3p^2v^2}{r^2} \overset{(7.58)}{=} 3\left(\frac{v\sin\phi}{1}\right)^2 = 3\left(\frac{\text{transversal velocity}}{1}\right)^2. \tag{7.94}$$

In Einstein's theory the ratio of *transversal velocity* to unit velocity (this is there the velocity of light) is of the order of 10^{-8}. Thus Einstein's theory changes Newton's equation only by a tiny correction term. In dominant (or

---

[§]The parameter μ will later in Einstein's theory be identified as GM/c^2, M the mass of the sun. See Eq. (16.100) and thereafter.

[¶]We had $\sin\phi = r d\theta/ds, v = ds/dt$. $\therefore$ $\sin\phi = r(d\theta/dt)(dt/ds), v\sin\phi = r\dot\theta$, cf. Eq. (7.12) and Example 5.1.

zeroth) order, the solution of Eq. (7.93) is therefore again given by Eq. (7.92). In order to obtain an improved approximation which takes Einstein's term into account, we replace in Eq. (7.93) u by the $u_0 + u_1$, where u_0 is the solution of the equation without the Einstein term. Then

$$\frac{d^2(u_0 + u_1)}{d\theta^2} + (u_0 + u_1) = \frac{m^2}{l^2}\mu + 3\mu(u_0 + u_1)^2, \tag{7.95}$$

so that (dropping the zeroth order terms)

$$\begin{aligned}
\frac{d^2 u_1}{d\theta^2} + u_1 &= 3\mu(u_0^2 + 2u_0 u_1 + u_1^2) \simeq 3\mu u_0^2 \\
&= 3\mu \left[\mu \frac{m^2}{l^2} \{1 + e\cos(\theta - \alpha)\} \right]^2 \\
&\simeq 3\mu^3 \left(\frac{m^2}{l^2} \right)^2 + 6\mu^3 \left(\frac{m^2}{l^2} \right)^2 e\cos(\theta - \alpha) + O(e^2). \tag{7.96}
\end{aligned}$$

The first and constant term on the right can be taken care of by setting $u_1' = u_1 - 3\mu^3(m^2/l^2)^2$. A particular solution of the equation

$$\frac{d^2 u}{d\theta^2} + u = A\cos(\theta - \alpha)$$

is, as one can check by differentiation,[||]

$$u = \frac{1}{2}A\theta\sin(\theta - \alpha).$$

We can therefore obtain the solution of Eq. (7.95) by adding to u_0 the contribution

$$3\mu^3 \left(\frac{m^2}{l^2} \right)^2 \theta e \sin(\theta - \alpha).$$

We obtain therefore

$$u = \mu\left(\frac{m^2}{l^2} \right)[1 + e\cos(\theta - \alpha)] + 3\mu^3\left(\frac{m^2}{l^2} \right)^2 \theta e\sin(\theta - \alpha) + 3\mu^3\left(\frac{m^2}{l^2} \right)^2. \tag{7.97}$$

This solution can be rewritten:

$$\begin{aligned}
u &= \mu\left(\frac{m^2}{l^2} \right)[1 + e\cos(\theta - \alpha - \delta\alpha)] + 3\mu^3\left(\frac{m^2}{l^2} \right)^2 \\
&= \mu\left(\frac{m^2}{l^2} \right)[1 + e\cos(\theta - \alpha)\cos\delta\alpha + e\sin(\theta - \alpha)\sin\delta\alpha] + 3\mu^3\left(\frac{m^2}{l^2} \right)^2 \\
&\simeq \mu\left(\frac{m^2}{l^2} \right)[1 + e\cos(\theta - \alpha) + e\sin(\theta - \alpha)\delta\alpha] + 3\mu^3\left(\frac{m^2}{l^2} \right)^2, \tag{7.98}
\end{aligned}$$

[||] $du/d\theta = (1/2)A\sin(\theta - \alpha) + (1/2)A\theta\cos(\theta - \alpha)$, $d^2u/d\theta^2 = (2/2)A\cos(\theta - \alpha) - (1/2)A\theta\sin(\theta - \alpha) = A\cos(\theta - \alpha) - u$.

since $\delta\alpha$ is small. Comparison with Eq. (7.97) implies that

$$\delta\alpha = 3\mu^2 \left(\frac{m^2}{l^2}\right)\theta. \tag{7.99}$$

For completion of one orbit, *i.e.* for $\theta = 2\pi$, one obtains a *constant apsis rotation* or *perihelion anomaly* (p.a.) of

$$\text{p.a.} = 3\mu^2 \left(\frac{m^2}{l^2}\right)2\pi = \frac{6\pi\mu^2 m^2}{l^2}. \tag{7.100}$$

The Special Theory of Relativity (see later) yields for this effect $\pi\mu^2 m^2/l^2$, *i.e.* without the factor of 6. For the ratio l/m we had earlier (*cf.* Eq. (7.74) and remarks after Eq. (7.41))

$$\frac{l}{m} = b\sqrt{\frac{\mu}{a}} \quad \text{with} \quad b^2 = a^2(1 - e^2),$$

so that the perihelion anomaly becomes

$$\text{p.a.} = \frac{6\pi\mu^2 a}{b^2\mu} = \frac{6\pi\mu a}{b^2} = \frac{6\pi\mu}{a(1 - e^2)}. \tag{7.101}$$

We see that the anomaly is the more pronounced, the smaller a is, or the bigger the eccentricity e. For this reason the anomaly is best observable in

Table 7.1 Secular Perihelion Anomalies

Planet	calculated $(\delta\alpha)_{100}$		observed $(\delta\alpha)_{100}$
	B.S.	C.	B.S., C.
Mercury	43″.15	43″.03	42″.56 ± 0″.94,
			43″.11 ± 0″.45
Venus	—	8″.60	—
Earth	3″.84	3″.80	4″.6 ± 2″.7
Mars	1″.35	1″.35	—

the case of the planet Mercury. The above Table 7.1 gives the calculated and observed values of the "*secular perihelion anomaly*", *i.e.* the apsidal shift which accumulates over 100 years, as cited in the literature. In view of more precise measurements of the constants involved in the calculated values, and more precise measurements of the observed quantities, the values cited in the literature differ minutely. In Table 7.1 the values cited by Bucerius and Schneider are indicated by the initials B.S., the more recent values cited by Carmeli are indicated by C.**

We recall from Fig. 7.7 that α is the angular coordinate of the perihelion apsis and θ is the corresponding angular coordinate of the planet. The first line of Eq. (7.98) shows that the orbit of the planet is an ellipse, however, with a gradual displacement, since the angle $\alpha + \delta\alpha$ which determines the position of the apsides, varies with θ. With respect to θ this variation is

$$\frac{\delta\alpha}{\theta} = 3\mu^2\left(\frac{m^2}{l^2}\right). \tag{7.102}$$

The motion thus calculated and illustrated in Fig. 7.15 agrees very well with the data of observations. Above we solved Eq. (7.93) in the dominant plus next order approximation. In Appendix A we show that the solution can also be obtained in a closed form in terms of Jacobian elliptic functions.

7.11 Stability of Circular Orbits

It is interesting to explore in the present context the *stability of an orbit*. The trajectory of a particle is said to be *stable* if its trajectory after a small disturbance proceeds in the vicinity of the original path. Every attractive central force permits a circular orbit. We had Eq. (7.25), *i.e.*

$$\frac{d^2u}{d\theta^2} + u = -\left(\frac{m^2}{l^2}\right)\frac{f}{m}\frac{1}{u^2}, \quad f = f(u) \equiv f(1/r). \tag{7.103}$$

For $r = a, u = 1/a$, we obtain

$$\frac{1}{a} = -\left(\frac{m^2}{l^2}\right)\frac{f(r=a)}{m}a^2,$$

i.e. (in the case of the circle $p = a$)

$$\left(\frac{l^2}{m^2}\right) = -a^3\left(\frac{f}{m}\right)_{r=a} = a^2v_0^2, \quad \text{with} \quad \frac{l}{m}\overset{(7.28)}{=}pv, \tag{7.104}$$

if v_0 is the initial velocity of the particle on the circular orbit. It follows that

$$v_0 = \left(-\frac{af}{m}\right)^{1/2}_{r=a} = \text{const.}$$

**H. Bucerius and M. Schneider [7], p. 239; M. Carmeli [10], p. 55.

Setting in Eq. (7.103) $u = 1/a + \xi$, where the fluctuation ξ is the new variable and is assumed to be small compared with $1/a$, we obtain (using Eq. (7.104))

$$\frac{d^2}{d\theta^2}\left(\frac{1}{a}+\xi\right) + \left(\frac{1}{a}+\xi\right) = \frac{m}{a^3 f(1/a)}\frac{f(u)}{mu^2} = \frac{f(u)}{f(1/a)u^2 a^3},$$

$$f(u)|_{u=1/a} = f\left(\frac{1}{a}\right).$$

Thus we obtain the *fluctuation equation*

$$\frac{d^2\xi}{d\theta^2} + \xi = \frac{f(u)}{f(1/a)u^2 a^3} - \frac{1}{a}. \qquad (7.105)$$

Expanding the right hand side around $u = 1/a$, we have

$$f(u) = f\left(\frac{1}{a}+\xi\right) \simeq f\left(\frac{1}{a}\right) + \xi f'\left(\frac{1}{a}\right),$$

and (expanding $(1/a + \xi)^2$ in the denominator)

$$\frac{d^2\xi}{d\theta^2} + \xi = \frac{f(1/a) + \xi f'(1/a)}{f(1/a)a^3(1/a+\xi)^2} - \frac{1}{a}$$

$$\simeq \frac{1}{a}\frac{f(1/a)+\xi f'(1/a)}{f(1/a)}(1-2a\xi) - \frac{1}{a}$$

$$\simeq \frac{1}{a}\xi\frac{f'(1/a)}{f(1/a)} - 2\xi. \qquad (7.106)$$

Hence

$$\frac{d^2\xi}{d\theta^2} + \xi\left[3 - \frac{f'(1/a)}{af(1/a)}\right] = 0. \qquad (7.107)$$

We set

$$n^2 \equiv 3 - \frac{f'(1/a)}{af(1/a)}. \qquad (7.108)$$

We then have the equation

$$\frac{d^2\xi}{d\theta^2} + n^2\xi = 0. \qquad (7.109)$$

The solution of this equation is

$$\xi = A\cos(n\theta + \alpha). \qquad (7.110)$$

We see that as long as $n^2 > 0$, the solution ξ is periodic in θ and oscillates, so that the orbit, *i.e.* $u = 1/a + \xi$, around the unperturbed circular path, is stable. In order to realize clearly the consequences of the *stability condition* $n^2 > 0$, we also look at the case $n^2 \equiv \nu^2 < 0$. In this case

$$\xi = Ae^{\nu\theta} + Be^{-\nu\theta} \quad \text{and} \quad u = \frac{1}{r} = \frac{1}{a} + Ae^{\nu\theta} + Be^{-\nu\theta}. \qquad (7.111)$$

Clearly, with $A \neq 0, \theta > 0$ this solution spirals away from the circle, and for $A = 0, \theta > 0$ it collapses toward the circle and thus dies off. For a force $f(u) = -m\mu u^N$ (recall the Kepler case $f_{\text{Kepler}} = -m\mu u^2$ has $N = 2$) we have $f'(u) = -m\mu N u^{N-1}$ and

$$n^2 = 3 - \frac{m\mu N (1/a)^{N-1}}{am\mu(1/a)^N} = 3 - \frac{Na}{a} = 3 - N.$$

This is positive for $N < 3$, *i.e.* $f \propto u^{N<3}, V \propto r^{-N+1}$. It will be observed later in the context of General Relativity that the number 3 appearing here results from the three spatial dimensions.

7.12 Scattering in Central Force Fields

The topic of *scattering* that we now come to is very different from what we considered above. Moreover, in the following we jump to a different physical situation. Above we had planets in mind, and their orbits around the sun, and the relevant potential was Newton's gravitational potential. What permits us now to jump to microscopic physics is the analogy between Newton's potential and the Coulomb potential acting between electrically charged particles. In the latter case, of course, we wish to consider a beam of such particles, since this would be what is required in an experiment in

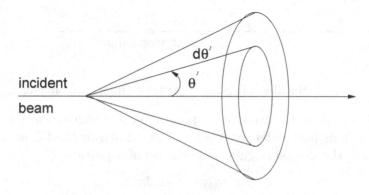

Fig. 7.16 Scattering in the direction θ'.

atomic physics. This is the new aspect we therefore encounter here. In other respects, if we consider a single particle and its trajectory in space (in the presence of the Coulomb potential), the treatment is analogous to that of a *comet* in the gravitational field of a star since our considerations here are purely classical, *i.e. deterministic*, meaning the equation of motion and initial conditions determine the trajectory and whereabouts of the body or particle at all later times.

We consider a homogeneous beam of particles, all of which have the same mass m. The incident beam is characterized by its *intensity I*, here also called *current density*, *i.e.* the number of particles that pass in unit time through unit area perpendicular to the beam. The so-called *cross section* $\sigma(\Omega)$ for scattering in a given direction of solid angle Ω is defined by

$$d\sigma \equiv \sigma(\Omega)d\Omega$$

= the number of particles which are scattered per unit time into the solid angle element $d\Omega$ divided by the incident intensity I. The quantity $\sigma(\Omega)$ is also described as the *differential cross section*. For central forces the scattering is completely symmetric about the axis of the incident beam, as indicated in Fig. 7.16, where θ' is the angle between the scattering direction and the direction of the incident beam. Then, on a unit sphere,

$$d\Omega = 2\pi \sin \theta' d\theta'.$$

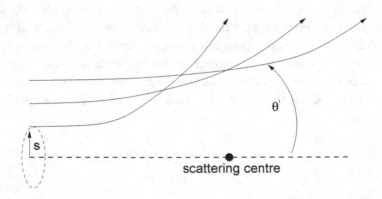

Fig. 7.17 Smaller deflection θ' for larger impact parameter s.

It is convenient to introduce a parameter s which is known as *impact parameter*. This parameter, *cf.* Fig. 7.17, is defined by the following relation in which l is the orbital angular momentum of a particle,[††]

$$l = mv_0 s = s\sqrt{2mE}, \tag{7.112}$$

[††]The angular momentum vector l is, of course, orthogonal to the trajectory of the individual

where v_0 is the incident velocity (far away from the scattering centre so that the potential is effectively zero and therefore $E \simeq$ kinetic energy). We have:

Number of outgoing particles with scattering angle between θ' and $\theta' + d\theta'$
= number of incident particles with impact parameter between s and $s + ds$
= number of incident particles with angular momentum between $l(s)$
= mvs and $l(s + ds) = mv(s + ds)$.
We thus have:

Number of particles scattered per unit time into the solid angle element $d\Omega$
= number of incident particles with impact parameter between s and $s + ds$,
i.e. ($2\pi s$ is the length of the circumference of the circle, and ds the width shown in Fig. 7.17)

$$-\sigma(\Omega)d\Omega I = 2\pi s ds I = -2\pi \sin \theta' d\theta' \sigma(\theta')I.$$

The minus-sign is introduced because an increase in the impact parameter s by an amount ds implies a weaker force acting on the particle, and this again implies a diminished scattering angle by an amount $d\theta'$, as can also be

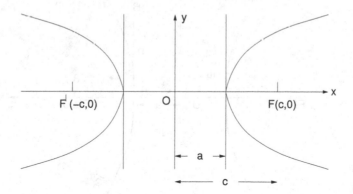

Fig. 7.18 The hyperbola with foci F and F'.

seen from Fig. 7.17. We thus obtain

$$\sigma(\theta') = -\frac{s ds}{\sin \theta' d\theta'}. \tag{7.113}$$

In the following we want to calculate $s ds$ from the central force.

---

particle. The corresponding cyclic variable is an angle ψ of the rotation about the axis perpendicular to the trajectory. The cyclic variable ψ must not be confused with the angle ϕ of rotation around the z-axis, of which $\sigma(\theta)$ is independent, and which appears here only in connection with a mean of particles.

In the case of the *Coulomb potential* between a fixed charge $-Ze$ and the charge $-Z'e$ of the incident particles, the force is

$$f = \frac{ZZ'e^2}{r^2}, \tag{7.114}$$

and hence a *repelling Kepler force*. We can take over our earlier results provided we replace there, *i.e.* in the expression $f = -m\mu/r^2$ (*cf.* Eq. (7.35)) the quantity $m\mu$ by $-ZZ'e^2$. Hence we can set

$$m\mu = -ZZ'e^2. \tag{7.115}$$

In the case of the *trajectory with pole at the focus* we then have (from now on we use ϵ as symbol for the *eccentricity* of the ellipse, in order to avoid

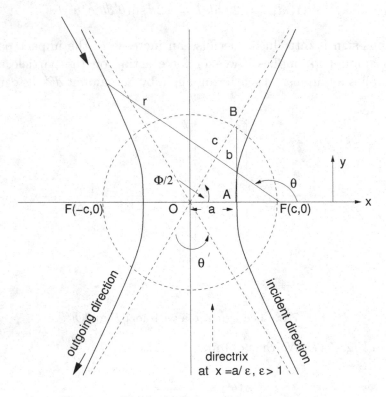

Fig. 7.19 The hyperbola and its asymptotes.

confusion with the charge e), *cf.* Eq. (7.42) and recall Eq. (7.49) for the relation $l' = l^2/\mu m^2$,

$$\frac{1}{r} = \frac{1}{l'}[1 + \epsilon \cos(\theta - \alpha')] = -\frac{mZZ'e^2}{l^2}[1 + \epsilon \cos \theta]. \tag{7.116}$$

Here α' is a suitably chosen zero point of the angle θ: We choose $\alpha' = 0$. Moreover, as we saw:

$$l' \stackrel{(7.49)}{=} \frac{l^2}{\mu m^2} \stackrel{\text{here}}{=} -\frac{l^2}{mZZ'e^2}. \tag{7.117}$$

We had also derived the following expression for the eccentricity ϵ:

$$\epsilon \stackrel{(7.49)}{=} \sqrt{1 + \frac{2El^2}{\mu^2 m^3}} \stackrel{(7.115)}{=} \sqrt{1 + \frac{2El^2}{m(ZZ'e^2)^2}} \stackrel{(7.112)}{=} \sqrt{1 + \left(\frac{2Es}{ZZ'e^2}\right)^2}. \tag{7.118}$$

We see that for $\epsilon > 1$ the trajectory of the particle is a hyperbola (*cf.* Eq. (7.50)). However, to ensure that r in Eq. (7.116) is positive, we must have

$$\epsilon \cos\theta < -1 \quad \Rightarrow \quad \frac{\pi}{2} < \theta < \frac{3\pi}{2},$$

In the case of a hyperbola, as in Figs. 7.18, 7.19, we have $\epsilon = c/a$, where c is the abscissa of the focus F, and a is the length of the semi-major axis.

The *eccentricity of the hyperbola* is defined[‡‡] in the same way as that of the ellipse (*cf.* Eq. (7.41)), except that b^2 is replaced by $-b^2$, and is then given by the following expression (see Fig. 7.19, and obtain from the footnote after Eq. (7.41) that in the present case of the hyperbola in Fig. 7.19 $\epsilon = OF/OA$, whereas in the case of the ellipse of Fig. 7.7 $e = OS/ON, OS = ae$):

$$\epsilon = \frac{c}{a}, \quad i.e. \quad \epsilon = \frac{1}{\cos(\Phi/2)}, \tag{7.119}$$

where Φ is the angle between the asymptotes of the hyperbola, as indicated in Fig. 7.19.[*] Since in our case here, the particle comes from infinity, is then scattered or deflected in the domain of the force, and then drifts off to infinity, we have

$$\Phi = \pi - \theta',$$

where θ' is the *scattering angle* as indicated in Fig. 7.19. Since[†]

$$\frac{\pi}{2} < \theta < \frac{3\pi}{2},$$

[‡‡]See *e.g.* D. M. Y. Sommerville [46], p. 66.

[*]The equation of the tangent at A of Fig. 7.19 is $x - a = 0$. The asymptotes to the hyperbola $x^2/a^2 - y^2/b^2 = 1$ are obtained by letting $x, y \to \pm\infty$. Then $y^2/x^2 = b^2/a^2$, and the asymptotes are given by the straight line equations $\pm y/x = b/a, \pm ay = bx$. Thus the asymptote $ay - bx = 0$ and the tangent $x - a = 0$ intersect at $y = b$, as indicated in Fig. 7.19.

[†]This condition on the polar angle θ does not imply that for any ϵ the entire angular domain from $\pi/2$ to $3\pi/2$ is possible. The boundary values $\pi/2, 3\pi/2$ are reached only for $\epsilon = 0$, *i.e.* for the focus F at the origin O.

the focus or pole lies without the hyperbola, as indicated Fig. 7.19. Then

$$\cos\left(\frac{\Phi}{2}\right) = \frac{1}{\epsilon} = \sin\left(\frac{\theta'}{2}\right),$$

and hence (using $\cot^2 x = \text{cosec}^2 x - 1$)

$$\left(\cot\frac{\theta'}{2}\right)^2 = \left(\sin\frac{\theta'}{2}\right)^{-2} - 1 = \epsilon^2 - 1 \overset{(7.118)}{=} \left(\frac{2Es}{ZZ'e^2}\right)^2,$$

$$\text{or } \cot\frac{\theta'}{2} = \frac{2Es}{ZZ'e^2}. \tag{7.120}$$

From this we deduce (using $d\cot x/dx = -\text{cosec}^2 x$):

$$s = \frac{ZZ'e^2}{2E}\cot\frac{\theta'}{2} \quad \Rightarrow \quad ds = -\frac{ZZ'e^2 d\theta'}{4E\sin^2(\theta'/2)^2}.$$

Hence with Eq. (7.113) we obtain (with $\sin\theta' = 2\sin(\theta'/2)\cos(\theta'/2)$ in the second step)

$$\sigma(\theta') = \frac{1}{2}\left(\frac{ZZ'e^2}{2E}\right)^2 \frac{\cot(\theta'/2)}{\sin\theta'(\sin(\theta'/2))^2}$$

$$= \frac{1}{4}\left(\frac{ZZ'e^2}{2E}\right)^2 \frac{1}{\sin^4(\theta'/2)}. \tag{7.121}$$

This expression for the cross section $\sigma(\theta')$ is the *Rutherford formula*, which Rutherford derived originally for the scattering of α-particles from atomic nuclei. Quantum mechanics in its nonrelativistic limit leads to an identical expression. This is therefore a very basic result.

The *total cross section* σ_{tot} is obtained by integrating over all angles Ω, and hence is defined by the relation

$$\sigma_{\text{tot}} = \int \sigma(\Omega)d\Omega = 2\pi\int_0^\pi \sigma(\theta')\sin\theta'd\theta'. \tag{7.122}$$

Inserting here the result of Eq. (7.121), we obtain infinity. Physically this consequence can be understood as follows. The Coulomb potential has an infinite range, *i.e.* any particle incident with an impact parameter s in the range $0 < s < \infty$ is scattered. Only when this infinite range is truncated, *e.g.* by a suitable exponentially decreasing multiplicative factor in the potential — this is known as *screening* — does the scattering cross section assume a finite value. In Nature such screening effects are provided by other effects, like those of the electron shells.

We deduce from Eq. (7.121) and the second of the relations preceding Eq. (7.121) that $\sigma(\theta')$ can also be written

$$\sigma(\theta') = \left(\frac{ds}{d\theta'}\right)^2. \tag{7.123}$$

Thus when $d\theta'/ds = 0$, singularities appear in $\sigma(\theta')$, *i.e.* when

$$\sin^2\frac{\theta'}{2} = 0, \quad \frac{1}{2}(1 - \cos\theta') = 0, \quad \theta' = 2\pi n, n = 0, \pm 1, \dots.$$

The singularities at $0, 2\pi$ are described as *forward and backward glories*. Generally, also in the case of potentials different from the Coulomb potential discussed here, one describes as *rainbow singularities* those angles θ' or their respective impact parameters s, for which

$$\frac{d\theta'}{ds} = 0.$$

In the case of the Coulomb potential this relation yields only the forward and backward singularities.

7.13 Miscellaneous Examples

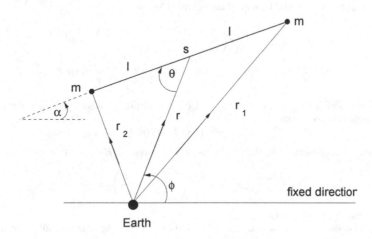

Fig. 7.20   The dumb-bell satellite with masses m and connecting rod of length $2l$.

Example 7.7: The dumb-bell satellite

A dumb-bell satellite consists of two equal point masses m which are connected by a rigid massless rod of length $2l$. Derive the Euler–Lagrange equations of motion of this satellite in a plane in the gravitational field of the Earth (using polar coordinates r and ϕ with respect to the mass centre

and the angle θ as angle of inclination of the dumb-bell axis to the position vector of the centre of mass). What is the law of conservation of energy? Investigate in particular the following special solutions for $l/r_0 < 1$:
(a) $r = r_0 = $ const., $\dot\phi = \dot\phi_0 = $ const., $\theta = 0$, ("*spoke orientation*"),
(b) $r = r_0 = $ const., $\dot\phi = \dot\phi_0 = $ const., $\theta = \pi/2$, ("*spear orientation*").
In which case is the orbital velocity faster or slower than that of a masspoint? Finally show that the motion of a sufficiently small dumb-bell satellite (l/r small) subdivides into the Kepler motion of the mass centre and a rotational motion of the satellite about its centre of mass.

Solution: We choose coordinates as in Fig. 7.20. Let v be the velocity of the centre of mass; this velocity consists of the radial velocity $\dot r$ and the transversal velocity $r\dot\theta$ (*cf.* Example 5.1). The kinetic energy T of the satellite consists of the kinetic energy of the centre of mass and the kinetic energy of the motion around the centre of mass. The kinetic energy of the centre of mass is $(2m)v^2/2$, that of the motion around it is $m[l(\dot\theta - \dot\phi)]^2$.[‡] We thus obtain

$$T = mv^2 + ml^2(\dot\theta - \dot\phi)^2 = m\dot r^2 + mr^2\dot\phi^2 + ml^2(\dot\theta - \dot\phi)^2.$$

The potential energy V is (applying the cosine theorem to the θ-obtuse-angled and θ-acute-angled triangles in Fig. 7.20)

$$V = -m\left(\frac{\mu}{r_1} + \frac{\mu}{r_2}\right) = -m\mu[(r^2 + l^2 + 2rl\cos\theta)^{-1/2} + (r^2 + l^2 - 2rl\cos\theta)^{-1/2}], \qquad (7.124a)$$

where $\mu = GM_{\text{Earth}} = 3986 \times 10^2 \text{km}^3\text{s}^{-2}$. The Lagrangian L is therefore

$$L = T - V = m\dot r^2 + mr^2\dot\phi^2 + ml^2(\dot\theta - \dot\phi)^2 + m\mu[(r^2 + l^2 + 2rl\cos\theta)^{-1/2} + (r^2 + l^2 - 2rl\cos\theta)^{-1/2}].$$

The system is conservative. Hence the Euler–Lagrange equations are

$$\frac{d}{dt}\left(\frac{\partial L}{\partial\dot q_k}\right) - \frac{\partial L}{\partial q_k} = 0,$$

where $q_k = r, \phi, \theta$. We obtain the equations (divided by $m/2$):

$$\ddot r - r\dot\phi^2 = -\frac{\mu}{2}\left[r\left(\frac{1}{r_1^3} + \frac{1}{r_2^3}\right) + l\left(\frac{1}{r_1^3} - \frac{1}{r_2^3}\right)\cos\theta\right],$$

$$\frac{d}{dt}[r^2\dot\phi - l^2(\dot\theta - \dot\phi)] = 0, \quad \ddot\theta - \frac{\mu}{2}\frac{r}{l}\left(\frac{1}{r_1^3} - \frac{1}{r_2^3}\right)\sin\theta = \ddot\phi. \qquad (7.124b)$$

Here r and ϕ describe the orbital motion, and θ the rotation. These motions are coupled. From the second equation we obtain

$$r^2\dot\phi - l^2(\dot\theta - \dot\phi) = \text{const}.$$

Since the system is conservative, the total energy E is given by $E = T + V = $ const. It follows that

$$\dot r^2 + r^2\dot\phi^2 + l^2(\dot\theta - \dot\phi)^2 - \mu\left(\frac{1}{r_1} + \frac{1}{r_2}\right) = \text{const}.$$

We consider two types of solutions:
(a) $r = r_0 = $ const., $\dot\phi = \dot\phi_0 = $ const., $\theta = 0$. This is the "spoke orientation". From the first of Eqs. (7.124b) we obtain

$$-r_0\dot\phi_0^2 = -\frac{\mu}{2}\left[r_0\left\{\frac{1}{(r_0 + l)^3} + \frac{1}{(r_0 - l)^3}\right\} + l\left\{\frac{1}{(r_0 + l)^3} - \frac{1}{(r_0 - l)^3}\right\}\right].$$

[‡] Extending the connecting rod of the satellite by a line to meet the fixed direction at an angle α, we have $\phi = \alpha + \theta, \alpha = \phi - \theta$.

For small values of l/r_0 we obtain from this equation:

$$\dot{\phi}_0^2 \simeq \frac{\mu}{r_0^3}\left\{1 + 3\left(\frac{l}{r_0}\right)^2\right\} \equiv \Omega^2\left\{1 + 3\left(\frac{l}{r_0}\right)^2\right\} > \Omega^2,$$

if we consider $\Omega = (\mu/r_0^3)^{1/2}$ as the angular velocity of a point-like mass (on a circular orbit around the centre of mass).
(b) $r = r_0 = \text{const.}, \dot{\phi} = \dot{\phi}_0 = \text{const.}, \theta = \pi/2$. This is the "arrow orientation". From the first of Eqs. (7.124b) we obtain

$$-r_0\dot{\phi}_0^2 = -\frac{\mu}{2}\left[r_0\frac{2}{(r_0^2+l^2)^{3/2}}\right], \quad i.e. \quad \dot{\phi}_0^2 = \frac{\mu}{r_0^3}\left\{1 + \left(\frac{l}{r_0}\right)^2\right\}^{-3/2} \simeq \frac{\mu}{r_0^3}\left\{1 - \frac{3}{2}\left(\frac{l}{r_0}\right)^2\right\} < \Omega^2.$$
$$(7.124c)$$

We see that in this arrow orientation the dumb-bell is slower than a mass point, and the latter is slower than the dumb-bell in the spoke-orientation.
Finally, when $l/r \ll 1$, we can expand r_1, r_2:

$$\frac{1}{r_1} = (r^2 + l^2 + 2rl\cos\theta)^{-1/2} = \frac{1}{r}\left(1 + 2\frac{l}{r}\cos\theta + \frac{l^2}{r^2}\right)^{-1/2}$$

$$= \frac{1}{r}\left\{1 - \frac{l}{r}\cos\theta + \left(\frac{3}{2}\cos^2\theta - \frac{1}{2}\right)\left(\frac{l}{r}\right)^2 + \cdots\right\},$$

$$\frac{1}{r_2} = \frac{1}{r}\left\{1 + \frac{l}{r}\cos\theta + \left(\frac{3}{2}\cos^2\theta - \frac{1}{2}\right)\left(\frac{l}{r}\right)^2 + \cdots\right\}. \quad (7.124d)$$

We obtain therefore:

$$\frac{1}{r_1^3} + \frac{1}{r_2^3} = \frac{1}{r^3}\left\{2 + (15\cos^2\theta - 3)\left(\frac{l}{r}\right)^2 + \cdots\right\}, \quad \frac{1}{r_1^3} - \frac{1}{r_2^3} = \frac{1}{r^3}\left\{-6\frac{l}{r}\cos\theta + \cdots\right\}.$$

Inserting these expressions into Eqs. (7.124b), we obtain

$$\ddot{r} - r\dot{\phi}^2 = -\frac{\mu}{r^2}\left[1 + \frac{3}{2}(3\cos^2\theta - 1)\left(\frac{l}{r}\right)^2 + \cdots\right], \quad \frac{d}{dt}\left[r^2\left\{\dot{\phi} + \left(\frac{l}{r}\right)^2(\dot{\phi} - \dot{\theta})\right\}\right] = 0,$$

$$\ddot{\theta} + \frac{3\mu}{2r^3}\sin 2\theta + \left(\frac{l}{r}\right)^2(\text{const.} + \cdots) = \ddot{\phi}. \quad (7.124e)$$

We observe that the first two of these equations which describe the orbital motion involve θ only in terms of the order of l^2/r^2. If we ignore these contributions, the orbital and rotational motions decouple and we obtain:

$$\ddot{r} - r\dot{\phi}^2 = -\frac{\mu}{r^2}, \quad \frac{d}{dt}(r^2\dot{\phi}) = 0, \quad \ddot{\theta} + \frac{3\mu}{2r^3}\sin 2\theta = 0. \quad (7.124f)$$

The first two of these equations describe the Kepler motion, and the last equation an independent rotational motion.

Example 7.8: Lifting a satellite into a higher orbit
A satellite encircling the Earth on a circular orbit of radius r_1 is to be lifted into a concentric circular orbit of radius $r_2 > r_1$ by means of two rocket impulses. The required amount of fuel is optimized if — as indicated in Fig. 7.21 — the transition is made via an elliptic orbit with the Earth at one of the foci. Compute as a function of r_1 and r_2 the required velocity increment for the transition, *i.e.* the quantity

$$\Delta v = (v_{\max} - v_{r_1}) + (v_{r_2} - v_{\min}),$$

where v_{r_1}, v_{r_2} are the velocities on the circular orbits, and $v_{\max}$ and $v_{\min}$ are the maximum and minimum velocities on the elliptic orbit.

Solution: The velocities of the circular orbits are (see explanation below)

$$v_{r_1} = \sqrt{\frac{\mu}{r_1}}, \qquad v_{r_2} = \sqrt{\frac{\mu}{r_2}}.$$

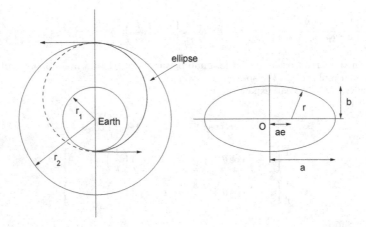

Fig. 7.21 Transition from one circular orbit to another via an elliptic orbit.

Here μ is defined such that Newton's gravitational law is $f = -m\mu/r^2$. For the Earth as a gravitating mass the value of μ is $\mu = GM_{\text{Earth}} = 3986 \times 10^2 \text{km}^3\text{s}^{-2}$. The velocities given above are special cases of the relation (7.76), *i.e.*

$$v^2 = \frac{2\mu}{r} - \frac{\mu}{a}$$

for the velocity of a particle at point r on an elliptic orbit. This equation follows from the law of conservation of energy with the condition that the orbit does not end at infinity. The velocity v is respectively a maximum or minimum at an apsis when $r \to r_1 = a(1 - e)$ and $r \to r_2 = a(1 + e)$ respectively where e is the eccentricity[§] of the ellipse (*cf.* Fig. 7.21 and the footnote after Eq. (7.41)). Then

$$v_{\max}^2 = \frac{2\mu}{a(1 - e)} - \frac{\mu}{a} = \frac{\mu}{a}\frac{1 + e}{1 - e}, \qquad v_{\min}^2 = \frac{2\mu}{a(1 + e)} - \frac{\mu}{a} = \frac{\mu}{a}\frac{1 - e}{1 + e}.$$

Thus for a circular orbit of radius r (eccentricity $e = 0$, *cf* Eq. (7.50)) we have $v_{\max} = v_{\min} = \sqrt{\mu/r}$. But (*cf.* Fig. 7.21)

$$\frac{r_{\min}}{r_{\max}} = \frac{1 - e}{1 + e} = \frac{r_1}{r_2}, \quad \text{and} \quad r_1 + r_2 = 2a, \quad a = \frac{1}{2}(r_1 + r_2),$$

so that

$$v_{\max} = \sqrt{\frac{\mu(1 + e)}{a(1 - e)}} = \sqrt{\frac{\mu}{a}\frac{r_2}{r_1}} = \sqrt{\frac{2\mu r_2}{(r_1 + r_2)r_1}}, \qquad v_{\min} = \sqrt{\frac{2\mu r_1}{(r_1 + r_2)r_2}},$$

[§]Whenever there is no confusion with a charge e, we use again e instead of ϵ to denote the eccentricity.

and hence

$$\triangle v = \sqrt{\frac{2\mu r_2}{(r_1 + r_2)r_1}} - \sqrt{\frac{\mu}{r_1}} + \sqrt{\frac{\mu}{r_2}} - \sqrt{\frac{2\mu r_1}{(r_1 + r_2)r_2}}.$$

Example 7.9: Calculation of the lengths of the seasons

The Earth moves on an elliptic orbit with the sun at a focus. Calculate the time t it takes the Earth to travel from the polar angle $\theta = 0$ at the perihelion to an angle θ. Express the parameters appearing in the equation in terms of the eccentricity e and the frequency n (equal to $2\pi/T, T$ = period). Without complete evaluation of the resulting integral for t, obtain an approximation for small values of e. In the case of the Earth $e = 1/60$ and $T = 365.24$ days. Assuming that the Earth is on March 21 at angular position $\theta = 78°47''$, calculate the time-length of spring to a precision of hours. By reversal of the expression for t, obtain a formula for the angle θ which is valid for small eccentricities. Up to the next angular minute calculate the largest angular motion in one day.

Solution: Our starting point is the polar equation of an ellipse, *i.e.*

$$\frac{l'}{r} = 1 + e\cos\theta, \quad \text{where} \quad l' \stackrel{(7.83)}{=} a(1 - e^2) \stackrel{(7.49)}{=} \frac{l^2}{\mu m^2}. \tag{7.125}$$

Here l' is the *semi-latus rectum*, e is the eccentricity, a, b are the lengths of the x, y semi-axes with $a > b$, l is the orbital angular momentum, and μ is the gravitational constant multiplied by the mass of the Earth. We have (see Eq. (7.83))

$$r^2\dot{\theta} = \frac{l}{m} = \sqrt{\mu l'}. \tag{7.126a}$$

For the frequency n we have (*cf.* Kepler's third law, Eq. (7.84))

$$n = \left(\frac{\mu}{a^3}\right)^{1/2}. \tag{7.126b}$$

Eliminating r from Eqs. (7.125) and (7.126a) we obtain

$$\dot{\theta} = \frac{d\theta}{dt} = \frac{\sqrt{\mu l'}}{r^2} = \frac{\sqrt{\mu}}{l'^{3/2}}(1 + e\cos\theta)^2, \quad dt = \frac{l'^{3/2}}{\sqrt{\mu}}\frac{d\theta}{(1 + e\cos\theta)^2}. \tag{7.127}$$

It follows that (using $l'/a = (1 - e^2)$)

$$ndt = (1 - e^2)^{3/2}\frac{d\theta}{(1 + e\cos\theta)^2}, \quad \int_0^t dt = \frac{l'^{3/2}}{\mu^{1/2}}\int_0^\theta \frac{d\theta}{(1 + e\cos\theta)^2}. \tag{7.128}$$

The integral can be looked up in Tables of Integrals, or else one can evaluate it oneself as follows:

$$1 + e\cos\theta = [\sin^2(\theta/2) + \cos^2(\theta/2)] + e[\cos^2(\theta/2) - \sin^2(\theta/2)] = (1+e)\cos^2(\theta/2) + (1-e)\sin^2(\theta/2).$$

Hence

$$\frac{1}{1 + e\cos\theta} = \frac{\sec^2(\theta/2)}{(1+e) + (1-e)\tan^2(\theta/2)}, \quad \int \frac{d\theta}{(1+e\cos\theta)^2} = \int \frac{\sec^4(\theta/2)d\theta}{[(1+e) + (1-e)\tan^2(\theta/2)]^2}.$$

We now set

$$\tan\phi \equiv \sqrt{\frac{1-e}{1+e}}\tan\frac{\theta}{2}, \quad \therefore \quad \tan\frac{\theta}{2} = \sqrt{\frac{1+e}{1-e}}\tan\phi.$$

Differentiation yields

$$\sec^2\frac{\theta}{2}d\theta = 2\sqrt{\frac{1+e}{1-e}}\sec^2\phi d\phi.$$

Furthermore we have

$$(1+e) + (1-e)\tan^2\frac{\theta}{2} = (1+e) + (1-e)\left(\frac{1+e}{1-e}\right)\tan^2\phi = (1+e)(1+\tan^2\phi) = (1+e)\sec^2\phi.$$

Using $\sec^2(\theta/2) = 1 + \tan^2(\theta/2) = 1 + \tan^2\phi(1+e)/(1-e)$, the integral becomes

$$\int \frac{d\theta}{(1+e\cos\theta)^2} = \int \frac{2\sqrt{(1+e)/(1-e)}\sec^2\phi\,d\phi}{(1+e)^2\sec^4\phi}\left\{1 + \frac{1+e}{1-e}\tan^2\phi\right\}$$

$$= \frac{2}{(1-e^2)^{3/2}}\int \frac{(1-e)+(1+e)\tan^2\phi}{\sec^2\phi}d\phi$$

$$= \frac{2}{(1-e^2)^{3/2}}\int [(1-e)\cos^2\phi + (1+e)\sin^2\phi]d\phi$$

$$= \int \frac{2(1-e\cos 2\phi)d\phi}{(1-e^2)^{3/2}} = \frac{2}{(1-e^2)^{3/2}}\left[\phi - \frac{e}{2}\sin 2\phi\right]. \qquad (7.129a)$$

We now have to re-express ϕ again in terms of θ:

$$\sin 2\phi = \frac{2\tan\phi}{1+\tan^2\phi} = \frac{2[(1-e)/(1+e)]^{1/2}\tan(\theta/2)}{1+[(1-e)/(1+e)]\tan^2(\theta/2)}.$$

With the help of the half-angle formulas

$$\sin\theta = \frac{2\tan(\theta/2)}{1+\tan^2(\theta/2)}, \qquad \cos\theta = \frac{1-\tan^2(\theta(2)}{1+\tan^2(\theta/2)},$$

we obtain *e.g.* by *componendo et dividendo*

$$\tan^2\frac{\theta}{2} = \frac{1-\cos\theta}{1+\cos\theta}, \quad \text{and} \quad 2\tan\frac{\theta}{2} = \sin\theta\left(\frac{2}{1+\cos\theta}\right).$$

We obtain therefore (with $(1+e)(1+\cos\theta) + (1-e)(1-\cos\theta) = 2(1+e\cos\theta)$):

$$\sin 2\phi = \frac{\sqrt{(1-e)/(1+e)}\sin\theta[2/(1+\cos\theta)]}{1+[(1-e)/(1+e)](1-\cos\theta)/(1+\cos\theta)} = \frac{(1-e^2)^{1/2}\sin\theta}{(1+e\cos\theta)}.$$

Hence we can now write the integral:

$$\int \frac{d\theta}{(1+e\cos\theta)^2} = \frac{1}{(1-e^2)^{3/2}}\left[2\tan^{-1}\left\{\sqrt{\frac{1-e}{1+e}}\tan\frac{\theta}{2}\right\} - \frac{e\sqrt{1-e^2}\sin\theta}{1+e\cos\theta}\right]. \qquad (7.129b)$$

Returning to Eq. (7.128) we obtain (with Eq. (7.125)

$$t = \frac{a^{3/2}(1-e^2)^{3/2}}{\sqrt{\mu}}\int \frac{d\theta}{(1+e\cos\theta)^2} = \frac{a^{3/2}}{\sqrt{\mu}}\left[2\tan^{-1}\left\{\sqrt{\frac{1-e}{1+e}}\tan\frac{\theta}{2}\right\} - \frac{e\sqrt{1-e^2}\sin\theta}{1+e\cos\theta}\right],$$

or (*cf.* Eq. (7.126b))

$$nt = 2\tan^{-1}\left\{\sqrt{\frac{1-e}{1+e}}\tan\frac{\theta}{2}\right\} - \frac{e\sqrt{1-e^2}\sin\theta}{1+e\cos\theta}. \qquad (7.130)$$

Case of small e: We return to Eq. (7.128) and consider:

$$ndt = (1-e^2)^{3/2}(1+e\cos\theta)^{-2}d\theta = \left(1 - \frac{3}{2}e^2 - \cdots\right)(1 - 2e\cos\theta + 3e^2\cos^2\theta - \cdots)d\theta$$

$$= \left(1 - 2e\cos\theta + \frac{3}{2}e^2\cos 2\theta + \cdots\right)d\theta.$$

We integrate this expression termwise with the initial condition $\theta = 0$ at time $t = 0$ and obtain:

$$nt = \theta - 2e\sin\theta + \frac{3}{4}e^2\sin 2\theta + \cdots, \quad \text{or} \quad \theta = nt + 2e\sin\theta - \frac{3}{4}e^2\sin 2\theta - \cdots.$$

We now iterate this expression:

$$
\begin{aligned}
\theta &\simeq nt + 2e\sin\{nt + 2e\sin(nt)\} - \frac{3}{4}e^2\sin 2(nt + 2e\sin nt) - \cdots \\
&= nt + 2e[\sin nt\cos(2e\sin nt) + \cos nt\sin(2e\sin nt)] \\
&\quad - \frac{3}{4}e^2[\sin 2nt\cos(4e\sin nt) + \cos 2nt\sin(4e\sin nt)] \\
&= nt + 2e\sin nt + 4e^2\cos nt\sin nt - \frac{3}{4}e^2\sin 2nt + O(e^3) \\
&= nt + 2e\sin nt + \frac{5}{4}e^2\sin 2nt + O(e^3).
\end{aligned}
\tag{7.131}
$$

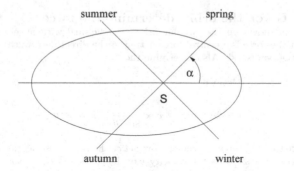

Fig. 7.22 Calculation of the lengths of the seasons.

We are now in a position to calculate the lengths of the seasons. On the 21st of March the Earth is at a position on the Kepler orbit where $\theta = \alpha = 78°47'$ (called *spring equinox*). We use Eq. (7.131):

$$nt \simeq \theta - 2e\sin\theta, \tag{7.132}$$

where $n = 2\pi/T$, T being the period. Let t_{spsu} (spring to summer) be the duration of the spring season (on the northern hemisphere). Then (*cf.* Fig. 7.22)

$$
\begin{aligned}
nt_{\text{spsu}} &= nt_{\text{summer}} - nt_{\text{spring}} = (\theta_{\text{summer}} - 2e\sin\theta_{\text{summer}}) - (\theta_{\text{spring}} - 2e\sin\theta_{\text{spring}}) \\
&= \frac{\pi}{2} - 2e[\sin\theta_{\text{summer}} - \sin\theta_{\text{spring}}] = \frac{\pi}{2} - 2e\left[\sin\left(\alpha + \frac{\pi}{2}\right) - \sin\alpha\right] \\
&= \frac{\pi}{2} - 2e(\cos\alpha - \sin\alpha).
\end{aligned}
\tag{7.133}
$$

Replacing n by $2\pi/T$ (where $T = 365.24$ days), we obtain (in days and given that $e = 1/60$ and $\alpha = 78°47'$, $\cos\alpha = 0.194$, $\sin\alpha = 0.981$) and multiplying through by $T/2\pi$:

$$t_{\text{spsu}} = \frac{T}{4} - \frac{eT}{\pi}(\cos\alpha - \sin\alpha) = \frac{365.24}{4} - 0.376 + 1.910. \tag{7.134}$$

This means, the length of spring is $t_{\text{spsu}} = 92.84$ days. Similarly one obtains for the lengths of summer: 93.60 days, autumn: 89.78 days, winter: 89.02 days.

In the next step we use Eq. (7.131), $i.e.$

$$\theta = nt + 2e\sin nt + \frac{5}{4}e^2 \sin 2nt + O(e^3). \tag{7.135}$$

The largest angular motion in a day is obtained by setting in this equation $t = 1$ day, $i.e.$

$$nt = \frac{2\pi \times \text{one day}}{365.24 \text{ days}}, \quad e = \frac{1}{60}.$$

Then

$$
\begin{aligned}
\theta &\simeq \frac{2\pi}{365} + \frac{2\times 1}{60}\sin\frac{2\pi}{365} + \frac{5}{4}\left(\frac{1}{60}\right)^2 \sin\frac{4\pi}{365} \\
&= 0.0172 + 0.0333\sin(0.0172) + 0.0003\sin(0.344) \\
&= 0.0172 + 0.00057 + 0.00010 = 0.01779 \text{ radian} = 1°1'.
\end{aligned}
\tag{7.136}
$$

This result is what one would expect approximately for a total of 360 degrees in 365 days.

Example 7.10: Given the orbit, determine the force

A particle of mass m moves on the circular orbit $r = 2a\sin\theta$ with a velocity proportional to $1/\sin^2\theta$. Show that the force is a central force acting in the direction towards $r = 0$ ($i.e.$ that the transverse acceleration vanishes). Also determine the force.

Solution: We are given the relations

$$r = 2a\sin\theta, \quad v \propto \frac{1}{\sin^2\theta} = k\text{cosec}^2\theta.$$

We first demonstrate that the force is a central force ($i.e.$ the transverse acceleration is zero), and so $r^2\dot\theta = \text{const}$. From the expression for the velocity v of the particle, $i.e.$ $v^2 = \dot r^2 + r^2\dot\theta^2 = k^2\text{cosec}^4\theta$, we obtain therefore

$$(2a\dot\theta\cos\theta)^2 + r^2\dot\theta^2 = k^2\left(\frac{2a}{r}\right)^4,$$

and since $r^2 = 4a^2\sin^2\theta$,

$$4a^2\dot\theta^2(\cos^2\theta + \sin^2\theta) = \frac{16a^4 k^2}{r^4}, \quad \dot\theta^2 = \frac{4a^2 k^2}{r^4}, \quad \dot\theta = \frac{2ak}{r^2}, \quad r^2\dot\theta = 2ak = \text{const}.$$

The transverse component of the acceleration is ($cf.$ Example 5.1)

$$\frac{1}{r}\frac{d}{dt}\left(r^2\dot\theta\right).$$

It follows that in the present case the transverse component of the acceleration vanishes, $i.e.$ the force acts radially and is therefore a central force.

Next we determine the force. From $r = 2a \sin \theta$, we obtain

$$\dot{r} = 2a\dot{\theta}\cos\theta = 2a\cos\theta \frac{2ak}{r^2} = \frac{4a^2 k \cos\theta}{r^2}.$$

Differentiating this expression with respect to t, we obtain:

$$\ddot{r} = 4a^2 k \left\{ -\frac{2\cos\theta}{r^3}\dot{r} + \frac{1}{r^2}(-\dot{\theta}\sin\theta) \right\} = -4a^2 k \left\{ \frac{2}{r^3} 2a\dot{\theta}\cos^2\theta + \frac{\dot{\theta}}{r^2}\sin\theta \right\}$$

$$= -4a^2 k \left\{ \frac{2}{r^3} 2a\cos^2\theta + \frac{1}{r^2}\sin\theta \right\} \frac{2ak}{r^2}. \qquad (7.137)$$

Also

$$r\dot{\theta}^2 = r \frac{4a^2 k^2}{r^4} = \frac{4a^2 k^2}{r^3}.$$

Equation (7.137) can therefore be written:

$$\ddot{r} - r\dot{\theta}^2 = -\frac{8a^3 k^2}{r^2} \left\{ \frac{4a\cos^2\theta}{r^3} + \frac{1}{r^2}\sin\theta \right\} - \frac{4a^2 k^2}{r^3}$$

$$= -\frac{8a^3 k^2}{r^2} \left\{ \frac{4a(1 - \sin^2\theta)}{r^3} + \frac{1}{r^2}\sin\theta \right\} - \frac{4a^2 k^2}{r^3}. \qquad (7.138)$$

With $r = 2a\sin\theta$ we obtain

$$\ddot{r} - r\dot{\theta}^2 = -\frac{8a^3 k^2}{r^2} \left\{ \frac{4a}{r^3} - \frac{4a}{r^3}\frac{r^2}{4a^2} + \frac{r}{2a}\frac{1}{r^2} \right\} - \frac{4a^2 k^2}{r^3} = -\frac{8a^3 k^2}{r^2} \left\{ \frac{4a}{r^3} - \frac{1}{ra} + \frac{1}{2ar} \right\} - \frac{4a^2 k^2}{r^3}$$

$$= -\frac{8a^3 k^2}{r^2} \left\{ \frac{4a}{r^3} - \frac{1}{2ar} \right\} - \frac{4a^2 k^2}{r^3} = -\frac{32a^4 k^2}{r^5}, \qquad (7.130)$$

Thus the force is attractive and proportional to $1/r^5$. The problem shows how the law of force can be found if the orbit is given.

The force can also be determined by an alternative method. Let p be the length of the perpendicular from the pole to the tangent to the orbit at the point r. The angular momentum of the particle is then $l = mpv$, and for the force we have Eq. (7.66), i.e.

$$f = -\frac{l^2}{mp^3}\frac{dp}{dr}, \quad i.e. \quad f = -\frac{m^2 p^2 v^2}{mp^3}\frac{dp}{dr} = -\frac{mv^2}{p}\frac{dp}{dr}.$$

Next we require the polar equation of a circle expressed in terms of p and r. This is given by Eq. (7.59b) as $r^2 = 2ap$ and gives us p as a function of r, i.e. $p = r^2/2a$. Differentiating this expression we obtain

$$\frac{dp}{dr} = \frac{r}{a}.$$

With this relation we can now determine the force f:

$$f = -\frac{mv^2}{p}\frac{r}{a} = -\frac{mv^2}{r^2/2a}\frac{r}{a} = -\frac{2mv^2}{r}.$$

But we are given (see beginning) that $v = k/\sin^2\theta = k(2a/r)^2$. The force is therefore given by

$$f = -\frac{2m}{r}k^2 \left(\frac{2a}{r} \right)^4 = -\frac{32ma^4 k^2}{r^5},$$

which is in agreement with the acceleration of the first derivation.

Example 7.11: Given the force, determine the orbit

A particle of mass m moves in the field of the attractive central force per unit mass $f/m = -k^2[2(a^2 + b^2)u^5 - 3a^2b^2u^7], u = 1/r$. Here a, b, k are constants. At an apsis, distance a away from the centre of the force, the particle was shot into orbit with a velocity $v = k/a$. Show that the orbit is given by

$$r^2 = a^2 \cos^2 \theta + b^2 \sin^2 \theta.$$

Solution: As in the text we have, with the given initial condition,

$$\frac{l}{m} = r^2 \dot\theta = rv = \text{const.} = k,$$

where l is the angular momentum. We obtain the orbit with Eq. (7.25), *i.e.*

$$\frac{d^2u}{d\theta^2} + u = -\frac{mf}{l^2u^2} = 2(a^2 + b^2)u^3 - 3a^2b^2u^5.$$

We integrate this equation with the help of the relation

$$\frac{d^2u}{d\theta^2} = \frac{d}{du}\left[\frac{1}{2}\left(\frac{du}{d\theta}\right)^2\right],$$

and obtain

$$\frac{1}{2}\left(\frac{du}{d\theta}\right)^2 + \frac{u^2}{2} = \frac{1}{2}(a^2 + b^2)u^4 - \frac{1}{2}a^2b^2u^6 + \text{const.}$$

Initially $u = 1/a$ at an apsis, where $du/d\theta = 0$. This implies the relation

$$\frac{1}{a^2} = \frac{a^2 + b^2}{a^4} - \frac{b^2}{a^4} + 2(\text{const.}), \quad \therefore \text{ const.} = 0.$$

Next we set $u^2 = 1/q, q = r^2$. Then

$$2u\frac{du}{d\theta} = -\frac{1}{q^2}\frac{dq}{d\theta}, \quad \therefore \left(\frac{du}{d\theta}\right)^2 = \frac{1}{4q^3}\left(\frac{dq}{d\theta}\right)^2.$$

With this and the earlier result we obtain

$$\frac{dq}{d\theta} = \pm 2q^{3/2}\left(\frac{du}{d\theta}\right) = \pm 2q^{3/2}\left[\frac{a^2 + b^2}{q^2} - \frac{a^2b^2}{q^3} - \frac{1}{q}\right]^{1/2}$$
$$= \pm 2[q(a^2 + b^2) - a^2b^2 - q^2]^{1/2} = \pm 2[(a^2 - q)(q - b^2)]^{1/2}.$$

Instead of integrating this expression, we proceed as follows. We consider the equation (observe we write ϕ here, not θ)

$$q = \frac{1}{u^2} = r^2 = a^2 \cos^2 \phi + b^2 \sin^2 \phi.$$

Differentiation with respect to ϕ yields:

$$\frac{dq}{d\phi} = -2a^2 \cos\phi \sin\phi + 2b^2 \sin\phi \cos\phi = 2(b^2 - a^2)\cos\phi \sin\phi.$$

However, $q = a^2(1 - \sin^2\phi) + b^2 \sin^2\phi = a^2 + (b^2 - a^2)\sin^2\phi$, or

$$q = a^2 \cos^2 \phi + b^2 \sin^2 \phi = a^2 \cos^2 \phi + b^2(1 - \cos^2\phi) = b^2 + (a^2 - b^2)\cos^2\phi.$$

Hence $dq/d\phi$ becomes:

$$\frac{dq}{d\phi} = 2(b^2 - a^2)\sqrt{\frac{q - b^2}{a^2 - b^2}}\sqrt{\frac{q - a^2}{b^2 - a^2}} = \pm 2\sqrt{(a^2 - q)(q - b^2)}.$$

Comparison with the expression we had for $dq/d\theta$ above now shows that the orbit is given by the equation

$$r^2 = a^2 \cos^2\theta + b^2 \sin^2\theta.$$

This problem demonstrates how the orbit can be found if the force is given.

Example 7.12: A steamer encircling a lighthouse

A steamer is cruising around a lighthouse, thereby its velocity v relative to the water is always perpendicular to the line connecting it to the lighthouse. The velocity $u < v$ is the velocity of an additional current in the seawater. Determine the orbit of the steamer.

Solution: We choose the angle θ and the geometry as in Fig. 7.23. Then (in the usual notation)

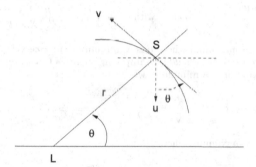

Fig. 7.23 The lighthouse at L and the steamer at S.

$$\dot{r} = -u\sin\theta, \quad r\dot\theta = v - u\cos\theta.$$

It follows that

$$-\frac{u\sin\theta}{v - u\cos\theta} = \frac{\dot{r}}{r\dot\theta} = \frac{1}{r}\frac{dr}{d\theta},$$

so that

$$\int \frac{dr}{r} = -\int \frac{u\sin\theta}{v - u\cos\theta}\,d\theta, \quad \ln r = -\ln(v - u\cos\theta) + \text{const.}, \quad r = \frac{\text{const.}}{v - u\cos\theta} = \frac{\text{const.}/v}{1 - u\cos\theta/v}.$$

Comparison with the polar equation of a conical section, *i.e.* $r = k/(1 - e\cos\theta)$, shows that the orbit is an ellipse with eccentricity $e = u/v < 1$.

Example 7.13: Apsidal precessions of planetary orbits

A planet moves on an orbit which is determined by a corrected central force per unit mass f/m of the form

$$\frac{f}{m} = -\left(\frac{\mu}{r^2} + \frac{\lambda}{r^3}\right), \quad \lambda \text{ small.}$$

Show that the correction term λ/r^3 leads to a precession of the apsides. What is under the given law of force the condition for the stability of the planet on a circular orbit of radius a? Assuming this condition to be satisfied, determine the resulting orbit and the angles of the apsides.

Solution: With the substitution $u = 1/r$, Eq. (7.24), the equation of motion of mass m, i.e. $m\ddot{r} - mr^2\dot{\theta} = f$, assumes — with $l/m = r^2\dot{\theta} = \text{const.}$ and $h^2 = l^2/m^2$ — the form (cf. Eq. (7.25)):

$$\frac{d^2u}{d\theta^2} + u = -\frac{mf}{l^2}\frac{1}{u^2} \equiv -\frac{f/m}{h^2u^2}, \qquad \frac{f}{m} = -(\mu u^2 + \lambda u^3),$$

and hence

$$\frac{d^2u}{d\theta^2} + u = \frac{1}{h^2}(\mu + \lambda u) \qquad \text{or} \qquad \frac{d^2u}{d\theta^2} + u\left(1 - \frac{\lambda}{h^2}\right) = \frac{\mu}{h^2}. \qquad (7.140)$$

The solution of this equation can be written (by first setting $u' = u - \mu/(h^2 - \lambda)$ and changing from θ to $\theta' = \theta\sqrt{1 - \lambda/h^2}$)

$$u = A\cos n(\theta - \alpha) + \frac{\mu}{h^2 - \lambda}, \qquad \text{where} \qquad n^2 = 1 - \frac{\lambda}{h^2}, \qquad (7.141)$$

and A and α are constants to be determined by boundary conditions. Choosing θ such that $\alpha = 0$, we can rewrite the equation as

$$\frac{l'}{r} = 1 + e\cos n\theta \qquad \text{with} \qquad A = \frac{e}{l'}, \qquad \frac{\mu}{h^2 - \lambda} = \frac{1}{l'}. \qquad (7.142)$$

The radial distance r is a maximum where u is a minimum and *vice versa*. At these points which are the apsides, we have therefore $du/d\theta = 0$, implying $\sin n\theta = 0$, i.e. $n\theta = 0, \pi, 2\pi, \dots$. Thus (cf. Eq. (7.141)) u is a maximum, r a minimum, where

$$\theta = 0, \quad \frac{2\pi}{\sqrt{1 - \lambda/h^2}}, \quad \frac{4\pi}{\sqrt{1 - \lambda/h^2}}, \quad \dots, \qquad (7.143a)$$

and u is a minimum, r a maximum, where

$$\theta = \frac{\pi}{\sqrt{1 - \lambda/h^2}}, \quad \frac{3\pi}{\sqrt{1 - \lambda/h^2}}, \quad \dots. \qquad (7.143b)$$

We expand the denominators of these expressions for $\lambda/h^2 < 1$. Then:

$$\frac{\pi}{\sqrt{1 - \lambda/h^2}} \simeq \pi\left(1 + \frac{1}{2}\frac{\lambda}{h^2}\right).$$

We observe that with every *complete* rotation the apsides shift by an angular amount of $\pi\lambda/h^2$; the associated motion is the *precession*. Thus we obtain as orbit or trajectory of the planet a continuously rotating ellipse. The angular velocity $\omega = 2\pi\nu$, ν frequency, is in terms of the period $T = 1/\nu$:

$$\omega = \frac{1}{T}\left(2\pi + \frac{2\pi}{\sqrt{1 - \lambda/h^2}}\right).$$

We now consider the conditions under which the planet is stable on a circular orbit of radius a. For this circular orbit $r = 1/u = a$. We insert this constant value into Eq. (7.140) and obtain

$$\frac{1}{a} = \frac{m^2}{l^2}\left(\mu + \frac{\lambda}{a}\right), \qquad \text{i.e.} \qquad \frac{m^2}{l^2} = \frac{1}{a\mu + \lambda}.$$

For the perturbed circular orbit we set, with χ, ξ, δ small compared with $a, 1/a, 1$ respectively:

$$r = a + \chi, \qquad u = \frac{1}{a} + \xi, \qquad \text{and} \qquad h^2 \equiv \frac{l^2}{m^2} = (a\mu + \lambda)(1 + \delta).$$

Here δ is fixed since l is constant, but χ and ξ are variables. Inserting these expressions into Eq. (7.140), we obtain:

$$\frac{d^2}{d\theta^2}\left[\frac{1}{a}+\xi\right]+\left[\frac{1}{a}+\xi\right]=\frac{\mu+\lambda(1/a+\xi)}{(a\mu+\lambda)(1+\delta)}=\frac{a\mu+\lambda+a\lambda\xi}{a(a\mu+\lambda)(1+\delta)},$$

or

$$\frac{d^2\xi}{d\theta^2}+\xi\simeq\left[\frac{a\mu+\lambda+a\lambda\xi}{a(a\mu+\lambda)}\right](1-\delta)-\frac{1}{a}=\frac{\lambda\xi}{a\mu+\lambda}-\left[\frac{a\mu+\lambda+a\lambda\xi}{a(a\mu+\lambda)}\right]\delta.$$

Thus (neglecting $a\lambda\xi$)

$$\left(\frac{a\mu+\lambda}{a\mu}\right)\frac{d^2\xi}{d\theta^2}+\xi\simeq-\frac{\delta}{a}\left(\frac{a\mu+\lambda}{a\mu}\right). \tag{7.144}$$

Setting

$$\theta'\equiv\theta\left(\frac{a\mu}{a\mu+\lambda}\right)^{1/2},\qquad\xi'=\xi+\frac{\delta}{a}\left(\frac{a\mu+\lambda}{a\mu}\right),$$

the equation can be rewritten as

$$\frac{d^2\xi'}{d\theta'^2}+\xi'=0. \tag{7.145}$$

The solution of this equation is

$$\xi'=A\sin(\theta'+\alpha),\qquad A,\alpha\quad\text{const.}$$

Expressed in terms of ξ and θ, this solution is

$$\xi=-\frac{\delta}{a}\left(\frac{a\mu+\lambda}{a\mu}\right)+A\sin\left(\sqrt{\frac{a\mu}{a\mu+\lambda}}\theta+\alpha\right),\qquad\text{provided}\qquad\frac{a\mu}{a\mu+\lambda}>0. \tag{7.146}$$

The inequality is the condition for the *stability of the circular orbit* (*i.e.* for the bounded trigonometric solution instead of an unbounded exponential solution), since the planet oscillates about the circular orbit with the frequency $\sqrt{a\mu/(a\mu+\lambda)}$. If $a\mu/(a\mu+\lambda)$ were negative, ξ could be a linear combination of $\exp(-c\theta),\exp(+c\theta)$ where $c=|a\mu/(a\mu+\lambda)|$. In such a case the orbit would not remain close to the circular path.

We return to expressions in terms of r, taking into account the inequality of Eq. (7.146). Then

$$r=\frac{1}{u}=\frac{1}{1/a+\xi}=\frac{a}{1+a\xi}\simeq a(1-a\xi).$$

With the solution of Eq. (7.146) we obtain therefore

$$r\simeq a-a^2\xi=a-a^2\left[-\frac{\delta}{a}\left(\frac{a\mu+\lambda}{a\mu}\right)+A\sin\left(\sqrt{\frac{a\mu}{a\mu+\lambda}}\theta+\alpha\right)\right],$$

i.e. we obtain the orbit:

$$r=a+\delta\frac{a\mu+\lambda}{\mu}-a^2A\sin\left(\sqrt{\frac{a\mu}{a\mu+\lambda}}\theta+\alpha\right). \tag{7.147}$$

We see now that r oscillates about $a+\delta(a\mu+\lambda)/\mu$ with the frequency $\sqrt{a\mu/(a\mu+\lambda)}$. The apsides are obtained from the condition:

$$\frac{dr}{d\theta}=0,\qquad\theta=\theta_{\text{apsis}}.$$

From Eq. (7.147) we obtain:

$$\frac{dr}{d\theta}=-a^2A\sqrt{\frac{a\mu}{a\mu+\lambda}}\cos\left(\sqrt{\frac{a\mu}{a\mu+\lambda}}\theta+\alpha\right).$$

This expression vanishes for

$$
\sqrt{\frac{a\mu}{a\mu + \lambda}}\,\theta_{\text{apsis}} + \alpha = \pm\frac{1}{2}(2n+1)\pi, \quad n = 1, 2, 3, \ldots
$$

Choosing $\alpha = 0$, the angular positions of the apsides are given by

$$
\theta_{\text{apsis}} = \mp\sqrt{\frac{a\mu + \lambda}{a\mu}}\left(n + \frac{1}{2}\right)\pi. \tag{7.148}
$$

Example 7.14: Elliptic orbits and old quantum mechanics

The *Bohr–Sommerfeld–Wilson quantization conditions* per degree of freedom[¶] in the present case of elliptic orbits given by the polar equation $1/r = c_1 + c_2\cos\theta$ and of eccentricity e are given by the following integrals ("quantum conditions" involving integral multiples of the Planck constant h):

$$
\oint p_\theta\, d\theta = n_\theta h, \qquad \oint p_r\, dr = n_r h, \tag{7.149}
$$

where $n_\theta, n_r = 1, 2, \ldots$, and p_θ, p_r are the momenta associated with the generalized coordinates θ, r.

(a) If $n = n_\theta + n_r$ and b and a are the lengths of the semi-major and -minor axes of the elliptic orbit of an electron around an atomic nucleus, also given that $b/a = (1 - e^2)^{1/2}$ (*cf.* Eq. (7.83)), show that $n_\theta/n = b/a$.

(b) In the case of the Coulomb potential $V(r) = -e_0^2 Z/r$, show that together with the total energy $W = T + V = $ const. (a conservative system and hence W independent of θ) the above quantization conditions lead to the *Bohr formula*

$$
W = -\frac{2\pi^2 m Z^2 e_0^4}{n^2 h^2} \qquad \text{together with} \qquad a = \frac{h^2 n^2}{4\pi^2 m e_0^2 Z}. \tag{7.150}
$$

Solution: (a) The most general equation of an ellipse in polar coordinates r, θ is $1/r = c_1 + c_2\cos\theta$, where c_1, c_2 are constants. Since $b = a(1 - e^2)^{1/2}$, we have (see Example 7.8) at the perihelion $(\theta = 0)$ $r = a(1 - e)$ and at the aphelion $(\theta = \pi)$ $r = a(1 + e)$. Hence

$$
\frac{1}{a(1 - e)} = c_1 + c_2 \qquad \text{and} \qquad \frac{1}{a(1 + e)} = c_1 - c_2.
$$

It follows that (in agreement with Eqs. (7.42), (7.85))

$$
c_1 = \frac{1}{a(1 - e^2)}, \qquad c_2 = \frac{e}{a(1 - e^2)}, \qquad \text{so that} \qquad \frac{1}{r} = \frac{1 + e\cos\theta}{a(1 - e^2)}. \tag{7.151}
$$

We can take the logarithm of the last expression and differentiate. Then

$$
\ln r = -\ln\frac{1 + e\cos\theta}{a(1 - e^2)}, \quad \frac{d}{d\theta}\ln r = -\frac{d}{d\theta}\ln\frac{1 + e\cos\theta}{a(1 - e^2)} = -\frac{a(1 - e^2)}{1 + e\cos\theta}\frac{(-e\sin\theta)}{a(1 - e^2)} = \frac{e\sin\theta}{1 + e\cos\theta},
$$

[¶]The Bohr theory of the atom of around 1913 — effectively Newtonian mechanics with discretized energy, often described as "old quantum theory" — is, of course, wrong and was corrected by the quantum mechanics developed by Heisenberg, Dirac and others in 1925 and thereafter. Nonetheless the Bohr–Sommerfeld–Wilson quantization conditions can for many cases be derived from correct quantum mechanical wave mechanics. However, there are cases where they do not apply. For the corrected version see H. J. W. Müller–Kirsten [35], p. 297.

and

$$\frac{d}{d\theta}\ln r = \frac{d}{dr}\ln r\,\frac{dr}{d\theta} = \frac{1}{r}\frac{dr}{d\theta} = \frac{e\sin\theta}{1+e\cos\theta}, \qquad \frac{dr}{d\theta} = r\,\frac{e\sin\theta}{1+e\cos\theta}.$$

With this result we obtain the generalized momenta p_r, p_θ:

$$p_r = m\dot{r} = m\frac{dr}{d\theta}\dot{\theta} = mr^2\dot{\theta}\frac{1}{r^2}\frac{dr}{d\theta} = p_\theta\frac{1}{r^2}\frac{dr}{d\theta} = \frac{p_\theta}{r}\frac{e\sin\theta}{1+e\cos\theta}.$$

Hence

$$p_r\,dr = p_\theta\left(\frac{e\sin\theta}{1+e\cos\theta}\right)\frac{1}{r}\frac{dr}{d\theta}d\theta = p_\theta\left(\frac{e\sin\theta}{1+e\cos\theta}\right)^2 d\theta. \tag{7.152}$$

One phase space integral can now be written:

$$n_r h = \oint p_r\,dr = \oint m\dot{r}\,dr = \oint p_\theta\left(\frac{e\sin\theta}{1+e\cos\theta}\right)^2 d\theta. \tag{7.153}$$

But we know p_θ from the other phase space integral, where $l = mr^2\dot{\theta} = $ const.:

$$n_\theta h = \oint p_\theta\,d\theta = \oint mr^2\dot{\theta}\,d\theta = \oint l\,d\theta = 2\pi p_\theta. \tag{7.154}$$

Hence (the value of the integral on the far right being derived at the end)

$$n_r h = \oint p_r\,dr = \frac{n_\theta h}{2\pi}\oint\left(\frac{e\sin\theta}{1+e\cos\theta}\right)^2 d\theta,$$

$$\frac{n_r}{n_\theta} = \frac{1}{2\pi}\int_0^{2\pi}\left(\frac{e\sin\theta}{1+e\cos\theta}\right)^2 d\theta = \frac{1}{\sqrt{1-e^2}} - 1. \tag{7.155}$$

It follows that (cf. the given equation $b/a = (1-e^2)^{1/2}$):

$$1 - e^2 = \frac{n_\theta^2}{(n_r+n_\theta)^2} = \frac{n_\theta^2}{n^2} = \frac{b^2}{a^2}. \tag{7.156}$$

(b) We have:

$$W = T + V = \frac{1}{2}m\dot{r}^2 + \frac{1}{2}mr^2\dot{\theta}^2 - \frac{e_0^2 Z}{r} = \frac{p_r^2}{2m} + \frac{p_\theta^2}{2mr^2} - \frac{e_0^2 Z}{r}$$

$$= \frac{1}{2mr^2}p_\theta^2\left[1 + \left(\frac{e\sin\theta}{1+e\cos\theta}\right)^2\right] - \frac{e_0^2 Z}{r}. \tag{7.157}$$

From the beginning of (a) we can use for $1/r$ the replacement:

$$\frac{1}{r} = \frac{1+e\cos\theta}{a(1-e^2)},$$

so that

$$W = \frac{p_\theta^2}{2m}\left\{\frac{1+e\cos\theta}{a(1-e^2)}\right\}^2\left[1 + \left(\frac{e\sin\theta}{1+e\cos\theta}\right)^2\right] - \frac{e_0^2 Z(1+e\cos\theta)}{a(1-e^2)}$$

$$= \frac{p_\theta^2}{ma^2(1-e^2)^2}\left[\frac{1+e^2}{2} + e\cos\theta\right] - \frac{e_0^2 Z(1+e\cos\theta)}{a(1-e^2)}. \tag{7.158}$$

For this expression to be independent of θ, the coefficient of $\cos\theta$ must vanish. This means we must have

$$\frac{p_\theta^2 e}{ma^2(1-e^2)^2} = \frac{e_0^2 Z e}{a(1-e^2)}, \quad i.e. \quad a = \frac{p_\theta^2}{me_0^2 Z(1-e^2)}, \tag{7.159}$$

and inserting the latter expression for a into the terms remaining in W, this becomes

$$
\begin{aligned}
W &= \frac{p_\theta^2(1+e^2)}{2ma^2(1-e^2)^2} - \frac{e_0^2 Z}{a(1-e^2)} = \frac{p_\theta^2(1+e^2)m^2 e_0^4 Z^2(1-e^2)^2}{2m(1-e^2)^2 p_\theta^4} - \frac{e_0^2 Z me_0^2 Z(1-e^2)}{(1-e^2)p_\theta^2} \\
&= \frac{e_0^4 Z^2 m}{p_\theta^2}\left[\frac{1+e^2}{2} - 1\right] = -\frac{e_0^4 Z^2(1-e^2)m}{2p_\theta^2}.
\end{aligned}
\tag{7.160}
$$

But above we had (see Eqs. (7.156), (7.154))

$$1 - e^2 = \frac{n_\theta^2}{n^2} \quad \text{and} \quad p_\theta = \frac{n_\theta h}{2\pi}. \tag{7.161}$$

It follows that

$$W = -\frac{2\pi^2 m Z^2 e_0^4}{h^2}\frac{1}{n^2}. \tag{7.162}$$

We note here that this expression is identical with that of Bohr, although we assumed an elliptic orbit (Kepler–Newton elliptic orbit) of the electron here (not a circular orbit as Bohr). For a we obtain (this is the *Bohr radius* of the hydrogen-like atom)

$$a = \frac{p_\theta^2}{me_0^2 Z(1-e^2)} = \frac{h^2 n^2}{4\pi^2 me_0^2 Z}. \tag{7.163}$$

Evaluation of the integral: The following integral can be found in Tables:[||]

$$\int_0^\pi \frac{\sin^2 x}{p+q\cos x}dx = \frac{p\pi}{q^2}\left[1 - \sqrt{1-\frac{q^2}{p^2}}\right]. \tag{7.164}$$

Differentiation with respect to p yields

$$-\int_0^\pi \frac{\sin^2 x}{(p+q\cos x)^2}dx = \frac{\pi}{q^2}\left[1 - \sqrt{1-\frac{q^2}{p^2}}\right] - \frac{p\pi}{q^2}\frac{1}{2}\frac{(-q^2)(-2)}{\sqrt{1-q^2/p^2}}\frac{1}{p^3}.$$

For $p = 1$ this implies

$$\int_0^\pi \frac{\sin^2 x}{(1+q\cos x)^2}dx = -\frac{\pi}{q^2}[1-\sqrt{1-q^2}] + \frac{\pi}{\sqrt{1-q^2}}.$$

But

$$\int_\pi^{2\pi} \frac{\sin^2 x}{(p+q\cos x)^2}dx \overset{x\to x+\pi}{=} \int_0^\pi \frac{\sin^2 x}{(1-q\cos x)^2}dx.$$

Hence

$$
\begin{aligned}
\int_0^{2\pi} \frac{\sin^2 x}{(1+q\cos x)^2}dx &= -\frac{2\pi}{q^2}[1-\sqrt{1-q^2}] + \frac{2\pi}{\sqrt{1-q^2}} \\
&= \frac{2\pi}{q^2\sqrt{1-q^2}}[1-\sqrt{1-q^2}] = \frac{2\pi}{q^2}\left[\frac{1}{\sqrt{1-q^2}} - 1\right].
\end{aligned}
$$

Hence, as claimed in Eq. (7.155),

$$\frac{n_r}{n_\theta} = \frac{1}{2\pi}\int_0^{2\pi}\left(\frac{e\sin\theta}{1+e\cos\theta}\right)^2 d\theta = \frac{1}{\sqrt{1-e^2}} - 1. \tag{7.165}$$

Again the reader can find interesting additional discussion in the book of Penrose.[**]

[||]I. S. Gradshteyn and I. N. Ryzhik [20], p. 379.
[**]R. Penrose [38], pp. 572 - 573.

Chapter 8

Rigid Body Dynamics

8.1   Introductory Remarks

The theoretical description of the motion of rigid bodies is, of course, the ultimate aim of classical mechanics. Rigid bodies are those conglomerates of masspoints which are such that the distance between any two of their masspoints remains fixed. Since such bodies may appear in any arbitrary shape, it is clear that a theory describing their motion cannot be simple. In particular the shape of such objects has to be taken into account. One expects, of course, that bodies with a reasonably simple, and symmetric shape, are easier to handle. This will, in fact, turn out to be the case. The quantities describing finite volumes of a given shape which are filled with masspoints are known as *moments of inertia*. These rotational analogues of mass play an important role in the formulation of the equations of motion. The even more difficult part of the problem of deriving the appropriate equations of motion is concerned with the distinction between the relevant inertial frame and a reference frame fixed in the body, and how these are related in the rigid body's linear and rotational motion. Naturally one exploits the presence of symmetry axes of the body, known as its *principal axes*. The transformation to these is known as the problem of *transformation to principal axes* and requires the solution of an eigenvalue problem. Since this, in particular an understanding of its eigenvalues and eigensolutions, is the most difficult aspect, we consider this also in detail in illustrative examples. We shall see that the transformation to principal axes has a simple geometrical interpretation. Finally we consider important aspects of rotating frames, in particular the Coriolis force and various cases where this appears. Throughout a number of diverse examples is given in order to provide illustrations, as well as clarifying aspects of the general theoretical background.

257

8.2   Moments of Inertia

We noted earlier (*cf.* Sec. 5.4.1) that the orientation of a rigid body can be described by one translation together with one rotation. If we choose the origin of the reference frame fixed in the rigid body such that this coincides with its *centre of mass*, then kinetic energy and total angular momentum split up into parts derived from the translation of the centre of mass, and from rotation around this centre of mass. In many cases the potential energy can be similarly split up; thus for instance the potential energy of gravitation depends only on the Cartesian vertical coordinate of the *centre of gravity*, which in the case of a *homogeneous gravitational field* coincides with the centre of mass.*

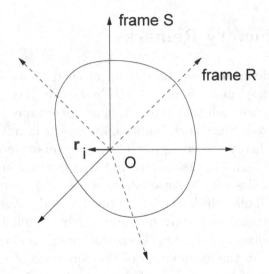

Fig. 8.1 Spatial frame S and rigid body frame R with same origin O.

In each of these cases we have for the Lagrange function L:

$$L = L_{\text{translation}} + L_{\text{rotation}}.$$

It is therefore necessary to obtain expressions for the energy and the angular momentum of the motion of a rigid body with respect to its centre of mass. The angular momentum $\mathbf{L}$ of a rigid body with respect to an arbitrarily

---

*For concreteness recall that centre of mass $\mathbf{R}$ and centre of gravity $\mathbf{R}_G$ are defined by

$$\mathbf{R} = \sum_i m_i r_i / \sum_i m_i \quad \text{and} \quad \mathbf{R}_G = \sum_i m_i g_K(r_i) r_i / \sum_i m_i g_K(r_i),$$

where $g_K(r_i)$ is the spatially dependent acceleration due to gravity.

chosen point in that body is given by, *cf.* Fig. 8.1,

$$\mathbf{L} = \sum_i m_i(\mathbf{r}_i \times \mathbf{v}_i). \tag{8.1}$$

Since the body is rigid the velocity of m_i with respect to any other point of the body vanishes, and we have $(\dot{\mathbf{r}}_i)_{\text{body}} = 0$. Consequently Eq. (5.55), *i.e.*

$$\left(\frac{d\mathbf{G}}{dt}\right)_S = \left(\frac{d\mathbf{G}}{dt}\right)_R + \boldsymbol{\omega} \times \mathbf{G},$$

where S stands for *spatial frame* and R for *rigid body frame*, implies, for $\mathbf{G}$ replaced by $\mathbf{r}_i$,

$$\left(\frac{d\mathbf{r}_i}{dt}\right)_S = \mathbf{v}_i = \boldsymbol{\omega} \times \mathbf{r}_i \tag{8.2}$$

for a spatially fixed reference frame S whose origin here coincides with that of the frame R fixed in the body. Hence referred to the spatial frame S with the same origin as the rigid body frame R, the angular momentum $\mathbf{L}$ is[†]

$$\mathbf{L} = \sum_i m_i \mathbf{r}_i \times (\boldsymbol{\omega} \times \mathbf{r}_i) = \sum_i m_i[\boldsymbol{\omega} r_i^2 - (\mathbf{r}_i \cdot \boldsymbol{\omega})\mathbf{r}_i], \tag{8.3}$$

and hence, for instance with $\mathbf{r}_i = (x_i, y_i, z_i) = (r_{i1}, r_{i2}, r_{i3})$, one has

$$\begin{aligned} L_x &= \sum_i m_i[\omega_x r_i^2 - \sum_j r_{ij}\omega_j x_i] \\ &= \sum_i m_i(\omega_x r_i^2 - \omega_x x_i^2 - \omega_y x_i y_i - \omega_z z_i x_i) \\ &= \omega_x \sum_i m_i(r_i^2 - x_i^2) - \omega_y \sum_i m_i x_i y_i - \omega_z \sum_i m_i z_i x_i. \end{aligned} \tag{8.4}$$

We see therefore, that $\mathbf{L}$ can be rewritten in the following matrix form:

$$\begin{pmatrix} L_x \\ L_y \\ L_z \end{pmatrix} = \begin{pmatrix} I_{xx} & I_{xy} & I_{xz} \\ I_{yx} & I_{yy} & I_{yz} \\ I_{zx} & I_{zy} & I_{zz} \end{pmatrix} \begin{pmatrix} \omega_x \\ \omega_y \\ \omega_z \end{pmatrix}, \tag{8.5a}$$

or

$$L_i = I_{ij}\omega_j, \tag{8.5b}$$

[†]Recall that a scalar triple product has the property: $(\mathbf{A} \times \mathbf{B}) \cdot \mathbf{C} = \mathbf{A} \cdot (\mathbf{B} \times \mathbf{C})$, and a vector triple product the property: $\mathbf{A} \times (\mathbf{B} \times \mathbf{C}) = (\mathbf{A} \cdot \mathbf{C})\mathbf{B} - (\mathbf{A} \cdot \mathbf{B})\mathbf{C}$.

where[‡]

$$I_{xx} = \sum_i m_i(r_i^2 - x_i^2), \quad I_{xy} = -\sum_i m_i x_i y_i = I_{yx}, \quad \text{etc.} \tag{8.6}$$

Obviously the matrix I is symmetric. We deduce from the relation (8.5a) that **L** and $\boldsymbol{\omega}$ are connected by a *linear transformation*. Each of the elements of the matrix I depends on the masses m_i, and thus on inertia. The matrix I is therefore known as *tensorial representation of the moment of inertia* of the rigid body, the diagonal elements I_{xx}, I_{yy}, I_{zz} are known as *coefficients of the moment of inertia*, and the off-diagonal elements I_{xy}, I_{yz}, I_{zx} are known as *products of inertia*. If we do not have a system of particles, but instead a continuous mass distribution of density $\rho(r)$, the sums in the above expressions become integrals. For instance:

$$I_{xx} = \int_V \rho(r) dV (r^2 - x^2). \tag{8.7}$$

The quantity or 3×3 matrix I is a second rank tensor. A vector is a tensor of rank 1, which — as we saw earlier — means that it transforms under orthogonal transformations A of the group $SO(3)$ according to the law (this, in fact, *defines* the vector in the three-dimensional space $\mathbb{R}^3$):

$$T_i' = \sum_j a_{ij} T_j, \quad A = (a_{ij}), \quad AA^T = \mathbb{1}. \tag{8.8}$$

The *tensor of rank* 2, *i.e.* its components, transform correspondingly, *i.e.*

$$T_{ij}' = \sum_{k,l} a_{ik} a_{jl} T_{kl}. \tag{8.9}$$

For the kinetic energy of the motion of a rigid body around a point fixed in the body we now have (note we use the discrete sum $\sum_i$ instead of the integral $\int_V dV$ over a continuous mass distribution)

$$
\begin{aligned}
T &= \frac{1}{2}\sum_i m_i v_i^2 = \frac{1}{2}\sum_i m_i \mathbf{v}_i \cdot \mathbf{v}_i \overset{(8.2)}{=} \frac{1}{2}\sum_i m_i \mathbf{v}_i \cdot (\boldsymbol{\omega} \times \mathbf{r}_i) \\
&= \frac{1}{2}\sum_i m_i (\boldsymbol{\omega} \times \mathbf{r}_i) \cdot \mathbf{v}_i = \frac{1}{2}\sum_i m_i \boldsymbol{\omega} \cdot (\mathbf{r}_i \times \mathbf{v}_i) \\
&\overset{(8.1)}{=} \frac{1}{2}\boldsymbol{\omega} \cdot \mathbf{L} \overset{(8.5b)}{=} \frac{1}{2}\sum_{i,j} \omega_i I_{ij} \omega_j.
\end{aligned}
\tag{8.10}
$$

[‡]Note that like H. Goldstein [19] we have $I_{xy} = -\sum_i m_i x_i y_i$. Other authors define this product with a plus sign. The elements of the inertial tensor depend on the choice of the origin O, but not on the direction of the axis of rotation. The moment of inertia I, of course, depends on both.

If **n** is a unit vector in the direction of $\boldsymbol{\omega}$, so that $\boldsymbol{\omega} = \omega\mathbf{n}$, then

$$T = \frac{\omega^2}{2} \sum_{i,j} n_i I_{ij} n_j = \frac{1}{2} I \omega^2, \tag{8.11}$$

where (see below)

$$I = \sum_{i,j} n_i I_{ij} n_j = \sum_i m_i [r_i^2 - (\mathbf{r}_i \cdot \mathbf{n})^2]. \tag{8.12}$$

We see this, by multiplying Eq. (8.3) by $\boldsymbol{\omega}$, because then

$$\frac{1}{\omega^2} \boldsymbol{\omega} \cdot \mathbf{L} \overset{(8.3)}{=} \sum_i m_i \left\{ r_i^2 - \left(\frac{\mathbf{r}_i \cdot \boldsymbol{\omega}}{\omega} \right)^2 \right\} = \sum_i m_i [r_i^2 - (\mathbf{r}_i \cdot \mathbf{n})^2],$$

but also

$$\frac{1}{\omega^2} \boldsymbol{\omega} \cdot \mathbf{L} \overset{(8.5b)}{=} \sum_{i,j} n_i I_{ij} n_j.$$

Thus Eq. (8.12) follows by comparison. The quantity I is called *moment of inertia* around the axis of rotation. We next show that we can rewrite the result Eq. (8.12) in the following form:

$$I = \sum_i m_i (\mathbf{r}_i \times \mathbf{n}) \cdot (\mathbf{r}_i \times \mathbf{n}). \tag{8.13}$$

We can see this as follows. We have:[§]

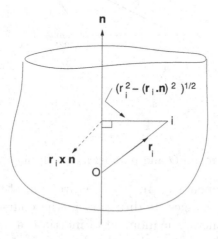

Fig. 8.2 The unit vector **n** in relation to the vector to vector $\mathbf{r}_i$ of mass m_i.

[§] For properties of scalar triple products and vector triple product see earlier footnote.

$$\sum_i m_i (\mathbf{r}_i \times \mathbf{n}) \cdot (\mathbf{r}_i \times \mathbf{n}) = \sum_i m_i (\mathbf{r}_i \times \mathbf{n}) \times \mathbf{r}_i \cdot \mathbf{n}$$

$$= -\sum_i m_i [\mathbf{r}_i \times (\mathbf{r}_i \times \mathbf{n})] \cdot \mathbf{n} = -\sum_i m_i [(\mathbf{r}_i \cdot \mathbf{n})\mathbf{r}_i - r_i^2 \mathbf{n}] \cdot \mathbf{n}$$

$$= \sum_i m_i [r_i^2 \mathbf{n}^2 - (\mathbf{r}_i \cdot \mathbf{n})^2],$$

where $\mathbf{n}^2 = \alpha^2 + \beta^2 + \gamma^2 = 1$ (the sum of the squares of the appropriate *direction cosines*). This is the expression of Eq. (8.12). Equation (8.13) says: The moment of inertia around the axis of rotation is equal to the sum of the products of the particle masses m_i and the squares of the perpendicular distances of these mass points from the axis as indicated in Fig. 8.2. We note here for later reference, that if the unit vector $\mathbf{n}$ points along the z-axis, or the x-axis or the y-axis, the result (8.12) implies that $I \to I_{zz}, I_{xx}, I_{yy}$ with

$$I_{zz} = \sum_i m_i [x_i^2 + y_i^2], \quad I_{xx} = \sum_i m_i [y_i^2 + z_i^2], \quad I_{yy} = \sum_i m_i [z_i^2 + x_i^2]. \quad (8.14)$$

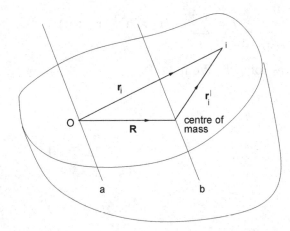

Fig. 8.3 Axis a through O and parallel axis b through the centre of mass.

We observe that I depends on $\mathbf{n}$, *i.e.* the direction of the axis of rotation. If the axis of rotation changes its direction in the course of time, then I also changes. Thus, in general the moment of inertia is a function of time. In the latter case, $\boldsymbol{\omega}$ must be expressed as the derivative of an angle.

We also see from the above that the moment of inertia I depends on $\mathbf{r}_i$, in other words also on the specific choice of the position of the reference frame. We enquire therefore how the moment of inertia with axis a through an arbitrarily fixed point O differs from that about a parallel axis b which passes through the centre of mass of the rigid body, as indicated in Fig. 8.3. With position vectors defined as in Fig. 8.3 we have

$$\mathbf{r}_i = \mathbf{R} + \mathbf{r}_i'.$$

The moment of inertia I_a about axis a is

$$I_a \overset{(8.13)}{=} \sum_i m_i(\mathbf{r}_i \times \mathbf{n})^2 = \sum_i m_i[(\mathbf{R} + \mathbf{r}_i') \times \mathbf{n}]^2, \tag{8.15}$$

and that around the parallel axis b through the centre of mass is

$$I_{\text{cm}} \equiv I_b = \sum_i m_i(\mathbf{r}_i' \times \mathbf{n})^2. \tag{8.16}$$

With $M = \sum_i m_i$ and

$$\mathbf{R} = \frac{1}{M}\sum_i m_i\mathbf{r}_i, \quad i.e. \quad \sum_i m_i(\mathbf{r}_i - \mathbf{R}) = 0, \quad \text{and hence} \quad \sum_i m_i\mathbf{r}_i' = 0,$$

it follows that

$$
\begin{aligned}
I_a &= I_b + \sum_i m_i(\mathbf{R} \times \mathbf{n})^2 + 2\sum_i m_i(\mathbf{R} \times \mathbf{n}) \cdot (\mathbf{r}_i' \times \mathbf{n}) \\
&= I_b + \sum_i m_i(\mathbf{R} \times \mathbf{n})^2 + 2(\mathbf{R} \times \mathbf{n}) \cdot [\underbrace{(\sum_i m_i\mathbf{r}_i')}_{0} \times \mathbf{n}] \\
&= I_b + M(\mathbf{R} \times \mathbf{n})^2, \quad M = \sum_i m_i. \tag{8.17}
\end{aligned}
$$

We have therefore

$$I_a = I_{\text{cm}} + M(\mathbf{R} \times \mathbf{n})^2. \tag{8.18}$$

The quantity $|\mathbf{R} \times \mathbf{n}|$ is the length of the perpendicular from the centre of mass to the axis through the point O. The result (8.18) is known as *Steiner's theorem*. The subdivision of the moment of inertia which it implies is very similar to that of the momentum, the energy and the angular momentum of particle systems, that we dealt with earlier.

Example 8.1: The rolling inhomogeneous wheel

A solid wheel made of *inhomogeneous* material has mass M and radius R, and its centre of mass a distance a away from the axis of the wheel. The wheel's moment of inertia with respect to an axis

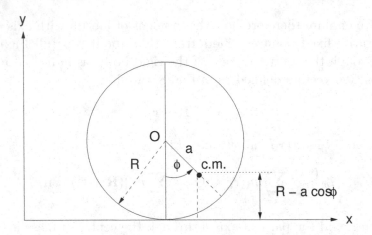

Fig. 8.4 The wheel with centre of mass distance a away from O.

through the centre of mass is I_0. The wheel rolls under gravity on a horizontal plane with initial values of the angle ϕ indicated in Fig. 8.4, and the angular velocity given by $\phi(0) = 0$ and $\dot{\phi}(0) = \omega$. Derive the Euler–Lagrange equation of the motion of the wheel and solve this for small values of a, *i.e.* $\phi(t) = \phi_{a=0}(t) + a\eta(t)$.

Solution: With *Steiner's theorem* the Lagrangian is

$$L(\phi, \dot{\phi}) = \frac{1}{2}(I_0 + Ma^2)\dot{\phi}^2 - Mg(R - a\cos\phi).$$

The Euler–Lagrange equation is

$$\frac{d}{dt}\left(\frac{\partial L}{\partial \dot{\phi}}\right) - \frac{\partial L}{\partial \phi} = 0, \qquad \frac{d}{dt}[(I_0 + Ma^2)\dot{\phi}] + Mga\sin\phi = 0.$$

Thus we have (*cf.* Eq. (3.101) of Example 3.14)

$$\ddot{\phi} + \Omega^2 \sin\phi = 0, \qquad \Omega^2 = \frac{Mga}{I_0 + Ma^2}.$$

For $a = 0$: $\Omega^2 = 0, \ddot{\phi}_{a=0} = 0, \dot{\phi}_{a=0}(t) = \omega, \phi_{a=0}(t) = \omega t$. With the given ansatz for small values of a we obtain:

$$\phi(t) = \omega t + a\eta(t), \quad \dot{\phi}(t) = \omega + a\dot{\eta}(t), \quad \ddot{\phi}(t) = a\ddot{\eta}(t).$$

$$\therefore \ a\ddot{\eta}(t) + \Omega^2\underbrace{\sin[\omega t + a\eta(t)]}_{\simeq \sin[\omega t] + a\eta(t)\cos[\omega t]} = 0,$$

where we approximated the argument of the sine for small values of a. Inserting the expression for Ω^2 and again retaining only terms linear in a, we obtain

$$a\ddot{\eta}(t) + \left(\frac{Mga}{I_0}\right)\sin\omega t = O(a^2) \simeq 0.$$

It follows that $\eta(t) = A\sin(\omega t + \alpha)$. The constants can be determined by subsitution. Inserting $\eta(t)$ into $\phi(t)$, we obtain

$$\phi(t) = \omega t + aA\sin(\omega t + \alpha).$$

Thus the ususal periodic behaviour (*i.e.* that investigated in Example 3.14) is superimposed by a periodic oscillation.

8.3   Diagonalization and Principal Axes

8.3.1   The ellipsoid of inertia

By definition,

$$I_{xy} = I_{yx},$$

i.e. the inertial tensor is symmetric. Since, in addition, the components are real, it follows that the 3×3 matrix I is selfadjoint or *Hermitian*, and thus possesses only six independent elements. The *eigenvalues of I are therefore real*, as we verify in Example 8.3.¶

The question arises: Is it possible in the case of a given origin of the reference frame fixed in the rigid body to choose the orientation of the axes of this system in such a way, that the inertial tensor is diagonal? The answer is yes, as we shall see, since — as we saw in the previous section — with a rotation we can bring the three-dimensional reference frame into any desired position. The *diagonalized inertial tensor* is then the matrix

$$\begin{pmatrix} I_1 & 0 & 0 \\ 0 & I_2 & 0 \\ 0 & 0 & I_3 \end{pmatrix}. \tag{8.19}$$

The eigenvalues $I_i, i = 1, 2, 3$ are known as *principal moments of inertia*. The angular momentum vector $\mathbf{L}$ would then correspondingly (*cf.* Eq. (8.5b)) be given by the components

$$L_x = I_1 \omega_x, \quad L_y = I_2 \omega_y, \quad L_z = I_3 \omega_z, \tag{8.20}$$

and the kinetic energy T by the expression

$$T = \frac{1}{2} I_1 \omega_x^2 + \frac{1}{2} I_2 \omega_y^2 + \frac{1}{2} I_3 \omega_z^2. \tag{8.21}$$

We consider in the first place a quantity known as the *ellipsoid of inertia*, also known as the *momental ellipsoid*.‖

We let $\alpha_a, \beta_a, \gamma_a$ be the x, y, z components of the unit vector $\mathbf{n}_a$ along

---

¶We repeat: The *Hermitian* matrix has real eigenvalues. Thus this is not the case for an *orthogonal* matrix $A \in SO(3)$, whose eigenvalues $\lambda_i, i = 1, 2, 3$, can be complex. However, the modulus of any λ_i must be 1, as one can conclude from the fact that the length of the rotated vector $\mathbf{R}' = A\mathbf{R} = \lambda_i \mathbf{R}$ remains unchanged as a result of the orthogonality condition.

‖A. S. Ramsey [41], p. 192.

the axis of rotation (hence index 'a'), *i.e.*[**]

$$\mathbf{n}_a = \begin{pmatrix} \alpha_a \\ \beta_a \\ \gamma_a \end{pmatrix}, \quad \alpha_a, \beta_a, \gamma_a \text{ (direction cosines)}, \quad \alpha_a^2 + \beta_a^2 + \gamma_a^2 = 1. \quad (8.22)$$

Then (*cf.* Eq. (8.12)) the moment of inertia I around the rotation axis with direction that of $\mathbf{n}_a$ is

$$\begin{aligned} I &= \sum_{i,j} n_{ai} I_{ij} n_{aj} \\ &= (\alpha_a \ \beta_a \ \gamma_a) \begin{pmatrix} I_{xx} & I_{xy} & I_{xz} \\ I_{yx} & I_{yy} & I_{yz} \\ I_{zx} & I_{zy} & I_{zz} \end{pmatrix} \begin{pmatrix} \alpha_a \\ \beta_a \\ \gamma_a \end{pmatrix} \\ &= I_{xx}\alpha_a^2 + I_{yy}\beta_a^2 + I_{zz}\gamma_a^2 + 2I_{xy}\alpha_a\beta_a \\ &\quad + 2I_{yz}\beta_a\gamma_a + 2I_{zx}\gamma_a\alpha_a, \end{aligned} \quad (8.23)$$

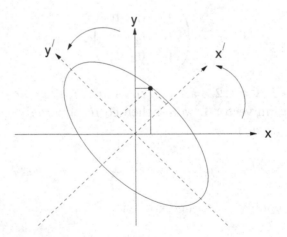

Fig. 8.5 Rotation of frame (x, y) to (x', y').

or if we introduce the vector

$$\boldsymbol{\rho}_a = \frac{\mathbf{n}_a}{\sqrt{I}}, \quad \rho_a^2 = \frac{1}{I}, \quad (8.24)$$

[**]The direction cosines are the cosines of the angles $\delta_1, \delta_2, \delta_3$, which an oriented line makes with the positive $x-, y-, z-$axes at unit distance from the origin. Thus in a two-dimensional Cartesian system in polar representation (polar angle θ) we have $\alpha = \cos\delta_1 = \cos\theta, \beta = \cos\delta_2 = \cos(\pi/2 - \theta) = \sin\theta$. These direction cosines are the elements of a row or column of a matrix decribing the rotation around the third axis. In three dimensions with spherical polar angles θ, ϕ, one has $\alpha = \cos\delta_1 = \cos\phi\sin\theta, \beta = \cos\delta_2 = \sin\phi\sin\theta, \gamma = \cos\delta_3 = \cos\theta$.

i.e.

$$\rho_{a1} = \frac{\alpha_a}{\sqrt{I}}, \quad \rho_{a2} = \frac{\beta_a}{\sqrt{I}}, \quad \rho_{a3} = \frac{\gamma_a}{\sqrt{I}}, \tag{8.25}$$

we can rewrite Eq. (8.23) as (note on the left 1 instead of I)

$$1 = I_{xx}\rho_{a1}^2 + I_{yy}\rho_{a2}^2 + I_{zz}\rho_{a3}^2 + 2I_{xy}\rho_{a1}\rho_{a2} + 2I_{yz}\rho_{a2}\rho_{a3} + 2I_{zx}\rho_{a3}\rho_{u1}. \tag{8.26}$$

This expression is the equation of an *ellipsoid* with the $x-, y-, z-$coordinates of a point on its surface given by $\rho_{a1}, \rho_{a2}, \rho_{a3}$. With a transformation (rotation) to *principal axes* (rotating either the object with axes kept fixed, or the $x-, y-, z-$axes in the opposite sense with the object untouched) the expression (8.26) can be brought into the following form called "*normal form*" with new direction cosines $\alpha'_a, \beta'_a, \gamma'_a$ and coordinates $\rho'_{a1}, \rho'_{a2}, \rho'_{a3}$ of the axis of rotation

$$1 = I_1 {\rho'_{a1}}^2 + I_2 {\rho'_{a2}}^2 + I_3 {\rho'_{a3}}^2, \tag{8.27}$$

as indicated in Fig. 8.5 for the respective direction cosines with $\sqrt{I}\rho'_a = \mathbf{n}'_a = (\alpha'_a, \beta'_a, \gamma'_a)$. Comparing this equation with the standard (normal form) equation of an ellipsoid in three-dimensional Cartesian coordinates, *i.e.*

$$\frac{x'^2}{a^2} + \frac{y'^2}{b^2} + \frac{z'^2}{c^2} = 1,$$

we see that the lengths of the semi-axes of the ellipsoid are given by

$$\frac{1}{\sqrt{I_1}}, \quad \frac{1}{\sqrt{I_2}}, \quad \frac{1}{\sqrt{I_3}}.$$

If, for instance $I_1 = I_2$, the ellipsoid is an *ellipsoid of revolution*. In the case of $I_1 = I_2 = I_3$ the ellipsoid reduces to a sphere with radius $1/\sqrt{I_1} = 1/\sqrt{I_2} = 1/\sqrt{I_3}$. Phrased differently, the *principal moments of inertia* I_1, I_2, I_3 determine the lengths of the principal semi-axes of the ellipsoid. As emphasized by Goldstein,[††] the use of the direction cosines has definite advantages. An important aspect is that the *elements of a rotation matrix* $A \in SO(3)$ represent such direction cosines. Thus in Eq. (8.23) $\alpha_a, \beta_a, \gamma_a$, the direction cosines of the axis of rotation with respect to axes x, y, z, are contained as row or column elements in a matrix $A \in SO(3)$ which brings each of the three axes into the position of the axis of rotation or, *vice versa* (in each such case of rotation to principal axes, one of the resulting row or column elements being 1, the other two 0, as will be seen below, *cf.* Eq. (8.35)). For a better understanding of the role of direction cosines in later contexts, we consider these in more detail in Example 8.2 for the simple case of the circle.

[††]H. Goldstein [19], Sec. 4.1.

Example 8.2: Direction cosines

In the simple case of a circle obtain the direction cosines α, β (a) from a radius considered as a line $\lambda = r$, and (b) from the normal to the circumference (more generally surface) of $\phi = \pi r^2$.

Solution: For (a) $\lambda = r = \sqrt{x^2 + y^2}$ and (b) $\phi = \pi r^2 = \pi(x^2 + y^2)$ we obtain

$$(a) \qquad \frac{\partial \lambda}{\partial x} = \frac{x}{r} = \cos\theta, \qquad \frac{\partial \lambda}{\partial y} = \frac{y}{r} = \sin\theta,$$

$$\therefore \qquad \alpha := \frac{(\partial\lambda/\partial x)}{\sqrt{(\frac{\partial\lambda}{\partial x})^2 + (\frac{\partial\lambda}{\partial y})^2}} = \cos\theta, \qquad \beta := \frac{(\partial\lambda/\partial y)}{\sqrt{(\frac{\partial\lambda}{\partial x})^2 + (\frac{\partial\lambda}{\partial y})^2}} = \sin\theta;$$

$$(b) \qquad \frac{\partial\phi}{\partial x} = 2\pi r \frac{\partial r}{\partial x} = 2\pi x = 2\pi r \cos\theta, \qquad \frac{\partial\phi}{\partial y} = 2\pi y = 2\pi r \sin\theta,$$

$$\therefore \qquad \alpha := \frac{(\partial\phi/\partial x)}{\sqrt{(\frac{\partial\phi}{\partial x})^2 + (\frac{\partial\phi}{\partial y})^2}} = \frac{2\pi r \cos\theta}{2\pi r \sqrt{\cos^2\theta + \sin^2\theta}} = \cos\theta,$$

$$\beta := \frac{(\partial\phi/\partial y)}{\sqrt{(\frac{\partial\phi}{\partial x})^2 + (\frac{\partial\phi}{\partial y})^2}} = \frac{2\pi r \sin\theta}{2\pi r \sqrt{\cos^2\theta + \sin^2\theta}} = \sin\theta.$$

8.3.2   Transformation to principal axes

We have a Cartesian (x, y, z)-reference frame fixed in the rigid body. The axis of rotation of the body has direction cosines α, β, γ with respect to this reference frame. We now wish to bring this axis into a position with direction cosines α', β', γ', which are such that the *inertial ellipsoid* assumes its *normal form*. Alternately we can rotate the reference frame at the same origin but into a different orientation.

The principal axes are axes fixed in the rigid body; moreover, they are *symmetry axes* of the body.[‡‡] Thus we wish to find a frame of coordinates, called the system of principal axes, in which the inertial tensor I_{ij} is diagonal. To this end we consider a rotation described by a nonsingular matrix $A^{-1} \in SO(3)$, say, for which

$$(A_{ij}^{-1} I_{jk} A_{kl}) = \begin{pmatrix} I_1 & 0 & 0 \\ 0 & I_2 & 0 \\ 0 & 0 & I_3 \end{pmatrix}, \tag{8.28}$$

or, multiplying from the left by A,

$$(I_{ij}A_{jk}) \equiv \begin{pmatrix} I_{xx} & I_{xy} & I_{xz} \\ I_{yx} & I_{yy} & I_{yz} \\ I_{zx} & I_{zy} & I_{zz} \end{pmatrix} \begin{pmatrix} A_{11} & A_{12} & A_{13} \\ A_{21} & A_{22} & A_{23} \\ A_{31} & A_{32} & A_{33} \end{pmatrix} \overset{(8.28)}{=} A \begin{pmatrix} I_1 & 0 & 0 \\ 0 & I_2 & 0 \\ 0 & 0 & I_3 \end{pmatrix},$$

[‡‡]For instance if $I_{xy} = -\sum_i m_i x_i y_i = 0$, the object can have as many positive x_i as negative for fixed y_i, and therefore possesses symmetry.

or explicitly,

$$(I_{ij}A_{jk}) \equiv \begin{pmatrix} A_{11} & A_{12} & A_{13} \\ A_{21} & A_{22} & A_{23} \\ A_{31} & A_{32} & A_{33} \end{pmatrix} \begin{pmatrix} I_1 & 0 & 0 \\ 0 & I_2 & 0 \\ 0 & 0 & I_3 \end{pmatrix}$$

$$= \begin{pmatrix} A_{11}I_1 & A_{12}I_2 & A_{13}I_3 \\ A_{21}I_1 & A_{22}I_2 & A_{23}I_3 \\ A_{31}I_1 & A_{32}I_2 & A_{33}I_3 \end{pmatrix}. \tag{8.29}$$

Thus the matrix equation separates into three separate homogeneous equations:

$$\begin{pmatrix} I_{xx} & I_{xy} & I_{xz} \\ I_{yx} & I_{yy} & I_{yz} \\ I_{zx} & I_{zy} & I_{zz} \end{pmatrix} \begin{pmatrix} A_{11} \\ A_{21} \\ A_{31} \end{pmatrix} = \begin{pmatrix} A_{11}I_1 \\ A_{21}I_1 \\ A_{31}I_1 \end{pmatrix} = I_1 \begin{pmatrix} A_{11} \\ A_{21} \\ A_{31} \end{pmatrix},$$

$$\begin{pmatrix} I_{xx} & I_{xy} & I_{xz} \\ I_{yx} & I_{yy} & I_{yz} \\ I_{zx} & I_{zy} & I_{zz} \end{pmatrix} \begin{pmatrix} A_{12} \\ A_{22} \\ A_{32} \end{pmatrix} = \begin{pmatrix} A_{12}I_2 \\ A_{22}I_2 \\ A_{32}I_2 \end{pmatrix} = I_2 \begin{pmatrix} A_{12} \\ A_{22} \\ A_{32} \end{pmatrix},$$

$$\begin{pmatrix} I_{xx} & I_{xy} & I_{xz} \\ I_{yx} & I_{yy} & I_{yz} \\ I_{zx} & I_{zy} & I_{zz} \end{pmatrix} \begin{pmatrix} A_{13} \\ A_{23} \\ A_{33} \end{pmatrix} = \begin{pmatrix} A_{13}I_3 \\ A_{23}I_3 \\ A_{33}I_3 \end{pmatrix} = I_3 \begin{pmatrix} A_{13} \\ A_{23} \\ A_{33} \end{pmatrix}. \tag{8.30}$$

We see therefore that the problem of *diagonalization of the inertial tensor* is equivalent to the following *eigenvalue problem*:

$$(I_{jk})\chi_{E_i} = I_i\chi_{E_i}, \quad i = 1, 2, 3, \quad \text{with} \quad \chi_{E_i} = \begin{pmatrix} A_{1i} \\ A_{2i} \\ A_{3i} \end{pmatrix} \equiv \begin{pmatrix} \alpha_i \\ \beta_i \\ \gamma_i \end{pmatrix}. \tag{8.31a}$$

Equation (8.31a) therefore combines all three equations. The number I_i in Eq. (8.31a) is called *eigenvalue of* (I_{jk}) with *eigenvector* χ_{E_i}. Since (I_{ij}) is Hermitian, the eigenvalues I_i are real, as we verify in Example 8.3. The eigenvalues $I_i, i = 1, 2, 3$, are called *principal moments of inertia*, and the axes of the new reference frame fixed in the rigid body, with respect to which (I_{jk}) is diagonal, are called *principal axes*.

Example 8.3: The eigenvalues of Hermitian matrix I are real
Show that the inertial tensor I has real eigenvalues.

Solution: We have the equation $I\chi_E = I_E\chi_E$. Hence with Hermitian conjugation and E replaced by E': $\chi_{E'}^\dagger I^\dagger = I_{E'}^* \chi_{E'}^\dagger$. Multiplying the first equation from the left by $\chi_{E'}^\dagger$ and the second from the right by χ_E, we obtain the equations

$$\chi_{E'}^\dagger I\chi_E = \chi_{E'}^\dagger I_E\chi_E, \quad \chi_{E'}^\dagger I^\dagger\chi_E = I_{E'}^*\chi_{E'}^\dagger\chi_E.$$

Since $I = I^\dagger$ (Hermitian), it follows by subtraction of one equation from the other that:

$$0 = (I_E - I_{E'}^*)\chi_{E'}^\dagger \chi_E.$$

Since $\chi_E \neq 0$, it follows for $E = E'$ that $I_E = I_E^*$, i.e. real, or when $E \neq E'$ we have: $\chi_{E'}^\dagger \chi_E = 0$. Thus in these latter cases χ_E and $\chi_{E'}$ are orthogonal to each other. In the present context this means that the principal axes are orthogonal to each other. More generally, since the extension of the concept of orthogonality of real matrices to those with complex elements is effected by defining a unitary matrix U by the relation $UU^\dagger = 1, U^\dagger = U^{*T}$, if a matrix is Hermitian it can be diagonalized by a unitary similarity transformation (*cf.* Sec. 5.4.5, Example 8.5, and Aitken [1], p. 58).

In order to understand the significance of the eigenvectors χ_{E_i}, we recall that the components A_{1i}, A_{2i}, A_{3i} of the matrix A are the elements of column i of a rotation matrix (with $A^{-1} \in SO(3)$ the matrix A is also an element of the group $SO(3)$), as one can verify in the case of examples. These elements represent the *direction cosines* of the principal axis i with respect to the given original axes. This follows from the fact that Eq. (8.31a) reproduces Eq. (8.23) (now with respect to an axis 'i') by multiplying from the left by $\chi_{E_i}^T$ (T meaning transposition):

$$\chi_{E_i}^T (I_{jk}) \chi_{E_i} = I_i (\chi_{E_i}^T \chi_{E_i}) = I_i \qquad (8.31b)$$

with (recall that the components of χ_{E_i} are direction cosines)

$$\chi_{E_i}^T \chi_{E_i} = 1, \quad \text{and} \quad \chi_{E_i} = \begin{pmatrix} \alpha_i \\ \beta_i \\ \gamma_i \end{pmatrix} \equiv \mathbf{n}_i.$$

As we have seen, the eigenvalue equation (8.31a) implies the rotation to a new reference frame and is equivalent to the equation

$$(\chi_{E_i}^T \overbrace{A)(A^{-1}}^{1} I \overbrace{A)(A^{-1}}^{1} \chi_{E_i}) = I_i (\chi_{E_i}^T \overbrace{AA^{-1}}^{1} \chi_{E_i}). \qquad (8.32)$$

$$A \begin{pmatrix} I_1 & 0 & 0 \\ 0 & I_2 & 0 \\ 0 & 0 & I_3 \end{pmatrix} A^{-1}$$

We set

$$(A^{-1}\chi_{E_i}) = \begin{pmatrix} \alpha_i' \\ \beta_i' \\ \gamma_i' \end{pmatrix}, \qquad (8.33)$$

then Eq. (8.32) is analogous to Eq. (8.23) and yields on multiplying out

$$\alpha_i'^2 I_1 + \beta_i'^2 I_2 + \gamma_i'^2 I_3 = I_i. \qquad (8.34)$$

It follows that:

$$
\begin{aligned}
\text{for } i = 1: \quad &\alpha_1' = 1, \quad &\beta_1' = 0, \quad &\gamma_1' = 0; \\
\text{for } i = 2: \quad &\alpha_2' = 0, \quad &\beta_2' = 1, \quad &\gamma_2' = 0; \\
\text{for } i = 3: \quad &\alpha_3' = 0, \quad &\alpha_3' = 0, \quad &\gamma_3' = 1.
\end{aligned}
\tag{8.35}
$$

The vectors $\mathbf{n}_i'$ are therefore unit vectors along the new axes.

In order to obtain the principal moments of inertia I_i explicitly, we have to determine the eigenvalues of Eq. (8.31a). We write this equation now:

$$
[(I_{jk}) - I_i \mathbf{1}]\chi_{E_i} = 0.
\tag{8.36a}
$$

This is a system of homogeneous equations in the three components of the column vector χ_{E_i}. We see the crucial aspects of this matrix equation by writing it out in detail. Thus with $\chi_{E_i} = (\chi_{E_i 1}, \chi_{E_i 2}, \chi_{E_i 3})$ we have

$$
\begin{pmatrix}
I_{xx} - I_i & I_{xy} & I_{zx} \\
I_{xy} & I_{yy} - I_i & I_{yz} \\
I_{zx} & I_{yz} & I_{zz} - I_i
\end{pmatrix}
\begin{pmatrix}
\chi_{E_i 1} \\
\chi_{E_i 2} \\
\chi_{E_i 3}
\end{pmatrix} = 0,
$$

or

$$
\begin{aligned}
(I_{xx} - I_i)\chi_{E_i 1} + I_{xy}\chi_{E_i 2} + I_{xz}\chi_{E_i 3} &= 0, \\
I_{yx}\chi_{E_i 1} + (I_{yy} - I_i)\chi_{E_i 2} + I_{yz}\chi_{E_i 3} &= 0, \\
I_{zx}\chi_{E_i 1} + I_{zy}\chi_{E_i 2} + (I_{zz} - I_i)\chi_{E_i 3} &= 0.
\end{aligned}
\tag{8.36b}
$$

We see that these equations imply a linear dependence of the column vectors $((I_{xx} - I_i), I_{yx}, I_{zx}), (I_{xy}, (I_{yy} - I_i), I_{zy}), (I_{xz}, I_{yz}, (I_{zz} - I_i))$ with coefficients $\chi_{E_i 1}, \chi_{E_i 2}, \chi_{E_i 3}$. We observe that — without normalization to unity as for direction cosines — the equations depend really only on two ratios like $\chi_{E_i 1}/\chi_{E_i 3}, \chi_{E_i 2}/\chi_{E_i 3}$, which means that the coefficient of a particular column in the dependence relation is unity. It is a theorem[*] in matrix algebra that if the rows or columns of a square matrix A are linearly dependent, then $|A| = 0$. The reason is that there exists an operation on rows (or equivalently columns) which — given the relation of linear dependence — replaces all the elements of some row (or column) by zeros, without altering the value of $|A|$. This vanishing of $|A|$ is illustrated in the case of an example in Example 8.4. Thus for sensible solutions in the present case, we must have

$$
\begin{vmatrix}
I_{xx} - I_i & I_{xy} & I_{zx} \\
I_{xy} & I_{yy} - I_i & I_{yz} \\
I_{zx} & I_{yz} & I_{zz} - I_i
\end{vmatrix} = 0.
\tag{8.37}
$$

---

[*]See, for instance, A. C. Aitken [1], p. 62.

This equation is called *secular equation* or *characteristic equation*. Equation (8.37), a cubic equation in I_i, can be solved for each of the three roots of this equation; the complete solution of this equation then determines the lengths of the principal axes of the ellipsoid.[†]

The last considerations point the way to the most suitable procedure for solving problems in practice. In establishing the Lagrange function of a system, the kinetic energy T can be expressed as the sum of a translational part of the centre of mass, $Mv^2/2$, and a rotational part around the centre of mass, $I\omega^2/2$. The latter part is most conveniently expressed if one resorts to principal axes, because then we have, as we saw,

$$T = \frac{1}{2}Mv^2 + \frac{1}{2}I_1\omega_x^2 + \frac{1}{2}I_2\omega_y^2 + \frac{1}{2}I_3\omega_z^2, \qquad (8.38)$$

where further simplifications are possible as a result of some rotational symmetry. The components of ω can then be expressed in terms of the Euler angles, as we saw. (We had considered in Eq. (2.39) the separation of the energy into a translational part and a rotational part only with respect to the centre of mass).

Example 8.4: Vanishing of $|A|$ if columns of A linearly dependent
Assuming the following linear dependence

$$\begin{pmatrix} I_{xx} - I_i \\ I_{yx} \\ I_{zx} \end{pmatrix} - 2 \begin{pmatrix} I_{xy} \\ I_{yy} - I_i \\ I_{xy} \end{pmatrix} - 3 \begin{pmatrix} I_{xz} \\ I_{yz} \\ I_{zz} - I_i \end{pmatrix} = 0$$

of the columns of the matrix $[(I_{jk}) - I_i\mathbb{1}]$ and $(\chi_{E1}, \chi_{E2}, \chi_{E3}) = (1, -2, -3)$, show that $|(I_{jk}) - I_i\mathbb{1}| = 0$.

Solution: We use the property that a determinant is unaltered in value when to any row or column is added a constant multiple of any other row or column.[‡] We apply the operation

$$\text{column}_1 \longrightarrow \text{column}_1 - 2\,\text{column}_2 - 3\,\text{column}_3$$

to the determinent of the matrix $[(I_{jk}) - I_i\mathbb{1}]$. Then

$$|(I_{jk}) - I_i\mathbb{1}| = \begin{vmatrix} (I_{xx} - I_i) & I_{xy} & I_{xz} \\ I_{yx} & (I_{yy} - I_i) & I_{yz} \\ I_{zx} & I_{zy} & (I_{zz} - I_i) \end{vmatrix}$$

$$= \begin{vmatrix} (I_{xx} - I_i) - 2I_{xy} - 3I_{xz} & I_{xy} & I_{xz} \\ I_{yx} - 2(I_{yy} - I_i) - 3I_{yz} & (I_{yy} - I_i) & I_{yz} \\ I_{zx} - 2I_{zy} - 3(I_{zz} - I_i) & I_{zy} & (I_{zz} - I_i) \end{vmatrix}$$

$$= \begin{vmatrix} 0 & I_{xy} & I_{xz} \\ 0 & (I_{yy} - I_i) & I_{yz} \\ 0 & I_{zy} & (I_{zz} - I_i) \end{vmatrix} = 0.$$

[†] The solutions of the cubic with (I_{ij}) replaced by the $SO(3)$ rotation matrix (A_{ij}) are considered in great detail by H. Goldstein [19], Sec. 4.6, in particular with regard to the theorem of Euler, which says that in that case one eigenvalue must be 1.

[‡] See, for instance, A. C. Aitken [1], p. 40.

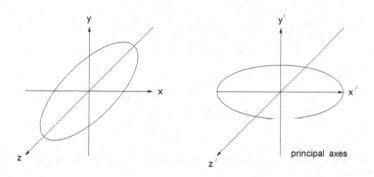

Fig. 8.6 Transformation to principal axes.

Example 8.5: Calculation of eigenvalues and eigenvectors

Determine the eigenvalues and eigenvectors of the real symmetric (hence Hermitian) matrix

$$
A = \begin{pmatrix} 2 & 0 & 0 \\ 0 & 2 & \sqrt{2} \\ 0 & \sqrt{2} & 3 \end{pmatrix}.
$$

Let $R = (x, y, z)$, and $F(R) = R^T A R = 1$. Investigate the transformation of F with respect to that particular transformation which diagonalizes the matrix A (set $R' = UR$).

Solution: We have seen (*cf.* Example 8.3) that a quadratic matrix A can be diagonalized by a unitary similarity transformation if and only if the matrix A is *Hermitian*, also called *normal*, *i.e.* if $AA^\dagger = A^\dagger A$, where the superscript $\dagger$ means complex conjugation as well as transposition. Thus the problem is to find a unitary matrix U which is such that

$$
UAU^{-1} = A_d \equiv \begin{pmatrix} \lambda_1 & 0 & 0 \\ 0 & \lambda_2 & 0 \\ 0 & 0 & \lambda_3 \end{pmatrix}, \quad i.e. \quad UA = A_d U.
$$

We obtain the eigenvalues $\lambda = \lambda_i, i = 1, 2, 3$, of matrix A from the solutions of the cubic secular equation (*cf.* Eq. (8.36a))

$$
|A - \lambda \mathbb{1}_{3\times3}| = 0, \qquad \begin{vmatrix} 2 - \lambda & 0 & 0 \\ 0 & 2 - \lambda & \sqrt{2} \\ 0 & \sqrt{2} & 3 - \lambda \end{vmatrix} = 0,
$$

i.e. $(2 - \lambda)[(2 - \lambda)(3 - \lambda) - 2] = 0$, or $(\lambda - 2)(\lambda - 1)(\lambda - 4) = 0$. Thus the eigenvalues are $\lambda_1 = 2, \lambda_2 = 1, \lambda_3 = 4$. Next we wish to determine the matrix U with the following property

$$
U = \begin{pmatrix} a & b & c \\ d & e & f \\ g & h & j \end{pmatrix}, \quad \begin{pmatrix} a & b & c \\ d & e & f \\ g & h & j \end{pmatrix} \begin{pmatrix} 2 & 0 & 0 \\ 0 & 2 & \sqrt{2} \\ 0 & \sqrt{2} & 3 \end{pmatrix} = \begin{pmatrix} 2 & 0 & 0 \\ 0 & 1 & 0 \\ 0 & 0 & 4 \end{pmatrix} \begin{pmatrix} a & b & c \\ d & e & f \\ g & h & j \end{pmatrix}.
$$

Multiplying out, we obtain the following equations

$$
(1; a, b, c): \; 2a = 2a, \quad 2d = d, \quad 2g = 4g,
$$
$$
(2; a, b, c): \; 2b + \sqrt{2}c = 2b, \quad 2e + \sqrt{2}f = e, \quad 2h + \sqrt{2}j = 4h,
$$
$$
(3; a, b, c): \; \sqrt{2}b + 3c = 2c, \quad \sqrt{2}e + 3f = f, \quad \sqrt{2}h + 3j = 4j.
$$

We deduce from Eqs. (1b), (1c), (2a), (3a), that $d = 0, g = 0, c = 0, b = 0$. The equations remaining are

$$e + \sqrt{2}f = 0, \qquad 2h - \sqrt{2}j = 0,$$

Thus the matrix U has the form

$$U = \begin{pmatrix} a & 0 & 0 \\ 0 & e & -\frac{1}{\sqrt{2}}e \\ 0 & h & \sqrt{2}h \end{pmatrix}.$$

Since U is unitary (a real unitary matrix is an orthogonal matrix, see Sec. 5.4.5), we must have $h = -e/\sqrt{2}$, *i.e.*

$$U = \begin{pmatrix} a & 0 & 0 \\ 0 & e & -\frac{1}{\sqrt{2}}e \\ 0 & -\frac{1}{\sqrt{2}}e & -e \end{pmatrix}.$$

Since $UU^\dagger = \mathbb{1}$, *i.e.* $U^2 = \mathbb{1}$, we have $a^2 = 1, 3e^2/2 = 1$, and hence $a = \pm 1, e = \pm\sqrt{2/3}$. Choosing the upper signs, we obtain

$$U = \begin{pmatrix} 1 & 0 & 0 \\ 0 & \sqrt{\frac{2}{3}} & -\sqrt{\frac{1}{3}} \\ 0 & -\sqrt{\frac{1}{3}} & -\sqrt{\frac{2}{3}} \end{pmatrix}.$$

The eigenvectors v_1, v_2, v_3 belonging respectively to the eigenvalues $\lambda_1, \lambda_2, \lambda_3$ are the columns of the matrix U. Hence:

$$v_1 = \begin{pmatrix} 1 \\ 0 \\ 0 \end{pmatrix}, \quad v_2 = \begin{pmatrix} 0 \\ \sqrt{\frac{2}{3}} \\ -\sqrt{\frac{1}{3}} \end{pmatrix}, \quad v_3 = \begin{pmatrix} 0 \\ -\sqrt{\frac{1}{3}} \\ -\sqrt{\frac{2}{3}} \end{pmatrix}.$$

One can verify now that these vectors are orthogonal and are eigenvectors, *i.e.* that

$$v_i^T v_j = \delta_{ij}, \qquad Av_j = \lambda_j v_j.$$

The latter of these equations shows that the eigenvectors are those vectors whose *directions* are not affected by the application of A. However, their moduli can change. Let R be the vector (x, y, z) and consider as stated in the problem the scalar quantity $F(R) = R^T A R = 1$. The unitary matrix U has the following effects:

$$R \longrightarrow R' = UR, \quad A \longrightarrow A_d = UAU^{-1} = \text{diagonal}.$$

Then (since U is unitary, $U^\dagger U = \mathbb{1}$ and with $R' = (x', y', z')$):

$$1 = F(R) = R^T U^\dagger UAU^{-1}UR = R'^T A_d R', \qquad A_d = \begin{pmatrix} \lambda_1 & 0 & 0 \\ 0 & \lambda_2 & 0 \\ 0 & 0 & \lambda_3 \end{pmatrix}.$$

Written out in detail the still untransformed equation is:

$$\begin{aligned} 1 = F(R) &= R^T A R = (x, y, z) \begin{pmatrix} 2 & 0 & 0 \\ 0 & 2 & \sqrt{2} \\ 0 & \sqrt{2} & 3 \end{pmatrix} \begin{pmatrix} x \\ y \\ z \end{pmatrix} = (x, y, z) \begin{pmatrix} 2x \\ 2y + \sqrt{2}z \\ \sqrt{2}y + 3z \end{pmatrix} \\ &= 2x^2 + (2y + \sqrt{2}z)y + (\sqrt{2}y + 3z)z = 2x^2 + 2y^2 + 3z^2 + 2\sqrt{2}yz. \end{aligned}$$

Alternatively, the transformed equation is (recall the eigenvalues)

$$F(R) = R'^T A_d R' = 2x'^2 + y'^2 + 4z'^2.$$

We see that the transformation U transforms the equation of the *ellipsoid*

$$2x^2 + 2y^2 + 3z^2 + 2\sqrt{2}yz = 1 \quad \text{into} \quad 2x'^2 + y'^2 + 4z'^2 = 1.$$

This means the axes have been shitfted into the position of the *principal axes* or *symmetry axes* of the ellipsoid, as indicated in Fig. 8.6.

8.4   The Equations of Motion

The equations of motion of rigid bodies are again obtained as the Euler–Lagrange equations derived from a Lagrange function. However, if a point of the rigid body is kept fixed, it is convenient to employ a different set of equations, known as *Euler's equations of motion*. In the case of conservative forces the Lagrange function is then

$$L = T - V = \frac{1}{2}(I_1\omega_x^2 + I_2\omega_y^2 + I_3\omega_z^2) - V(\theta, \phi, \psi). \qquad (8.39)$$

(Observe: No translational part because, as stated, one point of the rigid body is kept fixed). Here I_1, I_2, I_3 are the principal moments of inertia for rotation about the fixed point. If the orientation of the system of principal axes with respect to the spatial reference frame is determined by the *Euler angles* θ, ϕ, ψ, then the angular velocities $\omega_x, \omega_y, \omega_z$ (defined in Sec. 5.4.5) are given by the expressions (5.61) (note that previously we had x', y', z', where we now write x, y, z). Thus with x, y, z now the principal axes and θ, ϕ, ψ their orientation in the spatially fixed frame, we have

$$
\begin{aligned}
\omega_x &= \dot{\phi}\sin\theta\sin\psi + \dot{\theta}\cos\psi, \\
\omega_y &= \dot{\phi}\sin\theta\cos\psi - \dot{\theta}\sin\psi, \\
\omega_z &= \dot{\phi}\cos\theta + \dot{\psi}.
\end{aligned}
\qquad (8.40)
$$

In this way the Lagrange function L can be written as a function of the rotation angles and their time derivatives:

$$L = L(\theta, \phi, \psi, \dot{\theta}, \dot{\phi}, \dot{\psi}).$$

We recall that the Euler–Lagrange equations are expressed in terms of the generalized coordinates. If there are no constraints, and — as in the present case — one point of the rigid body is held fixed, three possible generalized coordinates are the Euler angles. The problem is now, to compute the *generalized forces* Q_j. We recall that Q_j was defined in Eq. (3.14) by the following relation in terms of the *applied force* $\mathbf{F}_i \equiv \mathbf{F}_i^{(a)}$,

$$Q_j = \sum_i \mathbf{F}_i \cdot \frac{\partial \mathbf{r}_i}{\partial q_j}, \qquad \mathbf{F}_i \equiv \mathbf{F}_i^{(a)}, \tag{8.41}$$

with the Euler–Lagrange equation (3.27),

$$\frac{d}{dt}\left(\frac{\partial L}{\partial \dot{q}_j}\right) - \frac{\partial L}{\partial q_j} = Q_j. \tag{8.42}$$

In the present case dq_j corresponds to an infinitesimal rotation of the vector $\mathbf{r}_i$ around a given axis.

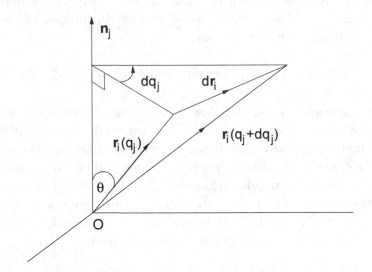

Fig. 8.7 Vector **n** a unit vector along the axis of rotation.

We have, from Eq. (5.50), see also Fig. 8.7,

$$d\mathbf{r}_i = \mathbf{r}_i \times d\boldsymbol{\Omega}, \tag{8.43}$$

so that with

$$d\boldsymbol{\Omega} = -dq_j \mathbf{n}_j, \qquad d\mathbf{r}_i = dq_j \mathbf{n}_j \times \mathbf{r}_i. \tag{8.44}$$

Hence

$$\frac{\partial \mathbf{r}_i}{\partial q_j} = \mathbf{n}_j \times \mathbf{r}_i, \tag{8.45}$$

and

$$Q_j = \sum_i \mathbf{F}_i \cdot \mathbf{n}_j \times \mathbf{r}_i = \sum_i \mathbf{n}_j \cdot \mathbf{r}_i \times \mathbf{F}_i = \mathbf{n}_j \cdot \sum_i \mathbf{r}_i \times \mathbf{F}_i = \mathbf{n}_j \cdot \mathbf{N}, \quad (8.46)$$

where $\mathbf{N}$ is the total *torque*,

$$\mathbf{N} = \sum_i \mathbf{r}_i \times \mathbf{F}_i. \quad (8.47)$$

It follows that the *generalized force* corresponding to a rotational coordinate, *i.e.* an angle, is the component of the torque along the axis of rotation. The generalized forces associated with the angles θ, ϕ, ψ are therefore the components of the torque **not** along the principal axes but along the axes belonging to the angles θ, ϕ, ψ.

The Euler–Lagrange equation in the angle ψ is therefore (*cf.* Fig. 5.4)

$$\frac{d}{dt}\left(\frac{\partial T}{\partial \dot{\psi}}\right) - \frac{\partial T}{\partial \psi} = \mathbf{n}_\psi \cdot \mathbf{N} = N_z. \quad (8.48)$$

Here N_z refers to the z-axis of the frame fixed in the rigid body. With Eqs. (8.40) we obtain

$$\frac{\partial}{\partial \dot{\psi}}(\omega_x, \omega_y) = 0, \quad \frac{\partial \omega_z}{\partial \dot{\psi}} = 1, \quad \frac{\partial \omega_x}{\partial \psi} = \omega_y, \quad \frac{\partial \omega_y}{\partial \psi} = -\omega_x,$$

$$\frac{\partial \omega_z}{\partial \psi} = 0. \quad (8.49)$$

Since

$$T = \frac{1}{2}I_1\omega_x^2 + \frac{1}{2}I_2\omega_y^2 + \frac{1}{2}I_3\omega_z^2,$$

we obtain

$$\frac{\partial T}{\partial \dot{\psi}} = \frac{\partial T}{\partial \omega_x}\frac{\partial \omega_x}{\partial \dot{\psi}} + \frac{\partial T}{\partial \omega_y}\frac{\partial \omega_y}{\partial \dot{\psi}} + \frac{\partial T}{\partial \omega_z}\frac{\partial \omega_z}{\partial \dot{\psi}} = I_3\omega_z,$$

$$\frac{\partial T}{\partial \psi} = \frac{\partial T}{\partial \omega_x}\frac{\partial \omega_x}{\partial \psi} + \frac{\partial T}{\partial \omega_y}\frac{\partial \omega_y}{\partial \psi} + \frac{\partial T}{\partial \omega_z}\frac{\partial \omega_z}{\partial \psi} = -I_2\omega_y\omega_x + I_1\omega_x\omega_y. \quad (8.50)$$

Equation (8.48) thus becomes (remembering that I_1, I_2, I_3 are constants in time)

$$I_3\dot{\omega}_z - (I_1 - I_2)\omega_x\omega_y = N_z. \quad (8.51a)$$

Here we have identified the principal axis z with the axis of rotation of ψ. By cyclic permutation of indices the remaining equations follow (the time derivatives with respect to the rotating axes):

$$I_1\dot{\omega}_x - (I_2 - I_3)\omega_y\omega_z = N_x, \quad (8.51b)$$

$$I_2\dot{\omega}_y - (I_3 - I_1)\omega_z\omega_x = N_y. \tag{8.51c}$$

Equations (8.51a), (8.51b), (8.51c) are known as *Euler's equations of motion*. Equations (8.51b), (8.51c) are not the Euler–Lagrange equations in θ, ϕ.[§]

The Euler equations of motion can be obtained more easily from the relation (2.5),

$$\left(\frac{d\mathbf{L}}{dt}\right)_{\text{space}} = \mathbf{N}, \tag{8.52}$$

which, by Newton's second law, applies to any inertial (spatially fixed) frame. Since, as we saw before, we have for $\mathbf{L} = I\omega$ instead of $\mathbf{G}$ in Eq. (5.57):

$$\left(\frac{d\mathbf{L}}{dt}\right)_{\text{space}} = \left(\frac{d\mathbf{L}}{dt}\right)_{\text{rigid body}} + \omega \times \mathbf{L} \tag{8.53}$$

considered with respect to a frame whose axes are parallel to those of the spatially fixed frame, but with origin as the frame fixed in the rigid body. Then in the direction x of the rigid body frame:

$$N_x = \frac{dL_x}{dt} + \omega_y L_z - \omega_z L_y. \tag{8.54}$$

However, in the frame of principal axes x, y, z of the rigid body:

$$L_x = I_1\omega_x, \quad L_y = I_2\omega_y, \quad L_z = I_3\omega_z, \tag{8.55}$$

so that

$$\frac{d}{dt}L_x = I_1\dot{\omega}_x. \tag{8.56}$$

It follows that

$$I_1\dot{\omega}_x + \omega_y I_3\omega_z - \omega_z I_2\omega_y = N_x \quad \text{or} \quad I_1\dot{\omega}_x - \omega_y\omega_z(I_2 - I_3) = N_x. \tag{8.57}$$

In this case we can make cyclic replacements of x, y, z, because all reference to the Euler angles and their axes has been lost. In this way we reproduce the above Eqs. (8.51a) to (8.51c).

[§]In elementary physics courses a formula like $I_1\omega_x\omega_y = -N_z$ is taught with angular momentum component $L_x = I_1\omega_x$, and a 3-finger rule is given to determine the axes, see *e.g.* E. H. Booth and P. M. Nicol [6], p. 74. Such a formula can be seen to result from the Euler equations (8.51a) to (8.51c) as a special case by assuming ω to be constant, $\omega_z = 0$ and $I_2 = I_3 = 0$. Then ω_x is the angular frequency of rotation and ω_y that of precession, with the torque $-N_z$ in the third direction.

8.5   Miscellaneous Examples I

In the following we consider a variety of examples illustrating different points related to the preceding theoretical considerations. The first example illustrates the calculation of the position of the centre of mass of a rigid body (centres of gravity have been defined earlier), and the others illustrate the calculation of moments of inertia in various cases, the results of several of which will be required later in other contexts. An important aim of Example 8.13 is to exhibit the significance of the ellipsoid of inertia in the context of a nontrivial application. In particular we show that with the help of the ellipsoid and its associated tangential planes one obtains a geometrical picture of the diagonalization of a symmetric 3×3 matrix, and hence of the problem of principal axes.

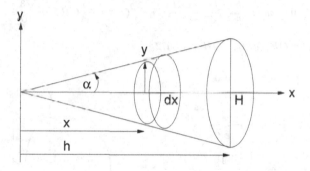

Fig. 8.8 The cone (solid or hollow).

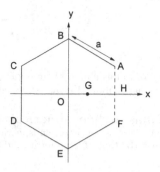

Fig. 8.9 The hexagonal wire with one link missing.

Example 8.6: (a) A cone and (b) a hexagonal wire less one link

Determine the centre of mass of the cone shown in Fig. 8.8, either solid or planar, and that of the uniform hexagonal wire shown in Fig. 8.9.¶

¶K. E. Bullen [8], p. 173.

Solution: (a) Using the formula $\mathbf{R} = \sum_i m_i \mathbf{r}_i / \sum_i m_i$, and assuming a uniform mass density ρ, the centre of mass of the cone as in Fig. 8.8 is given by the x-coordinate ($y = x \tan \alpha$)

$$\bar{x} = \frac{\int_0^h x \pi y^2 dx \rho}{\int_0^h \pi y^2 dx \rho} = \frac{\int_0^h x^3 \pi \tan^2 \alpha \rho dx}{\int_0^h x^2 \pi \tan^2 \alpha \rho dx} = \frac{\int_0^h x^3 dx}{\int_0^h x^2 dx} = \frac{3}{4} h.$$

We observe that this location is independent of the angle α. If the cone were a shell of mass per unit area λ, the centre of mass would be given by

$$\bar{x} = \frac{\int_0^h x 2\pi y dx \lambda}{\int_0^h 2\pi y dx \lambda} = \frac{\int_0^h x^2 dx}{\int_0^h x dx} = \frac{h^3}{3} \frac{2}{h^2} = \frac{2}{3} h.$$

(b) Considering the wire frame shown in Fig. 8.9, we assume a mass k and length a of each of the five links. In Fig. 8.9 the origin O is the centre of mass of the hexagon with all six sides, and G that of the structure with one side removed. The point H is the centre of mass of the single side AF. The centre of mass of the complete six-component structure is

$$0 = \bar{x}_6 = \frac{5k.OG + k.OH}{6k}, \quad i.e. \quad OG = -\frac{1}{5}OH = -\frac{\sqrt{3}}{10} a,$$

since $\sqrt{3} = \tan 60° = OH/(a/2), \therefore OH = \sqrt{3}a/2$.

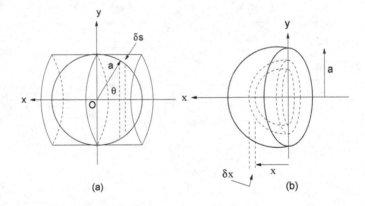

Fig. 8.10 (a) A sphere of radius a and its enveloping cylinder of length $2a$,
(b) the solid hemisphere subdivided into thin hemispherical shells.

Example 8.7: Centres of mass of hemispherical shell and solid hemisphere
Determine the position of the centre of mass of (a) a hemispherical shell (*i.e.* a surface with a thin mass distribution) of radius a, and (b) that of a uniform solid hemisphere of radius a and mass density ρ.[‖]

Solution: (a) The surface area of a sphere of radius a is $4\pi a^2$. We can write this as $2\pi a \times 2a$, and recognize the area of the sphere as equal to the area of its enveloping circular cylinder of base radius a and length $2a$, as illustrated in Fig. 8.10(a). There the area of the cylindrical slice of the sphere (between two planes perpendicular to the axis of the cylinder) of arc length or width δs at angle θ is $\delta A = 2\pi a \sin \theta \times \delta s$, where $\delta s = a \delta \theta$ and $\delta x = \delta(-a \cos \theta) = a \sin \theta \delta \theta$. Hence $\delta A \simeq 2\pi a \delta x$. Thus the area of every segment of length dx of the cylinder (with x measured along its axis) is

---

[‖] This case has also been considered by K. E. Bullen [8], p. 171.

equal to the area of the corresponding slice of the surface of the sphere. Since the centre of mass of half of the entire enveloping cylinder is at its midpoint, the centre of mass of the hemispherical shell of radius a is also at $x = a/2$ along the axis of the cylinder and measured from the origin of the sphere.

(b) Thus we know that the centre of mass of a shell of radius x is at $x/2$ along the axis of the enveloping cylinder, and hence, subdividing the solid hemisphere of mass density ρ and total mass $(2/3)\pi a^3 \rho$ into thin shells of thickness dx, as illustrated in Fig. 8.10(b), the position of its centre of mass along the axis of the enveloping cylinder is given by

$$\bar{x} = \frac{\sum_i \text{area}_i \times \text{width}_i \times \text{mass density} \times \text{position}_i}{\text{total mass}}$$

$$= \frac{\int_0^a 2\pi x^2 dx \rho (x/2)}{(2/3)\pi a^3 \rho} = \frac{\pi a^4 \rho}{4(2/3)\pi a^3 \rho} = \frac{3}{8}a.$$

This result will be required in Example 9.6.

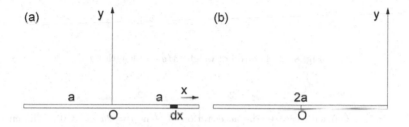

Fig. 8.11 The rod with axis of rotation through (a) its centre, and (b) an end.

Example 8.8: Moments of inertia of a massive rod
Determine the moments of inertia of the massive rod shown in Figs. 8.11(a) and 8.11(b) for a vertical axis of rotation (a) through its centre, and (b) through an end.

Solution: We assume a mass per unit length of the rod given by λ, and a total mass $M = 2a\lambda$. Then in case (a) the moment of inertia I is given by

$$I = \int_{-a}^{a} x^2 (\lambda dx) = \left[\lambda \frac{x^3}{3}\right]_{-a}^{a} = \frac{2}{3}\lambda a^3 = \frac{1}{3}Ma^2, \qquad M = 2a\lambda.$$

In case (b) we use the theorem of Steiner and obtain for the moment of inertia

$$I = Ma^2 + \frac{1}{3}Ma^2 = \frac{4}{3}Ma^2,$$

Example 8.9: The rigid body (compound) pendulum
One end of a thin rod of mass M and length $2a$ is fixed at a point C a vertical distance h above the surface of the Earth (acceleration due to gravity g). What is the Lagrange function L in terms of the angle ϕ between the vertical and the rod? What is the Hamilton function H? The rod is allowed to fall from a position with $\phi = \pi/2, \dot{\phi} = 0$. What is the acceleration of the free end A of the rod when released?

Solution: With the geometry as in Fig. 8.12 we have

$$L = T - V = \frac{1}{2}I_C \dot{\phi}^2 - V(\phi), \quad V(\phi) = Mg(h - a\cos\phi),$$

where from Example 8.7 $I_C = 4Ma^2/3$. The coordinates and velocities of the free end A of the rod are:

$$x = -2a\sin\phi, \quad y = h - 2a\cos\phi, \quad \dot{x} = -2a\cos\phi\dot{\phi}, \quad \dot{y} = 2a\sin\phi\dot{\phi},$$
$$\therefore \quad \ddot{x} = 2a\sin\phi\dot{\phi}^2 - 2a\cos\phi\ddot{\phi}, \quad \ddot{y} = 2a\cos\phi\dot{\phi}^2 + 2a\sin\phi\ddot{\phi}.$$

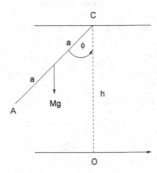

Fig. 8.12 The rod of weight Mg and length $2a$.

Hence

$$\ddot{x}_{\phi=\pi/2} = 2a\dot{\phi}^2, \quad \ddot{y}_{\phi=\pi/2} = 2a\ddot{\phi},$$

and at $\phi = \pi/2, t = 0$ with $\dot{\phi} = 0$, the acceleration of A is $\ddot{y} = 2a\ddot{\phi}(t = 0)$. The energy E is conserved, i.e.

$$E = \frac{1}{2}\frac{4}{3}Ma^2\dot{\phi}^2 + Mg(h - a\cos\phi) = \text{const.}, \quad \therefore \frac{dE}{dt} = 0, \quad 0 = \frac{4}{6}Ma^2.2\dot{\phi}\ddot{\phi} + Mga\sin\phi\dot{\phi}.$$

It follows that $\ddot{\phi}_{\phi=\pi/2}(t = 0) = -3g/4a$, and therefore the requested acceleration of A is $2a\ddot{\phi}(t = 0) = -3g/2$. The Hamilton function is E with the angular velocity $\dot{\phi}$ replaced by momentum $p_\phi = \partial L/\partial\dot{\phi} = I_C\dot{\phi}$. A rigid body like the rod which is free to rotate about a horizontal axis on which the centre of mass (or gravity) does *not* lie can oscillate about that axis, and constitutes what is called a *compound pendulum*.

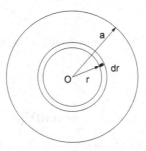

Fig. 8.13 The circular disc of radius a.

Example 8.10: Moment of inertia of a circular disc

Determine the moment of inertia of the circular disc of mass per unit area σ in the plane (x, y) shown in Fig. 8.13 for (a) a vertical axis of rotation through its centre (i.e. along the z-axis), and (b) an axis along a diameter (i.e. in the plane (x, y)).

Solution: In case (a) we obtain for the moment of inertia with the help of Fig. 8.13:

$$I = \int_0^a r^2 dm(r) = \int_0^a 2\pi r dr \sigma r^2 = \frac{1}{2}\pi a^4 \sigma = \frac{1}{2}Ma^2 \equiv I_z, \qquad M = \pi a^2 \sigma.$$

In case (b) we have a system of masspoints in the plane (x, y). Recall first that

$$I = \sum_i m_i [\mathbf{r}_i^2 - (\mathbf{r}_i \cdot \mathbf{n})^2], \quad \text{so that} \quad I_y = \sum_i m_i(\mathbf{r}_i^2 - y_i^2) = \sum_i m_i x_i^2, \quad \text{since } z_i = 0.$$

Thus in this case the moment of inertia of the disc with axis along the z-axis (as in case (a)) is

$$I_z \equiv I_{\text{along } z-\text{axis}} = \sum_i m_i(x_i^2 + y_i^2) = I_y + I_x.$$

Since $I_x = I_y$, we have $I_z = 2I_x$, and with the result of case (a) we obtain $I_x = Ma^2/4$.

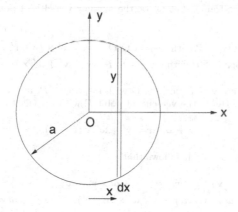

Fig. 8.14   The solid sphere, axis along a diameter.

Example 8.11: Moment of inertia of a solid sphere about a diameter

Determine the moment of inertia of the sphere of mass per unit volume ρ and radius a as shown in Fig. 8.14 for an axis of rotation through its centre.

Solution: The mass of the disc-like element shown in Fig. 8.14 is $\pi y^2 \delta x \rho$. Hence the moment of inertia of this disc about OX is (*cf.* Example 8.10(a))

$$\frac{1}{2} \times \text{mass} \times (\text{radius})^2 = \frac{1}{2}\pi y^2 \delta x \rho \times y^2 = \frac{1}{2}\pi(a^2 - x^2)^2 \rho \delta x.$$

The desired moment of inertia I is therefore obtained by summing over all such discs, so that

$$I = \frac{1}{2}\pi\rho \int_{-a}^a (a^2 - x^2)^2 dx = \frac{8}{15}\pi a^5 \rho = \frac{2}{5}Ma^2, \qquad M = \frac{4}{3}\pi a^3 \rho.$$

Example 8.12: Independence of $\boldsymbol{\omega}$ of choice of origin of rigid body frame

Show that the angular velocity $\boldsymbol{\omega}$ is independent of the choice of the location of the origin O of the rigid body frame.

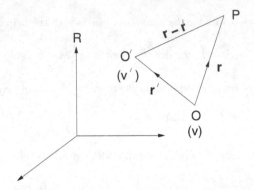

Fig. 8.15 Velocities $\mathbf{v}$ and $\mathbf{v}'$ with respect to frame R.

Solution: Let O, O' and P be points of a rigid body. Let $\mathbf{v}'$ be the velocity of O' in the inertial or spatially fixed reference frame. Let $\boldsymbol{\omega}'$ be the angular velocity of point P with respect to O'. Then, *cf.* Fig. 8.15,

$$\begin{aligned}
\text{velocity of } P \text{ with respect to } O' &= \boldsymbol{\omega}' \times (\mathbf{r} - \mathbf{r}'), \\
\text{velocity of } P \text{ with respect to } R &= \mathbf{v}' + \boldsymbol{\omega}' \times (\mathbf{r} - \mathbf{r}').
\end{aligned}$$

But $\mathbf{v}' = \mathbf{v} + \boldsymbol{\omega} \times \mathbf{r}'$, where $\mathbf{v} = $ velocity of O with respect to R, and $\boldsymbol{\omega} = $ angular velocity of O' and P with respect to O. Hence the velocity of point P with respect to R is

$$\mathbf{v} + \boldsymbol{\omega} \times \mathbf{r} = \mathbf{v}' + \boldsymbol{\omega}' \times (\mathbf{r} - \mathbf{r}').$$

With, as stated, $\mathbf{v}' = \mathbf{v} + \boldsymbol{\omega} \times \mathbf{r}'$, it follows that

$$\begin{aligned}
\mathbf{v} + \boldsymbol{\omega} \times \mathbf{r} &= \mathbf{v} + \boldsymbol{\omega} \times \mathbf{r}' + \boldsymbol{\omega}' \times (\mathbf{r} - \mathbf{r}'), \\
\textit{i.e. } \boldsymbol{\omega} \times (\mathbf{r} - \mathbf{r}') &= \boldsymbol{\omega}' \times (\mathbf{r} - \mathbf{r}'), \quad \text{and} \quad \therefore \ \boldsymbol{\omega} = \boldsymbol{\omega}'.
\end{aligned}$$

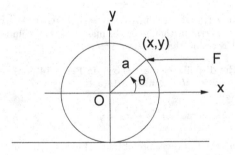

Fig. 8.16 Force F hitting the solid sphere.

Example 8.13: An impulsive force applied to a solid sphere

A uniform solid sphere of radius a and mass M as shown in Fig. 8.16 is to be given an impulse so that it rolls on a flat frictionless plane. At what height above the table does the force F have to be applied to achieve this rolling?[**]

[**]A similar problem may be found in K. E. Bullen [8], p. 241.

Solution: *Rolling* means velocity $v = a\omega \equiv a\dot{\theta}$. In general, for a force F acting on a particle of mass m in the short time interval δt implies the equation:

$$F\delta t = m\ddot{x}\delta t = m\frac{\delta\dot{x}}{\delta t}\delta t = \delta(m\dot{x}).$$

In the present case and for rolling this means $F\delta t = m\delta(a\dot{\theta})$ or $F = md(a\dot{\theta})/dt$. Inserting this expression as the external force in the Euler equation (2.5) or (8.52) we obtain (*cf.* Fig. 8.16)

$$\frac{d}{dt}(I\dot{\theta}) = Fy = M\frac{da(\dot{\theta})}{dt}a\sin\theta, \quad \therefore\ I = Ma^2\sin\theta.$$

Thus, with $I = 2Ma^2/5$ from Example 8.10 we have $y/a = \sin\theta = I/Ma^2 = 2/5 = 0.4$, *i.e.* $y = 0.4a$.

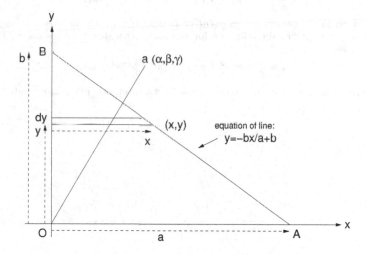

Fig. 8.17 The triangular plane.

Example 8.14: Moment of inertia of a triangular plane

Determine the moment of inertia of a triangular plane about axes of various orientations.[††]

Solution: We assume a mass per unit area λ. For the geometry we refer to Fig. 8.17. The line AB there is given by the equation $y = -bx/a + b$. We let dm be the mass of an element of area. The moment of inertia of the triangular plane about the x-axis is by definition

$$I_{xx} \overset{(8.6)}{=} \int_{y=0}^{y=b} y^2\,dm = \int_{y=0}^{y=b} y^2(xdy\lambda) = \lambda\int_0^b y^2(b-y)\frac{a}{b}dy = \frac{1}{6}Mb^2, \quad M = \frac{1}{2}ab\lambda.$$

Similarly we obtain (since $z = 0$)

$$I_{yy} = \frac{1}{6}Ma^2, \quad I_{zz} \overset{(8.14)}{=} I_{xx} + I_{yy} = \frac{1}{6}M(a^2 + b^2).$$

[††]The triangular plane is also treated in great detail in D. A. Wells [51].

Next we calculate the inertial products around an axis through O. Thus by definition

$$-I_{xy} \overset{(8.6)}{=} \int xy \, dm = \int\int \lambda xy \, dx \, dy = \lambda \int_{x=0}^{a} \int_{y=0}^{b-bx/a} xy \, dx \, dy = \lambda \int_{x=0}^{a} x \, dx \frac{1}{2}\left(b - \frac{b}{a}x\right)^2$$

$$= \frac{\lambda}{2}\int_0^a \left(b^2 - \frac{2b^2}{a}x + \frac{b^2}{a^2}x^2\right)x \, dx = \frac{\lambda}{2}\left[b^2\frac{a^2}{2} - \frac{2}{3}b^2 a^2 + \frac{b^2 a^2}{4}\right]$$

$$= \frac{1}{24}\lambda a^2 b^2 = \frac{1}{12}Mab, \quad M = \frac{1}{2}\lambda ab; \quad \text{also} \quad I_{xz} = 0 = I_{yz}.$$

For the *ellipsoid of inertia* with respect to an axis passing through O and perpendicular to the plane of the triangle, we obtain therefore from Eq. (8.26),

$$1 = I_{xx}\rho_1^2 + I_{yy}\rho_2^2 + I_{zz}\rho_3^2 + 2I_{xy}\rho_1\rho_2 + 2I_{yz}\rho_2\rho_3 + 2I_{zx}\rho_3\rho_1,$$

the expression

$$1 = \frac{1}{6}M[b^2\rho_1^2 + a^2\rho_2^2 + (a^2 + b^2)\rho_3^2 - ab\rho_1\rho_2].$$

With this last result we obtain the *moment of inertia* I_{Oa} about an axis Oa (as indicated in Fig. 8.17) with an arbitrary direction, as for instance, with direction cosines α, β, γ and $\rho = (\alpha, \beta, \gamma)/\sqrt{I_{Oa}}$ as:

$$I_{Oa} = \frac{1}{6}M[b^2\alpha^2 + a^2\beta^2 + (a^2 + b^2)\gamma^2 - ab\alpha\beta].$$

Next we determine for later use the centre of mass G of a triangular plane as indicated in Fig. 8.18.

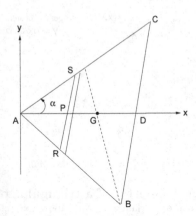

Fig. 8.18 The triangular plane with centre of mass G.

We assume again a mass λ per unit area. In Fig. 8.18 the point D is the midpoint of the line BC. We can then consider the strip RS as a rod, whose centre of mass is at $P(x, 0)$. The corresponding element of mass is $\lambda . \delta x . RS$. We can re-express this in terms of x with a little geometry. Considering the congruent triangles RAS and BAC, we have

$$\frac{AP}{RS} = \frac{AD}{BC} = \text{const.} \equiv \frac{1}{c}.$$

Thus now the element of mass becomes $\delta m = \lambda . \delta x . RS = \lambda . \delta x . c . AP = \lambda . \delta x . c . x$. The x-coordinate of the centre of mass of the triangle, $\bar{x}$, is then given by

$$\bar{x} = \frac{\int x \, dm}{\int dm} = \frac{\int_0^{AD} x\lambda \, dx \, cx}{\int_0^{AD} \lambda \, dx \, cx} = \frac{AD^3/3}{AD^2/2} = \frac{2}{3}AD.$$

Correspondingly we have the simpler case of the y-coordinate:

$$\bar{y} = \frac{\int_0^{AD} 0.x dx}{\int_0^{AD} x dx} = 0.$$

Hence the centre of mass G of triangle ABC is at the position $AG = 2AD/3$ along the line AD.[‡‡]
   We now return to the triangle itself. We wish to calculate its moments of inertia, more precisely first the coefficients or elements $\bar{I}_{xx}, \bar{I}_{yy}, \bar{I}_{zz}, \bar{I}_{xy}$ of its as yet undiagonalized inertial tensor, about a vertical axis to its plane and through its centre of mass. Thereafter we consider the diagonalization of the tensor.

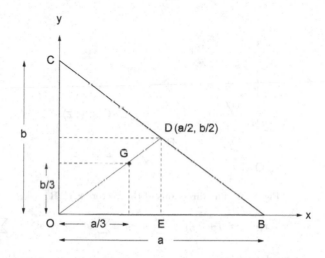

Fig. 8.19 The triangular plane with sides along Cartesian axes.

In Fig. 8.19 DBE, CBO are congruent triangles and from the preceding calculations we know that this centre of mass G has the coordinates of $2/3$ of the coordinates of the point D (the midpoint of the line BC, and this has the coordinates $(a/2, b/2)$), so that the coordinates of G are $(a/3, b/3)$. We use the *theorem of Steiner*, in order to obtain $\bar{I}_{xx}, \bar{I}_{yy}, \bar{I}_{zz}$ from I_{xx}, I_{yy}, I_{zz} about the origin O, and this means we have

$$\bar{I}_{xx} \overset{(8.18)}{=} I_{xx} - M\bar{y}^2 = \frac{1}{6}Mb^2 - M\left(\frac{b}{3}\right)^2 = \frac{1}{18}Mb^2.$$

Correspondingly one obtains

$$\bar{I}_{yy} = I_{yy} - M\bar{x}^2 = \frac{1}{6}Ma^2 - M\left(\frac{a}{3}\right)^2 = \frac{1}{18}Ma^2,$$

$$\bar{I}_{zz} = \bar{I}_{xx} + \bar{I}_{yy} = \frac{1}{18}M(a^2 + b^2).$$

[‡‡]This result is actually one contained in a theorem already taught in High School Euclidean geometry. There the theorem is enunciated as follows: The medians of a triangle are concurrent, the point of intersection (the centroid of the triangle) being one third of the way along each median from the base to the opposite vertex like GD in Fig. 8.18, so that G is distant $2/3$ of the way from A, or from O in Fig. 8.19. This means, again with reference to Fig. 8.18, that the coordinates of G are $2/3$ of the coordinates of D, and are therefore $(2/3)(a/2, b/2) = (a/3, b/3)$.

For the inertial product coefficients similar formulas apply as for the moments of inertia; for instance (we explain this below),

$$\bar{I}_{xy} = I_{xy} + M\bar{x}\bar{y}.$$

For the explanation we recall that with the vectors $\mathbf{r}, \mathbf{r}', \mathbf{R}$ as in Fig. 8.20:

$$-\sum_i m_i x_i' y_i' = \bar{I}_{xy}, \qquad -\sum_i m_i x_i y_i = I_{xy}$$

and (for the present calculations in the case of the triangle in the (x, y)-plane all z-components are zero)

$$\mathbf{r}_i = \mathbf{R} + \mathbf{r}_i', \quad \therefore \quad x_i = \bar{x} + x_i', \ y_i = \bar{y} + y_i',$$

and therefore

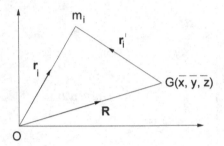

Fig. 8.20 The directions of the vectors $\mathbf{r}, \mathbf{r}', \mathbf{R}$.

$$\sum_i m_i x_i y_i = \sum_i m_i [\bar{x}\bar{y} + x_i' y_i' + x_i' \bar{y} + \bar{x} y_i'] = \sum_i m_i \bar{x}\bar{y} + \sum_i m_i x_i' y_i', \quad \text{since} \quad \sum_i m_i \mathbf{r}_i' \overset{(8.17)}{=} 0.$$

It follows, as above claimed, that $-I_{xy} = M\bar{x}\bar{y} - \bar{I}_{xy}$. As a consequence we obtain (with $I_{xy} = -Mab/12$ from above)

$$\bar{I}_{xy} = I_{xy} + M\bar{x}\bar{y} = -\frac{1}{12}Mab + M\frac{a}{3}\frac{b}{3} = \frac{1}{36}Mab, \qquad \bar{I}_{xz} = 0 = \bar{I}_{yz}.$$

The *ellipsoid of inertia* about the centre of mass G is now given by (*cf.* Eq. (8.26))[*]

$$1 = \bar{I}_{xx}x'^2 + \bar{I}_{yy}y'^2 + \bar{I}_{zz}z'^2 + 2\bar{I}_{xy}x'y' + 2\bar{I}_{yz}y'z' + 2\bar{I}_{zx}z'x',$$

$$\text{or} \qquad 1 = \frac{1}{18}M[b^2 x'^2 + a^2 y'^2 + (a^2 + b^2)z'^2 + abx'y']. \tag{8.58}$$

With this result we obtain the moment of inertia of the triangular plane about any axis GG' through the centre of mass and with direction cosines α', β', γ'. Thus with (*cf.* Eq. (8.24))

$$\mathbf{r}' = (x', y', z') = \frac{\mathbf{n}'}{\sqrt{I_{GG'}}}, \qquad \mathbf{n}' = \alpha'\mathbf{e}_{x'} + \beta'\mathbf{e}_{y'} + \gamma'\mathbf{e}_{z'},$$

this is given by

$$I_{GG'} = \bar{I}_{xx}\alpha'^2 + \bar{I}_{yy}\beta'^2 + \bar{I}_{zz}\gamma'^2 + 2\bar{I}_{xy}\alpha'\beta' + 2\bar{I}_{yz}\beta'\gamma' + 2\bar{I}_{zx}\gamma'\alpha',$$

$$\text{or} \quad I_{GG'} = \frac{1}{18}M[b^2\alpha'^2 + a^2\beta'^2 + (a^2 + b^2)\gamma'^2 + ab\alpha'\beta']. \tag{8.59}$$

---

[*]Note that it might be more appropriate to write $\bar{I}_{xx} \to I_{x'x'}$, etc.

Example 8.15: Principal moments of inertia of a triangular plane

Determine the principal moments of inertia of the triangular plane of Example 8.14.

Solution: Our aim now is to derive the principal moments of inertia I^p of the triangular plane of Example 8.13 about an axis through its centre of mass. To this end we have to solve the equation (*cf.* Eq. (8.37))

$$\begin{vmatrix} \bar{I}_{xx} - I^p & \bar{I}_{xy} & \bar{I}_{xz} \\ \bar{I}_{xy} & \bar{I}_{yy} - I^p & \bar{I}_{yz} \\ \bar{I}_{xz} & \bar{I}_{yz} & \bar{I}_{zz} - I^p \end{vmatrix} = 0.$$

Since $\bar{I}_{xz} = 0 = \bar{I}_{yz}$, we have

$$\begin{vmatrix} \bar{I}_{xx} - I^p & \bar{I}_{xy} & 0 \\ \bar{I}_{xy} & \bar{I}_{yy} - I^p & 0 \\ 0 & 0 & \bar{I}_{zz} - I^p \end{vmatrix} = 0,$$

and hence we obtain the equation

$$(\bar{I}_{zz} - I^p)[(\bar{I}_{xx} - I^p)(\bar{I}_{yy} - I^p) - \bar{I}_{xy}^2] = 0. \tag{8.60}$$

We deduce from one root of this equation that

$$I_3^p = \bar{I}_{zz} \overset{(8.14)}{=} \bar{I}_{xx} + \bar{I}_{yy}.$$

The other two roots are given by the quadratic equation

$$I^{p2} - I^p(\bar{I}_{xx} + \bar{I}_{yy}) + \bar{I}_{xx}\bar{I}_{yy} - \bar{I}_{xy}^2 = 0$$

with roots

$$I_{1,2}^p = \frac{1}{2}\left[\bar{I}_{xx} + \bar{I}_{yy} \pm \sqrt{(\bar{I}_{xx} + \bar{I}_{yy})^2 - 4(\bar{I}_{xx}\bar{I}_{yy} - \bar{I}_{xy}^2)}\right] = \frac{1}{2}\left[\bar{I}_{xx} + \bar{I}_{yy} \pm \sqrt{(\bar{I}_{xx} - \bar{I}_{yy})^2 + 4\bar{I}_{xy}^2}\right].$$

We now insert here the expressions previously obtained for $\bar{I}_{xx}, \bar{I}_{yy}, \bar{I}_{xy}$, and obtain

$$\begin{aligned} I_{1,2}^p &= \frac{1}{2}\left[\frac{1}{18}M(b^2 + a^2) \pm \sqrt{\left(\frac{1}{18}\right)^2 M^2(b^2 - a^2)^2 + 4\left(\frac{1}{36}Mab\right)^2}\right] \\ &= \frac{1}{36}M\left[b^2 + a^2 \pm \sqrt{(b^2 - a^2)^2 + a^2b^2}\right] = \frac{1}{36}M\left[a^2 + b^2 \pm \sqrt{a^4 + b^4 - a^2b^2}\right], \\ I_3^p &= \frac{1}{18}M(a^2 + b^2). \end{aligned} \tag{8.61}$$

Example 8.16: Directions of the principal axes of a triangular plane

Determine the directions of the principal axes of a triangular plane with respect to a given set of axes through the centre of mass.

Solution: We wish to determine the directions of the principal axes with the centre of mass G as origin and with respect to the axes (x', y', z') there. The ellipsoid of inertia about G is given by Eq. (8.58), or

$$\begin{aligned} 1 = \phi(x', y', z') &= \bar{I}_{xx}x'^2 + \bar{I}_{yy}y'^2 + \bar{I}_{zz}z'^2 + 2\bar{I}_{xy}x'y' + 2\bar{I}_{yz}y'z' + 2\bar{I}_{zx}z'x', \\ \text{or} \qquad 1 &= \frac{1}{18}M[b^2x'^2 + a^2y'^2 + (a^2 + b^2)z'^2 + abx'y']. \tag{8.62} \end{aligned}$$

This equation represents the surface of the ellipsoid (if ϕ represented a potential, this would be an equipotential surface with the constant value 1 everywhere). In view of the constant on one side of the equation, the gradient of $\phi(x', y', z')$ has the component zero in every direction tangent to the surface $\phi(x', y', z') - 1 = 0$, and thus has components only along the normal there, and is therefore perpendicular to the surface.[†] Thus the vector (*cf.* Example 8.2)

$$\mathbf{N}(\mathbf{r}') \equiv \left(\frac{\partial \phi}{\partial x'}, \frac{\partial \phi}{\partial y'}, \frac{\partial \phi}{\partial z'} \right), \qquad N = \sqrt{\left(\frac{\partial \phi}{\partial x'} \right)^2 + \left(\frac{\partial \phi}{\partial y'} \right)^2 + \left(\frac{\partial \phi}{\partial z'} \right)^2} \qquad (8.63)$$

is a perpendicular on the surface $\phi(x', y', z') - 1 = 0$ at the point (x', y', z') (or equivalently the normal to the tangential plane at this point) with its direction cosines given by the expressions[‡]

$$l = \cos \delta_1 = \frac{1}{N} \frac{\partial \phi}{\partial x'}, \qquad m = \cos \delta_2 = \frac{1}{N} \frac{\partial \phi}{\partial y'}, \qquad n = \cos \delta_3 = \frac{1}{N} \frac{\partial \phi}{\partial z'}.$$

Equating the derivatives involved here to their respective direction cosines multiplied by a constant k, we obtain

$$\begin{aligned} \frac{\partial \phi}{\partial x'} &= 2[\bar{I}_{xx} x' + \bar{I}_{xy} y' + \bar{I}_{zx} z'] = kl, \\ \frac{\partial \phi}{\partial y'} &= 2[\bar{I}_{yy} y' + \bar{I}_{xy} x' + \bar{I}_{yz} z'] = km, \\ \frac{\partial \phi}{\partial z'} &= 2[\bar{I}_{zz} z' + \bar{I}_{zx} x' + \bar{I}_{zy} y'] = kn. \end{aligned} \qquad (8.64)$$

In the next Example 8.17 we illustrate the derivation of the equation of the tangential plane at a specific point of an ellipsoid. The principal axes are orthogonal to each other (*cf.* Example 8.6) and (*cf.* Eq. (8.27)) to the surface of the ellipsoid. Such an axis cuts the surface of the ellipsoid at

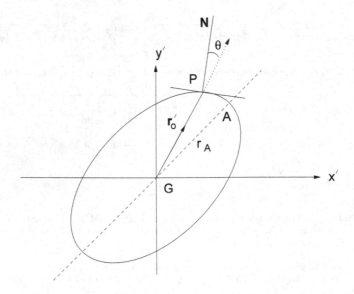

Fig. 8.21 Alignment of $\mathbf{r}_0'$ and $\mathbf{N}$ as P approaches A.

[†] For further discussions see, for instance, R. Courant [12], Vol. II, p. 90.
[‡] For further discussions of these see, for instance, R. Courant [12], Vol. II, p. 130.

a point A (say) like the perihelion in Fig. 8.21 a distance $r'_A = \sqrt{{x'_A}^2 + {y'_A}^2 + {z'_A}^2}$ from the origin G. Suppose a radial vector to the surface at a point P (coordinates (x'_0, y'_0, z'_0)) and the normal to the surface there meet with an inclination angle θ to each other, as shown in Fig. 8.21. Clearly, as the point P moves to the perihelion position A, the angle θ decreases to zero, and the radial vector and the normal align to a principal axis of the ellipsoid. Thus the condition for location of a principal axis is that the direction cosines of the radius to a point P on the surface and those of the normal to the surface there, *i.e.* (l, m, n), must be proportional to each other. The direction cosines of the radial vector $\mathbf{r}'_0$ to the point P on the surface in the reference frame (x', y', z') are given by the projections

$$\frac{x'_0}{r'_0}, \quad \frac{y'_0}{r'_0}, \quad \frac{z'_0}{r'_0}. \tag{8.65}$$

Thus, considering Eqs. (8.64) now at x'_0, y'_0, z'_0 with

$$x'_0 = r'_0 l, \quad y'_0 = r'_0 m, \quad z'_0 = r'_0 n,$$

we obtain the equations

$$\bar{I}_{xx} l + \bar{I}_{xy} m + \bar{I}_{xz} n = \frac{kl}{2r'_0},$$

$$\bar{I}_{yy} m + \bar{I}_{yx} l + \bar{I}_{yz} n = \frac{km}{2r'_0},$$

$$\bar{I}_{zz} n + \bar{I}_{zx} l + \bar{I}_{zy} m = \frac{kn}{2r'_0}. \tag{8.66}$$

We now multiply the first of these equations by l, the second by m, the third by n, and add them. This gives

$$\frac{k}{2r'_0} = \bar{I}_{xx} l^2 + \bar{I}_{yy} m^2 + \bar{I}_{zz} n^2 + 2\bar{I}_{xy} lm + 2\bar{I}_{yz} mn + 2\bar{I}_{zx} ln. \tag{8.67}$$

This is precisely (*cf.* Eq. (8.59)) $I_{GG'} = k/2r'_0$ with $r_0 \to r_A$ (the length of the semi-principal axis) for the specific values l, m, n of α', β', γ'. Thus we can write Eqs. (8.66)

$$(\bar{I}_{xx} - I^p)l + \bar{I}_{xy} m + \bar{I}_{zx} n = 0,$$

$$\bar{I}_{xy} l + (\bar{I}_{yy} - I^p)m + \bar{I}_{yz} n = 0, \tag{8.68}$$

$$\bar{I}_{zx} l + \bar{I}_{zy} m + (\bar{I}_{zz} - I^p)n = 0,$$

where I^p is the principal moment of inertia associated with the principal axis under consideration. Hence, given the ellipsoid of inertia in any reference frame with respect to the centre of mass as origin (this implies $\bar{I}_{xx} \equiv I_{x'x'}, \ldots$), then one obtains the direction cosines l, m, n of a principal axis by solving these equations for the principal moment of inertia I^p of this axis. In effect, we had observed this earlier, since Eqs. (8.68) are seen to be identical with our earlier Eqs. (8.36b). We also observed that we can divide the equations by (*e.g.*) n, so that — without the additional condition $l^2 + m^2 + n^2 = 1$ — they really determine only two ratios like l/m and m/n. We have thus achieved a geometrical understanding (as illustrated in Fig. 8.21) of the diagonalization of a matrix or, as here, of the inertial tensor. We can now proceed to the actual calculations.

In the present case we have (with the plane of the triangle in the (x', y')-plane, implying for this $z' = 0$, but not for the axes)

$$\bar{I}_{xz} = 0 = \bar{I}_{yz},$$

so that the equations become

$$(\bar{I}_{xx} - I^p)l + \bar{I}_{xy} m = 0, \quad \bar{I}_{xy} l + (\bar{I}_{yy} - I^p)m = 0, \quad (\bar{I}_{zz} - I^p)n = 0. \tag{8.69}$$

(a)   In the case of $I^p \equiv I_3^p$ — and we know from Eq. (8.60) that $I_3^p = \bar{I}_{zz}$ — we obtain

$$(\bar{I}_{xx} - I_3^p)l_3 + \bar{I}_{xy}m_3 = 0 \qquad \text{implying} \qquad l_3 = \frac{\bar{I}_{xy}m_3}{I_3^p - \bar{I}_{xx}},$$

$$\bar{I}_{xy}l_3 + (\bar{I}_{yy} - I_3^p)m_3 = 0 \qquad \text{implying} \qquad 0 = \left(\underbrace{\frac{\bar{I}_{xy}^2}{I_3^p - \bar{I}_{xx}}}_{\bar{I}_{xy}l_3} + \bar{I}_{yy} - I_3^p \right)m_3 = 0,$$

$$(\bar{I}_{zz} - I_3^p)n_3 \qquad = \qquad 0.$$

It follows that $m_3 = 0$, and hence $l_3 = 0$. Then from the relation $l_3^2 + m_3^2 + n_3^2 = 1$, we obtain the result that $n_3 = 1$. Thus we have

$$(l_3, m_3, n_3) = (0, 0, 1).$$

(b)   In the cases of $I^p \equiv I_1^p$ and I_2^p, we obtain immediately from Eqs. (8.69) that $n_1 = 0, n_2 = 0$ (since $I_1^p, I_2^p \neq \bar{I}_{zz}$). The components l_1, m_1 follow from

$$(\bar{I}_{xx} - I_1^p)l_1 + \bar{I}_{xy}m_1 = 0, \qquad \bar{I}_{xy}l_1 + (\bar{I}_{yy} - I_1^p)m_1 = 0,$$

and l_2, m_2 from

$$(\bar{I}_{xx} - I_2^p)l_2 + \bar{I}_{xy}m_2 = 0, \qquad \bar{I}_{xy}l_2 + (\bar{I}_{yy} - I_2^p)m_2 = 0,$$

together with $l_1^2 + m_1^2 = 1, l_2^2 + m_2^2 = 1$. Eliminating l_1 from the first two equations, we obtain

$$[-(\bar{I}_{xx} - I_1^p)(\bar{I}_{yy} - I_1^p) + \bar{I}_{xy}^2]m_1 = 0.$$

Equation (8.60) requires that the expression in square brackets vanish. Thus we cannot conclude from the last equation that m_1 would be zero. In order to determine m_1, we replace in the first equation above l_1 by $l_1 = \sqrt{1 - m_1^2}$ and solve for m_1. We obtain

$$(\bar{I}_{xx} - I_1^p)\sqrt{1 - m_1^2} + \bar{I}_{xy}m_1 = 0, \quad \text{or by squaring} \quad (\bar{I}_{xx} - I_1^p)^2(1 - m_1^2) = \bar{I}_{xy}^2 m_1^2.$$

Hence

$$(\bar{I}_{xx} - I_1^p)^2 = m_1^2[\bar{I}_{xy}^2 + (\bar{I}_{xx} - I_1^p)^2], \quad \text{or} \quad m_1^2 = \frac{(\bar{I}_{xx} - I_1^p)^2}{[\bar{I}_{xy}^2 + (\bar{I}_{xx} - I_1^p)^2]}.$$

Here we can insert the explicit expressions we obtained earlier for $\bar{I}_{xx}, I_1^p, \bar{I}_{xy}$. The other coefficients can then be determined correspondingly together with the relation $l_{1,2}^2 + m_{1,2}^2 + n_{1,2}^2 = 1$.

Example 8.17: Tangential planes to ellipsoids

Given the equation of an ellipse in normal form — and correspondingly that of an ellipsoid — derive the equation of the tangent — correspondingly tangential plane — at a point (x_0, y_0) — correspondingly at (x_0, y_0, z_0). The result is referred to in Examples 7.1 and 8.16.

Solution: In the case of an ellipse we differentiate the equation

$$\frac{x^2}{a^2} + \frac{y^2}{b^2} = 1 \quad \text{and obtain} \quad \frac{2x}{a^2} + \frac{2y}{b^2}\frac{dy}{dx} = 0.$$

The equation of the tangent at a point (x_0, y_0) on the ellipse is therefore given by

$$\left(\frac{dy}{dx}\right)_{\substack{x \to x_0 \\ y \to y_0}} = \frac{y - y_0}{x - x_0} = -\frac{b^2}{a^2}\frac{x_0}{y_0}, \quad \text{or} \quad \frac{x_0}{a^2}x + \frac{y_0}{b^2}y = \frac{x_0^2}{a^2} + \frac{y_0^2}{b^2} = 1.$$

In the case of the ellipsoid, we differentiate

$$\frac{x^2}{a^2} + \frac{y^2}{b^2} + \frac{z^2}{c^2} = 1 \quad \text{and obtain} \quad \frac{2x}{a^2} + \frac{2y}{b^2}\frac{dy}{dx} + \frac{2z}{c^2}\frac{dz}{dx} = 0.$$

The equation of the tangential plane to the ellipsoid at the point (x_0, y_0, z_0) on its surface is therefore

$$\frac{2x_0}{a^2} + \frac{2y_0}{b^2}\left(\frac{y - y_0}{x - x_0}\right) + \frac{2z_0}{c^2}\left(\frac{z - z_0}{x - x_0}\right) = 0, \quad \text{or} \quad \frac{x_0}{a^2}x + \frac{y_0}{b^2}y + \frac{z_0}{c^2}z = \frac{x_0^2}{a^2} + \frac{y_0^2}{b^2} + \frac{z_0^2}{c^2} = 1.$$

8.6 Force-free Motion

We now consider the dynamics of rigid bodies that are not subjected to external forces or torques, whose energy is, however, nonzero. Such situations are possible, for instance, when a rigid body is given a certain initial velocity. It is convenient in these cases to choose the centre of mass of the body as the origin of the body-fixed reference frame. The *Euler equations* (8.51a), (8.51b), (8.51c) then become[§]

$$\begin{aligned}
I_1\dot{\omega}_x &= \omega_y\omega_z(I_2 - I_3), \\
I_2\dot{\omega}_y &= \omega_z\omega_x(I_3 - I_1), \\
I_3\dot{\omega}_z &= \omega_x\omega_y(I_1 - I_2).
\end{aligned} \tag{8.70}$$

These equations can be integrated, in fact their solutions can immediately be read off the derivatives of the three basic *Jacobian elliptic functions* of argument u, namely $\mathrm{sn}[u], \mathrm{cn}[u], \mathrm{dn}[u]$, as will be shown in Example 8.18. Here we pursue a simpler procedure. Two integration constants can be taken as the *total energy* (here kinetic energy) and *total angular momentum*. The motion of the rigid body in the force-free case can also be understood geometrically, and — in fact — with the following construction. We choose the axes of the reference frame along the principal axes of the rigid body. Then

$$I = \sum_{i,j} n_i I_{ij} n_j, \quad \mathbf{n} = \sqrt{I}\rho, \quad \rho = \frac{\omega}{\omega\sqrt{I}}, \quad |\boldsymbol{\omega}| = \omega, \tag{8.71}$$

and hence the surface of the ellipsoid of inertia is given by

$$F(\rho) \equiv \sum_{i,j}(\rho_i I_{ij} \rho_j) = 1. \tag{8.72}$$

As indicated in Fig. 8.22, the unit vector $\mathbf{n}$ points in the direction of the axis of rotation.

[§]For readers also interested in modern developments, we add that these equations are symmetric under parity (P) and time (T) inversions, and permit complex solutions, which have been studied recently by C. M. Bender, D. D. Holm and D. W. Hook [3].

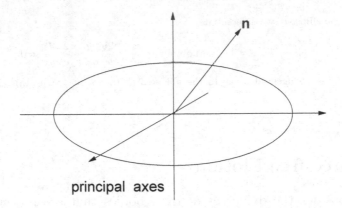

principal axes

Fig. 8.22 The ellipsoid of inertia, the principal axes and the axis of rotation.

Since, by assumption, the inertial tensor I_{ij} has been diagonalized, we can rewrite Eq. (8.72) as

$$F(\boldsymbol{\rho}) \equiv \begin{pmatrix} \rho_1 & \rho_2 & \rho_3 \end{pmatrix} \begin{pmatrix} I_1 & 0 & 0 \\ 0 & I_2 & 0 \\ 0 & 0 & I_3 \end{pmatrix} \begin{pmatrix} \rho_1 \\ \rho_2 \\ \rho_3 \end{pmatrix} = \rho_1^2 I_1 + \rho_2^2 I_2 + \rho_3^2 I_3.$$

$$(8.73)$$

It follows — by recalling from Eq. (8.5b) that $L_i = \sum_j I_{ij}\omega_j$ — that

$$\frac{\partial F}{\partial \rho_1} \equiv (\boldsymbol{\nabla} F)_1 = 2I_1\rho_1 = \frac{2\omega_1 I_1}{\omega\sqrt{I}} = \frac{2L_x}{\omega\sqrt{I}}, \quad \text{or} \quad \boldsymbol{\nabla} F = \frac{2}{\omega\sqrt{I}}\mathbf{L}. \qquad (8.74)$$

The gradient of F is a vector which is perpendicular to the surface of the ellipsoid at a point (ρ_1, ρ_2, ρ_3), but is not parallel to $\boldsymbol{\rho}$, hence also not parallel to $\boldsymbol{\omega}$. We see therefore, that as $\boldsymbol{\omega}$ varies, the perpendicular to the ellipsoid always points in the direction of the angular momentum. In view of the relation

$$\frac{d\mathbf{L}}{dt} = \mathbf{N} = 0 \quad (\text{force−free motion}), \quad \mathbf{L} = \text{const.},$$

the angular momentum $\mathbf{L}$ is conserved, including its direction in space. Thus the motion of the ellipsoid of inertia in space is such that the direction of the angular momentum $\mathbf{L}$, and hence the gradient of F, remains constant. One defines as an *invariable plane* one whose normal points in the constant direction of $\mathbf{L}$. In particular, the plane that touches the ellipsoid is referred to as the *tangential plane*.

We consider now a symmetric rigid body, for which $I_1 = I_2$, which implies that the z-axis is a symmetry axis of the body like the NS–polar axis of the Earth with ω_z ist angular frequency around this axis which is 2π per day.

The Euler equations (8.70) are then

$$I_1\dot\omega_x = \omega_y\omega_z(I_1 - I_3), \tag{8.75a}$$

$$I_1\dot\omega_y = -\omega_z\omega_x(I_1 - I_3), \tag{8.75b}$$

$$I_3\dot\omega_z = 0. \tag{8.75c}$$

Equation (8.75c) implies that $\omega_z = $ const. Differentiation of Eq. (8.75a) implies therefore

$$I_1\ddot\omega_x = (I_1 - I_3)\omega_z\dot\omega_y = -\frac{[(I_1 - I_3)\omega_z]^2}{I_1}\omega_x,$$

or

$$\ddot\omega_x + \Omega^2\omega_x = 0, \quad \text{with} \quad \Omega^2 = \left\{\frac{(I_1 - I_3)\omega_z}{I_1}\right\}^2 \neq 0 \text{ for } I_1 \neq I_3 \tag{8.76}$$

We thus obtain as the simplest solution of the harmonic motion the result:

$$\omega_x = A\sin\Omega t, \quad A = \text{const.} \tag{8.77}$$

The constant A is determined by an initial condition (like a given initial frequency or kinetic energy). From Eq. (8.75a) we obtain now

$$\omega_y = \frac{I_1\dot\omega_x}{(I_1 - I_3)\omega_z} = \frac{\dot\omega_x}{\Omega} = \frac{1}{\Omega}A\Omega\cos\Omega t = A\cos\Omega t. \tag{8.78}$$

We see therefore, that the vector

$$\omega_x\mathbf{i} + \omega_y\mathbf{j}, \quad \omega_x^2 + \omega_y^2 = A^2,$$

lies in the (x, y)-plane and has the constant modulus A, and rotates with the frequency Ω around the z-axis. This is illustrated in Fig. 8.23. It follows that, since $\omega_z = $ const.,

$$|\boldsymbol\omega| = \sqrt{\omega_x^2 + \omega_y^2 + \omega_z^2} = \text{const.}$$

Since the axis of the rotation (in space) is parallel to $\boldsymbol\omega$, the result implies: The axis of rotation *precesses* with angular velocity Ω around the z-axis of the reference frame fixed in the rigid body. From Eqs. (8.75a), (8.75b) follows that with $I_1 \to I_3$, *i.e.* in the spherically symmetric case ($I_1 = I_2 = I_3$), all components $\omega_x, \omega_y, \omega_z$ are constant, and this implies that there is then no precession of the axis of rotation, *i.e.* of the vector $\boldsymbol\omega$, around the z-axis (or any other axis). The *precession* of the nonspherical rigid body thus results

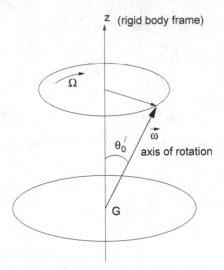

Fig. 8.23 The vector $\boldsymbol{\omega}$ rotating with frequency Ω about the z-axis.

from the fact that the axis of rotation does not coincide with the rigid body's symmetry axis, as illustrated in Fig. 8.23. The angle θ_0' between the axis of rotation (and hence $\boldsymbol{\omega}$) and the z-axis of the rigid body frame (as indicated in Fig. 8.23) is given by

$$\tan \theta_0' = \frac{(\omega_x^2 + \omega_y^2)^{1/2}}{\omega_z} = \frac{A}{\omega_z}. \tag{8.79}$$

The constant A and ω_z can be re-expressed in terms of the constants T (kinetic energy) and $\mathbf{L}^2$ (angular momentum squared):

$$\begin{aligned}
T &= \frac{1}{2}I_1\omega_x^2 + \frac{1}{2}I_2\omega_y^2 + \frac{1}{2}\omega_z^2 = \frac{1}{2}I_1 A^2 + \frac{1}{2}I_3\omega_z^2, \\
\mathbf{L}^2 &= (I\boldsymbol{\omega})^2 = (I_1\omega_x)^2 + (I_1\omega_y)^2 + (I_3\omega_z)^2 = I_1^2 A^2 + I_3^2\omega_z^2, \\
\therefore\ 2T &= I_1 A^2 + I_3\omega_z^2, \qquad 2T I_3 - L^2 = (I_1 I_3 - I_1^2)A^2, \tag{8.80}
\end{aligned}$$

and it follows that

$$A^2 = \frac{2T I_3 - L^2}{I_1(I_3 - I_1)}, \qquad \omega_z^2 = \frac{L^2 - I_1^2 A^2}{I_3^2} = \frac{2T I_1 - L^2}{I_3(I_1 - I_3)}. \tag{8.81}$$

The vector shown in Fig. 8.24 as "*axis of rotation*" is the vector resulting from the vector-addition of the angular velocities,

$$\boldsymbol{\omega} = \dot{\boldsymbol{\phi}} + \dot{\boldsymbol{\psi}} + \dot{\boldsymbol{\theta}}, \tag{8.82}$$

where ϕ, ψ, θ are the Euler angles of Sec. 5.4.5 defined with respect to an

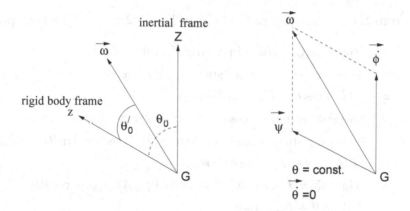

Fig. 8.24 The orientation of the axis of rotation.

inertial frame. One should note that this vector is not parallel to the Z axis of the inertial system, as we explain now with reference to Fig. 8.24. The angles θ_0 (this will be determined below) and θ_0' in Fig. 8.24 are given by

$$\theta_0 = \tan^{-1}\left(\frac{I_1 A}{I_3 \omega_z}\right), \quad \theta_0' = \tan^{-1}\left(\frac{A}{\omega_z}\right). \tag{8.83}$$

We observed earlier, that in the case of force-free motion the angular momentum is conserved. We want to verify this now in the case of the above example. To this end we remind ourselves of the following relations:

$$\left.\begin{array}{l} L_x = I_1 \omega_x = I_1 A \sin \Omega t, \\[4pt] L_y = I_1 \omega_y = I_1 A \cos \Omega t, \\[4pt] L_z = I_3 \omega_z = \text{const.} \end{array}\right\} \tag{8.84}$$

The expressions in the middle of these relations demonstrate already that **L** *is not parallel to* $\boldsymbol{\omega}$ for $I_1 \neq I_3$. The relations (8.84) provide the components of the angular momentum along the axes fixed in the rigid body. This does not lead to any confusion since **L** is a (constant) vector in the inertial frame, and we can consider its components at any time t along any arbitrary axes, hence also along the axes X, Y, Z of the inertial frame, by an appropriate rotation of the vector with components L_x, L_y, L_z. As with the rotation A of Eq. (5.21a), we then have ($A^{-1} = A^T$):

$$\begin{pmatrix} L_x \\ L_y \\ L_z \end{pmatrix} = (A) \begin{pmatrix} L_X \\ L_Y \\ L_Z \end{pmatrix}, \quad \begin{pmatrix} L_X \\ L_Y \\ L_Z \end{pmatrix} = (A^T) \begin{pmatrix} L_x \\ L_y \\ L_z \end{pmatrix}.$$

Then from the explicit expression of A in Eq. (5.21b), and its transposition,

$$
\begin{aligned}
L_X &= (\cos\psi\cos\phi - \cos\theta\sin\phi\sin\psi)I_1 A\sin\psi \\
&\quad +(-\sin\psi\cos\phi - \cos\theta\sin\phi\cos\psi)I_1 A\cos\psi + (\sin\theta\sin\phi)I_3\omega_z \\
&= -(I_1 A\cos\theta - I_3\omega_z\sin\theta)\sin\phi, \\
L_Y &= (\cos\psi\sin\phi + \cos\theta\cos\phi\sin\psi)I_1 A\sin\psi \\
&\quad +(-\sin\psi\sin\phi + \cos\theta\cos\phi\cos\psi)I_1 A\cos\psi - \sin\theta\cos\phi I_3\omega_z \\
&= (I_1 A\cos\theta - I_3\omega_z\sin\theta)\cos\phi, \\
L_Z &= (\sin\psi\sin\theta)I_1 A\sin\psi + (\cos\psi\sin\theta)I_1 A\cos\psi + \cos\theta I_3\omega_z \\
&= I_1 A\sin\theta + I_3\omega_z\cos\theta.
\end{aligned} \tag{8.85}
$$

We see that, putting $\mathbf{L}$ along the Z-axis, L_X, L_Y vanish if

$$
I_1 A\cos\theta - I_3\omega_z\sin\theta = 0,
$$

which implies that

$$
\tan\theta \to \tan\theta_0 = \frac{I_1 A}{I_3\omega_z} = \frac{I_1}{I_3}\tan\theta_0' \equiv \frac{I_1}{I_3}\frac{\sqrt{\omega_x^2 + \omega_y^2}}{\omega_z},
$$

which determines the angle between the Z-axis of the inertial frame and the symmetry axis (which is the z-axis) of the rigid body (in the frame of principal axes), *i.e.* θ_0. Differentiating L_Z, we obtain

$$
\frac{dL_Z}{dt} = (I_1 A\cos\theta - I_3\omega_z\sin\theta)\dot{\theta} = 0. \tag{8.86}
$$

Thus the angular momentum $\mathbf{L}$ is conserved; it has the constant direction along the $Z-$axis of the inertial system. With these considerations the physical contents of Eqs. (8.84) becomes clearer, if we look at the motion of the rigid body from the viewpoint of the inertial frame. To this end we have to re-express $\omega_x, \omega_y, \omega_z$ in terms of $\theta, \phi, \psi, \dot{\theta}, \dot{\phi}, \dot{\psi}$, since the Euler angles were defined with respect to an inertial frame (X, Y, Z) with ϕ the angle of rotation about the Z-axis. This means, we have to replace $\omega_x, \omega_y, \omega_z$ by the relations (8.40) which we used several times before:

$$
\begin{aligned}
\omega_x &= \dot{\phi}\sin\theta\sin\psi + \dot{\theta}\cos\psi, \\
\omega_y &= \dot{\phi}\sin\theta\cos\psi - \dot{\theta}\sin\psi, \\
\omega_z &= \dot{\phi}\cos\theta + \dot{\psi}.
\end{aligned} \tag{8.87}
$$

For ease of reading we reproduce in Fig. 8.25 our earlier Fig. 5.4, but with the axes renamed as required here (*i.e.* z in Fig. 5.4 corresponds to Z here,

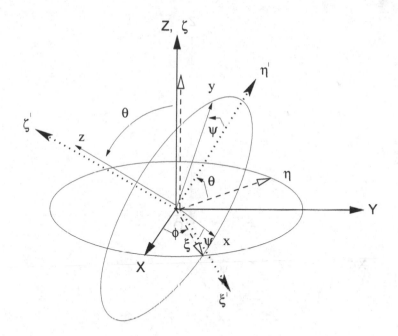

Fig. 8.25 Euler angles, principal axes x', y', z' and inertial axes X, Y, Z.

and z' in Fig. 5.4 to z here). Equations (8.87) can then be read off directly from Fig. 8.26. In the present case

$$\theta = \theta_0 = \text{const.}, \quad \theta_0 \neq 0, \frac{\pi}{2},$$

hence also $\dot{\theta} = 0$, and with $\omega_z = \text{const.}$ and $\omega_x^2 + \omega_y^2 = A^2$ from above, we have

$$\left.\begin{array}{l} \omega_x = \dot{\phi}\sin\theta_0\sin\psi \\ \omega_y = \dot{\phi}\sin\theta_0\cos\psi \end{array}\right\} \omega_x^2 + \omega_y^2 = \dot{\phi}^2\sin^2\theta_0 \equiv A^2 = \text{const.},$$

$$\omega_z = \dot{\phi}\cos\theta_0 + \dot{\psi} = \text{const.} \tag{8.88}$$

The z-axis, *i.e.* the symmetry axis of the rigid body, rotates at an angle θ around the Z-axis of the inertial system. The angular velocity of this rotation is not simply $\dot{\phi}$, since the Euler angles were not defined with respect to mutually orthogonal rotations. The relevant angular velocity is, as we deduce from Fig. 8.26 by considering components along the Z-axis

$$\dot{\phi} + \dot{\psi}\cos\theta_0, \quad (\dot{\theta} = 0). \tag{8.89}$$

Here $\dot{\psi}$ is that part of the angular velocity along the symmetry axis which originates solely from the rotation around the symmetry axis as illustrated in Fig. 8.26. In the case of the Earth we have

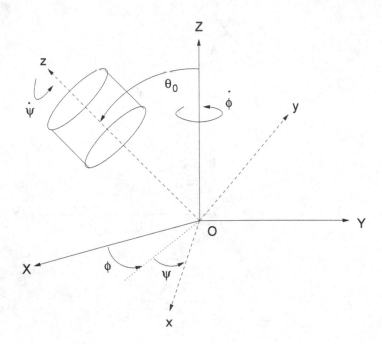

Fig. 8.26 Expressing $\omega_x, \omega_y, \omega_z$ in terms of $\dot{\phi}, \dot{\psi}$.

$$\omega_z = \dot{\phi}\cos\theta_0 + \dot{\psi} = 2\pi \text{ per day}, \tag{8.90}$$

i.e. not $\dot{\psi} = 2\pi$ per day. In the inertial frame we see that the symmetry axis rotates around the Z-axis. In the rigid body frame with principal axes, however, we see the Z-axis rotating around the symmetry axis.

Example 8.18: Force-free precession of the Earth

Calculate the period of precession of the Earth on the assumption that the radius b at the equator is larger than the radius a at the poles, $b > a$.

Solution: We assume the Earth to be symmetric around its polar axis, and slightly flattened at the poles, as indicated in Fig. 8.27. Since in the principal moment of inertia I_3 the maximum value of the distance r_i of any point-mass m_i from the z-axis is $b > a$, but in the principal moment of inertia I_1 this is $a < b$, we conclude that $I_3 > I_1$. Literature cites the approximate value

$$\frac{I_3 - I_1}{I_1} = 0.0033 \simeq \frac{1}{300}.$$

Using Eq. (8.76), we obtain the angular frequency of precession as

$$|\Omega| \simeq \frac{I_3 - I_1}{I_1}\omega_z = \frac{\omega_z}{300}.$$

The period of precession is therefore

$$P_p = \frac{2\pi}{\Omega} = \left(300\frac{2\pi}{\omega_z}\right)_{\omega_z = 2\pi \text{ per day}} \simeq 300 \text{ days} \simeq 10 \text{ months}.$$

Thus an observer on Earth ought to be able to observe a precession of the axis of rotation of the Earth around the North pole, this means, that the axis of rotation should perform a circular orbit around the North pole in the course of 10 months. A phenomenon of this sort is actually observed, however, with a period of 427 days. The deviation from the above value is attributed to the fact that the Earth is not a completely rigid object.

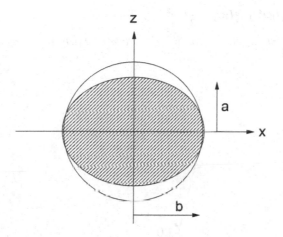

Fig. 8.27 The Earth, flattened at the poles.

Example 8.19: Complete integration of the force-free Euler equations

Demonstrate by citing the derivatives of the three basic Jacobian elliptic functions of argument u and elliptic modulus k, namely $\mathrm{sn}[u, k], \mathrm{cn}[u, k], \mathrm{dn}[u, k]$, that the solutions of the three coupled Euler equations (8.70) are given by the three basic Jacobian elliptic functions.

Solution: Comparing the coupled Eqs. (8.70) with the derivatives[¶]

$$\frac{d}{du}\mathrm{sn}[u, k] = \mathrm{cn}[u, k]\mathrm{dn}[u, k], \quad \frac{d}{du}\mathrm{cn}[u, k] = -\mathrm{sn}[u, k]\mathrm{dn}[u, k], \quad \frac{d}{du}\mathrm{dn}[u, k] = -k^2\mathrm{sn}[u, k]\mathrm{cn}[u, k],$$

we see immediately that — apart from constants a, t_0, l, m, n — the solutions of Eqs. (8.70) are

$$\omega_x = l\mathrm{sn}[a(t - t_0)], \quad \omega_y = m\mathrm{cn}[a(t - t_0)], \quad \omega_x = n\mathrm{dn}[a(t - t_0)].$$

The constants can be determined by substitution of these expressions in the differential equations. Thus, with $u = at$, we obtain

$$I_1\frac{d\omega_x}{du} = \frac{\omega_y\omega_z}{a}(I_2 - I_3), \quad I_2\frac{d\omega_y}{du} = \frac{\omega_z\omega_x}{a}(I_3 - I_1), \quad I_3\frac{d\omega_z}{du} = \frac{\omega_x\omega_y}{a}(I_1 - I_2),$$

or

$$(a)\ \ \frac{I_1 l}{mn} = \frac{I_2 - I_3}{a}, \qquad (b)\ \ \frac{I_2 m}{ln} = -\frac{I_3 - I_1}{a}, \qquad (c)\ \ -\frac{I_3 n k^2}{lm} = \frac{I_1 - I_2}{a}.$$

Multiplying the left hand sides and right hand sides of (a) and (b), and dividing sides of (a) and (c), and (b) and (c), we obtain three relations connecting the constants:

$$a^2 = \frac{(I_2 - I_3)(I_1 - I_3)}{I_1 I_2}n^2, \qquad k^2 = \frac{l^2}{n^2}\frac{I_1}{I_3}\left(\frac{I_2 - I_1}{I_2 - I_3}\right), \qquad \frac{m^2}{l^2} = \frac{I_1}{I_2}\left(\frac{I_1 - I_3}{I_2 - I_3}\right).$$

[¶]L. M. Milne–Thomson [30], p. 22.

One can now investigate the complete solution of the problem further together with the constant quantities of angular momentum squared L^2 and kinetic energy T:

$$
\begin{aligned}
L^2 &= (I_1\omega_x)^2 + (I_2\omega_y)^2 + (I_3\omega_z)^2 \\
&= I_1^2 l^2 \mathrm{sn}^2[a(t-t_0),k] + I_2^2 m^2 \mathrm{cn}^2[a(t-t_0),k] + I_3^2 n^2 \mathrm{dn}^2[a(t-t_0),k], \\
T &= \frac{1}{2}I_1\omega_x^2 + \frac{1}{2}I_2\omega_y^2 + \frac{1}{2}I_3\omega_z^2 \\
&= \frac{1}{2}l^2 I_1 \mathrm{sn}^2[a(t-t_0),k] + \frac{1}{2}m^2 I_2 \mathrm{cn}^2[a(t-t_0),k] + \frac{1}{2}n^2 I_3 \mathrm{dn}^2[a(t-t_0),k].
\end{aligned}
$$

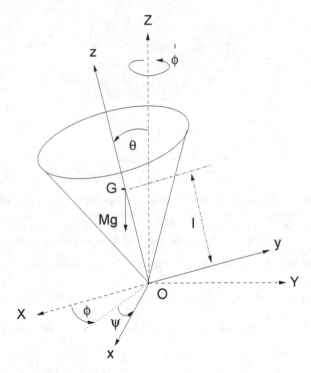

Fig. 8.28 The spinning top in the gravitational field.

8.7 The Spinning Top in the Gravitational Field

After consideration of force-free motion in the preceding section, we now consider the motion of a rigid body in the field of a force. We consider a solid conical top of mass M with its symmetry axis the principal z-axis of the (x, y, z)-frame fixed in the top, with the origin O at its only stationary point, as illustrated in Fig. 8.28 with the centre of mass a distance l above O. In view of this arrangement two principal moments of inertia are equal,

$I_1 = I_2$. The kinetic energy T of the top is therefore given by the relation

$$T = \frac{1}{2}I_1(\omega_x^2 + \omega_y^2) + \frac{1}{2}I_3\omega_z^2, \tag{8.91}$$

where $\omega_x, \omega_y, \omega_z$ are the components of its angular frequency. These components are related to the Euler angles θ, ϕ, ψ defined with respect to a spatially fixed reference frame (inertial frame) as in Eq. (8.87), i.e.

$$\left.\begin{aligned}
\omega_x &= \dot{\phi}\sin\theta\sin\psi + \dot{\theta}\cos\psi, \\
\omega_y &= \dot{\phi}\sin\theta\cos\psi - \dot{\theta}\sin\psi, \\
\omega_z &= \dot{\phi}\cos\theta + \dot{\psi}.
\end{aligned}\right\} \tag{8.92}$$

The kinetic energy T then becomes

$$T - \frac{1}{2}I_1(\dot{\theta}^2 + \dot{\phi}^2\sin^2\theta) + \frac{1}{2}I_3(\dot{\psi} + \dot{\phi}\cos\theta)^2. \tag{8.93}$$

The symmetry axis is a principal axis. Points on this axis have the coordinates $(0, 0, z)$. It follows that for the inertial products

$$I_{xz} = -\sum_i m_i x_i z_i = 0, \qquad I_{yz} = -\sum_i m_i y_i z_i = 0, \tag{8.94}$$

in view of the symmetry of the top (for every m_i with $y_i > 0$ there is an m_i with $y_i < 0$). In order to verify that the symmetry axis is a principal axis, we have to demonstrate that the unit vector along the symmetry axis, i.e. the z-axis, is eigenvector of the inertial tensor with eigenvalue $I_3 = I_{zz}$. Hence, we take $I_{xz} = 0, I_{yz} = 0$ and $I_3 = I_{zz}$ and let the direction cosines of the z-axis be given by

$$(l_3, m_3, n_3) \quad \text{with} \quad l_3^2 + m_3^2 + n_3^2 = 1. \tag{8.95}$$

Then I_3 follows from the solution of the *determinantel secular equation*

$$\begin{vmatrix} I_{xx} - I_3 & I_{yx} & 0 \\ I_{xy} & I_{yy} - I_3 & 0 \\ 0 & 0 & I_{zz} - I_3 \end{vmatrix} = 0,$$

and — as observed in connection with Eqs. (8.31a), (8.31b) and Eq. (8.36b) — l_3, m_3, n_3 follow from the set of equations:

$$\begin{pmatrix} I_{xx} - I_3 & I_{yx} & 0 \\ I_{xy} & I_{yy} - I_3 & 0 \\ 0 & 0 & I_{zz} - I_3 \end{pmatrix} \begin{pmatrix} l_3 \\ m_3 \\ n_3 \end{pmatrix} = 0, \tag{8.96}$$

and hence from

$$(I_{xx} - I_3)l_3 + I_{xy}m_3 = 0, \\ I_{xy}l_3 + (I_{yy} - I_3)m_3 = 0, \Bigg\} \qquad (8.97a)$$

$$(I_{zz} - I_3)n_3 = 0. \qquad (8.97b)$$

Since (note: for I_3) $(I_{xx} - I_3)(I_{yy} - I_3) - I_{xy}^2 \neq 0$, it follows as in Example 8.14 that $m_3 = 0, l_3 = 0$, and we have $I_3 = I_{zz}$, and hence from Eqs. (8.95) and (8.97b) that $n_3 = 1$, as had to be shown.

The potential energy of the top is given by

$$V = Mgl\cos\theta.$$

Thus the Lagrange function L is given by

$$\begin{aligned} L &= T - V \\ &= \frac{1}{2}I_1(\dot{\theta}^2 + \dot{\phi}^2\sin^2\theta) + \frac{1}{2}I_3(\dot{\psi} + \dot{\phi}\cos\theta)^2 - Mgl\cos\theta. \quad (8.98) \end{aligned}$$

The *Euler–Lagrange equation* is for $q = \theta, \phi, \psi$:

$$\frac{d}{dt}\left(\frac{\partial L}{\partial \dot{q}}\right) - \frac{\partial L}{\partial q} = 0.$$

Since ϕ and ψ do not appear explicitly in L, we have:

$$\frac{\partial L}{\partial \psi} = 0, \qquad \frac{\partial L}{\partial \phi} = 0,$$

and hence

$$\frac{d}{dt}p_\psi \equiv \frac{d}{dt}\left(\frac{\partial L}{\partial \dot{\psi}}\right) = 0, \qquad \frac{d}{dt}p_\phi \equiv \frac{d}{dt}\left(\frac{\partial L}{\partial \dot{\phi}}\right) = 0.$$

The generalized momenta p_ψ, p_ϕ are therefore constants in time. What are they explicitly? From the Lagrangian L we obtain

$$p_\psi = \frac{\partial L}{\partial \dot{\psi}} = I_3(\dot{\psi} + \dot{\phi}\cos\theta) = I_3\omega_z \equiv I_1 a, \qquad a = \frac{I_3}{I_1}\omega_z. \qquad (8.99)$$

Hence $\omega_z = I_1 a/I_3$ is also a constant. Similarly

$$\begin{aligned} p_\phi &= \frac{\partial L}{\partial \dot{\phi}} = I_3(\dot{\psi} + \dot{\phi}\cos\theta)\cos\theta + I_1\dot{\phi}\sin^2\theta \\ &= I_3\dot{\psi}\cos\theta + \dot{\phi}(I_1\sin^2\theta + I_3\cos^2\theta) \equiv I_1 b = \text{const.} \\ &= I_1\dot{\phi}\sin^2\theta + I_3\cos\theta(\dot{\psi} + \dot{\phi}\cos\theta). \qquad (8.100) \end{aligned}$$

The constants arising here are determined by initial conditions. Since the system is conservative, we also have

$$
\begin{aligned}
E &= T + V = \text{const.} \\
&= \frac{1}{2}I_1(\dot{\theta}^2 + \dot{\phi}^2 \sin^2\theta) + \frac{1}{2}I_3(\dot{\psi} + \dot{\phi}\cos\theta)^2 + Mgl\cos\theta. \quad (8.101)
\end{aligned}
$$

We have thus determined three constants of motion. We now eliminate $\dot{\psi}$ from the equations. Multiplying Eq. (8.99) by $\cos\theta$, we obtain

$$
\cos\theta I_3(\dot{\psi} + \dot{\phi}\cos\theta) = I_1 a \cos\theta. \quad (8.102)
$$

Inserting this relation into Eq. (8.100) we obtain:

$$
I_1\dot{\phi}\sin^2\theta + I_1 a \cos\theta = I_1 b,
$$

or

$$
\dot{\phi}(\theta) = \frac{b - a\cos\theta}{\sin^2\theta}, \quad \theta = \theta(t). \quad (8.103)
$$

If $\theta(t)$ were known, this equation could be integrated immediately. We insert the result (8.103) into Eq. (8.102) and obtain

$$
\dot{\psi} = \frac{I_1 a}{I_3} - \frac{b - a\cos\theta}{\sin^2\theta}\cos\theta. \quad (8.104)
$$

This means $\dot{\psi}$ would be known if $\theta(t)$ were known. We now insert the expressions for $\dot{\phi}$ and $\dot{\psi}$ into Eq. (8.101) for E, and thus obtain an equation determining θ:

$$
\begin{aligned}
E &= Mgl\cos\theta + \frac{1}{2}I_1\left[\dot{\theta}^2 + \frac{(b - a\cos\theta)^2}{\sin^2\theta}\right] \\
&\quad + \frac{1}{2}I_3\left[\frac{I_1 a}{I_3} - \frac{(b - a\cos\theta)\cos\theta}{\sin^2\theta} + \frac{\cos\theta(b - a\cos\theta)}{\sin^2\theta}\right]^2 \\
&= Mgl\cos\theta + \frac{1}{2}I_1\left[\dot{\theta}^2 + \frac{(b - a\cos\theta)^2}{\sin^2\theta}\right] + \frac{1}{2}\frac{I_1^2 a^2}{I_3}. \quad (8.105)
\end{aligned}
$$

With this equation we have a differential equation that involves only θ. However, we can also proceed in a different way. Since

$$
\dot{\psi} + \dot{\phi}\cos\theta = \omega_z = \text{const.},
$$

it follows from Eq. (8.101) that

$$
\text{const.} = E' \equiv E - \frac{1}{2}I_3\omega_z^2 = \frac{1}{2}I_1(\dot{\theta}^2 + \dot{\phi}^2\sin^2\theta) + Mgl\cos\theta. \quad (8.106)
$$

Inserting here Eq. (8.103), we obtain

$$E' = \frac{1}{2}I_1\left[\dot{\theta}^2 + \left(\frac{b - a\cos\theta}{\sin\theta}\right)^2\right] + Mgl\cos\theta,$$

or

$$\dot{\theta}^2\sin^2\theta = \sin^2\theta\frac{(E' - Mgl\cos\theta)}{I_1/2} - (b - a\cos\theta)^2. \tag{8.107}$$

Setting

$$\alpha = \frac{2E'}{I_1}, \quad \beta = \frac{2Mgl}{I_1}, \tag{8.108}$$

the equation becomes

$$\dot{\theta}^2\sin^2\theta = \sin^2\theta(\alpha - \beta\cos\theta) - (b - a\cos\theta)^2. \tag{8.109}$$

Setting now

$$u = \cos\theta, \quad \dot{u} \equiv \frac{du}{dt} = -\dot{\theta}\sin\theta,$$

we obtain

$$\dot{u}^2 = (1 - u^2)(\alpha - \beta u) - (b - au)^2, \tag{8.110}$$

or

$$t = \int_{u(0)}^{u(t)} \frac{du}{\sqrt{(1 - u^2)(\alpha - \beta u) - (b - au)^2}}. \tag{8.111}$$

This equation expresses t as an elliptic integral and hence as a function of $u = \cos\theta$. In Example 8.19 we indicate how the integral can be evaluated analytically with the help of Tables of Elliptic Integrals that are now available. With reversal of Eq. (8.111), one obtains $\cos\theta$ as a function of t. This function can be inserted into Eqs. (8.103) and (8.104) and then permits the calculation of θ and ψ as functions of t. Consider Eq. (8.110),

$$\dot{u}^2 = f(u), \quad f(u) = (1 - u^2)(\alpha - \beta u) - (b - au)^2. \tag{8.112}$$

The function $f(u)$ is a cubic in $u = \cos\theta$ which has the general behaviour shown in Fig. 8.29. The zeros of $f(u)$ at u_1, u_2, u_3 provide the "turning points" of the "velocity" $\dot{u}$, and thus the "turning angles" of θ. At $u = \pm 1$, we have $f(u) = -(b \mp a)^2 < 0$, except for $f(u = \pm 1) = 0$ (when $u = \pm 1, \cos\theta = \pm 1$ implying $\theta = 0$, which implies the configuration of a vertically upright top, the angle π being unphysical). When $u = +1$, we have $f(u) < 0$, whereas for $u \to \infty$, the function $f(u)$ is positive, as indicated in Fig. 8.29. Thus there must be at least one root of the function between $u = 1$ and $u = \infty$, and hence in the domain $u > 1$. However, $u > 1$ is an unphysical domain of $\cos\theta$, since in this case θ does not correspond to a physical angle.

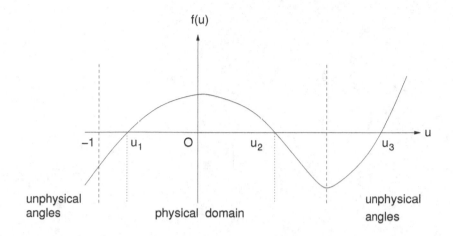

Fig. 8.29 The cubic function $f(u)$.

For a physically sensible motion the angle θ must lie between $-\pi$ and π, i.e. in the domain $-1 < u < +1$ with $\dot{u}^2 > 0$. Then there must exist two roots u_1, u_2 for the actual top. The top moves in such a way that $\cos\theta$ lies between -1 and $+1$, in other words between these two roots u_1, u_2, which are therefore known as "*turning points*". The spinning of the top around its fixed point can very conveniently be demonstrated with the help of the curve traced out by its axis on a unit sphere. This curve or trajectory is called the *locus of the axis of the top*. The Euler angles θ and ϕ are the polar coordinates of a point on this curve. With $u = \cos\theta$, the expression in Eq. (8.103) for the "*precession*" of the top, can be rewritten as

$$\dot{\phi}(u) = \frac{a}{1-u^2}(u' - u), \quad \text{where} \quad u' = \frac{b}{a}. \tag{8.113}$$

We see that, since $a > 0$, $\dot{\phi}(u)$ is positive for $u' > u$. Thus, if $u' > u$, the velocity $\dot{\phi}$ is always positive, which means the top does not execute backward steps in the sense of negative values of $\dot{\phi}$. The trajectory of the top's axis thus has the shape of the curve shown in Fig. 8.30(a), where

$$u' > u_2 \le u, \quad \theta_2 = \cos^{-1} u_2, \quad \theta_1 = \cos^{-1} u_1.$$

The difference

$$x_1 = u_2 - u_1$$

is a measure of the spinning or "*nutation*" of the top about the vertical axis. Thus the axis of the top precesses about this vertical axis, and its oscillations in the angle θ between θ_1 and θ_2 are what one describes as its nutations.

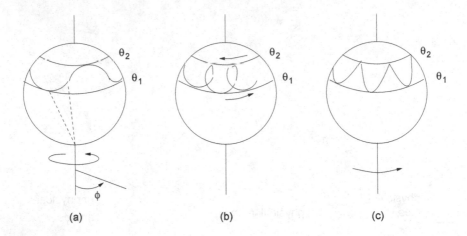

Fig. 8.30 The nutations of the spinning top.

It is evident from Eq. (8.113) that the angular velocity $\dot{\phi}$ can also be negative. Thus if $u_1 < u' < u_2$, then $\dot{\phi}$ is positive at $u = u_1$, but negative at u_2. This means the trajectory turns backward at the upper limiting circle and assumes the shape shown in Fig. 8.30(b). There is the further possibility that $u' = u_1$ or u_2, which are the turning points where $f(u) = 0$. In these cases the velocity $\dot{\phi} = 0$, also $\dot{\theta} = 0$, at the limiting circles, since with $f(u) = 0$, $i.e.$ $\dot{\theta} = 0$, and $\dot{\phi} = 0$ follows from Eq. (8.113). In the case of $u' = u_2$ the angular velocity $\dot{\phi}$ can not be negative, but can just reach zero. The trajectory then assumes the shape shown in Fig. 8.30(c). Such a motion to the right results from the following conditions:

$$t = 0, \quad \theta = \theta_0, \quad \therefore \dot{\theta} = 0, \quad \dot{\phi} = 0 \ (f(u) = 0), \quad u' = u_2 : \dot{\phi} = 0. \quad (8.114)$$

For motion to the left:

$$u' = u_1, \quad u' - u < 0, \quad \therefore \dot{\phi} < 0 \ \left(\theta_1 < \frac{\pi}{2}\right).$$

For $u' < u_1, 0 < u_1$, the velocity $\dot{\phi}$ is always negative. The top would now spin in the opposite direction and its axis would again precess about the vertical (u between u_1 and u_2).

Example 8.20: Analytical integration of the equation of the spinning top
Determine explicitly the turning points u_1, u_2, u_3 of the curve $f(u)$ and indicate how the complete integration of the equation of the spinning top may be achieved.

Solution: In order to determine the turning points, one has to find the roots of the cubic equation $f(u) = 0$. Cubic equations have a reputation of being difficult to solve. However, a cubic equation with real coefficients has always one real root, which when found, allows the cubic polynomial to be factorized in a linear factor and a quadratic one. The factorization can be achieved as follows.[||] We have, on rearranging terms in Eq. (8.112) in powers of u,

$$f(u) = \beta u^3 - (\alpha + a^2)u^2 + (2ab - \beta)u + (\alpha - b^2), \tag{8.115}$$

and set

$$u = y + \frac{\alpha + a^2}{3\beta}, \quad \text{so that} \quad f(u) = \beta[y^3 + py + q] = \beta(y - y_1)(y - y_2)(y - y_3), \tag{8.116}$$

where (as one finds with some algebra)

$$p = \frac{2ab - \beta}{\beta} - \frac{(\alpha + a^2)^2}{3\beta^2}, \quad q = \frac{(\alpha - b^2)}{\beta} + \frac{(2ab - \beta)(\alpha + a^2)}{3\beta^2} - \frac{2(\alpha + a^2)^3}{(3\beta)^3}. \tag{8.117}$$

The roots y_1, y_2, y_3 in terms of p and q are known.[**] Three real roots are obtained with

$$p < 0, \quad 27q^2 < 4p^2, \quad \text{and} \quad \cos(3\theta) \equiv \left(-\frac{27q^2}{4p^3}\right)^{1/2}, \tag{8.118}$$

as

$$y_1 = \mp\left(-\frac{4p}{3}\right)^{1/2}\cos\theta, \quad y_2 = \mp\left(-\frac{4p}{3}\right)^{1/2}\cos(\theta + 120°),$$

$$y_3 = \mp\left(-\frac{4p}{3}\right)^{1/2}\cos(\theta + 240°). \tag{8.119}$$

We do not pursue the calculations here any further. The reader who wants to get an impression of how the calculations have to proceed up to the explicit evaluation of an elliptic integral can consult the calculations in the very analogous case of the cubic potential in quantum mechanics given in the literature.[††] An analogous problem arises in relativity and is dealt with in Appendix A. That a complete solution is possible can be seen from the fact that extensive Tables of Elliptic Integrals are available today which permit the explicit and exact evaluation of an integral like that appearing here in Eq. (8.111), which is an integral of the form[‡‡]

$$\int_a^b \frac{dy}{\sqrt{(y - y_1)(y - y_2)(y - y_3)}}.$$

Example 8.21: The satellite with damped nutations
A satellite which is rotationally symmetric around its own z-axis contains — in order to damp its nutations — a mass m attached to two springs as illustrated in Fig. 8.31. The mass can move in the (x, z)-plane of the satellite frame of reference parallel to the x-axis at a distance $z = h$ from the origin. The mass m moves under the influence of a Hooke force $F_1 = -kx, k = \text{const.}$, as well as a velocity-dependent damping force $F_2 = -c\dot{x}, c = \text{const.}$ Establish the Euler equations of motion and show that $\omega_z = \text{const.}$, where $\boldsymbol{\omega} = (\omega_x, \omega_y, \omega_z)$ is the angular velocity of the satellite around its centre of mass O. Finally determine the acceleration $\mathbf{b}$ of the mass m relative to an inertial frame S with origin coincident with O.

Solution: It is easiest to obtain the Euler equations by recalling the law of angular momentum, *i.e.*

[||] L. M. Milne–Thomson [30], p. 36.
[**] See again L. M. Milne–Thomson [30], p. 37.
[††] See H. J. W. Müller–Kirsten [35], Example 14.6, p. 302.
[‡‡] P. F. Byrd and M. D. Friedman [9].

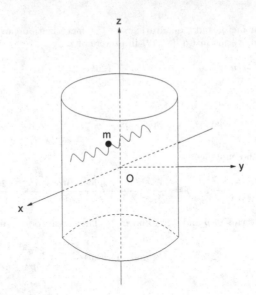

Fig. 8.31 The satellite with mass m attached to springs.

$$\left(\frac{d\mathbf{L}}{dt}\right)_S = \sum_i \mathbf{F}_i \times \mathbf{r},$$

where $\mathbf{L}$ is the angular momentum, $\mathbf{F}_i$ are the applied forces, and S refers to the inertial or spatial frame of reference, in accordance with Newtonian mechanics. If the axes of the rigid body, *i.e.* of the satellite-fixed frame R, are chosen to align with its symmmetry axes, the tensor of inertia is diagonalized and we have

$$\mathbf{L} = (I_1 \omega_x, I_2 \omega_y, I_3 \omega_z),$$

where $I_1 = I_2$ in view of the rotational symmetry of the satellite. Recalling Eq. (5.57) (or below Eq. (8.120)) which expresses the time-derivative relative to the inertial frame S as one relative to the rigid body frame R, we obtain from the above two equations (*cf.* also Eq. (8.54))

$$\sum_i \mathbf{F}_i \times \mathbf{r} = \left(\frac{d\mathbf{L}}{dt}\right)_R + \boldsymbol{\omega} \times \mathbf{L}$$
$$= (I_1\dot{\omega}_x, I_2\dot{\omega}_y, I_3\dot{\omega}_z) + \{(I_3 - I_2)\omega_y\omega_z, (I_1 - I_3)\omega_z\omega_x, (I_2 - I_1)\omega_x\omega_y)\},$$

where $I_1 = I_2$. From this equation we can read off the Euler equations (noting that here $\mathbf{F} \times \mathbf{r} = -F_x \sin(\pi/2)h\mathbf{e}_y$ is a vector in the direction of $\mathbf{e}_y$ as in Fig. 8.31):

$$I_1\dot{\omega}_x + (I_3 - I_2)\omega_y\omega_z = 0, \quad I_2\dot{\omega}_y + (I_1 - I_3)\omega_z\omega_x = h(kx + c\dot{x}), \quad I_3\dot{\omega}_z = 0.$$

The last equation tells us that $\omega_z = $ const. In the reference frame R fixed in the satellite we have

$$m\ddot{b}_x = -kx - c\dot{x}.$$

The position vector of the mass m is $\mathbf{r} = (x, 0, h)$. Again using Eq. (5.57), we have

$$\left(\frac{d\mathbf{r}}{dt}\right)_S = \left(\frac{d\mathbf{r}}{dt}\right)_R + \boldsymbol{\omega} \times \mathbf{r}, \quad \text{and} \quad \therefore \left(\frac{d}{dt}\right)_S = \left(\frac{d}{dt}\right)_R + \boldsymbol{\omega} \times .$$

It follows that the acceleration **b** is relative to frame S:

$$
\begin{aligned}
\mathbf{b} &= \left(\frac{d}{dt}\right)_S \left(\frac{d\mathbf{r}}{dt}\right)_S = \left(\frac{d^2\mathbf{r}}{dt^2}\right)_S = \left[\left(\frac{d}{dt}\right)_R + \boldsymbol{\omega}\times\right]\left[\left(\frac{d\mathbf{r}}{dt}\right)_R + \boldsymbol{\omega}\times\mathbf{r}\right] \\
&= \left(\frac{d^2\mathbf{r}}{dt^2}\right)_R + 2\boldsymbol{\omega}\times\left(\frac{d\mathbf{r}}{dt}\right)_R + \left(\frac{d\boldsymbol{\omega}}{dt}\right)_R \times\mathbf{r} + \boldsymbol{\omega}\times(\boldsymbol{\omega}\times\mathbf{r}) \\
&= \ddot{\mathbf{r}} + 2\boldsymbol{\omega}\times\dot{\mathbf{r}} + \dot{\boldsymbol{\omega}}\times\mathbf{r} + \boldsymbol{\omega}\times(\boldsymbol{\omega}\times\mathbf{r}).
\end{aligned}
$$

Since $\mathbf{r} = (x, 0, h)$, we have $\dot{\mathbf{r}} = (\dot{x}, 0, 0)$ and $\ddot{\mathbf{r}} = (\ddot{x}, 0, 0)$. Moreover,

$$
\boldsymbol{\omega}\times\dot{\mathbf{r}} = \begin{vmatrix} i & j & k \\ \omega_x & \omega_y & \omega_z \\ \dot{x} & 0 & 0 \end{vmatrix} = (0, \dot{x}\omega_z, -\dot{x}\omega_y), \quad \dot{\boldsymbol{\omega}}\times\mathbf{r} = \begin{vmatrix} i & j & k \\ \dot{\omega}_x & \dot{\omega}_y & 0 \\ x & 0 & h \end{vmatrix} = (h\dot{\omega}_y, -h\dot{\omega}_x, -x\dot{\omega}_y),
$$

and

$$
\boldsymbol{\omega}\times\mathbf{r} = \begin{vmatrix} i & j & k \\ \omega_x & \omega_y & \omega_z \\ x & 0 & h \end{vmatrix} = (h\omega_y, x\omega_z - h\omega_x, -x\omega_y),
$$

so that

$$
\begin{aligned}
\boldsymbol{\omega}\times(\boldsymbol{\omega}\times\mathbf{r}) &= \begin{vmatrix} i & j & k \\ \omega_x & \omega_y & \omega_z \\ h\omega_y & x\omega_z - h\omega_x & -x\omega_y \end{vmatrix} \\
&= (-x\omega_y^2 - x\omega_z^2 + h\omega_x\omega_z, h\omega_y\omega_z + x\omega_x\omega_y, x\omega_x\omega_z - h\omega_x^2 - h\omega_y^2).
\end{aligned}
$$

Thus we obtain for $\mathbf{b} = (b_x, b_y, b_z)$: $b_x = \ddot{x} + h\dot{\omega}_y + h\omega_x\omega_z - x(\omega_y^2 + \omega_z^2), b_y = 2\dot{x}\omega_z - h\dot{\omega}_x + h\omega_y\omega_z + x\omega_x\omega_y, \quad b_z = -2\dot{x}\omega_y - x\dot{\omega}_y + x\omega_x\omega_z - h(\omega_x^2 + \omega_y^2).$

8.8 Motion Relative to Rotations: Centrifugal and Coriolis Forces

In the following we consider a spatially fixed or inertial reference frame S, which has its origin at the centre of the Earth (for instance, at the latter's geometrical centre) and its orientation in space fixed with respect to some fixed star or the sun (this is the nearest one can come in practice to a spatially fixed reference frame). Apart from this frame, we assume an Earth-fixed reference frame R (rigid body frame) with its origin coincident with that of the frame S. Our starting point is Eq. (5.57), *i.e.* the equation

$$
\left(\frac{d\mathbf{G}}{dt}\right)_{\text{space},S} = \left(\frac{d\mathbf{G}}{dt}\right)_{\text{rigid body},R} + \boldsymbol{\omega}\times\mathbf{G}, \qquad \boldsymbol{\omega} = \frac{d\boldsymbol{\Omega}}{dt}. \qquad (8.120)
$$

We assume there is a particle of mass m at radius vector $\mathbf{r}$ away from the origin O. We write the velocity of this particle with respect to the spatially fixed reference frame $\mathbf{v}_S$. Then, replacing the arbitrary vector quantity $\mathbf{G}$ by

r (and using the following notation in spite of the fact that in the classical mechanics here there is only one time), we obtain the equation

$$\left(\frac{d\mathbf{r}}{dt}\right)_S \equiv \left(\frac{d}{dt}\right)_S \mathbf{r} = \mathbf{v}_S = \left[\left(\frac{d}{dt}\right)_R + \boldsymbol{\omega} \times\right]\mathbf{r} = \mathbf{v}_R + \boldsymbol{\omega} \times \mathbf{r}. \qquad (8.121)$$

Of course, we can also replace the vector **G** by the velocity $\mathbf{v}_S$ and obtain the corresponding "*absolute acceleration*" or acceleration relative to the inertial frame. Thus:

$$\begin{aligned}
\left(\frac{d}{dt}\right)_S \mathbf{v}_S &\equiv \mathbf{a}_S = \left(\frac{d}{dt}\right)_S (\mathbf{v}_R + \boldsymbol{\omega} \times \mathbf{r}) \\
&= \left(\frac{d}{dt}\right)_R (\mathbf{v}_R + \boldsymbol{\omega} \times \mathbf{r}) + \boldsymbol{\omega} \times [\mathbf{v}_R + \boldsymbol{\omega} \times \mathbf{r}] \\
&= \mathbf{a}_R + 2\boldsymbol{\omega} \times \mathbf{v}_R + \boldsymbol{\omega} \times (\boldsymbol{\omega} \times \mathbf{r}), \qquad (8.122)
\end{aligned}$$

where $\mathbf{a}_R$ is the acceleration relative to the rotating frame. The sign of the third term on the right hand side, *i.e.* of $\boldsymbol{\omega} \times (\boldsymbol{\omega} \times \mathbf{r})$, makes this term an inwardly directed (hence *centripetal*) acceleration, or multiplied by the mass, a centripetal force. Newton's second law of motion implies for a force **F** and in the direction of **r**, and with respect to the inertial frame S:

$$\mathbf{F} = m\mathbf{a}_S, \qquad (8.123)$$

so that

$$\mathbf{F} - m\{2\boldsymbol{\omega} \times \mathbf{v}_R + \boldsymbol{\omega} \times (\boldsymbol{\omega} \times \mathbf{r})\} = m\mathbf{a}_R. \qquad (8.124)$$

Thus to an observer on Earth it appears that the particle moves as if subjected to the *effective force*

$$\mathbf{F}_{\text{eff}} = \frac{d}{dt}(m\mathbf{v}_R) = \mathbf{F} - m\{2\boldsymbol{\omega} \times \mathbf{v}_R + \boldsymbol{\omega} \times (\boldsymbol{\omega} \times \mathbf{r})\}. \qquad (8.125)$$

Here the sign of the triple vector product is such that its direction is outwards and hence the term represents a *centrifugal force*. Since

$$\mathbf{a} \times (\mathbf{b} \times \mathbf{c}) = (\mathbf{a} \cdot \mathbf{c})\mathbf{b} - (\mathbf{a} \cdot \mathbf{b})\mathbf{c},$$

the effective force is

$$\mathbf{F}_{\text{eff}} = \frac{d}{dt}(m\mathbf{v}_R) = \mathbf{F} - 2m\boldsymbol{\omega} \times \mathbf{v}_R + m\omega^2\mathbf{r} - m(\boldsymbol{\omega} \cdot \mathbf{r})\boldsymbol{\omega}. \qquad (8.126)$$

Consider the terms appearing in this expression. The contribution $m\omega^2\mathbf{r}$ is the usual outwardly directed *centrifugal force*. If the particle is stationary on Earth, *i.e.* in the frame R attached to the Earth, then $\mathbf{v}_R = 0$, and apart from the term with $\boldsymbol{\omega} \cdot \mathbf{r}$, the centrifugal force is the only additional contribution of the effective force. The contribution with $\boldsymbol{\omega} \cdot \mathbf{r}$ vanishes when $\boldsymbol{\omega} \perp \mathbf{r} = 0$, and this is the case at the equator (or near the equator), if we take into account the *precession* of $\boldsymbol{\omega}$ around the symmetry axis of the Earth, which we dealt with in Secs. 8.6, 8.7. If the particle moves with respect to the reference frame R of the revolving Earth, the additional term, called "*Coriolis force*" and equal to mass times *Coriolis acceleration*,

$$\mathbf{F}_{\mathrm{Cor}} \equiv -2m\boldsymbol{\omega} \times \mathbf{v}_R, \tag{8.127}$$

comes into play (this is $-2m\omega v \sin \phi$ with ϕ the geographical *latitude, cf.* Example 10.5). In order to understand a crucial point of the Coriolis force consider Eq. (8.126) reduced to the main terms for this purpose, *i.e.* as $\mathbf{F}_{\mathrm{eff}} = \mathbf{F}_{\mathrm{Cor}}$. Recall the case of a stone attached to a string and swirled around on a planar circular orbit. The constant inwardly directed tension in the string (hence "*centripetal force*") deflects the stone from an otherwise rectilinear path. The magnitude of the stone's momentum is not changing with time, but its instantaneous direction. Hence Newton's second law requires a force to be present, the centripetal force, which is perpendicular to the instantaneous momentum. Similarly here (as the vector product shows) $\mathbf{F}_{\mathrm{Cor}}$ is perpendicular to $m\mathbf{v}_R$, and thus implies an orbit which at every instant curves away from a straight line path. This curving effect to an observer in a rotating frame of a path which is observed as a straight line by an observer in an inertial frame is explained further in Example 8.21.

The centrifugal acceleration $\omega^2 \mathbf{r}$ of a particle on the surface of the Earth at the equator can be calculated from a knowledge of the length of the radius of the Earth there, which is known to be approximately 6.37×10^8 cm. On the other hand, the frequency of rotation of the Earth is

$$\omega = \frac{2\pi}{1 \text{ day}} = \frac{2\pi}{24 \times 60 \times 60} = 7.26 \times 10^{-5} \mathrm{s}^{-1}.$$

Thus one obtains for the centrifugal acceleration (with $g \simeq 980$ cm/s^2)

$$\omega^2 r = 3.36 \text{ cm/s}^2 \simeq \frac{3.4}{1000} g \text{ (repulsive)}. \tag{8.128}$$

Away from the equator, the *centrifugal acceleration* is

$$\omega^2 \mathbf{r} - (\boldsymbol{\omega} \cdot \mathbf{r})\boldsymbol{\omega}. \tag{8.129}$$

We observe that this expression vanishes where $\boldsymbol{\omega}$ is parallel to $\mathbf{r}$, which is the case at the poles (also at the South pole, where the direction of $\mathbf{r}$ is opposite to that of $\boldsymbol{\omega}$). Thus only at the equator, where $\boldsymbol{\omega} \perp \mathbf{r}$, the centrifugal force is parallel to the radius vector, and is there also maximal. It is this difference between the magnitudes of the centrifugal forces at the equator and at the poles which explains the larger bulging out of the Earth in the equatorial region. In the above we have neglected the centrifugal force resulting from the orbiting of the Earth around the sun. The rotational frequency of this effect is smaller than that above by the factor $1/365 = 2.7 \times 10^{-3}$. Of course the radius r is now much bigger, in fact approximately by a factor $10^5/4$. Nonetheless, the ratio of the two centrifugal forces or accelerations to each other is small:

$$\frac{\text{centrifugal acceleration around sun}}{\text{centrifugal acceleration of rotation of Earth}} = \frac{10^5 \times (2.7 \times 10^{-3})^2}{4} = 0.2.$$

Example 8.22: Linear path in inertial frame seen curved in rotating frame
In an inertial frame S a particle of mass m is seen to travel with constant velocity $\mathbf{v}_S$ along a straight line. Determine the path of this particle observed from a circular disc which rotates with angular velocity ω_z around the axis of the disc (assumed perpendicular to the straight line path) by taking into account only the Coriolis effect.

straight line seen in inertial frame

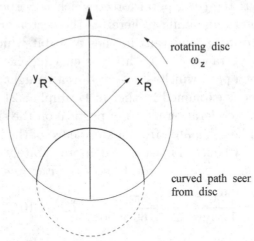

rotating disc
ω_z

y_R x_R

curved path seen
from disc

Fig. 8.32 Linear path in inertial frame seen curved from rotating frame.

Solution: Since $\mathbf{v}_S$ is constant, the acceleration $\mathbf{a}_S = 0$ and hence the force $\mathbf{F}$ in Eqs. (8.123) and (8.126) vanishes. Neglecting in Eq. (8.126) the terms which are not to be considered, the motion observed from the rotating disc is determined by the equation

$$m\dot{\mathbf{v}}_R = -2m\boldsymbol{\omega} \times \mathbf{v}_R.$$

Separating this equation into its components, we obtain the equations:

$$m\dot{v}_{Rx} = -2m(\omega_y v_{Rz} - \omega_z v_{Ry}),$$
$$m\dot{v}_{Ry} = -2m(\omega_z v_{Rx} - \omega_x v_{Rz}),$$
$$m\dot{v}_{Rz} = -2m(\omega_x v_{Ry} - \omega_y v_{Rz}).$$

Since $\omega_x, \omega_y, v_{Rz}$ are zero, the equations reduce to

$$m\dot{v}_{Rx} = +2m\omega_z v_{Ry}, \quad m\dot{v}_{Ry} = -2m\omega_z v_{Rx}.$$

Differentiating these equations with respect to time t, and inserting one into the other, we obtain the equations

$$\ddot{v}_{Rx} + 4\omega_z^2 v_{Rx} = 0, \quad \ddot{v}_{Ry} + 4\omega_z^2 v_{Ry} = 0.$$

The solutions are, with A and t_0 constants:

$$v_{Rx} = A\sin 2\omega_z(t - t_0), \quad v_{Ry} = A\cos 2\omega_z(t - t_0),$$

and hence (with appropriate initial conditions)

$$x_R - x_{R0} = -\frac{A}{2\omega_z}\cos 2\omega_z(t - t_0), \quad y_R - y_{R0} = \frac{A}{2\omega_z}\sin 2\omega_z(t - t_0).$$

Thus the path seen by an observer on the rotating disc is an arc of the circle $(x_R - x_{R0})^2 + (y_R - y_{R0})^2 = A^2/4\omega_z^2$, as illustrated in Fig. 8.32.

Example 8.23: A pendulum at rest in the Earth's gravitational field

Investigate the effect of the centrifugal force on the value of the apparent acceleration due to gravity at a *latitude* angle of 60°, and compare this value with that at the North pole.

Solution: We consider a simple pendulum at rest at a point P in the gravitational field of the Earth and just above the surface, as illustrated in Fig. 8.33. The pendulum bob has mass m. The pendulum is at rest in its equilibrium position, and there has $\mathbf{r} = $ const. This position of the pendulum bob is determined by the sum of all forces acting upon it (gravity and string tension F). Since $\mathbf{r} = $ const., the pendulum is at rest with respect to the Earth (rigid frame R) and therefore

$$\mathbf{v}_R = \left(\frac{d}{dt}\right)_R \mathbf{r} = 0.$$

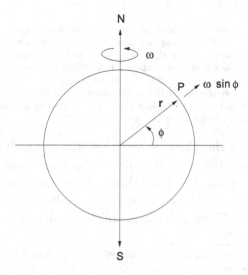

Fig. 8.33 A pendulum under gravity at P.

The pendulum string is therefore directed parallel to the centrifugal acceleration given by the vector (8.129). Thus the pendulum is not excatly parallel to the radial vector $\mathbf{r}$.

The *apparent* force of gravity acting on the pendulum, that appearing to an observer on Earth, is given by the effective force of Eq. (8.125) and thus defines the apparant acceleration due to gravity $\mathbf{g}_R$, *i.e.*

$$m\mathbf{g}_R = -\mathbf{F}_{\text{eff}}. \tag{8.130}$$

Since the actual force of gravity is $m\mathbf{g} = -\mathbf{F}$, where $\mathbf{F}$ is the tension in the string, we obtain from Eqs. (8.125), (8.126) the equation (with $\mathbf{v}_R = 0$)

$$\begin{aligned} -m\mathbf{g}_R &= -m\mathbf{g} + m\omega^2\mathbf{r} - m(\boldsymbol{\omega} \cdot \mathbf{r})\boldsymbol{\omega}, \\ m\mathbf{g} &= m\mathbf{g}_R + m\omega^2\mathbf{r} - m(\boldsymbol{\omega} \cdot \mathbf{r})\boldsymbol{\omega}. \end{aligned} \tag{8.131}$$

Here g is the *acceleration due to gravity* calculated from Newton's gravitational law:

$$m\mathbf{g} = -G_{\text{Newton}} \frac{mM_{\text{Earth}}}{r_{\text{Earth}}^2} \left(\frac{-\mathbf{r}_{\text{Earth}}}{r_{\text{Earth}}} \right). \tag{8.132}$$

Since the vector $m\omega^2\mathbf{r} - m(\boldsymbol{\omega}\cdot\mathbf{r})\boldsymbol{\omega} = m(\mathbf{g}-\mathbf{g}_R)$ is largest at the equator, the quantity g_R is smallest there. Hence g_R assumes its largest value at the poles, where $m\omega^2\mathbf{r} = m(\boldsymbol{\omega}\cdot\mathbf{r})\boldsymbol{\omega}$. Consider a point P at a latitude as shown in Fig. 8.33. We set $G_\phi \equiv g_R$. Then at the point P the apparent acceleration due to gravity is given by

$$\begin{aligned} g_R \equiv G_\phi &= g - [\omega^2 r - (\omega \cdot \mathbf{r})\omega \sin \phi] \\ &= g - \omega^2 r (1 - \sin^2 \phi) \\ &= g - \omega^2 r \cos^2 \phi. \end{aligned} \tag{8.133}$$

At the North pole: $\phi = 90°, G \equiv G_{\text{pole}} = g$. Hence

$$\begin{aligned} G_\phi &= G_{\text{pole}} - \omega^2 r \cos^2 \phi = G_{\text{pole}} \left(1 - \frac{\omega^2 r}{g} \cos^2 \phi \right) \\ &\overset{(8.128)}{=} G_{\text{pole}}(1 - 0.0034 \cos^2 \phi). \end{aligned} \tag{8.134}$$

In particular at angle $\phi = 60°$:

$$\frac{G_{60°}}{G_{\text{pole}}} \simeq 1 - 0.0008. \tag{8.135}$$

This result makes sense, since the centrifugal force (which vanishes at the pole) acts against the acceleration due to gravity. Thus the apparent acceleration due to gravity is largest at the pole and somewhat less away from it. Since the Earth really has the shape of a spheroid which bulges out at the equator, this also enhances the effect. Thus think of a merchant travelling around the world and selling coffee at a fixed price. If he carries a spring balance with him which measures the weight mg, he will make a better profit in a cold country than in one near the equator. However, with a beam balance (*i.e.* a beam resting on a fulcrum) and employing gauged masses on one side, he would not be able to make this extra profit.*

Next we consider the *Coriolis force*: $-2m\boldsymbol{\omega} \times \mathbf{v}_R$. This force is directed vertically to both $\boldsymbol{\omega}$ and $\mathbf{v}_R$ and leads to a deflection of a horizontally shot bullet to the right of $\mathbf{v}_R$ in the northern hemisphere, and to the left on the southern hemisphere; at the equator the Coriolis force vanishes for a horizontally shot bullet. These situations are shown diagramatically in Fig. 8.34.

---

*This is an amended formulation of problem 207 in E. H. Booth and P. M. Nicol [6], p. 60, concerning a merchant selling tea. As these authors remark: This is not intended to incite unlawful dealings. Rather, it is meant to entertain or amuse.

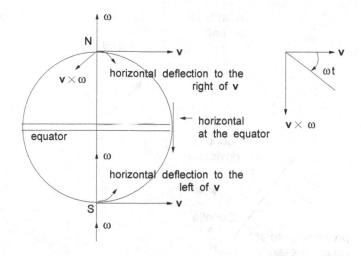

Fig. 8.34 The Coriolis force acting on a bullet.

We always have (recall from Eq. (8.128)): $\omega = 7.3 \times 10^{-5}\mathrm{s}^{-1}$, so that $2\omega \simeq 1.5 \times 10^{-4}\mathrm{s}^{-1}$,

$$|2\boldsymbol{\omega} \times \mathbf{v}_R| \lesssim 1.5 \times 10^{-4}v \sim 0.015v.$$

Thus in general the effect of the Coriolis force is neglibibly small.

Example 8.24: Trajectory of a bullet fired at the North pole

Determine the direction of the trajectory that a bullet executes when fired horizontally at the North pole and given there a velocity $\mathbf{v}_R = \mathbf{v}$ (implying $\mathbf{v} \perp \boldsymbol{\omega}$).

Solution: In an inertial frame S the path traced by the bullet is that of a straight line. This follows already from Newton's first law (recall that Newton's laws refer to an inertial system). However, we can understand this also as follows. The *Coriolis acceleration* is $2\omega v$. The deflecting velocity at time t is therefore equal to this acceleration times t, *i.e.* $2\omega vt$. The corresponding distance in the direction of this deflecting force is therefore at time t:

$$x_{\text{deflection}} = \text{velocity} \times t \equiv \int_0^t (2\omega vt)dt = \omega vt^2.$$

Consequently the corresponding angular deflection is, as indicated in Fig. 8.34:

$$\theta = \frac{x_{\text{deflection}}}{vt} = \frac{\omega vt^2}{vt} = \omega t.$$

Seen from the North pole, this is a deflection to the right, while the Earth rotates around the polar axis through the same angle to the left. Thus in the inertial frame, the trajectory of the bullet is that of a straight line.

Example 8.25: The Coriolis force and air drifts

Explain the approximate effect of Coriolis forces on air drifts in the northern hemisphere.

Solution: Before one can apply the previous one–particle consideration to a mass of air, Eq. (8.126) has to be rewritten in terms of the density ρ of air. We assume that all particles have the same

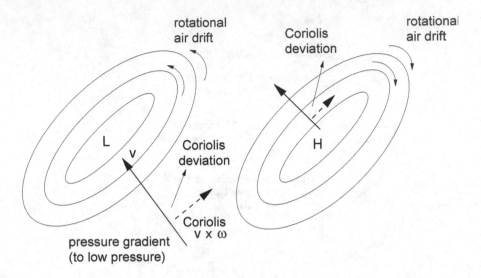

Fig. 8.35 The Coriolis force and High and Low pressure domains
with directions as for northern hemisphere.

mass and can be treated as identical. Also we ignore the weaker terms $m\omega^2\mathbf{r}$, $-m(\boldsymbol{\omega} \cdot \mathbf{r})\boldsymbol{\omega}$ in Eq. (8.126), as well as other effects, such as friction. Attaching a name index k to one–particle quantities and summing over all particles, we obtain from Eq. (8.126) the equation

$$\sum_k \mathbf{F}_{\text{eff},k} = \sum_k \mathbf{F}_k - \sum_k 2m\boldsymbol{\omega} \times \mathbf{v}_{R,k} = \sum_k \mathbf{F}_k - 2\boldsymbol{\omega} \times \sum_k m\mathbf{v}_{R,k}. \qquad (8.136)$$

Wind is a moving mass of air. The density of the appropriate current over a distance x and per second through an area A is the *wind current density*

$$\mathbf{j}_{R,\text{wind}} = \frac{1}{Ax} \sum_k m\mathbf{v}_{R,k} \equiv \rho\mathbf{v}_{R,\text{wind}}. \qquad (8.137)$$

Dividing Eq. (8.136) by Ax (or arguing on the basis of infinitesimal quantities), we have for the applied force

$$\frac{1}{Ax} \sum_k \mathbf{F}_k = \boldsymbol{\nabla}_x P,$$

where P is the pressure on the area A, and (since the force is d/dt of momentum)

$$\frac{1}{Ax} \sum_k \mathbf{F}_{\text{eff},k} = \frac{d}{dt}\left[\frac{1}{Ax} \sum_k m\mathbf{v}_{R,k}\right] \equiv \frac{d}{dt}\mathbf{j}_{R,\text{wind}}.$$

Inserting these expressions into Eq. (8.136) divided by Ax, we obtain

$$\frac{d}{dt}\mathbf{j}_{R,\text{wind}} = \boldsymbol{\nabla}_x P - 2\boldsymbol{\omega} \times \rho\mathbf{v}_{R,\text{wind}}. \qquad (8.138)$$

For a reasonably steady wind or approximate equilibrium, the wind current is approximately constant and hence the left hand side vanishes. The result

$$\boldsymbol{\nabla}_x P = -[-2\boldsymbol{\omega} \times \rho\mathbf{v}_{R,\text{wind}}] \qquad (8.139)$$

expresses a balancing between the pressure gradient in one direction and the Coriolis forces in the opposite direction. Since the Coriolis force has a curving effect, the result is a wind–drift configuration known as a *cyclone*, as is illustrated in Fig. 8.35. As a matter of general interest we add the following comments. A low pressure domain L in the atmosphere can be visualized as a funnel with a deficiency of air in its trough. A high pressure domain H on the other hand can be visualized as a mountain with too much air at its top. Nature aims at equilibrium, and the most straightforward way to achieve this in the present case would be with surplus air of H to flow into L. However, it is precisely the Coriolis force which prevents this simple way. The resulting cyclone-type of domains prevent the immediate passage of air from one type to another. In an L domain with a deficiency of air, the air can expand and therefore cools, the result being cloud formation and rain. In an H domain the air is compressed and the temperature rises, and rain drops evaporate.

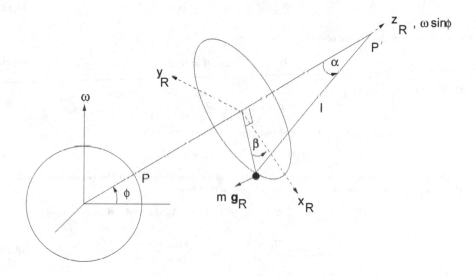

Fig. 8.36 The Foucault pendulum at latitude ϕ, colatitude $\pi/2 - \phi$.

Example 8.26: The Foucault pendulum

A pendulum bob of mass m is attached to a string of a long length l and is allowed to oscillate freely at a *latitude* ϕ in the gravitational field of the Earth. Explore the effect of the *Coriolis force*, *i.e.* the effect of the term (8.127) in Eq. (8.126), on the oscillation of the pendulum.

Solution: The pendulum is held at a point P' along the z_R-axis of the Earth-fixed reference frame R with coordinates x_R, y_R, z_R above the point P with latitude ϕ, as indicated in Fig. 8.36. From the geometry we deduce that the coordinates of the pendulum bob are given by

$$\frac{x_R}{l} = \cos\beta\sin\alpha, \qquad \frac{y_R}{l} = \sin\beta\sin\alpha, \qquad \frac{z_R}{l} = \cos\alpha. \qquad (8.140)$$

The acceleration due to gravity observed in frame R, namely $\mathbf{g}_R$, now has components in all directions. These are:

$$g_{Rx} = g_R\cos\beta\sin\alpha = g_R\frac{x_R}{l}, \quad g_{Ry} = g_R\sin\beta\sin\alpha = g_R\frac{y_R}{l}, \quad g_{Rz} = g_R\cos\alpha = g_R\frac{z_R}{l}, \quad (8.141)$$

or

$$\mathbf{g}_R = g_R\frac{\mathbf{r}_R}{l}. \qquad (8.142)$$

Inserting $\mathbf{F} = -mg_R = -mg_R\mathbf{r}_R/l$ into Eq. (8.126) (and ignoring the centrifugal terms), we obtain

$$m\ddot{\mathbf{r}}_R = -mg_R\frac{\mathbf{r}_R}{l} - 2m\boldsymbol{\omega}\times\dot{\mathbf{r}}_R. \tag{8.143}$$

In components and with $\omega_z = \omega\sin\phi, \dot{r}_{Rz} \sim 0$, this equation implies

$$\ddot{x}_R = -g_R\frac{x_R}{l} - 2(-\omega_z\dot{y}_R), \quad \ddot{y}_R = -g_R\frac{y_R}{l} - 2(\omega_z\dot{x}_R), \tag{8.144}$$

as well as the third equation $\ddot{z}_R = -g_Rz_R/l - 2(\omega_x\dot{y}_R - \omega_y\dot{x}_R)$. Since for l very long, the oscillations are practically restricted to the (x_R, y_R)-plane, the last equation can be ignored. The two remaining equations constitute a pair of coupled equations. It is convenient to set $Z = x_R + iy_R$, and hence to add i times the second equation to the first. Then

$$\ddot{Z} + \frac{g_R}{l}Z + 2i\omega_z\dot{Z} = 0. \tag{8.145}$$

Setting now $Z = Ae^{iBt}$, we obtain:

$$B^2 - \frac{g_R}{l} + 2\omega_z B = 0, \tag{8.146}$$

with solutions

$$B_\pm = -\omega_z \pm \sqrt{\omega_z^2 + \frac{g_R}{l}}. \tag{8.147}$$

Hence the general solution $Z(t)$ is (with integration constants $A_\pm$)

$$Z(t) = A_+e^{iB_+t} + A_-e^{-iB_-t}. \tag{8.148}$$

In order to fix the integration constants, we assume that at time $t = 0$ the pendulum is released with velocity $\dot{Z}(0) = ib$ from position $x_R = a, y_R = 0$. Then

$$Z(0) = a = A_+ + A_-, \quad \dot{Z}(0) = ib = i(A_+B_+ - A_-B_-),$$

with solutions

$$A_+ = \frac{aB_- + b}{B_+ + B_-}, \quad A_- = \frac{aB_+ - b}{B_+ + B_-}. \tag{8.149}$$

For $A_+ = A_- = a/2$, we find $b = a(B_+ - B_-)/2$. Then

$$Z(t) = \frac{a}{2}[e^{iB_+t} + e^{-iB_-t}] = \frac{a}{2}e^{i\sqrt{\omega_z^2 + g_R/l}}[e^{-i\omega_z t} + e^{i\omega_z t}] = a\cos\omega_z t\, e^{i\sqrt{\omega_z^2 + g_R/l}}.$$

It follows from $Z = x_R + iy_R$ that

$$x_R(t) = a\cos\omega_z t\cos\sqrt{\omega_z^2 + g_R/l}t, \quad y_R(t) = a\cos\omega_z t\sin\sqrt{\omega_z^2 + g_R/l}t. \tag{8.150}$$

The angular velocity $\omega_z = \omega\sin\phi$ of the Earth is small compared with g_R/l. Hence the extrema between which the pendulum oscillates on the approximately circular orbit

$$x_R^2(t) + y_R^2(t) = a^2\cos^2\omega_z t$$

are approximately

$$\pm(\cos\sqrt{g_R/l}t, \sin\sqrt{g_R/l}t).$$

The period of oscillation is

$$T_{\text{Foucault}} = \frac{2\pi}{\omega_z} = \frac{2\pi}{\omega\sin\phi}.$$

It is this oscillation that one observes in the various science museums which exhibit the Foucault pendulum. At the equator the period is infinite, at the pole ($\phi = \pi/2$) it is $2\pi/\omega$. A similar discussion of the Foucault pendulum can be found in the book of Kibble and Berkshire [25].[†]

[†]T. W. B. Kibble and F. H. Berkshire [25], p. 117.

8.9   Miscellaneous Examples II

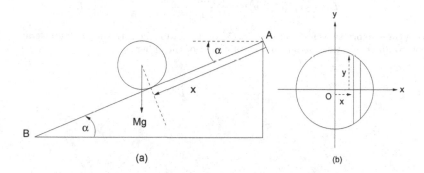

Fig. 8.37 Dynamical identification of a sphere.

Example 8.27: Two identically looking spheres on an inclined plane

Two spheres A and B of different materials (*i.e.* densities) have the same mass, the same radius and the same appearance. It is known that one of the spheres has no cavities, whereas the other sphere has a concentric hollow spherical cavity. Describe a dynamical method to determine which sphere is which.

Solution: We let the spheres roll down an inclined plane with angle of inclination α as shown in Fig. 8.37 and determine their accelerations. Let A be the initial point of contact of a sphere with the inclined plane, from where it starts off with zero velocity at time $t = 0$, and let B be its point of contact at time t later. Let $AB = x$, M the mass of a sphere, and a its radius. We then have the constraint $x = a\theta$, where $\dot{\theta}$ is the angular velocity of the sphere. We use the law of conservation of energy, which equates the work done by the external forces to the kinetic energy of the rolling sphere. Then, if I is the moment of inertia of the sphere about a diameter, the potential energy lost by the sphere in rolling down the distance x is equal to the kinetic energy gained, and hence

$$Mgx\sin\alpha = \frac{1}{2}I\dot{\theta}^2 = \frac{1}{2}I\frac{\dot{x}^2}{a^2}.$$

Differentiating this equation with respect to t and using the relation

$$\ddot{x} = \frac{d}{dx}\left(\frac{1}{2}\dot{x}^2\right), \quad \text{we obtain} \quad Mg\sin\alpha = \frac{I\ddot{x}}{a^2}, \quad \text{and} \quad \therefore \ddot{x} = \frac{Ma^2}{I}g\sin\alpha. \tag{8.151a}$$

Next we determine the moments of inertia of (a) a solid sphere, and of (b) a spherical shell of a definite thickness. (a) A sphere of radius a and density ρ has mass $M = 4\pi a^3\rho/3$. Using diameters as axes, the mass δM of a slice of the sphere of thickness δx and its moment of inertia δI about the axis Ox are given by (*cf.* Example 8.10 for δI)

$$\delta M = \pi y^2 \delta x \rho \quad \text{and} \quad \delta I = \frac{1}{2}\delta My^2 = \frac{1}{2}\pi(a^2 - x^2)^2\rho\delta x.$$

It follows that for the solid sphere the moment of inertia I is given by (the result of Example 8.11)

$$I = \frac{1}{2}\pi\rho\int_{-a}^{a}(a^2 - x^2)^2 dx = \frac{8}{15}\pi a^5\rho = \frac{2}{5}Ma^2. \tag{8.151b}$$

(b) In the case of the spherical shell we consider radii a and b with $a > b$, and density ρ'. The mass of the hollow sphere is then $M = 4\pi\rho'(a^3 - b^3)/3$. Again using a diameter as axis, and recalling that the moment of inertia I is defined by $I = \sum_i m_i r_i^2$, we obtain for this case from the expression in the middle of Eq. (8.151b)

$$I = \frac{8}{15}\pi\rho'(a^5 - b^5) = \frac{8}{15}\frac{3M}{4}\frac{a^5 - b^5}{a^3 - b^3} = \frac{2}{5}\frac{a^5 - b^5}{a^3 - b^3}M.$$

Inserting these expressions for I into Eq. (8.151a), we obtain the corresponding accelerations, which are seen to differ and thus permit an identification of the spheres.

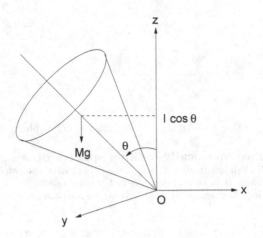

Fig. 8.38 The spinning top.

Example 8.28: Nutations of the spinning top

The motion of a spinning top like that used as a toy can be studied with the help of the locus of the top's axis on a unit sphere around the only fixed point of the spinning top. The polar coordinates of a point on this locus are θ and ϕ. What is the equation of motion in θ? Assuming the nutations of the spinning top are small, *i.e.* θ close to a constant value α, calculate the period of its nutations.

Solution: We let M be the mass of the top and l the distance of its centre of mass from the fixed point O as indicated in Fig. 8.38. We begin with Eq. (8.105) for the energy E:

$$E = Mgl\cos\theta + \frac{1}{2}I_1\left[\dot{\theta}^2 + \frac{(b - a\cos\theta)^2}{\sin^2\theta}\right] + \frac{1}{2}\frac{I_1^2 a^2}{I_3}. \tag{8.152}$$

The constants a and b are given by the following expressions obtained from Eqs. (8.99) and (8.100):

$$\begin{aligned} p_\psi &= I_3(\dot{\psi} + \dot{\phi}\cos\theta) = I_3\omega_z = I_1 a, \\ p_\phi &= I_3(\dot{\psi} + \dot{\phi}\cos\theta)\cos\theta + \dot{\phi}I_1\sin^2\theta = I_1 a\cos\theta + \dot{\phi}I_1\sin^2\theta = I_1 b. \end{aligned} \tag{8.153}$$

The angles θ, ϕ, ψ are *Euler angles*. To simplify the present problem we consider the case of $\psi = \text{const.}, \dot{\psi} = 0$. This condition will be used later in Eq. (8.160). Note in the definition of the constants a and b in Eqs. (8.153) the change from I_3 to I_1. We differentiate Eq. (8.152) with respect to θ. Using the relation

$$\ddot{\theta} = \frac{d}{d\theta}\left(\frac{1}{2}\dot{\theta}^2\right),$$

we obtain the equation

$$I_1\ddot{\theta} - Mgl\sin\theta - \frac{I_1(b - a\cos\theta)^2}{\sin^3\theta}\cos\theta + \frac{I_1(b - a\cos\theta)a}{\sin\theta} = 0. \tag{8.154}$$

The nutations are oscillations in the angle θ. If these are small, we can set $\theta(t) = \alpha + x(t)$, where $\alpha = $ const. is the mean value of $\theta(t)$. From Eq. (8.154) we can then compute the period of the nutations. First, for $|x| \ll \alpha$, we have:

$$
\begin{aligned}
\sin\theta &= \sin(\alpha + x) = \sin\alpha\cos x + \cos\alpha\sin x \simeq \sin\alpha + x\cos\alpha, \\
\cos\theta &= \cos(\alpha + x) = \cos\alpha\cos x - \sin\alpha\sin x \simeq \cos\alpha - x\sin\alpha.
\end{aligned}
$$

Inserting these expressions into Eq. (8.154), we obtain.

$$
\begin{aligned}
& I_1\ddot{x} - Mgl(\sin\alpha + x\cos\alpha) \\
& - I_1\frac{[b - a(\cos\alpha - x\sin\alpha)]^2[\cos\alpha - x\sin\alpha]}{(\sin\alpha + x\cos\alpha)^3} + I_1\frac{a[b - a(\cos\alpha - x\sin\alpha)]}{(\sin\alpha + x\cos\alpha)} = 0.
\end{aligned}
$$

Retaining only terms which are linear in x, this leaves

$$
\begin{aligned}
& I_1\ddot{x} - Mgl(\sin\alpha + x\cos\alpha) \\
& - I_1\frac{[b - a\cos\alpha + ax\sin\alpha]^2[\cos\alpha - x\sin\alpha]}{(\sin\alpha + x\cos\alpha)^3} + I_1\frac{a[b - a(\cos\alpha - x\sin\alpha)]}{(\sin\alpha + x\cos\alpha)} \simeq 0,
\end{aligned}
$$

or

$$
\begin{aligned}
I_1\ddot{x} \quad & - \quad Mgl(\sin\alpha + x\cos\alpha) \\
& - I_1\frac{(1 - 3x\cos\alpha/\sin\alpha)(b - a\cos\alpha)^2}{\sin^3\alpha}\left(1 + 2\frac{ax\sin\alpha}{b - a\cos\alpha}\right)\cos\alpha\left(1 - x\frac{\sin\alpha}{\cos\alpha}\right) \\
& + I_1 a\frac{[(b - a\cos\alpha) + ax\sin\alpha]}{\sin\alpha}\left(1 - x\frac{\cos\alpha}{\sin\alpha}\right) \simeq 0. \qquad (8.155)
\end{aligned}
$$

Considering first the x-independent terms we have (as also follows from Eq. (8.154) if we set there $\theta = \alpha$):

$$
-Mgl\sin\alpha - I_1\frac{(b - a\cos\alpha)^2\cos\alpha}{\sin^3\alpha} + I_1\frac{a(b - a\cos\alpha)}{\sin\alpha} = 0. \qquad (8.156)
$$

This is the equation determining the angle α. The equation remaining determines the nutation $x(t)$. This equation is:

$$
\begin{aligned}
& I_1\ddot{x} + x\left[-Mgl\cos\alpha - I_1\frac{(b - a\cos\alpha)^2\cos\alpha}{\sin^3\alpha}\left\{-\frac{3\cos\alpha}{\sin\alpha} + \frac{2a\sin\alpha}{b - a\cos\alpha} - \frac{\sin\alpha}{\cos\alpha}\right\}\right. \\
& \left.+ I_1\frac{a(b - a\cos\alpha)}{\sin\alpha}\left\{\frac{a\sin\alpha}{b - a\cos\alpha} - \frac{\cos\alpha}{\sin\alpha}\right\}\right] \simeq 0,
\end{aligned}
$$

or

$$
\begin{aligned}
& I_1\ddot{x} + x\left[-Mgl\cos\alpha - 3I_1\frac{a(b - a\cos\alpha)\cos\alpha}{\sin^2\alpha} + I_1 a^2\right. \\
& \left.+ I_1\frac{(b - a\cos\alpha)^2\cos\alpha}{\sin^3\alpha}\left(\frac{3\cos^2\alpha + \sin^2\alpha}{\sin\alpha\cos\alpha}\right)\right] \simeq 0. \qquad (8.157)
\end{aligned}
$$

With the help of Eq. (8.153) we can make here the replacements:

$$
I_1(b - a\cos\alpha) = \Omega I_1\sin^2\alpha, \quad \Omega = \frac{b - a\cos\alpha}{\sin^2\alpha}, \quad \text{with} \quad \dot{\phi} \equiv \Omega = \text{const.} \qquad (8.158)
$$

Then Eq. (8.157) becomes

$$
I_1\ddot{x} + x[-Mgl\cos\alpha - 3a\cos\alpha\,\Omega I_1 + I_1 a^2 + I_1\Omega^2(1 + 2\cos^2\alpha)] \simeq 0. \qquad (8.159)
$$

The first of Eqs. (8.153) yields for $\dot{\psi} = 0$, the specialization referred to there (so that we deal only with two of the three Euler angles):

$$I_3 \Omega \cos \alpha = I_1 a. \tag{8.160}$$

Inserting this expression for Ω into Eq. (8.159) we obtain:

$$I_1 \ddot{x} + x \left[- Mgl \cos \alpha - \frac{3a^2 I_1^2}{I_3} + I_1 a^2 + \frac{I_1^3 a^2}{I_3^2} \left(2 + \frac{1}{\cos^2 \alpha} \right) \right] \simeq 0. \tag{8.161}$$

Thus the motion is *stable* (meaning periodic and not exponentially increasing or decreasing), if

$$\frac{I_1^3 a^2}{I_3^2} \left(2 + \frac{1}{\cos^2 \alpha} \right) + I_1 a^2 > Mgl \cos \alpha + \frac{3a^2 I_1^2}{I_3}. \tag{8.162}$$

In order to see whether this condition is satisfied, we return to Eq. (8.154) and set there $\theta = \alpha = $ const. Then, in terms of Ω, we obtain:

$$\sin \alpha (-I_1 \Omega^2 \cos \alpha + I_1 a \Omega - Mgl) = 0 \quad \text{or} \quad \Omega^2 - \frac{a}{\cos \alpha} \Omega + \frac{Mgl}{I_1 \cos \alpha} = 0. \tag{8.163}$$

The roots of this equation are

$$\Omega_{\pm} = \frac{a}{2 \cos \alpha} \pm \frac{a}{2 \cos \alpha} \sqrt{1 - \frac{4 Mgl \cos \alpha}{I_1 a^2}}.$$

Thus Eq. (8.163) possesses *real roots* $\Omega_{\pm}$, and this means physically sensible roots, if the discriminant is positive, *i.e.*

$$I_1 a^2 - 4 Mgl \cos \alpha > 0, \qquad a^2 > \frac{4 Mgl \cos \alpha}{I_1}. \tag{8.164}$$

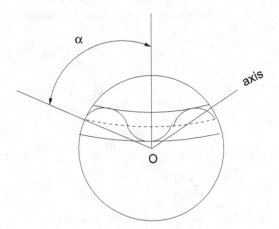

Fig. 8.39 The nutations about angle α.

Thus this inequality must be satisfied. We now compare this condition with the condition (8.162):

$$4 \frac{I_1^3 a^2}{I_3^2} \left(2 + \frac{1}{\cos^2 \alpha} \right) + 4 I_1 a^2 - \frac{12 a^2 I_1^2}{I_3} > 4 Mgl \cos \alpha. \tag{8.165}$$

For $I_1 \sim I_3$ this inequality is seen to be satisfied in view of Eq. (8.164), i.e. Eq. (8.165) is roughly

$$4a^2 \left(2 + \frac{1}{\cos^2 \alpha} \right) + 4a^2 - 12a^2 > \frac{4Mgl \cos \alpha}{I_1}, \quad \frac{4a^2}{\cos^2 \alpha} > a^2 > \frac{4Mgl \cos \alpha}{I_1}.$$

We set, thereby defining a parameter P^2,

$$I_1 P^2 \equiv -Mgl \cos \alpha - \frac{3a^2 I_1^2}{I_3} + I_1 a^2 + \frac{I_1^3 a^2}{I_3^2} \left(2 + \frac{1}{\cos^2 \alpha} \right). \tag{8.166}$$

Equation (8.161) now becomes

$$\ddot{x} + P^2 x = 0. \tag{8.167}$$

It follows that, with integration constants A and B,

$$x = A \sin(Pt + B). \tag{8.168}$$

We obtain therefore for the angle $\theta(t)$ the expression

$$\theta(t) = \alpha + x(t) = \alpha + A \sin(Pt + B). \tag{8.169}$$

Thus P is the period of the nutations of the spinning top, as indicated in Fig. 8.39.

Example 8.29: A beetle crawling on a rotating disc

A uniform, circular disc of radius a and mass M can rotate in a horizontal plane around a frictionless axis vertically through its centre. A beetle of mass nM is sitting on the disc at a point very close to the axis. When the disc is given a constant angular velocity Ω the beetle begins to crawl with velocity V (relative to the disc) along a radius drawn on the disc. What is the angle through which the disc has rotated at the moment when the beetle reaches the edge of the disc? Assume that the pressure on the disc due to the weight of the beetle can be neglected.

Solution: Naively one might argue the beetle reaches the edge of the disc in time a/V and hence the angle through which the disc has rotated is $a\Omega/V$. However, this argument ignores that the mass of the beetle is an extra mass point which enters the effective or instantaneous moment of inertia of the rotating system. We use the Lagrange formalism. The kinetic energy T of the system "disc plus beetle" is made up of the kinetic energies of both the disc and the beetle. The kinetic energy of the disc is $I\dot{\theta}^2/2$ where $I = Ma^2/2$ is the moment of inertia of the disc with respect to an axis vertically through its centre (Example 8.9). For the computation of the kinetic energy of the beetle we require the relation (8.120) which expresses the velocity of the beetle relative to an inertial frame in terms of that relative to the disc, i.e. the relation

$$\left(\frac{d\mathbf{r}}{dt} \right)_{\text{inertial}} = \left(\frac{d\mathbf{r}}{dt} \right)_{\text{disc}} + \frac{d\boldsymbol{\theta}}{dt} \times \mathbf{r}.$$

Here $\mathbf{r}$ is the radius vector of the beetle. Since $\boldsymbol{\theta}$ is a vector parallel to the axis of rotation, we have $\boldsymbol{\theta} \perp \mathbf{r}$. Further, since

$$\left(\frac{d\mathbf{r}}{dt} \right)_{\text{disc}} = \dot{\mathbf{r}},$$

it follows that $\dot{\mathbf{r}} \perp \dot{\boldsymbol{\theta}} \times \mathbf{r}$. The square of $(d\mathbf{r}/dt)_{\text{inertial}}$ is therefore

$$\left(\frac{d\mathbf{r}}{dt} \right)_{\text{inertial}}^2 = \dot{r}^2 + r^2 \dot{\theta}^2.$$

Hence we obtain for the kinetic energy T:

$$T = \frac{1}{2} I \dot{\theta}^2 + \frac{1}{2} nM (\dot{r}^2 + r^2 \dot{\theta}^2).$$

We ignore the effect of the weight of the beetle on the disc, and this means a potential energy V. Thus $L = T$. The general equations of motion are, where Q_{q_j} are the generalized forces,

$$\frac{d}{dt}\left(\frac{\partial L}{\partial \dot{q}_j}\right) - \frac{\partial L}{\partial q_j} = Q_j.$$

With $q_j = \theta, r$, we obtain the equations:

$$\frac{d}{dt}(I\dot{\theta} + nMr^2\dot{\theta}) = Q_\theta, \qquad \frac{d}{dt}(nM\dot{r}) - nMr\dot{\theta}^2 = Q_r. \qquad (8.170)$$

In these equations $Q_\theta = 0$, and Q_r (the force responsible for the motion of the beetle) is unknown. The first equation expresses what the law of conservation of angular momentum says, namely

$$\frac{d\mathbf{L}}{dt} = \mathbf{N} = 0,$$

where $\mathbf{N}$ is the moment of external forces, here ignoring the effect of the weight of the beetle. The given constraint $\dot{r} = V = $ const. implies here that $r = Vt$ (with initial condition $r = 0$ at time $t = 0$). The first of Eqs. (8.170) implies (defining Ω)

$$I\dot{\theta} + nMr^2\dot{\theta} = \text{const.} \equiv I\Omega,$$

where $I = Ma^2/2, M = 2I/a^2$. We observe here the addition of nMr^2 to I to yield the instantaneous moment of inertia. Hence

$$\dot{\theta} = \frac{\Omega}{1 + 2nr^2/a^2} = \frac{d\theta}{dt}. \qquad (8.171)$$

But

$$\frac{d\theta}{dt} = V\frac{d\theta}{d(Vt)} = V\frac{d\theta}{dr}.$$

Thus Eq. (8.171) can be rewritten as

$$d\theta = \frac{1}{V}\frac{\Omega dr}{1 + 2nr^2/a^2},$$

or

$$\theta = \frac{a\Omega}{V\sqrt{2n}}\int_0^r \frac{ad(\sqrt{2n}r)}{a^2 + (\sqrt{2n}r)^2} = \frac{a\Omega}{V\sqrt{2n}}\tan^{-1}\left(\frac{\sqrt{2n}r}{a}\right) + \text{const.}$$

With $\theta = 0$ at $r = 0, t = 0$, the constant is zero. It follows that when $r = a$ the angle θ is

$$\theta_a = \frac{a\Omega}{V\sqrt{2n}}\tan^{-1}(\sqrt{2n}).$$

We see that for $2n \to 0$, i.e. a negligible mass of the beetle, the angle approaches the naively expected value.

Example 8.30: A sphere rolling on a rotating inclined plane

A solid sphere of mass M and radius a rolls on a plane which itself rotates with a uniform angular velocity Ω around an axis perpendicular to the plane. The plane is not horizontal. Establish the equations of motion and show that the motion of the point of contact of the sphere with the plane can be considered as the motion of a rotation around a point which moves with a uniform velocity in a horizontal direction.

Solution: Let O be the origin of a spatially fixed, i.e. inertial, reference frame, and Ω the angular velocity of the plane around the axis Oz as indicated in Fig. 8.40. Let $\boldsymbol{\omega}$ be the angular velocity of the sphere (around its own axis). Also, let $\mathbf{i}, \mathbf{j}, \mathbf{k}$ be unit vectors along the x, y, z-axes

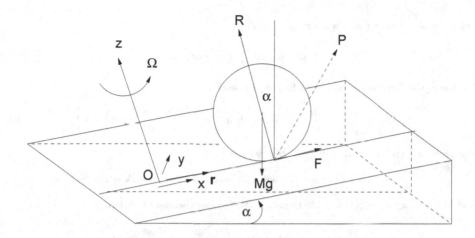

Fig. 8.40 A sphere rolling down a rotating inclined plane.

of the inertial frame. Further, let $\mathbf{P}$ be the vector whose components are the reaction $\mathbf{R}$ and the friction $\mathbf{F}$, and let $\mathbf{n}$ be the unit vector $\mathbf{n} = \mathbf{k}\cos\alpha + \mathbf{i}\sin\alpha$. Let $\mathbf{r}$ be the position vector of the point of contact of the sphere with the plane relative to the origin O. The equation of motion of the sphere, or rather this point of contact, is then

$$M\ddot{\mathbf{r}} = \mathbf{P} - Mg\mathbf{n}, \quad \text{or} \quad M\ddot{r} = F - Mg\sin\alpha, \quad 0 = R - Mg\cos\alpha. \qquad (8.172)$$

The law "Time rate of change of angular momentum about an axis =moment of the external forces about this axis" (Eqs. (2.32), (8.52)) yields the equation

$$I\dot{\omega} = Fa, \quad \text{or} \quad I\dot{\omega} = -a\mathbf{k} \times \mathbf{P}, \qquad (8.173)$$

where I is the moment of inertia of the sphere about its diameter. The constraint "*rolling*" implies: Velocity of the point of contact relative to the rigid body frame, *i.e.* that of the sphere, is given by

$$\left(\frac{d\mathbf{r}}{dt}\right)_{\text{rigid body}} = a\omega, \quad \text{or} \quad \left(\frac{d\mathbf{r}}{dt}\right)_{\text{rigid body}} = a\boldsymbol{\omega} \times \mathbf{k}.$$

The velocity $d\mathbf{r}/dt$ relative to the inertial reference frame is therefore:

$$\left(\frac{d\mathbf{r}}{dt}\right)_{\text{inertial}} = \left(\frac{d\mathbf{r}}{dt}\right)_{\text{rigid body}} + \boldsymbol{\Omega} \times \mathbf{r} = a\boldsymbol{\omega} \times \mathbf{k} + \boldsymbol{\Omega} \times \mathbf{r}. \qquad (8.174)$$

With the help of the three equations (8.172), (8.173), (8.174) the three unknown quatities $\mathbf{r}, \mathbf{P}, \boldsymbol{\omega}$ can now be determined. We eliminate $\mathbf{P}$ from Eqs. (8.172) and (8.173):

$$I\dot{\boldsymbol{\omega}} = -a\mathbf{k} \times \mathbf{P} = -a\mathbf{k} \times [M\ddot{\mathbf{r}} + Mg\mathbf{n}]. \qquad (8.175)$$

Scalar multiplication by $\mathbf{k}$ yields (because the resulting right hand side contains $\mathbf{k} \times \mathbf{k}$ of a scalar triple product):

$$\mathbf{k} \cdot \dot{\boldsymbol{\omega}} = 0, \quad \text{and} \quad \therefore \ \mathbf{k} \cdot \boldsymbol{\omega} = \text{const.} \qquad (8.176)$$

Integrating Eq. (8.175) we obtain

$$I\boldsymbol{\omega} = -a\mathbf{k} \times [M\dot{\mathbf{r}} + Mg\mathbf{n}t] + \text{const.}$$

Let $\boldsymbol{\omega}$ be $\boldsymbol{\omega}_0$ at time $t = 0$, when $\dot{\mathbf{r}}_{t=0} = 0$. Then

$$\boldsymbol{\omega} - \boldsymbol{\omega}_0 = \frac{Ma}{I}(\dot{\mathbf{r}} + gn t) \times \mathbf{k}. \tag{8.177}$$

We insert this expression for $\boldsymbol{\omega}$ into Eq. (8.174) and obtain:

$$\left(\frac{d\mathbf{r}}{dt}\right)_{\text{inertial}} \equiv \frac{d\mathbf{r}}{dt} = \boldsymbol{\Omega} \times \mathbf{r} - a\mathbf{k} \times \left\{\boldsymbol{\omega}_0 + \frac{Ma}{I}\left(\frac{d\mathbf{r}}{dt} + gn t\right) \times \mathbf{k}\right\}. \tag{8.178}$$

But since $\mathbf{a} \times (\mathbf{b} \times \mathbf{c}) = (\mathbf{a} \cdot \mathbf{c})\mathbf{b} - (\mathbf{a} \cdot \mathbf{b})\mathbf{c}$, we have

$$\mathbf{k} \times (\mathbf{n} \times \mathbf{k}) = \mathbf{n} - (\mathbf{n} \cdot \mathbf{k})\mathbf{k} = \mathbf{k}\cos\alpha + \mathbf{i}\sin\alpha - \cos\alpha\mathbf{k} = \mathbf{i}\sin\alpha,$$

and $\mathbf{k} \times (\mathbf{r} \times \mathbf{k}) = \mathbf{r}$ (since $\mathbf{k} \perp \mathbf{r}$). Equation (8.178) can therefore be written:

$$\frac{d\mathbf{r}}{dt} = \boldsymbol{\Omega} \times \mathbf{r} - a\mathbf{k} \times \boldsymbol{\omega}_0 - \frac{Ma^2}{I}\frac{d\mathbf{r}}{dt} - a\frac{Magt}{I}\sin\alpha\mathbf{i},$$

or, since $\boldsymbol{\Omega}$ is parallel to $\mathbf{k}$:

$$\left(1 + \frac{Ma^2}{I}\right)\frac{d\mathbf{r}}{dt} = \boldsymbol{\Omega} \times \mathbf{r} - a\mathbf{k} \times \boldsymbol{\omega}_0 - \frac{Ma^2}{I}gt\sin\alpha\mathbf{i}$$

$$= \boldsymbol{\Omega} \times \left[\mathbf{r} - \frac{Ma^2 g}{I\Omega}\sin\alpha t\mathbf{j} - \frac{a\boldsymbol{\omega}_0}{\Omega}\right]. \tag{8.179}$$

We observe that this equation is of the form

$$M'\frac{d\mathbf{r}}{dt} = \boldsymbol{\Omega} \times [\mathbf{r} - \mathbf{v}t - \mathbf{r}_0]. \tag{8.180}$$

This equation describes the motion of a point $\mathbf{r} - \mathbf{r}_0$, which moves with the velocity $\mathbf{v}$ in the direction of the vector $\mathbf{j}$, and this means in a direction perpendicular to the vectors $\mathbf{i}, \mathbf{k}$.

Example 8.31: A mass attached to one end of a string around a wheel

The quantity I is the moment of inertia of a wheel with an axle about a common, fixed horizontal axis. A light (meaning weightless) inelastic string is wound around the axle of radius a, and a mass M is attached to its other free end as indicated in Fig. 8.41. A constant frictional torque G acts against the rotation. The mass M is allowed to fall from a vertical position. What is the angular velocity of the axle when the mass M has fallen through a distance h?

Solution: The torque of the weight is Mga, and the frictional torque is G. Hence the entire torque acting on the system is $Mga - G$. Let l be the angular momentum and I the moment of inertia of the wheel. The time rate of change of angular momentum (Eq. (8.52)) then implies the equation:

$$Ta - G = \frac{dl}{dt} = I\dot{\omega}, \tag{8.181}$$

where T is the tension in the string. We also have the equation of motion

$$M\ddot{x} = Mg - T, \tag{8.182}$$

and the constraint (where $\omega = \dot{\theta}$ is the angular velocity) $a\theta = x$, $a\dot{\theta} = \dot{x}$, $a\ddot{\theta} = \ddot{x}$. We now solve the problem with the help of the Lagrange formalism. Let T_k be the kinetic energy of the system, L the Lagrange function, and V the potential energy of the mass M. Then

$$L = T_k - V, \qquad T_k = \frac{1}{2}I\omega^2 + \frac{1}{2}M\dot{x}^2, \qquad V = -Mgx. \tag{8.183}$$

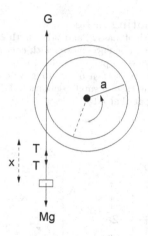

Fig. 8.41 Mass M attached to one end of a string around a wheel.

In addition the system is subjected to the nonconservative force $K = G/a$. The Euler–Lagrange equation is

$$\frac{d}{dt}\left(\frac{\partial L}{\partial \dot{x}}\right) - \frac{\partial L}{\partial x} = Q_x, \quad \text{where} \quad Q_x = -K = -\frac{G}{a}.$$

Together with the constraint $a\omega = \dot{x}, \omega = \dot{x}/a$, we obtain

$$\frac{d}{dt}\frac{\partial}{\partial \dot{x}}\left(\frac{1}{2}I\frac{\dot{x}^2}{a^2} + \frac{1}{2}M\dot{x}^2\right) - \frac{\partial}{\partial x}(Mgx) = -\frac{G}{a}, \quad \text{or} \quad \frac{d}{dt}\left(I\frac{\dot{x}}{a^2} + M\dot{x}\right) - Mg + \frac{G}{a} = 0,$$

or

$$I\frac{\ddot{x}}{a^2} + M\ddot{x} - Mg + \frac{G}{a} = 0, \quad \text{or} \quad \ddot{x} = \frac{Mg - G/a}{I/a^2 + M}. \tag{8.184}$$

Integration yields

$$\dot{x} = \frac{(Mga - G)a}{I + Ma^2}t + c. \tag{8.185}$$

With $\dot{x} = 0$ at time $t = 0$ the integration constant c vanishes. We obtain then

$$\omega = \frac{\dot{x}}{a} = \frac{Mga - G}{I + Ma^2}t.$$

The next integration yields

$$x = \frac{1}{2}\frac{(Mga - G)a}{I + Ma^2}t^2. \tag{8.186}$$

The integration constant here vanishes for $x = 0$ at $t = 0$. From the last two equations we obtain

$$\omega(x) = \frac{Mga - G}{I + Ma^2}\left[\frac{2x(I + Ma^2)}{(Mga - G)a}\right]^{1/2} = \left[\frac{2x(Mga - G)}{(I + Ma^2)a}\right]^{1/2}. \tag{8.187}$$

For $x = h$ we obtain

$$\omega(h) = \left[\frac{2h(Mga - G)}{(I + Ma^2)a}\right]^{1/2}.$$

In order to obtain the equation of motion we could also have used Eq. (8.181). Together with the constraint $\ddot{x} = a\ddot{\theta}$, this yields in agreement with Eq. (8.184):

$$I\frac{\ddot{x}}{a} + Ma\ddot{x} = Mga - G, \quad i.e. \quad \ddot{x} = \frac{(Mga - G)a}{I + Ma^2}.$$

Example 8.32: The oscillating cube

A cube consist of 6 thin plates, each of mass M and side length 2a. Determine the moment of inertia of the cube about one of its edges. The cube moves with constant linear velocity u. Suddenly the cube is stopped abruptly along an edge perpendicular to u. Determine the angular velocity ω of the cube about this edge. The cube is then hung along an edge in the gravitational field of the Earth. What is its period T in the case of small oscillations? Establish equations which allow the determination of the force acting at that edge?

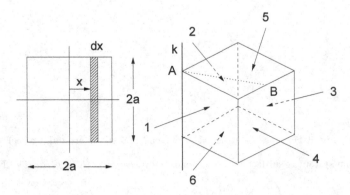

Fig. 8.42 The cube.

Solution: We calculate first the moment of inertia of a quadratic plate of mass $M = 4a^2\rho$, of side length 2a and mass per unit area ρ about several axes through its centre of mass. (a) Axis in the plane of the plate and parallel to an edge. In this case the moment of inertia is ($cf.$ Fig. 8.42)

$$I_1 = \int_{-a}^{a} 2adx\rho x^2 = 2a\rho \int_{-a}^{a} x^2 dx = \frac{2}{3}\rho a[x^3]_{-a}^{a} = \frac{4}{3}\rho a^4 = \frac{1}{3}Ma^2. \qquad (8.188)$$

(b) The moment of inertia about an axis in the plane but perpendicular to that in case (a) is the same, $i.e.$ $I_2 = Ma^2/3$. (c) The moment of inertia about an axis through the centre of mass but perpendicular to the plate is

$$I_3 = \sum_i m_i r_i^2 = \sum_i m_i(x_i^2 + y_i^2) = I_1 + I_2 = \frac{2}{3}Ma^2. \qquad (8.189)$$

We obtain the moment of inertia of the entire cube about one of its edges k as in Fig. 8.42 with the help of the *theorem of Steiner*. We assign numbers to the 6 faces as in Fig.8.42. The moment of inertia of each of faces 1 and 2 about k is

$$I_1 + Ma^2 = \frac{4}{3}Ma^2.$$

The moment of inertia of each of faces 5 and 6 about the axis k is

$$I_3 + M(a^2 + a^2) = \frac{8}{3}Ma^2.$$

The moment of inertia of each of faces 3 and 4 about k is (see line AB in Fig. 8.42)

$$I_1 + M(4a^2 + a^2) = \frac{16}{3}Ma^2.$$

Adding we obtain the moment of inertia of the entire cube about one of its edges as

$$I_k = 2\left(\frac{4}{3} + \frac{8}{3} + \frac{16}{3}\right)Ma^2 = \frac{56}{3}Ma^2 = 18\frac{2}{3}Ma^2. \qquad (8.190)$$

Next we apply the law of angular momentum at the edge k, *i.e.* angular momentum immediately after the abrupt fixture of the cube - the moment of its momentum just before $= 0$. Then

$$I_k \omega = Mua, \quad \omega = \frac{Mua}{I_k} = \frac{3Mua}{56Ma^2} = \frac{3u}{56a}. \tag{8.191}$$

Let θ be the angle of oscillation of the cube, *i.e.* the angle between the vertical from k and the line from there through the centre of gravity. Then d/dt of angular momentum about axis $k =$ moment of the gravitational force (*cf.* Fig. 8.43),

$$\frac{d}{dt}(I_k \dot{\theta}) = -Mg2a \sin\theta \cos\left(\frac{\pi}{4}\right), \quad i.e. \quad \ddot{\theta} + \frac{\sqrt{2}Mga}{I_k}\theta \simeq 0, \tag{8.192}$$

with $\sin\theta \simeq \theta$. It follows that the period T is

$$T = 2\pi\sqrt{\frac{I_k}{\sqrt{2}Mga}} = 2\pi\sqrt{\frac{56a}{3\sqrt{2}g}}.$$

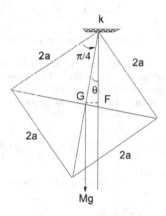

Fig. 8.43 The oscillating cube.

If X, Y are the components of the force acting at the edge, the equation of motion there separated in components yields respectively in transversal and radial directions the equations (*cf.* Example 5.1):

$$X - Mg\sin\theta = 2aM\cos\left(\frac{\pi}{4}\right)\ddot{\theta}, \quad Y - Mg\cos\theta = 2aM\cos\left(\frac{\pi}{4}\right)\dot{\theta}^2.$$

Together with Eq. (8.192) we thus have three equations from which X, Y and θ can be obtained.

Example 8.33: The precessions of a boomerang
Explain briefly how the precessions of a boomerang come about.

Solution: It is precisely the somewhat crooked construction of the boomerang with its twisted arms which enables it to perform its remarkable flight. We assume the boomerang is thrown in a horizontal direction with the two arms vertical. The resultant of the air pressure opposing the horizontal start of the bomerang acts at a point of the boomerang which is some distance away from the centre of mass. As a result a couple is given and the boomerang begins to rotate about the other horizontal direction. If one places the boomerang on a flat table one notices that it itself is not perfectly flat — the reason being that its two arms do not lie in the same plane. Consequently the boomerang executes a precession about the vertical axis, and the torque about this causes a further precession about the direction of projection.

Example 8.34: Moment of inertia of a torus

Determine the moment of inertia of a solid toroidal ring of central radius R, cross sectional radius a, and mass M about its axis, as shown in Fig. 8.44, where $x = \cos\theta$.

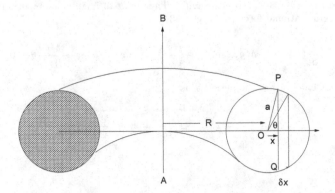

Fig. 8.44 Cross section through the torus.

Solution: A cross section of the ring is depicted in Fig. 8.44. With mass density m, the mass M of the ring is given by $M = m.\pi a^2.2\pi R = 2\pi^2 mRa^2$. The mass of the cylindrical slice of thickness $PQ.\delta x$ indicated in Fig. 8.44 is given by $2\pi(R+x).2\sqrt{a^2-x^2}\delta x.m$. The moment of inertia of the toroidal ring about the axis AB is therefore given by (note:[‡] $\cos 4\theta = 8\cos^4\theta - 8\cos^2\theta + 1$, and $\therefore\ \cos^2\theta\sin^2\theta = (1-\cos 4\theta)/8$)

$$
\begin{aligned}
I_{AB} &= 4\pi m \int_{-a}^{a} (R+x)^3\sqrt{a^2-x^2}\,dx = -4\pi ma^2\int_{\pi}^{0}(R+a\cos\theta)^3\sin^2\theta\,d\theta \\
&= 4\pi ma^2\int_0^\pi \left[\frac{R^3}{2}(1-\cos 2\theta) + 3R^2 a\cos\theta\sin^2\theta + \frac{3Ra^2}{8}(1-\cos 4\theta)\right. \\
&\quad \left. + a^3\cos^3\theta\sin^2\theta\right]d\theta \\
&= 4\pi ma^2\left[\frac{R^3\theta}{2} - \frac{R^3\sin 2\theta}{4} + R^2 a\sin^3\theta + \frac{3Ra^2\theta}{8} - \frac{3Ra^2\sin 4\theta}{32}\right. \\
&\quad \left. + \frac{a^3\sin^3\theta}{3} - \frac{a^3\sin^5\theta}{5}\right]_0^\pi \\
&= \frac{1}{2}\pi^2 mRa^2(4R^2+3a^2) = \frac{1}{4}M(4R^2+3a^2).
\end{aligned}
$$

[‡]H. B. Dwight [13], formula 403.24, p. 76.

Chapter 9

Small Oscillations and Stability

9.1 Introductory Remarks

The topic of small oscillations about some equilibrium situation is evidently of universal importance and is closely related to the topic of stability. If a system is in a configuration of stable equilibrium it can execute small oscillations about this configuration and always return to it. If, on the other hand, a configuration is that of instability, an oscillation there will fling the system into a different state. We encountered such cases already in earlier chapters. Thus in the case of the simple pendulum we made the approximation $\sin\theta \simeq \theta$ in order to confine the motion of the bob to small oscillations about a configuration which minimizes the energy. Similarly, in the case of planetary motion, we observed that small oscillations about a stable elliptic orbit led to the vital perihelian anomaly of the planet Mercury. Criteria for the stability of a system are therefore intimately linked to the study of small oscillations of a system. A deeper study of these topics is therefore the object of this chapter. Here we use the Lagrangian formalism. One can also use equivalently the Hamiltonian method, as is done, for instance, in the book of Penrose.[*]

9.2 Resonance Frequencies and Normal Modes

In comments to literature supplementing his chapter on small vibrations, Goldstein [19] emphasizes the use of examples and says, that relatively simple

---

[*]R. Penrose [38], pp. 478 – 483.

systems give an idea of the meaning of terms like *normal modes of vibration, resonances* and others, which is frequently lost in the abstractness of a general theory. Nonetheless he develops the theory first in general terms and therein specializes at one point to the case when the generalized coordinates are the Cartesian coordinates. Here we prefer to plunge immediately into a nontrivial case, that of the double pendulum,[†] and learn concurrently the new aspects arising therein. An important aspect is the simultaneous diagonalization of two quadratic forms (kinetic energy T and potential energy V).

Example 9.1: The double pendulum

A double pendulum consisting of two weightless rods of equal lengths l and attached bobs of masses m_1, m_2 is oscillating in a vertical plane in the gravitational field of the Earth. Determine its resonating frequencies and normal modes of oscillation.

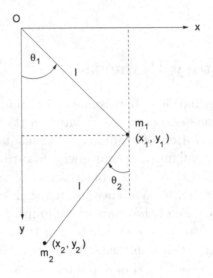

Fig. 9.1 The double pendulum.

Solution:

(1) Equations of motion

We use coordinates as indicated in Fig. 9.1. The coordinates $(x_1, y_1), (x_2, y_2)$ and velocities v_1, v_2 of the masses m_1, m_2 are given by the following expressions in which a dot means differentiation with respect to time t:

$$
\begin{aligned}
(x_1, y_1) &= (l \sin \theta_1, l \cos \theta_1), \quad v_1^2 = \dot{x}_1^2 + \dot{y}_1^2 = (l \cos \theta_1 \dot{\theta}_1)^2 + (-l \sin \theta_1 \dot{\theta}_1)^2 = l^2 \dot{\theta}_1^2, \\
(x_2, y_2) &= (x_1 - l \sin \theta_2, y_1 + l \cos \theta_2), \quad v_2^2 = \dot{x}_2^2 + \dot{y}_2^2 \\
v_2^2 &= (l \cos \theta_1 \dot{\theta}_1 - l \cos \theta_2 \dot{\theta}_2)^2 + (-l \sin \theta_1 \dot{\theta}_1 - l \sin \theta_2 \dot{\theta}_2)^2 \\
&= l^2 \dot{\theta}_1^2 + l^2 \dot{\theta}_2^2 - 2l^2 \dot{\theta}_1 \dot{\theta}_2 \cos(\theta_1 + \theta_2). \quad\quad (9.1)
\end{aligned}
$$

[†]This is a problem left unsolved by H. Goldstein [19], see his Chapter I, problem 10, Chapter X, problem 1. The double pendulum is treated extensively (also with different lengths and masses) by T. W. B. Kibble and F. H. Berkshire [25], pp. 255 – 261, and in *The Physics Coaching Class* [48], problems 2035, 2036, 2054, pp. 537, 539, 584.

The kinetic energy T of the system is therefore given by

$$T = \frac{1}{2}m_1 v_1^2 + \frac{1}{2}m_2 v_2^2 = \frac{1}{2}(m_1 + m_2)l^2\dot{\theta}_1^2 + \frac{1}{2}m_2 l^2\dot{\theta}_2^2 - m_2 l^2\dot{\theta}_1\dot{\theta}_2 \cos(\theta_1 + \theta_2). \tag{9.2}$$

For small oscillations we can approximate this expression by

$$T - \frac{1}{2}(m_1 + m_2)l^2\dot{\theta}_1^2 + \frac{1}{2}m_2 l^2\dot{\theta}_2^2 - m_2 l^2\dot{\theta}_1\dot{\theta}_2. \tag{9.3}$$

Inserting the approximation of small angles θ_1, θ_2 into y_1, y_2, we see that this approximates these to constant values $l, 2l$, and the motion is effectively that of two oscillators restricted to the direction of x. With $g \equiv l\omega_0^2$ the acceleration due to gravity, the potential energy V of the system is given by (again for small angles)

$$
\begin{aligned}
V &= -m_1 gl\cos\theta_1 - m_2 g(l\cos\theta_1 + l\cos\theta_2) = -(m_1 + m_2)gl\cos\theta_1 - m_2 gl\cos\theta_2 \\
&= -(m_1 + m_2)gl\left[1 - \frac{1}{2}\theta_1^2\right] - m_2 gl\left[1 - \frac{1}{2}\theta_2^2\right] \\
&= V_0 + \frac{1}{2}(m_1 + m_2)l^2\omega_0^2\theta_1^2 + \frac{1}{2}m_2 l^2\omega_0^2\theta_2^2,
\end{aligned}
\tag{0.4}
$$

where

$$V_0 = -(m_1 + m_2)gl - m_2 gl, \quad g \equiv l\omega_0^2. \tag{9.5}$$

We thus obtain the following Lagrange function

$$
\begin{aligned}
L(\theta_1, \theta_2, \dot{\theta}_1, \dot{\theta}_2) = T - V &= \frac{1}{2}(m_1 + m_2)l^2\dot{\theta}_1^2 + \frac{1}{2}m_2 l^2\dot{\theta}_2^2 - m_2 l^2\dot{\theta}_1\dot{\theta}_2 \\
&\quad -V_0 - \frac{1}{2}(m_1 + m_2)l^2\omega_0^2\theta_1^2 - \frac{1}{2}m_2 l^2\omega_0^2\theta_2^2.
\end{aligned}
\tag{9.6}
$$

The *Euler–Lagrange equations*,

$$\frac{d}{dt}\left(\frac{\partial L}{\partial \dot{\theta}_i}\right) - \frac{\partial L}{\partial \theta_i} = 0, \quad i = 1, 2,$$

therefore imply the following two coupled equations of motion in which we dropped the factor l^2:

$$(a) \quad (m_1 + m_2)\ddot{\theta}_1 - m_2\ddot{\theta}_2 + (m_1 + m_2)\omega_0^2\theta_1 = 0, \quad (b) \quad m_2\ddot{\theta}_2 - m_2\ddot{\theta}_1 + m_2\omega_0^2\theta_2 = 0. \tag{9.7}$$

(2) Solution of the *secular equation*
With an assumed periodic time dependence $\theta_1, \theta_2 \propto \exp(-i\omega t)$, and rewriting the equations in matrix form, we obtain

$$\sum_{j=1,2} M_{ij}\theta_j \equiv \begin{pmatrix} (m_1 + m_2)(\omega_0^2 - \omega^2) & m_2\omega^2 \\ m_2\omega^2 & m_2(\omega_0^2 - \omega^2) \end{pmatrix}\begin{pmatrix} \theta_1 \\ \theta_2 \end{pmatrix} = 0. \tag{9.8}$$

We observe that the matrix is symmetric. As explained in Sec. 8.3.2 (Eq. (8.36a)), two coupled equations like this have solutions only for a vanishing of the determinant of M, *i.e.*

$$\begin{vmatrix} (m_1 + m_2)(\omega_0^2 - \omega^2) & m_2\omega^2 \\ m_2\omega^2 & m_2(\omega_0^2 - \omega^2) \end{vmatrix} = 0.$$

Thus

$$m_2(m_1 + m_2)(\omega_0^2 - \omega^2)^2 - m_2^2\omega^4 = 0, \quad (\omega_0^2 - \omega^2)^2 - \frac{m_2}{m_1 + m_2}\omega^4 = 0,$$

or

$$\left(\omega_0^2 - \omega^2 - \sqrt{\frac{m_2}{m_1 + m_2}}\omega^2\right)\left(\omega_0^2 - \omega^2 + \sqrt{\frac{m_2}{m_1 + m_2}}\omega^2\right) = 0.$$

The roots ω_+, ω_- are therefore given by

$$\omega_\pm^2 = \frac{\omega_0^2}{1 \pm \sqrt{m_2/(m_1 + m_2)}}, \quad \omega_+^2\omega_-^2 = \left(\frac{m_1 + m_2}{m_1}\right)\omega_0^4, \quad (\omega_0^2 - \omega_\pm^2) = \pm\omega_\pm^2\sqrt{\frac{m_2}{m_1 + m_2}}. \tag{9.9}$$

We observe that if the lower mass m_2 is very small the two resonance frequencies $\omega_\pm$ are almost equal. By solving the equations (9.8) for each of $\omega = \omega_+, \omega_-$ we obtain the ratio of the amplitudes θ_1/θ_2. Consider the case of $\omega = \omega_+$. In this case we have the two equations

$$(m_1 + m_2)(\omega_0^2 - \omega_+^2)\theta_1 + m_2\omega_+^2\theta_2 = 0, \quad m_2\omega_+^2\theta_1 + m_2(\omega_0^2 - \omega_+^2)\theta_2 = 0.$$

Thus we have two equations for one and the same ratio. We now check that these ratios are the same. The first equation yields the following ratio into which we insert the third of Eqs. (9.9). Then

$$\omega^2 = \omega_+^2 : \quad \frac{\theta_1}{\theta_2} = -\frac{m_2\omega_+^2}{(m_1 + m_2)(\omega_0^2 - \omega_+^2)} \overset{(9.9)}{=} -\frac{m_2\sqrt{m_1 + m_2}}{(m_1 + m_2)\sqrt{m_2}} = -\sqrt{\frac{m_2}{m_1 + m_2}}. \tag{9.10}$$

Similarly we obtain from the second equation

$$\frac{\theta_1}{\theta_2} = -\frac{m_2(\omega_0^2 - \omega_+^2)}{m_2\omega_+^2} \overset{(9.9)}{=} -\sqrt{\frac{m_2}{m_1 + m_2}}.$$

For $\omega^2 = \omega_-^2$ we proceed similarly and obtain

$$\omega^2 = \omega_-^2 : \quad \frac{\theta_1}{\theta_2} = +\sqrt{\frac{m_2}{m_1 + m_2}}. \tag{9.11}$$

We now return to Eqs. (9.7) (re-inserting the factor l^2 we dropped there), and rewrite these in matrix form, thereby separating the kinetic energy part T and the potential energy part V with symmetric matrices $(T_{ij}), (V_{ij})$. Then

$$\sum_{j=1,2} T_{ij}\ddot{\theta}_j + \sum_{j=1,2} V_{ij}\theta_j = 0, \tag{9.12}$$

$$T\ddot{\theta} = \begin{pmatrix} (m_1 + m_2)l^2 & -m_2l^2 \\ -m_2l^2 & m_2l^2 \end{pmatrix}\begin{pmatrix} \ddot{\theta}_1 \\ \ddot{\theta}_2 \end{pmatrix}, \quad V\theta = \begin{pmatrix} (m_1 + m_2)l^2\omega_0^2 & 0 \\ 0 & m_2l^2\omega_0^2 \end{pmatrix}\begin{pmatrix} \theta_1 \\ \theta_2 \end{pmatrix}. \tag{9.13}$$

Our aim is to obtain the *normal modes of oscillation* (or vibration). The crucial point to realize in the present example is that the diagonalization of two quadratic forms required here necessitates a somewhat different treatment of T and V. This is why we separate these first. Now inserting into Eq. (9.12) (with an additional normalization factor C)

$$\theta_j = Ca_j e^{-i\omega t}, \tag{9.14}$$

(only the real part proportional to $\cos \omega t$ being physically meaningful), we obtain (with the determinant of the linear operator vanishing as for Eq. (9.8))

$$\sum_j (V_{ij}a_j - \omega^2 T_{ij}a_j) = 0, \quad \text{or} \quad \sum_j (V_{ij} - \omega^2 T_{ij})a_j = 0. \tag{9.15}$$

We now consider first the kinetic energy T. This is given by the expression

$$
T = \frac{1}{2}(\dot{\theta}_1, \dot{\theta}_2) \begin{pmatrix} (m_1 + m_2)l^2 & -m_2 l^2 \\ -m_2 l^2 & m_2 l^2 \end{pmatrix} \begin{pmatrix} \dot{\theta}_1 \\ \dot{\theta}_2 \end{pmatrix}
$$

$$
= \frac{1}{2}(m_1 + m_2)l^2 \dot{\theta}_1^2 + \frac{1}{2} m_2 l^2 \dot{\theta}_2^2 - m_2 l^2 \dot{\theta}_1 \dot{\theta}_2. \tag{9.16}
$$

We write the second of Eqs. (9.15) in the following form which demonstrates its form as an *eigenvalue equation* with eigenvalues $\omega_\pm^2 \equiv \lambda_\pm$:

$$
\sum_j V_{ij} a_j = \lambda \sum_j T_{ij} a_j, \qquad Va = \lambda Ta, \qquad \lambda \equiv \omega^2. \tag{9.17}
$$

Here V is real and symmetrical, from which — in a general case — one can conclude (*cf.* below or see Goldstein [19], Sec. 10.2) that the eigenvalues λ_k are real. We make the following vital observation: Different from ordinary eigenvalue problems, the action of V on a does not produce a multiple of a, but that of T applied to a. The matrix $A = (a_{ij})$ will be shown to diagonalize T as well as V, however in a different sense: T is diagonalized to a unit matrix, and V to a matrix whose diagonal elements are ω_+^2 and ω_-^2. To this end we first attach to the two-component vector a a superscript to form a^+ or a^-, depending on whether $\lambda = \lambda_+$ or λ_- respectively. Then

$$
\sum_{j=1,2} V_{ij} a_j^+ = \lambda_+ \sum_{j=1,2} T_{ij} a_j^+, \quad \text{and} \quad \sum_{j=1,2} V_{ij} a_j^- = \lambda_- \sum_{j=1,2} T_{ij} a_j^-. \tag{9.18}
$$

(3) Diagonalization of T and V

We add to a_j^+ another index k to form a matrix A consisting of the two column vectors a_j^+, a_j^-, i.e.

$$
A \equiv (a_{ij}) = \begin{pmatrix} a_1^+ & a_1^- \\ a_2^+ & a_2^- \end{pmatrix} \equiv \begin{pmatrix} a_{11} & a_{12} \\ a_{21} & a_{22} \end{pmatrix}. \tag{9.19}
$$

We can thus combine the equations of (9.18) in the single equation

$$
\sum_{m=1,2} V_{im} a_{mk} = \lambda_k \sum_{m=1,2} T_{im} a_{mk}, \qquad \lambda_{k=1,2} = \lambda_\pm, \tag{9.20}
$$

so that

$$
\sum_m V_{im} a_{m1} = \lambda_1 \sum_m T_{im} a_{m1}, \qquad \sum_j V_{ij} \underbrace{a_{j1}}_{a_j^+} = \lambda_+ \sum_j T_{ij} \underbrace{a_{j1}}_{a_j^+},
$$

and correspondingly for $k = 2$. The complex conjugate of Eq. (9.20) is (with $k \to l$ and recalling that T_{ij} and V_{ij} are symmetric)

$$
\sum_{m=1,2} V_{im} a_{ml}^* = \lambda_l^* \sum_{m=1,2} T_{im} a_{ml}^* \quad \text{or} \quad \sum_m a_{lm}^{*T} V_{mi} = \lambda_l^* \sum_m a_{lm}^{*T} T_{mi}. \tag{9.21}
$$

Multiplying Eq. (9.20) from the left by by a_{li}^{*T} and summing over i and Eq. (9.21) from the right by a_{ik} and summing over i, and then subtracting the latter equation from the former, we obtain

$$
0 = (\lambda_k - \lambda_l^*) \sum_{i,m} a_{lm}^{*T} T_{mi} a_{ik}, \qquad (T_{ij}) \overset{(9.13)}{=} l^2 \begin{pmatrix} m_1 + m_2 & -m_2 \\ -m_2 & m_2 \end{pmatrix}. \tag{9.22}
$$

We fix the eigenvectors θ (recall the normalization factor C in Eq. (9.14)) by demanding that

$$
\tilde{T}_{lk} \equiv \sum_{i,j} a_{li}^{*T} T_{ij} a_{jk} = \delta_{lk}, \qquad \tilde{T} = A^{*T} T A = \mathbb{1}_{2\times2}. \tag{9.23}
$$

This equation implies the diagonalization of the kinetic energy matrix (T_{ij}) to a unit matrix and concurrently provides a normalization of the vectors with components a_{jk}; the equation also implies that the eigenvalues λ_k are real. From Eqs. (9.10), (9.11) we deduce for the ratio of the elements $a_{1,2}^{\pm}$:

$$\frac{\theta_1^{\pm}}{\theta_2^{\pm}} = \mp\sqrt{\frac{m_2}{m_1 + m_2}} = \frac{a_1^{\pm}}{a_2^{\pm}}. \tag{9.24}$$

We add here parenthetically that if one of the equations of motion (9.7) is inserted into the other, one obtains

$$\ddot{\theta}_1 + \left[\frac{m_1 + m_2}{m_1} + \frac{m_2}{m_1}\left(\frac{\theta_2}{\theta_1}\right)\right]\omega_0^2\theta_1 = 0, \quad \ddot{\theta}_2 + \left[\frac{m_1 + m_2}{m_1}\left(1 + \frac{\theta_1}{\theta_2}\right)\right]\omega_0^2\theta_2 = 0.$$

With θ_1/θ_2 as in Eq. (9.24) one obtains from these equations $\theta_1, \theta_2 \propto \exp[\pm i\omega_{\pm}t]$.

With the help of Eq. (9.23) we obtain the normalized expressions of $a_{1,2}^{\pm}$ individually. From Eqs. (9.10), (9.11) we obtain for the ratios of $a_i^{\pm}$:

$$\frac{a_1^+}{a_2^+} = -\sqrt{\frac{m_2}{m_1 + m_2}}, \quad \frac{a_1^-}{a_2^-} = +\sqrt{\frac{m_2}{m_1 + m_2}}. \tag{9.25}$$

The matrix A of Eq. (9.19) therefore becomes

$$A \equiv (a_{ij}) = \begin{pmatrix} a_1^+ & a_1^- \\ a_2^+ & a_2^- \end{pmatrix} = \begin{pmatrix} a_1^+ & a_1^- \\ -\sqrt{\frac{m_1+m_2}{m_2}}a_1^+ & \sqrt{\frac{m_1+m_2}{m_2}}a_1^- \end{pmatrix}. \tag{9.26}$$

We evaluate Eq. (9.23) for the four possible cases, inserting for T_{ij} the elements given in Eq. (9.22). In the first case with $l = k = 1$ we obtain

$$1/l^2 = a_{11}^{*T}(m_1 + m_2)a_{11} + a_{12}^{*T}(-m_2)a_{11} + a_{11}^{*T}(-m_2)a_{21} + a_{12}^{*T}(m_2)a_{21}.$$

Inserting the expressions for a_{ij} given in Eq. (9.26), we obtain

$$1/l^2 = (a_1^+)^2(m_1 + m_2) - \sqrt{\frac{m_1 + m_2}{m_2}}a_1^+(-m_2)a_1^+ + a_1^+(-m_2)\left(-\sqrt{\frac{m_1 + m_2}{m_2}}\right)a_1^+$$
$$+ \left(-\sqrt{\frac{m_1 + m_2}{m_2}}a_1^+\right)m_2\left(-\sqrt{\frac{m_1 + m_2}{m_2}}a_1^+\right)a_1^+,$$
$$1/l^2 = 2[(m_1 + m_2) + \sqrt{m_2(m_1 + m_2)}](a_1^+)^2$$
$$(a_1^+)^2 = \frac{1}{2m_1l^2}\left[1 - \sqrt{\frac{m_2}{m_1 + m_2}}\right] \stackrel{(9.9)}{=} \frac{\omega_0^2}{2m_1l^2\omega_-^2} \stackrel{(9.9)}{=} \frac{1}{2l^2}\frac{\omega_+^2}{(m_1 + m_2)\omega_0^2}. \tag{9.27}$$

In the next case with $l = 1, k = 2$ we verify similarly that the sum vanishes. Inserting these values into the sum of Eq. (9.23) we obtain:

$$\sum_{i,j} a_{1i}^{*T}T_{ij}a_{j2} = a_{11}^{*T}T_{11}a_{12} + a_{12}^{*T}T_{21}a_{12} + a_{11}^{*T}T_{12}a_{22} + a_{12}^{*T}T_{22}a_{22}$$

$$= l^2\left[a_1^+(m_1 + m_2)a_1^- + \left(-\sqrt{\frac{m_1 + m_2}{m_2}}a_1^+\right)(-m_2)a_1^-\right.$$
$$+ a_1^+(-m_2)\left(\sqrt{\frac{m_1 + m_2}{m_2}}a_1^-\right)$$
$$+ \left.\left(-\sqrt{\frac{m_1 + m_2}{m_2}}a_1^+\right)m_2\sqrt{\frac{m_1 + m_2}{m_2}}a_1^-\right] = 0.$$

The same result follows for $l = 2, k = 1$. Similarly for $l = k = 2$:

$$\sum_{i,j} a_{2i}^{*T} T_{ij} a_{j2} = 2l^2 [(m_1 + m_2) - \sqrt{m_2(m_1 + m_2)}](a_1^-)^2,$$

$$(a_1^-)^2 = \frac{1}{2m_1 l^2} \left[1 + \sqrt{\frac{m_2}{m_1 + m_2}} \right] = \frac{\omega_0^2}{2m_1 l^2 \omega_+^2}. \tag{9.28}$$

With Eq. (9.24) we obtain

$$(a_2^\pm)^2 = \frac{m_1 + m_2}{m_2} (a_1^\pm)^2 = \frac{m_1 + m_2}{m_2} \frac{\omega_0^2}{2m_1 l^2 \omega_\mp^2} \overset{(9.9)}{=} \frac{\omega_\pm^2}{2m_2 l^2 \omega_0^2}. \tag{9.29}$$

The next step is the diagonalization of the potential, *i.e.* the evaluation of the left hand side of Eq. (9.17) when multiplied from the left by A^{*T}, or the evaluation of

$$\tilde{V} := A^{*T} V A, \quad \tilde{V}_{lk} = \sum_{i,j} a_{li}^{*T} V_{ij} a_{jk}, \quad V = \begin{pmatrix} (m_1 + m_2) l^2 \omega_0^2 & 0 \\ 0 & m_2 l^2 \omega_0^2 \end{pmatrix}. \tag{9.30}$$

Again we show how the evaluation proceeds. We begin with the case of $l = k = 1$. In this case we have

$$\tilde{V}_{11} = a_{11}^{*T} V_{11} a_{11} + a_{12}^{*T} V_{22} a_{21} \overset{(9.26)}{=} V_{11} (a_1^+)^2 + \frac{m_1 + m_2}{m_2} V_{22} (a_1^+)^2$$

$$= 2(m_1 + m_2) l^2 \omega_0^2 (a_1^+)^2 \overset{(9.27)}{=} \omega_+^2. \tag{9.31}$$

Similarly we have

$$\tilde{V}_{22} = a_{21}^{*T} V_{11} a_{12} + a_{22}^{*T} V_{22} a_{22} = V_{11} (a_1^-)^2 + \frac{m_1 + m_2}{m_2} V_{22} (a_1^-)^2$$

$$\overset{(9.28)}{=} \frac{\omega_0^2 (m_1 + m_2) l^2 \omega_0^2}{2m_1 l^2 \omega_+^2} + \left(\frac{m_1 + m_2}{m_2} \right) \frac{m_2 l^2 \omega_0^4}{2m_1 l^2 \omega_+^2}$$

$$\overset{(9.27)}{=} \frac{1}{2} \omega_-^2 + \frac{1}{2} \omega_-^2 = \omega_-^2. \tag{9.32}$$

We next verify our results. Equation (9.20) multiplied from the left by a_{li}^{*T} and summed over i implies the equation

$$\sum_{i,m} a_{li}^{*T} V_{im} a_{mk} = \lambda_k \sum_{i,m} a_{li}^{*T} T_{im} a_{mk}. \tag{9.33}$$

In matrix form, *i.e.* for $l, k = 1, 2$, this equation is seen to be the expected identity

$$\tilde{V} = \begin{pmatrix} \omega_+^2 & 0 \\ 0 & \omega_-^2 \end{pmatrix} = \begin{pmatrix} \lambda_1 = \omega_+^2 & 0 \\ 0 & \lambda_2 = \omega_-^2 \end{pmatrix}. \tag{9.34}$$

(4)   Reduction of L to *normal form* (*i.e.* principal axes)

We now come to what we set out to achieve: To express the Lagrangian L in *normal form*, *i.e.* without mixed terms. This transformation of L corresponds precisely to those steps in our treatment of rigid bodies which lead from the original *ellipsoid of inertia* to that in normal form, *i.e.* the steps from Eq. (8.26) to Eq. (8.27), and is the procedure of transformation to principal axes. We rewrite the Lagrangian $L = T - V$ of Eq. (9.6) (ignoring the constant V_0) in matrix form as

$$L = \frac{1}{2} (\dot{\theta}_1, \dot{\theta}_2) \begin{pmatrix} (m_1 + m_2) l^2 & -m_2 l^2 \\ -m_2 l^2 & m_2 l^2 \end{pmatrix} \begin{pmatrix} \dot{\theta}_1 \\ \dot{\theta}_2 \end{pmatrix}$$

$$- \frac{1}{2} (\theta_1, \theta_2) \begin{pmatrix} (m_1 + m_2) l^2 \omega_0^2 & 0 \\ 0 & m_2 l^2 \omega_0^2 \end{pmatrix} \begin{pmatrix} \theta_1 \\ \theta_2 \end{pmatrix}. \tag{9.35}$$

This Lagrangian contains the mixed or product terms $\dot{\theta}_1\dot{\theta}_2$. We use the matrix A defined by Eq. (9.19) and insert at matrix junctions unit matrix factors AA^{-1}, and define

$$A^{-1}\begin{pmatrix} \theta_1 \\ \theta_2 \end{pmatrix} \equiv \begin{pmatrix} \theta'_1 \\ \theta'_2 \end{pmatrix} \equiv \theta', \quad (\theta_1, \theta_2)(A^{-1})^T = (\theta'_1, \theta'_2) = \theta'^T. \tag{9.36}$$

The Lagrangian L therefore becomes ($\tilde{T}$ given by Eq. (9.23))

$$
\begin{aligned}
L &= \frac{1}{2}(\dot{\theta}_1, \dot{\theta}_2)(A^T)^{-1}A^T \begin{pmatrix} (m_1+m_2)l^2 & -m_2 l^2 \\ -m_2 l^2 & m_2 l^2 \end{pmatrix} AA^{-1}\begin{pmatrix} \dot{\theta}_1 \\ \dot{\theta}_2 \end{pmatrix} \\
&\quad - \frac{1}{2}(\theta_1, \theta_2)(A^T)^{-1}A^T \begin{pmatrix} (m_1+m_2)l^2\omega_0^2 & 0 \\ 0 & m_2 l^2\omega_0^2 \end{pmatrix} AA^{-1}\begin{pmatrix} \theta_1 \\ \theta_2 \end{pmatrix} \\
&= \frac{1}{2}(\dot{\theta}_1, \dot{\theta}_2)(A^T)^{-1}\tilde{T}A^{-1}\begin{pmatrix} \dot{\theta}_1 \\ \dot{\theta}_2 \end{pmatrix} - \frac{1}{2}(\theta_1, \theta_2)(A^T)^{-1}\tilde{V}A^{-1}\begin{pmatrix} \theta_1 \\ \theta_2 \end{pmatrix} \\
&= \frac{1}{2}\dot{\theta}'^T\tilde{T}\dot{\theta}' - \frac{1}{2}\theta'^T\tilde{V}\theta' \\
&= \frac{1}{2}[(\dot{\theta}'_1)^2 + (\dot{\theta}'_2)^2] - \frac{1}{2}[\omega_1^2(\theta'_1)^2 + \omega_2^2(\theta'_2)^2]. \tag{9.37}
\end{aligned}
$$

Here $\omega_1 = \omega_+, \omega_2 = \omega_-$. This is now the Lagrange function of two independent harmonic oscillators, or the sum of the Lagrangians of these. We obtain their equations of motion from $L(\theta'_1, \theta'_2, \dot{\theta}'_1, \dot{\theta}'_2)$ as

$$\frac{d}{dt}\left(\frac{\partial L}{\partial \dot{\theta}'_i}\right) - \frac{\partial L}{\partial \theta'_i} = 0, \quad \ddot{\theta}'_i + \omega_i^2\theta'_i = 0, \quad i = 1, 2, \text{ with } \theta'_i = c_i e^{-i\omega_i t}. \tag{9.38}$$

Each of these new coordinates is therefore a simple periodic function of only one specific *"resonance frequency"* ω_i. The coordinates θ'_i are therefore referred to as *"normal coordinates"* corresponding to a specific normal mode of oscillation of the system.

Finally we observe that the total energy E is given by the sum of positive quadratic forms:

$$E = \frac{1}{2}[(\dot{\theta}'_1)^2 + (\dot{\theta}'_2)^2] + \frac{1}{2}[\omega_1^2(\theta'_1)^2 + \omega_2^2(\theta'_2)^2]. \tag{9.39}$$

Thus for E to be conserved, the minimum of any term here requires an appropriate maximum of the other terms. This explains the continual periodic shift of oscillatory motion of one mass to that of the other and back (since no friction is assumed).

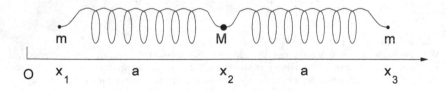

Fig. 9.2 The linear molecule consisting of three atoms.

Example 9.2: The linear molecule consisting of three atoms

A linear molecule like CO_2 consisting of three atoms as depicted in Fig. 9.2 has two atoms of the same mass m arranged symmetrically on either side of the third atom of mass M. In its state of equilibrium the atoms of mass m are a distance a away from the central atom. Considering only longitudinal motion along the axis and forces given by *Hooke's law* (shown as springs in Fig. 9.2, with spring constant k), determine the resonance frequencies of the molecule and its normal modes of vibration.[‡] How does this case differ from one with the central atom attached to a similar spring whose other end is fixed at the origin?

Solution: With coordinates x_1, x_2, x_3 of the atoms as shown in Fig. 9.2, the nonequilibrium extensions of the springs are $(x_2 - x_1 - a)$ and $(x_3 - x_2 - a)$ and hence the potential is given by

$$V = \frac{k}{2}(x_2 - x_1 - a)^2 + \frac{k}{2}(x_3 - x_2 - a)^2.$$

This is the expression for the case when the central atom is free to move; in the case when the latter is attached to a spring and this to the origin, a contribution $k(x_2 - x_{02})^2$ has to be added, where x_{02} is the equilibrium value of x_2. In the following we continue with the first case and point out changes in the case of the latter. In equilibrium $a \equiv x_{02} - x_{01} = x_{03} - x_{02}$. We set $\eta_i = x_i - x_{0i}$, which defines the coordinates of the atoms with respect to their equilibrium positions. Then (for mass M fixed a contribution $k\eta_2^2/2$ has to be added)

$$
\begin{aligned}
V &= \frac{k}{2}(\eta_2 - \eta_1)^2 + \frac{k}{2}(\eta_3 - \eta_2)^2 = \frac{k}{2}(\eta_1^2 + 2\eta_2^2 + \eta_3^2 - 2\eta_1\eta_2 - 2\eta_2\eta_3) \\
&- \frac{1}{2}(\eta_1 \ \eta_2 \ \eta_3)\begin{pmatrix} k & -k & 0 \\ -k & 2k & -k \\ 0 & -k & k \end{pmatrix}\begin{pmatrix} \eta_1 \\ \eta_2 \\ \eta_3 \end{pmatrix}.
\end{aligned}
\tag{9.40}
$$

The kinetic energy T is, written similarly,

$$T = \frac{m}{2}(\dot{\eta}_1^2 + \dot{\eta}_3^2) + \frac{M}{2}\dot{\eta}_2^2 = \frac{1}{2}(\dot{\eta}_1 \ \dot{\eta}_2 \ \dot{\eta}_3)\begin{pmatrix} m & 0 & 0 \\ 0 & M & 0 \\ 0 & 0 & m \end{pmatrix}\begin{pmatrix} \dot{\eta}_1 \\ \dot{\eta}_2 \\ \dot{\eta}_3 \end{pmatrix} \equiv \frac{1}{2}\dot{\eta}^T \tilde{T}\dot{\eta}. \tag{9.41}$$

The Lagrangian is $L = T - V$ and the equations of motion, assuming a time dependence $\eta_i \propto \exp(-i\omega t)$, are

$$m\ddot{\eta}_1 + k(\eta_1 - \eta_2) = 0, \quad M\ddot{\eta}_2 + k(2\eta_2 - \eta_1 - \eta_3) = 0, \quad m\ddot{\eta}_3 + k(\eta_3 - \eta_2) = 0, \quad \text{or}$$

$$M(\omega)\eta = 0, \quad M(\omega) = \begin{pmatrix} k - m\omega^2 & -k & 0 \\ -k & 2k - M\omega^2 & -k \\ 0 & -k & k - m\omega^2 \end{pmatrix}, \tag{9.42}$$

where in the case of mass M fixed the central element is $3k - M\omega^2$. As in the preceding example the *secular equation* of the equations of motion is the determinant of the matrix M, *i.e.*[§]

$$0 = \det(M) = (k - m\omega^2)[(2k - M\omega^2)(k - m\omega^2) - 2k^2] = \omega^2(k - m\omega^2)[mM\omega^2 - k(M + 2m)]. \tag{9.43}$$

The roots of this equation are

$$\omega_1 = 0, \quad \omega_2 = \sqrt{\frac{k}{m}}, \quad \omega_3 = \sqrt{\frac{k}{m}\left(1 + \frac{2m}{M}\right)}, \tag{9.44}$$

[‡]This is the example discussed by H. Goldstein [19] in Sec. 10.4 of his book and in his Example 3.
[§]Recall the signs arising in the evaluation of a determinant:

$$\begin{vmatrix} a_{11} & a_{12} & a_{13} \\ a_{21} & a_{22} & a_{23} \\ a_{31} & a_{32} & a_{33} \end{vmatrix} = a_{11}\begin{vmatrix} a_{22} & a_{23} \\ a_{32} & a_{33} \end{vmatrix} - a_{12}\begin{vmatrix} a_{21} & a_{23} \\ a_{31} & a_{33} \end{vmatrix} + a_{13}\begin{vmatrix} a_{21} & a_{22} \\ a_{31} & a_{32} \end{vmatrix}.$$

and in the case of mass M attached to the origin the roots are

$$\omega_1 = \sqrt{\frac{k}{m}}, \quad \omega_{2,3} = \frac{2k}{(3m + M) \pm \sqrt{9m^2 + M^2 + 2mM}}.$$

The zero eigenvalue with eigenfunction called 'zero mode' does not arise unexpectedly. If we take another look at the Lagrangian $L(\eta_i, \dot\eta_i)$ and replace there η_i by $\eta_i - b$, where b is a constant translational shift, we see that L remains unchanged, *i.e.* is invariant under such a shift, like also the equations of motion. This means under such a shift the energy remains unchanged. Zero modes of similar types occur in many areas of physics, *e.g.* in quantum mechanics and field theory. We see that the zero mode does not arise once the translational invariance is violated, as in the case of the mass M attached to the origin.

Next we determine the eigenvectors η_i of the equation $M(\omega_i)\eta_i = 0$ for each of the three eigenvalues $\omega_1, \omega_2, \omega_3$. This means we set $\eta_i = (a_{1i}, a_{2i}, a_{3i})$ and thus have to solve the set of simultaneous equations

$$\begin{pmatrix} k - m\omega_i^2 & -k & 0 \\ -k & 2k - M\omega_i^2 & -k \\ 0 & -k & k - m\omega_i^2 \end{pmatrix} \begin{pmatrix} a_{1i} \\ a_{2i} \\ a_{3i} \end{pmatrix} = 0.$$

The vectors η_i are again normalized as in Eq. (9.23) with the matrix $\tilde T$ in T, *i.e.* by

$$1 = \eta^T \tilde T \eta = (a_{1i}\ a_{2i}\ a_{3i}) \begin{pmatrix} m & 0 & 0 \\ 0 & M & 0 \\ 0 & 0 & m \end{pmatrix} \begin{pmatrix} a_{1i} \\ a_{2i} \\ a_{3i} \end{pmatrix} = m(a_{1i}^2 + a_{3i}^2) + M a_{2i}^2.$$

Inserting the first eigenvalue $\omega_1 = 0$ in these equations one obtains $a_{11} = a_{21} = a_{31}$ and hence

$$a_{11} = a_{21} = a_{31} = \frac{1}{\sqrt{2m + M}}.$$

In the other two cases the corresponding calculations yield:

$$a_{12} = \frac{1}{\sqrt{2m}}, \quad a_{22} = 0, \quad a_{32} = -\frac{1}{\sqrt{2m}},$$

where the equation $a_{22} = 0$ implies that the central atom remains at rest, and

$$a_{13} = \left[2m \left(1 + \frac{2m}{M} \right) \right]^{-1/2} = a_{33}, \quad a_{23} = -\frac{2}{\sqrt{2M(2 + M/m)}}.$$

In this case the two external atoms oscillate with the same amplitude (as expected in view of the symmetry of the molecule), whereas the central atom has a different amplitude. Further discussion of this example can be found in the book of Goldstein [19], Sec. 10.4.

9.3   Stability

The topic of stability is of considerable importance as one realizes immediately by thinking, for instance, of a bus driving along a sloping roadway in the mountains. Clearly it is vital to construct the bus such that it does not topple. One distinguishes between *stable*, *unstable* and *neutral* equilibrium, as illustrated in the case of a solid cone in Fig. 9.3. In each case the cone

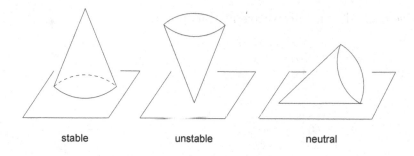

stable   unstable   neutral

Fig. 9.3 The three types of equilibrium.

is minimally displaced from the shown configuration. Stability means the
cone returns to this original configuration, instability means it does not return to that configuration, and neutral equilibrium means it does neither. In
the case of several degrees of freedom instability is given as soon as this is
provided by a single degree of freedom; this case is illustrated by the *saddle
configuration* in Fig. 9.4 compared with the absolutely stable configuration.

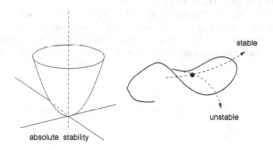

stable

unstable

absolute stability

Fig. 9.4 Instability in the case of more than one degree of freedom.

The principle of virtual work described in Sec. 3.3 equates to zero the
net work done by the acting forces under an infinitesimal displacement of the
system. If there is only one degree of freedom and if the force is conservative,
i.e. expressible as the first derivative of a potential, then the vanishing of this
multiplied by the displacement implies the vanishing of this first derivative, so
that the potential $V(x)$ (say) is an extremum there (note that x can also be an
angle). Thus if $dV(x)/dx = 0$ at $x_1, x_2, \ldots$, these positions define equilibrium
configurations of the system. We then form d^2V/dx^2. If this expression is
positive at $x = x_1$, then this value of x is a stable equilibrium position
or configuration, if it is negative an unstable position. This is indicated
graphically in Fig. 9.5. In the case of several variables $x, y, \ldots$, the potential
is $V(x, y, \ldots)$, and partial derivatives have to be formed. A configuration will
then be stable and define the *absolute minimum*, only if there is no single

negative second order derivative at that point.

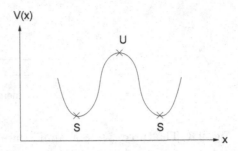

Fig. 9.5 Positions of stable (S) and unstable (U) equilibrium.

One can summarize the conditions of stability in the two statements: There must be no resultant force acting on the body, otherwise there would ensue translational motion, and there must be no resulting torque acting on the body, otherwise there would be rotation.

Example 9.3: Stability of a bus

A double-decker bus has a wheel span of $2b = 1.5$ meters. A school class is boarding the bus and fills the upper deck, leaving the lower deck empty. We assume one can treat this case as a symmetrical rigid body, and that the centre of mass of the bus is $l = 1.5$ meters above the ground. Ignoring other effects, find the greatest inclination θ of the road tolerable for a safe driving of the bus.

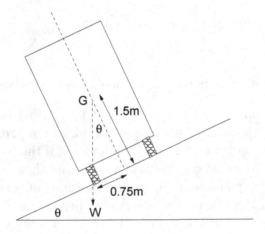

Fig. 9.6 A bus on a sloping roadway.

Solution: For a safe driving the axles of the bus have to be parallel to the road as indicated in Fig. 9.6 (the bus is on the point of tilting when the vertical line through the centre of gravity just passes through the left-hand edge of the left-hand tyre). From Fig. 9.6 we deduce:

$$\tan\theta = \frac{b}{l} = \frac{0.75}{1.5} = \frac{3 \times 2}{4 \times 3} = \frac{1}{2}, \quad \theta \simeq 26.5°.$$

This is approximately the angle for London busses. That this is the critical angle θ_c can be determined by considering the height $\bar{y}$ of the centre of mass G above the ground, here above the lowest point of the left-hand tyre:

$$\bar{y} = l\cos\theta + b\sin\theta, \quad \bar{y}' = -l\sin\theta + b\cos\theta, \quad \bar{y}'(\theta_c) = 0, \quad \tan\theta_c = \frac{b}{l}.$$

Example 9.4: The solid cone in the hole of a table

A solid cone with circular base and of height h and angle 2α at its top as indicated in Fig. 9.7 is placed with its apex into the frictionless circular hole of the plate of a table. The hole has radius a and $a < 3h\sin(2\alpha)/16$. Determine the angle between the symmetry axis of the cone and the plate of the table in the position of stable equilibrium of the cone.

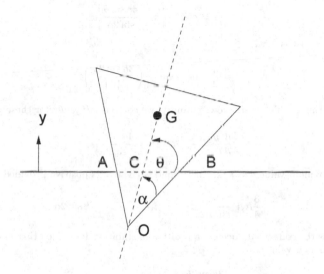

Fig. 9.7 The solid cone in the hole of a table.

Solution: The location of the centre of mass G of the cone was determined in Example 8.6(a) and was found to be at distance $3h/4$ from its top. The equilibrium position is determined by the location of the maxima and minima of the potential. A position of equilibrium is stable if after a small displacement the object returns to its original position. The position is unstable if after the small displacement the object assumes a different position. In Fig. 9.5 the curve depicts a potential in the gravitational field of the Earth. Points U are positions of unstable equilibrium, points S are points of stable equilibrium.

We have to determine the height $\bar{y}$ of the centre of mass G above the plate of the table. For this reason we have to determine first the distance OC in Fig. 9.7. In view of the circular symmetry of the cone and the hole, we can treat it by considering its planar cross section. Thus in Fig. 9.7 AOB is a planar triangle. We recall the theorem from High School Euclidean geometry which says that the ratio of two sides of a triangle is equal to the ratio of the sines of their opposite angles.¶ Using this theorem we obtain from Fig. 9.7, where $AB = 2a$:

$$\frac{OC}{AB} = \frac{OC}{OA}\frac{OA}{AB} = \frac{\sin(\theta+\alpha)}{\sin\theta}\frac{\sin(\theta-\alpha)}{\sin(2\alpha)},$$

¶The reader who wants to reassure himself of this theorem is advised to draw the perpendicular from one vertex of the triangle to its opposite side and then to express the ratio of two sides of the triangle in terms of sines and tangents of angles.

since we can deduce from the geometry of Fig. 9.7 that

$$\angle CAO = \pi - \alpha - \theta, \quad \angle ABO = \pi - 2\alpha - (\pi - \alpha - \theta) = \theta - \alpha.$$

Thus we obtain

$$OC = \frac{2a \sin(\theta + \alpha) \sin(\theta - \alpha)}{\sin \theta . \sin(2\alpha)} = \frac{a(\cos(2\alpha) - \cos(2\theta))}{\sin(2\alpha) \sin \theta}.$$

It follows that the cone's centre of mass is above the table at

$$\bar{y} = (OG - OC) \sin \theta = \left[\frac{3}{4} h - \frac{a(\cos(2\alpha) - \cos(2\theta))}{\sin(2\alpha) \sin \theta} \right] \sin \theta = \frac{3}{4} h \sin \theta + \frac{a \cos 2\theta}{\sin 2\alpha} - a \cot 2\alpha.$$

The cone's potential energy is proportional to $\bar{y}$. Therefore we differentiate this expression with respect to θ and obtain:

$$\frac{d\bar{y}}{d\theta} = \cos \theta \left[\frac{3}{4} h - \frac{4a \sin \theta}{\sin 2\alpha} \right].$$

This expression vanishes when

$$\theta = \frac{\pi}{2} \quad \text{or} \quad \sin^{-1} \left(\frac{3h \sin 2\alpha}{16a} \right).$$

In order to determine which case allows stability, we compute $d^2\bar{y}/d\theta^2$ for these values:

$$\frac{d^2\bar{y}}{d\theta^2} = -\frac{3}{4} h \sin \theta - \frac{4a \cos 2\theta}{\sin 2\alpha}.$$

For the first extremum value $\theta = \pi/2$, this second derivative is positive provided

$$-\frac{3}{4} h + \frac{4a}{\sin 2\alpha} > 0, \quad i.e. \quad a > \frac{3}{16} h \sin 2\alpha.$$

Thus in this case the cone would occupy a position of stability. If on the other hand, $\sin \theta$ assumes the second extremum value, $i.e.$

$$\sin \theta = \frac{3h \sin 2\alpha}{16a} \leq 1, \quad \text{since} \quad \sin \theta \leq 1,$$

then $d^2\bar{y}/d\theta^2$ is positive when (recall $\cos 2\theta = 1 - 2 \sin^2 \theta$)

$$-\frac{3h \sin \theta}{4} - \frac{4a \cos 2\theta}{\sin 2\alpha} = -\frac{9h^2 \sin 2\alpha}{64a} - \frac{4a}{\sin 2\alpha} \left[1 - \frac{9h^2 \sin^2 2\alpha}{128a^2} \right] > 0,$$

$$i.e. \quad \frac{9h^2 \sin 2\alpha}{64a} > \frac{4a}{\sin 2\alpha}, \quad \text{or} \quad \frac{3h \sin 2\alpha}{16a} > 1.$$

Since this contradicts the condition $\sin \theta \leq 1$ above, stable equilibrium is possible only for $\theta = \pi/2$; other positions are unstable.

Example 9.5: Stability of a toy top

A toy top of uniform density ρ consists of a circular cone of height h and base radius a, and the hemisphere of a solid sphere of radius a attached to its base as depicted in Fig. 9.8. Determine the location of the centre of mass along its symmetry axis. What is the condition of stability of the top with its hemispherical part placed on a plane? For what ratio h/a is this satisfied?

Solution: The base of a cross section of the cone subtends an angle 2α at its apex with half of this angle given by $\tan \alpha = a/h$. We consider a slice of thickness dx of the toy top at a distance x away from the origin O. The cone was already considered in this way in Example 8.6(a), from

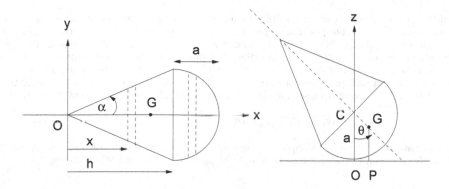

Fig. 9.8 The toy top.

where we obtain the position of its centre of mass as $\bar{x}_{\text{cone}} = 3h/4$. The mass of this conical part of the top is

$$m_{\text{cone}} = \int_0^h \pi y^2 \rho dx = \pi \rho \int_0^h (x \tan \alpha)^2 dx = (\tan \alpha)^2 \pi \rho \frac{h^3}{3} = \frac{\pi \rho h a^2}{3}. \tag{9.45}$$

In dealing with the hemispherical part, we have to keep in mind that the moment of a mass point, and hence the location of the centre of mass of the hemisphere, depends on the location of the origin of coordinates, but the mass of the hemisphere does not. We observe this also in calculating the latter first. Since in the hemispherical region $y^2 = a^2 - (x - h)^2$, we obtain for this

$$m_{\text{hemis}} = \int_{x=h}^{h+a} \pi \rho [a^2 - (x-h)^2] dx = \int_{z=x-h=0}^a \pi \rho [a^2 - z^2] dz = \pi \rho \left[a^2 z - \frac{z^3}{3} \right]_0^a = \frac{2\pi \rho a^3}{3}. \tag{9.46}$$

This is, of course, the result expected as half that of a sphere, and is independent of h. The moment, however, depends on h and we have for this

$$\begin{aligned}
\left(\sum_i m_i x_i \right)_{\text{hemis}} &= \int_{x=h}^{h+a} \pi \rho x [a^2 - (x-h)^2] dx = \int_{z=x-h=0}^a \pi \rho (z+h)[a^2 - z^2] dz \\
&= \pi \rho \left[\frac{a^2 z^2}{2} - \frac{z^4}{4} \right]_0^a + \pi \rho h \left[a^2 z - \frac{z^3}{3} \right]_0^a = \frac{\pi \rho a^4}{4} + \frac{2\pi \rho h a^3}{3}. \tag{9.47}
\end{aligned}$$

The required coordinate of the centre of mass of the entire top is now obtained as (using the result $3a/8$ of Example 8.6(b))

$$\begin{aligned}
\bar{x} &= \frac{\bar{x}_{\text{cone}} m_{\text{cone}} + \bar{x}_{\text{hemis}} m_{\text{hemis}}}{m_{\text{cone}} + m_{\text{hemis}}} \\
&= \frac{(3/4)h(\pi \rho h a^2/3) + (h + 3a/8)(2\pi \rho a^3/3)}{(\pi \rho h a^2/3) + (2\pi \rho a^3/3)} \\
&= \frac{3h^2 + 3a^2 + 8ah}{4(h + 2a)}. \tag{9.48}
\end{aligned}$$

To determine the stability of the top in the gravitational field of the Earth, we consider its potential energy or equivalently again the height of its centre of mass G at $\bar{x}$ above the plane of the table. We infer from Fig. 9.8 that $\bar{z} = a - CG \cos \theta$. The derivative of $\bar{z}$ with respect to θ vanishes for $\theta = 0$. The second derivative at this value of θ is then positive, provided CG is positive. This means, for stability of the top its centre of mass must be located in the hemispherical part. This condition $\bar{x} > h$ then implies $3a^2 > h^2$, or $h < \sqrt{3}a$. We can verify this result, by evaluating the condition on the direction of the resultant moment of forces about the point of support O to turn the top back into a stable position. This time we consider the two parts of the top separately, *i.e.* their weights $W_{\text{hemis}}, W_{\text{cone}}$. Taking the difference of the moments of these weights about the point of support O in Fig. 9.8, and demanding that this difference be positive in the sense to push the top into its position of stability, we obtain (CG_{hemis} and CG_{cone} being the distances of the centres of mass of each part from the point C in Fig. 9.8, the line OC necessarily pasing through the centre of the circular cross section since it is the perpendicular to a tangent to the circle)

$$W_{\text{hemis}} CG_{\text{hemis}} \sin \theta - W_{\text{cone}} CG_{\text{cone}} \sin \theta > 0.$$

Here we have to let the angle θ approach zero, but we can cancel it out just before we do so. Then with CG_{hemis} from Example 8.7 as $3a/8$ and $CG_{\text{cone}} = h - 3h/4 = h/4$, we obtain:

$$m_{\text{cone}} \frac{h}{4} - m_{\text{hemis}} \frac{3a}{8} < 0, \quad \text{or} \quad \frac{m_{\text{cone}}}{m_{\text{hemis}}} < \frac{3a}{2h}.$$

We know from the above that the ratio of the masses is $h/2a$. Hence the condition for stability is $h/2a < 3a/2h$, or $h^2 < 3a^2$. One can construct many similar examples; the case of a circular cylinder attached to the hemisphere is a worked problem for instance in the book of Bullen [8].||

9.4   Miscellaneous Examples

Example 9.6: Equivalence of E and entropy S equilibrium principles

Assuming E is the total energy of a system A in contact with a heat bath R, and S the entropy, show that the equilibrium values of the internal parameters (*e.g.* temperature T) can be determined from either the *minimum energy principle* or the *maximum entropy principle, i.e.*

$$\delta E = 0 \quad \text{with} \quad \delta^2 E > 0, \qquad \text{or} \qquad \delta S = 0 \quad \text{with} \quad \delta^2 S < 0. \tag{9.49}$$

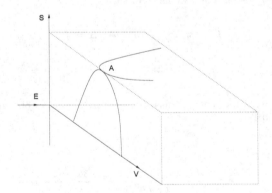

Fig. 9.9   Maximum S, minimum E of equilibrium state A.

Solution: We establish the equivalence of the two principles by showing that if the energy is not minimal, then at equilibrium the entropy can not be maximal. Thus we begin by assuming that at a given maximal entropy the energy E is not minimal. We consider the *second law of thermodynamics,*

$$T dS = dQ = dE + dW. \tag{9.50}$$

|| K. E. Bullen [8], Example 15.82, p. 279.

We keep the entropy constant, so that $dQ = 0$. Since the energy is not that of a minimum, we can lower the energy by an amount dE by allowing the system to perform the amount of work dW (with no change in the entropy — for instance by allowing a piston to slowly push the gas outside, quasistatically and adiabatically). Thereafter we re-establish the original energy of the system by adding the amount of heat dQ. Then the energy of the system is the same as before. However, we have increased the entropy by the amount dQ/T. Also we assumed that the system occupied a state of maximal entropy in its state of equilibrium. Hence our assumption must be wrong, *i.e.* the energy of the system in equilibrium at maximum entropy must be minimal. Hence the equilibrium state follows from both principles. This is what one would expect, as a state of equilibrium is usually connected with a minimum of energy. This equivalence is illustrated in Fig. 9.9 in which the point A represents a state of equilibrium with in one case S maximal at $E = \text{const.}$, and in the other case E minimal at $S = \text{const.}$

Example 9.7: Equation of stability about an extremum

By varying the action $S = \int L(x, \dot{x})dt$ up to the second order obtain the condition for stability around the solution $x_c(t)$ of Newton's equation. Does the conditional equation possess a 'zero mode'?**

Solution: We consider the following Lagrangian of the theory for a one-dimensional configuration space

$$L(x, \dot{x}) \equiv L(x) = \frac{1}{2}m\dot{x}^2 - V(x), \tag{9.51}$$

and consider the variation

$$
\begin{aligned}
\delta L &= L(x + \delta x) - L(x) = \frac{1}{2}m(\dot{x} + \delta\dot{x})^2 - \frac{1}{2}m\dot{x}^2 - V(x + \delta x) + V(x) \\
&\simeq m\dot{x}\frac{d}{dt}\delta x - V'(x)\delta x.
\end{aligned}
$$

Setting this first variation equal to zero, we obtain in the usual way (by shifting d/dt in front of $\dot{x}$ with partial integration in $S = \int dt\, L$) the equation of motion with solution $x(t) \to x_c(t)$,

$$0 = \frac{\delta L}{\delta x}\bigg|_{x_c(t)} = -\frac{d}{dt}m\dot{x}_c(t) - V'(x_c(t)). \tag{9.52}$$

Now varying the following equation further,

$$\frac{\delta L}{\delta x} = -\frac{d}{dt}m\dot{x} - V'(x),$$

we obtain

$$
\begin{aligned}
\frac{\delta^2 L}{\delta x(t')\delta x(t)} &= -\frac{d}{dt}\left(m\frac{\delta}{\delta x(t')}\dot{x}(t)\right) - \frac{\delta V'(x(t))}{\delta x(t')} = -\frac{d}{dt}\left[m\frac{d}{dt}\delta(t - t')\right] - V''(x(t))\delta(t - t') \\
&= -\left[m\frac{d^2}{dt^2} + V''(x(t))\right]\delta(t - t') \equiv M\delta(t - t').
\end{aligned}
$$

Thus

$$\delta^2 L = \delta x(t')M\delta(t - t')\delta x(t), \qquad M(x_c(t)) = -m\frac{d^2}{dt^2} - V''(x_c(t)). \tag{9.53}$$

The equation

$$M\psi_i(t) = \omega_i^2\psi(t), \quad \text{with completeness relation } \delta(t - t') = \sum_i \psi_i(t)\psi_i(t')$$

**This example assumes familiarity with Dirac's delta function.

is the required *equation of stability* with eigenvalues ω_i^2 and eigenfunctions $\psi_i(t)$. Thus

$$\left.\frac{\delta^2 L}{\delta x(t')\delta x(t)}\right|_{x_c(t)} = M(x_c(t))\delta(t-t') = M(x_c(t))\sum_i \psi_i(t)\psi_i(t') = \sum_i \omega_i^2\psi_i(t)\psi_i(t'). \qquad (9.54)$$

This is positive, implying stability with the action a minimum at $x_c(t)$, if $\omega_i^2 > 0$ for all i. Differentiating Newton's equation, *i.e.* the Euler–Lagrange equation (9.52), from the left by t, we obtain

$$0 = -\frac{d}{dt}\left[-\frac{d}{dt}m\dot{x}_c - V'(x_c)\right] = \left[-m\frac{d^2}{dt^2} - V''(x_c)\right]\dot{x}_c(t) = 0. \qquad (9.55)$$

Thus the time derivative of $x_c(t)$ is the eigenfunction of the differential operator $M(x_c(t))$ with eigenvalue $\omega_0^2 = 0$. Thus, apart from a normalization constant, the derivative of the solution of Newton's equation is the zero mode. Put differently this means the *generator of time translations*, *i.e.* d/dt applied to the solution $x_c(t)$, is the zero mode. This zero mode results, as one can show quite generally, as a consequence of the time-translation invariance (*i.e.* under replacements $t \to t + t_0$) of the Lagrangian L and Newton's equation of motion. The *fluctuation equation* or *stability equation*, $(M - \omega_i^2)\psi_i = 0$, is, in general, a complicated equation. It has the form of a standard equation of mathematical physics or of a Schrödinger equation in quantum mechanics. The set of eigenvalues $\{\omega_i^2\}$ would usually also include a continuous set, and thus its further detailed study is beyond the scope of the present topic. A single negative eigenvalue ω_i^2 implies instability and implies a *saddle point* structure of eigenfunctions around $x_c(t)$, as in Fig. 9.4. A zero mode is usually easy to detect in view of its relation to symmetry properties of the Lagrangian. However, if one wants to investigate a problem in more detail, one has to exclude it by the introduction of some constraint. The study of this is again beyond our objectives here.[††]

Example 9.8: Three particles held together by three springs
Three particles of the same mass m are held together in a plane by identical springs of the same spring constant k, and when in equilibrium are each located symmetrically at a distance l away from the origin O. Determine the equations of motion of the particles and the frequencies of their *normal modes* of vibration.[‡‡]

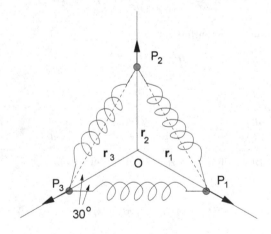

Fig. 9.10 Three particles held together by three springs in a plane.

[††]The reader interested in more details and in specific examples can find these in literature on pseudoparticle configurations, as in H. J. W. Müller–Kirsten [35].

[‡‡]For an analogous problem see *The Physics Coaching Class* [48], problem 1123, p. 206.

Solution: We consider the particles at positions r_1, r_2, r_3 as sketched in Fig. 9.10. In their equilibrium the length of a spring is (with $\cos 30° = \sqrt{3}/2$) $r_{32} = P_3 P_2 = l\sqrt{3}$, and correspondingly for the other 2 springs. Applying *Hooke's law* the potential of particle 3, for instance, is given by

$$V_3 = \frac{1}{2}k(r_3 - r_2 - \sqrt{3}le_{32})^2 + \frac{1}{2}k(r_3 - r_1 - \sqrt{3}le_{31})^2. \tag{9.56}$$

In equilibrium we assume the positions of the three particles to be at $r_1^{(0)}, r_2^{(0)}, r_3^{(0)}$. Their deviations from these equilibrium positions are then given by

$$R_i = r_i - r_i^{(0)}, \quad \text{with} \quad i = 1, 2, 3.$$

It follows that *e.g.*

$$r_3 - r_2 - \sqrt{3}le_{32} = (R_3 - R_2) + \underbrace{(r_3^{(0)} - r_2^{(0)} - \sqrt{3}le_{32})}_{0}.$$

The potential V of the system can therefore be written

$$V = \frac{1}{2}k[(R_3 - R_2)^2 + (R_2 - R_1)^2 + (R_1 - R_3)^2]. \tag{9.57}$$

The kinetic energy T is correspondingly given by

$$T = \frac{1}{2}m[\dot{R}_1^2 + \dot{R}_2^2 + \dot{R}_3^2], \tag{9.58}$$

and the Lagrangian L therefore by

$$L = T - V.$$

The Euler–Lagrange equations or Newton equations are then

$$m\ddot{R}_1 + k(2R_1 - R_2 - R_3) = 0,$$
$$m\ddot{R}_2 + k(2R_2 - R_3 - R_1) = 0,$$
$$m\ddot{R}_3 + k(2R_3 - R_1 - R_3) = 0. \tag{9.59}$$

Assuming a periodic time dependence $R_I \propto \exp(i\omega t)$, the set of these three coupled equations is in matrix form:

$$(M)\begin{pmatrix} R_1 \\ R_2 \\ R_3 \end{pmatrix} \equiv \begin{pmatrix} 2k - m\omega^2 & -k & -k \\ -k & 2k - m\omega^2 & -k \\ -k & -k & 2k - m\omega^2 \end{pmatrix}\begin{pmatrix} R_1 \\ R_2 \\ R_3 \end{pmatrix} = 0. \tag{9.60}$$

The determinant of the matrix M is

$$\begin{aligned} \det(M) &= (2k - m\omega^2)[(2k - m\omega^2)^2 - k^2] + k[k(m\omega^2 - 2k) - k^2] - k[k^2 + k(2k - m\omega^2)] \\ &= m\omega^2[(2k - m\omega^2) + k][m\omega^2 - 3k]. \end{aligned}$$

This determinant is seen to vanish for the values

$$m\omega^2 = 0, 3k, 3k. \tag{9.61}$$

Thus two eigenvalues are equal and hence imply *degeneracy* (different eigenvectors with the same eigenvalue). The eigenvalue zero is that of the eigenvector called *zero mode*, and results from the translational invariance of the Lagrangian and the equations of motion. Inserting $m\omega^2 = 3k$ into the matrix equation, we see that all three equations imply one and the same relation,

$$\mathbf{R}_1 + \mathbf{R}_2 + \mathbf{R}_3 = 0,$$

so that with $\mathbf{R}_1 = 0$ one has $\mathbf{R}_2 = -\mathbf{R}_3$, and with $\mathbf{R}_1 = \mathbf{c}$ one has $\mathbf{R}_2 = \mathbf{R}_3 = -\mathbf{c}/2$. In the case of the zero mode, the matrix equation yields the set

$$\mathbf{R}_1 = \frac{1}{2}(\mathbf{R}_2 + \mathbf{R}_3), \quad \mathbf{R}_2 = \frac{1}{2}(\mathbf{R}_1 + \mathbf{R}_3), \quad \mathbf{R}_3 = \frac{1}{2}(\mathbf{R}_2 + \mathbf{R}_1),$$

implying $\mathbf{R}_1 = \mathbf{R}_2 = \mathbf{R}_3$.

Example 9.9: Four masses on a circle and constrained by springs

Four masses m are connected by identical springs and are constrained to move on a frictionless circle of radius a. Establish the Lagrangian L for small oscillations of the particles about their equilibrium configuration depicted in Fig. 9.11.

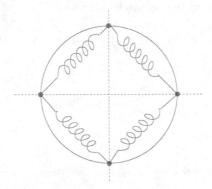

Fig. 9.11 Four particles on a circle and held together by springs.

Solution: If ϕ is the angle subtended by a chord AB at the centre of the circle, the length of this chord is $z = 2a\sin(\phi/2) \simeq a\phi$ for small angles ϕ. When in equilibrium the four masses are located at angular intervals of $\pi/2$. Hence in the case of small oscillations the distance between masses i and j is $z_{ji} = 2a\sin[(\phi_i - \phi_j)/2] \simeq 2a\sin(\pi/4) \simeq a[(\phi_i - \phi_j) - \sqrt{2}]$. Applying *Hooke's law*, it follows that the potential V constraining the masses is given by

$$\begin{aligned} V &= \frac{1}{2}ka^2[(\phi_1 - \phi_4)^2 - 2\sqrt{2}(\phi_1 - \phi_4) + 2 + (\phi_2 - \phi_1)^2 - 2\sqrt{2}(\phi_2 - \phi_1) + 2 \\ &\quad + (\phi_3 - \phi_2)^2 - 2\sqrt{2}(\phi_3 - \phi_2) + 2 + (\phi_4 - \phi_3)^2 - 2\sqrt{2}(\phi_4 - \phi_3) + 2] \\ &= ka^2[\phi_1^2 + \phi_2^2 + \phi_3^2 + \phi_4^2 - \phi_1\phi_2 - \phi_2\phi_3 - \phi_3\phi_4 - \phi_4\phi_1 + 4]. \end{aligned} \tag{9.62}$$

The kinetic energy T is correspondingly given by

$$T = \frac{1}{2}ma^2[\dot{\phi}_1^2 + \dot{\phi}_2^2 + \dot{\phi}_3^2 + \dot{\phi}_4^2].$$

Hence the Lagrangian is $L = T - V$. The problem can now be investigated along the lines of Examples 9.1 and 9.8. The solution can also be looked up in the literature.*

---

*See *The Physics Coaching Class* [48], problem 2034, p. 535.

Chapter 10

Motivation of the Theory of Relativity

10.1 Introductory Remarks

As explained at the beginning, we attempt here a treatment of classical mechanics linked with an introduction into the theory of relativity, so that the entire development of the field, spanning from Newton's laws and Newton's gravitational law through Lagrangian and Hamiltonian mechanics to Einstein's Special and General Theories of Relativity is covered, and the reader can see how the various stages of the development are related. Consequently we shall try to link mathematical steps in the next few chapters, such as coordinate transformations and considerations of action principles, to considerations in the preceding chapters in the hope that this simplifies an appreciation of some basic aspects of relativity, which a sophisticated mathematical treatment might obscure. In the compilation of the following chapters we utilized, of course, only selected works of the vast amount of literature which has meanwhile accumulated. We name already here some references that we found particularly useful for our present purposes. These are the huge, encyclopaedic text on *Gravitation* by Misner, Thorne and Wheeler [33], the similarly huge text entitled *The Road to Reality* by Penrose [38] (recommendable particularly for its in-depth-going discussion), the text on *Gravitation and Cosmology* by Weinberg [50], a recently published text on *Cosmological Relativity* by Carmeli [10], and the book on *Tensor Calculus and Relativity* by Lawden [27]. Very illuminating and helpful we found the collection of articles of Wilczek [52] and internationally distributed lecture notes on General Relativity by Leite–Lopes [28], Regge [42], Carroll [11] and Williams [53]. Other sources and references will be cited in the running text.

The first problem that many people encounter when confronted with the theory of relativity is that they have only a fuzzy feeling of some frequently used terms which many writers hardly care to explain, or explain very briefly or in different ways — *e.g.* inertia or inertial mass, the principle of equivalence in its weak and strong forms, later the distinction between covariance and invariance, affinity and others. Also, a novice in the topic may be surprised to learn that what he had assumed as natural from his days at school is here questioned and subjected to an experimental test, namely whether the mass of a body as it arises in Newton's equation of motion (called *inertial mass*) is actually the same as that appearing in Newton's law of gravitation (there called the *gravitational mass*). These concepts, however, are very important and their clear understanding can help a long way to appreciate the fundamental features even of General Relativity. We begin therefore with a nonmathematical discussion of some of these, and in particular consider the questions that motivated the General Theory of Relativity. In this connection we recapitulate aspects of the Special Theory of Relativity, not all of which will be considered here in great detail, since this topic is usually included in electrodynamics.

10.2   The Weak Equivalence Principle

Newton's equation is conveniently considered with respect to an assumed absolutely fixed space (which, of course, does not exist), and hence with respect to an unaccelerated, nonrotating three-dimensional reference frame with coordinates $\{x_i\}$ (say). This absolute reference frame has no reference to anything outside of it (*i.e.* gravitation). A point-like particle in this space which is given an applied force $\mathbf{F}_a$ is accelerated according to Newton's second law to acceleration $\ddot{\mathbf{x}} \equiv \ddot{\mathbf{r}}$ given by $\mathbf{F}_a \propto \ddot{\mathbf{r}}$. Thus

$$\mathbf{F}_a = m_i \ddot{\mathbf{r}}, \tag{10.1}$$

and the constant m_i is called the *inertial mass* of the particle. This inertial mass is a measure of the "inertia" of the particle or body, *i.e.* a measure of its reluctance to be set into motion, as we explained already in Chapter 2. The "fixed" reference frame is therefore also called the *inertial frame* (but see below!).

On the other hand the *gravitational mass* m_g of the particle or body is deduced from Newton's law of gravitation. This law says that if there is another body of (gravitational) mass M (*e.g.* the Earth), then the body is attracted to it with a force given by

$$\mathbf{F}_g = -G\frac{M m_g}{r^3}\mathbf{r}, \tag{10.2}$$

where r is the separation of their centres of mass and G is the universal gravitational constant,

$$G = 6.668 \times 10^{-8} \mathrm{g}^{-1} \mathrm{cm}^3 \mathrm{s}^{-2}, \qquad \frac{G}{c^2} = 7.425 \times 10^{-29} \mathrm{cm\,g}^{-1}.$$

Taking this gravitational force as the force $\mathbf{F}_a$ applied to the body, we have

$$\mathbf{F}_g = \mathbf{F}_a, \quad i.e. \quad -G\frac{Mm_g}{r^3}\mathbf{r} = m_i\ddot{\mathbf{r}}, \tag{10.3}$$

(this is a case of so-called *"free fall"*). Thus *if* we argue that *empirically* we have $m_i \propto m_g$ or $m_i = m_g$ (choosing the constant to be 1), m_i and M_g cancel and we have

$$-G\frac{M}{r^3}\mathbf{r} = \ddot{\mathbf{r}},$$

i.e. the gravitational acceleration is the same for all masses along the same trajectory in Newton's space with $M \neq 0$. Galilei argued the other way. He selected bodies, in fact balls of copper, lead and other materials, with different masses and observed that the acceleration of the falling bodies (falling meaning free motion in a gravitational field) is the same. This is the *"universality of free fall"* for which Newtonian gravity offers no explanation. Galilei thus arrived at the equivalence*

$$m_i = m_g, \tag{10.4}$$

which is now known as (Galilei's) *weak equivalence principle.*[†] Newton is said to have been aware of this. Thus local experiments on freely falling bodies cannot detect the presence of gravity, and all bodies move inertially irrespective of the presence of gravity.

In order to obtain a better understanding of Eq. (10.4), we think of a body of gravitational mass m_g on the surface of the Earth (assumed homogeneous, perfectly spherical, *etc.*). This body has a weight which is the force opposite to the thrust against the body which holds it where it sits and is, say, $\mathbf{F}_g$. For the particle placed in a homogeneous and uniform gravitational field (observe $\mathbf{r}$ is taken as directed outward from the centre of the Earth)

$$\mathbf{F}_g = +m_g\mathbf{g}, \quad \mathbf{g} = -G\frac{M\mathbf{r}}{r^3} \tag{10.5}$$

(thus weight is the pull of gravity, and one has weightlessness without this pull).

---

*This means $m_i/m_g =$ const. and the constant was chosen to be 1; below a consideration is described to *prove* that the constant is 1.

[†]As in Sec. 2.6 we refer again to the discussion of R. Penrose [38], pp. 390 – 299.

We now also take into consideration the motion, *i.e.* rotation of the spherically assumed Earth around its axis with angular velocity $\boldsymbol{\omega}$ with respect to Newton's (say, assumed fixed) reference frame (assumed originally coincident with that of the Earth). The centrifugal force is then given by the following expression (*cf.* Eq. (8.129)) with *centrifugal acceleration*

$$F_c = m_i[\omega^2 \mathbf{r} - (\boldsymbol{\omega} \cdot \mathbf{r})\boldsymbol{\omega}]. \qquad (10.6)$$

At the equator (where $\boldsymbol{\omega} \perp \mathbf{r}$) this is $m_i\omega^2 \mathbf{r}$, at the North pole it is zero.

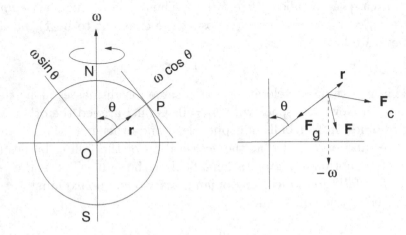

Fig. 10.1 The rotating Earth.

We let θ be the angle of inclination of $\mathbf{r} \equiv \mathbf{x}$ to the direction of $\boldsymbol{\omega}$ or the North pole, as indicated in Fig. 10.1. The total force acting on the body at P is the resultant

$$F = F_g + F_c = m_g\left[\mathbf{g} + \frac{m_i}{m_g}\{\omega^2 \mathbf{r} - (\boldsymbol{\omega} \cdot \mathbf{r})\boldsymbol{\omega}\}\right]. \qquad (10.7)$$

The first experiment to test the equivalence or identity of m_i and m_g was that of Eötvös[‡] in Budapest towards the end of the nineteenth century. At Budapest the colatitude angle θ is 42.5°, *i.e.* there roughly $\theta \simeq 45°$ and $\cos\theta = \sin\theta = 1/\sqrt{2}$. The vertical component of F_c/m_i, *i.e.* parallel to $-\mathbf{g}$, is obtained by taking the scalar product of F_c and $\mathbf{r}/r$, *i.e.*

$$F_c \cdot \frac{\mathbf{r}}{r} = m_i[\omega^2 \mathbf{r} - (\boldsymbol{\omega} \cdot \mathbf{r})\boldsymbol{\omega}] \cdot \frac{\mathbf{r}}{r} = m_i\omega^2 r(1 - \cos^2\theta) = m_i\omega^2 r \sin^2\theta,$$

and thus the vertical component of F_c/m_i is $\omega^2 r \sin^2\theta$. This is too small compared with $|\mathbf{g}|$ to be noteworthy. It is therefore the horizontal, *i.e.* tangential

---

[‡]R. V. Eötvös [15].

component which is of interest to us here and this means the component perpendicular to $\mathbf{r}$, *i.e.* (*cf.* the second term on the right of Eq. (10.6))

$$(\mathbf{F}_c)_{\text{horiz.}} = -m_i(\boldsymbol{\omega} \cdot \mathbf{r})\left(\boldsymbol{\omega} \times \frac{\mathbf{r}}{r}\right) = m_i\omega^2 r \cos\theta \sin\theta.$$

Thus the fibre or thread attached to the body hanging like a pendulum bob at P (*cf.* Fig. 10.1) will — as a result of the rotation of the Earth — experience an angular deflection (from the straight line joining the centre of the Earth with the point of support) and point in the direction of the resultant force, *i.e.* the small angular deflection is[§]

$$
\begin{aligned}
\delta\theta &= \frac{\text{horizontal centrifugal force}}{\text{vertical gravitational force}} = \frac{m_i\omega^2 r \sin\theta \cos\theta}{m_g g} \\
&= \frac{m_i}{m_g} \times \frac{3.4 \text{ cm s}^{-2}}{980 \text{ cm s}^{-2}} \sin\theta \cos\theta \\
&= \frac{m_i}{m_g} \times 1.7 \times 10^{-3} \text{ radian at } \theta = 45°.
\end{aligned}
\tag{10.8}
$$

Now the idea was to take two objects, *i.e.* pendulum bobs, with the same gravitational mass (*i.e.* the same weight at the same point), *i.e.* $m_g = m_g'$, but of very different materials like wood and platinum, and to measure the deflection in both cases. Eötvös [15] reported $\delta\theta = 1.7 \times 10^{-3}$ with an error $\leq 10^{-9}$, so that from Eq. (10.8) $m_i/m_g = 1$. Above we outlined only the principle of the experiment. In actual fact the experiment was designed as a torsion balance experiment involving a beam hanging from some support with masses attached to the ends with fibres. The equality of m_i and m_g, or rather the independence of the acceleration due to gravity of the materials of the falling bodies has more recently been re-investigated with refined experiments, foremost among these being that of Roll, Krotkov and Dicke at Princeton in 1964.[¶] Descriptions of this more complicated experiment can be found in standard literature.[‖] The possibility of testing the weak equivalence principle with artificial Earth satellites (and there reaching a precision level of 10^{-15} to 10^{-18}) has been discussed in a recent paper by Iorio [23]. The experiment is one of the few very important null experiments of physics. Maybe one can say that the simple relation (10.3) hints already at a deep connection between acceleration and gravity. In Example 10.1 we ask: Does the photon obey the (weak form of the) principle of equivalence?

[§]See Eq. (8.128) for the numerical value of $\omega^2 r$ that we insert here.

[¶]P. G. Roll, R. Krotkov and R. H. Dicke [44].

[‖]C. W. Misner, K. S. Thorne and J. A. Wheeler [33], pp. 13 – 14. See also M. Carmeli [10], pp. 39 – 41.

10.3   Inertial Frames

Newton's mechanics can conveniently be based on the assumption of an absolute space with a fixed reference frame, and one can visualize this roughly as one whose centre lies in a fixed star or in the sun or in something else with a seemingly fixed appearance. But such an assumed absolute space does not exist, *i.e.* is not observable. Einstein made the concept more precise by emphasizing the significance of a *local frame of reference* — the frame in which one is weightless, *i.e.* in which the effect of gravity is negligible (thus "inertial motion" is motion under no forces). In such a frame

$$m_i \ddot{\mathbf{x}} = 0, \qquad i.e. \qquad \frac{\dot{\mathbf{x}}}{m_i} = \text{const.}$$

Thus in such a frame the free object moves in a straight line with constant velocity. The motion in this case is particularly simple. The weightlessness of the motion is indicative of the (local) inertial frame of reference — evidenced by thousands of experiments, *e.g.* in particle physics where gravity plays no role whatsoever. The "constant velocity" of this local motion brings us immediately to the Special Theory of Relativity and to its difference from the General Theory of Relativity (the Special form being that in which gravity is negligible). We can imagine various local reference frames moving with different velocities since one is moving with respect to another, but we cannot imagine something like an absolute velocity. For a quantitative description we define local reference frames, say frames K and K', with space and time coordinates $(\mathbf{x}, t)$ and $(\mathbf{x}', t')$ respectively. In the Special Theory of Relativity Einstein added the postulate of the constancy of the velocity of light (that it is independent of the motion of the source and is the same in vacuum in the presence of electromagnetic fields for all observers in inertial frames). This constancy expresses, in fact, the *Lorentz invariance* of the line element squared,

$$ds^2 = c^2 dt^2 - d\mathbf{x}^2,$$

under transformation from one local inertial frame to a translated one (we come to rotations later!). We can rewrite ds^2 in the form

$$ds^2 = (cdt, d\mathbf{x})(g) \begin{pmatrix} cdt \\ d\mathbf{x} \end{pmatrix} \quad \text{with} \quad (g) = \begin{pmatrix} 1 & 0 & 0 & 0 \\ 0 & -1 & 0 & 0 \\ 0 & 0 & -1 & 0 \\ 0 & 0 & 0 & -1 \end{pmatrix}.$$

This matrix g with elements $g_{\mu\nu}, \mu, \nu = 0, 1, 2, 3$, is called a *"metric"*. Its significance is that it describes the geometry of the space of the local inertial frame. In view of the constancy of the diagonal elements it implies a so-called *"flat space"*. Thus the local inertial reference frame is that of a flat space. Therefore *locally* we deal with a flat space (flat spacetime as well as flat position space). Deformation, *i.e.* curvature or (which means the same) warping of space, arises at huge distances, *i.e.* in nonlocal frames.

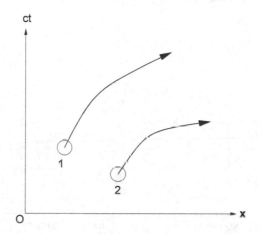

Fig. 10.2 Events 1 and 2 and their worldlines in spacetime.

The space defined by the 4 coordinates $\mathbf{x}, t$ is called *Minkowski space* or *spacetime* and is denoted by $\mathbb{M}_4$. The introduction of this space as a single concept was a revolutionary idea in physics. A point in this space is called an *event*, and the path traced out by it in spacetime is called a *worldline*, as indicated in Fig. 10.2. The position part is obviously Euclidean. Before the advent of Einstein's theory of relativity this space was assumed to be Euclidean to any distance. But we don't have to go to these distances to see gravity. The above example of the Eötvös experiment showed that with rotation an acceleration appears (*i.e.* a frame accelerated relative to an inertial frame), and this acceleration implies a fictitious force, in fact a gravitational force. We consider this in more detail in the next section.

10.4 The Strong Principle of Equivalence

We consider now the principle known as the *strong form* of the principle of equivalence which implies that an observer cannot distinguish between the effect of a uniform gravitational field and that of a constant acceleration. It is said that Einstein found the weak form of the principle of equivalence that we considered in Sec. 10.2 very amazing and expanded it into the so-called strong form, and thus into a basic principle of physics. Einstein considered so-called thought-experiments in which the gravitational field[**] is mimicked by going to an accelerating frame, and then extended the principle into one of unrestricted validity. The idea is to construct situations which make it impossible for the observer to distinguish the effect of acceleration (like that of a rocket) and that of gravity. The thought-experiment may be considered as consisting of four stages, as illustrated in Figs. 10.3 and 10.4[††]

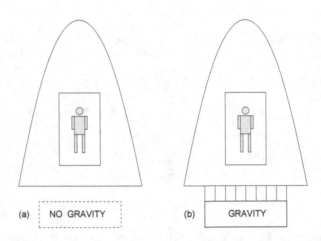

Fig. 10.3 Can the observer distinguish between (a) and (b)?

(a) *No gravity, no engine (this is the case of a stationary, i.e. unaccelerated rocket*: In this case the rocket drifts with uniform motion, *i.e.* constant velocity, through space. The observer inside checks that free objects (inside, where the experiment takes place) move with no acceleration.

(b) *With gravity (free fall), and again no engine*: The entire rocket cabin with the observer inside is allowed to fall freely (free fall is motion solely under gravity and inertia) in a uniform gravitational field.[*] The cabin as well as objects inside all fall with the same acceleration of gravity g (weak equivalence), so that the observer has no way to distinguish this case of constant acceleration from that of case (a) with constant velocity, since with $m_i \ddot{x} = m_g g, m_i = m_g$ (weak equivalence) one now has $\ddot{x} = g, \dot{x} = gt + \text{const}.$

(c) *No gravity, but with an engine accelerating the cabin*: In this case the rocket imparts to the cabin (or shell) the uniform acceleration $-\mathbf{a}$. Visualizing a huge space inside the rocket, all objects inside then appear accelerated to an acceleration $\mathbf{a}$, *i.e.* opposite to that of the cabin.

[**]The idea of a "field" which spreads through the whole of space became the accepted view as a result of the Michelson and Morley experiment.

[††]These figures are motivated by corresponding figures in the work of J. Leite-Lopes [28].

[*]In C. W. Misner, K. S. Thorne and J. A. Wheeler [33], p. 13, this case is described as that of weightlessness.

(d) *Gravity, but no engine*: Finally we place the cabin at rest relative to the Earth in the gravitational field with $\mathbf{g} = \mathbf{a}$. Then the observer will not observe a difference to case (c).

Thus an observer cannot distinguish between cases (a) and (b) or between cases (c) and (d). Strictly speaking we see the equivalence of (a) and (b) only from a local point of view — but this is what Einstein extended into a principle of unrestricted validity.

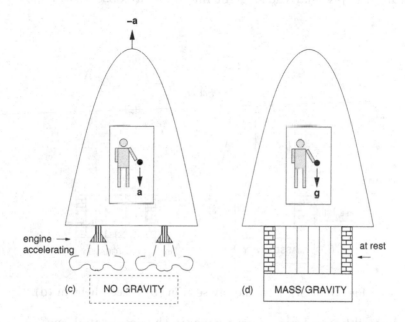

Fig. 10.4 Can the observer distinguish between (c) and (d)?

In summary: What do we learn from cases (a) and (b)? We see that an observer cannot distinguish between inertial motion, *i.e.* motion under no forces, and motion in a uniform gravitational field. Put differently: In *"free fall"* all bodies move inertially whether or not gravity is present. Thus inertial motion in one frame does not have to be inertial motion in another frame. What do we learn from cases (c) and (d)? An observer in a frame which is not an inertial frame cannot distinguish between the effect of a constant acceleration (of the rocket) and the effect of a corresponding gravitational field. The statement that observations in an accelerated frame of reference are indistinguishable from those in a uniform gravitational field is known as the *strong principle of equivalence*. As a consequence of the above equivalences we can get an idea of how the *deflection* or *bending of light rays* comes about, *i.e.* that the strong equivalence principle leads to a bending of light rays. The observer of case (a) (no gravity) will see light rays travelling in straight lines across his cabin. Thus the observer of case (b) will also observe these as straight while falling freely (actually not quite! Only if light has no mass! Since light carries energy $\sim$ mass $\times c^2$ this observer will also, strictly speaking, see a bending of the light ray). However, the observer of case (d) will regard the light ray (which the observer of case (c) sees) as being bent down because (c) is accelerated relative to (a) (the floor of the cabin in (c) moves up). This is illustrated in Fig. 10.5. Why does the relative acceleration imply that the straight line light-ray observed in the

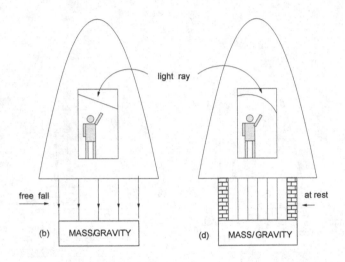

Fig. 10.5 Straight line ray seen in (b) is seen bent in (d).

unaccelerated frame appears bent down in the accelerated one? An intuitive answer is the following. The strong equivalence implies that observations in the accelerated frame are indistinguishable from those in a uniform gravitational field. Thus imagine a gravitational field in space. This can be looked at as making the vacuum a dense medium in which the light-ray is refracted, *i.e.* bent down. We give an illustration of this later (for a simplified view see Sec. 11.3.2). The observer of this bent light-ray will argue that spacetime is not flat.

We conclude from the above that to be able to cover observations of any observer, we need to go to curved spacetimes (*i.e.* accelerating frames). Of course, gravitational fields are usually nonuniform which makes it also imperative to consider curved spacetimes. In the following we shall employ some simple considerations in order to establish a linkage between acceleration and curvature.

In summary: The weak principle of equivalence says that $m_{\text{inertial}} = m_{\text{gravitational}}$, and the strong principle that a case with acceleration **a** can-

not be distinguished from one under gravity with **g** = **a**. Hence the word
"*equivalence*": A uniform gravitational field (*i.e.* that of **g**) is *equivalent* to
an acceleration **a** over a small region, *i.e.* to the use of a frame which is
accelerated with respect to the local inertial frame ("*free fall*" without the
presence of gravity). For an example see Sec. 12.7.

Example 10.1: Does the photon obey the principle of equivalence?

A photon of frequency ν is falling through a distance x in the gravitational field of the Earth. Does
the photon obey the principle of equivalence?[†]

Solution: The principle of equivalence in the weak form states that gravitational and inertial
masses are equal. In the case of the photon with mass zero (the graviton is another particle with
mass zero) the effective inertial mass m is given in terms of its energy E and frequency ν by[‡]

$$m = \frac{E}{c^2} = \frac{h\nu}{c^2},$$

where h is Planck's constant and c the velocity of light. The photon falling through a distance x
in the gravitational field of the Earth gains potential energy mgx (m being now the gravitational
mass) resulting in an increase of its frequency from ν to ν':

$$h\nu \longrightarrow h\nu' = h\nu + mgx = h\nu\left(1 + \frac{mgx}{h\nu}\right) = h\nu\left(1 + \frac{gx}{c^2}\right).$$

For $\nu' = \nu + \triangle\nu$, this implies

$$\frac{\triangle\nu}{\nu} = \frac{gx}{c^2},$$

and a shift to the blue part of the spectrum of light, which is therefore called a *blueshift*. Since
this shift is observed, the photon can be said to obey the principle of equivalence.

10.5 The Fundamental Postulate

The fundamental postulate of general relativity is that the *gravitational field
is described by the (nonflat) metric tensor* or *metric field* $g_{\mu\nu}(x)$ (more precisely, the graviton field is $g_{\mu\nu}(x)$ with the flat space metric subtracted out
as in Eq. (15.107)). Before we explore the concept of a curved spacetime
we consider briefly a case of rotation and how this alters the metric, and we
attempt to understand its meaning.

We consider first an *inertial system S* with flat spacetime line-element ds
given by

$$ds^2 = (cdt)^2 - d\mathbf{x}^2 \equiv g^{(0)}_{\mu\nu}(x)dx^\mu dx^\nu \tag{10.9}$$

[†]See also The Physics Coaching Class [48], problem 3051, p. 744.

[‡]Here we are using without prior derivation the wellknown Einstein formula of energy = mass
× (velocity of light)^2, and the formula $E = h\nu$ from elementary quantum mechanics. Einstein's
famous formula is derived in Sec. 12.8, *cf.* Eq. (12.48). For a non-mathematical discussion of
"*Mass without Mass*" see F. Wilczek [52], pp. 72 – 80.

with constant metric tensor

$$g_{\mu\nu}^{(0)} = \begin{pmatrix} 1 & 0 & 0 & 0 \\ 0 & -1 & 0 & 0 \\ 0 & 0 & -1 & 0 \\ 0 & 0 & 0 & -1 \end{pmatrix}$$

(ds^2 is also written $dx_\mu dx^\mu$ as we recapitulate later). Here the metric (tensor) has the Lorentz or Minkowski form. We now want to pass to a frame S' which undergoes uniform rotation about the z-axis as described by the following transformation:

$$x \to x'\cos\omega t - y'\sin\omega t, \quad y \to x'\sin\omega t + y'\cos\omega t, \quad z \to z', \quad t \to t'. \tag{10.10}$$

Thus S is the inertial frame relative to which a particle moves with constant velocity $\dot{\mathbf{x}}$. Seen from the noninertial frame S', *i.e.* in its coordinates, the particle experiences fictitious forces, here centrifugal and Coriolis forces or appropriately the metric of a different space. Thus an observer on S', who would not observe the particle as moving uniformly, would not attribute this noninertial motion to the fact that his frame is not inertial but to inertial forces, *i.e.* here the fictitious forces, the corresponding accelerations being independent of the mass as the gravitational acceleration. In a spinning but non-accelerated rocket the observer of a free particle in the cabin can attribute this noninertial motion to the existence of a gravitational field which is such as to account for the centrifugal and Coriolis forces (see also Sec. 11.7). We have[*]

$$d\mathbf{x}^2 = (d\mathbf{x}')^2 + \omega^2 dt^2 [x'^2 + y'^2] + 2\omega dt[x'dy' - y'dx'], \tag{10.11}$$

and hence

$$\begin{aligned} ds'^2 &= \left[1 - \frac{\omega^2}{c^2}(x'^2 + y'^2)\right](cdt)^2 - (d\mathbf{x}')^2 - 2\frac{\omega}{c}(x'dy' - y'dx')cdt \\ &= g_{\mu\nu}(\mathbf{x}', t)dx^\mu dx^\nu, \end{aligned}$$

where

$$g_{\mu\nu}(\mathbf{x}', t) = \begin{pmatrix} 1 - \frac{\omega^2}{c^2}(x'^2 + y'^2) & 2\frac{\omega}{c}y' & -2\frac{\omega}{c}x' & 0 \\ 2\frac{\omega}{c}y' & -1 & 0 & 0 \\ -2\frac{\omega}{c}x' & 0 & -1 & 0 \\ 0 & 0 & 0 & -1 \end{pmatrix}. \tag{10.12}$$

---

[*] $dx = (dx' - \omega y'dt)\cos\omega t - (dy' + \omega x'dt)\sin\omega t, dy = (dx' - \omega y'dt)\sin\omega t + (dy' + \omega x'dt)\cos\omega t.$

Here we could now drop the primes. We observe that the metric changed in the course of the rotation. The tensor $g_{\mu\nu}$ defines the metric of the physical space. Its nonflat form hints at fictitious forces.[†] Einstein says the forces are real and generate the gravitational field. The change of metric of the space implies a change of the geometry of the space. Einstein postulates: The gravitational field is described by $g_{\mu\nu}(x)$. This quantity is a symmetric second rank tensor. (It may be helpful to recall that the electromagnetic field is described by the vector potential $A_\mu(x)$; in fact $g_{\mu\nu}(x)$ corresponds to that in the case of gravity, although the forces are very different). One should note that the coefficients $g_{\mu\nu}(x)$ are not uniquely determined by the gravitational field since transformations

$$x^\mu = f^\mu(x^\nu)$$

leave the line element invariant (recall gauge invariance in electrodynamics does not permit a unique electromagnetic vector potential $A_\mu(x)$ to be determined).

10.6 Curvature

In *Lorentz geometry*, *i.e.* in a spacetime with the flat Minkowski metric, particles move in straight lines. It is fairly evident that this is always the case if one considers a sufficiently small part of space, *i.e.* a local frame even in a spacetime which is curved in the large. We have seen that acceleration leads to nonflatness of spacetime and that acceleration implies the gravitational field and *vice versa*. Since the gravitational field is also associated with masses (yet another aspect which has to be made clearer later, in particular when we extract Newton's equation from that of a particle in a gravitational field), these masses can be looked at as being responsible for an observed curvature of spacetime. Put differently: A space traveller far away from any masses will fancy himself travelling in a flat Minkowski space. This is the key point of relativity. Thus if one goes to a sufficiently large part of physical space, accelerations will become evident, and, in fact, a good example is provided by the *tides produced on Earth*. Before we touch this topic briefly after Eq. (10.19), we dwell a little longer on the idea of curvature and the fact that a local or infinitesimal part of a curved space may be considered as flat.

The simplest curved space to visualize is a sphere of radius R in 3 space dimensions. If we let $R \to \infty$ even a large part of its surface will appear as

---

[†]One can say the fictitious forces owe their existence to the choice of the coordinate system in the above example. If we return to the inertial frame, these fictitious forces disappear. Thus the space is really flat.

flat, *i.e.* as a Euclidean space. The 11th axiom of Euclid says that given any straight line AB and a point P (not on the line), one can draw a line parallel to AB as in Fig. 10.6. This 11th axiom establishes the familiar result concerning the sum of the angles of a planar triangle, *i.e.* that

$$\alpha + \beta + \gamma = \pi. \tag{10.13}$$

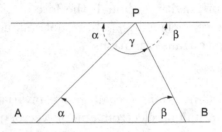

Fig. 10.6 Euclid's 11th axiom.

The equivalent of a straight line on a sphere is a *geodesic, i.e.* an arc whose length is the shortest (or longest) distance between two points on a sphere. A triangle on the sphere corresponding to a planar triangle whose sides are straight lines is a *spherical triangle* whose sides are arcs of great circles. If

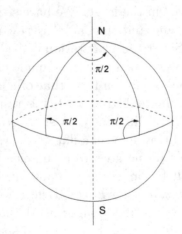

Fig. 10.7 An octant of the surface of a sphere.

α, β, γ are the interior angles of such a triangle, we obviously expect a relation

$$\alpha + \beta + \gamma - \pi = \left(\frac{1}{R}\right),$$

which reduces to the planar result for $R \to \infty$. In actual fact one has

$$\alpha + \beta + \gamma - \pi = \left(\frac{A}{R^2} \right), \qquad (10.14a)$$

where A is the area of the spherical triangle. One can verify this result in simple cases, *e.g.* in the case of an octant for which $\alpha = \beta = \gamma = \pi/2$ as in Fig. 10.7, and $A = (1/8) \times 4\pi R^2$, so that the left hand side of Eq. (10.14a) is $= 3\pi/2 - \pi = \pi/2$, and the right hand side $= (1/2) \times \pi R^2/R^2 = \pi/2$. Thus both sides are equal. We observe that now

$$\alpha + \beta + \gamma = \pi + \frac{A}{R^2} > \pi.$$

We leave the proof of formula (10.14a) to Example 10.2. One requires for this the area of a *"lune"*. A lune is the area of the spherical part of a segment of a sphere like that of an orange. This is the area between two great circles. The area between two great circles is determined by the angle between the tangents at their common point as indicated in Fig. 10.8.

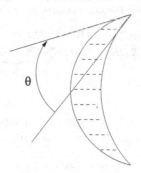

θ

Fig. 10.8 The area $\triangle_\theta$ of a lune of angle θ.

If the angle is θ, and the radius of the sphere is R, the area of the lune is, as we demonstrate immediately, $2R^2\theta$. In order to obtain this result, we suppose the angle $\theta = 2\pi/n$, n an integer. Then n identical lunes cover the sphere and the area of each lune is $4\pi R^2/n$, where $n = 2\pi/\theta$. Hence the area $\triangle_\theta$ of each lune is $4\pi R^2/n = 4\pi R^2/(2\pi/\theta) = 2R^2\theta$. From this simple result we obtain the above formula for the area of the spherical triangle to be derived in Example 10.2, *i.e.*

$$\triangle_\theta = 2\theta R^2 \quad \text{and} \quad \triangle ABC = R^2(A + B + C - \pi). \qquad (10.14b)$$

In spherical geometry one denotes an angle by the symbol of the vertex, like A, B, C in Fig. 10.9.

Example 10.2: Area of a spherical triangle

Show that the area of a spherical triangle ABC on a sphere of radius R is given by

$$^\circ\!\!\triangle = R^2(A + B + C - \pi).$$

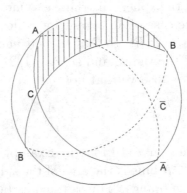

Fig. 10.9 The spherical triangle ABC.

Solution: We consider the lunes formed by the vertices of the spherical triangle as indicated in Fig. 10.9. Then (*e.g.* the area of the lune $AC\overline{A}BA$ around vertex A of the triangle under consideration is $2AR^2$, similarly for the others considering the vertices of the triangle ABC)

$$2\,AR^2 = {}^\circ\!\!\triangle ABC + {}^\circ\!\!\triangle\overline{A}BC, \qquad 2\,BR^2 = {}^\circ\!\!\triangle ABC + {}^\circ\!\!\triangle A\overline{B}C,$$
$$2\,CR^2 = {}^\circ\!\!\triangle ABC + \underbrace{{}^\circ\!\!\triangle AB\overline{C}}_{={}^\circ\!\!\triangle\overline{A}\overline{B}C}.$$

Hence

$$2(A + B + C)R^2 = 3\,{}^\circ\!\!\triangle ABC + {}^\circ\!\!\triangle\overline{A}BC + {}^\circ\!\!\triangle A\overline{B}C + {}^\circ\!\!\triangle\overline{A}\overline{B}C = 2\,{}^\circ\!\!\triangle ABC + 2\pi R^2,$$
$$\therefore \quad {}^\circ\!\!\triangle ABC = (A + B + C - \pi)R^2.$$

(Not clear? Remember the hemisphere of the sphere facing you in Fig. 10.9 is a lune of angle π — to be inserted in $\triangle_\theta$ of Eq. (10.14b) — and consists of 4 spherical triangles). Another, maybe smarter, way to write the result is (now with angles α, β, γ and area A)

$$\kappa = \frac{1}{R^2} = \lim_{A \to 0} \frac{\alpha + \beta + \gamma - \pi}{A}, \qquad A \text{ now area.} \tag{10.15}$$

The parameter κ is called *Gaussian curvature*. One should note that κ has dimensions of an inverse area and that this formula refers to a triangle on the curved spherical surface.

The parameter κ defined in Example 10.2, Eq. (10.15), appears as a measure of the violation of Euclid's 11th axiom, and thus as a measure of curvature. However, this definition is not so easy to handle. We observe that κ can even be negative if

$$\alpha + \beta + \gamma < \pi.$$

How then can we distinguish a curved surface from a flat one in a simple way (*i.e.* without using tensor quantities that will be introduced much later)?

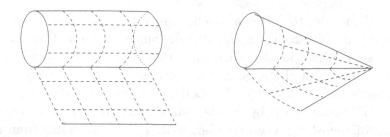

Fig. 10.10 The circular cylinder and cone have flat surfaces.

One way to test for curvature of a surface, *i.e.* for κ unequal to zero, is to roll the surface on a plane — if a coordinate system printed on the surface can thereby be transferred onto the plane, one can say the surface is flat. Examples of such surfaces are those of a circular cylinder and a circular cone.[‡] These can clearly be rolled on a plane, as indicated in Fig. 10.10. With a sphere or a torus this will not be possible. Not quite so obvious is

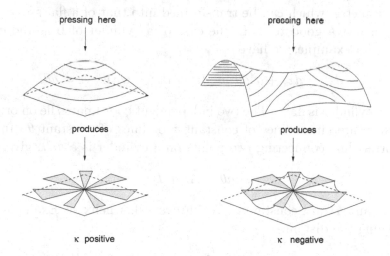

Fig. 10.11 Examples illustrating positive and negative curvature as evident from the angles less than 2π and greater than 2π around the points of squashing on nonplanar surfaces.

the following procedure of trying to flatten the given surface, but this procedure — different from the triangle consideration above — demonstrates the difference between surfaces of positive curvature and those of negative curvature. If we try to flatten a hemispherical or spherically-like hilly surface the result is characterized by angular gaps (a very similar result would be

---

[‡]At the end of Example 3.11 it was shown that the sum of the angles of a triangle on the surface of a circular cylinder is π, as in the case of a plane.

obtained if one tries to flatten a cone in this way by pressing its apex instead
of unrolling its surface as in Fig. 10.10). Since we know from Eq. (10.15)
that the sphere has positive curvature, this hilly surface also has positive
curvature, *i.e.* with the angular gaps around the point of squashing, the
surface-angle there is less than 2π. Correspondingly a surface with negative
curvature is characterized by angular parts which overlap (the angle around
the point of squashing now being larger than 2π) and result from a saddle
point structure, as one can verify with a piece of paper. These effects are
illustrated in Fig. 10.11 (see also Sec. 12.7). Mathematically one can see
these effects by looking at the geodesics associated with the surface: Two
such curves which start off parallel (*cf.* Fig. 10.18) will aim to converge if
the curvature of the surface is positive, and will diverge when this is negative
(for an illustration see Example 10.3).

How can we arrive at a more precise definition of curvature (still without using tensor calculus)? Since a plane has a flat metric (Euclidean or
Minkowskian, depending on the space) a surface with zero curvature ought
to have a metric which can be transformed into that of a flat space globally.
This is in fact a good test. In the case of a cylinder of base radius a and
height z, for example, we have

$$ds^2 = dz^2 + a^2 d\theta^2 = dz^2 + dx^2 \quad \text{with} \quad x = a\theta.$$

Thus the cylinder is flat. The two independent coordinates lie on orthogonal
axes which are either lines of constant z or lines of constant θ. In general
the shortest line connecting two points on a cylinder is a *helix* given by

$$a\theta = Az + B, \tag{10.16}$$

where A and B are constants. We derived this helix in Example 3.11 by
extremizing the distance

$$S_{12} := \int_1^2 ds = \int_1^2 dz \left[1 + a^2 \left(\frac{d\theta}{dz} \right)^2 \right]^{1/2}. \tag{10.17}$$

In Example 3.11 we have also shown that the sum of the angles of a triangle on
the surface of the circular cylinder is π — again substantiating the planarity
of the surface. Since we will again be considering action integrals later to
derive equations of motion, we might as well use these already at this stage
in the discussion of geometry. Thus the line corresponding to a geodesic on
the sphere is a helix on the cylinder.

The metric of a cone of height h is similarly derived and found to be

$$ds^2 = a^2 dh^2 + b^2 h^2 d\theta^2, \tag{10.18}$$

where a^2, b^2 are constants depending on the conical angle. Thus with

$$dx^2 = ds^2|_{\theta=\text{const.}} = a^2 dh^2, \qquad dy^2 = ds^2|_{h=\text{const.}} = b^2 h^2 d\theta^2,$$

we again have a flat metric: $ds^2 = dx^2 + dy^2$. Of course, one can also introduce orthogonal coordinates in the case of a sphere, *i.e.* angles θ, ϕ specifying *latitude* (θ : angular distance of a point on a meridian (great circle through Greenwich) north or south of the equator and given by $\pi/2 - \theta$) and *longitude* (ϕ : the angular distance of a point on the sphere east or west from the standard meridian). So this is not the criterion. The real criterion is, as stated, whether or not the metric tensor, *i.e.* the metric as matrix, can be *globally* transformed into Euclidean form or not. In the case of the sphere of radius a we have

$$ds^2 = a^2 d\theta^2 + a^2 \sin^2 \theta d\phi^2.$$

Can we transform this as required? *Locally* the space is flat, but globally? In Example 10.4 we consider this case in analogy to that of the circular cylinder treated in Example 3.12, and show that the equation of the geodesic can not be mapped onto a Euclidean plane. Thus the sphere is not flat!

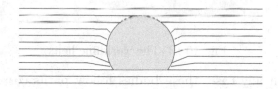

Fig. 10.12 The massive ball distorting the lattice of Euclidean space around it.

In general it is not possible to see at a glance whether a given surface (*i.e.* that defined by the metric tensor $g_{\mu\nu}$) is flat or not. One has to evaluate its *Riemann curvature tensor* to be defined later, and if this is zero, the surface is flat. We shall see later that curvature κ is given by a quantity R, immediately below written R_{Ricci}, known as the *Ricci curvature scalar*. We can extract its rough meaning from dimensional considerations. Since we expect the gravitational constant G to come in, we expect

$$\kappa = R_{\text{Ricci}} \propto G.$$

In natural units this would imply $c = 1$. If we take

$$R_{\text{Ricci}} = \frac{G}{c^2}\rho, \qquad \frac{G}{c^2} = 7.425 \times 10^{-29} \frac{\text{cm}}{g},$$

and since we know from Eq. (10.15) that κ has dimension (length)$^{-2}$, we see that ρ has dimensions of $R_{\text{Ricci}}g/\text{cm}$ and so of mass/volume. Thus the density of matter is the source of curvature (this is what Einstein's equation (15.86) implies). If one wants to have an analogy with electrodynamics, one could consider this type of "ordinary gravity" as playing a role analogous to that of electrostatics. Some people illustrate this with pictures like that in Fig. 10.12 in which the solid ball is seen to distort the lattice of the Euclidean space around it, thus curving or warping it. Thus massive bodies like the sun and the planets produce curvature — in space and more generally in spacetime. This curvature again is reflected in the motion of

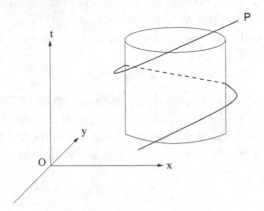

Fig. 10.13 The worldline helix.

these bodies. In any case, these bodies follow paths corresponding to the shortest distance in spacetime. Above we pictured only a shortest distance in a Euclidean configuration or position space. Let us now consider paths

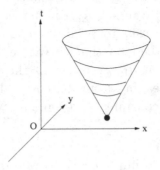

Fig. 10.14 The conical spreading of a disturbance in a pond.

in spacetime. A planet in (let us assume) circular motion around the sun in the (x, y)-plane traces out a helix in spacetime, its so-called *worldline*, as Fig. 10.13 illustrates. We can also picture something similar to the so-called *light cone* by constructing the cone of spreading circular ripples produced by throwing a stone into a pond, as illustrated in Fig. 10.14.

This brings us already to the Special Theory of Relativity which deals with flat spacetimes and uniform (unaccelerated) motion of inertial frames. The so-called *"light cone"* is defined by the condition $ds^2 = 0$, where $ds^2 = c^2 dt^2 - d\mathbf{x}^2$. The future light cone ($t > 0$) of the event at O is shown schematically in Fig. 10.15.

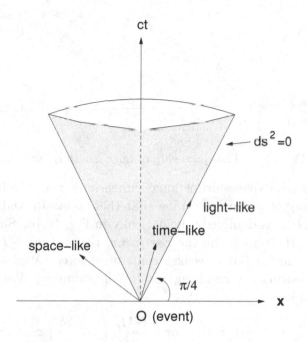

Fig. 10.15 The future light cone of event O.

Before we conclude these first considerations of curvature it is, of course, an interesting question to ask over what sort of distances one can confidently use Euclidean geometry. Again we are here only interested in a simple-minded estimate. We can get an idea by considering a space ship on a circular orbit around the Earth (of mass M_E) or the sun, or by considering the Earth as a space ship orbiting around the sun. In the former case we have at the equator a balance of gravitational and centrifugal forces,

$$\frac{GM_E}{r^2} = \omega^2 r. \tag{10.19}$$

The left hand side of the equation represents gravity, this means the gravitational attraction, and the right hand side an acceleration, so that the equation expresses the equivalence principle (*i.e.* the equal gravitational and inertial masses of the spaceship have cancelled out, and the acceleration is indistinguishable from the gravitational acceleration, *cf.* Sec. 10.4). The radius r is the distance from the centre of the Earth (or the sun) to the centre of mass of the space ship.

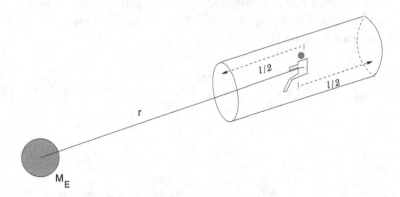

Fig. 10.16 The space ship of huge length l, but $l \ll r$.

We now imagine a spaceship of huge dimensions, specifically in the radial direction — say of width $l = 2\delta$, so that these opposite walls are at radial distances $r \pm l/2$, as depicted schematically in Fig. 10.16. Since l is huge in spite of $l \ll r$, the wall facing the Earth, *i.e.* the one at $r - l/2$, experiences a stronger attraction but a weaker centrifugal force. We can estimate the effect by computing the resulting *residual acceleration*. We have, together with Eq. (10.19),

$$\frac{GM_E}{(r \pm \delta)^2} = \omega^2(r \pm \delta), \quad \text{or} \quad \frac{GM_E}{r^2}\left(1 \mp \frac{2\delta}{r}\right) \simeq \omega^2(r \pm \delta).$$

Replacing in the second relation $\delta\omega^2$ by δ times the dominant term $\omega^2 = GM_E/r^3$, we can write this

$$\frac{GM_E}{r^2} \mp 3\delta\left(\frac{GM_E}{r^3}\right) = \omega^2 r.$$

The term in δ represents a *residual gravitational acceleration*, also called "*tidal acceleration*". In the case of an observer falling into a black hole this tidal acceleration will at some point begin to cause pain for him.[§] Now,

$$\text{dimension of} \quad \left(\frac{GM_E}{r^3}\right) \quad \text{is that of } \text{time}^{-2},$$

$$\text{or dimension of} \quad \left(\frac{GM_E}{c^2 r^3}\right) \quad \text{is that of } \frac{1}{\text{area}}.$$

[§]D. Raine and E. Thomas [40], p. 36.

The latter suggests that the residual acceleration can be related to curvature. We can consider residual accelerations of masses M as those of *tidal forces* $\mp 3\delta c^2 M/$area. Relating tidal forces to a quantity with dimension $1/$area means relating them to (Gaussian) curvature as in Eq. (10.15). In this way one can relate physics directly to curvature of space. We consider a more sophisticated case demonstrating tidal effects in Example 10.3, which is an example outlined verbally in the book of Penrose.[¶]

We could also consider Newtonian *linear motion along the radial direction* and picture the two opposite walls of the space ship as particles subjected to the residual accelerations. If $l = 2\delta$ represents the separation at time t, the residual acceleration is given by

$$\left|\frac{d^2 l}{dt^2}\right| = 2\frac{GM_E}{r^3}l, \tag{10.20}$$

where the factor 2 comes from the difference of $(\mp 2\delta/r)$. How does this equation exhibit a curvature of space? In order to obtain a glimpse of the answer we return to the spherical space, *i.e.* a sphere of radius a, considered earlier, and perform the exercise in Example 10.4 which is a brief version of Example 3.12. In Example 10.4, Eq. (10.39), the difference ξ corresponds to l in Eq. (10.20). The geodesics shown in Fig. 10.18 are originally parallel with separation ξ_0, but are no longer parallel after a distance L. In fact they bend towards each other — which is the case of positive (Gaussian) curvature (two geodesics starting off parallel from the left of a saddle as in Fig. 10.11 bend away from each other — which is the case of negative curvature). When the radius a of the sphere is allowed to go to infinity, of course, $\xi \rightarrow \xi_0$, and the paths are parallel. This is the limit in which we have Euclidean geometry. We see therefore: The geodesics in curved space correspond to straight lines in Euclidean space. Comparing Eq. (10.20) with Eq. (10.39), the latter multiplied by c^2, we see that indeed the residual gravitational acceleration can be related to curvature. From here we could proceed and allow a twisting of the separation as it varies with L. We would then arrive at a picture of the *Riemann curvature tensor* which enters the definition of the curvature scalar mentioned earlier. But we shall not do this here.

Before we consider in some detail aspects of the Special Theory of Relativity, we consider in the next chapter a *simplified* or naive way of looking at the main spectacular results of the General Theory, so that we obtain a rough idea of where we are heading, and what the theory will eventually explain.

[¶]R. Penrose [38], p. 397.

10.7   Miscellaneous Examples

Example 10.3: The tidal effect

A mass M, like that of an astronaut, at a distance R away from the centre of the Earth of mass M_E is surrounded by a circle (three-dimensionally sphere) of particles, each of mass m, originally at rest at a distance r away from mass M, as indicated in Fig. 10.17(a). Calculate the difference between the accelerations $\mathbf{a}_m$ and $\mathbf{a}_M$ of a particle and the mass M, and show that this forms an ellipse (three-dimensionally a prolate ellipsoid of revolution), thus demonstrating the tidal effects on the particles due to attraction to the Earth. Also show that the area of this ellipse (volume of the ellipsoid) is equal to the area (volume) of the circle (sphere) originally. A detailed verbal explanation of this problem can be found in the book of Penrose.[||]

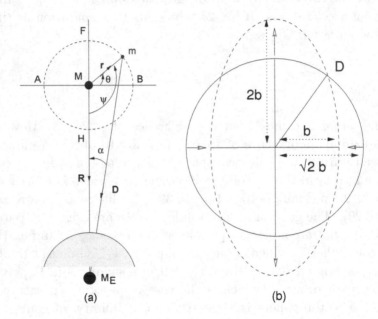

Fig. 10.17 (a) The mass M (*e.g.* of an astronaut) surrounded by a circular cloud of
particles of mass m far above the surface of the Earth.
Fig. 10.17 (b) Displacement of the particles (relative to their initial configuration) due
to the attraction of the Earth; this is the tidal effect.

Solution: From the weak equivalence principle we know that the masses m, M do not affect the accelerations. We consider angles θ, ψ and α as indicated in Fig. 10.17(a). Since the distance from a mass m at location $\theta = -\pi/2$ (*i.e.* from H in Fig. 10.17(a)) to the centre of the Earth is less than the distance of the mass M to the centre of the Earth, Newton's law of gravitation tells us that the acceleration $\mathbf{a}_m$ of the particle towards the Earth will be larger than that of M, whereas for a particle's acceleration at location $\theta = +\pi/2$ this will be the opposite, as indicated by arrows in Fig. 10.17(b). In both cases of a mass m at $\theta = 0$ or π, the mass M will be closer to the centre of the Earth, and hence the arrows point inward. We can verify these considerations by explicit calculation. We set

$$\cos\psi = \cos\left(\frac{\pi}{2} + \theta\right) = -\sin\theta, \quad \mathbf{D} = \mathbf{R} - \mathbf{r}, \quad \mathbf{r}\cdot\mathbf{R} = rR\cos\psi = -rR\sin\theta. \tag{10.21}$$

From Newton's law we obtain for the difference of the accelerations (G being Newton's gravitational

---

[||]R. Penrose [38], p. 397.

constant)

$$\mathbf{a}_m - \mathbf{a}_M \;=\; GM_E \frac{(\mathbf{R} - \mathbf{r})}{|\mathbf{R} - \mathbf{r}|^3} - GM_E \frac{\mathbf{R}}{R^3} = GM_E \frac{(\mathbf{R} - \mathbf{r})}{[R^2 + r^2 - 2\mathbf{r}\cdot\mathbf{R}]^{3/2}} - GM_E \frac{\mathbf{R}}{R^3}$$

$$=\; GM_E \frac{(\mathbf{R} - \mathbf{r})}{R^3 [1 + (r^2 - 2\mathbf{r}\cdot\mathbf{R})/R^2]^{3/2}} - GM_E \frac{\mathbf{R}}{R^3}$$

$$=\; GM_E \frac{(\mathbf{R} - \mathbf{r})}{R^3}\left[1 - \frac{3}{2}\frac{(r^2 - 2\mathbf{r}\cdot\mathbf{R})}{R^2} + \frac{15}{8}\left(\frac{r^2 - 2\mathbf{r}\cdot\mathbf{R}}{R^2}\right)^2 + \cdots\right] - GM_E \frac{\mathbf{R}}{R^3}.$$

$$(10.22)$$

Collecting terms which dominate for $R \gg r$, but also retaining terms in r in case the condition $\theta = 0$ removes higher order contributions, we obtain

$$\mathbf{a}_m - \mathbf{a}_M \;=\; GM_E \frac{(\mathbf{R} - \mathbf{r})}{R^3}\left[1 - \frac{1}{2R^2}\left\{3(r^2 - 2\mathbf{r}\cdot\mathbf{R}) - 15\frac{(\mathbf{r}\cdot\mathbf{R})^2}{R^2}\right\} + \cdots\right] - GM_E \frac{\mathbf{R}}{R^3}$$

$$=\; GM_E \frac{(\mathbf{R} - \mathbf{r})}{R^3}\left[1 - \frac{1}{2R^2}\{3r^2 + 6rR\sin\theta - 15r^2\sin^2\theta\} + \cdots\right] - GM_E \frac{\mathbf{R}}{R^3}.$$

$$(10.23)$$

Before we continue with the general case, we consider the special cases of $\theta = 0, \pi/2$ and verify the behaviour of the relative accelerations expected above.
(a) The case $\theta = 0$.
In this case we consider the component of the relative acceleration in the direction of $\mathbf{r}$, *i.e.*

$$(\mathbf{a}_m - \mathbf{a}_M)\cdot\mathbf{e}_r \;=\; GM_E \frac{(-r)}{R^3}\left[1 - \frac{1}{2R^2}\{3r^2 + 6rR\sin\theta - 15r^2\sin^2\theta\}\right]_{\theta=0}$$

$$\simeq\; -GM_E \frac{r}{R^3}\left[1 - \frac{3r^2}{2R^2}\right].$$

$$(10.24)$$

Hence

$$\mathbf{a}_m\cdot\mathbf{e}_r \simeq \underbrace{\mathbf{a}_M\cdot\mathbf{e}_r}_{0} - GM_E \frac{r}{R^3}.$$

$$(10.25)$$

Thus the component of the particle's acceleration at B in Fig. 10.17(a) in the direction of $\mathbf{e}_r$ is negative as indicated in Fig. 10.17(b). For symmetry reasons the same must be the case at $\theta = \pi$, *i.e.* for the particle at point A.
(b) The case $\theta = \pi/2$.
In this case we have $\mathbf{R} - \mathbf{r} = \mathbf{D}$ with (for $\theta = \pi/2$: $\mathbf{e}_D = \mathbf{e}_R$)

$$\mathbf{D}\cdot\mathbf{e}_D = R\cos 0 - r\cos\pi = (R + r), \quad \mathbf{R}\cdot\mathbf{e}_D = R, \quad \mathbf{r}\cdot\mathbf{e}_D = -r. \qquad (10.26)$$

It follows that

$$(\mathbf{a}_m - \mathbf{a}_M)\cdot\mathbf{e}_D \;=\; GM_E \frac{(R + r)}{R^3}\left[1 - \frac{3}{2R^2}\{r^2 + 2Rr - 5r^2\} + \cdots\right] - GM_E \frac{R}{R^3}$$

$$=\; GM_E \frac{1}{R^2}\left[1 - \frac{3}{2R^2}\{2Rr - 4r^2\} + \cdots\right] - GM_E \frac{1}{R^2}$$

$$+ GM_E \frac{r}{R^3}\left[1 - \frac{3}{2R^2}\{2Rr - 4r^2\} + \cdots\right]$$

$$\simeq\; GM_E \frac{1}{R^4}[-3(Rr - 2r^2)] + GM_E \frac{r}{R^3} - GM_E \frac{3r^2}{R^4}$$

$$\simeq\; -GM_E \frac{2r}{R^3}.$$

$$(10.27)$$

Since the result is predominantly negative we have at $\theta = \pi/2$ as expected (since $R + r > R$) that $\mathbf{a}_m \cdot \mathbf{e}_D < \mathbf{a}_M \cdot \mathbf{e}_D$. We observe that the magnitudes of the dominant result of Eq. (10.27) is twice that of the dominant result of Eq. (10.25), thus suggesting in the general case an ellipse with major axis twice the length of the minor axis.

Returning to the general case, Eq. (10.23), we have in dominant approximation

$$\mathbf{a}_m - \mathbf{a}_M = GM_E \frac{(-\mathbf{r})}{R^3} - GM_E \frac{3(\mathbf{R} - \mathbf{r})}{2R^5} [r^2 + 2Rr \sin\theta - 5r^2 \sin^2\theta]. \qquad (10.28)$$

Squaring this expression and again selecting the dominant terms, we have

$$\frac{(\mathbf{a}_m - \mathbf{a}_M)^2}{(GM_E/R^3)^2} = r^2 + \frac{3}{R^2}(\underline{\mathbf{R} \cdot \mathbf{r}} - r^2)\{r^2 + 2rR\sin\theta - 5r^2\sin^2\theta\}$$

$$+ \frac{9}{4R^4}(\underline{R^2} + r^2 - 2\mathbf{R} \cdot \mathbf{r})\{r^2 + 2rR\sin\theta - 5r^2\sin^2\theta\}^2. \qquad (10.29)$$

The products of the underlined terms (in the second case the latter is squared) yield the dominant contributions. We obtain thus

$$\frac{(\mathbf{a}_m - \mathbf{a}_M)^2}{(GM_E/R^3)^2} \simeq r^2 - 6r^2\sin^2\theta + 9r^2\sin^2\theta = r^2 + 3r^2\sin^2\theta + O\left(\frac{1}{R}\right). \qquad (10.30)$$

We now show that this equation is the polar equation of an ellipse with the pole at its centre. The general form of this equation is that of Eq. (7.63a) or (7.63b). Thus, considering the equation (note that ρ is the radius of the ellipse)

$$\frac{1}{\rho^2} = \frac{\cos^2\theta}{a^2} + \frac{\sin^2\theta}{b^2}, \qquad (10.31)$$

with (in our case here as observed after Eq. (10.27)) $a = 2b$, we obtain

$$\gamma^2 \equiv \frac{b^2}{\rho^2} = \frac{1}{4}\cos^2\theta + \sin^2\theta = \frac{1}{4}(1 - \sin^2\theta) + \sin^2\theta = \frac{1}{4}(1 + 3\sin^2\theta). \qquad (10.32)$$

We see that this is an equation like that in Eq. (10.30) with

$$\frac{(\mathbf{a}_m - \mathbf{a}_M)^2}{r^2(GM_E/R^3)^2} = \frac{4b^2}{\rho^2} = 4\gamma^2. \qquad (10.33)$$

We can also determine the radius of the circle by going to the point D in Fig. 10.17(b). At this point the elliptic fluctuation about the circle vanishes. From Eq. (10.28) we obtain in dominant approximation with $r \ll R$

$$\mathbf{a}_m - \mathbf{a}_M = \frac{GM_E(-\mathbf{r})}{R^3} - GM_E \frac{3\mathbf{R}}{2R^5}(2Rr \sin\theta) + O\left(\frac{r^2}{R^3}, \frac{1}{R^4}\right),$$

and hence in the direction of point D (with $\mathbf{R} \cdot \mathbf{r} = -Rr\sin\theta$)

$$0 = \mathbf{r} \cdot \left[-\mathbf{r} - \frac{3}{2R^2} 2Rr \sin\theta \mathbf{R} \right] = -r^2 + \frac{3}{2R^2} 2r^2 R^2 \sin^2\theta, \quad i.e. \quad \sin^2\theta = \frac{1}{3}.$$

Thus (from Eq. (10.30))

$$\sqrt{\frac{(\mathbf{a}_m - \mathbf{a}_M)^2}{(GM_E/R^3)^2}} = \sqrt{2}r = 2\gamma r,$$

and hence $2\gamma^2 = 1$ and $\rho = \sqrt{2}b$. On the other hand for $\theta = 0, \pi$ we obtain from Eq. (10.32) $\rho = 2b$, and for $\theta = \pm\pi/2$ we get $\rho = b$. These are the lengths of the semi-major and minor axes of the ellipse as indicated in Fig. 10.17(b). Thus the area of the ellipse (known to be given by π times the product of the lengths of the semi-major and -minor axes) is $2\pi b^2$, and is equal to that of the (initial) circle $\pi(\sqrt{2}b)^2 = 2\pi b^2$.

Example 10.4: Geodesics as periodic solutions of a differential equation

Show that the geodesics, providing the shortest or longest distance between two points on a sphere of radius a, are given by the periodic solutions of a differential equation. (See also Examples 3.12 and 10.5).

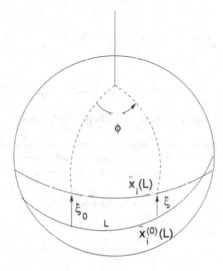

Fig. 10.18 A geodesic $\bar{x}_i(L)$ and a reference geodesic $\bar{x}_i^{(0)}(L)$.

Solution: The distance $L(s)$ between two points on a sphere at positions $s = 0, s$ is given by the expression

$$L(s) = \int_0^s ds\sqrt{\dot{\mathbf{x}}^2(s)}, \quad \text{with} \quad \mathbf{x}^2(s) - a^2 = 0. \tag{10.34}$$

Thus the geodesics are obtained by extremizing (since we use L for length, we use here $\mathcal{L}$ for the Lagrangian)

$$I = \int ds\left[\sqrt{\dot{\mathbf{x}}^2(s)} + \lambda(s)(\mathbf{x}^2(s) - a^2)\right] \equiv \int ds\mathcal{L}(x_i(s), \lambda(s), \dot{x}_i(s), \dot{\lambda}(s)), \tag{10.35}$$

where $\lambda(s)$ is a Lagrange multiplier. The Euler–Lagrange equations are

$$\mathbf{x}^2 - a^2 = 0, \quad \frac{d}{ds}\left(\frac{\dot{x}_i}{\sqrt{\dot{\mathbf{x}}^2}}\right) - 2\lambda x_i = 0.$$

Here

$$\sqrt{\dot{\mathbf{x}}^2(s)} = \frac{dL}{ds} \equiv v, \quad \frac{d}{ds} = \frac{dL}{ds}\frac{d}{dL} = v\frac{d}{dL}.$$

We set (the prime denoting differentiation with respect to L, the dot with respect to s)

$$\bar{x}_i(L) = x_i(s(L)), \quad \bar{x}_i' = \frac{d}{dL}x_i(s(l)) = \frac{\dot{x}_i}{v}. \tag{10.36}$$

The second Euler–Lagrange equation now becomes

$$\bar{x}_i'' - \mu\bar{x}_i = 0, \qquad \mu = \frac{2\bar{\lambda}(L)}{v},$$

so that

$$\sum_i \bar{x}_i(\bar{x}_i'' - \mu\bar{x}_i) = 0.$$

But

$$\frac{d^2}{dL^2}\left(\sum \bar{x}_i^2 - a^2\right) = 0, \qquad \therefore \quad \sum_i (\bar{x}_i'^2 + \bar{x}_i\bar{x}_i'') = 0,$$

and hence

$$\sum_i \bar{x}_i\bar{x}_i'' = -\sum_i \bar{x}_i'^2 = -\sum_i \left(\frac{\dot{x}_i}{v}\right)^2 = -1,$$

and therefore

$$\mu\sum_i \bar{x}_i^2 = -1, \quad \mu = -\frac{1}{a^2}, \quad \bar{x}_i'' + \frac{1}{a^2}\bar{x}_i = 0. \tag{10.37}$$

This is an equation with periodic solutions, these being the geodesics in Fig. 10.18, where $\phi/2\pi = L/2\pi a$, so that $\phi = L/a$.

Thus the period T of revolution is determined by the radius a of the sphere — $T = 2\pi a$ — a large radius requiring a large period. Suppose now we take a reference or fiducial geodesic with coordinates $\bar{x}_i^{(0)}(L)$ and define the separation of the two nearby geodesics as

$$\xi(L) := \bar{x}_i(L) - \bar{x}_i^{(0)}(L). \tag{10.38}$$

Then

$$\frac{d^2\xi(L)}{dL^2} + \frac{1}{a^2}\xi(L) = 0, \tag{10.39}$$

and a solution is the periodic function

$$\xi = \xi_0 \cos\left(\frac{L}{a}\right) = \xi_0 \cos\phi. \tag{10.40}$$

For further discussion we refer to Misner, Thorne and Wheeler [33], p. 31.

Example 10.5: Geodesics in terms of spherical angles
Determine the equation of the geodesics on a sphere of radius a in terms of polar angle θ and azimuthal angle ϕ, and verify the equivalence with the Cartesian equations of Example 10.4.

Solution: The infinitesimal length ds of a curve on the sphere of radius a is given by $ds^2 = dx^2 + dy^2 + dz^2$, which with spherical polar coordinates θ and ϕ as in Fig. 10.19. given by

$$x = a\cos\phi\sin\theta, \quad y = a\sin\phi\sin\theta, \quad z = a\cos\theta, \tag{10.41}$$

or alternatively with Pythagoras' theorem, becomes

$$ds^2 = a^2\sin^2\theta d\phi^2 + a^2 d\theta^2. \tag{10.42}$$

The length s of the distance between two points 1 and 2 on the sphere is given by the integral

$$I(\theta,\phi) = \int_1^2 ds = a\int_1^2 dt L(\theta,\phi), \quad L(\theta,\phi) = \sqrt{\sin^2\theta\dot{\phi}^2 + \dot{\theta}^2}, \quad \dot{\phi} = \frac{d\phi}{dt}, \dot{\theta} = \frac{d\theta}{dt}, \tag{10.43}$$

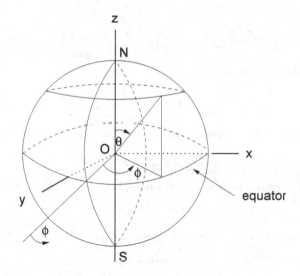

Fig. 10.19 The spherical angles ϕ and θ:
ϕ: The angular distance of a point on the sphere east or west
from the standard meridian (great circle through Greenwich) is called *longitude*,
θ: the angular distance on a meridian north or south
of the equator ($\pi/2 - \theta$) is called its *latitude*, θ its *colatitude*.

where l parametrizes the curve. The shortest or longest distance between the two points on the
sphere is obtained by extremizing $I(\theta, \phi)$, *i.e.* from $\delta I(\theta, \phi) = 0$, or the Euler–Lagrange equations

$$\frac{d}{dt}\left(\frac{\partial L}{\partial \dot{\gamma}}\right) - \frac{\partial L}{\partial \gamma} = 0, \quad \gamma = \theta, \phi.$$

The equations for θ and ϕ are:

$$\frac{d}{dt}\left[\frac{\dot{\theta}}{\sqrt{\sin^2 \theta \dot{\phi}^2 + \dot{\theta}^2}}\right] - \frac{\sin \theta \cos \theta \dot{\phi}^2}{\sqrt{\sin^2 \theta \dot{\phi}^2 + \dot{\theta}^2}} = 0, \tag{10.44}$$

$$\frac{d}{dt}\left[\frac{\dot{\phi}\sin^2 \theta}{\sqrt{\sin^2 \theta \dot{\phi}^2 + \dot{\theta}^2}}\right] = 0. \tag{10.45}$$

From the last equation we obtain

$$\frac{\dot{\phi}\sin^2 \theta}{\sqrt{\sin^2 \theta \dot{\phi}^2 + \dot{\theta}^2}} = A = \text{constant}. \tag{10.46}$$

With $\dot{\theta} = 0$ at $\theta = \pi/2$ and $t = 0$ we have $A = 1$. With $A = 1$ Eq. (10.44) becomes

$$\frac{d}{dt}\left[\frac{\dot{\theta}}{\dot{\phi}\sin^2 \theta}\right] - \frac{\cos \theta \dot{\phi}}{\sin \theta} = 0. \tag{10.47}$$

Extracting $\dot{\phi}/\dot{\theta}$ from Eq. (10.46), we obtain

$$\dot{\phi}^2 \sin^4 \theta = (\sin^2 \theta \dot{\phi}^2 + \dot{\theta}^2), \quad \dot{\phi}^2(\sin^4 \theta - \sin^2 \theta) = \dot{\theta}^2, \quad \dot{\phi} = \pm\frac{\dot{\theta}}{\sin \theta \sqrt{\sin^2 \theta - 1}}.$$

Hence

$$\frac{d}{dt}\left[\pm\frac{\sqrt{\sin^2\theta-1}}{\sin\theta}\right]=\frac{\cos\theta}{\sin\theta}\dot{\phi}, \qquad \pm i\frac{d}{dt}\left(\frac{\cos\theta}{\sin\theta}\right)=\left(\frac{\cos\theta}{\sin\theta}\right)\dot{\phi}. \tag{10.48}$$

Thus we end up with the equation

$$\dot{\phi}\pm i\frac{\dot{\theta}}{\cos\theta\sin\theta}=0, \qquad \text{or} \qquad \sin\theta\dot{\phi}\pm i\frac{\dot{\theta}}{\cos\theta}=0. \tag{10.49}$$

We use the second form of this equation later. Differentiating the first form with respect to t we obtain

$$\ddot{\phi}\pm\frac{2i}{\sin^2 2\theta}[\ddot{\theta}\sin 2\theta-2\dot{\theta}^2\cos 2\theta]=0. \tag{10.50}$$

Equating real and imaginary parts of the left hand side to zero (on the right), we obtain

$$\ddot{\phi}=0, \qquad \ddot{\theta}=2\dot{\theta}^2\cot 2\theta.$$

Integration of the first equation yields, with constants ϕ_0,ϕ_1:

$$\phi=\phi_0+\phi_1 t. \tag{10.51}$$

Considering now the second equation, and using the relation $\ddot{\theta}=d(\dot{\theta}^2/2)/d\theta$, we obtain with $g(\theta)=\dot{\theta}^2/2$:

$$\frac{d}{d\theta}g(\theta)=4g(\theta)\cot 2\theta, \quad \int\frac{dg(\theta)}{g(\theta)}=4\int\cot 2\theta d\theta, \quad \ln g(\theta)=2\ln|\sin 2\theta|, \quad g(\theta)\propto|\sin 2\theta|^2, \tag{10.52}$$

where we used a formula from Tables of Integrals.** It follows that

$$\dot{\theta}^2\propto 2|\sin 2\theta|^2, \quad \int\frac{d\theta}{\sin 2\theta}\propto\int dt, \quad \frac{1}{2}\ln|\tan\theta|\propto t, \quad \tan\theta=Ke^{2t}, \tag{10.53}$$

where we used again a formula from Tables of Integrals.†† Setting $\theta=\theta_0$ at $t=0$, we obtain $K=\tan\theta_0$ and

$$\frac{\tan\theta}{\tan\theta_0}=e^{2t}, \qquad t=\frac{1}{2}\ln\left|\frac{\tan\theta}{\tan\theta_0}\right|. \tag{10.54}$$

Now eliminating the parameter t from Eqs. (10.51) and (10.54), we obtain

$$\phi=\phi_0+\phi_1 t=\phi_0+\frac{1}{2}\phi_1\ln\left|\frac{\tan\theta}{\tan\theta_0}\right|. \tag{10.55}$$

We observe that this is a nonlinear relation between the angles θ and ϕ (in comparison with the case of the circular cylinder considered in Example 3.11).

In order to see the equivalence of this result with the set of three equations (10.37) obtained in Example 10.4, we consider the second of the two forms of Eq. (10.49), *i.e.*

$$\sin\theta\dot{\phi}+i\frac{\dot{\theta}}{\cos\theta}=0, \tag{10.56}$$

and re-express the angles here in terms of x,y,z. Thus

$$\phi=\tan^{-1}\left(\frac{y}{x}\right), \qquad \dot{\phi}=\frac{d}{dt}\left[\tan^{-1}\left(\frac{y}{x}\right)\right]=\frac{x\dot{y}-y\dot{x}}{x^2+y^2},$$

$$\theta=\cos^{-1}\left(\frac{z}{a}\right), \qquad \dot{\theta}=-\frac{\dot{z}}{\sqrt{a^2-z^2}}. \tag{10.57}$$

** H. B. Dwight, [13], formula 453.11, p. 104.
†† H. B. Dwight [13], formula 432.10, p. 90.

Thus Eq. (10.56) becomes ($\sin\theta = \sqrt{x^2+y^2}/a, \cos\theta = z/a$):

$$\frac{\sqrt{x^2+y^2}}{a}\frac{x\dot{y}-y\dot{x}}{x^2+y^2} - i\frac{\dot{z}}{\sqrt{a^2-z^2}}\frac{a}{z} = 0. \tag{10.58}$$

Since $x^2+y^2+z^2 = a^2$, this is

$$x\dot{y} - y\dot{x} - ia^2\frac{\dot{z}}{z} = 0. \tag{10.59}$$

Differentiating with respect to t, this implies

$$x\ddot{y} - y\ddot{x} - ia^2\frac{d}{dt}\left(\frac{\dot{z}}{z}\right) = 0. \tag{10.60}$$

Equating real and imaginary parts on both sides, we obtain the equations:

$$x\ddot{y} - y\ddot{x} = 0, \qquad \frac{d}{dt}\left(\frac{\dot{z}}{z}\right) = 0. \tag{10.61}$$

The first equation implies

$$\frac{\ddot{y}}{y} = \frac{\ddot{x}}{x} = \text{const.} \equiv -\alpha^2, \tag{10.62}$$

and the second

$$\dot{z} - \text{const.}z, \qquad \ddot{z} = \text{const.}\dot{z} = \text{const.}^2 z \equiv -\beta^2 z. \tag{10.63}$$

In view of the spherical symmetry we must have $\alpha^2 = \beta^2$, and we obtain:

$$\ddot{x} + \alpha^2 x = 0, \qquad \ddot{y} + \alpha^2 y = 0, \qquad \ddot{z} + \alpha^2 z = 0. \tag{10.64}$$

These are periodic equations with e.g.

$$x = \text{const.}\cos(\alpha t + \alpha_0) = \cos[\alpha(t+2\pi a) + \alpha_0].$$

Thus $\cos\alpha 2\pi a = 1$, and $\alpha = 1/a$, as in Examples 10.4 (Eq. (10.37)) and 3.12 (with unit radius).

Example 10.6: Comparison of tidal effects of the sun and the moon

As seen on the sky, the angular diameters of the sun and the moon are roughly equal. Using this observation and assuming that the height h_m of the tide raised by the moon on Earth is about twice that raised by the sun, h_s, i.e. $h_m = 2h_s$, relate the ratio of these to the mass densities ρ_m and ρ_s of the moon and the sun respectively.[‡‡]

Solution: Let R_E, R_m, R_s and M_E, M_m, M_s be respectively the radii and masses of the Earth, the moon, and the sun, and d_m and d_s the distances from the centres of the moon and the sun to the centre of the Earth. Consider first only the Earth and the moon, and think of a mass m at each of the two ends of a line which is approximately a diameter of the Earth. The system of these two point-like masses on the surface of the Earth in the gravitational field of the moon is analogous to the dumb-bell satellite in the field of the Earth considered in Example 7.7. From Eqs. (7.124a) and (7.124d) there we obtain as the potential energy of the masses m separated approximately by a diameter $2R_E$ on Earth the expression

$$V \simeq -mGM_m\left[\frac{2}{d_m} + \frac{(3\cos^2\theta - 1)R_E^2}{d_m^3}\right], \tag{10.65}$$

where θ is the angle defined in the geometry of Fig. 7.20. The first part of V accounts for the Kepler motion of the masses m (in the present case, these are fixed on the surface of the Earth).

[‡‡] Cf. *The Physics Coaching Class* [48], problem 1263, p. 439.

Here we take the surface of the Earth as a reference level with potential zero, so that this term plays no role. The second part distorts the spherical symmetry of the first part and hence may be considered as a surface distortion (*i.e.* of size h_m) resulting from the potential of the moon, half of that term at each mass m. The gravitational force of the Earth thus performs there the work

$$w = -mgh_m, \qquad g = \frac{GM_E}{R_E^2}, \tag{10.66}$$

which in the static case here has to equal the particle's energy there in the field of the moon. Thus we equate:

$$mgh_m \equiv m\frac{GM_E}{R_E^2}h_m = mgM_m\frac{(1 - 3\cos^2\theta)R_E^2}{d_m^3}. \tag{10.67}$$

Analogously we consider the effect of the sun's gravitational field at the same point with orientation θ. Then taking the ratio, we obtain (with $M = 4\pi\rho R^3/3$)

$$\frac{h_m}{h_s} = \left(\frac{d_s}{d_m}\right)^3\frac{M_m}{M_s} = \left(\frac{d_s}{d_m}\right)^3\left(\frac{R_m}{R_s}\right)^3\frac{\rho_m}{\rho_s}. \tag{10.68}$$

Since the moon and the sun as seen from the Earth have roughly identical angular diameters $\phi_m = \phi_s \equiv \phi$, it follows that

$$R_m = d_m\phi, \quad R_s = d_s\phi, \quad \therefore \quad \frac{R_m}{d_m} = \frac{R_s}{d_s}.$$

Hence we obtain for the requested ratio:

$$2 = \frac{h_m}{h_s} = \frac{\rho_m}{\rho_s}, \tag{10.69}$$

and thus $\rho_m = 2\rho_s$.

Chapter 11

A Simple Look at Phenomenological Consequences

11.1 Introductory Remarks

Relativity theory has the reputation of being hard to grasp, and as a consequence it is frequently thought that experimental consequences are just as difficult to comprehend. In spite of these prejudices, Einstein's relativity theory fascinates and attracts the minds of countless people, and even laymen attempt to derive a glimpse of insight, and are curious to understand what the theory is known to predict. It is no surprise therefore that attempts were made, also by serious researchers, to find some roundabout way to one of the spectacular results which — though not scientifically sound and rigorous — at least provides a sketchy view in that direction. Thus it has been shown that one can devise some interesting *"trick calculations"* in order to get the flavour of what the theory predicts. These simple considerations — following Rowlands* — are no substitute for the full theory and, in fact, pose further questions as to what fraction of the prediction can be attributed to effects of the Special Theory of Relativity, and which parts are classical in the sense of Newtonian gravity. Nonetheless these simplified considerations without the heavy load of tensor calculus are useful for didactic purposes and point to what is to be derived in the full theory, and may thus help to open one's eyes for this goal. In this chapter we shall not attempt to enter into such considerations in detail. Here we assume a first familiarity with

---

*P. Rowlands [45]. This Chapter 11 is based on this paper.

385

the Special Theory as presented *e.g.* in connection with electrodynamics. In the next chapter we recapitulate the results of the Special Theory from the point of view of its use and significance in connection with the General Theory. Hence the Special Theory will be revised to a certain extent in the following chapter. But in concluding this introduction with a simple and naive look at the phenomenological consequences of the General Theory we require already the general main results of the Special Theory. These are therefore first summarized in the next section.

11.2 Results of the Special Theory Summarized

The most famous result is *Einstein's formula*, *i.e.* the relation $E = mc^2$, expressing mass m as a form of energy E. The other general results concern measurements of length, time and mass of the same event in two different inertial reference frames moving with constant velocity v relative to another. A rod of initial length l_0 which moves towards or away from us with velocity v is seemingly contracted from l_0 to l, *i.e.*[†]

$$l_0 \longrightarrow l = l_0\sqrt{1 - \frac{v^2}{c^2}} \simeq l_0\left(1 - \frac{v^2}{2c^2}\right). \tag{11.1}$$

This is the *FitzGerald–Lorentz contraction* which applies to bulk matter; it is an apparent, and not a real contraction of the object. The FitzGerald–Lorentz contraction will be further discussed in Chapter 12 (*cf.* Example 12.4).

The time interval measured by a clock moving like the rod is increased from t_0 to t, *i.e.*

$$t_0 \longrightarrow t = \frac{t_0}{\sqrt{1 - v^2/c^2}}, \qquad \frac{1}{t} = \frac{1}{t_0}\sqrt{1 - \frac{v^2}{c^2}}, \tag{11.2}$$

and a mass m_0, in obsolete usage called *rest mass*, increases from m_0 to m, where m is in corresponding usage called *relativistic mass*, *i.e.*

$$m_0 \longrightarrow m = \frac{m_0}{\sqrt{1 - v^2/c^2}}. \tag{11.3}$$

(*cf.* Eqs. (12.44) to (12.45)). These observational, *i.e.* measurement effects are, of course, related. There are three main tests of general relativity already investigated by Einstein. These are briefly referred to in the subsections of the next section.

[†]Note that "length" is not to be regarded as a property of the rod, but as the result of a measurement operation.

11.3 Main Tests of General Relativity

11.3.1 The gravitational redshift

The gravitational redshift is probably the simplest of the three effects. The trick calculation proceeds as follows. We know the photon is massless like the graviton (*cf.* Eq. (15.107)), but using $E = mc^2$, we can attribute an effective inertial mass to the photon. Thus if the photon has frequency ν, we can define the effective mass by setting $E = h\nu$, where h is Planck's constant. In the gravitational field of the sun, say at the edge of the sun (of mass M, radius R), the energy of the photon will be modified by the gravitational potential to

$$h\nu' = h\nu - G\frac{Mm}{R}, \qquad m = \frac{h\nu}{c^2}.$$

Hence

$$h\nu' = h\nu\left(1 - \frac{GM}{Rc^2}\right), \qquad t' = \frac{t}{(1 - GM/Rc^2)}. \qquad (11.4)$$

Thus the effective (*i.e.* measured) frequency of the photon, ν', is reduced or "*redshifted*" by the factor in brackets. Since ν has dimension of (time)$^{-1}$, one interprets $1 - GM/Rc^2$ as the relativistic factor $\sqrt{1 - v^2/c^2} \simeq (1 - v^2/2c^2)$ in comparison with Eq. (11.2). Thus one identifies

$$v^2 \Longleftrightarrow \frac{2GM}{R}. \qquad (11.5)$$

Recalling the considerations of Example 2.3, we see that v is the *velocity of escape* from the massive body. This velocity is a characteristic quantity in the theory of gravitation. The velocity appears also in the Schwarzschild solution, as we shall see later (see discussion after Eq. (16.60)). The redshift here should be compared with the *blueshift* considered in Example 10.1, the differences arise from opposite signs of the gravitational potential. A more convincing derivation is given in Sec. 13.5.

11.3.2 The gravitational deflection of light

A velocity u (distance/time) measured in a gravitational field with a measuring rod which is contracted by the factor $(1 - GM/Rc^2)$, and a clock with time dilated by the factor $(1 - GM/Rc^2)$ (as in Eq. (11.4)) is reduced to

$$u\left(1 - \frac{GM}{Rc^2}\right)^2 \simeq u\left(1 - \frac{2GM}{Rc^2}\right).$$

In optics and electromagnetic theory the refractive index n is defined as the ratio:[‡]

$$n := \frac{\text{absolute velocity of light } (c)}{\text{measured velocity of light in gravitational field } (c')}.$$

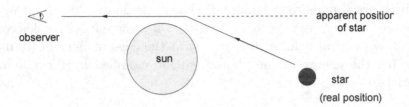

observer

sun

apparent position of star

star
(real position)

Fig. 11.1 The deflection of a ray of light in the field of the sun.

Thus we have:

$$n = \frac{c}{c(1 - 2GM/Rc^2)} \simeq 1 + \frac{2GM}{Rc^2}.$$

Hence gravity (G) has a refractive effect, as illustrated in Fig. 11.1 (see also the discussion at the end of Sec. 10.4). The gravitational deflection of a light ray as a consequence of Einstein's equation is treated in Sec. 16.6.

11.3.3   The precession of the planet Mercury's perihelion

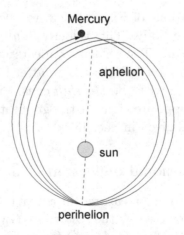

Mercury

aphelion

sun

perihelion

Fig. 11.2 Precession of Mercury's perihelion.

[‡]See books on electrodynamics, e.g. H. J. W. Müller–Kirsten [34], Sec. 11.4.2, Eq. (11.46).

The precession of the perihelion of Mercury (point of closest approach to the sun) means that its elliptic orbit is not closed and its perihelion precesses by a measurable amount, as illustrated in Fig. 11.2. It is said that the effect was already conceived by Newton. In 1859 Leverrier of France calculated that Mercury precessed by 574 seconds of arc per century. Of these 531 can be accounted for by perturbations due to other planets, not however, the remaining 43 seconds. It is these 43 seconds which Einstein was able to account for in 1915 with his spectacular calculation.

The trick calculation which allows a glimpse of how this comes about goes as follows. One starts from the attractive gravitational potential

$$V(r) = -\frac{GM}{r},$$

where r is the radius of the orbit of the planet with pole of polar coordinates at the focus, *i.e.* the sun. Assuming the relativistic contraction in the radial direction, this becomes

$$V(r) = -\frac{GM}{r(1 - GM/rc^2)}. \tag{11.6}$$

As in the Kepler problem considered in great detail in Chapter 7, we have for the *tangential velocity* $v = ds/dt$ Eq. (7.33), which is equivalent to the equation expressing conservation of energy, *i.e.*

$$\frac{1}{2}mv^2 + \text{const.} = \int^r f(r)dr, \quad f(r) = -\frac{mMG}{r^2}. \tag{11.7}$$

In order to avoid confusion, we verify in Example 11.1 that the tangential kinetic energy here is equal to the total polar kinetic energy. Thus, dividing the relation by the mass m and performing the integration, one obtains the relation

$$\frac{1}{2}v^2 = \frac{GM}{r} + \text{const.} \tag{11.8}$$

The idea is to insert this relation into Eq. (11.6), so that (ignoring the constant in Eq. (11.8) or setting this equal to zero)

$$V(r) = -\frac{GM}{r(1 - v^2/2c^2)} = -\frac{GM}{r} - \frac{GMv^2}{2rc^2}. \tag{11.9}$$

We now recall from Eqs. (7.28) and (7.29) the following relations in terms of the tangential coordinates p (the length of the perpendicular) and ϕ (angle between the tangent and the vector to the pole):

$$pv = \frac{l}{m}, \quad p = r\sin\phi.$$

Thus

$$v = \frac{l}{mr \sin \phi} \simeq \frac{l}{mr},\qquad(11.10)$$

where the approximation assumes a small eccentricity, or deviation of the orbit from a circle. Inserting this expression for v in Eq. (11.9), we obtain

$$V(r) = -\frac{GM}{r} - \frac{GMl^2}{2m^2c^2r^3}.\qquad(11.11)$$

Hence with $\alpha = l/m = \text{const.}$ we have

$$V = -\frac{GM}{r} - \frac{GM\alpha^2}{2r^3c^2},$$

and

$$F = -\frac{\partial V}{\partial r} = -\frac{GM}{r^2} - \frac{3GM\alpha^2}{2r^4c^2}.$$

We know from Chapter 7, Eq. (7.93), that this expression with the characteristic factor 3 is the force which yields the perihelion precession (the new contribution involving $u = 1/r$ with two powers higher than the Newton term).

The above calculations are not to be taken really seriously, but in presenting these here in a very brief form they may help to appreciate what some of the great achievements of the General Theory of Einstein — to be developed in the following — turn out to be.

Example 11.1: Tangential kinetic energy equals total polar kinetic energy

Show that the tangential kinetic energy of a particle of mass m in a central force field is equal to the polar kinetic energy, i.e.

$$\frac{1}{2}mv^2 = \frac{1}{2}m(\dot{r}^2 + r^2\dot{\theta}^2).$$

Solution: The tangential velocity of a particle was defined in Sec. 7.4, Eq. (7.28), as

$$v = \frac{ds}{dt},\qquad \text{with}\quad pv = \frac{l}{m} = \text{const.}$$

Here $\delta s \to ds$ is the element of arc of the orbit, l is the angular momentum, and p the length of the perpendicular from the pole to the tangent. We also had the relation (7.29): $p = r \sin \phi$. We thus have

$$v^2 = \frac{l^2}{m^2p^2} = \frac{l^2}{m^2r^2 \sin^2 \phi} = \frac{l^2}{m^2r^2}\operatorname{cosec}^2\phi.$$

With the relation (7.30), i.e. $\tan\phi = r(d\theta/dr)$, and the relation $\operatorname{cosec}^2\phi = 1 + 1/\tan^2\phi$, we can re-express v^2 as:

$$\begin{aligned}
v^2 &= \frac{l^2}{m^2r^2}\left[1 + \frac{1}{\tan^2\phi}\right] = \frac{l^2}{m^2r^2}\left[1 + \frac{1}{r^2(d\theta/dr)^2}\right] = \frac{l^2}{m^2r^2}\left[1 + \frac{\dot{r}^2}{r^2\dot{\theta}^2}\right]\\
&= \frac{l^2}{m^2r^2}[r^2\dot{\theta}^2 + \dot{r}^2]\frac{1}{r^2\dot{\theta}^2} = \frac{l^2}{m^2}[r^2\dot{\theta}^2 + \dot{r}^2]\frac{1}{r^4\dot{\theta}^2}\\
&= r^2\dot{\theta}^2 + \dot{r}^2,
\end{aligned}$$

since $l = mr^2\dot{\theta}$, as in Eq. (7.23).

Chapter 12

Aspects of Special Relativity

12.1 Introductory Remarks

The *Special Theory of Relativity* has already been frequently referred to. We repeat: What makes this theory *special* is that it considers motion far away from any gravitational effects (and hence excludes acceleration). Without acceleration the motion is that of inertial frames. But also in a gravitational field it is always possible to define a frame relative to which the field vanishes over a limited region of space and thus behaves like an inertial frame. We consider here the basic aspects (specifically Lorentz transformations) and vital consequences (as in electrodynamics), practically all of which were obtained by Einstein in his famous paper of 1905.* Our intention is mainly (1) to appreciate motion in inertial frames of reference, (2) to acquire familiarity with Lorentz indices and their manipulation, and (3) to obtain a glimpse of the significance of Lorentz transformations in electrodynamics. In electrodynamics the field is a 1-form, but the field strength a 2-form. This case is therefore simpler than the *General Theory of Relativity* with the gravitational field which is a 2-form. Thus we do not recapitulate every aspect of the Special Theory, in particular we do not rederive here the general form of the magnetic field observed in a reference frame which moves with a constant velocity relative to the rest frame of a charge e; the derivation of this relationship can be found in most books on electrodynamics.[†] The appearance of curvature is illustrated by appealing to examples. For a somewhat deeper appreciation of light cones associated with metrics we digress a little and consider briefly also the Schwarzschild metric, which, of course, is a topic of General Relativity, and will be treated in detail in Chapter 16.

---

*A. Einstein [14].

[†]See, for instance, J. D. Jackson [24] or H. J. W. Müller–Kirsten [34].

12.2   Basics and Physical Motivation of the Lorentz Transformation

The basic postulates of Einstein's Special Theory of Relativity are the following, of which the first is the most revolutionary postulate, and on the basis of previous classical mechanics the most difficult to comprehend (because the mind does not distinguish immediately between massive and massless particles):

(1) The velocity of light in vacuum has the same value c in all directions in all inertial frames, and is thus independent of the observer or source, and

(2) all inertial frames are equivalent.

An inertial frame is, as we saw, a frame of reference in which any free mass-point called particle moves with constant velocity, and the coordinates can be called inertial coordinates. In electrodynamics a constant was found to be important in linking units of electrostatics with those of electromagnetism, and it was found that this constant is precisely the velocity of light, c. The idea of a stationary ether was introduced solely for the purpose of understanding the propagation of electromagnetic waves in space, and this incorrect idea suggested that the velocity of light in a reference frame moving with velocity u relative to ether (c') would differ from that (c) in a frame stationary as the ether, $i.e.$ $c' = c - u.$[‡] In that case Maxwell's equations would have to be different in different reference frames, in order to allow for different values of c. The resolution of this apparent contradiction, between the constancy of the velocity of light (with its wavelike nature in Maxwell's electrodynamics but also pictureable as quanta of rest mass zero) on the one hand, and the equivalence of inertial frames (for particles of nonzero mass) on the other hand, was achieved with the union of space and time into the new concept of a *spacetime* with its own symmetries; this then embodies the constancy of the velocity of light in all directions (see below after Eq. (12.4b)), and with the resulting light cone the causal distinction between past and future. These connections were realized by Einstein and subsequently made clearer by Minkowski's introduction of the concept of spacetime and Poincaré's insight into the relevant invariance groups. Today it is clear that the photon of light, which in view of its masslessness does not have a rest frame, cannot be treated like a particle with inertia.

It is historically wellknown that the interferometer experiments of Michelson and Morley were planned to check the idea of an absolute velocity v of the Earth in the ether, which would imply an observable etherwind in the

---

[‡]See Sec. 5.3 on the Galilei transformation.

opposite direction. We can mimick[§] the experiment in the following way by considering a swimmer with velocity c in a stationary water (*i.e.* with stationary current) who swims in different directions in a river of constant velocity v.

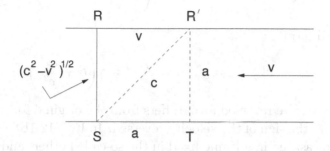

Fig. 12.1 The swimmer in a river of current velocity v.

We let $SR = ST = a$ as indicated in Fig. 12.1. Then the time taken by the swimmer in swimming from S to T and back is

$$t_l = \frac{a}{c-v} + \frac{a}{c+v} = \frac{2ac}{c^2 - v^2}.$$

In order to reach R from S the swimmer must swim to R'. The time taken by him to reach R from S is the same as the time taken to reach S from R. Thus the time taken by him to swim from S to R and back is

$$t_\perp = 2\frac{a}{\sqrt{c^2 - v^2}}.$$

Michelson and Morley observed in their experiment to high precision the equivalent of (here) $t_l = t_\perp$ for $v \neq 0$, which implied

$$t_\perp = \frac{2a}{\sqrt{c^2 - v^2}} \stackrel{?}{=} \frac{2ac}{c^2 - v^2} = t_l.$$

Lorentz observed that these expressions could be made compatible if parallel to v matter shrinks; this is the Lorentz contraction hypothesis, earlier put forward by FitzGerald and H. A. Lorentz and therefore also known as the *FitzGerald–Lorentz contraction*,[¶] meaning that in t_l

$$a \longrightarrow \frac{a\sqrt{c^2 - v^2}}{c} < a.$$

[§] See L. R. Lieber [29], or H. J. W. Müller–Kirsten [34], p. 383.
[¶] *Cf.* J. D. Jackson [24], p. 506.

This was rejected in particular by Poincaré (1904) who undertook a new look at the transformations now known as *Lorentz transformations* (but introduced earlier for other reasons by Voigt in 1887), *i.e.* the transformations

$$x' = \gamma(x-vt), \ \ y' = y, \ \ z' = z, \ \ t' = \gamma\left(t-\frac{vx}{c^2}\right), \ \ \gamma = \frac{1}{\sqrt{1-v^2/c^2}}, \quad (12.1a)$$

with reversed form

$$x = \gamma(x' + vt'), \ \ y = y', \ \ z = z', \ \ t = \gamma\left(t' + \frac{vx'}{c^2}\right), \quad (12.1b)$$

We observe that the reversed form differs from the original form in having — as expected — the sign of the velocity reversed. In Eq. (12.1b) x would be the coordinate measured in a frame fixed in the so-called ether, and x' that fixed on Earth, c being the same in both. Of course, we know today that the ether hypothesis was wrong (one of the many conclusions of Einstein in 1905); instead one considers fields in space. Also we know that the *FitzGerald–Lorentz contraction* is not a real contraction of an object, but only an apparent shrinkage which appears as the difference of measurements of observers in different inertial frames; the question of the *visibility* of the contraction is considered further in Example 12.4 where — following the arguments of Terrell and Penrose‖ — it is shown that by considering a body moving at high speed, the effect, *i.e.* the shrinkage $\sqrt{1-v^2/c^2}x' = (x-vt)_{\text{rest frame}}$, is not directly observable.** In view of the importance and later use of the Lorentz transformation we express it also in matrix form. Thus, now in the order ct, x, y, z, and with

$$(x'_\mu) = \begin{pmatrix} ct' \\ -x' \\ -y' \\ -z' \end{pmatrix}, \quad (x_\mu) = \begin{pmatrix} ct \\ -x \\ -y \\ -z \end{pmatrix}, \quad (12.1c)$$

$$x'_\mu = l_\mu{}^\nu x_\nu, \quad (l_\mu{}^\nu) = \begin{pmatrix} \gamma & \gamma\beta & 0 & 0 \\ \gamma\beta & \gamma & 0 & 0 \\ 0 & 0 & 1 & 0 \\ 0 & 0 & 0 & 1 \end{pmatrix}, \quad \beta = \frac{v}{c}, \quad \gamma = \frac{1}{\sqrt{1-\beta^2}}. \quad (12.1d)$$

‖See also J. Terrell [47] and R. Penrose [39].

**A lot of confusion can arise from an imprecise use of words. J. Terrell [47] therefore distinguishes very clearly between "observing" and "seeing", but concludes by saying that none of his "statements should be construed as casting any doubt on either the observability or the reality of the Lorentz contraction, as all the results given are derived from the special theory of relativity".

This is therefore how a covariant 4-vector $x_\mu = (x_0, x_i) = (ct, -x, -y, -z) \equiv (ct, -\mathbf{x})$ (with lower indices) transforms in Minkowski space $\mathbb{M}_4$, and other covariant vectors transform accordingly. The transformation describes the relation between inertial frames K and K' with coordinates x_μ, x'_μ respectively, and the $x-, y-, z-$axes parallel with frame K' moving with constant velocity $v = \beta c$ along the x-direction. Correspondingly one has for contravariant vectors $x^\mu = (x^0, x^i) = (ct, x, y, z) \equiv (ct, +\mathbf{x})$ (with upper indices), and

$$(x'^\mu) = \begin{pmatrix} ct' \\ x' \\ y' \\ z' \end{pmatrix}, \qquad (x^\mu) = \begin{pmatrix} ct \\ x \\ y \\ z \end{pmatrix}, \qquad (12.1e)$$

$$x'^\mu = l^\mu{}_\nu x_\nu, \quad (l^\mu{}_\nu) = \begin{pmatrix} \gamma & -\gamma\beta & 0 & 0 \\ \gamma\beta & \gamma & 0 & 0 \\ 0 & 0 & 1 & 0 \\ 0 & 0 & 0 & 1 \end{pmatrix}, \quad \beta = \frac{v}{c}, \qquad \gamma = \frac{1}{\sqrt{1-\beta^2}}.$$

$$(12.1f)$$

Covariant and *contravariant vectors* are explained in detail in Sec. 12.5. As examples of vectors of relevance later we mention the 4-momentum vectors

$$p^\mu = \left(\frac{E}{c}, \mathbf{p}\right), \quad p_\mu = \left(\frac{E}{c}, -\mathbf{p}\right),$$

and the wave 4-vector

$$k^\mu = \left(\frac{\omega}{c}, \mathbf{k}\right), \quad k_\mu = \left(\frac{\omega}{c}, -\mathbf{k}\right).$$

Thus these transform like x^μ of Eq. (12.1f).

The diagram in Fig. 12.2 shows an event P with coordinates (t, x, y, z) in one frame (that of observer A, say) and (t', x', y', z') in another (that of observer B, say). The proper understanding was achieved by Einstein with his interpretation of t and t' as measurements of clocks in different inertial frames.

12.3 Active and Passive Transformations

The transformation

$$x' = x\cos\theta + y\sin\theta, \qquad y' = -x\sin\theta + y\cos\theta \qquad (12.2)$$

describes the passage from coordinate frame K with coordinates (x, y) to a coordinate frame K' with coordinates (x', y'). Thus for $\theta = 0$ both frames coincide. K' is then rotated through angle θ so that the coordinates x, y in K become x', y' in K'.

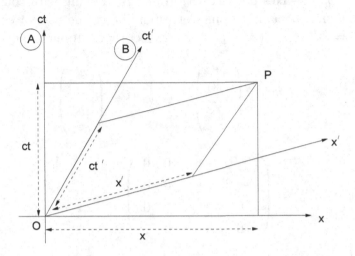

Fig. 12.2 The one event P seen by observers in two different inertial frames. Axes x', ct' are orthogonal even though they do not appear so.

The point $P(x, y)$ is not rotated. Such a rotation or transformation is described as *passive*. One also considers a so-called *active transformation*, as illustrated in Fig. 12.3. In the above example this is a rotation of the point P with the frame K' so that with respect to K' the coordinates remain unchanged (similarly later in an active Lorentz transformation length and time measurements are preserved).

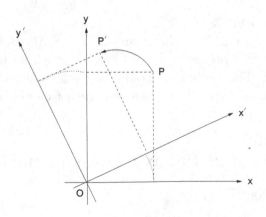

Fig. 12.3 Active rotation of P to P'.

Since the rotation is characterized by invariance of $x^2 + y^2$, i.e.

$$x^2 + y^2 = x'^2 + y'^2 = \text{const.} = a^2, \qquad (12.3)$$

a point subjected to a series of successive active rotations will appear in K to be moving in a circle as in Fig. 12.4. In this diagram lines of constant $x^2 + y^2$ are circles.

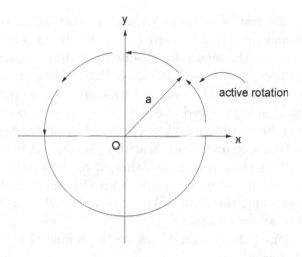

Fig. 12.4 The active rotation tracing a circle.

A *Lorentz transformation* is more complicated than the rotation in the plane, but we can consider it along similar lines as already briefly mentioned in Sec. 6.2. The distance s from the origin to the event point (t, x, y, z) in Minkowski space M_4 is given by

$$s^2 = c^2 t^2 - x^2 - y^2 - z^2, \qquad (12.4a)$$

or infinitesimally by

$$ds^2 = c^2 dt^2 - dx^2 - dy^2 - dz^2. \qquad (12.4b)$$

Corresponding to the rotational invariance of the expression (12.3), *i.e.* under transformations of the group $SO(2)$, so now the spacetime path-length squared s^2 is invariant under Lorentz transformations, *i.e.* those of the group $SO(3,1)$, with the velocity of light c remaining untouched, *i.e.* the same for all inertial frames related by a Lorentz transformation. We thus see how the equivalence of inertial frames (inside the future light cone) are related to the constancy of the velocity of light in all such frames. Since this is one of the hardest points to swallow for people entering the topic, and we are not addressing the great experts on relativity here, we repeat the argument. The transformation from one inertial frame of reference K with coordinates (ct, x, y, z) to another inertial frame K' with coordinates (ct', x', y', z') is given by a Lorentz transformation which is an element of the group $SO(3, 1)$ or $SO(1, 3)$, and this is defined as that for which

$$(ct)^2 - x^2 - y^2 - z^2 = (ct')^2 - x'^2 - y'^2 - z'^2 = \cdots = \text{invariant.}$$

Inertial frames are frames of particles with inertia, of mass m_i (say), and that means of mass unequal to zero ($m_i \neq 0$), which are travelling with constant speed (v) — the reason for this being that gravitational fields (bigg masses) are far away (*i.e.* negligible), so that (think of Newton's second law) $m_i \times$ acceleration $= 0$ (0 on the right meaning the gravitational force is zero). It follows that $m_i \times$ speed $=$ constant, or speed $=$ constant$/m_i$. Thus the mass cannot be zero (a massless particle has no inertia and hence no inertial or rest frame of reference), and the photon travelling with velocity c is excluded! This velocity c is not that of an inertial frame. Thus the vanishing mass conspires, so to speak, with the maximization of velocity to demand a constant, the finite velocity of light, which is the same in all directions and in all such frames.

The expressions (12.4a), (12.4b) are to be compared with the Cartesian equation of a hyperbola, *i.e.*

$$\frac{x^2}{a^2} - \frac{y^2}{b^2} = 1. \tag{12.5}$$

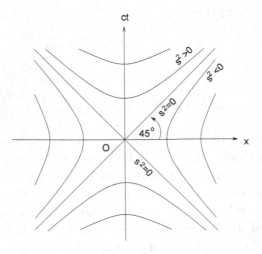

Fig. 12.5 Trajectories of constant $s^2 = c^2 t^2 - x^2$.

Lines of constant s are seen to have hyperbolic shape as expected (observe the minus sign in Eq. (12.5) which makes the difference to an ellipse). Just as the length of the radius a in Eq. (12.3) remains unchanged by the rotation (defined as an orthogonal transformation, *i.e.* an element of the group $SO(2)$), so the distance s remains unchanged, *i.e.* invariant under a Lorentz transformation which is an element of the noncompact group $SO(3,1)$. The diagram in Fig. 12.5 shows lines (or surfaces if $x^2 \to x^2 + y^2 + z^2$) of constant s^2 which are clearly hyperbolas. The coordinates of any point on the hyperbola may be expressed by means of the equations

$$x = \pm a \cosh \varphi, \qquad y = \pm b \sinh \varphi, \tag{12.6}$$

with φ varying from $-\infty$ to $+\infty$. We can analogously re-express the Lorentz transformation in terms of hyperbolic functions. To this end it is convenient to introduce a parameter ζ defined by

$$\tanh \zeta \equiv \beta = \frac{v}{c}, \quad \gamma \beta = \sinh \zeta, \quad \cosh \zeta \equiv \gamma = \frac{1}{\sqrt{1 - \beta^2}}. \tag{12.7a}$$

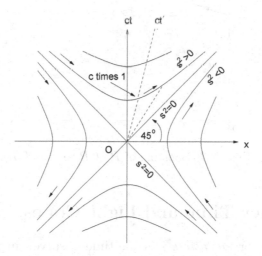

Fig. 12.6 The dotted lines indicate two unit clock measurements in K' at different velocities relative to K.

Then the transformation (for convenience we give it in two forms)

$$ct' = ct \cosh \zeta - x \sinh \zeta, \quad x' = -ct \sinh \zeta + x \cosh \zeta, \quad y' = y, \quad z' = z,$$

or

$$\begin{pmatrix} ct' \\ x' \\ y' \\ z' \end{pmatrix} = \begin{pmatrix} \cosh \zeta & -\sinh \zeta & 0 & 0 \\ -\sinh \zeta & \cosh \zeta & 0 & 0 \\ 0 & 0 & 1 & 0 \\ 0 & 0 & 0 & 1 \end{pmatrix} \begin{pmatrix} ct \\ x \\ y \\ z \end{pmatrix}, \tag{12.7b}$$

implies the invariance

$$s^2 = c^2 t^2 - \mathbf{x}^2 = c^2 t'^2 - \mathbf{x}'^2.$$

The minus sign here permits the distinction between "spacelike" directions ($\mathbf{x}^2 > c^2 t^2$), "timelike" directions ($c^2 t^2 > \mathbf{x}^2$), and "null" or "lightlike" directions ($c^2 t^2 = \mathbf{x}^2$).

An *active Lorentz transformation* as depicted in Fig. 12.6 is known as a *boost*. It helps to visualize this also with respect to the moving frame K' with coordinates $t', \mathbf{x}'$, in which the event point P retains its coordinates because P is dragged along with K' (*i.e.* length and time measurements are preserved under this *active* Lorentz transformation which can be looked at as the inverse of the Lorentz transformation (12.7a)). This is shown in Fig. 12.7 in which P, Q, R are *boosted* (as one says) to P', Q', R'.

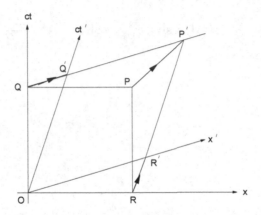

Fig. 12.7 Boosting of P, Q, R to P', Q', R'.

12.4   Proper Time and Light Cones

The *proper time* or *eigentime* τ is the time observed in the observer's rest frame, *i.e.* where $d\mathbf{x}^2 = 0$ (or $d\mathbf{x}^2 = 0$) ,

$$ds^2 = +c^2 d\tau^2.$$

The two dotted lines in Fig. 12.6 thus show unit clock measurements of observers moving in K' at different velocities relative to K. Similarly unit spatial measurements can be considered with respect to the right hand branch of the hyperbola. Thus these hyperbolas of $s^2 = \text{const.}$ represent curves (or surfaces) of unit time measurements in frames moving with different velocities v.

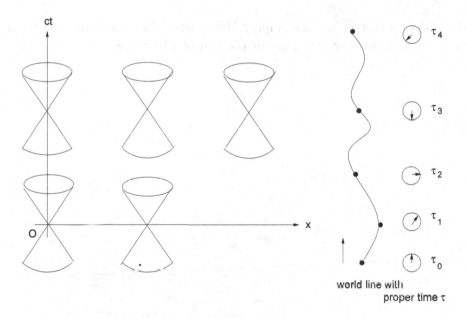

Fig. 12.8 The flat (or null) cones in Minkowski space,
parallel and with boundary at 45°.

This brings us to the topic of light cones and metric coefficients. In the flat Minkowskian metric the coefficients are constant, resulting in identical and parallel light cones or null cones at each spacetime point determined by $ds^2 = 0$. This is illustrated in Fig. 12.8 which shows the flat light cones (or *null cones*, 'null' referring to $s^2 = 0$, *i.e.* a vector of length zero, in Eq. (12.4a)). Each light cone at an event point P consists of two parts — the *past cone* representing the history of a flash of light imploding at P and the *future cone* of a flash of light emanating from P. In terms of proper time τ, we have

$$ds^2 = c^2 d\tau^2, \quad \tau = \int \frac{ds}{c}, \quad \overset{(12.4b)}{\tau_a} = \int_1^2 dt\sqrt{1 - \frac{v_a^2}{c^2}}, \quad \tau_b = \int_{1'}^{2'} dt\sqrt{1 - \frac{v_b^2}{c^2}}, \dots .$$

This is the time measurement of a clock moving along a time-like world-line between consecutive events (say) 1 and 2. Here causality comes in because the motion of light rays is given by $ds^2 = 0$, and no velocity can exceed c. In the General Theory of Relativity we shall encounter very different and nonflat metrics. For purposes of illustration we consider a space with the different (and imagined) metric (observe the factor t^2 in the coefficient of dx^2 in this illustration)

$$ds^2 = c^2 dt^2 - a^2 t^2 (dx^2 + dy^2 + dz^2), \quad a^2 = \text{const}.$$

What does a metric like this imply? We observe that the light cone given by $ds^2 = 0$ now changes its shape in the course of time as shown in Fig. 12.9 for $t = t_1 < t_2 < t_3$.

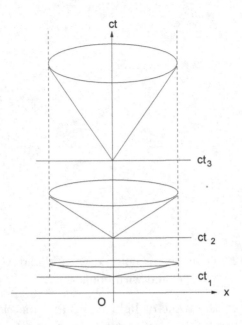

Fig. 12.9 The cones $cdt = atdx$, $dt/dx \propto t$, for $t_1 < t_2 < t_3$.

We see that t again measures the proper time $(c^2 d\tau^2 = c^2 dt^2)$, but along *e.g.* a curve with t, y, z constant:

$$ds^2 = -a^2 t^2 dx^2 \quad (t, y, z \text{ const.})$$

Thus $ds^2 = -a^2 t^2 dx^2$ measures a distance for different constant values of t. This case is called that of an *expanding universe*. We mention here only briefly as a matter of interest that later in General Relativity (Chapter 16) we shall come across other metrics, such as the *Schwarzschild metric* in spherical coordinates, *i.e.* (here m, called *"geometric mass"* — with dimension of a length, thus in SI units in meters — is an integration constant, as we shall see in Chapter 16, where (*cf.* Eq. (16.60)) it will be identified as $m = 2GM/c^2$, the spherical mass M being the source of the gravitational field):

$$ds^2 = -\left(1 - \frac{2m}{r}\right)c^2 dt^2 + \left(1 - \frac{2m}{r}\right)^{-1} dr^2 + r^2(d\theta^2 + \sin^2\theta d\phi^2), \quad (12.8)$$

which implies a more complicated light cone behaviour as Fig. 12.10 shows. We do not enter into a derivation and deeper discussion of the Schwarzschild

spacetime here (this will be done in Chapter 16), but point out a few properties and indicate how one arrives at the picture of *tilting light cones* (which
for $r \to \infty$ become again those for a flat spacetime) as compared to the 45°
upright light or null cones in Special Relativity.

We observe first that at $r = 2m$ the Schwarzschild metric becomes singular owing to the factor in front of dr^2, but is otherwise — as we shall see in
Chapter 16 — on either side an acceptable solution of Einstein's equation.
In order to recognize $r = 2m$ as a singularity note that at $r = 2m$, as in
Fig. 12.10, ct runs through all possible values — which is analogous to what
happens at the pole $r = 0$ of polar coordinates (r, θ) when r approaches the
fixed value zero: A circle with $-\pi \le \theta \le \pi$ collapses to a point.* This singularity at $r = 2m$ is a result of the coordinates used and is therefore not a
physical singularity. With different coordinates the singularity at $r = 2m$ can
be avoided. One such procedure is the following. One sets — this expression
is known as *Eddington–Finkelstein coordinate* —

$$v := ct + r + 2m \ln \left| \frac{r}{2m} - 1 \right|, \qquad (12.9a)$$

so that after transformation to v (replacing t) in the metric, this becomes

$$ds^2 = -\left(1 - \frac{2m}{r}\right) dv^2 + 2dvdr + r^2(d\theta^2 + \sin^2 d\phi^2). \qquad (12.9b)$$

In this transformation

$$
\begin{aligned}
dv &= \left(\frac{\partial v}{\partial t}\right) dt + \left(\frac{\partial v}{\partial r}\right) dr = cdt + \left[1 + \frac{2m}{(r - 2m)}\right] dr \\
&= cdt + \left(\frac{r}{r - 2m}\right) dr = cdt + \left(1 - \frac{2m}{r}\right)^{-1} dr,
\end{aligned}
$$

so that

$$\left(1 - \frac{2m}{r}\right)^2 c^2 dt^2 = \left(1 - \frac{2m}{r}\right)^2 dv^2 + dr^2 - 2drdv \left(1 - \frac{2m}{r}\right),$$

and thus yields Eq. (12.9b):

$$
\begin{aligned}
\therefore ds^2 \overset{(12.8)}{=} & -\left(1 - \frac{2m}{r}\right) dv^2 - \left(1 - \frac{2m}{r}\right)^{-1} dr^2 + 2rdrdv \\
& + \left(1 - \frac{2m}{r}\right)^{-1} dr^2 + r^2(d\theta^2 + \sin^2 \theta d\phi^2).
\end{aligned}
$$

---

*See also the comparison with the squashing of an egg crate to zero volume in C. W. Misner,
K. S. Thorne and J. A. Wheeler [33], p. 11 and pp. 822 – 825.

From $ds^2 = 0$ (which determines the null or light cone) for $d\theta = d\phi = 0$ *i.e.* θ, ϕ constant), we obtain from Eq. (12.9b)

$$dv\left[2dr - \left(1 - \frac{2m}{r}\right)dv\right] = 0.$$

The solutions of this equation are (observe we are considering this as effectively two-dimensional)

$$v(r) = \text{const.} \quad \text{and} \quad \frac{dv}{dr} = \frac{2r}{r - 2m} \equiv \frac{2}{1 - 2m/r}. \qquad (12.9c)$$

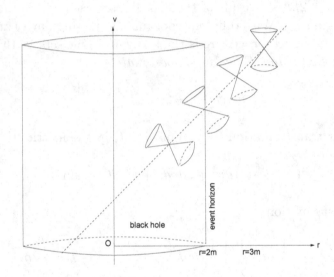

Fig. 12.10 Tilting of light cones towards spatial origin in Schwarzschild metric.

We now draw $v = \text{const.}$ at $45°$ to the ct-axis and consider the slope of the opposite arm of the cone for $r = \infty, 3m, 2m, m$. The boundary $r = 2m$ is called *event horizon* ($r = 0$ is clearly a singular point). This horizon is a trap: Ingoing matter and radiation, *i.e.* that which crosses the horizon into the interior, the so-called *black hole*, can never get out again. This can be inferred from the tilting of the light cones towards the singularity $r = 0$ as indicated in Fig. 12.10 and the fact, that propagation of anything is possible only into the (interior of the) forward light cone (propagation into an outside region would require a velocity larger than that of light, c). The null cones are tangent to the horizon, as we see there. In Fig. 12.11 we illustrate the approach to the event horizon with a larger view. One should observe that the tilting of the instantaneous light cones is such that motion of matter is always confined to the interior, radiation to the hull, and is

directed towards the central singularity. At the event horizon one arm of the
light cone (sketched two-dimensionally) is tangential to the horizon and once
the forward part enters the black hole, there is no going back, and there is no
situation with the backward part inside the black hole and the forward part
outside. Thus matter can only go in, but never come out. These directions
are dictated by the slope dv/dr of Eq. (12.9c). Thus the tilting is such that
once one has entered the black hole domain, *e.g.* with a spaceship, there is
no escape, although (with a moderate solar mass) the crossing of the horizon
would barely be noted. However, from then on tidal forces (*cf.* Example 10.3)
become effective and eventually the journey becomes a catastrophe.[†]

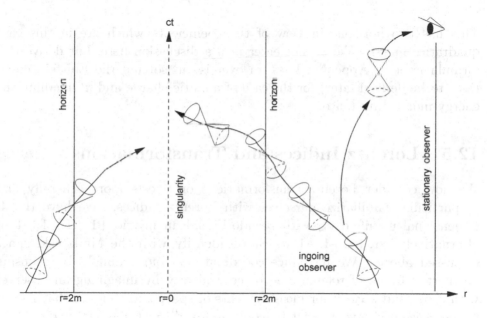

Fig. 12.11 A stationary observer and an observer passing the horizon
into the black hole.

It is now of interest — also in comparison with Special Relativity — to
explore the meaning of proper time in the Schwarzschild spacetime geometry.
An increment $d\tau$ of the proper time τ of a stationary clock at a radius r has
$dr = d\theta = d\phi = 0$, and

$$ds^2 \rightarrow -c^2 d\tau^2 = -c^2 \left(1 - \frac{2m}{r} \right) dt^2, \quad d\tau = \left(1 - \frac{2m}{r} \right)^{1/2} dt.$$

Thus t becomes proper time τ at $r = \infty$. It follows that t is not the time
measured by a clock outside of the gravitating mass M. But clocks at the

---

[†]See also R. Penrose [38], p. 713.

same radius run at the same rate. If τ_∞ is the proper time measured by a clock far away, we have $d\tau_r/d\tau_\infty \simeq (1-m/r) < 1, d\tau_\infty > d\tau_r$. Thus a clock at r runs slower than a clock at infinity. This effect is called *"gravitational time dilation"*. Correspondingly the radial increment of proper distance obtained by setting $dt = d\theta = d\phi = 0$ is not dr but $dr/(1 - 2m/r)^{1/2}$.

We end with a comment: A generalization of the Schwarzschild metric (containing electromagnetic charge Q) is the *Reissner–Nordstrom metric* given by

$$ds_4^2 = -\left(1 - \frac{2m}{r} + \frac{Q^2}{r^2}\right)c^2dt^2 + \left(1 - \frac{2m}{r} + \frac{Q^2}{r^2}\right)^{-1}dr^2 + d\Omega_2^2 r^2.$$

This has two horizons in view of the coefficients which are in this case quadratics in $1/r$. We do not enter into a discussion here, but derive this formula later in Appendix B as an exercise in solving the Einstein equation (to be derived later) for the case of a static charge and in handling the energy-momentum tensor.

12.5   Lorentz Indices and Transformations

We now consider Lorentz transformations or boosts more generally and in particular familiarize ourselves with Lorentz indices. We have the 4-dimensional manifold M_4 with pseudo-Euclidean metric $+1, -1, -1, -1$ or alternatively $-1, +1, +1, +1$ which we identify with the Minkowski space discussed above. We introduce coordinates in our manifold by choosing an inertial frame of reference — or equivalently by imagining an observer with a rod and a clock for measurements of space and time, so that he can characterize each event point by coordinates $x^0 = ct$ and x^1, x^2, x^3. These 4 coordinates together are written x^μ where μ is a *Lorentz index* and the vector is called a *contravariant vector*. A 4-dimensional vector space S_4 in which a metric is defined, *i.e.* an interval ds between two neighbouring points $x^\mu, x^\mu + dx^\mu$ given by

$$\mathbb{R}_4 : \qquad ds^2 = \sum_{\mu,\nu} g_{\mu\nu} dx^\mu dx^\nu,$$

with coefficients $g_{\mu\nu}$ specified in some coordinate frame at every point $x^\mu \in \mathsf{S}_4$, is called a *4-dimensional Riemann space* $\mathbb{R}_4$. Clearly the 4-dimensional Minkowski space M_4 is a particular case of $\mathbb{R}_4$ with metric given by

$$\mathsf{M}_4 : \qquad ds^2 = dx^{0^2} - d\mathbf{x}^2, \quad x^0 = ct,$$

and the coefficients $g_{\mu\nu}$ given by $\eta_{\mu\nu}$, i.e.

$$
\eta_{\mu\nu} = \begin{pmatrix} +1 & 0 & 0 & 0 \\ 0 & -1 & 0 & 0 \\ 0 & 0 & -1 & 0 \\ 0 & 0 & 0 & -1 \end{pmatrix}.
$$

12.5.1 Contravariant vectors and covariant vectors

We assume $\{x^\mu\}$ are the coordinates of a point P in $\mathbb{M}_4$ relative to some given coordinate frame which we do not need to specify further. We now let x'^μ be the coordinates of the same point with respect to another frame (in Special Relativity these will both be inertial frames) and we assume that the two systems of coordinates are related by equations

$$
x'^\mu = x'^\mu(x^p).
$$

The coordinates of P in the first frame are x^μ, and those of a neighbouring point P' are $x^\mu + dx^\mu$, and in the second frame the latter are

$$
x'^\mu + dx'^\mu = x'^\mu(x^\mu + dx^\mu),
$$

or (without $\sum_\nu$ summation over repeated indices being understood)

$$
dx'^\mu = \sum_\nu \frac{\partial x'^\mu}{\partial x^\nu} dx^\nu \equiv \frac{\partial x'^\mu}{\partial x^\nu} dx^\nu. \tag{12.10}
$$

We use this relation to define quite generally a *contravariant vector* A^μ by saying that $\{A^\mu\}$ are the components of this contravariant vector at the point x^ν if the components in the primed frame are given by

$$
A'^\mu(x') = \frac{\partial x'^\mu}{\partial x^\nu} A^\nu(x). \tag{12.11}
$$

In the *Special Theory of Relativity* one considers transformations from one inertial frame to another described by Lorentz boosts in the three possible space directions. Of course, rotations around these three axes are also possible and are contained in the transformations called *homogeneous Lorentz transformations* given by:

$$
\mathsf{L}_6^+ : \qquad x^\mu \longrightarrow x'^\mu = l^\mu{}_\nu x^\nu. \tag{12.12}
$$

The matrices $(l^\mu{}_\nu)$ define a 4×4 representation of the noncompact Lorentz group $SO(3,1)$ with 6 independent parameters corresponding to 3 angles and 3 velocities. The "+" on $\mathbb{L}_6^+$ indicates that the transformations are those which evolve continuously from the identity. If we add spacetime translations a^μ to $\mathbb{L}_6^+$, *i.e.* $\mathbb{T}_4$, we have the 10-parameter *Poincaré group* of transformations given by

$$\mathbb{L}_6^+ \oplus \mathbb{T}_4 : \qquad x^\mu \longrightarrow x'^\mu = l^\mu_{\ \nu} x^\nu + a^\mu. \tag{12.13}$$

Thus in Special Relativity Eq. (12.11) becomes

$$A'^\mu(x') = l^\mu_{\ \nu} A^\nu(x). \tag{12.14}$$

We now consider a second vector B^ν at a point $y^\mu \in \mathsf{M}_4$ in the system of unprimed coordinates. Then in the primed frame

$$B'^\mu(y') = l^\mu_{\ \nu} B^\nu(y). \tag{12.15}$$

Here the sum of two vectors is still a vector, *i.e.*

$$A'^\mu(x') + B'^\mu(y') = l^\mu_{\ \nu} [A^\nu(x) + B^\nu(y)]. \tag{12.16}$$

But in the general case of Eq. (12.11) this is not the case, *i.e.*

$$A'^\mu(x') + B'^\mu(y') \neq \frac{\partial x'^\mu}{\partial x^\nu} [A^\nu(x) + B^\nu(y)].$$

This point becomes *a crucial issue* in the case of tensors and is treated in detail in Sec. 14.5. One can say, a contravariant vector can be defined only at a single point $x^\mu \in \mathbb{R}_4$.

A function F which remains unaltered in value when the reference frame is changed is called a *scalar* or *invariant* of M_4, *i.e.*

$$F'(x') = F(x). \tag{12.17}$$

But

$$dF'(x') = \frac{\partial F'(x')}{\partial x'^\lambda} dx'^\lambda = dF(x) = \frac{\partial F(x)}{\partial x^\lambda} dx^\lambda, \tag{12.18}$$

or

$$\frac{\partial F'(x')}{\partial x'^\lambda} \frac{\partial x'^\lambda}{\partial x^\alpha} dx^\alpha = \frac{\partial F(x)}{\partial x^\alpha} dx^\alpha.$$

From Eq. (12.18) we obtain the transformation of the gradient, *i.e.*

$$\frac{\partial F'(x')}{\partial x'^\mu} = \frac{\partial x^\beta}{\partial x'^\mu} \frac{\partial F(x)}{\partial x^\beta}. \tag{12.19}$$

Comparing Eq. (12.19) with Eq. (12.11), we see that the gradient $\partial F'/\partial x'^\mu$ does not transform as a contravariant vector. The gradient is taken to be the prototype of a different type of vector called *covariant vector*. Thus one defines a *covariant vector* $B_\nu(x)$ at x^μ by the transformation property

$$B'_\mu(x') = B_\nu(x)\frac{\partial x^\nu}{\partial x'^\mu}. \tag{12.20}$$

In particular we have

$$x'_\mu = x_\nu(x)\frac{\partial x^\nu}{\partial x'^\mu}. \tag{12.21}$$

In the case of Special Relativity we can invert the Lorentz transformation (12.12) and obtain

$$x^\nu = l^\nu{}_\mu x'^\mu \tag{12.22a}$$

with

$$l^\nu{}_\mu l^\mu{}_\rho = \delta^\nu_\rho. \tag{12.22b}$$

With the latter we can also verify that

$$x'^\mu x'_\mu = x^\nu x_\nu.$$

In the case of a Lorentz boost along the x-axis we have, for instance (*cf.* Eq. (12.1f)),

$$(l^\nu{}_\mu) = \begin{pmatrix} \gamma & -\gamma\beta & 0 & 0 \\ -\gamma\beta & \gamma & 0 & 0 \\ 0 & 0 & 1 & 0 \\ 0 & 0 & 0 & 1 \end{pmatrix}, \qquad \beta = \frac{v}{c}, \qquad \gamma = \frac{1}{\sqrt{1-\beta^2}}. \tag{12.22c}$$

12.5.2 Tensors

If $A^\mu(x)B^\nu(x)$ is the product of components of two contravariant vectors, the $4 \times 4 = 16$ such quantities are taken as the components of a contravariant tensor T of rank two with transformation:

$$A'^\mu(x')B'^\nu(x') = \frac{\partial x'^\mu}{\partial x^\alpha}\frac{\partial x'^\nu}{\partial x^\beta}A^\alpha(x)B^\beta(x),$$

or simply

$$T'^{\mu\nu}(x') = \frac{\partial x'^\mu}{\partial x^\alpha}\frac{\partial x'^\nu}{\partial x^\beta}T^{\alpha\beta}(x). \tag{12.23}$$

A quantity like $T^\mu{}_\nu$ is called a *mixed tensor*, the *covariant tensor* being defined appropriately, *i.e.* by

$$S'_{\mu\nu}(x') = S_{\alpha\beta}(x)\frac{\partial x^\alpha}{\partial x'^\mu}\frac{\partial x^\beta}{\partial x'^\nu}. \tag{12.24}$$

The mixed tensor is defined later explicitly in Sec. 14.2.

12.6 Lorentz Boosts in Electrodynamics

The best and clearest application of the previous and some earlier considerations is that to electrodynamics, *i.e.* the case of a static electric charge in
frame K. Then if an initially coinciding frame K' (not with the charge) is
subjected to a Lorentz boost with velocity $\mathbf{v}$ in the x-direction, this charge
appears in K' to move with velocity $\mathbf{v}$ in the opposite direction. The observer
in the transformed or boosted frame K' then observes also a magnetic field
due to the motion of the charge. This can readily be verified by considering
the appropriate transformation of the electromagnetic field tensor $F^{\mu\nu}$. In
electrodynamics one learns that the fields $\mathbf{E}$ and $\mathbf{B}$ are gauge-invariant observables given by components of $F^{\mu\nu}$. Thus in general the electromagnetic
field tensor is (the upper and lower indices being field component indices)

$$F^{\mu\nu}(x) = \begin{pmatrix} 0 & \frac{E^1}{c} & \frac{E^2}{c} & \frac{E^3}{c} \\ -\frac{E^1}{c} & 0 & -B_3 & B_2 \\ -\frac{E^2}{c} & B_3 & 0 & -B_1 \\ -\frac{E^3}{c} & -B_2 & B_1 & 0 \end{pmatrix}. \tag{12.25}$$

In the case of a static charge we have $B_i = 0, i = 1, 2, 3$. If we now subject
the frame K' to a boost along the $x-$axis, the field tensor observed in K'
becomes:

$$F'^{\mu\nu}(x') = \frac{\partial x'^{\mu}}{\partial x^{\alpha}} \frac{\partial x'^{\nu}}{\partial x^{\beta}} F^{\alpha\beta}(x) = l^{\mu}{}_{\alpha} l^{\nu}{}_{\beta} F^{\alpha\beta}(x) = l^{\mu}{}_{\alpha} F^{\alpha\beta} l^{T}{}_{\beta}{}^{\nu}. \tag{12.26}$$

For

$$F^{\alpha\beta}(x) = \begin{pmatrix} 0 & \frac{E^1}{c} & \frac{E^2}{c} & \frac{E^3}{c} \\ -\frac{E^1}{c} & 0 & 0 & 0 \\ -\frac{E^2}{c} & 0 & 0 & 0 \\ -\frac{E^3}{c} & 0 & 0 & 0 \end{pmatrix}$$

we find by multiplying out the product on the right hand side of Eq. (12.26):*

$$F^{\mu\nu}(x) = \begin{pmatrix} 0 & \frac{E^1}{c} & \gamma\frac{E^2}{c} & \gamma\frac{E^3}{c} \\ -\frac{E^1}{c} & 0 & -\gamma\beta\frac{E^2}{c} & -\gamma\beta\frac{E^3}{c} \\ -\gamma\frac{E^2}{c} & \gamma\beta\frac{E^2}{c} & 0 & 0 \\ -\gamma\frac{E^3}{c} & \gamma\beta\frac{E^3}{c} & 0 & 0 \end{pmatrix}.$$

---

*For the explicit multiplication see H. J. W. Müller–Kirsten [34], pp. 402 – 403; in Eq. (17.47)
there $-E_2/c$ should read $-\gamma E_2/c$.

Thus by comparison of the elements of this matrix with those of (12.25) with all field components there replaced by primed ones, we obtain the equations:

$$
\begin{array}{c|c}
E^{1\prime} = E^1 & B^{1\prime} = 0 \\[4pt]
E^{2\prime} = \gamma E^2 & B^{2\prime} = -\gamma\beta\frac{E^3}{c} \\[4pt]
E^{3\prime} = \gamma E^3 & B^{3\prime} = \gamma\beta\frac{E^2}{c}.
\end{array}
\tag{12.27}
$$

Thus the observer in K' observes also a magnetic field.

We close this digression into electrodynamcis by recalling that the action of the free electromagnetic field is given by

$$
S = \int \mathcal{L} d^4 x, \quad L = \int d^3 x \mathcal{L}, \quad \mathcal{L} = -\frac{1}{4} F^{\mu\nu} F_{\mu\nu}.
\tag{12.28}
$$

It is fairly clear that the Lagrange density $\mathcal{L}$ is a Lorentz invariant. One can also show that the volume element is Lorentz invariant, *i.e.* $d^4 x' = d^4 x$.

12.7 Curvature due to Lorentz Contraction

We consider an observer O' in an inertial frame with coordinates x', y', z', as shown in Fig. 12.12. A disc is allowed to rotate about the z'-axis with uniform angular velocity ω about the z'-axis. An observer at rest on the disc, *i.e.* in the noninertial frame with coordinates x, y, z, as shown, puts his measuring rod along radii to A, A', and measures lengths r. Thus the figure traced out by points $A, A', \ldots$ is a circle and observer O' agrees with his measurements.

Now observer O puts his rod in tangential directions perpendicular to a radius and measures the length of the circumference of the disc and finds a length L. Observer O' also measures, but obtains

$$
L\sqrt{1 - \frac{\omega^2 r^2}{c^2}}
$$

with FitzGerald–Lorentz contraction in the direction of motion, *i.e.* of velocity $v = \omega r$. Since O' is measuring in an inertial frame where the geometry is Euclidean, he will say

$$
2\pi r = L\sqrt{1 - \frac{\omega^2 r^2}{c^2}},
\tag{12.29}
$$

i.e.

$$
\frac{L}{r} = \frac{2\pi}{\sqrt{1 - \omega^2 r^2/c^2}} \simeq 2\pi\left(1 + \frac{\omega^2 r^2}{2c^2}\right).
\tag{12.30}
$$

Thus the figure or circle observed by O' subtends at the axis an angle larger than 2π, and consequently this observer (who is familiar with Fig. 10.11 and the discussion there) will claim the space to be curved. The larger r, the more pronounced the effect will be. Thus the observer O' can directly observe that space is curved.

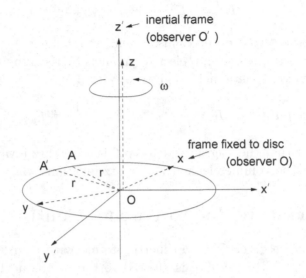

Fig. 12.12 Observation of curvature resulting from FitzGerald–Lorentz contraction.

12.8   Covariantization of Newton's Equation of a Charged Particle

Our objective now is to re-express the equation of motion of a Newtonian particle with charge q in the presence of an electromagnetic field (*i.e.* the Lorentz force) in covariant form, *i.e.* the equation we encountered several times earlier,

$$\frac{d\mathbf{p}}{dt} = q(\mathbf{E} + \mathbf{v} \times \mathbf{B}). \tag{12.31}$$

One should note that the left side of this equation consists of the vector $\mathbf{p}$ and the 0-component or d/dt of the 4-vector ∂_μ of a specific Lorentz frame. It is therefore necessary to change something in order to achieve the same transformation property on the left of the equation as on the right. This is done by replacing t by an invariant quantity. In order to generalize the *Newton equation*

$$\frac{d}{dt}\mathbf{p} = \mathbf{F}, \quad \frac{d}{dt}p^i = F^i,$$

relativististically, one introduces therefore the *proper time* or *eigentime* τ (introduced already in Sec. 12.4), defined by

$$(ds)^2 = c^2(dt)^2 - (d\mathbf{x})^2 = c^2(dt)^2\left[1 - \frac{1}{c^2}\left(\frac{d\mathbf{x}}{dt}\right)^2\right] = c^2(d\tau)^2. \qquad (12.32a)$$

The parameter τ is the time, measured by a clock fixed in the particle, *i.e.* in its *rest frame* (where $d\mathbf{x}^2 = 0$). This is the reason that τ is called the *eigentime of the particle*. With

$$\mathbf{v} = \frac{d\mathbf{x}}{dt}, \qquad v_i = \frac{dx_i}{dt} \qquad (12.32b)$$

(which is constant for inertial frames), this is the important relation which describes the so-called "*clock paradox*" (see Example 12.8), *i.e.*[†]

$$d\tau = \frac{dt}{\gamma} = \sqrt{1 - \beta^2}dt, \qquad \gamma^2 = \frac{1}{1 - \beta^2}. \qquad (12.33)$$

Since $d\tau > dt$, this relationship expresses, so to speak, that a stationary clock runs faster than a moving clock, a phenomenon known as "*time dilation*" (*cf.* the discussion in Jackson [24] with detailed explanation of experimental verifications; see also Example 12.8).[‡] Since $(ds)^2$ is Lorentz invariant, then also $(d\tau)^2$ or $\sqrt{(d\tau)^2} = d\tau$. This suggests to generalize Newton's equation as the following which also defines the *Minkowski force* on the right (with $\alpha = 0, 1, 2, 3$):

$$\frac{d}{d\tau}(mu_\alpha) = K_\alpha \equiv \text{Minkowski force.} \qquad (12.34a)$$

At this point we also introduce the mass m (frequently called "*rest mass*" m, in order to distinguish this from the "*relativistic mass*" $m/\sqrt{1 - \beta^2}$). The quantity u_α is the *4-velocity* (also to be understood as *tangential vector* with respect to the Minkowski manifold, *i.e.* tangential to the particle's worldline in the future light cone as explained in Sec. 12.9)

$$u_\alpha = \frac{dx_\alpha}{d\tau} = \frac{dx_\alpha}{\sqrt{1 - \beta^2}dt} \qquad (12.34b)$$

with (*cf.* (12.32b))

$$u_i := -\frac{1}{\sqrt{1 - \beta^2}}\frac{dx_i}{dt} = -\frac{v_i}{\sqrt{1 - \beta^2}}, \qquad (12.35)$$

[†]For curvilinear coordinate systems the corresponding equation is $d\tau = \sqrt{g^{\rho\kappa}dx_\rho dx_\kappa}/c$.
[‡]J. D. Jackson [24], pp. 520 – 532. There and elsewhere it is called 'dilatation'.

and

$$u_0 := \frac{dx_0}{d\tau} = \frac{cdt}{\sqrt{1-\beta^2}dt} = \frac{c}{\sqrt{1-\beta^2}}. \tag{12.36}$$

The 4-*form equation* (12.34a) can now be written

$$\frac{d}{d\tau}(mu_\alpha) = \frac{1}{\sqrt{1-\beta^2}}\frac{d(mu_\alpha)}{dt} = K_\alpha. \tag{12.37}$$

The spatial part implies (multiplied by $\sqrt{1-\beta^2}$)

$$-\frac{d}{dt}\left(m\frac{v_i}{\sqrt{1-\beta^2}}\right) = \sqrt{1-\beta^2}K_i.$$

Momentum conservation for $K_i = 0$ implies the identification

$$\text{momentum}: \quad p_i = \frac{mv_i}{\sqrt{1-\beta^2}}. \tag{12.38}$$

Hence also

$$F_i := -\sqrt{1-\beta^2}K_i. \tag{12.39}$$

The significance of the time component of the 4-form equation (12.37) can be seen as follows. We multiply the equation by u^α:

$$u^\alpha\frac{d}{d\tau}(mu_\alpha) = u^\alpha K_\alpha = \frac{d}{d\tau}\left(\frac{1}{2}mu^\alpha u_\alpha\right). \tag{12.40a}$$

However,

$$u^\alpha u_\alpha = (u_0)^2 - (u_i)^2 = \frac{c^2}{1-\beta^2} - \frac{v^2}{1-\beta^2} = \frac{c^2-v^2}{1-\beta^2} = c^2. \tag{12.40b}$$

Hence the right hand side of Eq. (12.40a) vanishes, and we have

$$0 = u^\alpha K_\alpha, \quad i.e. \quad u_0 K_0 = u_i K_i,$$

or with Eqs. (12.35) and (12.39):

$$\frac{c}{\sqrt{1-\beta^2}}K_0 = \frac{v_i}{\sqrt{1-\beta^2}}\frac{F_i}{\sqrt{1-\beta^2}},$$

i.e.

$$K_0 = \frac{1}{c}\frac{\mathbf{v}\cdot\mathbf{F}}{\sqrt{1-\beta^2}}. \tag{12.41}$$

Thus the fourth component of Eq. (12.37) becomes:

$$\frac{d(mc/\sqrt{1-\beta^2})}{\sqrt{1-\beta^2}dt} = \frac{1}{c}\frac{\mathbf{v}\cdot\mathbf{F}}{\sqrt{1-\beta^2}},$$

i.e.

$$\frac{d}{dt}\left(\frac{mc^2}{\sqrt{1-\beta^2}}\right) = \mathbf{v}\cdot\mathbf{F}, \tag{12.42}$$

i.e.

$$\frac{dT}{dt} = \mathbf{v}\cdot\mathbf{F}, \qquad T = \frac{mc^2}{\sqrt{1-\beta^2}}. \tag{12.43}$$

Since $\mathbf{v}\cdot\mathbf{F}$ = work per unit time (or *power*), T is the *total energy*, *i.e.*

$$T = \frac{mc^2}{\sqrt{1-\beta^2}} = mc^2 + \frac{1}{2}mv^2 + O\left(\frac{1}{c^2}\right). \tag{12.44}$$

We observe that for $v = 0$ (*i.e.* in the particle's rest frame) one has $T = mc^2$. This is the reason why the mass m is in common usage (see also remarks after Eq. (12.48)) described as the particle's *rest mass* as compared to its *relativistic mass* $m/\sqrt{1-\beta^2}$. Thus the particle's relativistic mass as observed in a frame (usually as good as an inertial frame) which is moving with velocity $\mathbf{v}$ relative to the particle is *larger than the mass in its rest frame*. Then

$$p_\alpha = mu_\alpha = \left(\frac{T}{c}, -p_i\right) = \left(\frac{T}{c}, -\frac{mv_i}{\sqrt{1-\beta^2}}\right), \tag{12.45}$$

with

$$p_0 = mu_0 = \frac{mc}{\sqrt{1-\beta^2}} = \frac{T}{c}. \tag{12.46}$$

We see that the zero-component of the 4-momentum p_α (or the energy-momentum 4-vector) is the total energy (apart from the factor c) and this total energy contains the energy of the mass m. Hence there is no separate conservation of mass — rather conservation of momentum, energy and mass, these together are all contained in the *law of conservation of 4-momentum*. A typical example to illustrate this important point is electron-positron annihilation into photons which is treated in Example 12.6. Thus in some reaction between particles the sum of the 4-momenta initially is equal to the sum of the 4-momenta finally. For $v \ll c$, Eq. (12.42) reproduces Newton's equation of motion if we insert the expansion (12.44) there:

$$\frac{d}{dt}\left(\frac{1}{2}mv^2\right) = \mathbf{v}\cdot\mathbf{F}, \qquad \text{or} \qquad \frac{d}{dt}(m\mathbf{v}) = \mathbf{F}.$$

From Eqs. (12.45) and (12.40b), we obtain in addition the following central and fully informative statement

$$p^\alpha p_\alpha = m^2 u^\alpha u_\alpha = m^2 c^2, \tag{12.47}$$

where the right hand side is the constant (in the corresponding representation) of the Lorentz group. In components the relation is

$$p_0^2 - \mathbf{p}^2 = m^2 c^2,$$

or

$$p_0^2 = \left(\frac{T}{c}\right)^2 = \mathbf{p}^2 + m^2 c^2. \tag{12.48}$$

For $\mathbf{p} = 0, T \equiv E$, this entails the wellknown *Einstein formula* $E = mc^2$ which identifies mass (in common — now obsolete — usage called rest mass) as the energy of the system in the rest frame divided by c^2.[§] But note again the difference with Eq. (12.46) in which $T = m_{\rm rel}.c^2, m_{\rm rel.} = m/\sqrt{1-\beta^2}$. We also observe that one can have a particle of zero rest mass, yet with energy $T = pc$. The typical example for this is the *photon, i.e.* the corpuscle of electromagnetic radiation.

Finally we consider the Minkowskian generalization of the Lorentz force. In comparison with Eq. (12.25) we note that with the metric $\eta_{\mu\nu}$ of Sec. 12.5 we have $E^i = -E_i$. Thus the Lorentz force generalized to four dimensions is

$$K^\alpha \equiv qF^{\alpha\beta}u_\beta$$

$$= q\begin{pmatrix} 0 & -\frac{E_1}{c} & -\frac{E_2}{c} & -\frac{E_3}{c} \\ \frac{E_1}{c} & 0 & -B_3 & B_2 \\ \frac{E_2}{c} & B_3 & 0 & -B_1 \\ \frac{E_3}{c} & -B_2 & B_1 & 0 \end{pmatrix}\begin{pmatrix} \frac{c}{\sqrt{1-\beta^2}} \\ \frac{-v_x}{\sqrt{1-\beta^2}} \\ \frac{-v_y}{\sqrt{1-\beta^2}} \\ \frac{-v_z}{\sqrt{1-\beta^2}} \end{pmatrix}$$

and so

$$K^\alpha = \frac{q}{\sqrt{1-\beta^2}}\begin{pmatrix} \frac{\mathbf{E}\cdot\mathbf{v}}{c} \\ E_1 + v_y B_3 - v_z B_2 \\ E_2 + v_z B_1 - v_x B_3 \\ E_3 + v_x B_2 - v_y B_1 \end{pmatrix}$$

$$= \frac{q}{\sqrt{1-\beta^2}}\begin{pmatrix} \frac{\mathbf{E}\cdot\mathbf{v}}{c} \\ \mathbf{E}+\mathbf{v}\times\mathbf{B} \end{pmatrix}. \tag{12.49}$$

[§]Thus saying "rest mass" in "rest frame" is like saying "snow is white", *i.e.* this is a tautology (saying the same thing twice). The only mass defined in modern physics is that in the invariant (12.47). Nonetheless, we shall use the term "rest mass" occasionally for reasons of clarity, *e.g.* in Example 12.5.

The equation

$$\frac{dp^\alpha}{d\tau} = \frac{d(mu^\alpha)}{d\tau} = K^\alpha$$

together with

$$\frac{dp^\alpha}{d\tau} = \frac{1}{\sqrt{1-\beta^2}}\frac{d}{dt}(p_0, \mathbf{p}) = \frac{1}{\sqrt{1-\beta^2}}\frac{d}{dt}\left(p_0, \frac{m\mathbf{v}}{\sqrt{1-\beta^2}}\right)$$

on the left hand side and (12.49) on the right, yields first for the vector part and then for the zero-component (and dividing out $\sqrt{1-\beta^2}$) the equations

$$\frac{d\mathbf{p}}{dt} \equiv \frac{d}{dt}\left(\frac{m\mathbf{v}}{\sqrt{1-\beta^2}}\right) = q(\mathbf{E} + \mathbf{v}\times\mathbf{B}),$$

$$\frac{dp_0}{dt} = q\frac{\mathbf{E}\cdot\mathbf{v}}{c}, \qquad i.e. \qquad \frac{d}{dt}\left(\frac{T}{c}\right) = q\frac{\mathbf{E}\cdot\mathbf{v}}{c},$$

and so

$$\underbrace{\frac{dT}{dt}}_{\text{power}} = q\underbrace{\frac{\mathbf{E}\cdot d\mathbf{x}}{dt}}_{\text{work/time}}. \tag{12.50}$$

We see that the 4-form

$$K^\alpha = qF^{\alpha\beta}u_\beta$$

is the Lorentz force supplemented by the *power* as the additional fourth component.

12.9 The Tangent Vector

The preceding considerations permit us to introduce briefly the first step towards a geometric description of the laws of physics. Relativity theory is pervaded today by the idea that every physical quantity must be describable by a geometric object, and that the laws of physics must be expressible as relationships between these. The main feature of these geometric objects in spacetime is their independence of a specific coordinate system or reference frame. The most elementary geometric object in spacetime is a point $\mathcal{P}$, generally called an "*event*". The geometric generalization of the vector linking two such points is the quantity known as the "*tangent vector*". Also the metric, tensors *etc.* can be looked at geometrically. Here we concentrate on the tangent vector using considerations of Sec. 12.8. How can a path be traced by an event $\mathcal{P}$ in spacetime in such a way that this is independent of any specific reference frame? We observed in Sec. 12.8 that proper time

τ is an invariant quantity. Thus it is suggestive to consider the event point $\mathcal{P}$ as $\mathcal{P}(\tau)$. How would this description be related to a specific Lorentzian coordinate $x_\mu(\tau)$? The Lorentzian coordinate $x_\mu(\tau)$ is the coordinate with respect to some reference frame with orthonormal basis vectors $\mathbf{e}^\mu$. Hence the world line is given by

$$\mathcal{P}(\tau) = x_\mu(\tau)\mathbf{e}^\mu.$$

The 4-velocity of a particle is now the spacetime displacement per unit of proper time along a straight-line approximation of the worldline, *i.e.*

$$u := \frac{d\mathcal{P}}{d\tau} = \frac{dx_\mu(\tau)}{d\tau}\mathbf{e}^\mu \equiv u_\mu(\tau)\mathbf{e}^\mu$$

with the components of ordinary velocity $\mathbf{v}$ given by

$$\mathbf{v} = \frac{d\mathbf{x}}{dt}, \qquad v_i = \frac{dx_i}{dt}.$$

as in Eq. (12.32b). We observe that the velocity $u = d\mathcal{P}(\tau)/d\tau$ is independent of any reference to a specific reference frame, and recalling Eq. (12.34b) we can identify u_μ as

$$u_\mu = \frac{dx_\mu}{d\tau}.$$

Our earlier Eq. (12.40b) tells us that $u_\mu u^\mu = c^2$. In order to relate this to u^2, one defines the symbol called "metric" g or "metric coefficients" $\eta_{\mu\nu}$ by the relation

$$\eta_{\mu\nu} \equiv g(\mathbf{e}_\mu, \mathbf{e}_\nu) = \mathbf{e}_\mu \cdot \mathbf{e}_\nu,$$

where in any Lorentz frame

$$(\eta_{\mu\nu}) = \begin{pmatrix} 1 & 0 & 0 & 0 \\ 0 & -1 & 0 & 0 \\ 0 & 0 & -1 & 0 \\ 0 & 0 & 0 & -1 \end{pmatrix}.$$

Then

$$u^2 = u^\mu(\tau)u^\nu(\tau)\mathbf{e}_\mu \cdot \mathbf{e}_\nu = u^\mu(\tau)\eta_{\mu\nu}u^\nu(\tau) = u^{0^2} - (u^i)^2 = u_\mu u^\mu = c^2.$$

Thus with $c = 1$ the velocity $u = d\mathcal{P}/d\tau$ is a unit vector. We can rewrite Eq. (12.45) now as the geometric (frame independent) "momentum vector"

$$p = mu, \qquad u = \frac{d\mathcal{P}}{d\tau}.$$

For further details we refer to Misner, Thorne and Wheeler [33], pp. 47 – 53.

12.10   Miscellaneous Examples

A large number of interesting and highly instructive examples of effects of Special Relativity are considered in detail and solved in the book *"The Physics Coaching Class in Mechanics"* [48]. We restrict ourselves therefore here to a few typical, and — in some cases — maybe, more involved examples.

Example 12.1: Law of addition of velocities

The reference frames K, K', K'' have parallel axes. Frame K' moves in the direction of the z-axes with velocity v away from K, and K'' parallel to these with velocity u away from K'. What is the velocity w of K'' relative to K?

Solution: We use the Lorentz transformations in the form $x' = Ax$, where $x = (x, y, z, ct)$ and

$$A \equiv A_v = \begin{pmatrix} 1 & 0 & 0 & 0 \\ 0 & 1 & 0 & 0 \\ 0 & 0 & \cosh\psi & -\sinh\phi \\ 0 & 0 & -\sinh\phi & \cosh\phi \end{pmatrix}, \tag{12.51}$$

where (see Eqs. (12.7a), (12.7b))

$$\cosh\phi = \frac{1}{\sqrt{1-\beta^2}}, \quad \sinh\phi = \frac{\beta}{\sqrt{1-\beta^2}}, \quad \beta = \frac{v}{c}. \tag{12.52}$$

Performing two Lorentz transformations one after the other, and using the relations $\cosh(x \pm y) = \cosh x \cosh y \pm \sinh x \sinh y$, $\sinh(x \pm y) = \sinh x \cosh y \pm \cosh x \sinh y$, we obtain

$$A_w = A_u A_v = \begin{pmatrix} 1 & 0 & 0 & 0 \\ 0 & 1 & 0 & 0 \\ 0 & 0 & \cosh(\phi+\phi') & -\sinh(\phi+\phi') \\ 0 & 0 & -\sinh(\phi+\phi') & \cosh(\phi+\phi') \end{pmatrix}$$

$$\equiv \begin{pmatrix} 1 & 0 & 0 & 0 \\ 0 & 1 & 0 & 0 \\ 0 & 0 & \cosh\overline{\phi} & -\sinh\overline{\phi} \\ 0 & 0 & -\sinh\overline{\phi} & \cosh\overline{\phi} \end{pmatrix}, \tag{12.53}$$

where $\phi, \phi', \overline{\phi}$ correspond to the velocities u, v, w. We obtain therefore $\overline{\phi} = \phi + \phi'$, where

$$\cosh\overline{\phi} = \frac{1}{\sqrt{1-w^2/c^2}}, \quad \cosh\phi = \frac{1}{\sqrt{1-v^2/c^2}}, \quad \cosh\phi' = \frac{1}{\sqrt{1-u^2/c^2}}.$$

With

$$\cosh\overline{\phi} \equiv \cosh(\phi+\phi') = \cosh\phi\cosh\phi' + \sinh\phi\sinh\phi',$$

and the relations (12.52), we obtain

$$\frac{1}{\sqrt{1-w^2/c^2}} = \frac{1}{\sqrt{1-v^2/c^2}}\frac{1}{\sqrt{1-u^2/c^2}} + \frac{v/c}{\sqrt{1-v^2/c^2}}\frac{u/c}{\sqrt{1-u^2/c^2}}$$

$$= \frac{1+(uv/c^2)}{\sqrt{1-u^2/c^2}\sqrt{1-v^2/c^2}}. \tag{12.54}$$

Squaring both sides and taking the reciprocal and then the square root we obtain

$$\frac{w^2}{c^2} = 1 - \frac{(1 - u^2/c^2)(1 - v^2/c^2)}{(1 + uv/c^2)^2} = \frac{(u/c + v/c)^2}{(1 + uv/c^2)^2},$$

i.e.

$$w = \frac{u + v}{1 + uv/c^2}. \tag{12.55}$$

This result was already given by Einstein [14].

Example 12.2: The relativistic aberration law

A transverse wave of frequency $\nu = \omega/2\pi$ propagates in a direction of angle α with the x−axis of inertial frame K of the source (in applications this source can, for example, be the end of a moving stick from which light is scattered). The source moves in the x−direction at speed $v < c$ towards

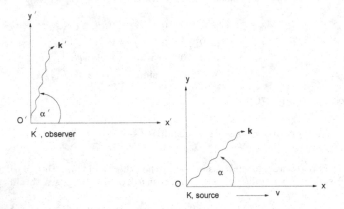

Fig. 12.13 Light leaving frame K which moves relative to frame K'.

an observer with rest frame K', the x'−axis being parallel to the x−axis, as indicated in Fig. 12.13. Determine the frequency ν' observed in K' and its angle of observation.

Solution: Let k^μ be the wave vector with components $(\omega/c, \mathbf{k})$. The Lorentz transformation of this vector is obtained from Eq. (12.1f), *i.e.*

$$\begin{pmatrix} \omega'/c \\ k'_x \\ k'_y \\ k'_z \end{pmatrix} = \begin{pmatrix} \gamma & -\gamma\beta & 0 & 0 \\ -\gamma\beta & \gamma & 0 & 0 \\ 0 & 0 & 1 & 0 \\ 0 & 0 & 0 & 1 \end{pmatrix} \begin{pmatrix} \omega/c \\ k_x \\ k_y \\ k_z \end{pmatrix},$$

so that

$$\omega' = \gamma(\omega - \beta k_x c), \quad k'_x = \gamma(-\beta\omega/c + k_x), \quad k'_y = k_y, \quad k'_z = k_z.$$

As given: $k_x = k\cos\alpha, k_y = k\sin\alpha, k_z = 0, \omega = kc$.

$$\therefore \quad \omega' = \gamma(\omega - \beta ck\cos\alpha) = \gamma(1 - \beta\cos\alpha)\omega = k'c,$$

and hence

$$\nu' = \frac{\omega'}{2\pi} = \gamma\frac{(1 - \beta\cos\alpha)}{2\pi}\omega = \gamma(1 - \beta\cos\alpha)\nu.$$

Also:

$$k'_x = \gamma(k_x - \beta\omega/c) = \gamma k(\cos\alpha - \beta).$$

Hence the angle α' which $\mathbf{k}'$ makes with the x'−axis is given by

$$\cos\alpha' = \frac{k_x'}{k'} = \frac{\gamma k(\cos\alpha - \beta)c}{\gamma kc(1 - \beta\cos\alpha)} = \frac{\cos\alpha - \beta}{1 - \beta\cos\alpha}. \tag{12.56}$$

This is the most general form of the *law of aberration* (*aberration* being an apparent straying from a path caused by an observer's motion), as Einstein remarks on p. 912 of his paper [14]. We observe that at an angle α or a velocity v so that $\cos\alpha = \beta, \sin\alpha = \sqrt{1 - \beta^2}$, the angle α' crosses $\pi/2$. Or when $\alpha = \pi/2$ we have $\cos\alpha' = -\beta, \sin\alpha' = \sqrt{1 - \beta^2}$, and the latter is the FitzGerald–Lorentz contraction factor. Thus in this special case the contraction can be related to the projection resulting from an apparent rotation through an angle $(\pi/2 - \alpha')$, the light ray being what can be registered on a photographic plate. Hence in either of these special situations the aberration implies a factor $\sqrt{1 - \beta^2}$ as in the FitzGerald–Lorentz contraction. This is an aspect which is elaborated on in Example 12.4.

Example 12.3: The Doppler effect
Light is emitted from a source in the inertial frame K' and observed in frame K, with frame K' moving at speed v away from K, the axes being parallel. ν is the frequency observed by the observer in frame K and ν' is the frequency of the light emitted by the source in frame K'. Derive the relation between ν and ν' when
(a) K, K' approach each other,
(b) K, K' recede from each other,
(c) K, K' pass each other in perpendicular positions, as indicated in Fig. 12.14.

Solution: We consider inertial frame K (frame of the observer), K' (frame of the light source) with axes parallel and K' moving along the x−direction with velocity v away from K. The 4-momentum $p^\mu = (E/c, \mathbf{p})$ transforms like x^μ in Eq. (12.1f), i.e. $(\beta = v/c, \gamma = 1/\sqrt{1 - \beta^2})$

$$\begin{pmatrix} E'/c \\ p_x' \\ p_y' \\ p_z' \end{pmatrix} = \begin{pmatrix} \gamma & -\gamma\beta & 0 & 0 \\ -\gamma\beta & \gamma & 0 & 0 \\ 0 & 0 & 1 & 0 \\ 0 & 0 & 0 & 1 \end{pmatrix} \begin{pmatrix} E/c \\ p_x \\ p_y \\ p_z \end{pmatrix} = \begin{pmatrix} \gamma E/c - \gamma\beta p_x \\ -\gamma\beta E/c + \gamma p_x \\ p_y \\ p_z \end{pmatrix}.$$

Similarly we have for the wave vector $k^\mu = (\omega/c, \mathbf{k})$:

$$\begin{pmatrix} \omega'/c \\ k_x' \\ k_y' \\ k_z' \end{pmatrix} = \begin{pmatrix} \gamma & -\gamma\beta & 0 & 0 \\ -\gamma\beta & \gamma & 0 & 0 \\ 0 & 0 & 1 & 0 \\ 0 & 0 & 0 & 1 \end{pmatrix} \begin{pmatrix} \omega/c \\ k_x \\ k_y \\ k_z \end{pmatrix} = \begin{pmatrix} \gamma\omega/c - \gamma\beta k_x \\ -\gamma\beta\omega/c + \gamma k_x \\ k_y \\ k_z \end{pmatrix}.$$

Thus with v the velocity of K' going away from K we obtain from the last equation:

$$\omega' = \gamma(\omega - \beta c k_x), \quad k_x' = \gamma(k_x - \beta\omega/c), \quad k_y' = k_y, \quad k_z' = k_z.$$

More generally with $\boldsymbol{\beta}\cdot\mathbf{k} = \beta k_x = \beta k\cos\theta$ the first expression is $\omega' = \gamma(\omega - c\boldsymbol{\beta}\cdot\mathbf{k})$ which reduces to the derived expression for $\mathbf{k} = k\mathbf{e}_x$. In view of γ there is a transverse Doppler shift even for $\theta = \pi/2$. The inverse transformation is

$$\omega = \gamma(\omega' + \beta c k_x'), \quad k_x = \gamma(k_x' + \beta\omega'/c), \quad k_y = k_y', \quad k_z = k_z', \quad \text{and} \quad \omega = kc, \quad \omega' = k'c. \tag{12.57}$$

(a) In the case of source and observer approaching each other we have:
Since K' is to approach K, its velocity relative to K is $v' = -v$, with $\beta' = -\beta, \gamma' = \gamma$ and $k_x' = -k' = -\omega'/c$, so that, also with $\omega = 2\pi\nu, \omega' = 2\pi\nu'$,

$$\omega = \gamma(\omega' - \beta' c k_x') = \gamma(\omega' + \beta'\omega') = \frac{\omega'(1 + \beta')}{\sqrt{1 - \beta'^2}} = \omega'\sqrt{\frac{1 + \beta'}{1 - \beta'}}, \quad \nu = \nu'\sqrt{\frac{1 + \beta'}{1 - \beta'}}.$$

(b) In the case of source and observer going in opposite directions we have $\beta \to -\beta', \gamma' = \gamma, k'_x = k' = \omega'/c$:

$$\omega = \gamma(\omega' - \beta'ck'_u) = \gamma(1 - \beta')\omega' = \omega'\sqrt{\frac{1-\beta'}{1+\beta'}}, \quad \nu = \nu'\sqrt{\frac{1-\beta'}{1+\beta'}}.$$

(c) In the case of source and observer passing each other at right angles we proceed as follows. We let the source in K' be at position $(0, y', 0)$ and the observer in K at $(0, 0, 0)$ when they pass each other at $t = t' = 0$ with $k'_x = 0, k'_y = -k, k'_z = 0$. Then Eq. (12.57) gives

$$\omega = \gamma(\omega' + \beta ck'_x) = \gamma\omega' = \frac{\omega'}{\sqrt{1-\beta^2}}, \quad \nu = \frac{\nu'}{\sqrt{1-\beta^2}}.$$

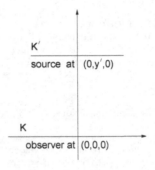

Fig. 12.14 Source and observer passing at right angles.

The relativistic Doppler shift has been observed spectroscopically with atoms in motion.[¶] Also note that the plane wave phase ϕ is an invariant, *i.e.* $\phi = \omega t - \mathbf{k} \cdot \mathbf{x} = \omega't' - \mathbf{k}' \cdot \mathbf{x}'$.

Example 12.4: FitzGerald–Lorentz contraction: Seeing vs. observing

Explain with the help of references of Terrell [47] and Penrose [39] (also [38], pp. 428 - 431) the difference between "seeing" and "observing" the FitzGerald–Lorentz contraction. Can the contraction be "seen"?[‖]

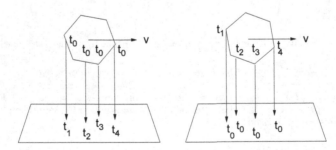

Fig. 12.15 The difference between observing and seeing a fast moving object.

[¶]See *e.g.* J. D. Jackson [24], p. 522.

[‖]The reader is cautioned that this is a rather difficult problem so that unless he has time to ponder over every word, he might prefer to skip it.

Solution: Following Terrell [47] we first distinguish between *"observing"* and *"seeing"*. In observing a moving object by means of light all light quanta leave the object at the same time t_0 as in Fig. 12.15 but arrive at the observer's position at different times (remember the velocity of light is finite, hence light requires a longer time to reach the eye from a point farther away) . In seeing the object by eye or on a photographic plate, all light quanta must arrive there simultaneously at time t_0, having left different points of the object at various earlier times. The Special Theory of Relativity implies that the FitzGerald–Lorentz contraction can be observed by a suitable experiment. Here we are concerned with the other issue of its visibility in view of the finiteness of the velocity of light — can it be seen (ignoring practical aspects)? We define angles θ, θ' as in Fig. 12.16. The relation between the angles θ and θ' is that of the relativistic aberration law, Eq. (12.56), of Example 12.2, which is for the angles defined in the present context and with $\beta = v/c$:

$$\cos\theta = \frac{\cos\theta' - \beta}{1 - \beta\cos\theta'}, \quad \sin\theta = \frac{\sqrt{1-\beta^2}\,\sin\theta'}{1 - \beta\cos\theta'}.$$

Recall the meaning of these equations. Light originating in frame K in direction θ is observed in frame K' (which moves with velocity v relative to K) in direction θ'. For $\theta = \pi/2$, at the distance of closest approach of the fast moving object to observer O, we have $\cos\theta' = \beta$, $\sin\theta' = \sqrt{1-\beta^2} = \cos(\theta - \theta')|_{\theta-\pi/2}$. Here (cf. Fig. 12.16)

$$(\theta - \theta') \quad = \quad \text{at same time, place :}$$

(known orientation θ of object — apparent orientation θ' of object).

Here θ is the angle at which the object (moving with velocity v parallel to $\theta = 0$ past the stationary observer at O) appears in frame K. θ' is the angle of the apparent direction in which (at the same time and place) an observer O' to whom the object appears stationary would see it.

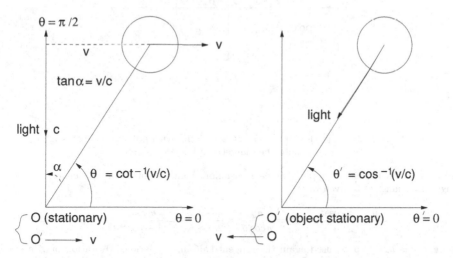

Fig. 12.16 Circular object as seen by observers O, O' momentarily at the same point (overlap graphs).

Since we are concerned with a visual problem here, *i.e.* light reaching our eye or the shutter of a camera, we are concerned with light rays travelling from various points of the rapidly moving object and reaching our eye simultaneously. The most straight-forward argument to demonstrate the invisibility of the FitzGerald–Lorentz contraction is that of Penrose [39]. The aberration formula is first re-expressed using the tangent-half-angle relation for a cosine, *i.e.*

$$\cos\theta = \frac{1 - t^2}{1 + t^2}, \quad \sin\theta = \frac{2t}{1 + t^2}, \quad t = \tan\left(\frac{\theta}{2}\right),$$

and then using *componendo et dividendo*. The aberration law then assumes the simple form given by Penrose, *i.e.***

$$\tan\left(\frac{\theta'}{2}\right) = \sqrt{\frac{1-\beta}{1+\beta}}\tan\left(\frac{\theta}{2}\right). \tag{12.58}$$

Penrose now considers the stereographic projection of a sphere (say of radius a) with centre at the point of observation O onto its equatorial plane from the pole at $\theta = \pi$. In the still simpler case of a circle of radius a projected on its diametrical line through O as depicted in Fig. 12.17 we have

$$x^2 + y^2 = a^2, \quad x = -a + r\cos\vartheta, \quad y = r\sin\vartheta,$$

so that with some algebra

$$y = a\sin 2\vartheta, \quad r = 2a\cos\vartheta,$$

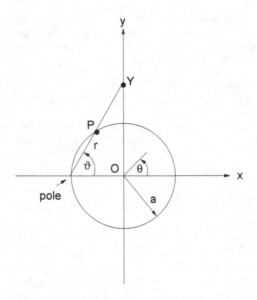

Fig. 12.17 Projection of a point P of a circle (with polar coordinates r, ϑ)
onto the diametrical line at Y.

and a point on the circle at polar angle ϑ is mapped into point Y on the y−axis which (from congruent triangles) is given by

$$\frac{y}{Y} = \frac{r\cos\vartheta}{a}, \quad Y = \frac{ay}{r\cos\vartheta} = \frac{a^2\sin 2\vartheta}{2a\cos^2\vartheta} = \frac{a\sin 2\vartheta}{1+\cos 2\vartheta} = a\tan\vartheta,$$

where in the last step we used again the tangent-half-angle relations. Rotation of Y around the x−axis then produces a circle. Using a circle of radius b with polar angle ϑ', one obtains for the point Y: $Y = b\tan\vartheta'$, so that

$$\tan\vartheta = \frac{b}{a}\tan\vartheta'.$$

***The corresponding expression for infinitesimal areas on the unit sphere, derivable from the aberration law, was given by J. Terrell [47], *i.e.*

$$\frac{d\theta'}{d\theta} = \frac{\sin\theta'}{\sin\theta} = \frac{\sqrt{1-\beta^2}}{1+\beta\cos\theta}, \quad d\phi' = d\phi.$$

Hence the three-dimensional version sends circles into circles and the factor b/a, or in the *aberration law* (12.58) the factor $\sqrt{(1-\beta)/(1+\beta)}$, merely expresses an expansion of the line or plane of projection. Thus a sphere is seen by both observers at O and O' as a circular object, though they see it of different size, and the FitzGerald–Lorentz contraction of the fast moving sphere into an ellipsoid is not visible.

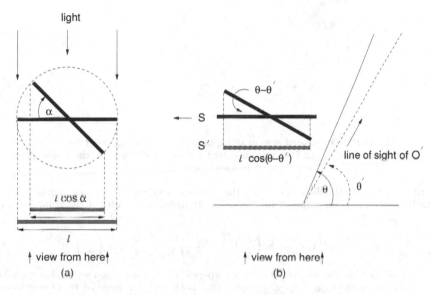

Fig. 12.18 (a) A rotation changing the apparent length of a stick from l to $l\cos\alpha$; (b) Observer O''s line of sight with length l of S (seen by observer O instantaneously coincident) aberrated to $l\cos(\theta - \theta')$.

The corresponding conclusion for the general case seems to have been inferred by Terrel [47], *i.e.* that both observers see identical pictures except for magnification, this conformality being a property of relativistic aberration. In a more detailed investigation of what actually happens, the shortening of the length of a stick as a consequence of an apparent rotation plays an important role. This (nonrelativistic) shortening of a stick (when viewed in a perpendicular direction from below against the direction of light) is illustrated in Fig. 12.18(a). We see that the broad-view length l of the stick observed from below (or on a photographic plate there) becomes $l\cos\alpha$ when the stick is rotated through an angle α. A similar projection or apparent rotation results in the relativistic case from multiplication by the factor $\cos(\theta - \theta')$ in the viewing of an object by the observer O' when instantaneously at the same time and point as observer O, as sketched in Fig. 12.18(b). This one-dimensional object, however, does not permit as clear a distinction between this apparent rotation and the FitzGerald–Lorentz contraction as the sphere that we consider for this reason in more detail.

Consider now again the case of the sphere, as this permits the clearest picture. In Fig. 12.19 the stationary sphere is shown with its lower hemisphere visible to observer O. When in motion the sphere is contracted in the direction of motion. This Lorentz contracted sphere of uncontracted diameter D is shown in Fig. 12.19 travelling to the right with velocity v relative to observer O. The sphere is viewed from O at $\theta = \pi/2$. The contracted diameter is $D\sqrt{1-\beta^2}$. However, viewed in motion (*i.e.* with light received from it) the length along the direction of motion is $D\sqrt{1-\beta^2}\cos(\theta - \theta') = D(1-\beta^2)$ owing to its apparent rotation (the effect of aberration). This is the distance along the direction of motion between the farthest simultaneously visible points A and B of the sphere: B is the farthest visible point because light leaving points on the sphere to the right of B are intercepted by the motion of the sphere. Correspondingly light emitted from A will not be stopped by the sphere which moves out of the path of the light.

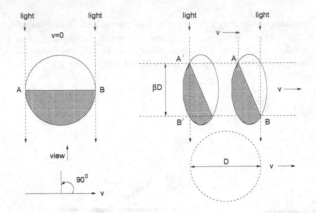

Fig. 12.19 The shaded area of the sphere shows its visible part for $v = 0$;
the shaded area of the Lorentz-contracted sphere, an ellipsoid, is the visible part
for $v \neq 0$ with AB tilted.

From the equation of an ellipse we obtain the distance between the farthest points A and B in Fig. 12.19 in the direction perpendicular to the motion. We have (with A the point (x, y) on the ellipse in Fig. 12.20)

$$\frac{x^2}{a^2} + \frac{y^2}{b^2} = 1, \quad x^2 = a^2\left[1 - \frac{y^2}{b^2}\right] = \left(\frac{D}{2}\right)^2\left[1 - \frac{D^2(1-\beta^2)^2 2^2}{2^2 D^2(1-\beta^2)}\right] = \left(\frac{D}{2}\right)^2\beta^2, \quad \therefore x = \beta\frac{D}{2}.$$

Thus the light reaching the observer O from A perpendicular to the motion leaves A at time $\beta D/c = Dv/c^2$ earlier than the light from B in order that both rays are received by O simultaneously. During this time the sphere moves the distance $Dv^2/c^2 = D\beta^2$. Adding this to the distance $D(1 - \beta^2)$, the overall distance *seen* by the observer is D, *i.e.* uncontracted.

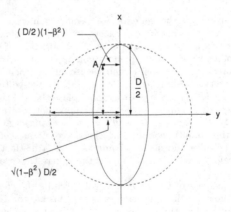

Fig. 12.20 The cross-sectional ellipse of semi-axes $a = D/2$ and $b = (D/2)\sqrt{1 - \beta^2}$.

Example 12.5: The decaying nucleus

A nucleus at rest with mass M decays into two parts with rest masses m_1 and m_2. Calculate E_1 and E_2, the energies of the decay parts expressed in terms of M, m_1, m_2 and, of course, c.

Solution: We have the relation (12.48):

$$E^2 \equiv p_0^2 c^2 = \mathbf{p}^2 c^2 + m^2 c^4. \tag{12.59}$$

Here E is the total energy of the particle, $\mathbf{p}$ its momentum and m its (rest) mass. In 4-dimensional notation (*cf.* Eq. (12.47) and note the choice of metric with diagonal elements having signs $+, -, -, -$; with signs taken as $-, +, +, +$ one has $p^2 = -m^2c^2, p^2 = -p_0^2 + \mathbf{p}^2$, which can lead to confusion):

$$p^2 \equiv p^\mu p_\mu = m^2c^2 = p_0^2 - \mathbf{p}^2. \tag{12.60}$$

Let P^μ be the 4-momentum of the decaying particle, and p_1^μ, p_2^μ the 4-momenta of the decay products with masses m_1, m_2 respectively. The law of conservation of energy and momentum yields (omitting Lorentz indices):

$$P^2 = (p_1 + p_2)^2 = p_1^2 + p_2^2 + 2p_1p_2. \tag{12.61}$$

However,

$$P^2 = M^2c^2, \quad p_1^2 = m_1^2c^2, \quad p_2^2 = m_2^2c^2,$$

so that

$$M^2c^2 = m_1^2c^2 + m_2^2c^2 + 2p_1p_2, \quad (M^2 - m_1^2 - m_2^2)c^2 = -2(\mathbf{p}_1 \cdot \mathbf{p}_2 - p_{10}p_{20}). \tag{12.62}$$

But $\mathbf{P} = \mathbf{p}_1 + \mathbf{p}_2 = 0$, *i.e.* $\mathbf{p}_1 = -\mathbf{p}_2 \equiv \mathbf{p}$. Hence Eq. (12.62) becomes

$$(M^2 - m_1^2 - m_2^2)c^2 = 2\mathbf{p}^2 + 2p_{10}p_{20}. \tag{12.63}$$

But here

$$cp_{10} = E_1 = (\mathbf{p}^2c^2 + m_1^2c^4)^{1/2}, \quad cp_{20} = E_2 = (\mathbf{p}^2c^2 + m_2^2c^4)^{1/2}. \tag{12.64}$$

Equation (12.63) can now be rewritten as

$$(M^2 - m_1^2 - m_2^2)c^2 = 2\mathbf{p}^2 + \frac{2}{c^2}(\mathbf{p}^2c^2 + m_1^2c^4)^{1/2}(\mathbf{p}^2c^2 + m_2^2c^4)^{1/2}. \tag{12.65}$$

This equation yields $\mathbf{p}^2$:

$$[(M^2 - m_1^2 - m_2^2)c^2 - 2\mathbf{p}^2]^2 = \frac{4}{c^4}(\mathbf{p}^2c^2 + m_1^2c^4)(\mathbf{p}^2c^2 + m_2^2c^4),$$

from which we obtain

$$\mathbf{p}^2 = \frac{c^2}{4M^2}[M^2 - (m_1 + m_2)^2][M^2 - (m_1 - m_2)^2]. \tag{12.66}$$

We can insert this expression for $\mathbf{p}^2$ into Eq. (12.64) in order to obtain E_1, E_2:

$$\begin{aligned} E_1^2 &= \mathbf{p}^2c^2 + m_1^2c^4 = \frac{c^4}{4M^2}\{[M^2 - (m_1 + m_2)^2][M^2 - (m_1 - m_2)^2] + 4m_1^2M^2\} \\ &= \frac{c^4}{4M^2}\{M^4 + (m_1^2 - m_2^2)^2 - M^2(-2m_1^2 + 2m_2^2)\} = \frac{c^4}{4M^2}(M^2 + m_1^2 - m_2^2)^2. \end{aligned}$$

Hence

$$E_1 = \frac{c^2}{2M}(M^2 + m_1^2 - m_2^2), \quad \text{and similarly} \quad E_2 = \frac{c^2}{2M}(M^2 + m_2^2 - m_1^2).$$

Example 12.6: Electron–positron annihilation

A positron e^+ is annihilated by colliding with an electron e^- which is almost at rest in the laboratory frame, *i.e.*

$$e^+ \text{ (fast)} + e^- \text{ (at rest)} \longrightarrow \text{ photons } \gamma.$$

(a) Consider the collision in a reference frame in which for the momenta of the initial particles

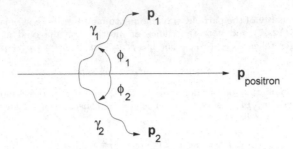

Fig. 12.21 Electron–positron annihilation with production of two photons.

$\mathbf{p}_{e+} + \mathbf{p}_{e-} = 0$. What is the smallest number of photons that must be created? (b) Considering the case $e^+ + e^- \longrightarrow \gamma_1 + \gamma_2$, derive an expression for the energy of one of the γ-rays γ_1, γ_2 in the laboratory frame ($\mathbf{p}_{e-} = 0$); the expression will involve the angle ϕ_1 between $\mathbf{p}_e$ and $\mathbf{p}_1$, the latter being the momentum of γ_1. (c) Finally obtain the largest and smallest γ-ray energies in the laboratory frame.

Solution: (a) Before annihilation $\mathbf{p}_{e+} + \mathbf{p}_{e-} = 0$. Since this is the total momentum which is conserved, the total momentum of the outgoing photons must also be zero. A single photon has a momentum $\mathbf{p} \neq 0$. Thus at least two photons must be produced, as illustrated in Fig. 12.21.
(b) With $m = m_{e+} = m_{e-}$ the law of conservation of energy implies: Energy E of e^+ plus energy of e^- (rest energy $\simeq mc^2$) = energy of 2 photons, *i.e.*

$$E + mc^2 = E_1 + E_2. \tag{12.67}$$

The law of conservation of momentum yields, with $\mathbf{p}_1, \mathbf{p}_2$ the momenta of the photons:

$$\mathbf{p}_{e+} = \mathbf{p}_1 + \mathbf{p}_2, \quad i.e. \quad p_{e+} = p_1 \cos\phi_1 + p_2 \cos\phi_2, \quad 0 = p_1 \sin\phi_1 - p_2 \sin\phi_2. \tag{12.68}$$

In the case of a photon the energy is $E_\gamma = pc$, so that $E_1 = p_1 c, E_2 = p_2 c$, and multiplying the relations (12.68) by c, we obtain:

$$cp_{e+} = E_1 \cos\phi_1 + E_2 \cos\phi_2, \quad 0 = E_1 \sin\phi_1 - E_2 \sin\phi_2. \tag{12.69}$$

We can eliminate the angle ϕ_2:

$$cp_{e+} = E_1 \cos\phi_1 + E_2 \sqrt{1 - \left(\frac{E_1}{E_2}\sin\phi_1\right)^2}. \tag{12.70}$$

Squaring the equation, we obtain using Eq. (12.67),

$$(cp_{e+} - E_1 \cos\phi_1)^2 = E_2^2 - E_1^2 \sin^2\phi_1 = (E + mc^2 - E_1)^2 - E_1^2 \sin^2\phi_1.$$

This equation now has to be solved for E_1:

$$
\begin{aligned}
(cp_{e+} - E_1 \cos\phi_1)^2 &= (mc^2 + E)^2 - 2(mc^2 + E)E_1 + E_1^2 \cos^2\phi_1, \\
(cp_{e+})^2 - 2cp_{e+}E_1 \cos\phi_1 &= (mc^2 + E)^2 - 2(mc^2 + E)E_1, \\
E_1 &= \frac{(mc^2 + E)^2 - c^2 p_{e+}^2}{2(mc^2 + E) - 2cp_{e+}\cos\phi_1}.
\end{aligned}
$$

But in the case of the incoming positron (*cf.* Einstein's formula Eq. (12.48))

$$E^2 = m^2 c^4 + p_{e+}^2 c^2. \tag{12.71}$$

Hence replacing $E^2 - c^2 p_{e+}^2$ by $m^2 c^4$:

$$E_1 = \frac{2m^2 c^4 + 2mc^2 E}{2(mc^2 + E) - 2cp_{e+}\cos\phi_1} = \frac{mc^2(mc^2 + E)}{E + mc^2 - p_{e+}c\cos\phi_1}. \qquad (12.72)$$

Setting T_0 equal to the kinetic energy of e^+, we can write (*cf.* Eq. (12.44)):

$$E = mc^2 + T_0, \quad p_{e+}^2 c^2 = E^2 - m^2 c^4 = (T_0 + mc^2)^2 - m^2 c^4 = T_0(T_0 + 2mc^2),$$

so that

$$E_1 = \frac{mc^2(2mc^2 + T_0)}{2mc^2 + T_0 - \sqrt{T_0^2 + 2mc^2 T_0}\cos\phi_1}. \qquad (12.73)$$

(c) For a given kinetic energy T_0 of the incoming positron, the energy E_1 is a maximum when $\cos\phi_1 = 1, \phi_1 = 0$, and a minimum when $\cos\phi_1 = -1, \phi_1 = \pi$, and hence (dividing by $T_0 + 2mc^2$):

$$E_{1,\text{max.}} = \frac{mc^2}{1 - (1 + 2mc^2/T_0)^{-1/2}}, \quad E_{1,\text{min.}} = \frac{mc^2}{1 + (1 + 2mc^2/T_0)^{-1/2}}. \qquad (12.74)$$

If T_0 is very small, *i.e.* $2mc^2/T_0 \gg 1$, then $E_{1,\text{max.}} \simeq mc^2 \simeq E_{1,\text{min.}}$. If T_0 is very large, $2mc^2/T_0 \ll 1$, then from Eq. (12.73)

$$E_{1,\text{max (min)}} \simeq \frac{mc^2(2mc^2 + T_0)}{2mc^2 + T_0 - (+)T_0\{1 + mc^2/T_0\}},$$

and hence $E_{1,\text{max.}} \approx T_0$ and $E_{1,\text{min.}} \approx mc^2/2$.

Example 12.7: The hurrying physicist

What are the Lorentz transformations for momentum and energy? A physicist in his car crossed traffic lights when red as indicated in Fig. 12.22, and was caught and put into jail. In the court hearing he admitted that he had approached the crossing at such a speed that instead of red he had seen green. The court imposed a penalty on the physicist which required him to pay one cent for every kilometer above the maximum permissible speed of 50 km per hour. What is the total amount of the penalty? (Note: $\lambda_{\text{green}} = 5300 A^\circ, \lambda_{\text{red}} = 6500 A^\circ$).

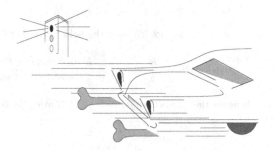

Fig. 12.22 The hurrying physicist.

Solution: Clearly this is again a problem based on the *aberration law*. Instead of simply referring to formula (12.56) we reconsider the aberration. The photon γ moves with momentum p in a direction with angle θ to the z-axis in the laboratory frame. Then

$$p_z = p\cos\theta, \quad p_y = p\sin\theta, \quad p_x = 0. \qquad (12.75)$$

In the case of uniform motion along parallel z-axes, the Lorentz transformations of momentum and energy in one frame to those of the other frame are (*cf.* Eq. (12.7b) and make there the replacements $ct \to E/c, x \to p_z, \zeta \to \phi$ for $p^\mu = (E/c, \mathbf{p})$)

$$
\begin{aligned}
p'_x &= p_x, \\
p'_y &= p_y, \\
p'_z &= p_z \cosh\phi - \frac{E}{c} \sinh\phi, \\
\frac{E'}{c} &= -p_z \sinh\phi + \frac{E}{c} \cosh\phi,
\end{aligned}
\tag{12.76}
$$

where (see Eq. (12.7a)) $\cosh\phi = 1/\sqrt{1-\beta^2}, \beta = v/c$. Let p' be the momentum and θ' the direction of the photon in the other uniformly moving reference frame. Then

$$
p'_z = p' \cos\theta', \quad p'_y = p' \sin\theta', \quad p'_x = 0.
\tag{12.77}
$$

From Eqs. (12.75) to (12.77) we obtain

$$
\begin{aligned}
\frac{E'}{c} &= -p \cos\theta \sinh\phi + \frac{E}{c} \cosh\phi, \\
p' \cos\theta' &= p \cos\theta \cosh\phi - \frac{E}{c} \sinh\phi, \\
p' \sin\theta' &= p \sin\theta.
\end{aligned}
\tag{12.78}
$$

But for a photon $E/c = p$, so that the first of these equations can be written:

$$
E' = E \cosh\phi \left(1 - \cos\theta \frac{\sinh\phi}{\cosh\phi} \right).
$$

But (*cf.* Eq. (12.7a))

$$
\sinh\phi = \frac{\beta}{\sqrt{1-\beta^2}}, \quad \cosh\phi = \frac{1}{\sqrt{1-\beta^2}}, \quad \tanh\phi = \beta,
$$

hence

$$
E' = E \cosh\phi (1 - \beta \cos\theta) = p'c.
\tag{12.79}
$$

We obtain $\cos\theta'$ from the second of Eqs. (12.78) with the help of Eq. (12.79):

$$
\cos\theta' = \frac{1}{p'} \left[p \cos\theta \cosh\phi - \frac{E}{c} \sinh\phi \right] = \frac{E \cos\theta \cosh\phi - E \sinh\phi}{E \cosh\phi (1 - \beta \cos\theta)}.
\tag{12.80}
$$

Thus (observe that this is again the *aberration formula* (12.56) already given by Einstein [14])

$$
\cos\theta' = \frac{\cos\theta - \beta}{1 - \beta \cos\theta}.
\tag{12.81}
$$

In approaching the crossing, the traffic lights send a signal in the direction of the driver, implying $\theta = \pi = \theta'$ (*cf.* Fig. 12.13). The relation between energy E and frequency ν is $E = h\nu$, where h is Planck's constant. From Eq. (12.79) we obtain therefore:

$$
\nu' = \nu \cosh\phi (1 + \beta) = \nu \frac{(1+\beta)}{\sqrt{1-\beta^2}} = \nu \left(\frac{1+\beta}{1-\beta} \right)^{1/2}.
$$

With $\nu = c/\lambda$ we obtain

$$\frac{\lambda'}{\lambda} = \left(\frac{1-\beta}{1+\beta}\right)^{1/2}, \quad \text{or} \quad \beta = \frac{1-(\lambda'/\lambda)^2}{1+(\lambda'/\lambda)^2}. \quad (12.82)$$

In the case of green light: $\lambda \to \lambda_g = 5300 \text{A}^\circ = \lambda'$, and for red light $\lambda \to \lambda_r = 6500 \text{A}^\circ = \lambda$, so that

$$\frac{\lambda'}{\lambda} = \frac{5300}{6500} = 0.81, \quad \left(\frac{\lambda'}{\lambda}\right)^2 = 0.66.$$

Here λ' is the wavelength of the light the driver claims to have received, and λ is that of the light sent out from the source, the traffic lights. For these values we obtain from the second relation of Eq. (12.82): $\beta = 0.34/1.66 = 0.20$. We thus obtain for the velocity v from this (since $\beta = v/c$):

$$\begin{aligned} v &= 0.20 \times c = 0.20 \times 3 \times 10^8 \text{ m/s} = 6 \times 10^7 \text{ m/s} \\ &= 6 \times 10^4 \text{ km/s} = 6 \times 10^4 \times 60 \times 60 \text{ km/hour} \\ &= 2.2 \times 10^8 \text{ km/hour}. \end{aligned} \quad (12.83)$$

Since 50 km/hour is the maximum permissible speed, the number of km/hour surpassing this is $= 2.2 \times 10^8 - 50 \simeq 2.2 \times 10^8$. Since the driver has to pay one cent for every such kilometer, his total penalty is $2.2 \times 10^8/100 = 2\,200\,000$ Euros.

Example 12.8: The twin problem, or clock paradox

A and B are twin brothers. On their 21st birthday A leaves his twin brother B on Earth and flies for 7 years in the direction of z (7 years measured on his own watch). The velocity v of A relative

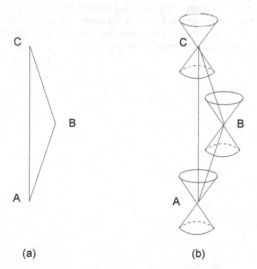

C   C

B   B

A   A

(a)   (b)

Fig. 12.23 (a) The Euclidean triangle with side-lenghts $AB + BC \geq AC$, (b) worldlines AC on Earth, ABC in space with clock measurements $AB + BC \leq AC$.

to that of the Earth is $\tilde{\gamma}c$, the fraction $\tilde{\gamma}$ of the velocity of light c. After 7 years A reverses his direction and returns to Earth with the same speed. How old are A and B then?

Solution: A is evidently $21 + 7 + 7 = 35$ years old according to his watch. We calculate the age of B (as measured by B on his watch) as follows. Reversing the Lorentz transformation of Eq. (12.7b), *i.e.* (with there $x \to z$ here, and there $\zeta \to \phi$ here)

$$x' = x, \quad y' = y, \quad z' = z\cosh\phi - ct\sinh\phi, \quad t' = -\frac{z}{c}\sinh\phi + t\cosh\phi, \quad (12.84)$$

we obtain (with $\cosh^2 \phi - \sinh^2 \phi = 1$)

$$x = x', \quad y = y', \quad z = z' \cosh \phi + ct' \sinh \phi, \quad t = \frac{z'}{c} \sinh \phi + t' \cosh \phi, \qquad (12.85)$$

where with $\beta = v/c$ we have $\cosh \phi = 1/\sqrt{1-\beta^2} = 1/\sqrt{1-\tilde{\gamma}^2} > 1$. We now use the last of Eqs. (12.85): $t = (z'/c) \sinh \phi + t' \cosh \phi$. Since A is at rest in his frame K' with primed coordinates, we have $z' = 0$. Then t, the time measured by B on Earth in frame K, is for $t' = 7$ years: $t = 7 \cosh \phi$ years, $i.e.$

$$t = \frac{7}{\sqrt{1-\tilde{\gamma}^2}} \quad \text{years.} \qquad (12.86)$$

The age of B is therefore at the time of return of A

$$= 21 + \frac{2 \times 7}{\sqrt{1-\tilde{\gamma}^2}} \quad \text{years.}$$

If $\tilde{\gamma} = 4/5$, the age of B on Earth follows as:

$$21 + \frac{2 \times 7}{\sqrt{1-(4/5)^2}} = 21 + \frac{2 \times 7 \times 5}{3} = 21 + \frac{70}{3} \simeq 44 \quad \text{years.}$$

Thus 35 years and 44 years are the time intervals which A and B measure on their respective watches. Hence in this way, as remarked after Eq. (12.33), the *stationary clock* appears to run faster than a *moving clock*. This difference in the clock measurements is illustrated in the book of Penrose[tt] by the reverse of the inequality satisfied by the lengths of the sides of a triangle ABC as in Fig. 12.23 (thus easy to remember), and is therefore referred to as the *clock paradox*. Note that in Fig. 12.23 the light cones are upright with their hulls or null surfaces at $45°$ to the vertical in the non-curved Minkowski space of Special Relativity.

[tt]R. Penrose [38], p. 421.

Chapter 13

Equation of Motion of a Particle in a Gravitational Field

13.1 Introductory Remarks

In the following we consider the variation of integrals which play the same role as action integrals in other areas of physics. The integrals are conceived as integrals determining the length of some path in a space with metric $g_{\mu\nu}(x)$, or some similar quantity, and we are interested in the extremum of such an integral as the condition for determining the corresponding *geodesic, i.e.* shortest or longest such quantity. The calculational steps are therefore those familiar from classical mechanics as explained in our earlier chapters. A main aim will be to uncover the relationship between the resulting equation and Newton's equation of motion for a particle in a gravitational field. We also point out some symmetry aspects of the chosen form of the action integral, consider again rotational motion as observed from an inertial frame, and finally consider the redshift in a gravitational field.

13.2 Equation of Motion

We start from the following relation for the line element ds in a curved (1+3)-dimensional spacetime, the *Riemann space* $\mathbb{R}_4$ (the more general space without metric being $\mathbb{S}_4$):

$$ds^2 = g_{\mu\nu}(x)dx^\mu dx^\nu, \quad g_{\mu\nu} = g_{\nu\mu}, \tag{13.1}$$

433

and define* (summation over repeated indices understood)

$$L(x^\alpha, \dot{x}^\alpha) \quad := \quad \left[g_{\mu\nu}(x) \frac{dx^\mu}{ds} \frac{dx^\nu}{ds} \right]^{1/2}$$

$$\equiv \quad [g_{\mu\nu}(x)\dot{x}^\mu(s)\dot{x}^\nu(s)]^{1/2}, \quad \dot{x}^\alpha = \frac{dx^\alpha}{ds}. \tag{13.2}$$

In the space defined by the line element ds of Eq. (13.1), the expression $\int_1^2 ds$ is the length of some path C in a *Riemann space* having this metric from event "1" to event "2", the quantity s being a parameter defined on C. The most direct such path is the *geodesic* determined by the extremum principle

$$\delta \int_1^2 ds = 0, \quad i.e. \quad \delta \int_1^2 ds L(x^\alpha, \dot{x}^\alpha) = 0. \tag{13.3}$$

Performing the variation, we obtain (with partial integration as in numerous earlier cases):

$$\delta \int_1^2 ds L(x^\alpha, \dot{x}^\alpha) \quad = \quad \int_1^2 ds \left[\frac{\partial L}{\partial x^\alpha} \delta x^\alpha + \frac{\partial L}{\partial \dot{x}^\alpha} \delta \dot{x}^\alpha \right]$$

$$= \quad \int_1^2 ds \left[\frac{\partial L}{\partial x^\alpha} - \frac{d}{ds}\left(\frac{\partial L}{\partial \dot{x}^\alpha} \right) \right] \delta x^\alpha,$$

so that the *Euler–Lagrange equation* of motion is found to be

$$\frac{\partial L}{\partial x^\alpha} - \frac{d}{ds}\left(\frac{\partial L}{\partial \dot{x}^\alpha} \right) = 0. \tag{13.4}$$

Here for the specific Lagrangian of Eq. (13.2) we have

$$\frac{\partial L}{\partial x^\alpha} = \frac{1}{2L} \frac{\partial g_{\mu\nu}}{\partial x^\alpha} \dot{x}^\mu \dot{x}^\nu, \qquad \frac{\partial L}{\partial \dot{x}^\alpha} = \frac{1}{L} g_{\alpha\nu}(x)\dot{x}^\nu. \tag{13.5}$$

Inserting these expressions into Eq. (13.4) we obtain

$$\frac{d}{ds}\left(\frac{\partial L}{\partial \dot{x}^\alpha} \right) \quad = \quad -\frac{1}{L^2} g_{\alpha\nu}(x)\dot{x}^\nu \frac{dL}{ds} + \frac{\partial g_{\alpha\nu}(x)}{\partial x^\beta} \frac{1}{L} \dot{x}^\beta \dot{x}^\nu + \frac{1}{L} g_{\alpha\nu} \ddot{x}^\nu$$

$$= \quad \frac{\partial L}{\partial x^\alpha} = \frac{1}{2L} \dot{x}^\mu \dot{x}^\nu \frac{dg_{\mu\nu}(x)}{dx^\alpha}, \tag{13.6}$$

---

*Thus (comparing with Eq. (13.1)) the value of L is 1, and we could — so one might argue — have chosen a different integral, as, for instance, one with L replaced by $L_2 = L^2$. The reason one prefers the case of Eq. (13.2) is that its action is invariant under a rescaling of the world-line parameter s, as is shown in Example 13.1.

or (multiplying through by L)

$$\frac{d}{ds}[g_{\alpha\nu}(x)\dot{x}^{\nu}] = \frac{1}{L}g_{\alpha\nu}(x)\dot{x}^{\nu}\frac{dL}{ds} + \frac{1}{2}\dot{x}^{\mu}\dot{x}^{\nu}\frac{\partial g_{\mu\nu}(x)}{\partial x^{\alpha}}. \tag{13.7}$$

Now from Eq. (13.1) we know that the magnitude of L is 1, so that $dL/ds = 0$. In L the expression dx^{μ}/ds is termed the *unit tangent* to the curve at a point P, its direction being that of the displacement dx^{μ} along the path as explained in Sec. 12.9. Thus[†]

$$\frac{d}{ds}[g_{\alpha\nu}(x)\dot{x}^{\nu}] = \frac{1}{2}\dot{x}^{\mu}\dot{x}^{\nu}\frac{\partial g_{\mu\nu}(x)}{\partial x^{\alpha}}. \tag{13.8}$$

The derivative $\partial g_{\mu\nu}/\partial x^{\alpha}$ can be looked at as the gradient of a potential. One therefore introduces the *Christoffel symbol* defined as the quantity:

$$\Gamma^{\mu}{}_{\alpha\beta}(x) := \frac{1}{2}g^{\mu\lambda}(x)\left[\frac{\partial g_{\alpha\lambda}}{\partial x^{\beta}} + \frac{\partial g_{\beta\lambda}}{\partial x^{\alpha}} - \frac{\partial g_{\alpha\beta}}{\partial x^{\lambda}}\right]. \tag{13.9}$$

In order to introduce this symbol into Eq. (13.8), we consider the latter, *i.e.* Eq. (13.8), written out as follows:

$$g_{\alpha\nu}(x)\frac{d^2x^{\nu}}{ds^2} + \frac{\partial g_{\alpha\nu}}{\partial x^{\lambda}}\frac{dx^{\lambda}}{ds}\frac{dx^{\nu}}{ds} = \frac{1}{2}\frac{\partial g_{\mu\nu}}{\partial x^{\alpha}}\frac{dx^{\mu}}{ds}\frac{dx^{\nu}}{ds},$$

$$\therefore g_{\alpha\nu}(x)\frac{d^2x^{\nu}}{ds^2} = \frac{1}{2}\frac{\partial g_{\mu\nu}}{\partial x^{\alpha}}\frac{dx^{\mu}}{ds}\frac{dx^{\nu}}{ds} - \frac{\partial g_{\alpha\nu}}{\partial x^{\lambda}}\frac{dx^{\lambda}}{ds}\frac{dx^{\nu}}{ds}.$$

Here λ is a dummy index which is summed over, and we can write

$$\begin{aligned}
g_{\alpha\nu}(x)\frac{d^2x^{\nu}}{ds^2} &= \frac{1}{2}\left(\frac{\partial g_{\mu\nu}}{\partial x^{\alpha}} - 2\frac{\partial g_{\alpha\nu}}{\partial x^{\mu}}\right)\frac{dx^{\mu}}{ds}\frac{dx^{\nu}}{ds} \\
&= \frac{1}{2}\left(\frac{\partial g_{\mu\nu}}{\partial x^{\alpha}} - \frac{\partial g_{\alpha\mu}}{\partial x^{\nu}} - \frac{\partial g_{\alpha\nu}}{\partial x^{\mu}}\right)\frac{dx^{\mu}}{ds}\frac{dx^{\nu}}{ds}.
\end{aligned} \tag{13.10}$$

With the property defining $g^{\mu\beta}$,

$$g^{\beta\alpha}(x)g_{\alpha\nu}(x) = \delta^{\beta}_{\nu}, \tag{13.11}$$

we obtain

$$\frac{d^2x^{\beta}}{ds^2} = \frac{1}{2}g^{\beta\alpha}\left(\frac{\partial g_{\mu\nu}}{\partial x^{\alpha}} - \frac{\partial g_{\alpha\mu}}{\partial x^{\nu}} - \frac{\partial g_{\alpha\nu}}{\partial x^{\mu}}\right)\dot{x}^{\mu}\dot{x}^{\nu}, \tag{13.12}$$

or

$$\frac{d^2x^{\beta}}{ds^2} + \Gamma^{\beta}{}_{\mu\nu}\frac{dx^{\mu}}{ds}\frac{dx^{\nu}}{ds} = 0. \tag{13.13}$$

[†]In Example 13.2 we show that the same equation follows also from an arbitrary power of L in the action integral.

This is the equation of a *geodesic in a Riemann space*, *i.e.* the equation of the most direct path between two points in the Riemann space. We shall later obtain a better understanding of the equation as the equation of motion of a test body by analyzing its geometry in some detail. In an *inertial frame* the metric coefficients are constants. Hence there the equation reduces as expected to

$$\frac{d^2 x^\beta}{ds^2} = 0, \quad \dot{x}^\beta = \text{const.}$$

In Eq. (13.13) we may therefore also look at the components of $\Gamma^\beta{}_{\mu\nu}$ (often called *affine connection*, see Eq. (14.48)) as potentials of fictitious forces, and if gravity is present it can appear here only through these quantities (and we shall uncover this in the following), and does not come in via an extra term on the right hand side of $d^2 x^\beta / ds^2 = 0$. Thus Eq. (13.13) is the equation of motion of a body whether or not gravity is present.[‡] One should note that the geodesic is the equation of a curve on a manifold in *arbitrary coordinates* so that the latter can be chosen in any possible form, *e.g.* as spherical coordinates or Cartesian or any other coordinates.

Example 13.1: Invariance of action S under rescaling

Show that the action S with L of Eq. (13.2) is invariant under a rescaling of the world line parameter s, $s \to \lambda = f(s)$.

Solution: The rescaling of s means the replacement $s \longrightarrow \lambda = f(s)$, $d\lambda = f'(s)ds$. Considering the action S associated with L of Eq. (13.2), now called S_1, we have

$$S_1 = \int \sqrt{g_{\mu\nu}(x) \frac{dx^\mu}{ds} \frac{dx^\nu}{ds}} ds = \int \sqrt{g_{\mu\nu}(x) \frac{dx^\mu}{d\lambda} \frac{dx^\nu}{d\lambda} (f'(s))^2} \frac{d\lambda}{f'(s)} = \int \sqrt{g_{\mu\nu}(x) \frac{dx^\mu}{d\lambda} \frac{dx^\nu}{d\lambda}} d\lambda.$$

But in the case of the following:

$$S_2 = \int g_{\mu\nu}(x) \frac{dx^\mu}{ds} \frac{dx^\nu}{ds} ds = \int g_{\mu\nu}(x) \frac{dx^\mu}{d\lambda} \frac{dx^\nu}{d\lambda} (f'(s))^2 \frac{d\lambda}{f'(s)} = \int g_{\mu\nu}(x) \frac{dx^\mu}{d\lambda} \frac{dx^\nu}{d\lambda} f'(s) d\lambda.$$

We see, however, that S_2 is invariant under the shift $s \to \lambda = s + \epsilon$ for ϵ infinitesimal.

Example 13.2: Different Lagrangians, same equation

Consider the action

$$S = \int ds\, L, \quad L = L_0^m, \quad L_0 = g_{\mu\nu}(x) \dot{x}^\mu \dot{x}^\nu,$$

and show that the Euler–Lagrange equation is the same as Eq. (13.8) with $L_0 = 1$.

Solution: We have for the first term on the left hand side of the Euler–Lagrange equation (13.4):

$$\frac{\partial L}{\partial x^\alpha} = m L_0^{m-1} \dot{x}^\mu \dot{x}^\nu \frac{\partial g_{\mu\nu}(x)}{\partial x^\alpha}.$$

[‡]For additional discussion of *"free fall"* see D. Raine and E. Thomas [40], p. 5.

On the other hand the second term of the Euler–Lagrange equation (13.4) implies

$$\frac{d}{ds}\left(\frac{\partial L}{\partial \dot{x}^\alpha}\right) = \frac{d}{ds}[2mL_0^{m-1}g_{\alpha\nu}(x)\dot{x}^\nu].$$

Equating both sides and dividing by $2m$ we obtain

$$\frac{1}{2}\dot{x}^\mu \dot{x}^\nu \frac{\partial g_{\mu\nu}(c)}{\partial x^\alpha}L_0^{m-1} = \frac{d}{ds}[g_{\alpha\nu}(x)\dot{x}^\nu L_0^{m-1}].$$

This equation agrees with Eq. (13.8) for $L_0 = 1$.

13.3   Reduction to Newton's Equation

We now show that Eq. (13.13) can be reduced to an equation having the form of Newton's equation. We do this by introducing the mass parameter m into L and setting

$$m\frac{d^2x^\beta}{ds^2} = F^\beta(x), \quad F^\beta(x) = -m\Gamma^\beta{}_{\mu\nu}\dot{x}^\mu\dot{x}^\nu. \tag{13.14}$$

Of course, this equation has four components, and Newton's equation involves time t, and not the parameter s. We expect Newton's gravitational potential to arise from the difference of the metric tensor $g_{\mu\nu}$ from that of the flat metric. We set therefore, for small ϵ,

$$g_{\mu\nu}(x) = g_{\mu\nu}^{(0)} + \epsilon\gamma_{\mu\nu}(x), \tag{13.15}$$

where

$$g_{\mu\nu}^{(0)} = \begin{pmatrix} 1 & 0 & 0 & 0 \\ 0 & -1 & 0 & 0 \\ 0 & 0 & -1 & 0 \\ 0 & 0 & 0 & -1 \end{pmatrix}.$$

The assumption of small values of ϵ will in Sec. 15.6 turn out to be the assumption of a weak gravitational field. Thus we ignore contributions of higher order in ϵ. Also, we assume that $\beta = v/c$, with $\mathbf{v} = d\mathbf{x}/dt$, is small and that higher order contributions in this parameter can also be neglected. With $x^0 = ct$ we have

$$ds^2 = g_{\mu\nu}(x)dx^\mu dx^\nu = (dx^0)^2 - (d\mathbf{x})^2 + \epsilon\gamma_{\mu\nu}(x)dx^\mu dx^\nu. \tag{13.16}$$

We make the additional assumptions that (a) $\gamma_{\mu\nu}(x)$ is independent of time t, and (b) that

$$\gamma_{\mu\nu}(\mathbf{x}) \xrightarrow{|\mathbf{x}|\to\infty} 0. \tag{13.17}$$

From Eq. (3.16) we obtain ($\mathbf{v} = d\mathbf{x}/dt$)

$$\left(\frac{ds}{dt}\right)^2 \simeq c^2 - v^2 + \epsilon\gamma_{\mu\nu}\frac{dx^\mu}{dt}\frac{dx^\nu}{dt} \simeq c^2\left[1 - \beta^2 + \epsilon\gamma_{\mu\nu}(\mathbf{x})\frac{dx^\mu}{dx^0}\frac{dx^\nu}{dx^0}\right],$$

or (see below)

$$\left(\frac{ds}{dt}\right)^2 \simeq c^2[1 + \epsilon\gamma_{00}(\mathbf{x})]. \tag{13.18}$$

Here we neglect contributions (implying here effectively $\gamma_{ik} \simeq 0, \gamma_{0k} \simeq 0$, and thus effectively also $g_{0k} \simeq 0$)

$$\epsilon\gamma_{ik}\frac{dx^i}{dx^0}\frac{dx^k}{dx^0} \sim \epsilon\beta^2, \quad \epsilon\gamma_{0k}\frac{dx^0}{dx^0}\frac{dx^k}{dx^0} \sim \epsilon\beta.$$

We now collect the nonnegligible terms of Eq. (13.13), *i.e.* of

$$\frac{d^2x^\alpha}{ds^2} + \Gamma^\alpha{}_{\mu\nu}(x)\frac{dx^\mu}{ds}\frac{dx^\nu}{ds} = 0.$$

These are given by (using Eq. (13.18))

$$\frac{d^2x^\alpha}{ds^2} + \Gamma^\alpha{}_{00}\left(\frac{dx^0}{ds}\right)^2 \simeq 0, \quad i.e. \quad \frac{d^2x^\alpha}{ds^2} + \Gamma^\alpha{}_{00}\frac{1}{1 + \epsilon\gamma_{00}} \simeq 0. \tag{13.19}$$

Next we evaluate the Christoffel symbols, *i.e.* here the *Newtonian affine connection.* We have (above we assumed $\gamma_{\mu\nu}$ to be independent of t)

$$\Gamma^\alpha{}_{00} \stackrel{(13.9)}{=} \frac{1}{2}g^{\alpha\lambda}\left[\frac{\partial g_{0\lambda}}{\partial x^0} + \frac{\partial g_{0\lambda}}{\partial x^0} - \frac{\partial g_{00}}{\partial x^\lambda}\right]$$

$$= \frac{1}{2}g^{\alpha 0}\left[2\frac{\partial g_{00}}{\partial x^0} - \frac{\partial g_{00}}{\partial x^0}\right] + \frac{1}{2}g^{\alpha i}\left[2\frac{\partial g_{0i}}{\partial x^0} - \frac{\partial g_{00}}{\partial x^i}\right]$$

$$= -\frac{1}{2}g^{\alpha i}\frac{\partial g_{00}}{\partial x^i}. \tag{13.20}$$

Thus

$$\Gamma^j{}_{00} = -\frac{1}{2}g^{ji}\frac{\partial g_{00}}{\partial x^i} = -\frac{1}{2}g^{(0)ji}\frac{\partial(\epsilon\gamma_{00}(\mathbf{x}))}{\partial x^i} = -\frac{1}{2}\partial^j(\epsilon\gamma_{00}(\mathbf{x})),$$

$$\Gamma^0{}_{00} = -\frac{1}{2}g^{0i}\frac{\partial g_{00}}{\partial x^i} = 0. \tag{13.21}$$

Hence from Eq. (13.19) to leading order in ϵ (and with $\partial^j = -\partial_j$):

$$\frac{d^2x^j}{ds^2} - \frac{1}{2}\epsilon\frac{\partial\gamma_{00}}{\partial x_j} \simeq 0, \quad \frac{d^2x^0}{ds^2} \simeq 0,$$

i.e. $dx^0/ds = $ const., $x^0 \equiv ct = as + b$, *i.e.* we can set $ct = s$. Hence

$$\frac{d^2x^j}{dt^2} + \frac{1}{2}c^2\epsilon\frac{\partial\gamma_{00}(\mathbf{x})}{\partial x^j} \simeq 0. \tag{13.22}$$

We have thus obtained *Newton's equation of motion, i.e.* (with $m = $ mass),

$$m\frac{d^2x^j}{dt^2} = -m\frac{\partial\phi}{\partial x^j}, \tag{13.23}$$

for a potential $V_{\text{Newton}} = m\phi(\mathbf{x})$ with

$$\phi(\mathbf{x}) = \frac{c^2}{2}\epsilon\gamma_{00}(\mathbf{x}). \tag{13.24}$$

In order to understand what we achieved we recall that we assumed a metric tensor with

$$g_{00}(x) = 1 + \epsilon\gamma_{00}(\mathbf{x}) = 1 + \frac{2}{c^2}\phi(\mathbf{x}). \tag{13.25}$$

This implies that for $\phi(\mathbf{x}) \neq 0$ the space is not flat (whereas for $\phi(\mathbf{x}) = 0$, Eq. (13.23) implies inertial motion in a flat space). This means $\phi(\mathbf{x})$ provides a curvature of space. Here $\phi(\mathbf{x})$ itself is not determined more specifically. The actual identification of $V_{\text{Newton}} = m\phi(\mathbf{x})$ with *Newton's gravitational potential* is achieved with *Einstein's equation of the gravitational field* later in Sec. 15.6.

Example 13.3: Christoffels for Minkowski space spherical coordinates
Compute the Christoffel symbols for the Minkowski space with spherical coordinates.

Solution: The Minkowski spacetime metric (for a change with an overall minus change compared with Eq. (13.16)) in spherical polar coordinates is given by

$$ds^2 = -c^2dt^2 + dr^2 + r^2(d\theta^2 + \sin^2\theta d\phi^2). \tag{13.26}$$

We assume that the action integral S to be extremized is given by

$$S = \int\left(\frac{ds}{d\tau}\right)^2 d\tau. \tag{13.27}$$

Thus we assume the Lagrangian to be

$$L\left(t, r, \theta, \phi, \frac{dt}{d\tau}, \frac{dr}{d\tau}, \frac{d\theta}{d\tau}, \frac{d\phi}{d\tau}\right) = -c^2\left(\frac{dt}{d\tau}\right)^2 + \left(\frac{dr}{d\tau}\right)^2 + r^2\left(\frac{d\theta}{d\tau}\right)^2 + r^2\sin^2\theta\left(\frac{d\phi}{d\tau}\right)^2. \tag{13.28}$$

The Euler–Lagrange equations in t, r, θ, ϕ are respectively (divided by 2):

$$\frac{d}{d\tau}\left(\frac{\partial L}{\partial(dt/d\tau)}\right) - \frac{\partial L}{\partial t} = 0 \longrightarrow \frac{d^2t}{d\tau^2} = 0,$$

$$\frac{d}{d\tau}\left(\frac{\partial L}{\partial(dr/d\tau)}\right) - \frac{\partial L}{\partial r} = 0 \longrightarrow \frac{d^2r}{d\tau^2} = r\left[\left(\frac{d\theta}{d\tau}\right)^2 + \sin^2\theta\left(\frac{d\phi}{d\tau}\right)^2\right],$$

$$\frac{d}{d\tau}\left(r^2\frac{d\theta}{d\tau}\right) = r^2\sin\theta\cos\theta\left(\frac{d\phi}{d\tau}\right)^2, \qquad \frac{d}{d\tau}\left(r^2\sin^2\theta\frac{d\phi}{d\tau}\right) = 0. \tag{13.29}$$

We can rewrite these equations as:

$$\frac{d^2t}{d\tau^2} = 0, \qquad \frac{d^2r}{d\tau^2} = r\left(\frac{d\theta}{d\tau}\right)^2 + r\sin^2\theta\left(\frac{d\phi}{d\tau}\right)^2,$$

$$\frac{d^2\theta}{d\tau^2} = -\frac{2}{r}\left(\frac{dr}{d\tau}\right)\left(\frac{d\theta}{d\tau}\right) + \sin\theta\cos\theta\left(\frac{d\phi}{d\tau}\right)^2,$$

$$\frac{d^2\phi}{d\tau^2} = -\frac{2}{r}\left(\frac{dr}{d\tau}\right)\left(\frac{d\phi}{d\tau}\right) - 2\cot\theta\left(\frac{d\theta}{d\tau}\right)\left(\frac{d\phi}{d\tau}\right). \tag{13.30}$$

These equations describe the motion of a free particle in space. We compare the equations with Eq. (13.13) in order to read off by comparison the expressions for the Christoffel symbols, *i.e.* we compare with

$$\frac{d^2x^\alpha}{d\tau^2} = -\Gamma^\alpha{}_{\mu\nu}\frac{dx^\mu}{d\tau}\frac{dx^\nu}{d\tau},$$

where we take $(x^\alpha) = (x^\tau, x^r, x^\theta, x^\phi)$. We obtain thus for the nonvanishing cases:

$$\Gamma^r{}_{\theta\theta} = -r, \quad \Gamma^r{}_{\phi\phi} = -r\sin^2\theta, \quad \Gamma^\theta{}_{r\theta} = \Gamma^\theta{}_{\theta r} = \frac{2}{r},$$

$$\Gamma^\theta{}_{\phi\phi} = -\sin\theta\cos\theta, \quad \Gamma^\phi{}_{r\phi} = \Gamma^\phi{}_{\phi r} = \frac{2}{r}, \quad \Gamma^\phi{}_{\theta\phi} = \Gamma^\phi{}_{\phi\theta} = 2\cot\theta. \tag{13.31}$$

13.4 Rotation Observed from an Inertial Frame

We have already touched rotational motion as observed from an inertial frame in Sec. 12.7. But we want to consider this now somewhat differently and in more detail. In an inertial reference frame S we have

$$ds^2 = dx^\mu dx_\mu = (dx^0)^2 - (d\mathbf{x})^2, \quad dx^0 = cdt. \tag{13.32}$$

We consider a frame $S'(x', y', z')$ which rotates with uniform angular velocity about the z-axis, as illustrated in Fig. 13.1. With notation as in Fig. 13.1. we have in cylindrical coordinates

$$x = r\cos\psi, \quad y = r\sin\psi, \quad z = z, \tag{13.33}$$

and obtain

$$d\mathbf{x}^2 = r^2 d\psi^2 + dr^2 + dz^2 \implies ds^2 = c^2 dt^2 - (dr^2 + r^2 d\psi^2 + dz^2). \tag{13.34}$$

From the geometry of Fig. 13.1 we deduce that

$$\psi = \phi + \omega t, \quad d\psi = d\phi + \omega dt. \tag{13.35}$$

Thus here $ds^2 = c^2 dt^2 - dr^2 - r^2(d\phi + \omega dt)^2 - dz^2$, or

$$ds^2 = (c^2 - \omega^2 r^2)dt^2 - (dr^2 + r^2 d\phi^2 + 2\omega r^2 d\phi dt + dz^2). \tag{13.36}$$

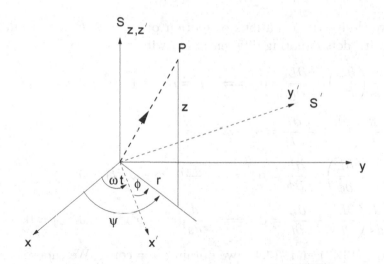

Fig. 13.1 Frame S' rotating about the z-axis of frame S.

Again

$$ds^2 = g_{\mu\nu}(x)dx^\mu dx^\nu. \tag{13.37}$$

The *proper time* interval $d\tau$ is given by $d\mathbf{x} = 0$, i.e.

$$d\tau = \frac{ds}{c}, \quad d\mathbf{x} = 0, \quad \Longrightarrow \quad d\tau = \sqrt{1 - \frac{\omega^2 r^2}{c^2}}dt. \tag{13.38}$$

We recall that dt is the time interval measured by an observer in S ($dr = 0, d\psi = 0, dz = 0$), and $d\tau$ is the time interval measured by an observer in S', i.e. in the rotating frame ($dr = 0, d\phi = 0, dz = 0$). This is what one expects since $v = \omega r$ and $d\tau = ds/c = \sqrt{1 - v^2/c^2}dt$, which implies the *time dilation*,

$$dt = \frac{d\tau}{\sqrt{1 - v^2/c^2}}. \tag{13.39}$$

We discussed earlier (*cf.* Sec. 12.7) that the geometry of the space S' is non-Euclidean, so we do not repeat these considerations here.

We now want to look at the *equations of motion* resulting from the world-line element of Eq. (13.37), and we choose again the action $s = \int dsL$ invariant only under the shift $s \longrightarrow s + a$, i.e. with L as

$$L = \underbrace{(c^2 - \omega^2 r^2)\left(\frac{dt}{ds}\right)^2}_{c^2(d\tau/ds)^2} - \left[\left(\frac{dr}{ds}\right)^2 + r^2\left(\frac{d\phi}{ds}\right)^2 + \left(\frac{dz}{ds}\right)^2 + 2\omega r^2\left(\frac{d\phi}{ds}\right)\left(\frac{dt}{ds}\right)\right].$$

$$\tag{13.40}$$

Again we derive the equations of motion of a point P seen from S', and obtain, with dots denoting differentiation with respect to s,

$$\frac{d}{ds}\left(\frac{\partial L}{\partial \dot{r}}\right) - \frac{\partial L}{\partial r} = 0 \implies \ddot{r} = r\omega^2 \dot{t}^2 + r\dot{\phi}^2 + 2\omega r \dot{\phi}\dot{t}, \qquad (13.41)$$

$$\frac{d}{ds}\left(\frac{\partial L}{\partial \dot{z}}\right) - \frac{\partial L}{\partial z} = 0 \implies \ddot{z} = 0, \qquad (13.42)$$

$$\frac{d}{ds}\left(\frac{\partial L}{\partial \dot{\phi}}\right) - \frac{\partial L}{\partial \phi} = 0 \implies \frac{d}{ds}(r^2\dot{\phi} + \omega r^2 \dot{t}) = 0, \qquad (13.43)$$

$$\frac{d}{ds}\left(\frac{\partial L}{\partial \dot{t}}\right) - \frac{\partial L}{\partial t} = 0 \implies \frac{d}{ds}[(c^2 - \omega^2 r^2)\dot{t} - \omega r^2 \dot{\phi}] = 0. \qquad (13.44)$$

From Eqs. (13.43) and (13.44) we obtain $c^2\dot{t} = $ const. We choose

$$c^2\dot{t} = c, \implies \dot{t} = \frac{1}{c}. \qquad (13.45)$$

We now consider a *radial motion* for which $\dot{\phi} = 0$. For this we obtain from Eq. (13.41) using Eq. (13.45):

$$\frac{d}{ds}\left(\frac{dr}{ds}\right) = \frac{d^2r}{ds^2} \equiv \ddot{r} = \frac{r\omega^2}{c^2}. \qquad (13.46)$$

But in the frame S' we have $d\tau = ds/c$, so that

$$\frac{d^2r}{d\tau^2} = r\omega^2, \qquad (13.47)$$

which is the *centrifugal acceleration*. We can also find the *Coriolis acceleration*. To obtain this, we consider Eq. (13.43), *i.e.*

$$\frac{d}{ds}(r^2\dot{\phi} + \omega r^2 \dot{t}) = 0, \quad \text{or} \quad r^2\ddot{\phi} + 2r\dot{r}\dot{\phi} + \omega r^2 \ddot{t} + 2\omega r \dot{r}\dot{t} = 0.$$

Here we set $\dot{t} = 1/c$ (*cf.* Eq. (13.45)). Then

$$r\frac{d^2\phi}{ds^2} + 2\frac{dr}{ds}\left(\frac{\omega}{c} + \frac{d\phi}{ds}\right) = 0,$$

or again with $d\tau = ds/c$:

$$r\frac{d^2\phi}{d\tau^2} = -2\omega\frac{dr}{d\tau} - 2\frac{dr}{d\tau}\frac{d\phi}{d\tau}, \qquad (13.48)$$

which links the velocity $\dot{r}$ with ω in $-2\omega dr/d\tau = -2|\boldsymbol{\omega} \times d\boldsymbol{r}/d\tau|$ as in the usual *Coriolis acceleration* (*cf.* Eq. (8.127)).

13.5   The Redshift

We are now, with Sec. 13.3, in a position to attempt a more serious calculation of the effect known as the *red-shift of the light* emitted by an atom
in the vicinity of the sun, *i.e.* to lower frequencies. We have seen that in
the presence of a gravitational field (and gravitational field here means an
x–dependent deviation of $g_{\mu\nu}(x)$ from the flat Minkowski spacetime tensor) the interval of proper time (*i.e.* that seen in a rest frame, *e.g.* in
a frame at rest on Earth or, differently, in a frame at rest on the sun) is
$d\tau = \sqrt{g_{00}(x)}dx^0/c = \sqrt{g_{00}(x)}dt$, where according to Eq. (13.25)

$$g_{00}(x) = 1 + \frac{2\phi(\mathbf{x})}{c^2}, \qquad (13.49)$$

and $\phi(\mathbf{x})$ is the (weak) gravitational potential. We now consider radiation
emitted by an atom in the sun and received on Earth. In this case we have
to distinguish between the potential ϕ_s at the surface of the sun and the
potential ϕ_E at the surface of the Earth. In the reference frames at rest
on the surface of the sun and correspondingly on the Earth we have proper
times given by

$$d\tau_s = \left(1 + \frac{2\phi_s}{c^2}\right)^{1/2} dt, \qquad d\tau_E = \left(1 + \frac{2\phi_E}{c^2}\right)^{1/2} dt, \qquad (13.50)$$

with the same interval dt. We assume that the frequency of the radiation
emitted by the atom on the sun is ν_s and that this is n waves per interval
$\Delta\tau_s$, *i.e.*

$$\nu_s = \frac{n}{\Delta\tau_s}, \qquad n = \nu_s\Delta\tau_s. \qquad (13.51)$$

The same number n of waves will be received on Earth, *i.e.* $n = \nu_E\Delta\tau_E$, so
that

$$\nu_s\Delta\tau_s = \nu_E\Delta\tau_E, \qquad \nu_E = \nu_s\frac{\Delta\tau_s}{\Delta\tau_E}. \qquad (13.52)$$

With Eq. (13.50) we obtain (with $d\tau_s \to \Delta\tau_s, d\tau_E \to \Delta\tau_E$)

$$\frac{\Delta\tau_s}{\Delta\tau_E} = \left(\frac{1 + 2\phi_s/c^2}{1 + 2\phi_E/c^2}\right)^{1/2}. \qquad (13.53)$$

Hence

$$\nu_E = \nu_s\left(\frac{1 + 2\phi_s/c^2}{1 + 2\phi_E/c^2}\right)^{1/2} \simeq \nu_s\left(\frac{1 + \phi_s/c^2}{1 + \phi_E/c^2}\right),$$

or (by *componendo et dividendo*)

$$\frac{\nu_E - \nu_s}{\nu_E + \nu_s} = \frac{\phi_s/c^2 - \phi_E/c^2}{2 + \phi_s/c^2 + \phi_E/c^2},$$

and with ν_E in the denominator re-expressed in terms of ν_s:

$$\frac{\nu_E - \nu_s}{\nu_s + \nu_s(1 + \phi_s/c^2)/(1 + \phi_E/c^2)} = \frac{\phi_s - \phi_E}{2c^2 + \phi_s + \phi_E},$$

and hence

$$\frac{\nu_E - \nu_s}{\nu_s} \simeq \frac{\phi_s - \phi_E}{c^2}. \tag{13.54}$$

Thus the gravitational field changes the frequency. The potential of the sun is larger than ϕ_E and negative, so that $|\phi_s| > |\phi_E|$ and $\phi_s < 0$. Hence

$$\nu_s > \nu_E, \quad \Delta\nu \equiv \nu_E - \nu_s < 0, \quad \nu_E \simeq \nu_s\left[1 + \frac{\phi_s - \phi_E}{c^2}\right] < \nu_s. \tag{13.55}$$

Thus the gravitational field leads to a decrease of the frequency (the frequency observed on Earth is less than that emitted in the sun), *i.e.* to a shift to lower frequencies, and this effect is called the *redshift*. The experimental test is difficult in view of other effects at the sun, like high temperatures, intense electromagnetic fields, high pressures, etc. However, a test with the Mössbauer effect is reported. Since the gravitational red-shift for the sun could also be explained without the use of Einstein's field equations it was frequently not considered to be a proof of Einstein's gravitational field equations. However, this view has changed as a result of other applications.[§]

We can obtain Eq. (13.54) also in the sense of the trick calculation of Sec. 11.3.1. There we used the Einstein formula $E = mc^2 = h\nu$ (h Planck's constant) to define an effective mass m of the photon, m_E its effective mass on Earth, and m_s its effective mass in the sun. Hence in the presence of a potential $\phi = V_{\text{Newton}}/m$ one has $mc^2 + m\phi = h\nu + m\phi$, and (with m replaced by m_s, m_E) conservation of energy of the photon implies

$$h\nu_s + m_s\phi_s = h\nu_E + m_E\phi_E, \quad \text{or} \quad h(\nu_E - \nu_s) = m_s\phi_s - m_E\phi_E. \tag{13.56}$$

At the sun $h\nu_s = m_sc^2$, so that (dividing Eq. (13.56) by $h\nu_s$)

$$\frac{\nu_E - \nu_s}{\nu_s} = \frac{m_s\phi_s - m_E\phi_E}{m_sc^2} = \frac{\phi_s - (m_E/m_s)\phi_E}{c^2}, \tag{13.57}$$

as in Eq. (13.54) and m_E/m_s to leading order 1.

[§]See M. Carmeli [10], pp. 51-52.

Chapter 14

Tensor Calculus for Riemann Spaces

14.1 Introductory Remarks

The concepts of scalars, vectors, and tensors (always defined with respect to some particular space like M_4 or $\mathbb{R}_4$ in Sec. 12.5) were introduced in Sec. 5.6 and were briefly referred to in Sec. 6.8. In these introductory sections we did not yet require the distinction between contravariant and covariant tensors. The concepts of contravariant and covariant vectors and tensors were subsequently introduced in Sec. 12.5. We now continue the study of tensor calculus and in particular study properties of the metric tensor, symmetric tensors, and other properties. The most important issue of this chapter, however, is the construction of covariant and contravariant derivatives of vectors and tensors of any type. This construction is achieved with the help of the method of *"parallel transport"* of vectors. The construction of these derivatives with a definite tensorial transformation property is an essential prerequisite for the derivation of tensorial equations which satisfy the fundamental requirement to be valid in any arbitrary reference frame which is a basic requirement of the theory of relativity. We shall also encounter quantities with several upper and lower indices called "mixed tensors". Mathematicians eschew such formulations if possible, *e.g.* in favour of differential geometry formulations, but for the uninitiated the explicit tensorial equations still provide the most direct and convincing way to a result, as also in electrodynamics. The latter also provides the most physical illustration of the concepts of antisymmetric and dual tensors which we encounter here. The comparison of the covariant derivatives here with those of the "minimal coupling" procedure in electrodynamics also helps to appreciate the physical significance of the former.

14.2   Tensors

The *mixed tensor* was referred to in Sec. 12.5.2 after introduction of con-
travariant and covariant tensors and is a quantity $T_{\alpha\beta\ldots}^{\mu\nu\ldots}(x), x \in \mathbb{R}_4$, with the
transformation property

$$T'^{\mu'\nu'\ldots}_{\alpha'\beta'\ldots}(x') = \frac{\partial x'^{\mu'}}{\partial x^\mu}\frac{\partial x'^{\nu'}}{\partial x^\nu}\cdots T_{\alpha\beta\ldots}^{\mu\nu\ldots}(x)\frac{\partial x^\alpha}{\partial x'^{\alpha'}}\frac{\partial x^\beta}{\partial x'^{\beta'}}\cdots . \tag{14.1}$$

The upper (lower) indices in the denominator play the role of lower (upper)
indices in the numerator. A tensor with p upper and q lower indices is said
to be a $[p, q]$-valent tensor. We see that the factor to the left of $T_{\alpha\beta\ldots}^{\mu\nu\ldots}(x)$
transforms the contravariant indices $\mu', \nu', \ldots$ and the factor to the right
the covariant indices. We now consider the particular symmetric covariant
metric tensor $g_{\mu\nu}(x), x \in \mathbb{R}_4$. We define

$$g = \det(g_{\mu\nu}) = \begin{vmatrix} g_{00} & g_{01} & g_{02} & g_{03} \\ g_{10} & g_{11} & g_{12} & g_{13} \\ g_{20} & g_{21} & g_{22} & g_{23} \\ g_{30} & g_{31} & g_{32} & g_{33} \end{vmatrix}. \tag{14.2}$$

We also let $\triangle^{\mu\nu}$ be the cofactor* of element $g_{\mu\nu}$ so that the expansion of the
determinant in terms of elements of row μ is

$$g = \sum_\lambda g_{\mu\lambda}\triangle^{\mu\lambda} \qquad \text{(no summation over } \mu\text{)}, \tag{14.3a}$$

and (see below)

$$\sum_\lambda g_{\nu\lambda}\triangle^{\mu\lambda} = 0 \quad \text{for} \quad \nu \neq \mu. \tag{14.3b}$$

The latter relation can be shown in general — *cf.* Example 14.1 — but is
also readily verified in a simple case; for instance for

$$g = \begin{vmatrix} g_{11} & g_{12} \\ g_{21} & g_{22} \end{vmatrix},$$

we have (expansion of the determinant in terms of elements of row 1)

$$g = \sum_i g_{1i}\triangle^{1i} = g_{11}\triangle^{11} + g_{12}\triangle^{12} = g_{11}g_{22} + g_{12}(-g_{21}),$$

---

*Recall that the *cofactor* of element $g_{\rho\kappa}$ in the determinant $g = |g_{\mu\nu}|$ is the determinant
obtained by suppressing the ρth row and the κth column of the matrix $(g_{\mu\nu})$ and giving it the sign
$(-1)^{\rho+\kappa}$. See *e.g.* A. C. Aitken [1], p. 39.

but

$$\sum_i g_{1i}\Delta^{2i} = g_{11}\Delta^{21} + g_{12}\Delta^{22} = g_{11}(-g_{12}) + g_{12}g_{11} = 0.$$

In this way we arrive at a *better definition* of the contravariant tensor $g^{\mu\nu}$ (recall that at the beginning $g_{\mu\nu}$ was assumed to be a covariant metric tensor). We now *define* the following quantity (still to be established to be the contravariant tensor, hence we use this notation)

$$g^{\mu\nu}(x) := \frac{\Delta^{\mu\nu}(x)}{g}, \qquad (14.4a)$$

because then (observe that δ^μ_ν is the fundamental mixed tensor,[†] as one can prove, *cf.* Eq. (14.68))

$$g^{\mu\lambda}(x)g_{\lambda\nu}(x) \overset{(14.4a)}{=} \frac{\Delta^{\mu\lambda}g_{\lambda\nu}}{g} \overset{(14.3a),(14.3b)}{=} \delta^\mu_\nu. \qquad (14.4b)$$

In Example 14.2 we convince ourselves that although the quantity $\Delta^{\mu\nu}$ is not a tensor, if $g \neq 0$ the quantity

$$\frac{\Delta^{\mu\nu}}{g} = g^{\mu\nu} \qquad (14.4c)$$

is a symmetric contravariant tensor, *i.e.* it transforms accordingly. Recall that the quantity $g_{\mu\nu}(x)$ was from the very beginning above assumed to be a covariant tensor.

Example 14.1: Determinant expansion in terms of alien cofactors is zero

The cofactors of elements in a row (or column) different from those of the elements of a given row (or column) in terms of which a determinant is expanded are known as *alien cofactors*. Show that expansions of a determinant in terms of alien cofactors vanish identically.[‡] This is the result (14.3b).

Solution: Substitute for the ith row of a determinant $|A| = |a_{ij}|$ a row of new elements b_{ij}. The cofactors of the b_{ij} in the new determinant are evidently the same as those of the corresponding a_{ij} in $|A|$, and so an expansion of the new determinant is (with $\Delta_{ij} \equiv |A_{ij}|$ denoting cofactors)

$$a_{i1}|A_{i1}| + a_{i2}|A_{i2}| + \cdots + a_{in}|A_{in}| \rightarrow b_{i1}|A_{i1}| + b_{i2}|A_{i2}| + \cdots + b_{in}|A_{in}|.$$

(A similar result is obtained when a new jth column is substituted in the matrix A). Now consider the case when the b_{ij} are elements of any row of A other than the ith. The expansion is then the expansion of a determinant with two rows identical, and hence gives zero. Thus as in Eq. (14.3b)

$$\sum_j a_{kj}|A_{ij}| = a_{k1}|A_{i1}| + a_{k2}|A_{i2}| + \cdots + a_{kn}|A_{in}| = 0 \quad \text{for} \quad k \neq i.$$

[†] This quantity, δ^μ_ν, is not written with displaced indices.
[‡] A. C. Aitken [1], p. 51.

Example 14.2: Proof that $g^{\mu\nu}(x)$ is a rank 2 contravariant tensor

Prove that $g^{\mu\nu}(x)$ is a rank 2 contravariant tensor.

Solution: We know from Eqs. (14.3a), (14.3b), that

$$\sum_\lambda \Delta^{\mu\lambda} g_{\nu\lambda} = g\delta^\mu_\nu \quad \text{and} \quad \sum_\lambda \Delta^{\lambda\mu} g_{\lambda\nu} = g\delta^\mu_\nu,$$

which correspond to expansion of the determinant either in terms of the cofactors of the elements of a row, or in terms of those of a column. Dividing the equations by $g \neq 0$ and using the definition (14.4a) of the quantity $g^{\mu\lambda}(x)$, we obtain (leaving summations understood)

$$g^{\mu\lambda} g_{\nu\lambda} = \delta^\mu_\nu, \quad g^{\lambda\mu} g_{\lambda\nu} = \delta^\mu_\nu. \tag{14.5}$$

We now let $A^\mu(x)$ be an arbitrary contravariant vector, and we define the covariant vector $B_\mu(x)$ by $B_\mu = g_{\mu\nu} A^\nu(x)$, $g_{\mu\nu}$ being by our earlier assumption a rank-2 covariant tensor. We consider the case $g \neq 0$, so that with the components B_μ chosen arbitrarily, the corresponding components of $A^\nu(x)$ can be calculated from the latter relation, *i.e.* from

$$g^{\mu\nu}(x) B_\mu(x) = g^{\mu\nu}(x) g_{\mu\rho}(x) A^\rho(x) = \delta^\nu_\rho A^\rho(x) = A^\nu(x),$$

where we used Eq. (14.5). We define with $\tilde{g}^{\mu\nu}(x')$ the actual components obtained when in $g^{\mu\nu}(x)$ the reference frame is changed from that of x to that of x', so that transforming to the x'-frame we have

$$\tilde{g}^{\mu\nu}(x') B'_\mu(x') = A'^\nu(x'). \tag{14.6a}$$

We now let $g'^{\mu\nu}(x')$ be the set of elements defined in the x'-frame by the transformation property of contravariant tensors, namely by (compare with Eq. (14.1))

$$g'^{\mu\nu}(x') := \sum_{\rho,\kappa} \frac{\partial x'^\mu}{\partial x^\rho} \frac{\partial x'^\nu}{\partial x^\kappa} g^{\rho\kappa}(x). \tag{14.6b}$$

Since this is a tensor transformation equation, we know that the elements satisfy the relation

$$g'^{\mu\nu}(x') B'_\mu(x') = A'^\nu(x'). \tag{14.7}$$

Hence by subtraction of Eq. (14.6a) from Eq. (14.7) we obtain:

$$[g'^{\mu\nu}(x') - \tilde{g}^{\mu\nu}(x')] B'_\mu(x') = 0.$$

Since $B'_\mu(x')$ has arbitrary components in the x-frame, its components in the x'-frame are also arbitrary. Thus assuming suitable values we obtain

$$g'^{\mu\nu}(x') - \tilde{g}^{\mu\nu}(x') = 0, \quad i.e. \quad \tilde{g}^{\mu\nu}(x') = g'^{\mu\nu}(x'),$$

i.e. $g^{\mu\nu}$ transforms as a rank 2 contravariant tensor.

From the metric relation $ds^2 = g_{\mu\nu} dx^\mu dx^\nu$ we know that $g_{\mu\nu} = g_{\nu\mu}$. Hence $g^{\mu\lambda} g_{\lambda\nu} = \delta^\mu_\nu$, $g_{\nu\lambda} g^{\lambda\mu} = \delta^\mu_\nu$, and so

$$g^{\mu\lambda} g_{\lambda\nu} - g_{\nu\lambda} g^{\lambda\mu} = 0, \quad \sum_\lambda g_{\lambda\nu}(g^{\mu\lambda} - g^{\lambda\mu}) = 0.$$

Since in general $g_{\lambda\nu} \neq 0$, we have for arbitrary $g_{\lambda\nu}$: $g^{\mu\lambda} = g^{\lambda\mu}$. The contravariant rank 2 tensor is also symmetric.

14.3   Symmetric and Antisymmetric Tensors

In handling tensors of various types, one also wants to know when and how indices, in particular those of mixed tensors, can be manipulated. This is the aspect we consider here, not all of which is required for later purposes, but may provide a better background, also in view of our frequent reference to electrodynamics. We assume we have a tensor $T^{\mu\nu}(x)$ at a point $x \in \mathbb{R}_4$. Then one can define

$$T^{\mu}{}_{\lambda}(x) := T^{\mu\nu}(x)g_{\nu\lambda}(x), \qquad (14.8a)$$

and we can write (with $g_{\lambda\nu} = g_{\nu\lambda}$ and summations understood)

$$T_{\lambda}{}^{\mu}(x) := g_{\lambda\nu}(x)T^{\nu\mu}(x) = T^{\nu\mu}(x)g_{\nu\lambda}(x). \qquad (14.8b)$$

Then, as in Eq. (14.1), and again with summations understood,

$$T'^{\mu}{}_{\lambda}(x') = \frac{\partial x''^{\mu}}{\partial x^{\alpha}}T^{\alpha}{}_{\beta}(x)\frac{\partial x^{\beta}}{\partial x'^{\lambda}} = \frac{\partial x'^{\mu}}{\partial x^{\alpha}}T^{\alpha\eta}(x)g_{\eta\beta}(x)\frac{\partial x^{\beta}}{\partial x'^{\lambda}}, \qquad (14.9a)$$

and

$$T'_{\lambda}{}^{\mu}(x') = \frac{\partial x'^{\mu}}{\partial x^{\alpha}}T_{\beta}{}^{\alpha}(x)\frac{\partial x^{\beta}}{\partial x'^{\lambda}} = \frac{\partial x'^{\mu}}{\partial x^{\alpha}}T^{\eta\alpha}(x)g_{\eta\beta}(x)\frac{\partial x^{\beta}}{\partial x'^{\lambda}}. \qquad (14.9b)$$

In general $T^{\mu\nu}(x) \neq T^{\nu\mu}(x)$ and $T^{\alpha}{}_{\beta}(x) \neq T_{\beta}{}^{\alpha}(x)$. But if

$$T^{\mu\nu}(x) = T^{\nu\mu}(x), \qquad (14.9c)$$

we obtain by rewriting respectively Eqs. (14.8b) and (14.8a) with different indices:

$$T_{\nu}{}^{\lambda}(x) = g_{\nu\mu}(x)T^{\mu\lambda}(x), \qquad T^{\lambda}{}_{\nu}(x) = T^{\lambda\mu}(x)g_{\mu\nu}(x),$$

and hence (using in the first equality the assumed symmetry $T^{\mu\nu} = T^{\nu\mu}$ and in the last equality the first of the two relations just written out)

$$T^{\lambda}{}_{\nu}(x) = T^{\lambda\mu}(x)g_{\mu\nu}(x) \overset{(14.9c)}{=} T^{\mu\lambda}(x)g_{\mu\nu}(x) = g_{\nu\mu}(x)T^{\mu\lambda}(x) = T_{\nu}{}^{\lambda}(x).$$

Thus if $T^{\mu\nu}(x) = T^{\nu\mu}(x)$ we have

$$T_{\lambda}{}^{\nu}(x) = T^{\nu}{}_{\lambda}(x) \equiv T^{\nu}_{\lambda}(x). \qquad (14.10)$$

If $T^{\alpha\beta}(x) = T^{\beta\alpha}(x)$, then

$$
\begin{aligned}
T'^{\mu\nu}(x') &= \frac{\partial x'^{\mu}}{\partial x^{\alpha}}\frac{\partial x'^{\nu}}{\partial x^{\beta}}T^{\alpha\beta}(x) = \frac{\partial x'^{\mu}}{\partial x^{\alpha}}\frac{\partial x'^{\nu}}{\partial x^{\beta}}T^{\beta\alpha}(x) \\
&= \frac{\partial x'^{\nu}}{\partial x^{\beta}}\frac{\partial x'^{\mu}}{\partial x^{\alpha}}T^{\beta\alpha}(x) = T'^{\nu\mu}(x'),
\end{aligned} \qquad (14.11)
$$

i.e. the transformed contravariant tensors are also symmetric. Corresponding considerations apply to antisymmetric tensors,

$$T^{\mu\nu}(x) = -T^{\nu\mu}(x), \qquad T_{\lambda}{}^{\nu}(x) = -T^{\nu}{}_{\lambda}(x).$$

14.4   Definition of Other Important Quantities

14.4.1   Transformation of the metric tensor

We encountered above, *e.g.* in Eq. (14.1), the derivatives

$$\alpha^{\mu}_{\nu}(x) = \frac{\partial x'^{\mu}}{\partial x^{\nu}}, \qquad \beta^{\mu}_{\nu}(x') = \frac{\partial x^{\mu}}{\partial x'^{\nu}} \tag{14.12}$$

with

$$\alpha^{\mu}_{\nu}(x)\beta^{\nu}_{\lambda}(x') = \delta^{\mu}_{\lambda}, \tag{14.13}$$

and hence

$$\det\alpha(x)\det\beta(x') = 1. \tag{14.14}$$

The transformation of the covariant metric tensor is

$$g'_{\mu\nu}(x') = g_{\alpha\beta}(x)\frac{\partial x^{\alpha}}{\partial x'^{\mu}}\frac{\partial x^{\beta}}{\partial x'^{\nu}}. \tag{14.15}$$

Thus

$$\det(g'_{\mu\nu}) \equiv g' = g(\det\beta)^2 = \frac{g}{(\det\alpha)^2}, \quad i.e. \quad g' = \frac{g}{(\det\alpha)^2}. \tag{14.16}$$

For a real number we take

$$\sqrt{-g'(x')} = \frac{\sqrt{-g(x)}}{|\det\alpha(x)|}. \tag{14.17}$$

For a manifold with Minkowski metric $g_{\mu\nu} = \eta_{\mu\nu}, \eta_{00} = -\eta_{11} = -\eta_{22} = -\eta_{33} = 1, \eta_{\mu\nu} \neq 0$ for $\mu \neq \nu$, the determinant of $\eta_{\mu\nu}$ is -1. One therefore imposes the condition $-g(x) > 0$ everywhere. Of course, in the case of a Euclidean metric the minus sign is left out.

14.4.2   Pseudo-tensors and duals

The following expression exhibits the transformation properties of a mixed tensor — except that the prefactor, $\det\alpha/|\det\alpha|$, allows for a $+$ or $-$ sign:

$$P'^{\mu'\nu'...}_{\alpha'\beta'...}(x') := \frac{\det\alpha}{|\det\alpha|}\frac{\partial x'^{\mu'}}{\partial x^{\mu}}\frac{\partial x'^{\nu'}}{\partial x^{\nu}} \cdots P^{\mu\nu...}_{\alpha\beta...}(x)\frac{\partial x^{\alpha}}{\partial x'^{\alpha'}}\frac{\partial x^{\beta}}{\partial x'^{\beta'}} \cdots . \tag{14.18}$$

If $\det\alpha/|\det\alpha| = +1$, the quantity on the left hand side transforms as a (mixed) *tensor*, if it is -1, the equation defines a (mixed) *pseudo-tensor*. We now introduce the *symbol* (summations understood)

$$\delta'_{\alpha'\beta'\gamma'\eta'} = (\det\alpha)\delta_{\alpha\beta\gamma\delta}\frac{\partial x^{\alpha}}{\partial x'^{\alpha'}}\frac{\partial x^{\beta}}{\partial x'^{\beta'}}\frac{\partial x^{\gamma}}{\partial x'^{\gamma'}}\frac{\partial x^{\eta}}{\partial x'^{\eta'}}, \tag{14.19}$$

where $\delta_{\alpha\beta\gamma\delta}$ is the *Levi–Civita symbol*[§] with $\delta_{1230} = 1 = -\delta_{2130}$, *etc.* as for the completely antisymmetrized quantity. One *defines* now the quantity

$$\epsilon_{\alpha\beta\gamma\eta}(x) := \sqrt{-g(x)}\,\delta_{\alpha\beta\gamma\eta}. \tag{14.20}$$

The transformation law of this quantity then follows from Eq. (14.19) as (note the difference between $\epsilon_{\alpha\beta\gamma\eta}$ and $\delta_{\alpha\beta\gamma\eta}$, we use the result (14.17): $|\det\alpha| = \sqrt{-g(x)}/\sqrt{-g'(x')}$)

$$\underbrace{\epsilon'_{\alpha'\beta'\gamma'\eta'}(x')}_{\sqrt{-g'}\delta'_{\alpha'\beta'\gamma'\eta'}} = \frac{\det\alpha}{|\det\alpha|}\underbrace{\epsilon_{\alpha\beta\gamma\eta}(x)}_{\sqrt{-g}\delta_{\alpha\beta\gamma\eta}}\frac{\partial x^\alpha}{\partial x'^{\alpha'}}\frac{\partial x^\beta}{\partial x'^{\beta'}}\frac{\partial x^\gamma}{\partial x'^{\gamma'}}\frac{\partial x^\eta}{\partial x'^{\eta'}}. \tag{14.21}$$

Here $\epsilon_{\alpha\beta\gamma\eta}(x)$ is — in agreement with the definition (14.18) — the *completely antisymmetric pseudo-tensor of order 4 of Levi–Civita*, also called the *Levi–Civita form* (note the difference between Levi–Civita *symbol* and Levi–Civita *form*).

We now introduce the concept of a *dual* or *dual map*. This is a useful concept since it allows us to link or identify forms of different ranks. Consider the following antisymmetric tensor or 2-form (like the field tensor of electrodynamics) on a 4-dimensional manifold with metric $g(x)$:

$$F^{\mu\nu}(x) = -F^{\nu\mu}(x).$$

We can define its *dual*[¶] as the 4-minus-2-form

$$^*F_{\alpha\beta}(x) = \frac{1}{2!}\epsilon_{\alpha\beta\mu\nu}(x)F^{\mu\nu}(x) \overset{(14.20)}{=} \frac{1}{2!}\sqrt{-g(x)}\,\delta_{\alpha\beta\mu\nu}F^{\mu\nu}(x). \tag{14.22a}$$

Here the factor $1/2!$ arises because, for instance, $^*F_{03}(x)$ (with summation over $\mu, \nu = 0, 1, 2, 3$) involves two relative permutations, the relative sign of which in the Levi-Civita symbol can be determined by counting the number of intersections of their connecting lines, as indicated in Fig. 14.1, an odd number of intersections implying a minus sign and an even number a plus sign. In this way we obtain

$$
\begin{aligned}
^*F_{03}(x) \quad &= \quad \frac{1}{2!}\epsilon_{0312}F^{12}(x) + \frac{1}{2!}\epsilon_{0321}F^{21}(x) \\
&= \quad \frac{1}{2}\epsilon_{0312}F^{12}(x) + \frac{1}{2}\epsilon_{0312}F^{12}(x) = \epsilon_{0312}F^{12}(x) \\
&\overset{(14.20)}{=} \quad \sqrt{-g(x)}\,\delta_{0312}F^{12}(x) = \sqrt{-g(x)}F^{12}(x). \tag{14.22b}
\end{aligned}
$$

[§] For extensive discussions see B. Felsager [16], p. 348.
[¶] See *e.g.* B. Felsager [16], p. 352.

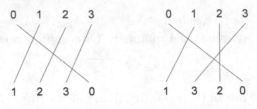

Fig. 14.1 Determination of sign of relative permutation.

From Eq. (14.22a) we obtain (where (a) results again from summation over $\mu, \nu = 0, 1, 2, 3$, and (b) from antisymmetry)

$$^*F_{12}(x) = \frac{1}{2}\sqrt{-g}\delta_{12\mu\nu}F^{\mu\nu}(x) \overset{(a)}{\equiv} -\sqrt{-g}F^{30}(x) \overset{(b)}{\equiv} \sqrt{-g}F^{03}(x),$$

$$^*F_{23}(x) = \frac{1}{2}\sqrt{-g}\delta_{23\mu\nu}F^{\mu\nu}(x) \overset{(a)}{\equiv} -\sqrt{-g}F^{10}(x) \overset{(b)}{\equiv} \sqrt{-g}F^{01}(x),$$

$$^*F_{31}(x) = \frac{1}{2}\sqrt{-g}\delta_{31\mu\nu}F^{\mu\nu}(x) \overset{(a)}{\equiv} -\sqrt{-g}F^{20}(x) \overset{(b)}{\equiv} \sqrt{-g}F^{02}(x),$$

$$^*F_{01}(x) \quad \equiv \quad \sqrt{-g}F^{23}(x),$$

$$^*F_{02}(x) \equiv \sqrt{-g}F^{31}(x), \quad ^*F_{03}(x) \quad \equiv \quad \sqrt{-g}F^{12}(x). \tag{14.23}$$

The equalities in the last line follow as Eq. (14.22b). Thus for example the dual of the electric field (made up of components F_{0i}) is the magnetic field, or that of the magnetic field (made up of components $F_{ij}, i, j \neq 0$) is the electric field. The electric field is a *polar vector*; the magnetic field is an *axial vector*.

14.4.3 Volume forms

We consider two contravariant vectors $A^\mu(x), B^\nu(x)$ on the manifold $\mathbb{R}_4$ and construct the antisymmetric quantity:

$$S^{\mu\nu}(x) := \begin{vmatrix} A^\mu & B^\mu \\ A^\nu & B^\nu \end{vmatrix} = A^\mu(x)B^\nu(x) - A^\nu(x)B^\mu(x) = -S^{\nu\mu}(x). \tag{14.24}$$

The expression reminds us of that for the area **f** of a parallelogram in a 2-dimensional subspace; *i.e.* if

$$\mathbf{a} = (a_1\mathbf{e}_1, a_2\mathbf{e}_2, 0), \qquad \mathbf{b} = (b_1\mathbf{e}_1, b_2\mathbf{e}_2, 0),$$

with $\mathbf{e}_1, \mathbf{e}_2, \mathbf{e}_3$ mutually orthogonal unit vectors, the area $\mathbf{f}$ of the parallelogram and its orientability (direction) are given by

$$\mathbf{f}: \quad = \quad \mathbf{a} \times \mathbf{b} = \mathbf{e}_3 ab \sin(\mathbf{a}, \mathbf{b}) = \mathbf{e}_3(a_1 b_2 - a_2 b_1),$$

$$f \quad = \quad \begin{vmatrix} a_1 & b_1 \\ a_2 & b_2 \end{vmatrix} = \sum_{j,k=1,2} \epsilon_{30jk} a_j b_k. \tag{14.25}$$

We observe that the expression is antisymmetric under interchange of indices 1 and 2, and implies a reversal of the vector $\mathbf{f}$.

From Eq. (14.22a) and insertion of the expression in Eq. (14.24), we obtain the dual of $S^{\mu\nu}(x)$:

$$\begin{aligned} {}^*S_{\alpha\beta} \quad &= \quad \frac{1}{2!} \epsilon_{\alpha\beta\mu\nu} S^{\mu\nu} = \frac{1}{2!} \sqrt{-g} \delta_{\alpha\beta\mu\nu} S^{\mu\nu} \\ &\quad - \frac{1}{2!} \sqrt{-g} \delta_{\alpha\beta\mu\nu}(A^\mu B^\nu - A^\nu B^\mu) \\ &= \quad \frac{1}{2!} \sqrt{-g} [\delta_{\alpha\beta\mu\nu} A^\mu B^\nu - \delta_{\alpha\beta\mu\nu} A^\nu B^\mu] \\ &= \quad \frac{1}{2!} \sqrt{-g} [\delta_{\alpha\beta\mu\nu} A^\mu B^\nu - \delta_{\alpha\beta\nu\mu} A^\mu B^\nu] \\ &= \quad \sqrt{-g} \delta_{\alpha\beta\mu\nu} A^\mu B^\nu. \tag{14.26} \end{aligned}$$

The intermediate steps here which demonstrate the cancellation of 2! are not repeated below in the higher order cases. For $\alpha = 0, \beta = 1$, we obtain

$$ {}^*S_{01} = \sqrt{-g} \delta_{01\mu\nu} A^\mu B^\nu = \sqrt{-g}[A^2 B^3 - A^3 B^2] = \sqrt{-g} \begin{vmatrix} A^2 & B^2 \\ A^3 & B^3 \end{vmatrix}.$$

Contracting indices we have:

$$\begin{aligned} {}^*S_{\alpha\beta} S^{\alpha\beta} \quad &= \quad \frac{1}{2!} \sqrt{-g} \delta_{\alpha\beta\mu\nu}(A^\mu B^\nu - A^\nu B^\mu)(A^\alpha B^\beta - A^\beta B^\alpha) \\ &= \quad (a) + (b) + (c) + (d) = 0, \tag{14.27} \end{aligned}$$

where the vanishing of the expression follows from:

(a) $\quad \delta_{\alpha\beta\mu\nu} A^\mu B^\nu A^\alpha B^\beta = \frac{1}{2}(\delta_{\alpha\beta\mu\nu} + \delta_{\mu\beta\alpha\nu}) A^\mu A^\alpha B^\nu B^\beta = 0,$

(b) $\quad \delta_{\alpha\beta\mu\nu} A^\mu B^\nu A^\beta B^\alpha \overset{\beta\leftrightarrow\mu}{=} \delta_{\alpha\mu\beta\nu} A^\beta B^\nu A^\mu B^\alpha = -\delta_{\alpha\beta\mu\nu} A^\mu B^\nu A^\beta B^\alpha = 0,$

and similarly for cases (c) and (d).

Correspondingly three vectors determine the volume of a parallelopiped:

$$V^{\mu\nu\lambda}(x) := \begin{vmatrix} A^\mu & B^\mu & C^\mu \\ A^\nu & B^\nu & C^\nu \\ A^\lambda & B^\lambda & C^\lambda \end{vmatrix}. \tag{14.28}$$

We define the dual of this quantity (on the 4-dimensional manifold with Minkowski metric) by the 1-form

$$^*V_\alpha(x) = \frac{1}{3!}\epsilon_{\alpha\beta\mu\nu}V^{\beta\mu\nu} = \frac{\sqrt{-g}}{3!}\delta_{\alpha\beta\mu\nu}V^{\beta\mu\nu} = \sqrt{-g}\delta_{\alpha\beta\mu\nu}A^\beta B^\mu C^\nu. \quad (14.29)$$

Here in the last equality the last expression follows from considerations analogous to those in the two-dimensional case.

In the case of four vectors we define

$$\Sigma^{\alpha\beta\lambda\eta}(x) := \begin{vmatrix} A^\alpha & B^\alpha & C^\alpha & D^\alpha \\ A^\beta & B^\beta & C^\beta & D^\beta \\ A^\lambda & B^\lambda & C^\lambda & D^\lambda \\ A^\eta & B^\eta & C^\eta & D^\eta \end{vmatrix} \quad (14.30)$$

with dual, where $\delta^{\alpha\beta\cdots}_{\gamma\rho\cdots} = \pm 1$ depending on whether $\gamma\rho\cdots$ is an even or odd permutation of $\alpha\beta\cdots$,

$$^*\Sigma := \frac{1}{4!}\epsilon_{\alpha\beta\lambda\eta}\Sigma^{\alpha\beta\lambda\eta} = \frac{\sqrt{-g}}{4!}\delta_{\alpha\beta\lambda\eta}\Sigma^{\alpha\beta\lambda\eta} \quad (14.31)$$

$$= \frac{\sqrt{-g}}{4!}\delta_{\alpha\beta\lambda\eta}\Sigma_{\gamma,\rho,\mu,\nu}\delta^{\alpha\beta\lambda\eta}_{\gamma\rho\mu\nu}A^\gamma B^\rho C^\mu D^\nu$$

$$= \sqrt{-g}\delta_{\alpha\beta\lambda\eta}A^\alpha B^\beta C^\lambda D^\eta \quad (14.32)$$

(for the cancellation of 4! see the analogous case of Eq. (14.26)), or

$$^*\Sigma = \sqrt{-g}\begin{vmatrix} A^0 & B^0 & C^0 & D^0 \\ A^1 & B^1 & C^1 & D^1 \\ A^2 & B^2 & C^2 & D^2 \\ A^3 & B^3 & C^3 & D^3 \end{vmatrix}. \quad (14.33)$$

In Example 14.3 we show that $^*\Sigma$ is a pseudoscalar.

Example 14.3: Proof that $^*\Sigma$, the dual of $\Sigma^{\alpha\beta\lambda\eta}$, is a pseudoscalar
Show that $^*\Sigma$, the dual of $\Sigma^{\alpha\beta\lambda\eta}$ as defined by Eq. (14.31), is a pseudoscalar.

Solution: Consider the expression

$$^*\Sigma'(x') \overset{(14.31)}{=} \frac{\sqrt{-g'(x')}}{4!}\delta'_{\alpha'\beta'\gamma'\eta'}\Sigma'^{\alpha'\beta'\gamma'\eta'}$$

$$\overset{(14.32)}{=} \sqrt{-g'(x')}\delta'_{\alpha'\beta'\gamma'\eta'}A'^{\alpha'}B'^{\beta'}C'^{\gamma'}D'^{\eta'}$$

$$\overset{(14.17),(14.19)}{=} \frac{\sqrt{-g(x)}}{|\det(\alpha)|}\det(\alpha)\delta_{abcd}\frac{\partial x^a}{\partial x'^{\alpha'}}\frac{\partial x^b}{\partial x'^{\beta'}}\frac{\partial x^c}{\partial x'^{\gamma'}}\frac{\partial x^d}{\partial x'^{\eta'}}$$

$$\times \frac{\partial x'^{\alpha'}}{\partial x^m}\frac{\partial x'^{\beta'}}{\partial x^n}\frac{\partial x'^{\gamma'}}{\partial x^p}\frac{\partial x'^{\eta'}}{\partial x^q}A^m B^n C^p D^q$$

$$= \frac{\sqrt{-g(x)}}{|\det(\alpha)|}\det(\alpha)\delta_{mnpq}A^m B^n C^p D^q \overset{(14.32)}{=} \frac{\det(\alpha)}{|\det(\alpha)|}{}^*\Sigma(x).$$

Thus

$$*\Sigma'(x') = \frac{\det(\alpha)}{|\det(\alpha)|} *\Sigma(x).$$ (14.34)

For a manifold with Minkowski metric, as mentioned after Eq. (14.17), $\det(\alpha) = -1$, implying that Σ is a pseudoscalar. The volume element of the space is correspondingly a pseudoscalar, *i.e.*

$$d\Sigma = \sqrt{-g}dx^0 dx^1 dx^2 dx^3,$$ (14.35)

i.e.

$$d\Sigma'(x') \quad := \quad \sqrt{-g'(x')}dx'^0 dx'^1 dx'^2 dx'^3 = \sqrt{-g'(x')}\left|\frac{\partial x'^\mu}{\partial x^\rho}\right| dx^0 dx^1 dx^2 dx^3$$

$$\overset{(14.17)}{=} \quad \frac{\sqrt{-g(x)}}{|\det(\alpha)|}\det(\alpha)dx^0 dx^1 dx^2 dx^3 = \frac{\det(\alpha)}{|\det(\alpha)|}d\Sigma(x).$$

14.5 Covariant Derivatives by the Method of Parallel Transport of a Vector

Tensor equations (like those we require here) are valid in every reference frame (and this is what we demand of equations describing the laws of nature). Thus the tensorial nature of any equation has to be ensured. In particular we require to this end tensorial derivatives, which are here called *"covariant derivatives"*. The construction of these covariant derivatives is therefore of basic importance, and is the topic of this section.[||] The construction utilizes a rectangular reference frame along with general (curved) frames, and a parallel translation or transport of a vector in this rectangular frame from one point in space to another point nearby. Writing out the appropriate transformation equations is in essence what leads to expressions which permit the definition of the covariant derivative as one which transforms like a tensor. An assumption we use throughout is that derivatives $\partial/\partial x^\alpha, \partial/\partial x^\beta$ commute, *i.e.* that

$$\frac{\partial^2}{\partial x^\alpha \partial x^\beta} = \frac{\partial^2}{\partial x^\beta \partial x^\alpha}.$$

This is the *"torsion-free"* case, which is also the usual case excepting some few specific theories which are of no relevance here.

We have the transformation of a vector field $F^\mu(x)$, *i.e.*

$$F'^\mu(x') = \sum_\alpha \frac{\partial x'^\mu}{\partial x^\alpha}F^\alpha(x) \equiv \frac{\partial x'^\mu}{\partial x^\alpha}F^\alpha(x),$$ (14.36)

[||]We make ample use here of D. F. Lawden [27], pp. 96 – 108. The reader can find additional and considerably detailed discussion of this topic in C. W. Misner, K. S. Thorne and J. A. Wheeler [33], pp. 258 – 264.

and we set

$$\frac{\partial}{\partial x^\lambda} F^\mu(x) := H^\mu{}_\lambda(x), \qquad \frac{\partial}{\partial x'^\lambda} F'^\mu(x') := H'^\mu{}_\lambda(x'). \qquad (14.37)$$

In a *flat space* we have (*cf.* Sec. 6.3.2):

$$x'^\mu = l^\mu_\alpha x^\alpha + a^\mu, \quad F'^\mu(x') = l^\mu{}_\alpha F^\alpha(x),$$

and

$$\frac{\partial F'^\mu(x')}{\partial x'^\lambda} = l^\mu{}_\alpha \frac{\partial F^\alpha(x)}{\partial x'^\lambda} = l^\mu{}_\alpha \frac{\partial x^\beta}{\partial x'^\lambda} \frac{\partial F^\alpha(x)}{\partial x^\beta},$$

where $l^\mu{}_\alpha = $ const. In our *nonflat case* we have:

$$\begin{aligned}
\frac{\partial}{\partial x'^\lambda} F'^\mu(x') &= \frac{\partial}{\partial x'^\lambda}\left[\frac{\partial x'^\mu}{\partial x^\alpha} F^\alpha(x)\right] = \frac{\partial}{\partial x^\eta}\left[\frac{\partial x'^\mu}{\partial x^\alpha} F^\alpha(x)\right]\frac{\partial x^\eta}{\partial x'^\lambda}\\
&= \frac{\partial x'^\mu}{\partial x^\alpha}\frac{\partial F^\alpha(x)}{\partial x^\eta}\frac{\partial x^\eta}{\partial x'^\lambda} + \frac{\partial^2 x'^\mu}{\partial x^\eta \partial x^\alpha} F^\alpha(x)\frac{\partial x^\eta}{\partial x'^\lambda}. \qquad (14.38)
\end{aligned}$$

We have seen that a tensor $T^\alpha{}_\nu(x)$ transforms as

$$T'^\mu{}_\lambda(x') = \frac{\partial x'^\mu}{\partial x^\alpha} T^\alpha{}_\nu(x) \frac{\partial x^\nu}{\partial x'^\lambda}. \qquad (14.39)$$

But

$$H'^\mu{}_\lambda(x') \stackrel{(14.37)}{\equiv} \frac{\partial}{\partial x'^\lambda} F'^\mu(x') \stackrel{(14.38)}{=} \frac{\partial x'^\mu}{\partial x^\alpha} H^\alpha{}_\eta(x) \frac{\partial x^\eta}{\partial x'^\lambda} + \frac{\partial^2 x'^\mu}{\partial x^\eta \partial x^\alpha} F^\alpha(x) \frac{\partial x^\eta}{\partial x'^\lambda}.$$
$$(14.40)$$

Thus $H'^\mu{}_\lambda(x') = \partial F'^\mu(x')/\partial x'^\lambda$ is not a tensor because it does not transform as such in view of the second term on the right (compare *e.g.* with the transformation of a mixed tensor, Eq. (14.1)). In order to achieve a transformation as a tensor another term is needed to cancel the latter. For the flat space Poincaré transformation $x'^\mu = l^\mu{}_\nu x^\nu + a^\mu$ we have

$$\frac{\partial^2 x'^\mu}{\partial x^\eta \partial x^\alpha} = 0.$$

We now look into the solution of this problem by the method of parallel transport or parallel displacement more closely. We consider the field $F^\mu(x)$ at points $P(x)$ and $P'(x')$. At the point $x' = x + dx$ its value is

$$F^\mu(x + dx) = F^\mu(x) + dF^\mu(x), \qquad dF^\mu(x) = \frac{\partial F^\mu(x)}{\partial x^\alpha} dx^\alpha. \qquad (14.41)$$

Since the tensor transformation law varies from point to point in $\mathbb{R}_4$ (or more generally, in the more general space without a metric, S_4), it follows that

$dF^\mu(x)$ is not a vector because $\partial F^\mu(x)/\partial x^\alpha$ is not a tensor (with an upper index μ and a lower index α), as we concluded from its transformation after Eq. (14.40). This is the difficulty we now face. It means that the difference $dF^\mu(x)$ of the values of the vector fields at two distinct but neighbouring points (*i.e.* $F^\mu(x), F^\mu(x+dx)$ in Eq. (14.41)) cannot lead to a tensorial $\partial F^\mu(x)/\partial x^\alpha$, just as the sum of tensors at different points is not a tensor (the algebra of tensors is defined only at a point). However, if the procedure could be replaced by one involving two vectors defined at the same point, the result can be expected to be a tensor equation. This is achieved with a new form of derivative, and this leads to the idea of parallel transport.

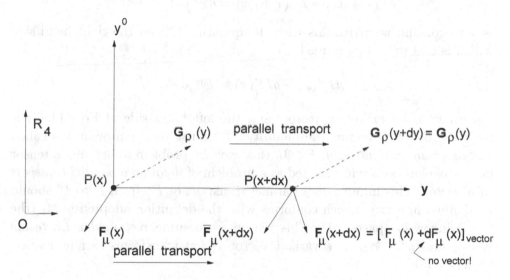

Fig. 14.2 Parallel transport of the vector $G_\rho(y)$ from $P(x)$ to $P'(x+dx)$.

We imagine the vector $F^\mu(x)$ "*transported*" from $P(x)$ (where it is defined) to the neighbouring point $P'(x+dx)$, so that it may be thought of as the same vector** defined at the other point as depicted in Fig. 14.2. The word "transported", of course, conveys at first an imprecise procedure, so that a precise method remains to be defined, which will be done below. In S_4 we have not defined "*magnitude*" or "*direction*" of a vector; in M_4 or a Euclidean space E_4 this would mean, the vector has the same components at $P'(x')$ as at $P(x)$. However, in general, if in this space curvilinear coordinates are used (*e.g.* spherical polar coordinates) the directions of the axes at P' will differ from those at P and hence also the components there of the vector $\overline{F}^\mu(x+dx)$ transported to this point (thus *e.g.* component $\overline{F}^2(x+dx)$ may be bigger than $F^2(x)$ and $\overline{F}^3(x+dx)$ smaller than $F^3(x)$, although the

---

**By vector $F^\mu(x)$ we mean, of course, the set of its components.

overall length of the vector consisting of all four components is the same, and therefore in general $\delta F^\mu \neq dF^\mu$). For this reason one introduces a new notation and writes in S_4:

$$F^\mu(x)|_{\text{transported from } P(x) \text{ to } P'(x+dx)} \equiv \overline{F}^\mu(x+dx) := F^\mu(x) + \delta F^\mu(x).$$
(14.42)

The vector $\overline{F}^\mu(x+dx)$ is a new vector which can now be compared with the field vector $F^\mu + dF^\mu$ at the same point $P'(x') = P'(x+dx)$. Since the two vectors are *defined at the same point*, the difference is a vector at this point, *i.e.*

$$F^\mu(x+dx) - \overline{F}^\mu(x+dx) = dF^\mu(x) - \delta F^\mu(x)$$

is a vector and we write this with the quantity $F^\mu_{;\alpha}$ on the right hand side which is still to be determined, *i.e.*

$$dF^\mu(x) - \delta F^\mu(x) = F^\mu_{;\alpha} dx^\alpha,$$
(14.43)

Since dx^α is an arbitrary vector, and the left hand side of Eq. (14.43) is by construction a vector, the quantity $F^\mu_{;\alpha}$ must be a tensor and is called the *covariant derivative of* F^μ. In this way the problem of defining a tensor derivative has been reformulated as a problem of *defining a parallel transport of a vector*. This infinitesimal parallel transport of F^μ from P to P' should be defined in a way, which conforms with the definition adopted in M_4 (the flat Minkowski space). For this reason we assume *rectangular Cartesian coordinates* $y^\rho \in M_4$. A covariant vector $B_\beta(x)$ transforms as (summation understood)

$$B'_\mu(x') = \frac{\partial x^\beta}{\partial x'^\mu} B_\beta(x),$$

as we have seen. Thus in our case the transformation from the frame with rectangular coordinates to the frame with general coordinates is

$$F_\mu(x) = \frac{\partial y^\rho}{\partial x^\mu} G_\rho(y) \quad \text{or} \quad G_\rho(y) = \frac{\partial x^\mu}{\partial y^\rho} F_\mu(x)$$
(14.44)

(this is a coordinate transformation which implies that $G_\rho(y)$ are the components of the vector F_μ with respect to the rectangular axes). If parallel transport of $F_\mu(x)$ is now carried out, as depicted in Fig. 14.2,

$$F_\mu(x) \longrightarrow \overline{F}_\mu(x+dx),$$

its Cartesian components $G_\rho(y)$ will not change under $y \longrightarrow y + dy$, *i.e.* $\delta G_\rho(y) = 0$ (this is the crucial point: no change of this vector in its transport to $y + dy$), however, $F_\mu(x)$ changes by δF_μ, and this is given by

$$\delta F_\mu(x) \quad = \quad \delta \left\{ \frac{\partial y^\rho}{\partial x^\mu} G_\rho(y) \right\} \tag{14.45}$$

$$= \quad \delta \left\{ \frac{\partial y^\rho}{\partial x^\mu} \right\} G_\rho(y) = \frac{\partial^2 y^\rho}{\partial x^\kappa \partial x^\mu} dx^\kappa G_\rho(y)$$

$$\stackrel{(14.44)}{=} \frac{\partial^2 y^\rho}{\partial x^\kappa \partial x^\mu} dx^\kappa \frac{\partial x^\nu}{\partial y^\rho} F_\nu(x). \tag{14.46}$$

Thus (we use already here the Christoffel symbol but will *still have to show* — by specializing from S_4 to $\mathbb{R}_4$, *i.e.* to a space with metric tensor $g_{\mu\nu}$ — that this is, in fact, the quantity we defined earlier in Eq. (13.9))

$$\delta F_\mu(x) = \Gamma^\nu_{\mu\kappa} F_\nu(x) dx^\kappa, \tag{14.47}$$

where (observe that this quantity is not a tensor of rank 3)

$$\Gamma^\nu_{\mu\kappa} = \frac{\partial^2 y^\rho}{\partial x^\kappa \partial x^\mu} \frac{\partial x^\nu}{\partial y^\rho} = \Gamma^\nu_{\kappa\mu}. \tag{14.48}$$

In S_4 we *define* accordingly the quantity $\delta F_\mu(x)$ by Eq. (14.47), determining the coefficients $\Gamma^\nu_{\mu\kappa}$ arbitrarily at every point of S_4 (subject to requirements of continuity) but once determined with respect to one frame they are then fixed with respect to any other frame by transformation laws. The set of coefficients $\Gamma^\nu_{\mu\kappa}$ is called an *affinity* and specifies an *affine connection between points* P, *i.e.* x, and points P', *i.e.* $x + dx$ of S_4, which are therefore said to be *affinely connected.*[††] Now we can write for the covariant quantities

$$dF_\mu(x) - \delta F_\mu(x) \stackrel{(14.41),(14.47)}{=} \frac{\partial F^\mu(x)}{\partial x^\alpha} dx^\alpha - \Gamma^\nu_{\mu\kappa} F_\nu(x) dx^\kappa$$

$$= \left[\frac{\partial F_\mu(x)}{\partial x^\alpha} - \Gamma^\nu_{\mu\alpha} F_\nu(x) \right] dx^\alpha. \tag{14.49}$$

We complete therefore Eq. (14.43), *i.e.*

$$dF_\mu(x) - \delta F_\mu(x) = F_{\mu;\alpha} dx^\alpha, \tag{14.50}$$

with the identification

$$F_{\mu;\alpha}(x) = \frac{\partial F_\mu(x)}{\partial x^\alpha} - \Gamma^\nu_{\mu\alpha}(x) F_\nu(x). \tag{14.51}$$

[††]The word affinity is derived from the Latin word *affinitas* and means relationship. It thus refers to a kind of mapping of domains or spaces which maintain certain geometrical properties. In the present context it refers to maintaining the same geometrical figure of the vector.

Since the left hand side of Eq. (14.50) is a vector for arbitrary dx^α, the quantity $F_{\mu;\alpha}$ must be a covariant tensor of rank 2 and therefore transforms accordingly (like a rank 2 covariant tensor). It is called the *covariant derivative of F_μ*. In tensor equations, which are to be valid in every frame, only covariant derivatives may appear. From Eq. (14.51) the transformation law of the affinity, which is not a tensor of rank 3, can be deduced, since

$$F'_{\mu;\alpha}(x') = \frac{\partial F'_\mu(x')}{\partial x'^\alpha} - \Gamma'^\nu_{\mu\alpha}(x')F'_\nu(x'), \tag{14.52}$$

and the transformation laws of $F_\mu(x)$ and the covariant tensor $F'_{\mu;\alpha}$ are known, *i.e.*

$$F'_\mu(x') = \frac{\partial x^\rho}{\partial x'^\mu}F_\rho(x), \qquad F'_{\mu;\alpha}(x') = \frac{\partial x^\rho}{\partial x'^\mu}\frac{\partial x^\kappa}{\partial x'^\alpha}F_{\rho;\kappa}(x). \tag{14.53}$$

Substituting these into Eq. (14.52), we obtain

$$\frac{\partial x^\rho}{\partial x'^\mu}\frac{\partial x^\kappa}{\partial x'^\alpha}F_{\rho;\kappa}(x) = \frac{\partial}{\partial x'^\alpha}\left(\frac{\partial x^\rho}{\partial x'^\mu}F_\rho(x)\right) - \Gamma'^\nu_{\mu\alpha}(x')\frac{\partial x^\rho}{\partial x'^\nu}F_\rho(x). \tag{14.54}$$

With Eq. (14.51) this is

$$\underbrace{\frac{\partial x^\rho}{\partial x'^\mu}\frac{\partial x^\kappa}{\partial x'^\alpha}\left(\frac{\partial F_\rho(x)}{\partial x^\kappa} - \Gamma^{\tilde\rho}_{\rho\kappa}(x)F_{\tilde\rho}(x)\right)}$$

$$= \frac{\partial^2 x^\rho}{\partial x'^\alpha \partial x'^\mu}F_\rho(x) + \underbrace{\frac{\partial x^\rho}{\partial x'^\mu}\frac{\partial F_\rho(x)}{\partial x^\kappa}\frac{\partial x^\kappa}{\partial x'^\alpha}} - \Gamma'^\nu_{\mu\alpha}(x')\frac{\partial x^\rho}{\partial x'^\nu}F_\rho(x). \tag{14.55}$$

The contributions underbraced cancel. Since $F_\rho(x)$ is an arbitrary vector, we can equate coefficients of $F_\rho(x)$ on both sides and obtain

$$\Gamma'^\nu_{\mu\alpha}(x')\frac{\partial x^\rho}{\partial x'^\nu} = \frac{\partial x^{\tilde\rho}}{\partial x'^\mu}\frac{\partial x^\kappa}{\partial x'^\alpha}\Gamma^\rho_{\tilde\rho\kappa}(x) + \frac{\partial^2 x^\rho}{\partial x'^\alpha \partial x'^\mu}. \tag{14.56}$$

Multiplying both sides of this equation by $\partial x'^\nu/\partial x^\rho$ and using

$$\frac{\partial x'^\lambda}{\partial x^\rho}\frac{\partial x^\rho}{\partial x'^\nu} = \frac{\partial x'^\lambda}{\partial x'^\nu} = \delta^\lambda_\nu, \tag{14.57}$$

we obtain the *transformation law of the affinity*, *i.e.*

$$\Gamma'^\nu_{\mu\alpha}(x') = \frac{\partial x^\epsilon}{\partial x'^\mu}\frac{\partial x^\kappa}{\partial x'^\alpha}\frac{\partial x'^\nu}{\partial x^\rho}\Gamma^\rho_{\epsilon\kappa}(x) + \frac{\partial x'^\nu}{\partial x^\rho}\frac{\partial^2 x^\rho}{\partial x'^\alpha \partial x'^\mu}. \tag{14.58}$$

One may note that the presence of the second term on the right implies that the expression is not a tensor of rank 3.

The expression (14.51) is the covariant derivative of the covariant vector $F_\mu(x)$. We now want to find the covariant derivative (lower index) of the contravariant vector $F^\mu(x)$. For this purpose we consider the scalar product of $F^\mu(x)$ with an arbitrary covariant vector $A_\mu(x)$. Then $A_\mu(x)F^\mu(x)$ is invariant and remains unchanged in value under parallel transport from P to P'. Thus

$$\left.\begin{array}{l} \delta(A_\mu F^\mu) = 0, \\ A_\mu(\delta F^\mu) + (\delta A_\mu)F^\mu = 0, \end{array}\right\} \tag{14.59}$$

and hence with Eq. (14.47) for the variation of a covariant vector

$$A_\mu(x)\delta F^\mu(x) = -\delta A_\mu(x)F^\mu(x) = -\Gamma^\nu_{\mu\kappa}A_\nu(x)dx^\kappa F^\mu(x). \tag{14.60}$$

Since $A_\mu(x)$ is arbitrary, their coefficients can be equated. Hence

$$\delta F^\mu(x) = -\Gamma^\mu_{\nu\kappa}F^\nu(r)dx^\kappa. \tag{14.61}$$

This equation defines the parallel transport of a contravariant vector. The covariant derivative (note: derivative index in lower position, since $\partial/\partial x^\kappa \equiv \partial_\kappa$) is now deduced as before, *i.e.* from

$$dF^\mu(x) - \delta F^\mu(x) = \left(\frac{\partial F^\mu(x)}{\partial x^\kappa} + \Gamma^\mu_{\nu\kappa}F^\nu(x)\right)dx^\kappa. \tag{14.62}$$

Since dx^κ is an arbitrary vector and $dF^\mu - \delta F^\mu$ is known to be a vector, the expression

$$F^\mu_{;\kappa}(x) = \frac{\partial F^\mu(x)}{\partial x^\kappa} + \Gamma^\mu_{\nu\kappa}(x)F^\nu(x) \tag{14.63}$$

is a tensor and is called *covariant derivative of $F^\mu(x)$*.

We can continue this process and consider the invariant

$$\delta(H_\mu{}^\nu(x)F_\nu(x)A^\mu(x)) = 0, \tag{14.64}$$

to deduce for arbitrary $A^\mu(x)$ that

$$\delta H_\mu{}^\nu = \Gamma^\lambda_{\mu\kappa}H_\lambda{}^\nu dx^\kappa - \Gamma^\nu_{\lambda\kappa}H_\mu^\lambda dx^\kappa, \tag{14.65}$$

and the covariant derivative of the tensor $H_\mu{}^\nu$ is:

$$H_{\mu;\kappa}{}^\nu = \frac{\partial H_\mu{}^\nu}{\partial x^\kappa} + \Gamma^\nu_{\lambda\kappa}H_\mu{}^\lambda - \Gamma^\lambda_{\mu\kappa}H_\lambda{}^\nu. \tag{14.66}$$

Correspondingly from

$$\delta(H_{\mu\nu}F^\mu A^\nu) = 0,$$

we obtain:

$$H_{\mu\nu\,;\kappa} = \frac{\partial H_{\mu\nu}}{\partial x^{\kappa}} - \Gamma^{\rho}_{\mu\kappa} H_{\rho\nu} - \Gamma^{\rho}_{\nu\kappa} H_{\mu\rho}, \tag{14.67a}$$

and by similar procedures

$$T^{\alpha\beta}_{;\mu} = \frac{\partial T^{\alpha\beta}}{\partial x^{\mu}} + \Gamma^{\alpha}_{\lambda\mu} T^{\lambda\beta} + \Gamma^{\beta}_{\mu\lambda} T^{\alpha\lambda}, \tag{14.67b}$$

$$T^{\rho}_{\nu\mu\,;\lambda} = \frac{\partial T^{\rho}_{\nu\mu}}{\partial x^{\lambda}} + \Gamma^{\rho}_{a\lambda} T^{a}_{\nu\mu} - \Gamma^{\sigma}_{\nu\lambda} T^{\rho}_{\sigma\mu} - \Gamma^{\sigma}_{\mu\lambda} T^{\rho}_{\nu\sigma}, \tag{14.67c}$$

$$T^{\mu}_{\eta\alpha\lambda\,;\nu} = \frac{\partial T^{\mu}_{\eta\alpha\lambda}}{\partial x^{\nu}} + \Gamma^{\mu}_{a\nu} T^{a}_{\eta\alpha\lambda} - \Gamma^{m}_{\eta\nu} T^{\mu}_{ma\lambda} - \Gamma^{m}_{a\nu} T^{\mu}_{\eta m\lambda} - \Gamma^{m}_{\lambda\nu} T^{\mu}_{\eta a m}, \tag{14.67d}$$

(Eq. (14.67b) will be required in Example 14.5 and in Appendix B, Sec. B.7 for the case of an electrostatic field, and the last relation will be needed later for the derivative of the Riemann curvature tensor $R^{\mu}_{\eta\alpha\lambda}$, cf. Sec. 15.3) and in general

$$\begin{aligned} T^{\mu\nu\cdots}_{\alpha\beta\cdots\,;\lambda} &= \frac{\partial}{\partial x^{\lambda}} T^{\mu\nu\cdots}_{\alpha\beta\cdots} + \Gamma^{\mu}_{a\lambda} T^{a\nu\cdots}_{\alpha\beta\cdots} + \Gamma^{\nu}_{a\lambda} T^{\mu a\cdots}_{\alpha\beta\cdots} + \cdots \\ &\quad \cdots - \Gamma^{m}_{\alpha\lambda} T^{\mu\nu\cdots}_{m\beta\cdots} - \Gamma^{m}_{\beta\lambda} T^{\mu\nu\cdots}_{\alpha m\cdots}. \end{aligned} \tag{14.67e}$$

With these formulas — which reveal a clear systematics — we now have at our disposal the derivative of any tensor with the correct tensorial transformation properties. The "*Kronecker delta*" δ^{μ}_{ν}, also called the *fundamental mixed tensor*, is readily shown to be the same in all frames:

$$\delta^{\mu}_{\nu} = \frac{\partial x'^{\mu}}{\partial x'^{\nu}} = \frac{\partial x'^{\mu}}{\partial x^{\kappa}} \frac{\partial x^{\kappa}}{\partial x'^{\nu}} = \frac{\partial x'^{\mu}}{\partial x^{\kappa}} \frac{\partial x^{\lambda}}{\partial x'^{\nu}} \delta^{\kappa}_{\lambda} = \delta'^{\mu}_{\nu}. \tag{14.68}$$

This relation is seen to be an application of Eq. (14.1). Applying Eq. (14.66) to this tensor, we obtain its covariant derivative:

$$\delta^{\mu}_{\nu\,;\kappa} = -\Gamma^{\lambda}_{\nu\kappa} \delta^{\mu}_{\lambda} + \Gamma^{\mu}_{\lambda\kappa} \delta^{\lambda}_{\nu} = -\Gamma^{\mu}_{\nu\kappa} + \Gamma^{\mu}_{\nu\kappa} = 0. \tag{14.69}$$

Thus the fundamental tensor behaves like a constant under covariant differentiation.

14.6   Metric Affinity and Christoffel Symbols

The considerations of the previous section did not really require the space to have a metric. It can be shown by "diagonalization" that at any chosen point $P \in \mathbb{R}_4$ the homogeneous quadratic form can be transformed from the coordinates x^μ to coordinates y^μ, maybe with complex quantities, *i.e.*

$$g_{\mu\nu}dx^\mu dx^\nu = \pm(dy^0)^2 \mp (dy^1)^2 \mp (dy^2)^2 \mp (dy^3)^2. \qquad (14.70)$$

Thus in a small neighbourhood of P, the coordinates y^μ behave like pseudo-Euclidean coordinates. In such a neighbourhood the components of the affinity Γ are representable as in Eq. (14.48), *i.e.*

$$\Gamma^\nu_{\mu\kappa} = \frac{\partial^2 y^\rho}{\partial x^\kappa \partial x^\mu} \frac{\partial x^\nu}{\partial y^\rho} = \Gamma^\nu_{\kappa\mu}. \qquad (14.71)$$

Apart from this symmetry the expression is arbitrary. However, one can show* that to avoid confusion the affinity should be chosen so that the co-variant derivative of the metric tensor $g_{\mu\nu}$, also called *connection*, is zero, *i.e.* (*cf.* also Eq. (14.83) below in support of this)[†]

$$g_{\mu\nu;\kappa}(x) = 0. \qquad (14.72)$$

The affinity is then called "*metric affinity*". The argument is that the lowering of an index with $g_{\mu\nu}$ should commute with differentiation, *i.e.*

$$g_{\mu\nu}(x)F^\nu_{;\rho}(x) \overset{!}{=} (g_{\mu\nu}(x)F^\nu(x))_{;\rho},$$

or

$$g_{\mu\nu}(x)F^\nu_{;\rho}(x) = g_{\mu\nu;\rho}(x)F^\nu(x) + g_{\mu\nu}(x)F^\nu_{;\rho}(x).$$

Since two terms cancel and $F^\nu(x)$ is arbitrary, we end up with Eq. (14.72). Thus (*cf.* Eq. (14.67a))

$$\frac{\partial g_{\mu\nu}}{\partial x^\kappa} - \Gamma^\rho_{\mu\kappa} g_{\rho\nu} - \Gamma^\rho_{\nu\kappa} g_{\mu\rho} = 0 = g_{\mu\nu;\kappa},$$

$$\frac{\partial g_{\nu\kappa}}{\partial x^\mu} - \Gamma^\rho_{\nu\mu} g_{\rho\kappa} - \Gamma^\rho_{\kappa\mu} g_{\nu\rho} = 0 = g_{\nu\kappa;\mu},$$

$$\frac{\partial g_{\kappa\mu}}{\partial x^\nu} - \Gamma^\rho_{\kappa\nu} g_{\rho\mu} - \Gamma^\rho_{\mu\nu} g_{\kappa\rho} = 0 = g_{\kappa\mu;\nu}. \qquad (14.73)$$

Adding the last two equations and subtracting the first and remembering that the quantities $g_{\mu\nu}$ and $\Gamma^\rho_{\mu\nu}$ are symmetric in μ and ν, we obtain

$$g_{\kappa\rho}\Gamma^\rho_{\mu\nu} = \frac{1}{2}\left(\frac{\partial g_{\nu\kappa}}{\partial x^\mu} + \frac{\partial g_{\kappa\mu}}{\partial x^\nu} - \frac{\partial g_{\mu\nu}}{\partial x^\kappa}\right). \qquad (14.74)$$

---

*See for instance D. F. Lawden [27], p. 119.

[†]This relation is also written $\nabla g = 0$, and ∇ is called Riemannian or Christoffel connection, and the Christoffel symbols are also called connection quantities. See R. Penrose [38], p. 319.

Multiplying both sides by $g^{\sigma\kappa}$, it follows that

$$\Gamma^{\sigma}_{\mu\nu} = \frac{1}{2}g^{\sigma\kappa}\left(\frac{\partial g_{\nu\kappa}}{\partial x^{\mu}} + \frac{\partial g_{\kappa\mu}}{\partial x^{\nu}} - \frac{\partial g_{\mu\nu}}{\partial x^{\kappa}}\right). \tag{14.75}$$

We thus find that the affinity, here the *"metric affinity"*, is given by the *Christoffel symbol* (13.9). The affinity is also written

$$\Gamma^{\sigma}_{\mu\nu} \equiv \left\{ \begin{matrix} \sigma \\ \mu\ \nu \end{matrix} \right\}. \tag{14.76}$$

The symbol on the right is called the *Christoffel symbol of the second kind* (the symbol of the first kind being the quantity on the right of Eq. (14.75)).

14.7   Raising and Lowering of Indices

As stated earlier: If $A^{\mu}(x)$ is a contravariant vector defined at $x^{\mu} \in \mathbb{R}_4$, then $g_{\mu\nu}(x)A^{\nu}(x)$ is a covariant vector at the same point, and this is denoted by $A_{\mu}(x)$, *i.e.*

$$A_{\mu}(x) = g_{\mu\nu}(x)A^{\nu}(x). \tag{14.77}$$

Thus $A^{\mu}(x), A_{\nu}(x)$ are the contravariant and covariant components of one and the same vector relative to whichever coordinate frame is being used. The relation (14.77) is described as lowering the index. Analogously one has

$$A^{\mu}(x) = g^{\mu\nu}(x)A_{\nu}(x). \tag{14.78}$$

Then

$$\begin{aligned} A^{\mu}B_{\mu} &= g^{\mu\nu}A_{\nu}g_{\mu\rho}B^{\rho} = g^{\mu\nu}g_{\mu\rho}A_{\nu}B^{\rho} = g_{\rho\mu}g^{\mu\nu}A_{\nu}B^{\rho} \\ &= \delta^{\nu}_{\rho}A_{\nu}B^{\rho} = A_{\nu}B^{\nu}. \end{aligned} \tag{14.79}$$

Thus the dummy indices can be raised and lowered without affecting the result. Further (here the comma is not to be confused with a derivative):

$$\Gamma_{\mu,\alpha\beta}(x) := g_{\mu\nu}\Gamma^{\nu}_{\alpha\beta}(x) = \Gamma_{\mu,\beta\alpha}(x). \tag{14.80}$$

Also (using $g_{\nu\mu}g^{\mu\lambda} = \delta^{\lambda}_{\nu}, \delta^{\lambda}_{\nu}g_{\alpha\lambda} = g_{\alpha\nu}$):

$$\begin{aligned} \Gamma_{\nu,\alpha\beta}(x) &= g_{\nu\mu}\Gamma^{\mu}_{\alpha\beta}(x) \stackrel{(14.75)}{=} \frac{1}{2}g_{\nu\mu}g^{\mu\lambda}\left(\frac{\partial g_{\alpha\lambda}}{\partial x^{\beta}} + \frac{\partial g_{\beta\lambda}}{\partial x^{\alpha}} - \frac{\partial g_{\alpha\beta}}{\partial x^{\lambda}}\right) \\ &= \frac{1}{2}\left(\frac{\partial g_{\alpha\nu}}{\partial x^{\beta}} + \frac{\partial g_{\beta\nu}}{\partial x^{\alpha}} - \frac{\partial g_{\alpha\beta}}{\partial x^{\nu}}\right), \end{aligned} \tag{14.81}$$

and

$$\Gamma_{\alpha,\nu\beta}(x) = \frac{1}{2}\left(\frac{\partial g_{\alpha\nu}}{\partial x^\beta} + \frac{\partial g_{\beta\alpha}}{\partial x^\nu} - \frac{\partial g_{\nu\beta}}{\partial x^\alpha}\right), \tag{14.82}$$

so that

$$\Gamma_{\nu,\alpha\beta} + \Gamma_{\alpha,\nu\beta} = \frac{\partial g_{\alpha\nu}}{\partial x^\beta} \quad \text{or} \quad \frac{\partial g_{\alpha\nu}}{\partial x^\beta} - g_{\nu\mu}\Gamma^\mu_{\alpha\beta} - g_{\alpha\mu}\Gamma^\mu_{\nu\beta} = 0, \tag{14.83}$$

which verifies the condition (14.72), or, equivalently, any of the explicit relations (14.73). Analogously (*cf.* Eq. (14.67b))

$$\frac{\partial g^{\alpha\beta}}{\partial x^\lambda} + \Gamma^\alpha_{\lambda\eta}g^{\eta\beta} + \Gamma^\beta_{\lambda\eta}g^{\alpha\eta} = 0. \tag{14.84}$$

Thus in either case, the covariant derivative applied to the metric tensor yields a quantity which vanishes.

14.8 Rewriting Co- and Contravariant Derivatives

Having obtained the covariant and contravariant derivatives we can now consider applications. In this section and the following we therefore consider the divergence of a vector, the *D'Alembertian* in a Riemann space and Maxwell's equations in such a space.

We now rewrite Eq. (14.63) in the following form:

$$F^\alpha_{;\lambda}(x) := \partial_\lambda F^\alpha(x) + \Gamma^\alpha_{\lambda\beta}F^\beta(x) \equiv \mathcal{D}^\alpha_{\lambda\beta}F^\beta, \tag{14.85}$$

where

$$\mathcal{D}^\alpha_{\lambda\beta}(x) = \partial_\lambda\delta^\alpha_\beta + \Gamma^\alpha_{\lambda\beta}, \tag{14.86}$$

and Eq. (14.51) with $F_\mu(x)$ replaced by $G_\alpha(x)$ as

$$G_{\alpha;\lambda}(x) := \partial_\lambda G_\alpha(x) - \Gamma^\beta_{\alpha\lambda}G_\beta(x) \equiv \mathcal{D}^\beta_{\alpha\lambda}G_\beta(x), \tag{14.87}$$

where

$$\mathcal{D}^\beta_{\alpha\lambda}(x) = \partial_\lambda\delta^\beta_\alpha - \Gamma^\beta_{\alpha\lambda}. \tag{14.88}$$

Further, from Eqs. (14.67b) and (14.67a) respectively

$$\begin{aligned}
T^{\alpha\beta}_{;\lambda}(x) &= \partial_\lambda T^{\alpha\beta}(x) + \Gamma^\alpha_{m\lambda}T^{m\beta}(x) + \Gamma^\beta_{\lambda n}T^{\alpha n}(x) \equiv \mathcal{D}^{\alpha\beta}_{mn\lambda}(x)T^{mn}(x),\\
T_{\alpha\beta;\lambda}(x) &= \partial_\lambda T_{\alpha\beta}(x) - \Gamma^\eta_{\alpha\lambda}T_{\eta\beta}(x) - \Gamma^\eta_{\beta\lambda}T_{\alpha\eta}(x) \equiv \mathcal{D}^{mn}_{\alpha\beta\lambda}(x)T_{mn}(x),
\end{aligned} \tag{14.89}$$

where

$$
\begin{aligned}
\mathcal{D}^{\alpha\beta}_{mn\lambda}(x) &= \partial_\lambda \delta^\alpha_m \delta^\beta_n + \Gamma^\alpha_{m\lambda}\delta^\beta_n + \Gamma^\beta_{\lambda n}\delta^\alpha_m, \\
\mathcal{D}^{mn}_{\alpha\beta\lambda}(x) &= \partial_\lambda \delta^m_\alpha \delta^n_\beta - \Gamma^m_{\alpha\lambda}\delta^n_\beta - \Gamma^n_{\beta\lambda}\delta^m_\alpha.
\end{aligned}
\tag{14.90}
$$

Example 14.4: Leibnitz rule and covariant derivatives
Show by application to two examples that Leibnitz's rule holds for covariant derivatives.

Solution: (a) As the first example we derive the covariant derivative of the quantity $G_\mu := H^\nu_\mu F_\nu$. Since this expression is a covariant vector, we can apply Eq. (14.51) and obtain from this for its covariant derivative:

$$
G_{\mu;\kappa} = (H^\nu_\mu F_\nu)_{;\kappa} = \partial_\kappa (H^\nu_\mu F_\nu) - \Gamma^\lambda_{\mu\kappa}(H^\rho_\lambda F_\rho).
$$

Next we use Leibnitz's rule. Using Eqs. (14.66) and (14.51) we obtain:

$$
\begin{aligned}
G_{\mu;\kappa} &= H^\nu_{\mu;\kappa} F_\nu + H^\nu_\mu F_{\nu;\kappa} \\
&= [\partial_\kappa H^\nu_\mu + \Gamma^\nu_{\lambda\kappa}H^\lambda_\mu - \underline{\Gamma^\lambda_{\mu\kappa}H^\nu_\lambda}]F_\nu + H^\nu_\mu[\partial_\kappa F_\nu - \underline{\Gamma^\lambda_{\nu\kappa}F_\lambda}] \\
&= \partial_\kappa (H^\nu_\mu F_\nu) - \Gamma^\lambda_{\mu\kappa}(H^\nu_\lambda F_\nu).
\end{aligned}
$$

The underlined terms cancel each other. The result is seen to be in agreement with the above.
(b) As a second example we consider a tensor quantity $T_{\rho\nu} := H^\mu_\rho H_{\mu\nu}$. Applying Eq. (14.67a) we obtain for the covariant derivative of this quantity:

$$
(T_{\rho\nu})_{;\kappa} \equiv (H^\mu_\rho H_{\mu\nu})_{;\kappa} = \partial_\kappa (H^\mu_\rho H_{\mu\nu}) - \Gamma^\lambda_{\rho\kappa}(H^\mu_\lambda H_{\mu\nu}) - \Gamma^\lambda_{\nu\kappa}(H^\epsilon_\rho H_{\epsilon\lambda}).
$$

On the other hand, applying Leibnitz's rule and using Eqs. (14.66) and (14.67a), we obtain

$$
\begin{aligned}
(T_{\rho\nu})_{;\kappa} &\equiv (H^\mu_\rho H_{\mu\nu})_{;\kappa} = (H^\mu_{\rho;\kappa})H_{\mu\nu} + H^\mu_\rho (H_{\mu\nu;\kappa}) \\
&= [\partial_\kappa H^\mu_\rho + \Gamma^\mu_{\lambda\kappa}H^\lambda_\rho - \underline{\Gamma^\lambda_{\rho\kappa}H^\mu_\lambda}]H_{\mu\nu} + H^\mu_\rho[\partial_\kappa H_{\mu\nu} - \underline{\Gamma^\lambda_{\mu\kappa}H_{\lambda\nu}} - \Gamma^\lambda_{\nu\kappa}H_{\mu\lambda}] \\
&= \partial_\kappa (H^\mu_\rho H_{\mu\nu}) - \Gamma^\lambda_{\rho\kappa}(H^\mu_\lambda H_{\mu\nu}) - \Gamma^\lambda_{\nu\kappa}(H^\mu_\rho H_{\mu\lambda}).
\end{aligned}
$$

Again the underlined terms cancel (with replacement $\mu \leftrightarrow \lambda$ in the second of these). The result is seen to agree with that above.
     It is apparent that to write out a general proof is rather laborious.

14.9   Covariant Divergence, Rotation etc.

The covariant divergence (lower index) of the contravariant vector $F^\alpha(x)$ is given by Eq. (14.63),

$$
\mathrm{div}_\alpha F^\alpha(x) \equiv F^\alpha_{;\alpha}(x) = (\partial_\alpha + \Gamma^\lambda_{\lambda\alpha})F^\alpha(x).
\tag{14.91}
$$

One should note the contraction of two indices of Γ here. The covariant rotation of a vector G_μ is, using Eq. (14.51),

$$
\begin{aligned}
G_{\mu;\nu} - G_{\nu;\mu} &= (\partial_\nu G_\mu - \Gamma^\beta_{\mu\nu}G_\beta) - (\partial_\mu G_\nu - \Gamma^\beta_{\nu\mu}G_\beta) \\
&= \partial_\nu G_\mu - \partial_\mu G_\nu,
\end{aligned}
\tag{14.92}
$$

where we used the symmetry (14.48) of the affinity. The rotation therefore is the same as in the usual expression familiar from elementary vector analysis.

The consideration of a particle with charge e in an electromagnetic field by standard methods (*i.e.* by postulating the action integral and derivation of the Hamiltonian H) leads to the idea of "*minimal coupling*"* which says that the Hamiltonian is obtained from that in mechanics by the replacement

$$p^\mu \longrightarrow p^\mu - eA^\mu,$$

which in quantum mechanics implies

$$-i\hbar\partial^\mu \longrightarrow -i\hbar\partial^\mu - eA^\mu = -i\hbar\left(\partial^\mu - \frac{ie}{\hbar}A^\mu\right).$$

The quantity on the right is also called a contravariant derivative. Thus it is already simple *analogy* between the very different forces of electromagnetism and gravity which suggests the interpretation of Γ in Eq. (14.91) as something like a gravitational field that acts on the particle, and correspondingly that the former can be taken into account by a similar replacement of the ordinary derivative by the covariant derivative. From Riemann's point of view Γ takes into account the curvature of space, also described as a warping of space.

We now want to determine $\Gamma_{\lambda\alpha}^\lambda$. We have (*cf.* Eqs. (14.84) and (14.83))

$$g^{\alpha\beta}_{\ \ ;\lambda} = \partial_\lambda g^{\alpha\beta} + \Gamma_{\lambda\eta}^\alpha g^{\eta\beta} + \Gamma_{\lambda\eta}^\beta g^{\alpha\eta} = 0, \tag{14.93}$$

and

$$g_{\alpha\beta;\lambda} = \partial_\lambda g_{\alpha\beta} - \Gamma_{\alpha\lambda}^\eta g_{\eta\beta} - \Gamma_{\beta\lambda}^\eta g_{\alpha\eta} = 0. \tag{14.94}$$

Multiplying Eq. (14.93) by $g_{\alpha\beta}$, we obtain:

$$g_{\alpha\beta}\partial_\lambda g^{\alpha\beta} + \Gamma_{\lambda\eta}^\alpha g^{\eta\beta} g_{\alpha\beta} + \Gamma_{\lambda\eta}^\beta g^{\alpha\eta} g_{\alpha\beta} = 0,$$

i.e.

$$\Gamma_{\lambda\alpha}^\alpha + \Gamma_{\lambda\beta}^\beta + g_{\alpha\beta}(\partial_\lambda g^{\alpha\beta}) = 0, \quad \text{or} \quad \Gamma_{\lambda\alpha}^\alpha = -\frac{1}{2}g_{\alpha\beta}(\partial_\lambda g^{\alpha\beta}). \tag{14.95}$$

Now, we have

$$g^{\mu\nu}(x)g_{\nu\alpha}(x) = \delta_\alpha^\mu, \qquad g^{\mu\nu}(dg_{\nu\alpha}) + (dg^{\mu\nu})g_{\nu\alpha} = 0,$$

$$\frac{\partial g^{\mu\nu}}{\partial g_{\rho\sigma}}g_{\nu\alpha} + g^{\mu\nu}\delta_\nu^\rho\delta_\alpha^\sigma = 0, \qquad \frac{\partial g^{\mu\nu}}{\partial g_{\rho\sigma}}g_{\nu\alpha}g^{\alpha\lambda} + g^{\mu\nu}\delta_\nu^\rho\delta_\alpha^\sigma g^{\alpha\lambda} = 0,$$

$$\therefore \quad \frac{\partial g^{\mu\lambda}}{\partial g_{\rho\sigma}} = -g^{\mu\rho}g^{\sigma\lambda}, \tag{14.96}$$

---

*A corresponding procedure applies also to the quantization of gravity in spite of contrary statements. See F. Wilczek [52], pp. 134 – 135, 227.

and (*cf.* Eq. (14.3a))

$$g = \sum_\beta g_{\alpha\beta}\triangle^{\alpha\beta}, \qquad \frac{\partial g}{\partial g_{\alpha\beta}} = \triangle^{\alpha\beta} \overset{(14.4c)}{=} gg^{\alpha\beta}. \tag{14.97}$$

Hence (with the second of the five relations of Eq. (14.96))

$$dg = \frac{\partial g}{\partial g_{\alpha\beta}}dg_{\alpha\beta} \overset{(14.97)}{=} gg^{\alpha\beta}dg_{\alpha\beta} \overset{(14.96)}{=} -gg_{\alpha\beta}dg^{\alpha\beta}, \tag{14.98}$$

so that

$$\frac{\partial g}{\partial x^\lambda} = -gg_{\alpha\beta}\frac{\partial g^{\alpha\beta}}{\partial x^\lambda} \equiv -g(g_{\alpha\beta}\partial_\lambda g^{\alpha\beta}). \tag{14.99}$$

Hence (this expression will be required for the Schwarzschild solution, *cf.* Eq. (16.40))

$$\Gamma^\alpha_{\lambda\alpha} \overset{(14.95)}{=} -\frac{1}{2}g_{\alpha\beta}(\partial_\lambda g^{\alpha\beta}) \overset{(14.99)}{=} \frac{1}{2g}\frac{\partial g}{\partial x^\lambda} = \frac{1}{2\sqrt{-g}}\frac{-\partial g/\partial x^\lambda}{\sqrt{-g}}$$

$$= \frac{\partial}{\partial x^\lambda}(\ln\sqrt{-g(x)}). \tag{14.100}$$

It follows that

$$F^\alpha_{;\alpha} \overset{(14.91)}{=} (\partial_\alpha + \Gamma^\lambda_{\alpha\lambda})F^\alpha$$

$$= (\partial_\alpha + \partial_\alpha \ln\sqrt{-g})F^\alpha = \frac{1}{\sqrt{-g}}\partial_\alpha[F^\alpha(x)\sqrt{-g}]. \tag{14.101}$$

The *D'Alembert operator* in a Riemann space $\mathbb{R}_4$ is now defined as $\Box^2 \equiv$ div grad in the relation

$$\Box^2\psi := \left(g^{\alpha\lambda}\frac{\partial}{\partial x^\lambda}\psi\right)_{;\alpha} \equiv (\partial^\alpha\psi)_{;\alpha} \overset{(14.91)}{=} (\partial_\alpha + \Gamma^\rho_{\rho\alpha})\left(g^{\alpha\lambda}\frac{\partial\psi}{\partial x^\lambda}\right). \tag{14.102}$$

Note that this is also

$$\Box^2\psi = \frac{\partial}{\partial x^\alpha}\left(g^{\alpha\lambda}\frac{\partial\psi}{\partial x^\lambda}\right) + (\partial_\alpha \ln\sqrt{-g})g^{\alpha\lambda}\frac{\partial\psi}{\partial x^\lambda}$$

$$= \partial_\alpha\left(g^{\alpha\lambda}\frac{\partial\psi}{\partial x^\lambda}\right) + \Gamma^\rho_{\alpha\rho}g^{\alpha\lambda}\frac{\partial\psi}{\partial x^\lambda}. \tag{14.103}$$

As an application of the preceding considerations we consider in Example 14.3 Maxwell's equations in a Riemannian space. One can also prove the applicability of Gauss's and Stokes's theorems to covariant derivatives.

Example 14.5: Maxwell's equations in a Riemann space
Obtain Maxwell's equations in a Riemann space.

Solution: We have seen that (*cf.* Eq. (14.67b))

$$T^{\alpha\beta}_{\;\;;\lambda} = \partial_\lambda T^{\alpha\beta} + \Gamma^\alpha_{\lambda\mu}T^{\mu\beta} + \Gamma^\beta_{\mu\lambda}T^{\alpha\mu}. \tag{14.104}$$

Hence, with $\beta \to \lambda$,

$$T^{\alpha\lambda}_{\;\;;\lambda} = \partial_\lambda T^{\alpha\lambda} + \Gamma^\alpha_{\lambda\mu}T^{\mu\lambda} + \Gamma^\lambda_{\mu\lambda}T^{\alpha\mu} \stackrel{(14.100)}{=} \partial_\lambda T^{\alpha\lambda} + \Gamma^\alpha_{\lambda\mu}T^{\mu\lambda} + (\partial_\mu \ln\sqrt{-g})T^{\alpha\mu},$$

i.e.

$$T^{\alpha\lambda}_{\;\;;\lambda} = \frac{1}{\sqrt{-g}}\partial_\lambda(\sqrt{-g}T^{\alpha\lambda}) + \Gamma^\alpha_{\lambda\mu}T^{\mu\lambda}. \tag{14.105}$$

If we now assume that the tensor $T^{\alpha\beta}$ is antisymmetric, we have $T^{\mu\lambda} = -T^{\lambda\mu}$. In this case, since also $\Gamma^\alpha_{\lambda\mu} = \Gamma^\alpha_{\mu\lambda}$, and now writing F instead of T, we obtain with $T^{\mu\lambda} = (T^{\mu\lambda} - T^{\lambda\mu})/2$,

$$F^{\alpha\beta}_{\;\;;\beta} = \frac{1}{\sqrt{-g}}\partial_\lambda(\sqrt{-g}F^{\alpha\lambda}). \tag{14.106}$$

In a *flat space* Maxwell's equations are (*cf.* Example 6.9)

$$\partial_\nu F^{\mu\nu} = j^\mu, \qquad \partial_\lambda F_{\mu\nu} + \partial_\nu F_{\lambda\mu} + \partial_\mu F_{\nu\lambda} = 0. \tag{14.107}$$

In the general case of a *non flat space* the ordinary derivatives have to be replaced by covariant derivatives, *i.e.* the equations become, the second being the *Bianchi identity*,

$$F^{\mu\nu}_{\;\;;\nu} = j^\mu, \qquad F_{\mu\nu;\lambda} + F_{\lambda\mu;\nu} + F_{\nu\lambda;\mu} = 0, \tag{14.108}$$

with

$$F_{\mu\nu} = A_{\mu;\nu} - A_{\nu;\mu} \stackrel{(14.92)}{=} \partial_\nu A_\mu - \partial_\mu A_\nu. \tag{14.109}$$

Substitution of expression (14.106) yields

$$\frac{1}{\sqrt{-g}}\partial_\nu(\sqrt{-g}F^{\mu\nu}) = j^\mu, \qquad \partial_\nu(\sqrt{-g}F^{\mu\nu}) = \sqrt{-g}j^\mu, \tag{14.110}$$

and the *Bianchi identity* in electrodynamics as usual (*cf.* Eq. (14.92))

$$\partial_\lambda F_{\mu\nu} + \partial_\nu F_{\lambda\mu} + \partial_\mu F_{\nu\lambda} = 0, \qquad F_{\mu\nu} = \partial_\mu A_\nu - \partial_\nu A_\mu. \tag{14.111}$$

Two of the four Maxwell equations can be shown to follow from this.[†] We observe that

$$j^\mu_{\;;\mu} \stackrel{(14.63)}{=} (\partial_\mu + \Gamma^\lambda_{\lambda\mu})j^\mu \stackrel{(14.101)}{=} \frac{1}{\sqrt{-g}}\partial_\mu(\sqrt{-g}j^\mu) \stackrel{(14.110)}{=} \frac{1}{\sqrt{-g}}\partial_\mu\partial_\lambda(\sqrt{-g}F^{\mu\lambda}) = 0, \tag{14.112}$$

since $F^{\mu\nu} = -F^{\nu\mu}$. This result implies that the current j^μ is *conserved*. With Eq. (6.67) we can show that the charge is conserved, the charge being defined as the spatial integral of the 0-component of the current. In a flat space with a steady current through a volume V we have $\partial_i j_i = 0$ and

$$0 = \int_V d^3x\, \partial_\nu j^\nu = \int_V d^3x\, \partial_0 j^0, \quad \text{and} \quad \int_V d^3x\, j^0 = \text{constant}. \tag{14.113}$$

[†]See *e.g.* H. J. W. Müller–Kirsten [34], pp. 411 – 412. See also Example 6.9.

Thus the total charge is constant. On the other hand in the case of the non-flat space the steady current is defined by $\partial_i(\sqrt{-g}j^i) = 0$, so that (using the middle expression of (14.112))

$$
\begin{aligned}
0 &= \int_V \sqrt{-g}j^\mu{}_{;\mu}d^3x \overset{(14.112)}{=} \int_V \partial_\mu(\sqrt{-g}j^\mu)d^3x \\
&= \int_V \partial_0(\sqrt{-g}j^0)d^3x + \int_V \partial_i(\sqrt{-g}j^i)d^3x \\
&= \int_V \frac{\partial}{\partial x^0}(\sqrt{-g}j^0)d^3x = \frac{d}{dx^0}\int_V \sqrt{-g}j^0 d^3x,
\end{aligned}
\tag{14.114}
$$

so that $\int_V \sqrt{-g}j^0(x)d^3x = \text{const}$. Thus the gravitational force affects the charge distribution in space.

Example 14.6: Geodesic equation in terms of covariant derivative

Show that if $u^\nu = dx^\nu/d\tau$ is the tangent vector defined in Sec. 12.9, the equation

$$
u^\nu u^\mu{}_{;\nu} = 0 \quad \text{or} \quad p^\nu p^\mu{}_{;\nu} = 0,
$$

is the geodesic equation (13.13).

Solution: Using Eq. (14.63) we obtain

$$
\begin{aligned}
u^\nu u^\mu{}_{;\nu} &= u^\nu\left[\frac{\partial u^\mu}{\partial x^\nu} + \Gamma^\mu_{\epsilon\nu}u^\epsilon\right] = \frac{dx^\nu}{d\tau}\left[\frac{\partial}{\partial x^\nu}\left(\frac{dx^\mu}{d\tau}\right) + \Gamma^\mu_{\epsilon\nu}\frac{dx^\epsilon}{d\tau}\right] \\
&= \frac{d}{d\tau}\left(\frac{dx^\mu}{d\tau}\right) + \Gamma^\mu_{\epsilon\nu}\frac{dx^\nu}{d\tau}\frac{dx^\epsilon}{d\tau} = 0,
\end{aligned}
$$

which is the geodesic equation (13.13) which subdivides into 4 coupled second order differential equations.[*]

---

[*]See also C. W. Misner, K. S. Thorne and J. A. Wheeler [33], p. 657.

Chapter 15

Einstein's Equation of the Gravitational Field

15.1 Introductory Remarks

With a few more mathematical technicalities — in particular the definition of the *Riemann curvature tensor* $R^{\mu}_{\eta\alpha\lambda}$, the *Ricci tensor* $R_{\mu\nu}$, the *Ricci curvature scalar* R, and the *Einstein* or *Ricci–Einstein tensor* $G_{\alpha\beta}$ — we can write down Einstein's equation of the gravitational field. The main difficulty in the following is perhaps a patient and cumbersome keeping track of indices. However, after this we can proceed to investigate the properties and specific solutions of Einstein's equation.

15.2 The Riemann Curvature Tensor

In the case of a flat space the ordinary derivative is a vector and two such operations commute. In a space with nonzero curvature these properties are no longer valid.

Consider the *covariant derivative* of a vector F^{μ} in a curved space. This is a tensor $T^{\mu}{}_{\alpha}$, *i.e.* (note: In the following the comma denotes the ordinary derivative, that is ",$\alpha \equiv \partial_{\alpha}$", whereas the semi-colon denotes the covariant derivative),

$$
\begin{aligned}
T^{\mu}{}_{\alpha}(x) &\equiv F^{\mu}{}_{;\alpha}(x) \overset{(14.85)}{=} F^{\mu}{}_{,\alpha}(x) + \Gamma^{\mu}_{\alpha m}(x) F^{m}(x), \\
F^{\mu}{}_{,\alpha}(x) &\equiv \partial_{\alpha} F^{\mu}(x).
\end{aligned}
\tag{15.1}
$$

We now want to calculate the *covariant derivative of this tensor*. We have (with $T^{\mu}_{\alpha,\lambda} = \partial_{\lambda} T^{\mu}_{\alpha} = F^{\mu}{}_{,\alpha,\lambda} + \cdots$, and the underbraced symbols are meant to help the identification and/or cancellation of terms in Eq. (15.3)):

$$
F^{\mu}_{\ ;\alpha;\lambda} \ = \ T^{\mu}_{\ \alpha;\lambda} \overset{(14.66)}{=} T^{\mu}_{\ \alpha,\lambda} + \Gamma^{\mu}_{a\lambda} T^{a}_{\ \alpha} - \Gamma^{m}_{\alpha\lambda} T^{\mu}_{\ m}
$$

$$
\overset{(15.1)}{=} \ F^{\mu}_{\ ,\alpha,\lambda} + \underbrace{(\Gamma^{\mu}_{\alpha m})_{,\lambda} F^{m}}_{\cdots\cdots} + \underbrace{\Gamma^{\mu}_{\alpha m} F^{m}_{\ ,\lambda}}_{+++} + \Gamma^{\mu}_{a\lambda} (\underbrace{F^{a}_{\ ,\alpha}}_{---} + \underbrace{\Gamma^{a}_{\alpha\nu} F^{\nu}}_{\because\because})
$$

$$
- \Gamma^{m}_{\alpha\lambda} (\underbrace{F^{\mu}_{\ ,m}}_{\varnothing\varnothing\varnothing} + \underbrace{\Gamma^{\mu}_{m\eta} F^{\eta}}_{\diamond\diamond\diamond}). \tag{15.2a}
$$

All we did here was to replace T by F with the help of Eq. (15.1). Writing down the corresponding expression with α replaced by λ and λ by α we obtain

$$
F^{\mu}_{\ ;\lambda;\alpha} \ = \ F^{\mu}_{\ ,\lambda,\alpha} + \underbrace{(\Gamma^{\mu}_{\lambda m})_{,\alpha} F^{m}}_{***} + \underbrace{\Gamma^{\mu}_{\lambda m} F^{m}_{\ ,\alpha}}_{---} + \Gamma^{\mu}_{a\alpha} (\underbrace{F^{a}_{\ ,\lambda}}_{+++} + \underbrace{\Gamma^{a}_{\lambda\nu} F^{\nu}}_{\because\because})
$$

$$
- \Gamma^{m}_{\lambda\alpha} (\underbrace{F^{\mu}_{\ ,m}}_{\varnothing\varnothing\varnothing} + \underbrace{\Gamma^{\mu}_{m\eta} F^{\eta}}_{\diamond\diamond\diamond}). \tag{15.2b}
$$

Next we take the difference of Eqs. (15.2a) and (15.2b) and obtain the following relation, by renaming indices and using the symmetry property (14.71) of lower indices of $\Gamma^{\mu}_{\ \nu\rho}$,

$$
F^{\mu}_{\ ;\alpha;\lambda} - F^{\mu}_{\ ;\lambda;\alpha} = \left[\underbrace{(\Gamma^{\mu}_{\alpha\eta})_{,\lambda}}_{\cdots\cdots} - \underbrace{(\Gamma^{\mu}_{\lambda\eta})_{,\alpha}}_{***} + \underbrace{\Gamma^{\mu}_{a\lambda} \Gamma^{a}_{\alpha\eta}}_{\because\because} - \underbrace{\Gamma^{\mu}_{a\alpha} \Gamma^{a}_{\lambda\eta}}_{\because\because} \right] F^{\eta}. \tag{15.3}
$$

The expression in square braces, $[\cdots]$, is called the *Riemann curvature tensor, i.e.* the classical expression

$$
R^{\mu}_{\ \eta\alpha\lambda} := (\Gamma^{\mu}_{\alpha\eta})_{,\lambda} - (\Gamma^{\mu}_{\lambda\eta})_{,\alpha} + \Gamma^{\mu}_{a\lambda} \Gamma^{a}_{\alpha\eta} - \Gamma^{\mu}_{a\alpha} \Gamma^{a}_{\lambda\eta}. \tag{15.4}
$$

The most immediate property of $R^{\mu}_{\ \eta\alpha\lambda}$ that we can deduce from this definition is the following relation, sometimes described as the *first Bianchi identity* or as *Bianchi symmetry* (Penrose [38]):

$$
R^{\mu}_{\ \eta\alpha\lambda} + R^{\mu}_{\ \alpha\lambda\eta} + R^{\mu}_{\ \lambda\eta\alpha} = 0. \tag{15.5}
$$

We verify this relation in Example 15.1 below. One can show with somewhat laborious manipulations which we do not enter into here (since we do not use the result), that with operators $\mathcal{D}^{\alpha}_{\lambda\beta}, \mathcal{D}^{\alpha\beta}_{mn\lambda}$ defined by Eqs. (14.86) and (14.90), one has

$$
(\mathcal{D}^{\mu m}_{a\alpha\lambda} - \mathcal{D}^{\mu m}_{a\lambda\alpha}) \mathcal{D}^{a}_{\eta m} = R^{\mu}_{\ \eta\alpha\lambda}. \tag{15.6}
$$

We can now rewrite Eq. (15.3) in terms of the Riemann curvature tensor as

$$F^{\mu}_{;\alpha;\lambda} - F^{\mu}_{;\lambda;\alpha} = R^{\mu}_{\eta\alpha\lambda}F^{\eta}. \tag{15.7}$$

This result demonstrates, in fact, that the Riemann space is curved, since the covariant derivatives on the left do not commute as they would in the case of flat space covariant derivatives. A *flat space* can therefore be defined by the condition

$$R^{\mu}_{\eta\alpha\lambda}(x) = 0 \quad \text{at every } x \in \mathbb{R}_4. \tag{15.8}$$

We now note some further properties of the Riemann curvature tensor. The first property we deduce from Eq. (15.4) is:

$$R^{\mu}_{\eta\alpha\lambda} = -R^{\mu}_{\eta\lambda\alpha}. \tag{15.9}$$

Also since $R^{\mu}_{\eta\alpha\lambda}$ is a tensor, we can write (and obtain by using Eq. (15.9))

$$g_{\mu\eta}R^{\eta}_{\alpha\beta\gamma} = R_{\mu\alpha\beta\gamma}, \quad g_{\mu\eta}R^{\eta}_{\alpha\gamma\beta} = R_{\mu\alpha\gamma\beta}, \quad R_{\mu\alpha\beta\gamma} = -R_{\mu\alpha\gamma\beta}. \tag{15.10}$$

This implies antisymmetry in the last two indices. One can also prove antisymmetry in the first two indices, *i.e.*

$$R_{\mu\alpha\beta\gamma} = -R_{\alpha\mu\beta\gamma}. \tag{15.11}$$

The explicit proof of this relation is laborious, we leave it therefore to Example 15.2, as also the following symmetry property in the first pair of indices and the last, *i.e.*

$$R_{\mu\alpha\beta\gamma} = R_{\beta\gamma\mu\alpha}. \tag{15.12}$$

The symmetry relations reduce the number of independent components of the Riemann curvature tensor. Let's count them. To start, the tensor $R_{\mu\alpha\beta\gamma}$ has $4^4 = 256$ components. Instead of counting the number of dependent components and subtracting these from 256 to obtain the number of independent elements, we count the latter directly. The antisymmetry property (15.10) implies that independent of $\mu\alpha$, there are 6 independent components, *i.e.* as in the case of an antisymmetric 4×4 matrix $(R_{\beta\gamma})$ whose diagonal elements vanish. Similarly the antisymmetry of (15.11) implies that independent of the last two indices there are again 6 independent components. Thus altogether so far there are 36 independent components like those of a real 6×6 matrix, where a row index corresponds to a value of a $(\mu\alpha)$ pair and a column index to a value of a $(\beta\gamma)$ pair. Symmetry with respect to these pairs — expressed by Eq. (15.12) — implies independence of 6 diagonal elements and half the number of off-diagonal elements $= 6 + (36 - 6)/2 = 6 + 15 = 21$. With one remaining condition obtained from Eq. (15.5) (*e.g.* by multiplying by $g_{\rho\mu}$ and considering $R_{0123}+R_{0231}+R_{0312}=0$), the number of independent elements is found to be 20 (10 independent components in empty space, *i.e.* with no matter density). The formula is actually $n^2(n^2-1)/12$ which is 20 for $n=4$.*

Example 15.1: Bianchi symmetry relation
Verify the Bianchi symmetry relation (15.5), *i.e.*

$$R^{\mu}{}_{abc} + R^{\mu}{}_{bca} + R^{\mu}{}_{cab} = 0.$$

Solution: Using the definition (15.4) we have:

$$
\begin{aligned}
R^{\mu}{}_{abc} &= (\Gamma^{\mu}_{ba})_c - (\Gamma^{\mu}_{ca})_b + \Gamma^{\mu}_{\nu c}\Gamma^{\nu}_{ba} - \Gamma^{\mu}_{\nu b}\Gamma^{\nu}_{ca}, \\
R^{\mu}{}_{bca} &= (\Gamma^{\mu}_{cb})_a - (\Gamma^{\mu}_{ab})_c + \Gamma^{\mu}_{\nu a}\Gamma^{\nu}_{cb} - \Gamma^{\mu}_{\nu c}\Gamma^{\nu}_{ab}, \\
R^{\mu}{}_{cab} &= (\Gamma^{\mu}_{ac})_b - (\Gamma^{\mu}_{bc})_a + \Gamma^{\mu}_{\nu b}\Gamma^{\nu}_{ac} - \Gamma^{\mu}_{\nu a}\Gamma^{\nu}_{bc}.
\end{aligned}
$$

Recalling the relation (14.48), *i.e.* $\Gamma^{\mu}_{\nu\kappa}=\Gamma^{\mu}_{\kappa\nu}$, we see that the sum of these contributions vanishes, thus proving the above relation.

Example 15.2: Symmetry properties of Riemann curvature tensor
Verify the symmetry properties (15.10) (last relation there), (15.11) and (15.12) of the Riemann curvature tensor.

Solution: We employ the notation defined by Eq. (14.76), *i.e.*

$$\Gamma^{\sigma}_{\mu\nu} \equiv \left\{ \begin{matrix} \sigma \\ \mu\,\nu \end{matrix} \right\}.$$

We recall the first of the relations of Eq. (15.10) (here written in reverse order):

$$R_{\mu\alpha\beta\gamma} = g_{\mu\eta}R^{\eta}{}_{\alpha\beta\gamma} \overset{(15.4)}{=} g_{\mu\eta}\left[\Gamma^{\eta}_{\alpha\gamma}\Gamma^{a}_{\beta\alpha} - \Gamma^{\eta}_{\alpha\beta}\Gamma^{a}_{\gamma\alpha} + (\Gamma^{\eta}_{\beta\alpha})_{,\gamma} - (\Gamma^{\eta}_{\gamma\alpha})_{,\beta}\right],$$

i.e. in the bracket notation:

$$R_{ijkl} = g_{is}\left[\left\{ \begin{matrix} s \\ r\,l \end{matrix} \right\}\left\{ \begin{matrix} r \\ j\,k \end{matrix} \right\} - \left\{ \begin{matrix} s \\ r\,k \end{matrix} \right\}\left\{ \begin{matrix} r \\ j\,l \end{matrix} \right\} + \frac{\partial}{\partial x^l}\left\{ \begin{matrix} s \\ j\,k \end{matrix} \right\} - \frac{\partial}{\partial x^k}\left\{ \begin{matrix} s \\ j\,l \end{matrix} \right\}\right].$$

We know from Eq. (14.75) that

$$\left\{ \begin{matrix} \sigma \\ \mu\,\nu \end{matrix} \right\} := \frac{1}{2}g^{\sigma\kappa}\left(\frac{\partial g_{\nu\kappa}}{\partial x^{\mu}} + \frac{\partial g_{\kappa\mu}}{\partial x^{\nu}} - \frac{\partial g_{\mu\nu}}{\partial x^{\kappa}}\right).$$

Defining a square bracket $[ij,k]$ as the expression

$$[ij,k] := \frac{1}{2}\left(\frac{\partial g_{jk}}{\partial x^i} + \frac{\partial g_{ki}}{\partial x^j} - \frac{\partial g_{ij}}{\partial x^k}\right),$$

we have

$$
\begin{aligned}
g_{rs}\left\{ \begin{matrix} s \\ i\,j \end{matrix} \right\} &= g_{rs}\frac{1}{2}g^{sk}\left(\frac{\partial g_{jk}}{\partial x^i} + \frac{\partial g_{ki}}{\partial x^j} - \frac{\partial g_{ij}}{\partial x^k}\right) \\
&= \delta^k_r[ij,k] = [ij,r].
\end{aligned}
$$

---

*See *e.g.* R. Penrose [38], pp. 320, 459.

With these expressions we can rewrite R_{ijkl} as (note that additional terms are subtracted out)

$$
R_{ijkl} = [rl,i]\left\{\begin{matrix} r \\ jk \end{matrix}\right\} - [rk,i]\left\{\begin{matrix} r \\ jl \end{matrix}\right\}
$$
$$
+ \frac{\partial}{\partial x^l}\left[g_{is}\left\{\begin{matrix} s \\ jk \end{matrix}\right\}\right] - \frac{\partial}{\partial x^k}\left[g_{is}\left\{\begin{matrix} s \\ jl \end{matrix}\right\}\right] + \frac{\partial g_{is}}{\partial x^k}\left\{\begin{matrix} s \\ jl \end{matrix}\right\} - \frac{\partial g_{is}}{\partial x^l}\left\{\begin{matrix} s \\ jk \end{matrix}\right\}.
$$

From the definition of the square bracket above, we obtain

$$
[ik,s] + [sk,i] = \frac{1}{2}\left(\frac{\partial g_{ks}}{\partial x^i} + \frac{\partial g_{si}}{\partial x^k} - \frac{\partial g_{ik}}{\partial x^s}\right) + \frac{1}{2}\left(\frac{\partial g_{ki}}{\partial x^s} + \frac{\partial g_{is}}{\partial x^k} - \frac{\partial g_{sk}}{\partial x^i}\right) = \frac{\partial g_{is}}{\partial x^k}.
$$

Using this relation, we can rewrite R_{ijkl} as (similarly underbraced terms cancelling):

$$
R_{ijkl} = \underbrace{[rl,i]\left\{\begin{matrix} r \\ jk \end{matrix}\right\}}_{--} - \underbrace{[rk,i]\left\{\begin{matrix} r \\ jl \end{matrix}\right\}}_{++} + \frac{\partial}{\partial x^l}[jk,i] - \frac{\partial}{\partial x^k}[jl,i]
$$
$$
+\left\{\begin{matrix} s \\ jl \end{matrix}\right\}([ik,s] + \underbrace{[sk,i]}_{++}) - \left\{\begin{matrix} s \\ jk \end{matrix}\right\}([il,s] + \underbrace{[sl,i]}_{--})
$$
$$
= \frac{\partial}{\partial x^l}[jk,i] - \frac{\partial}{\partial x^k}[jl,i] + \left\{\begin{matrix} s \\ jl \end{matrix}\right\}[ik,s] - \left\{\begin{matrix} s \\ jk \end{matrix}\right\}[il,s]
$$
$$
= -\frac{1}{2}\left(\frac{\partial^2 g_{li}}{\partial x^j \partial x^k} + \frac{\partial^2 g_{jk}}{\partial x^i \partial x^l} - \frac{\partial^2 g_{jl}}{\partial x^i \partial x^k} - \frac{\partial^2 g_{ki}}{\partial x^j \partial x^l}\right)
$$
$$
- g_{sr}\left\{\begin{matrix} r \\ il \end{matrix}\right\}\left\{\begin{matrix} s \\ jk \end{matrix}\right\} + g_{sr}\left\{\begin{matrix} r \\ ik \end{matrix}\right\}\left\{\begin{matrix} s \\ jl \end{matrix}\right\}.
$$

Here we can read off the relations

$$
R_{ijkl} = -R_{jikl}, \qquad R_{ijkl} = -R_{ijlk}, \qquad R_{ijkl} = R_{klij}.
$$

15.3 Bianchi Identities and Ricci–Einstein Tensor

Our next step is to consider the covariant derivative of the 4-index Riemann curvature tensor of Eq. (15.4). Thus using Eq. (14.67d) the covariant derivative is given by

$$
R^{\mu}{}_{\eta\alpha\lambda;\nu} = \left\{\partial_\nu R^{\mu}{}_{\eta\alpha\lambda}\right\} + \Gamma^{\mu}_{\rho\nu}R^{\rho}{}_{\eta\alpha\lambda} - \Gamma^{\beta}_{\eta\nu}R^{\mu}{}_{\beta\alpha\lambda} - \Gamma^{\beta}_{\alpha\nu}R^{\mu}{}_{\eta\beta\lambda} - \Gamma^{\beta}_{\lambda\nu}R^{\mu}{}_{\eta\alpha\beta}
$$
$$
\overset{(15.4)}{=} \left\{(\Gamma^{\mu}_{\alpha\eta})_{,\lambda,\nu} - (\Gamma^{\mu}_{\lambda\eta})_{,\alpha,\nu} + \partial_\nu\left[\Gamma^{\mu}_{a\lambda}\Gamma^{a}_{\alpha\eta} - \Gamma^{\mu}_{a\alpha}\Gamma^{a}_{\lambda\eta}\right]\right\}
$$
$$
+ \Gamma^{\mu}_{\rho\nu}R^{\rho}{}_{\eta\alpha\lambda} - \Gamma^{\beta}_{\eta\nu}R^{\mu}{}_{\beta\alpha\lambda} - \Gamma^{\beta}_{\alpha\nu}R^{\mu}{}_{\eta\beta\lambda} - \Gamma^{\beta}_{\lambda\nu}R^{\mu}{}_{\eta\alpha\beta},
$$

and once again using Eq. (15.4) this becomes

$$
R^{\mu}_{\eta\alpha\lambda;\nu} \overset{(15.4)}{=} (\Gamma^{\mu}_{\alpha\eta}),_{\lambda,\nu} - (\Gamma^{\mu}_{\lambda\eta}),_{\alpha,\nu} + (\Gamma^{\mu}_{a\lambda}),_{\nu}\Gamma^{a}_{\alpha\eta} + \Gamma^{\mu}_{a\lambda}(\Gamma^{a}_{\alpha\eta}),_{\nu}
$$
$$
- (\Gamma^{\mu}_{a\alpha}),_{\nu}\Gamma^{a}_{\lambda\eta} - \Gamma^{\mu}_{a\alpha}(\Gamma^{a}_{\lambda\eta}),_{\nu}
$$
$$
+ \Gamma^{\mu}_{\rho\nu}\left[(\Gamma^{\rho}_{\alpha\eta}),_{\lambda} - (\Gamma^{\rho}_{\lambda\eta}),_{\alpha} + \Gamma^{\rho}_{a\lambda}\Gamma^{a}_{\alpha\eta} - \Gamma^{\rho}_{a\alpha}\Gamma^{a}_{\lambda\eta} \right]
$$
$$
- \Gamma^{\beta}_{\eta\nu}\left[(\Gamma^{\mu}_{\alpha\beta}),_{\lambda} - (\Gamma^{\mu}_{\lambda\beta}),_{\alpha} + \Gamma^{\mu}_{a\lambda}\Gamma^{a}_{\alpha\beta} - \Gamma^{\mu}_{a\alpha}\Gamma^{a}_{\lambda\beta} \right]
$$
$$
- \Gamma^{\beta}_{\alpha\nu}\left[(\Gamma^{\mu}_{\beta\eta}),_{\lambda} - (\Gamma^{\mu}_{\lambda\eta}),_{\beta} + \Gamma^{\mu}_{a\lambda}\Gamma^{a}_{\beta\eta} - \Gamma^{\mu}_{a\beta}\Gamma^{a}_{\lambda\eta} \right]
$$
$$
- \Gamma^{\beta}_{\lambda\nu}\left[(\Gamma^{\mu}_{\alpha\eta}),_{\beta} - (\Gamma^{\mu}_{\beta\eta}),_{\alpha} + \Gamma^{\mu}_{a\beta}\Gamma^{a}_{\alpha\eta} - \Gamma^{\mu}_{a\alpha}\Gamma^{a}_{\beta\eta} \right]. \qquad (15.13)
$$

We want to consider the following expression:

$$
B^{\mu}_{\eta\alpha\lambda\nu} := R^{\mu}_{\eta\alpha\lambda;\nu} + R^{\mu}_{\eta\nu\alpha;\lambda} + R^{\mu}_{\eta\lambda\nu;\alpha}. \qquad (15.14)
$$

We consider first the double-derivative contributions, *i.e.* those of the first two terms in Eq. (15.13), for each of the three terms in Eq. (15.14):

$$
B^{\mu}_{\eta\alpha\lambda\nu}{}^{(0)} := (\Gamma^{\mu}_{\alpha\eta}),_{\lambda,\nu} - \underline{(\Gamma^{\mu}_{\lambda\eta}),_{\alpha,\nu}} + \underbrace{(\Gamma^{\mu}_{\nu\eta}),_{\alpha,\lambda}}
$$
$$
- (\Gamma^{\mu}_{\alpha\eta}),_{\nu,\lambda} + \underline{(\Gamma^{\mu}_{\lambda\eta}),_{\nu,\alpha}} - \underbrace{(\Gamma^{\mu}_{\nu\eta}),_{\lambda,\alpha}} = 0. \qquad (15.15)
$$

The terms are seen to vanish in pairs as underlined or not, or underbraced. One can proceed similarly with the other expressions in Eq. (15.13) inserted into Eq. (15.14) and observe that they cancel. Thus finally:

$$
B^{\mu}_{\eta\alpha\lambda\nu} := R^{\mu}_{\eta\alpha\lambda;\nu} + R^{\mu}_{\eta\nu\alpha;\lambda} + R^{\mu}_{\eta\lambda\nu;\alpha} = 0. \qquad (15.16a)
$$

Multiplication of the entire equation (15.16a) from the left by $g^{\rho\eta}$ raises one index, *i.e.* (note the cyclic order of the lower indices)

$$
R^{\mu\rho}_{\alpha\lambda;\nu} + R^{\mu\rho}_{\nu\alpha;\lambda} + R^{\mu\rho}_{\lambda\nu;\alpha} = 0. \qquad (15.16b)
$$

These relations are valid in any coordinate frame and are called *Bianchi identities*. They correspond to Eq. (14.111) in electrodynamics.

We return to the Riemann tensor and contract two indices, *i.e.* we consider the following quantity known as the *Ricci tensor*:

$$
R_{\mu\nu} := R^{\alpha}_{\mu\alpha\nu}. \qquad (15.17)
$$

Here the indices μ and ν stand on either side of the lower index α. Our next essential step is to explore the meaning of cases where α stands on either side of the pair $\mu\nu$. The results are summarized below in Eq. (15.27). We consider the following properties. We have

$$R_{\mu\alpha\beta\gamma} = g_{\mu\eta} R^{\eta}{}_{\alpha\beta\gamma}.$$

Multiplying by $g^{\rho\mu}$ we obtain

$$g^{\rho\mu} R_{\mu\alpha\beta\gamma} = g^{\rho\mu} g_{\mu\eta} R^{\eta}{}_{\alpha\beta\gamma} = \delta^{\rho}_{\eta} R^{\eta}{}_{\alpha\beta\gamma} = R^{\rho}{}_{\alpha\beta\gamma}. \tag{15.18}$$

Setting $\rho = \alpha$ and then summing, we obtain (from the right to the left of Eq. (15.18)):

$$R^{\alpha}{}_{\alpha\beta\gamma} = g^{\alpha\mu} R_{\mu\alpha\beta\gamma} = g^{\mu\alpha} R_{\mu\alpha\beta\gamma}.$$

Renaming indices this is

$$R^{\alpha}{}_{\alpha\mu\nu} \overset{(a)}{\underset{(15.11),(b)}{=}} g^{\beta\lambda} R_{\beta\lambda\mu\nu} \overset{\lambda\leftrightarrow\beta}{=} g^{\lambda\beta} R_{\lambda\beta\mu\nu} = g^{\beta\lambda} R_{\lambda\beta\mu\nu}$$
$$-g^{\beta\lambda} R_{\beta\lambda\mu\nu}. \tag{15.19}$$

From the equalities (a) and (b) we deduce that

$$2g^{\beta\lambda} R_{\beta\lambda\mu\nu} = 0, \quad i.e. \quad R^{\alpha}{}_{\alpha\mu\nu} = 0. \tag{15.20}$$

From Eq. (15.18), i.e.

$$g^{\rho\mu} R_{\mu\alpha\beta\gamma} = R^{\rho}{}_{\alpha\beta\gamma},$$

we obtain by setting (1) $\rho = \alpha$:

$$R^{\alpha}{}_{\alpha\beta\gamma} = g^{\alpha\mu} R_{\mu\alpha\beta\gamma}, \tag{15.21}$$

and by setting (2) $\gamma = \rho$:

$$g^{\rho\mu} R_{\mu\alpha\beta\rho} = R^{\rho}{}_{\alpha\beta\rho}. \tag{15.22}$$

With Eq. (15.10) (antisymmetry in the last two indices) applied to the left hand side of this equation, we obtain

$$-g^{\rho\mu} R_{\mu\alpha\rho\beta} = R^{\rho}{}_{\alpha\beta\rho}. \tag{15.23}$$

Thus from this equation, (15.23), with β replaced by γ,

$$R^{\rho}{}_{\alpha\gamma\rho} = -g^{\rho\mu} R_{\mu\alpha\rho\gamma}. \tag{15.24}$$

Setting in Eq. (15.18) $\beta = \rho$, we obtain (from right to left):

$$R^{\rho}{}_{\alpha\rho\gamma} = g^{\rho\mu} R_{\mu\alpha\rho\gamma} \overset{(15.24)}{=} -R^{\rho}{}_{\alpha\gamma\rho}. \qquad (15.25)$$

Thus (renaming indices)

$$R^{\alpha}{}_{\mu\nu\alpha} = -R^{\alpha}{}_{\mu\alpha\nu} \overset{(15.17)}{=} -R_{\mu\nu}. \qquad (15.26)$$

For the sake of clarity we summarize the results:

$$R_{\mu\nu} \overset{(15.17)}{:=} R^{\alpha}{}_{\mu\alpha\nu} \overset{(15.26)}{=} -R^{\alpha}{}_{\mu\nu\alpha}, \qquad R^{\alpha}{}_{\alpha\mu\nu} \overset{(15.20)}{=} 0. \qquad (15.27)$$

The explicit form of $R_{\mu\nu}$ is

$$R_{\mu\nu} \overset{(15.4)}{=} (\Gamma^{\alpha}_{\alpha\mu})_{,\nu} - (\Gamma^{\alpha}_{\mu\nu})_{,\alpha} + \Gamma^{\alpha}_{a\nu}\Gamma^{a}_{\alpha\mu} - \Gamma^{\alpha}_{a\alpha}\Gamma^{a}_{\mu\nu}. \qquad (15.28)$$

We can now show that

$$R_{\mu\nu} = R_{\nu\mu}. \qquad (15.29)$$

We leave the explicit verification of this relation to Example 15.3.

The *Ricci curvature scalar* R is defined by contraction of $R_{\mu\nu}$, *i.e.*

$$R := g^{\mu\nu} R_{\mu\nu} \equiv R^{\mu}{}_{\mu}. \qquad (15.30)$$

It is this particular quantity which when nonzero distinguishes the Special Theory of Relativity (in which $R = 0$) from the General Theory. The *Ricci–Einstein tensor* $G_{\alpha}{}^{\mu}$ is (defined as) the tensor whose covariant divergence is zero, *i.e.*

$$G^{\mu}{}_{\alpha;\mu} = 0. \qquad (15.31)$$

We determine this tensor as follows. From the Bianchi identity (15.16b) we obtain (recall that the lower indices occur in cyclic order)

$$R^{\mu\alpha}{}_{\beta\gamma;\nu} + R^{\mu\alpha}{}_{\nu\beta;\gamma} + R^{\mu\alpha}{}_{\gamma\nu;\beta} = 0. \qquad (15.32)$$

The contraction of the indices μ and β gives

$$R^{\mu\alpha}{}_{\mu\gamma;\nu} + R^{\mu\alpha}{}_{\nu\mu;\gamma} + R^{\mu\alpha}{}_{\gamma\nu;\mu} = 0. \qquad (15.33)$$

Here we contract α and γ, and obtain

$$R^{\mu\alpha}{}_{\mu\alpha;\nu} + R^{\mu\alpha}{}_{\nu\mu;\alpha} + R^{\mu\alpha}{}_{\alpha\nu;\mu} = 0. \qquad (15.34)$$

We obtain the first term on the left without the covariant differentiation from

$$R \overset{(15.30)}{=} g^{\mu\nu} R_{\mu\nu} \overset{(15.17)}{=} g^{\mu\nu} R^{\alpha}{}_{\mu\alpha\nu} = R^{\alpha\nu}{}_{\alpha\nu}. \qquad (15.35)$$

We can obtain the second and third terms in Eq. (15.34), again without their covariant differentiations, as follows. We have

$$R^\rho{}_\nu = g^{\rho\mu}R_{\mu\nu} \overset{(15.17)}{=} g^{\rho\mu}R^\alpha{}_{\mu\alpha\nu} \overset{(15.9)}{=} -g^{\rho\mu}R^\alpha{}_{\mu\nu\alpha} = -R^{\alpha\rho}{}_{\nu\alpha} \equiv -R^{\mu\rho}{}_{\nu\mu}. \quad (15.36)$$

Similarly

$$R^\rho{}_\nu = g^{\rho\mu}R_{\mu\nu} = g^{\rho\mu}R^\alpha{}_{\mu\alpha\nu}, \quad i.e. \quad R^\rho{}_\nu = R^{\alpha\rho}{}_{\alpha\nu}, \quad (15.37a)$$

but also

$$R^\rho{}_\nu = R^{\alpha\rho}{}_{\alpha\nu} \overset{(15.11)}{=} -R^{\rho\alpha}{}_{\alpha\nu}. \quad (15.37b)$$

Inserting Eqs. (15.35), (15.36) and (15.37b) into Eq. (15.34), i.e. the results $R^{\mu\alpha}{}_{\mu\alpha} = R, R^{\mu\alpha}{}_{\nu\mu} = -R^\alpha{}_\nu, R^{\mu\alpha}{}_{\alpha\nu} = -R^\mu{}_\nu$, we obtain by taking the appropriate covariant derivatives

$$R_{;\nu} - R^\alpha{}_{\nu;\alpha} - R^\mu{}_{\nu;\mu} = 0, \quad i.e. \quad R_{;\nu} = 2R^\alpha{}_{\nu;\alpha}. \quad (15.38)$$

Now consider the quantity

$$\tilde{G}^\rho{}_\beta := R^\rho{}_\beta - \frac{1}{2}\delta^\rho_\beta R. \quad (15.39)$$

Then

$$g_{\alpha\rho}\tilde{G}^\rho{}_\beta = g_{\alpha\rho}R^\rho{}_\beta - \frac{1}{2}g_{\alpha\rho}\delta^\rho_\beta R, \quad i.e. \quad \tilde{G}_{\alpha\beta} = R_{\alpha\beta} - \frac{1}{2}g_{\alpha\beta}R, \quad (15.40)$$

and with covariant derivative:

$$
\begin{aligned}
\tilde{G}^\rho{}_{\beta;\rho} &= R^\rho{}_{\beta;\rho} - \frac{1}{2}\delta^\rho_\beta R_{;\rho} \overset{(15.38)}{=} R^\rho{}_{\beta;\rho} - \frac{1}{2}\delta^\rho_\beta(2R^\alpha{}_{\rho;\alpha}) \\
&= R^\rho{}_{\beta;\rho} - R^\alpha{}_{\beta;\alpha} = 0, \quad (15.41)
\end{aligned}
$$

and hence we can identify $\tilde{G} = G$ (G of Eq. (15.31)), and the Ricci–Einstein tensor

$$G_{\alpha\beta} = R_{\alpha\beta} - \frac{1}{2}g_{\alpha\beta}R = G_{\beta\alpha} \quad (15.42)$$

solves Eq. (15.31).

Example 15.3: Symmetry of the Ricci tensor $R_{\mu\nu}$
Verify that $R_{\mu\nu} = R_{\nu\mu}$.

Solution: The Ricci tensor $R_{\mu\nu}$ is given by the expression (15.28), i.e.

$$R_{\mu\nu} \overset{(15.4)}{=} (\Gamma^\alpha_{\alpha\mu})_{,\nu} - (\Gamma^\alpha_{\mu\nu})_{,\alpha} + \Gamma^\alpha_{\alpha\nu}\Gamma^a_{\alpha\mu} - \Gamma^\alpha_{\alpha\alpha}\Gamma^a_{\mu\nu}.$$

We introduce the following definitions:

$$A_{\mu\nu} \equiv (\Gamma^{\alpha}_{\alpha\mu})_{,\nu} = \partial_{\nu}\left[\frac{1}{2}g^{\alpha\kappa}(\partial_{\alpha}g_{\mu\kappa} + \partial_{\mu}g_{\kappa\alpha} - \partial_{\kappa}g_{\alpha\mu})\right], \tag{15.43}$$

$$B_{\mu\nu} \equiv (-\Gamma^{\alpha}_{\mu\nu})_{,\alpha} = -\partial_{\alpha}\left[\frac{1}{2}g^{\alpha\kappa}(\partial_{\mu}g_{\nu\kappa} + \partial_{\nu}g_{\kappa\mu} - \partial_{\kappa}g_{\mu\nu})\right]. \tag{15.44}$$

These replacements permit the Ricci tensor to be expressed as

$$R_{\mu\nu} = A_{\mu\nu} + B_{\mu\nu} + \frac{1}{4}C_{\mu\nu}, \quad \text{where} \quad C_{\mu\nu} = 4(\Gamma^{\alpha}_{a\nu}\Gamma^{a}_{\alpha\mu} - \Gamma^{\alpha}_{a\alpha}\Gamma^{a}_{\mu\nu}).$$

Hence, with similarly underbraced quantities cancelling each other,

$$\begin{aligned}
&2(A_{\mu\nu} + B_{\mu\nu} - A_{\nu\mu} - B_{\nu\mu}) \\
= \quad &\partial_{\nu}[g^{\alpha\kappa}(\partial_{\alpha}g_{\mu\kappa} + \partial_{\mu}g_{\kappa\alpha} - \partial_{\kappa}g_{\alpha\mu})] - \underbrace{\partial_{\alpha}[g^{\alpha\kappa}(\partial_{\mu}g_{\nu\kappa} + \partial_{\nu}g_{\kappa\mu} - \partial_{\kappa}g_{\mu\nu})]}_{\cdots\quad\cdots} \\
&- \partial_{\mu}[g^{\alpha\kappa}(\partial_{\alpha}g_{\nu\kappa} + \partial_{\nu}g_{\kappa\alpha} - \partial_{\kappa}g_{\alpha\nu})] + \underbrace{\partial_{\alpha}[g^{\alpha\kappa}(\partial_{\nu}g_{\mu\kappa} + \partial_{\mu}g_{\kappa\nu} - \partial_{\kappa}g_{\nu\mu})]}_{\cdots\quad\cdots},
\end{aligned}$$

and thus

$$\begin{aligned}
&2(A_{\mu\nu} + B_{\mu\nu} - A_{\nu\mu} - B_{\nu\mu}) \\
= \quad &(\partial_{\nu}g^{\alpha\kappa})(\partial_{\alpha}g_{\mu\kappa} + \partial_{\mu}g_{\kappa\alpha} - \partial_{\kappa}g_{\alpha\mu}) + g^{\alpha\kappa}(\partial_{\nu}\partial_{\alpha}g_{\mu\kappa} + \underbrace{\partial_{\nu}\partial_{\mu}g_{\kappa\alpha}}_{---} - \partial_{\nu}\partial_{\kappa}g_{\alpha\mu}) \\
&- g^{\alpha\kappa}(\partial_{\alpha}\partial_{\mu}g_{\nu\kappa} + \underbrace{\partial_{\mu}\partial_{\nu}g_{\kappa\alpha}}_{---} - \partial_{\mu}\partial_{\kappa}g_{\alpha\nu}) - (\partial_{\mu}g^{\alpha\kappa})(\partial_{\alpha}g_{\nu\kappa} + \partial_{\nu}g_{\kappa\alpha} - \partial_{\kappa}g_{\alpha\nu}) \\
= \quad &(\partial_{\nu}g^{\alpha\kappa})(\partial_{\alpha}g_{\mu\kappa} + \partial_{\mu}g_{\kappa\alpha} - \partial_{\kappa}g_{\alpha\mu}) + \partial_{\nu}\partial^{\kappa}g_{\mu\kappa} - \partial_{\nu}\partial^{\alpha}g_{\alpha\mu} \\
&- \partial^{\kappa}\partial_{\mu}g_{\nu\kappa} + \partial_{\mu}\partial^{\alpha}g_{\alpha\nu} - (\partial_{\mu}g^{\alpha\kappa})(\partial_{\alpha}g_{\nu\kappa} + \partial_{\nu}g_{\kappa\alpha} - \partial_{\kappa}g_{\alpha\nu}) \\
= \quad &(\partial_{\nu}g^{\alpha\kappa})(\partial_{\alpha}g_{\mu\kappa} + \partial_{\mu}g_{\kappa\alpha} - \partial_{\kappa}g_{\alpha\mu}) - (\partial_{\mu}g^{\alpha\kappa})(\partial_{\alpha}g_{\nu\kappa} + \partial_{\nu}g_{\kappa\alpha} - \partial_{\kappa}g_{\alpha\nu}). \tag{15.45}
\end{aligned}$$

Now consider the expression

$$\begin{aligned}
C_{\mu\nu} \quad \equiv \quad &4(\Gamma^{\alpha}_{a\nu}\Gamma^{a}_{\alpha\mu} - \Gamma^{\alpha}_{a\alpha}\Gamma^{a}_{\mu\nu}) \\
= \quad &g^{\alpha\kappa}(\partial_{a}g_{\nu\kappa} + \partial_{\nu}g_{\kappa a} - \partial_{\kappa}g_{a\nu})g^{a\rho}(\partial_{\alpha}g_{\mu\rho} + \partial_{\mu}g_{\rho\alpha} - \partial_{\rho}g_{\alpha\mu}) \\
&- g^{\alpha\kappa}(\partial_{a}g_{\alpha\kappa} + \partial_{\alpha}g_{\kappa a} - \partial_{\kappa}g_{a\alpha})g^{a\rho}(\partial_{\mu}g_{\nu\rho} + \partial_{\nu}g_{\rho\mu} - \partial_{\rho}g_{\mu\nu}). \tag{15.46}
\end{aligned}$$

Thus

$$\begin{aligned}
C_{\mu\nu} \quad = \quad &g^{\alpha\kappa}g^{a\rho}[(\partial_{a}g_{\nu\kappa} + \partial_{\nu}g_{\kappa a} - \partial_{\kappa}g_{a\nu})(\partial_{\alpha}g_{\mu\rho} + \partial_{\mu}g_{\rho\alpha} - \partial_{\rho}g_{\alpha\mu}) \\
&- (\partial_{a}g_{\alpha\kappa} + \partial_{\alpha}g_{\kappa a} - \partial_{\kappa}g_{a\alpha})(\partial_{\mu}g_{\nu\rho} + \partial_{\nu}g_{\rho\mu} - \partial_{\rho}g_{\mu\nu})].
\end{aligned}$$

We now have to write out all terms of the difference $C_{\mu\nu} - C_{\nu\mu}$. This is the following lengthy expression (for the cancellations make replacements $a \leftrightarrow \alpha, \rho \leftrightarrow \kappa$):

$$\begin{aligned}
C_{\mu\nu} - C_{\nu\mu} \quad = \quad &g^{\alpha\kappa}g^{a\rho}[\underbrace{\partial_{a}g_{\nu\kappa}\partial_{\alpha}g_{\mu\rho}}_{\cdot} + \underbrace{\partial_{a}g_{\nu\kappa}\partial_{\mu}g_{\rho\alpha}}_{\cdot\cdot} - \underbrace{\partial_{a}g_{\nu\kappa}\partial_{\rho}g_{\alpha\mu}}_{\cdots} + \underbrace{\partial_{\nu}g_{\kappa a}\partial_{\alpha}g_{\mu\rho}}_{*} + \underbrace{\partial_{\nu}g_{\kappa a}\partial_{\mu}g_{\rho\alpha}}_{**} \\
&- \underbrace{\partial_{\nu}g_{\kappa a}\partial_{\rho}g_{\alpha\mu}}_{***} - \underbrace{\partial_{\kappa}g_{a\nu}\partial_{\alpha}g_{\mu\rho}}_{+} - \underbrace{\partial_{\kappa}g_{a\nu}\partial_{\mu}g_{\rho\alpha}}_{++} + \underbrace{\partial_{\kappa}g_{a\nu}\partial_{\rho}g_{\alpha\mu}}_{+++} \\
&- \partial_{a}g_{\alpha\kappa}\partial_{\mu}g_{\nu\rho} - \partial_{a}g_{\alpha\kappa}\partial_{\nu}g_{\rho\mu} + \partial_{a}g_{\alpha\kappa}\partial_{\rho}g_{\mu\nu} - \partial_{a}g_{\kappa a}\partial_{\mu}g_{\nu\rho} - \partial_{a}g_{\kappa a}\partial_{\nu}g_{\rho\mu}
\end{aligned}$$

$$+\partial_\alpha g_{\kappa a}\partial_\rho g_{\mu\nu} + \partial_\kappa g_{a\alpha}\partial_\mu g_{\nu\rho} + \partial_\kappa g_{a\alpha}\partial_\nu g_{\rho\mu} - \partial_\kappa g_{a\alpha}\partial_\rho g_{\mu\nu}$$
$$- \underbrace{\partial_a g_{\mu\kappa}\partial_\alpha g_{\nu\rho}}_{*} - \partial_a g_{\mu\kappa}\partial_\nu g_{\rho\alpha} + \underbrace{\partial_a g_{\mu\kappa}\partial_\rho g_{a\nu}}_{+} - \underbrace{\partial_\mu g_{\kappa a}\partial_\alpha g_{\nu\rho}}_{..} - \underbrace{\partial_\mu g_{\kappa a}\partial_\nu g_{\rho\alpha}}_{**}$$
$$+ \underbrace{\partial_\mu g_{\kappa a}\partial_\rho g_{a\nu}}_{++} + \underbrace{\partial_\kappa g_{a\mu}\partial_a g_{\nu\rho}}_{...} + \underbrace{\partial_\kappa g_{a\mu}\partial_\nu g_{\rho\alpha}}_{***} - \underbrace{\partial_\kappa g_{a\mu}\partial_\rho g_{a\nu}}_{+++}$$
$$+ \partial_a g_{a\kappa}\partial_\nu g_{\mu\rho} + \partial_a g_{a\kappa}\partial_\mu g_{\rho\nu} - \partial_a g_{a\kappa}\partial_\rho g_{\nu\mu} + \partial_a g_{\kappa a}\partial_\nu g_{\mu\rho} + \partial_a g_{\kappa a}\partial_\mu g_{\rho\nu}$$
$$- \partial_a g_{\kappa a}\partial_\rho g_{\nu\mu} \quad \partial_\kappa g_{a\alpha}\partial_\nu y_{\mu\rho} - \partial_\kappa g_{a\alpha}\partial_\mu g_{\rho\nu} + \partial_\kappa g_{a\alpha}\partial_\rho g_{\nu\mu}] = 0. \tag{15.47}$$

Thus half of the number of terms cancel automatically (those not underbraced), the other half (underbraced) on replacement of $\rho \leftrightarrow \kappa$ with $a \leftrightarrow \alpha$. Collecting terms it follows that (again with cancellations as underbraced)

$$2(R_{\mu\nu} - R_{\nu\mu}) = (\partial_\nu g^{\alpha\kappa})(\partial_a g_{\mu\kappa} + \partial_\mu g_{\kappa a} - \partial_\kappa g_{a\mu}) - (\partial_\mu g^{\alpha\kappa})(\partial_a g_{\nu\kappa} + \partial_\nu g_{\kappa a} - \partial_\kappa g_{a\nu})$$
$$= \underbrace{(\partial_\nu g^{\alpha\kappa})(\partial_a g_{\mu\kappa})}_{--} + (\partial_\nu g^{\alpha\kappa})(\partial_\mu g_{\kappa a}) - \underbrace{(\partial_\nu g^{\alpha\kappa})(\partial_\kappa g_{a\mu})}_{--} - \underbrace{(\partial_\mu g^{\alpha\kappa})(\partial_a g_{\nu\kappa})}_{++}$$
$$- (\partial_\mu g^{\alpha\kappa})(\partial_\nu g_{\kappa a}) + \underbrace{(\partial_\mu g^{\alpha\kappa})(\partial_\kappa g_{a\nu})}_{++}$$
$$- (\partial_\nu g^{\alpha\kappa})(\partial_\mu g_{\kappa a}) - (\partial_\mu g^{\alpha\kappa})(\partial_\nu g_{\kappa a})$$
$$= (\partial_\nu g_{a\kappa})(\partial_\mu g^{\kappa\alpha}) - (\partial_\mu g^{\kappa\alpha})(\partial_\nu g_{\kappa a}) = 0. \tag{15.48}$$

This verifies the important property $R_{\mu\nu} = R_{\nu\mu}$.

15.4 The Energy–Momentum Tensor

15.4.1 The energy–momentum tensor in electrodynamics

We first familiarize ourselves with the energy–momentum tensor in (flat space) electrodynamics. There Newton's equation for a charge distribution $\rho(\mathbf{x}, t)$ in space is given by

$$\frac{d}{dt}(\text{mechanical } \mathbf{momentum} \text{ of } \rho) = \text{external } \mathbf{force}$$

$$= \int_V \rho(\mathbf{E} + \mathbf{v} \times \mathbf{B})dV. \tag{15.49}$$

With the help of Maxwell's equations[†] (*i.e.* $\rho = \nabla\cdot\mathbf{D}, \nabla\times\mathbf{H} = \mathbf{j}+\partial\mathbf{D}/\partial t, \mathbf{j} = \rho\mathbf{v}$) this equation can be re-expressed for component j as

$$\left[\frac{d}{dt}(\text{mechanical } \mathbf{momentum})\right]_j = -\frac{\partial}{\partial t}\int_V (\mathbf{D} \times \mathbf{B})_j dV + \int_V \frac{\partial}{\partial x_i}T_{ij}dV, \tag{15.50}$$

where[‡]

$$T_{ij} = (D_i E_j + B_i H_j) - \delta_{ij}\frac{1}{2}(\mathbf{D} \cdot \mathbf{E} + \mathbf{B} \cdot \mathbf{H}) = T_{ji} \tag{15.51}$$

[†] We recall a few equations. For details see for instance H. J. W. Müller–Kirsten [34], p. 413.
[‡] See *e.g.* H. J. W. Müller–Kirsten [34], p. 166.

is called the *Maxwell stress tensor.*§ With

$$\mathbf{P}_{\text{field}} = \int_V \mathbf{p}_{\text{field}} \, dV, \qquad \mathbf{p}_{\text{field}} = \mathbf{D} \times \mathbf{B}, \tag{15.52}$$

$\mathbf{p}_{\text{field}}$ is the spatial density of the field momentum. Using *Gauss' divergence theorem* we can rewrite Eq. (15.50) as

$$\frac{d}{dt}(\mathbf{P}_{\text{mech}} + \mathbf{P}_{\text{field}})_j - \int_{F(V)} T_{ij}(d\mathbf{F})_i = 0. \tag{15.53}$$

The quantity

$$K_j = T_{ij}(d\mathbf{F})_i$$

is the j-th component of a force acting on $d\mathbf{F}$, and

$$\frac{K_j n_j}{dF} = T_{ij} n_i n_j$$

(n_i being components of the unit vector) represents the radiation pressure on the surface element dF. We now want to be more general!

Each particle (name "n") carries a 4-momentum given by (recall from $E = mc^2$: dimension of E/c = dimension of mc)

$$P_n^\alpha(t) = \left(\frac{E_n(t)}{c}, \mathbf{p}_n(t) \right). \tag{15.54}$$

The total momentum $\mathbf{P}$ is the sum of the momenta of the individual particles, *i.e.* $\mathbf{P} = \sum_n \mathbf{p}_n(t)$, and the total energy divided by c is $P^0(t) \equiv E(t)/c = \sum_n E_n(t)/c \equiv \sum_n p_n^0(t)$, so that the three-dimensional or spatial density of energy and momentum is

$$T_{\text{mech}}^{\alpha 0}(t, \mathbf{x}) = \sum_n p_n^\alpha(t) \delta(\mathbf{x} - \mathbf{x}_n(t)), \tag{15.55}$$

which integrated over $d\mathbf{x}$ gives the total four-momentum. The current of momentum is obtained by replacing the particle density $\rho_n = \delta(\mathbf{x} - \mathbf{x}_n(t))$ or charge density $\rho_n = q_n \delta(\mathbf{x} - \mathbf{x}_n(t))$ by the current density $\rho_n d\mathbf{x}_n/dt$, and hence

$$T_{\text{mech}}^{\alpha j}(t, \mathbf{x}) = \sum_n p_n^\alpha(t) \delta(\mathbf{x} - \mathbf{x}_n(t)) \frac{dx_n^j}{dt}. \tag{15.56}$$

---

§This is a tensor only with respect to rectangular frames which are stationary with respect to the inertial frame used.

The last two equations together give in four dimensions with $x^0 = ct$

$$T^{\alpha\beta}_{\text{mech}}(t,\mathbf{x}) = \sum_n p^\alpha_n(t)\delta(\mathbf{x}-\mathbf{x}_n(t))\frac{dx^\beta_n(t)}{dt}$$

$$= \sum_n \int_{-\infty}^\infty p^\alpha_n(t_n)\delta(\mathbf{x}-\mathbf{x}_n(t_n))\frac{dx^\beta_n(t_n)}{dt_n}\delta(t_n - t)dt_n. \ (15.57)$$

The tensor has the following meaning (with $c = 1$ since dimensions of E and p differ):

$$T^{\alpha\beta}_{\text{mech}} = \begin{bmatrix} T^{00} & T^{0j} \\ T^{i0} & T^{ij} \end{bmatrix}$$

$$= \begin{bmatrix} \text{energy density} & \text{energy current density} \\ \text{momentum density} & \underbrace{\text{momentum current density}}_{\text{``stress tensor''}} \end{bmatrix} . \ (15.58)$$

Conservation of total energy and momentum are expressed by an equation of continuity for the entire tensor consisting of mechanical and electromagnetic (here abbreviated "el") contributions which we cite here without proof:[¶]

$$\partial_\beta T^{\alpha\beta} = 0, \qquad T^{\alpha\beta} = T^{\alpha\beta}_{\text{mech}} + T^{\alpha\beta}_{\text{el}}. \tag{15.59}$$

Thus we require the derivative of the tensor $T^{\alpha\beta}_{\text{mech}}(x^\mu)$ with respect to x^β. In Eq. (15.57) this x^β (note the difference between x^β and x^β_n) is contained in the delta functions (the product of four individual delta functions). To proceed it is most expedient to appeal to a *virial theorem, i.e.* an averaging relation.[‖] We define

$$G^{\alpha\beta}(t) := \sum_n p^\alpha_n(t)x^\beta_n(t), \quad \frac{d}{dt}G^{\alpha\beta}(t) = \sum_n\left[\frac{dp^\alpha_n(t)}{dt}x^\beta_n(t) + p^\alpha_n(t)\frac{dx^\beta_n(t)}{dt}\right].$$

Then, constructing the average over a time interval $2T$, we obtain

$$\overline{G^{\alpha\beta}(t)} = \frac{1}{2T}\int_{-T}^T \frac{dG^{\alpha\beta}(t)}{dt}dt = \frac{1}{2T}[G^{\alpha\beta}(T) - G^{\alpha\beta}(-T)]$$

$$= \overline{\sum_n\frac{dp^\alpha_n(t)}{dt}x^\beta_n(t) + \sum_n p^\alpha_n(t)\frac{dx^\beta_n(t)}{dt}}. \tag{15.60}$$

[¶]See *e.g.* H. J. W. Müller–Kirsten [34], p. 426. This equation is actually derived with the help of the Euler–Lagrange equation.

[‖]For a broader discussion see H. Goldstein [19], Sec. 3.4.

In a periodic case with $G^{\alpha\beta}(T) = G^{\alpha\beta}(-T)$ the expression vanishes. But the expression also approaches zero for $T \to \infty$ if the phase space coordinates assume only finite values. Thus assuming the latter, we obtain in an averaged sense, and omitting the overbars indicating this,

$$
\begin{aligned}
T^{\alpha\beta}_{\text{mech}}(t,\mathbf{x}) &\simeq \sum_n p^{\alpha}_n(t)\frac{dx^{\beta}_n(t)}{dt}\delta(\mathbf{x}-\mathbf{x}_n(t)) \\
&= -\sum_n \frac{dp^{\alpha}_n(t)}{dt}x^{\beta}_n(t)\delta(\mathbf{x}-\mathbf{x}_n(t)) \\
&= -\sum_n \frac{dp^{\alpha}_n(t)}{dt}x^{\beta}\delta(\mathbf{x}-\mathbf{x}_n(t))
\end{aligned}
\tag{15.61}
$$

Thus for

$$
T^{\alpha\beta} = T^{\alpha\beta}_{\text{mech}} + T^{\alpha\beta}_{\text{el}},
\tag{15.62a}
$$

conservation of energy and momentum requires

$$
\partial_{\beta}T^{\alpha\beta}_{\text{el}} = -\partial_{\beta}T^{\alpha\beta}_{\text{mech}},
\tag{15.62b}
$$

and so with the four-dimensional generalization of the better-known three-dimensional Lorentz force** $(F^{\alpha}_{\ \beta} = \partial^{\alpha}A_{\beta}(x) - \partial_{\beta}A^{\alpha}(x))$

$$
\frac{dp^{\alpha}_n(t)}{dt} = q_n F^{\alpha}_{\ \beta}\frac{dx^{\beta}_n(t)}{dt},
\tag{15.63}
$$

we have, in an averaged sense as above (and ignoring a lengthy discussion about the action of ∂_{β} on the delta function),

$$
\begin{aligned}
\partial_{\beta}T^{\alpha\beta}_{\text{el}} &= -\partial_{\beta}T^{\alpha\beta}_{\text{mech}} \overset{(15.61)}{=} \sum_n \frac{dp^{\alpha}_n(t)}{dt}\delta(\mathbf{x}-\mathbf{x}_n(t)) \\
&= \sum_n q_n F^{\alpha}_{\ \beta}\frac{dx^{\beta}_n}{dt}\delta(\mathbf{x}-\mathbf{x}_n(t)) \\
&= \text{source term} \equiv +F^{\alpha}_{\ \beta}j^{\beta}, \quad j^{\beta} : \text{the current density.} \tag{15.64}
\end{aligned}
$$

We observe clearly that the current density is made up of the charges multiplied by their velocities and the delta functions representing their instantaneous positions. In Example 15.4 we show that the source term of the electromagnetic energy-momentum tensor is given by

$$
F^{\alpha}_{\ \beta}j^{\beta} = \partial_{\beta}\left[F^{\alpha}_{\ \gamma}F^{\gamma\beta} + \frac{1}{4}\eta^{\alpha\beta}F^{\gamma\delta}F_{\gamma\delta}\right].
\tag{15.65}
$$

**For the derivation of this expression see H. J. W. Müller–Kirsten [34], pp. 414-417. The four-component expression contains in addition to the wellknown three-component force $d\mathbf{p}/dt = q(\mathbf{E} + \mathbf{v}\times\mathbf{B})$, an additional component expressing power = work/time = $q\mathbf{E}\cdot d\mathbf{x}/dt$.

Then from Eqs. (15.64) and (15.65) we can read off the result:

$$T_{el}^{\alpha\beta} = \left[F^{\alpha}_{\gamma} F^{\gamma\beta} + \frac{1}{4} \eta^{\alpha\beta} F^{\gamma\delta} F_{\gamma\delta} \right],$$ (15.66)

in agreement with the result of Example 15.5. Hence

$$\partial_{\beta} T_{el}^{\alpha\beta} = F^{\alpha}_{\beta} j^{\beta} \overset{(15.65)}{=} +\partial_{\beta}[F^{\alpha}_{\gamma} F^{\gamma\beta} + \frac{1}{4} \eta^{\alpha\beta} F^{\gamma\delta} F_{\gamma\delta}].$$ (15.67)

Example 15.4: Verification of Eq. (15.65), source term of $T_{el}^{\alpha\beta}$
Show that
$$F^{\alpha}_{\beta} j^{\beta} = \partial_{\beta} \left[F^{\alpha}_{\gamma} F^{\gamma\beta} + \frac{1}{4} \eta^{\alpha\beta} F^{\gamma\delta} F_{\gamma\delta} \right].$$

Solution: Using Eq. (14.107) for j^{β} we have

$$F^{\alpha}_{\beta} j^{\beta} = F^{\alpha}_{\beta} \partial_{\gamma} F^{\beta\gamma} = \partial_{\gamma}[F^{\alpha}_{\beta} F^{\beta\gamma}] - [\partial_{\gamma} F^{\alpha}_{\beta}] F^{\beta\gamma}.$$ (15.68)

We consider the last term here:

$$[\partial_{\gamma} F^{\alpha}_{\beta}] F^{\beta\gamma} = \frac{1}{2}(\partial^{\gamma} F^{\alpha\beta}) F_{\beta\gamma} + \frac{1}{2}(\partial^{\gamma} F^{\alpha\beta}) F_{\beta\gamma} = \frac{1}{2}(\partial^{\gamma} F^{\alpha\beta}) F_{\beta\gamma} + \frac{1}{2}(\partial^{\beta} F^{\alpha\gamma}) F_{\gamma\beta}$$

$$= \frac{1}{2}(\partial^{\gamma} F^{\alpha\beta}) F_{\beta\gamma} + \frac{1}{2}(\partial^{\beta} F^{\gamma\alpha}) F_{\beta\gamma} = \frac{1}{2}(\partial^{\gamma} F^{\alpha\beta} + \partial^{\beta} F^{\gamma\alpha}) F_{\beta\gamma}.$$

Using now the explicit form $F^{\alpha\beta} = \partial^{\alpha} A^{\beta} - \partial^{\beta} A^{\alpha}, \ldots$ or the *Bianchi identity* (14.111) (which says that $\partial^{\gamma} F^{\alpha\beta} + \partial^{\beta} F^{\gamma\alpha} = -\partial^{\alpha} F^{\beta\gamma}$ with the superscripts in cyclic order), this becomes

$$[\partial_{\gamma} F^{\alpha}_{\beta}] F^{\beta\gamma} = -\frac{1}{2}(\partial^{\alpha} F^{\beta\gamma}) F_{\beta\gamma} = -\frac{1}{4}\partial^{\alpha}[F^{\beta\gamma} F_{\beta\gamma}] = -\frac{1}{4}\partial_{\gamma}(\eta^{\alpha\gamma} F^{\beta\delta} F_{\beta\delta}).$$

Inserting this result into Eq. (15.68), we obtain (with $\gamma \leftrightarrow \beta$)

$$F^{\alpha}_{\beta} j^{\beta} = \partial_{\beta}\left[F^{\alpha}_{\gamma} F^{\gamma\beta} + \frac{1}{4} \eta^{\alpha\beta} F^{\gamma\delta} F_{\gamma\delta} \right].$$

This therefore verifies the relation (15.67).

15.4.2 The general case

Suppose we consider now the following action functional of a general form

$$S = \int L(g_{\mu\nu}, \phi) \sqrt{-g(x)} d^4 x.$$ (15.69)

In ordinary electrodynamics we have instead of the tensor field $g_{\mu\nu}$ the vector potential A_{μ}, and instead of the scalar field ϕ the field of a particle like that of the electron. For our arguments here we consider the scalar field ϕ as some arbitrary field (conveniently as a scalar since this is the easiest to

handle), and in addition we treat the metric tensor $g_{\mu\nu}$ as an equally important external field. This is the basic idea: To include the metric representing
the gravitational (tensor) field along with any physically relevant field and to
treat them all similarly. Typically an action consists of the interaction-free
parts plus the interacting parts, so that here we expect the free (matter) part
of $\phi(x)$, the free (gravitational) part of $g_{\mu\nu}(x)$ plus a matter-gravity interaction part, where "matter" can be mass or field energy (since the latter's
energy can again be linked to mass). In the usual way, we are interested in
the extremization of the action, and in the derivation of the Euler–Lagrange
equations for each of the fields contained in S. Thus the basic principle of
what we proceed to do is the same as in earlier chapters — always the same
as pointed out by Penrose (*cf.* Sec. 3.1). Naturally since we are here predominantly interested in gravity, our main interest focuses on the field $g_{\mu\nu}(x)$.
Below we derive an expression for the variation δS. We shall consider this
in a form which we can write in terms of a quantity $T^{\mu\nu}(x)$ to be specified
below, and actually identified as the energy–momentum tensor. Thus below
we shall write the variation of S in the form (observe the $1/2$ in front)

$$\delta S = -\frac{1}{2} \int T^{\mu\nu}(x) \delta g_{\mu\nu}(x) \sqrt{-g(x)} d^4 x + O(\delta\phi). \qquad (15.70)$$

We now consider δS of Eq. (15.69) ignoring thereby the variation $\delta\phi$ which
could be added. Then (assuming L contains also derivatives of $g_{\mu\nu}$)

$$\delta S \;-\; O(\delta\phi) = \int \left[\frac{\partial L}{\partial g_{\mu\nu}} \delta g_{\mu\nu} \sqrt{-g} + \left(\frac{\partial L}{\partial g_{\mu\nu,\lambda}} \right) \delta g_{\mu\nu,\lambda} \sqrt{-g} \right.$$
$$\left. + L \frac{\partial \sqrt{-g}}{\partial g_{\mu\nu}} \delta g_{\mu\nu} \right] d^4 x$$
$$= \int \left[\frac{\partial L}{\partial g_{\mu\nu}} - \left(\frac{\partial L}{\partial g_{\mu\nu,\lambda}} \right)_{,\lambda} + \frac{1}{2} g^{\mu\nu} L \right] \delta g_{\mu\nu} \sqrt{-g} d^4 x, \qquad (15.71)$$

since

$$\frac{\partial \sqrt{-g}}{\partial g_{\mu\nu}} = -\frac{1}{2\sqrt{-g}} \frac{\partial g}{\partial g_{\mu\nu}} \overset{(14.97)}{=} -\frac{1}{2\sqrt{-g}} g g^{\mu\nu} = \frac{1}{2} \sqrt{-g} g^{\mu\nu}.$$

Comparing Eqs. (15.70) and (15.71), we identify the quantity $T^{\mu\nu}$ as

$$-T^{\mu\nu} = 2 \frac{\partial L}{\partial g_{\mu\nu}} - 2 \left(\frac{\partial L}{\partial g_{\mu\nu,\lambda}} \right)_{,\lambda} + g^{\mu\nu} L. \qquad (15.72)$$

In Example 15.5 we verify that this formula for $T^{\mu\nu}$ reproduces the energy–
momentum tensor in electrodynamics.

Example 15.5: Identifying $T^{\mu\nu}$ as Maxwell energy–momentum tensor
Show that if Eq. (15.72) is applied to the Lagrangian of electrodynamics, the tensor $T^{\mu\nu}$ is identified as the energy–momentum tensor of electrodynamics.

Solution: In the simplest case of electrodynamics, *i.e.* that of the free electromagnetic field, the Lagrangian density L is given by (observe: no derivative of $g_{\mu\nu}$)

$$L(A_\mu(x), \partial_\mu A_\nu(x)) = -\frac{1}{4} F_{\mu\nu} F^{\mu\nu} = -\frac{1}{4} F_{\rho\sigma} F_{\alpha\beta} g^{\rho\alpha} g^{\sigma\beta}, \tag{15.73}$$

so that with Eq. (14.96), *i.e.*

$$\frac{\partial g^{\alpha\beta}}{\partial g_{\mu\nu}} = -g^{\alpha\mu} g^{\beta\nu}, \tag{15.74}$$

and

$$\frac{\partial L}{\partial g_{\mu\nu}} = -\frac{1}{4} F_{\rho\sigma} F_{\alpha\beta} [-g^{\mu\rho} g^{\nu\alpha} g^{\sigma\beta} - g^{\rho\alpha} g^{\mu\sigma} g^{\nu\beta}] = -\frac{1}{2} F^\mu_\alpha F^{\alpha\nu}, \quad (\text{using } F^{\alpha\nu} = -F^{\nu\alpha}). \tag{15.75}$$

Inserting this expression in Eq. (15.72), we obtain

$$-T^{\mu\nu} = -F^\mu_\alpha F^{\alpha\nu} - \frac{1}{4} g^{\mu\nu} F_{\rho\sigma} F^{\rho\sigma} \tag{15.76}$$

in agreement with Eq. (15.66) with Minkowski $\eta^{\mu\nu}$ there replaced by $g^{\mu\nu}$ here. One can show that $T^{\mu\nu}$ is conserved, *i.e.* $\partial_\nu T^{\mu\nu} = 0$, and that in a curved space the ordinary derivative becomes the covariant derivative (here indicated by a semi-colon), *i.e.* Eq. (15.59) becomes

$$T^{\mu\nu}_{;\nu} = 0. \tag{15.77}$$

However, one has to be careful: The energy–momentum tensor is not always symmetric.

15.5 Einstein's Equation of the Gravitational Field

We shall first write down Einstein's equation as he did originally, and there-after derive it from an action integral in the usual manner. In the preceding sections we arrived at the following expressions (*cf.* Eqs. (15.42), (15.30), (15.17), (14.75)):

$$G_{\alpha\beta} = R_{\alpha\beta} - \frac{1}{2} g_{\alpha\beta} R, \quad \text{Ricci} - \text{Einstein tensor}, \tag{15.78}$$

$$R = g^{\mu\nu} R_{\mu\nu}, \quad \text{Ricci curvature scalar}, \tag{15.79}$$

$$R_{\mu\nu} \;\;=\;\; R^\alpha_{\ \mu\alpha\nu}, \quad \text{Ricci tensor},$$
$$\overset{(15.28)}{=} (\Gamma^\alpha_{\alpha\mu})_{,\nu} - (\Gamma^\alpha_{\mu\nu})_{,\alpha} + \Gamma^\alpha_{a\nu} \Gamma^a_{\alpha\mu} - \Gamma^\alpha_{a\alpha} \Gamma^a_{\mu\nu}, \tag{15.80}$$

$$\Gamma^{\nu}_{\alpha\beta}(x) = \frac{1}{2}g^{\nu\lambda}(x)\left(\frac{\partial g_{\lambda\beta}}{\partial x^{\alpha}} + \frac{\partial g_{\alpha\lambda}}{\partial x^{\beta}} - \frac{\partial g_{\alpha\beta}}{\partial x^{\lambda}}\right), \quad \text{Christoffel symbol.} \quad (15.81)$$

Thus, making these replacements in $G_{\alpha\beta}$ of Eq. (15.78) we see that apart from nonlinear combinations of $g_{\mu\nu}$ and $g_{\mu\nu,\lambda}$, the expression $G_{\alpha\beta}$ contains second derivatives of the metric tensor or the gravitational field potential $g_{\mu\nu}(x)$ linearly. We saw that in a Riemann space (*cf.* Eq. (15.77))

$$T^{\mu\nu}_{\ ;\nu} = 0, \qquad T^{\mu}_{\nu;\mu} = 0. \tag{15.82}$$

We recall from Eq. (15.31) that the Ricci–Einstein tensor was defined as the tensor which satisfies the equation

$$G^{\mu}_{\alpha;\mu} = 0, \tag{15.83}$$

i.e. has covariant divergence zero. It is therefore plausible to postulate on the basis of the last two equations that the equations of the gravitational field are of the form*

$$G^{\alpha}_{\beta} = -\kappa T^{\alpha}_{\beta}, \tag{15.84}$$

where κ is a coupling constant. This is *the equation postulated by Einstein* in 1915 after several years of search for a generalization of the Poisson equation

$$\nabla^2 V(\mathbf{x}) = 4\pi G \rho_m(\mathbf{x}). \tag{15.85}$$

In the particular case of a weak gravitational field independent of time (which implies therefore only a small deviation from the metric tensor of Minkowski space) the comparison with Poisson's equation implies in terms of Newton's gravitational constant G (see below)

$$\kappa = \frac{8\pi G}{c^2}.$$

We can rewrite *Einstein's equation*, Eq. (15.84), the central equation of general relativity theory, as[†]

$$R_{\mu\nu} - \frac{1}{2}g_{\mu\nu}R = -\kappa T_{\mu\nu}, \quad \kappa > 0, \tag{15.86}$$

where $R_{\mu\nu}$ contains derivatives of the second order of the metric tensor. The energy–momentum tensor, of course, describes the distribution of matter in

---

*Note that field equations like this must involve second order derivatives — like a wave equation — so that the relevant equation would not follow from Eq. (14.72), *i.e.* with $g^{\mu}_{\alpha;\mu} = 0$.

[†]According to F. Wilczek [52], p. 73, "Einstein referred to the left hand side as a palace of gold, and to the right hand side as a hovel of wood". It is only on the right hand side that masses of particles occur. Note the agreement with the dimensional argument in Sec. 10.6.

space and acts as a source of the gravitational field. We can interpret the equation therefore as saying: the curvature of space (expressed by the left hand side) results from matter on the right hand side, or that the gravitational interaction is identified as the bending of spacetime by matter.

Naturally one would like to obtain Einstein's equation from the variation of an action integral. This can be done. Consider first the action functional (15.69), $i.e.$

$$S = \int L(g_{\mu\nu}, \phi)\sqrt{-g(x)}d^4x. \tag{15.87}$$

Above we focussed attention only on the $g_{\mu\nu}$ part. Including the variations with respect to some field which is here $\phi(x)$, we obtain as in all other earlier cases and in particular also in electrodynamics the variation

$$\delta S = \int d^4x\sqrt{-g}\left[\frac{\partial L}{\partial\phi}\delta\phi + \frac{\partial L}{\partial(\partial_\mu\phi)}\delta(\partial_\mu\phi)\right] - \frac{1}{2}\int d^4x\sqrt{-g}T_{\mu\nu}\delta y^{\mu\nu}, \tag{15.88}$$

where we inserted for the matter-gravity part derived from L the expression in Eq. (15.70). Now let's add to the action the purely gravitational contribution. We shall see that the correct choice is (with κ' a constant)

$$S = \int d^4x\sqrt{-g}\left[-\frac{1}{16\pi\kappa'}R + L\right], \tag{15.89}$$

where the *Ricci scalar* R depends only on the gravitational field $g_{\mu\nu}$ and thus without L the action would represent that of a fictitious "free" (sourceless) gravitational field; this part of the action is known as *Hilbert action* or *Hilbert–Einstein action*. Since $\kappa' \propto G$, and G is a small number, the factor $1/G$ in the Hilbert–Einstein action is large. Here we concentrate only on the variation with respect to $g^{\mu\nu}$ and we shall find *the result*:

$$\delta S = \int d^4x\sqrt{-g}\left[-\frac{1}{16\pi\kappa'}\left(R_{\mu\nu} - \frac{1}{2}g_{\mu\nu}R\right) - \frac{1}{2}T_{\mu\nu}\right]\delta g^{\mu\nu} + \text{the rest.} \tag{15.90}$$

Thus we are interested in the variation

$$\delta(\sqrt{-g}R) = \delta(\sqrt{-g}g^{\mu\nu}R_{\mu\nu}) = \sqrt{-g}g^{\mu\nu}\delta R_{\mu\nu} + \sqrt{-g}R_{\mu\nu}\delta g^{\mu\nu} + R\delta\sqrt{-g}. \tag{15.91}$$

Here the second term is already of the required form. So we look at the others. We recall that the Ricci tensor is $R_{\mu\nu} = R^\alpha_{\mu\alpha\nu}$ and that ($cf.$ Eq. (15.4))

$$R^\rho_{\mu\lambda\nu} = \partial_\nu\Gamma^\rho_{\lambda\mu} + \Gamma^\rho_{\alpha\nu}\Gamma^\alpha_{\lambda\mu} - (\lambda \leftrightarrow \nu). \tag{15.92}$$

Thus we would first have to vary the *connection quantity* Γ with respect to the metric $g^{\mu\nu}$. However, there is a smarter way. Consider arbitrary variations of the connection quantity or Christoffel symbol Γ by the replacement

$$\Gamma^\rho_{\lambda\mu} \longrightarrow \Gamma^\rho_{\lambda\mu} + \delta\Gamma^\rho_{\lambda\mu}. \tag{15.93}$$

Here $\delta\Gamma^\rho{}_{\lambda\mu}$ is the difference of two connection quantities and is therefore itself a tensor, as is evident from Eq. (14.58) — the contribution which prevents Γ from being a tensor then cancels out! Since $\delta\Gamma$ is therefore a tensor, we can take its covariant derivative, *i.e.* as in Eq. (14.67c)

$$(\delta\Gamma^\rho{}_{\nu\mu})_{;\lambda} = \partial_\lambda(\delta\Gamma^\rho{}_{\nu\mu}) + \Gamma^\rho{}_{a\lambda}(\delta\Gamma^a{}_{\nu\mu}) - \Gamma^\sigma{}_{\nu\lambda}(\delta\Gamma^\rho{}_{\sigma\mu}) - \Gamma^\sigma{}_{\mu\lambda}(\delta\Gamma^\rho{}_{\nu\sigma}). \quad (15.94)$$

Here we consider with the help of this difference the following quantity (the underbraced contributions cancelling out):

$$-(\delta\Gamma^\rho{}_{\nu\mu})_{;\lambda} + (\delta\Gamma^\rho{}_{\lambda\mu})_{;\nu}$$

$$= -\partial_\lambda(\delta\Gamma^\rho{}_{\nu\mu}) - \Gamma^\rho{}_{a\lambda}(\delta\Gamma^a{}_{\nu\mu}) + \underbrace{\Gamma^\sigma{}_{\nu\lambda}(\delta\Gamma^\rho{}_{\sigma\mu})} + \Gamma^\sigma{}_{\mu\lambda}(\delta\Gamma^\rho{}_{\nu\sigma})$$

$$\quad + \partial_\nu(\delta\Gamma^\rho{}_{\lambda\mu}) + \Gamma^\rho{}_{a\nu}(\delta\Gamma^a{}_{\lambda\mu}) - \underbrace{\Gamma^\sigma{}_{\lambda\nu}(\delta\Gamma^\rho{}_{\sigma\mu})} - \Gamma^\sigma{}_{\mu\nu}(\delta\Gamma^\rho{}_{\lambda\sigma})$$

$$= \partial_\nu(\delta\Gamma^\rho{}_{\lambda\mu}) + \delta(\Gamma^\rho{}_{a\nu}\Gamma^a{}_{\lambda\mu}) - \partial_\lambda(\delta\Gamma^\rho{}_{\nu\mu}) - \delta(\Gamma^\rho{}_{a\lambda}\Gamma^a{}_{\nu\mu})$$

$$\overset{(15.92)}{=} \delta R^\rho{}_{\mu\lambda\nu}, \quad (15.95)$$

which is what we need. Therefore the contribution of the first term on the right hand side of Eq. (15.91) to the action integral is with $R_{\mu\nu} = R^\rho{}_{\mu\rho\nu}$ and using Leibnitz's rule applied to the covariant derivatives:

$$\int d^4x\sqrt{-g}\,g^{\mu\nu}\delta R_{\mu\nu} = \int d^4x\sqrt{-g}\,g^{\mu\nu}[(\delta\Gamma^\rho{}_{\rho\mu})_{;\nu} - (\delta\Gamma^\rho{}_{\nu\mu})_{;\rho}]$$

$$\overset{(14.72)}{=} \int d^4x\sqrt{-g}[g^{\mu\nu}(\delta\Gamma^\rho{}_{\rho\mu}) - g^{\mu\sigma}(\delta\Gamma^\nu{}_{\sigma\mu})]_{;\nu}, \quad (15.96)$$

where in the last step we used $g^{\mu\nu}{}_{;\kappa} = 0$. Now we have the integral with respect to the volume element of the covariant divergence of a vector. One can show that Stokes's theorem can be applied to covariant derivatives; we omit the proof here. Applying *Stokes's theorem* to the volume integral over a covariant derivative here, this integral is equal to a boundary contribution at infinity which we can set to zero by making the variation vanish at infinity. Therefore this term does not contribute to the total variation.

The remaining variation in Eq. (15.91) to be calculated is $\delta\sqrt{-g}$. We obtain this from

$$\delta\sqrt{-g} = \frac{\partial\sqrt{-g}}{\partial g^{\mu\nu}}\delta g^{\mu\nu} = -\frac{1}{2\sqrt{-g}}\frac{\partial g}{\partial g^{\mu\nu}}\delta g^{\mu\nu}$$

$$\overset{(14.98)}{=} \frac{1}{2\sqrt{-g}}gg_{\mu\nu}\delta g^{\mu\nu} = -\frac{\sqrt{-g}}{2}g_{\mu\nu}\delta g^{\mu\nu}. \quad (15.97)$$

Thus finally we have (*cf.* Eq. (15.89))[‡]

$$\delta S = \int d^4x\sqrt{-g}\left[\left\{-\frac{1}{16\pi\kappa'}\left(R_{\mu\nu}-\frac{1}{2}g_{\mu\nu}R\right)-\frac{1}{2}T_{\mu\nu}\right\}\delta g^{\mu\nu}\right]+\cdots. \quad (15.98)$$

Independence of the elements of $\delta g^{\mu\nu}$ implies the equation of motion

$$R_{\mu\nu}-\frac{1}{2}g_{\mu\nu}R = -8\pi\kappa'T_{\mu\nu}, \quad (15.99)$$

i.e. Einstein's equation in agreement with Eq. (15.86) with $8\pi\kappa' = \kappa$.

15.6 Newton's Potential from Einstein's Equation

We now proceed to show that considerations of Einstein's equation for a weak field $g_{\mu\nu}$ lead to the Poisson equation. We start from Eq. (15.86):

$$R_{\mu\nu}-\frac{1}{2}g_{\mu\nu}R = -\kappa T_{\mu\nu},$$

and we multiply this equation by $g^{\mu\nu}$. Then, since with double summation $g_{\mu\nu}g^{\mu\nu} = \delta^\mu_\mu = \text{trace}(\mathbb{1}_{4\times4}) = 4$, we obtain

$$R - 2R = -\kappa T, \quad i.e. \quad R = \kappa T, \quad (15.100)$$

where

$$T = g^{\mu\nu}T_{\mu\nu} = T^\nu_\nu = \text{trace of } T_{\mu\nu}. \quad (15.101)$$

In electrodynamics the contracted $T = 0$, as we show below. Inserting the result (15.100) back into Eq. (15.86), we have

$$R_{\mu\nu}-\frac{1}{2}g_{\mu\nu}\kappa T = -\kappa T_{\mu\nu},$$

or

$$R_{\mu\nu} = -\kappa\left(T_{\mu\nu}-\frac{1}{2}g_{\mu\nu}T\right). \quad (15.102)$$

This equation is equivalent to Eq. (15.86) and demonstrates that $T_{\mu\nu}$ is the source of the gravitational field. For $T_{\mu\nu} = 0$ one has a space without a source and the equation becomes *Einstein's gravitational law*

$$R_{\mu\nu} = 0. \quad (15.103)$$

[‡]Observe that if we replace in Eq. (15.89) L by $L - 2\tilde{\Lambda}$, where $\tilde{\Lambda}$ is a constant, then in view of Eq. (15.97) the term $T_{\mu\nu}/2$ in Eq. (15.98) would be replaced by $T_{\mu\nu}/2 - \tilde{\Lambda}g_{\mu\nu}$, and would imply an additional term $\Lambda g_{\mu\nu}$ on the left hand side of Eq. (15.99) ($\Lambda = \text{const.}\tilde{\Lambda}$). This is the case of a cosmological constant and is discussed in Secs. 16.1 and 16.4.

A space with this property is said to be *Ricci flat*. In case the source is electromagnetic radiation in the absence of charges (the "free" electromagnetic field) the energy–momentum tensor is given by Eq. (15.76), *i.e.*

$$T_{\mu\nu} = \left(F_{\mu m} F^{m}_{\nu} + \frac{1}{4} g_{\mu\nu} F_{mn} F^{mn} \right). \tag{15.104}$$

We observe that the trace of $T_{\mu\nu}$ is zero, *i.e.*

$$T = T^{\mu}_{\mu} = g^{\mu\nu} T_{\mu\nu} = \left(\underbrace{F_{\mu m} F^{m\mu}}_{-F_{m\mu}} + \frac{1}{4} \underbrace{g_{\mu\nu} g^{\mu\nu}}_{4} F_{mn} F^{mn} \right) = 0. \tag{15.105}$$

In this case Eq. (15.102) becomes

$$R_{\mu\nu} = -\kappa T_{\mu\nu} \quad \text{(electromagnetism)}. \tag{15.106}$$

This equation constitutes the coupled *Einstein–Maxwell field equation*, or equations if the components are considered. These are differential equations of the second order in the components $g_{\mu\nu}$ and A_{μ} with a lot of coupling terms, including self-coupling ones. In the following we ignore the electromagnetic field and concentrate solely on the gravitational part. Thus we make a simplified ansatz for the relevant part of $T_{\mu\nu}$ in the Einstein equation (15.102).

Our aim is to extract Newton's gravitational potential from Einstein's equation. With this aim we consider an ansatz for $T_{\mu\nu}$ depending on the deviation of $g_{\mu\nu}$ from its flat Minkowski form. More precisely, we consider now the approximation of a *weak and time-independent gravitational field*, and retain only the dominant contribution for large values of c. Then, as earlier in Sec. 13.3, we expand the metric field around the flat space metric and write[§]

$$g_{\mu\nu}(x) = \tilde{g}_{\mu\nu}(x) + h_{\mu\nu}(x). \tag{15.107}$$

We saw previously that in this case of $\beta = v/c$ small (*cf.* Eq. (13.25))

$$g_{00}(x) = 1 + \frac{2}{c^2} \phi(x), \quad g_{0k} \simeq 0. \tag{15.108}$$

We consider this case now, *i.e.* for the source ansatz $T_{00} = \rho(x), T_{ij} = 0$, and therefore with trace of $T_{\mu\nu} = \rho(x)$ (weak field approximation). We recall

---

[§]For a non-mathematical discussion of this substitution see F. Wilczek [52], p. 355. Expanding the Einstein–Hilbert action (containg the factor $1/G$) around the flat space metric, one finds that the terms quadratic in an expansion parameter $\sqrt{G}$ attached to $h_{\mu\nu}$ give a properly normalized spin-2 *graviton field* $h_{\mu\nu}$.

from Eq. (15.58) that the 00-component of the energy momentum tensor represents an energy density. Thus, via Einstein's formula $E = mc^2$, this energy density, whatever its origin, also represents a mass density, and this is what we are considering now. The quantity $\phi(\mathbf{x})$ will turn out to be *Newton's gravitational potential*. In our approximation $\tilde{g}^{\mu\nu}(x)$ is the Minkowski metric with diagonal elements $+1, -1, -1, -1$. Thus our aim is to obtain from Eq. (15.102) the relation between the source $\rho(\mathbf{x})$ and the potential $\phi(\mathbf{x})$, this potential in Eq. (15.108) providing a deviation from the flat Minkowski metric.

We obtain from Eq. (15.102) with $T_{00} = \rho, T = \rho$ (weak case, $T_{ij} = 0$):

$$R_{00} \simeq -\kappa\left(\rho - \frac{1}{2}\rho\right) = -\frac{1}{2}\kappa\rho(\mathbf{x}), \quad R_{0k} = 0,$$

$$R_{jk} = \frac{1}{2}\kappa\tilde{g}_{jk}\rho(\mathbf{x}), \quad (T_{jk} = 0). \tag{15.109}$$

Evaluating R_{00} for the case of Eq. (15.108) we begin with the Christoffel symbols, Eq. (15.81), *i.e.*

$$\Gamma^\nu_{\alpha\beta}(x) = \frac{1}{2}g^{\nu\lambda}(x)\left(\frac{\partial g_{\lambda\beta}}{\partial x^\alpha} + \frac{\partial g_{\alpha\lambda}}{\partial x^\beta} - \frac{\partial g_{\alpha\beta}}{\partial x^\lambda}\right).$$

From this equation we obtain by inserting appropriate values:

$$\Gamma^\alpha_{\alpha 0} = \frac{1}{2}g^{\alpha\lambda}[\partial_\alpha g_{\lambda 0} + \underbrace{\partial_0 g_{\lambda\alpha}}_{0} - \partial_\lambda g_{\alpha 0}] = \frac{1}{2}(\partial^\lambda g_{\lambda 0} - \partial^\alpha g_{\alpha 0}) = 0,$$

$$\Gamma^\alpha_{0\nu} = \frac{1}{2}g^{\alpha\lambda}[\underbrace{\partial_0 g_{\lambda\nu}}_{0} + \partial_\nu g_{\lambda 0} - \partial_\lambda g_{0\nu}] \overset{g_{0k}\simeq 0}{=} \frac{1}{2}[g^{\alpha 0}\partial_\nu g_{00} - \partial^\alpha g_{0\nu}],$$

$$\Gamma^\alpha_{a0} = \frac{1}{2}g^{\alpha\lambda}(\partial_a g_{\lambda 0} + \partial_0 g_{a\lambda} - \partial_\lambda g_{a0}) = \frac{1}{2}[g^{\alpha 0}\partial_a g_{00} - \partial^\alpha g_{a0}],$$

$$\Gamma^\mu_{00} = \frac{1}{2}g^{\mu\lambda}[\underbrace{\partial_0 g_{\lambda 0}}_{0} + \underbrace{\partial_0 g_{\lambda 0}}_{0} - \partial_\lambda g_{00}] = -\frac{1}{2}\partial^\mu g_{00}. \tag{15.110}$$

Now, the *Ricci tensor* we require is given by

$$R_{\mu\nu} \overset{(15.80)}{=} (\Gamma^\alpha_{\alpha\mu})_{,\nu} - (\Gamma^\alpha_{\mu\nu})_{,\alpha} + \Gamma^\alpha_{\alpha\nu}\Gamma^\alpha_{\alpha\mu} - \Gamma^\alpha_{\alpha\alpha}\Gamma^\alpha_{\mu\nu}, \tag{15.111}$$

so that (again time-dependent derivatives are zero)

$$R_{00} = \underbrace{(\Gamma^\alpha_{\alpha 0})_{,0}}_{0} - (\Gamma^\alpha_{00})_{,\alpha} + \Gamma^\alpha_{a0}\Gamma^a_{\alpha 0} - \Gamma^\alpha_{a\alpha}\Gamma^a_{00},$$

and therefore

$$
\begin{aligned}
R_{00} &= -\left(-\frac{1}{2}\partial^\alpha g_{00}\right)_{,\alpha} + \frac{1}{2}[g^{a0}\partial_a g_{00} - \partial^\alpha g_{a0}]\frac{1}{2}[g^{a0}\partial_\alpha g_{00} - \partial^a g_{0\alpha}] \\
&\quad -\Gamma^\alpha_{a\alpha}\left(-\frac{1}{2}\partial^a g_{00}\right) \\
&= \frac{1}{2}\partial^\alpha\partial_\alpha g_{00} + \frac{1}{4}[g^{a0}(g^{a0}\partial_a g_{00})\partial_\alpha g_{00} + \partial^\alpha g_{a0}\partial^a g_{0\alpha} \\
&\quad \underbrace{-\partial_a g_{00}g^{a0}\partial^a g_{0\alpha}}_{--} \underbrace{-\partial^\alpha g_{a0}g^{a0}\partial_\alpha g_{00}}_{++}] + \frac{1}{2}(\partial^a g_{00})\Gamma^\alpha_{a\alpha}.
\end{aligned}
$$

With $\partial^\alpha\partial_\alpha \to -\nabla^2$ this is (note that again g_{0k} is of nonleading magnitude and g_{00} is independent of t)

$$
\begin{aligned}
R_{00} &= -\frac{1}{2}\nabla^2 g_{00} + \frac{1}{4}[-\underbrace{(g^{a0}\partial_a g_{00})(\partial^a g_{0\alpha})}_{--} - \underbrace{(\partial^\alpha g_{a0})(g^{a0}\partial_\alpha g_{00})}_{++} \\
&\quad + \underbrace{(\partial^0\cdots)}_{0}] + \frac{1}{2}\partial^a g_{00}\Gamma^\alpha_{a\alpha} + \cdots .
\end{aligned}
$$

In the underbraced terms we set $g^{a0} \to g^{00} \simeq 1$ and $g_{k0} \simeq 0$, so that these terms yield in this leading order the same contribution, and

$$
\begin{aligned}
R_{00} &= -\frac{1}{2}\nabla^2 g_{00} - \frac{1}{2}(\partial_\alpha g_{00})(\partial^\alpha g_{00}) \\
&\quad + \frac{1}{2}\partial^a g_{00}\frac{1}{2}g^{\alpha\lambda}(\partial_a g_{\lambda\alpha} + \partial_\alpha g_{a\lambda} - \partial_\lambda g_{a\lambda})_{\lambda\alpha\to00,a\lambda\to00,a\alpha\to00} + \cdots
\end{aligned}
$$

or

$$
\begin{aligned}
R_{00} &= -\frac{1}{2}\nabla^2 g_{00} - \frac{1}{2}(\partial_a g_{00})(\partial^a g_{00}) + \frac{1}{4}(\partial^a g_{00})(\partial_a g_{00}) + \cdots \\
&= -\frac{1}{2}\nabla^2 g_{00} - \frac{1}{4}(\partial_a g_{00})(\partial^a g_{00}) + \cdots ,
\end{aligned}
$$

and thus

$$
\begin{aligned}
R_{00} &\simeq -\frac{1}{2}\nabla^2 g_{00} - \frac{1}{4}\partial_a\left(1 + \frac{2}{c^2}\phi(\mathbf{x})\right)\partial^a\left(1 + \frac{2}{c^2}\phi(\mathbf{x})\right) + \cdots \\
&\overset{(15.108)}{=} -\frac{1}{c^2}\nabla^2\phi(\mathbf{x}) + O\left(\frac{1}{c^4},\cdots\right). \tag{15.112}
\end{aligned}
$$

Hence from Eq. (15.109) for R_{00}:

$$
\frac{1}{c^2}\nabla^2\phi(\mathbf{x}) = \frac{1}{2}\kappa\rho(\mathbf{x}), \qquad \nabla^2\phi(\mathbf{x}) = \frac{1}{2}c^2\kappa\rho(\mathbf{x}). \tag{15.113}
$$

We compare this equation with the *Poisson equation* for the Newtonian potential — this is solved in Example 15.6 — *i.e.*

$$\triangle\phi(\mathbf{x}) \equiv \nabla^2\phi(\mathbf{x}) = 4\pi G\rho_m(\mathbf{x}), \tag{15.114}$$

where $\rho_m(\mathbf{x})$ is the *mass density*, $|\mathbf{x}| = r$, and

$$\phi(\mathbf{x}) = -\frac{GM}{r},$$

where M is the mass of the Earth and G is *Newton's gravitational constant*. Hence, by comparison of Eqs. (15.113) and (15.114), as stated earlier,

$$\kappa = \frac{8\pi G}{c^2}. \tag{15.115}$$

Thus *Newton's gravitational potential* $\phi(\mathbf{x})$ appears as a non-Minkowskian contribution in the component g_{00} of the metric or gravitational field. In order to appreciate the significance of this result recall from Eq. (15.108) that $g_{00}(x) = 1 + (2/c^2)\phi(\mathbf{x})$. The deviation from 1 provided by Newton's potential $\phi(\mathbf{x})$ implies in the metric a deviation of the velocity of light c from its vacuum value of c^2 to $c'^2 = c^2[1+(2/c^2)\phi(\mathbf{x})]$ as conjectured in Sec. 11.3.2. Here c is — as in the first postulate of the Special Theory, *cf.* Sec. 12.2 — the velocity of light with the same value c in all directions in all inertial frames. Thus, on entering a gravitational region, gravity acts on a ray of light like a dielectric medium which the ray travels into from a vacuum.

Example 15.6: Solving the Poisson equation
Show that the solution of the Poisson equation (with $\mathbf{r} = \mathbf{x}$)

$$\triangle\phi(\mathbf{r}) = 4\pi G\rho_m(\mathbf{r}) \tag{15.116}$$

for a point mass M with $\rho(\mathbf{r}) \to \rho_m(\mathbf{r}) = M\delta(\mathbf{r})$ is given by

$$\phi(\mathbf{r}) = -G\frac{M}{|\mathbf{r}|}. \tag{15.117}$$

Solution: The standard method to solve an equation like Eq. (15.116) is with the help of the *Green's function* $\mathcal{G}(\mathbf{r})$ defined as an inhomogeneous solution of the equation

$$\triangle\mathcal{G}(\mathbf{r}) = \delta(\mathbf{r}), \qquad \delta(\mathbf{r}) = \begin{cases} 0 & \text{for} \quad \mathbf{r} \neq 0, \\ 1 & \text{for} \quad \mathbf{r} = 0. \end{cases} \tag{15.118}$$

The quantity $\delta(\mathbf{r})$ is known as the *Dirac delta function* (or distribution) and represents, effectively, the continuum version of the discrete Kronecker delta.[¶] Main properties of the delta function are as a consequence of its definition the relations

$$\int d\mathbf{r}\delta(\mathbf{r}) = 1, \qquad \int d\mathbf{r}'\delta(\mathbf{r} - \mathbf{r}')f(\mathbf{r}') = f(\mathbf{r}). \tag{15.119}$$

[¶]For simple introductions see in particular books on electrodynamics, *e.g.* H. J. W. Müller–Kirsten [34], p. 15.

In terms of the Green's function $\mathcal{G}(\mathbf{r})$ the solution $\phi(\mathbf{r})$ of the Poisson equation is given by, as we verify,

$$\phi(\mathbf{r}) = \int \mathcal{G}(\mathbf{r} - \mathbf{r}')4\pi G\rho_m(\mathbf{r}')d\mathbf{r}'. \tag{15.120}$$

Thus, applying the Laplacian $\triangle$ to this expression and using Eq. (15.119), we regain Eq. (15.117), i.e.

$$\triangle\phi(\mathbf{r}) = \int \triangle\mathcal{G}(\mathbf{r} - \mathbf{r}')4\pi G\rho_m(\mathbf{r}')d\mathbf{r}' = \int \delta(\mathbf{r} - \mathbf{r}')4\pi G\rho_m(\mathbf{r}')d\mathbf{r}' = 4\pi G\rho_m(\mathbf{r}).$$

In order to obtain the explicit form of the function $\phi(\mathbf{r})$ which will turn out to be *Newton's gravitational potential*, we require the explicit form of the Green's function $\mathcal{G}(\mathbf{r})$. We obtain this in two steps. In our first step we show that

$$\text{for } \mathbf{r} \neq 0: \qquad \triangle\frac{1}{|\mathbf{r}|} = 0. \tag{15.121}$$

This result is verified by the following equations using $r = \sqrt{x^2 + y^2 + z^2}, \partial r/\partial x = x/r$, and similarly for y and z:

$$\triangle = \frac{\partial^2}{\partial x^2} + \frac{\partial^2}{\partial y^2} + \frac{\partial^2}{\partial z^2}, \qquad \frac{\partial}{\partial x}\left(\frac{1}{r}\right) = -\frac{x}{r^3}, \qquad \frac{\partial^2}{\partial x^2}\left(\frac{1}{r}\right) = -\frac{1}{r^3} + \frac{3x^2}{r^5},$$

$$\therefore \quad \triangle\left(\frac{1}{r}\right) = -\frac{3}{r^3} + \frac{3r^2}{r^5} = 0.$$

In the second step we integrate and use *Stokes' theorem*. Thus

$$\int_V \triangle\frac{1}{r}d\mathbf{r} = \int_V \boldsymbol{\nabla}\cdot\boldsymbol{\nabla}\frac{1}{r}d\mathbf{r} = \int_{F(V)}\left(\boldsymbol{\nabla}\frac{1}{r}\right)\cdot d\mathbf{F} = -\int_F \frac{\mathbf{r}}{r^3}\cdot d\mathbf{F} = -\int_F \frac{dF}{r^2} = -\int d\Omega = -4\pi. \tag{15.122}$$

It follows therefore that we can write

$$\triangle\frac{1}{|\mathbf{r}|} = -4\pi\delta(\mathbf{r}), \quad \text{since} \quad \int_V \triangle\frac{1}{|\mathbf{r}|}d\mathbf{r} = -4\pi\int_V \delta(\mathbf{r})d\mathbf{r} = -4\pi. \tag{15.123}$$

It also follows from this Eq. (15.123) and Eq. (15.118) that the Green's function $\mathcal{G}$ is given by

$$\mathcal{G}(\mathbf{r} - \mathbf{r}') = -\frac{1}{4\pi|\mathbf{r} - \mathbf{r}'|}, \tag{15.124}$$

since

$$\triangle\mathcal{G}(\mathbf{r} - \mathbf{r}') = -\frac{1}{4\pi}\triangle\frac{1}{|\mathbf{r} - \mathbf{r}'|} = \delta(\mathbf{r} - \mathbf{r}').$$

We obtain therefore as solution of the Poisson equation from Eq. (15.120) the expression

$$\phi(\mathbf{r}) = \int \mathcal{G}(\mathbf{r} - \mathbf{r}')4\pi G\rho_m(\mathbf{r}')d\mathbf{r}' = -G\int \frac{\rho_m(\mathbf{r}')d\mathbf{r}'}{|\mathbf{r} - \mathbf{r}'|}. \tag{15.125}$$

For a point mass M at $\mathbf{r}'$ the mass density is $\rho_m(\mathbf{r}') = M\delta(\mathbf{r}')$, so that at a distance $\mathbf{r}$ away from it the function $\phi(\mathbf{r})$ is

$$\phi(\mathbf{r}) = -G\int \frac{M\delta(\mathbf{r}')d\mathbf{r}'}{|\mathbf{r} - \mathbf{r}'|} = -G\frac{M}{r}. \tag{15.126}$$

We see that this function is *Newton's gravitational potential* at a point which is a distance r away from a point mass M.

Chapter 16

The Schwarzschild Solution

16.1   Introductory Remarks

With Eq. (15.86) or (15.99) we now have *Einstein's equation*. The purely
gravitational part of the action integral, *i.e.* that with Lagrangian density
$L_H - \sqrt{-g}R$, is known as the *Hilbert action* or *Einstein–Hilbert action*, since
it was first considered by Hilbert. One can think of extensions of the above
action. Here we concentrate on one. In searching for the simplest possible
action for gravity, one realzes that any nontrivial scalar quantity has to be
at least of second order in derivatives of the metric like the *Ricci curvature
scalar R*; all one can do at lower order is to use a constant Λ. By itself the
constant does not lead to interesting dynamics, but when added to Hilbert's
action, it does. Thus if one considers

$$S = \int d^4 x \sqrt{-g}(R - 2\Lambda), \qquad (16.1)$$

the resulting field equations are, as we see from Eqs. (15.98), (15.99) (imagine
R there replaced by $R - 2\Lambda$),

$$R_{\mu\nu} - \frac{1}{2} R g_{\mu\nu} + \Lambda g_{\mu\nu} = 0. \qquad (16.2)$$

If the action of matter were included, there would be an energy-momentum
tensor on the right as in Eq. (15.99), *i.e.*

$$R_{\mu\nu} - \frac{1}{2} R g_{\mu\nu} + \Lambda g_{\mu\nu} = -\kappa T_{\mu\nu}. \qquad (16.3)$$

The constant Λ is known as the *cosmological constant* (which can be argued
to be proportional to the inverse of the square of the radius of the universe)

and plays an important role in cosmology.* It was introduced by Einstein after he had realized that his equation with matter content did not allow solutions representing a static universe (*i.e.* one which does not change for large times). However, after Hubble had demonstrated that the universe is expanding, the interest in static solutions faded and Einstein rejected his own suggestion. Thus General Relativity involves just two parameters: Newton's gravitational constant and the cosmological term. The latter, which had puzzled even Einstein, can be said to parameterize the energy density of empty space, but Wilczek [52] (p. 135) remarks:"The gaping hole in our understanding of Nature is the notorious problem of the cosmological term". For discussions of these points we refer in particular to Wilczek [52] and Carroll [11].

The first equation considered by Einstein was the one with the energy-momentum tensor set to zero, *i.e.*

$$T^{\mu\nu}_{;\nu} = 0, \tag{16.4}$$

so that (also ignoring Λ)

$$G^{\mu\nu}_{;\nu} \equiv \left(R^{\mu\nu} - \frac{1}{2} g^{\mu\nu} R \right)_{;\nu} = 0. \tag{16.5}$$

If $\Lambda g_{\mu\nu}$ were included in this case, its part disappears since (*cf.* Eq. (14.72))

$$g_{\mu\nu;\kappa} = 0.$$

In the following we consider first ds^2 in the case of spherical coordinates — and we shall finally obtain the result we cited and began with in Example 3.13. This result is the solution of Einstein's equation in spherical coordinates and is known as the *Schwarzschild solution*; this solution is the central topic of this chapter. For the solution given below we used the presentation of Leite–Lopes [28] as guideline. Thereafter we consider the inclusion of a cosmological term. Finally we consider the spectacular applications to the Kepler problem and the deflection of a ray of light in the field of a huge massive body like the sun. We shall not deal here at length with the Kepler problem as this application has been considered in detail in Chapter 7. It is evident that if matter fields are to be included, e.g. the electromagnetic field, one has to do this by taking into account the appropriate energy-momentum tensor in Eq. (16.3). For a static electric charge in the presence of gravity, the Einstein equation is considered in Appendix B.

---

*For a readable introduction and recent text see M. Carmeli [10], which, however, deals mainly with the variant called Cosmological Relativity.

16.2 The Spherical Solution Outside the Source

We assume there is a source like the sun which produces the gravitational field. More precisely we wish to consider the (vacuum) gravitational field produced by a spherical (nonrotating) mass M which we call the Sun (in fact we shall identify it as that by comparison with Newton's potential at a relevant stage; cf. Eq. (16.58b)). We shall encounter the Schwarzschild radius $r_S = 2GM/c^2$ which we identify with the radius of the interior of the body where the equations (or Schwarzschild coordinates) are invalid. The spherical radius r_S is called "*event horizon of the spherical black hole*".

We consider now the solution of Einstein's equations outside the source, and beyond the range of other fields, these being assumed to be of sufficiently short range so that all source effects are effectively zero and hence we can take $T_{\mu\nu} = 0$. Then, since $T = g^{\mu\nu}T_{\mu\nu}$ we also have $T = 0$ and hence (cf. Eq. (15.100)) $R = \kappa T = 0$. Thus Einstein's equation reduces to the purely gravitational equation

$$R_{\mu\nu} = 0. \tag{16.6}$$

This equation may be considered as the generalization of Newton's law of gravitation. Explicitly Eq. (16.6) implies (recall Eq. (15.80)):

$$(\Gamma^\alpha_{\alpha\nu})_{,\mu} - (\Gamma^\alpha_{\mu\nu})_{,\alpha} + \Gamma^\alpha_{\eta\mu}\Gamma^\eta_{\alpha\nu} - \Gamma^\alpha_{\eta\alpha}\Gamma^\eta_{\mu\nu} = 0. \tag{16.7}$$

This looks scary but can be handled, as we shall see. The *Schwarzschild solution* we shall consider is independent of time, is spherically symmetric, and requires the asymptotic behaviour:

$$g_{\mu\nu}(x) \overset{r\to\infty}{\longrightarrow} \overset{\circ}{g}_{\mu\nu} \quad \text{Lorentz}, \tag{16.8}$$

i.e. at infinity spacetime is flat. We have

$$ds^2 = g_{\mu\nu}(x)dx^\mu dx^\nu. \tag{16.9}$$

In spherical coordinates with

$$x_1 = r\cos\phi\sin\theta, \quad x_2 = r\sin\phi\sin\theta, \quad x_3 = r\cos\theta,$$

this means ds^2 involves $dt, dr, d\theta$ and $d\phi$, and we set

$$\begin{aligned} ds^2 = \ & Ac^2 dt^2 - (Bdr^2 + Cr^2 d\theta^2 + Dr^2\sin^2\theta\, d\phi^2) \\ & + adrd\theta + bd\theta d\phi + cdrd\phi + g_{0k}dx^0 dx^k, \end{aligned} \tag{16.10}$$

where A, B, C, D, a, b, c and g_{0k} are coefficients which have to be determined. At $r \to \infty$ we must have the Minkowski metric as in Example 13.3 (there with signature $-, +, +, +$).

$$ds^2|_\infty = c^2 dt^2 - (dr^2 + r^2 d\theta^2 + r^2\sin^2\theta d\phi^2). \tag{16.11}$$

Here r plays the role of the radial measure (actually only beyond r_S) and θ and ϕ are the usual polar and azimuthal angles. Next we exploit symmetry properties in order to remove some terms from Eq. (16.10). Thus if ds^2 is to be independent of t, then ds^2 does not change under the exchange $dx^0 \longrightarrow -dx^0$ (called inversion of time). Thus

$$g_{0k} = 0. \tag{16.12}$$

For spherical symmetry ds^2 is invariant under

$$d\phi \longrightarrow -d\phi, \quad \text{and} \quad d\theta \longrightarrow -d\theta,$$

so that

$$adrd\theta = 0, \quad bd\theta d\phi = 0, \quad cdrd\theta = 0. \tag{16.13}$$

Hence

$$ds^2 = Ac^2dt^2 - (Bdr^2 + Cr^2d\theta^2 + Dr^2\sin^2\theta d\phi^2), \tag{16.14}$$

where A, B, C, D are functions of r only. Spherical symmetry also implies invariance under the exchange $rd\theta \leftrightarrow r\sin\theta d\phi$ (or a shift of the north pole ($\theta = 0$) to the equator ($\theta = \pi/2$) which interchanges these infinitesimal lengths of the element of area on a sphere of radius r) implying $C = D$. Hence

$$ds^2 = Ac^2dt^2 - Bdr^2 - Cr^2(d\theta^2 + \sin^2\theta d\phi^2). \tag{16.15}$$

One can choose a coordinate system such that $C(r) = 1$. We can see this as follows. We introduce

$$\tilde{r} = \sqrt{C(r)}r, \quad C(r)r^2 = \tilde{r}^2, \tag{16.16}$$

so that by differentiation with respect to r:

$$2C(r)rdr + r^2C'(r)dr = 2\tilde{r}d\tilde{r}, \quad 2C(r)rdr\left[1 + \frac{r}{2C(r)}C'(r)\right] = 2\tilde{r}d\tilde{r},$$

or

$$C^2(r)r^2(dr)^2\left[1 + \frac{r}{2C(r)}C'(r)\right]^2 = \tilde{r}^2(d\tilde{r})^2 = C(r)r^2(d\tilde{r})^2. \tag{16.17}$$

This equation can be rewritten as (using $\tilde{r} = \sqrt{C(r)}r$)

$$(dr)^2 = \frac{1}{C(r)}\left[1 + \frac{rC'(r)}{2C(r)}\right]^{-2}(d\tilde{r})^2. \tag{16.18}$$

Multiplying by $B(r)$ we define a function $\tilde{B}(r)$:

$$B(r)(dr)^2 = \frac{B(r)}{C(r)}\left[1 + \frac{rC'(r)}{2C(r)}\right]^{-2}(d\tilde{r})^2 \equiv \tilde{B}(r)(d\tilde{r})^2. \tag{16.19}$$

Comparing this relation with Eq. (16.15), we can write (using $\tilde{r} = \sqrt{C(r)}r$)

$$ds^2 = Ac^2dt^2 - \tilde{B}(d\tilde{r})^2 - \tilde{r}^2(d\theta^2 + \sin^2\theta d\phi^2), \tag{16.20}$$

which, by comparison, is like choosing $C(r) = 1$ in Eq. (16.15). Thus we take

$$ds^2 = Ac^2dt^2 - Bdr^2 - r^2(d\theta^2 + \sin^2\theta d\phi^2). \tag{16.21}$$

16.3 The Schwarzschild Solution for $\Lambda = 0$

We consider first the case of a vanishing cosmological constant, *i.e.* $\Lambda = 0$. We now set

$$A(r) = e^{\nu(r)}, \quad B(r) = e^{\lambda(r)}. \tag{16.22}$$

Since the metric is to approach that of a flat space at $r \to \infty$, we expect (although this is here not really required to obtain the solution as emphasized by Carmeli[†])

$$ds^2 \xrightarrow{r\to\infty} ds^2_\infty,$$

and hence

$$A(r) \text{ and } B(r) \longrightarrow 1 \text{ for } r \to \infty, \quad \nu(r) \text{ and } \lambda(r) \longrightarrow 0 \text{ for } r \to \infty.$$

Thus with these conditions

$$ds^2 = e^{\nu(r)}c^2dt^2 - e^{\lambda(r)}dr^2 - r^2(d\theta^2 + \sin^2\theta d\phi^2). \tag{16.23}$$

We determine $\exp[\nu(r)]$ and $\exp[\lambda(r)]$ by solving — for the particular case we are considering — the gravitational equation without the source, Eq. (16.6), *i.e.*

$$R_{\mu\nu} = 0.$$

Thus we have to determine all the Christoffel symbols in this equation, altogether 40, since $\Gamma^\alpha_{\mu\nu}$ is symmetric in μ, ν, implying 6 (from the off-diagonal elements) + 4 (from diagonal elements) =10 different combinations multiplied by 4 for the 4 different values of α. Instead of evaluating these, we use a trick. We obtained earlier, *cf.* Eq. (13.13), the *equation of a geodesic, i.e.*

[†]M. Carmeli [10], p. 50.

the equation of the most direct curve between two points in as yet arbitrary coordinates, namely

$$\ddot{x}^\alpha + \Gamma^\alpha_{\beta\eta}\dot{x}^\beta\dot{x}^\eta = 0, \quad \dot{x}^\eta = \frac{dx^\eta}{ds}. \tag{16.24}$$

These equations involve all the Christoffel symbols. The variational principle yielding the geodesic is

$$\delta \int ds L(x^\alpha, \dot{x}^\alpha) = 0, \quad \dot{x}^\alpha = \frac{dx^\alpha}{ds},$$

where in the case of interest here we have to set $x^\alpha = (x^0, x^1 = r, x^2 = \theta, x^3 = \phi)$. Thus the variational principle boils down to

$$\delta \int ds L(x^0, r, \theta, \phi; \dot{x}^0, \dot{r}, \dot{\theta}, \dot{\phi}) = 0,$$

with the Lagrangian (Eq. (16.23) divided by ds^2, therefore of value 1)

$$L = e^{\nu(r)}(\dot{x}^0)^2 - (e^{\lambda(r)}\dot{r}^2 + r^2\dot{\theta}^2 + r^2\sin^2\theta\dot{\phi}^2), \tag{16.25}$$

and the Euler–Lagrange equations

$$\frac{d}{ds}\left(\frac{\partial L}{\partial \dot{x}^\alpha}\right) - \frac{\partial L}{\partial x^\alpha} = 0. \tag{16.26}$$

We leave the explicit derivation of the various equations and the evaluation of the Christoffel symbols involved to Example 16.1.

Example 16.1: Deriving Christoffel symbols from the geodesic equation

From the Lagrangian (16.25) and with the identification $x^\alpha \to x^0, x^1 = r, x^2 = \theta, x^3 = \phi$, derive the various Euler–Lagrange equations, and by comparison with the equation of the geodesic, Eq. (16.24), extract the expressions for the Christoffel symbols.

Solution:
(i) For $x^\alpha \equiv x^0$ one obtains the equation (Eq. (16.26) with L of Eq. (16.25)):

$$\frac{d}{ds}(2e^{\nu(r)}\dot{x}^0) = 0, \quad i.e. \quad \ddot{x}^0 + \nu'(r)\dot{r}\dot{x}^0 = 0. \tag{16.27}$$

We compare this equation with the corresponding component of the equation of the geodesic, Eq. (16.24):

$$\ddot{x}^0 + \Gamma^0_{\beta\eta}\dot{x}^\beta\dot{x}^\eta = 0, \tag{16.28}$$

where we can choose (since particular coordinates had not been chosen) $x^1 = r, x^2 = \theta, x^3 = \phi$, and hence obtain

$$\ddot{x}^0 + \Gamma^0_{00}(\dot{x}^0)^2 + \Gamma^0_{01}\dot{x}^0\underbrace{\dot{x}^1}_{\dot{r}} + \Gamma^0_{10}\underbrace{\dot{x}^1}_{\dot{r}}\dot{x}^0 + 2\Gamma^0_{20}\dot{x}^0\underbrace{\dot{x}^2}_{\dot{\theta}} + 2\Gamma^0_{30}\underbrace{\dot{x}^3}_{\dot{\phi}}\dot{x}^0 + 2\Gamma^0_{ik}\dot{x}^i\dot{x}^k = 0. \tag{16.29}$$

Thus by comparison with Eq. (16.27) we obtain

$$\Gamma^0_{01} = \Gamma^0_{10} = \frac{1}{2}\nu'(r), \quad \Gamma^0_{00} = 0. \tag{16.30}$$

The other Γ's are zero because ν and λ do not depend on θ and ϕ.
(ii)   For $x^\alpha \equiv x^1 = r$ one obtains from the variation principle the Euler–Lagrange equation (divided by $-2e^{\lambda(r)}$)

$$\ddot{r} + \frac{1}{2}\lambda'(r)\dot{r}^2 + \frac{1}{2}\nu'(r)e^{\nu-\lambda}(\dot{x}^0)^2 - e^{-\lambda}r\dot{\theta}^2 - r\sin^2\theta\dot{\phi}^2 e^{-\lambda} = 0. \tag{16.31}$$

Comparing with the equation obtained from Eq. (16.24),

$$\ddot{x}^1 + \Gamma^1_{\beta\eta}\dot{x}^\beta\dot{x}^\eta = 0, \tag{16.32}$$

one obtains

$$\Gamma^1_{00} = \frac{1}{2}\nu'e^{\nu-\lambda}, \quad \Gamma^1_{11} = \frac{1}{2}\lambda', \quad \Gamma^1_{22} = -e^{-\lambda}r, \quad \Gamma^1_{33} = -r\sin^2\theta e^{-\lambda}. \tag{16.33}$$

(iii)   For $x^\alpha \equiv x^2 = \theta$ we obtain from the variation principle the Euler–Lagrange equation (divided by $-2r^2$)

$$\ddot{\theta} + \frac{2}{r}\dot{\theta}\dot{r} - \sin\theta\cos\theta\dot{\phi}^2 = 0, \tag{16.34}$$

and compare this equation with (the superscript meaning "two", not squared)

$$\ddot{x}^2 + \Gamma^2_{\beta\eta}\dot{x}^\beta\dot{x}^\eta = 0. \tag{16.35}$$

The comparison yields the expressions (observe: No Γ^2_{00} and others)

$$\Gamma^2_{21} = \Gamma^2_{12} = \frac{1}{r}, \quad \Gamma^2_{33} = -\sin\theta\cos\theta. \tag{16.36}$$

(iv)   Finally for $x^\alpha \equiv x^3 = \phi$ we have the Euler–Lagrange equation (divided by $-2r^2\sin^2\theta$)

$$\ddot{\phi} + 2\cot\theta\dot{\phi}\dot{\theta} + \frac{2}{r}\dot{r}\dot{\phi} = 0, \tag{16.37}$$

which is to be compared with the equation

$$\ddot{x}^3 + \Gamma^3_{\beta\eta}\dot{x}^\beta\dot{x}^\eta = 0. \tag{16.38}$$

Comparison of the two equations implies

$$\Gamma^3_{23} = \Gamma^3_{32} = \cot\theta, \quad \Gamma^3_{13} = \Gamma^3_{31} = \frac{1}{r}. \tag{16.39}$$

We have thus determined the Christoffel symbols required for the evaluation of $R_{\mu\nu} = 0$.

We obtained previously (*cf.* Eq. (14.100))

$$\Gamma^\alpha_{m\alpha} = \frac{\partial}{\partial x^m}\ln\sqrt{-g(x)}. \tag{16.40}$$

The equation $R_{\mu\nu} = 0$, *i.e.* Eq. (16.6), Eq. (16.7), therefore becomes

$$R_{\mu\nu} = \frac{\partial^2}{\partial x^\mu \partial x^\nu} \ln \sqrt{-g(x)} - (\Gamma^\alpha_{\mu\nu})_{,\alpha} + \Gamma^\beta_{\eta\mu}\Gamma^\eta_{\beta\nu} - \Gamma^\eta_{\mu\nu}\frac{\partial}{\partial x^\eta} \ln \sqrt{-g(x)} = 0.$$
(16.41)

Now, in order to determine $\ln \sqrt{-g(x)}$ we consider:

$$ds^2 = e^{\nu(r)}c^2 dt^2 - e^{\lambda(r)}dr^2 - r^2(d\theta^2 + \sin^2\theta d\phi^2) \equiv g_{\mu\nu}(x)dx^\mu dx^\nu, \quad (16.42)$$

so that

$$g_{\mu\nu} = \begin{pmatrix} e^{\nu(r)} & 0 & 0 & 0 \\ 0 & -e^{\lambda(r)} & 0 & 0 \\ 0 & 0 & -r^2 & 0 \\ 0 & 0 & 0 & -r^2\sin^2\theta \end{pmatrix}$$
(16.43)

with

$$g(x) = \det(g_{\mu\nu}) = -r^4\sin^2\theta e^{\nu(r)+\lambda(r)}, \quad (16.44a)$$

and

$$\ln \sqrt{-g(x)} = \frac{1}{2}[\nu(r) + \lambda(r)] + 2\ln r + \ln|\sin\theta|. \quad (16.44b)$$

We leave the next step, which is the explicit evaluation of the elements R_{00} and R_{11}, to Example 16.2, and the step thereafter, *i.e.* the solution of the equations $R_{00} = 0, R_{11} = 0, R_{22} = 0, R_{33} = 0$, to Example 16.3.

Example 16.2: Evaluation of R_{00} and R_{11}
Evaluate R_{00} and R_{11} and hence obtain the equations $R_{00} = 0, R_{11} = 0$.

Solution:
(a) Evaluation of R_{00}: Setting in Eq. (16.41) $\mu = 0, \nu = 0$, we obtain

$$R_{00} = \frac{\partial^2}{\partial x^0 \partial x^0} \ln \sqrt{-g(x)} - (\Gamma^\alpha_{00})_{,\alpha} + \Gamma^\beta_{\eta 0}\Gamma^\eta_{\beta 0} - \Gamma^\eta_{00}\frac{\partial}{\partial x^\eta} \ln \sqrt{-g(x)} = 0. \quad (16.45)$$

Now, g is independent of x^0 and we need only consider Γ^1_{00} and Γ^0_{10}. Then

$$\begin{aligned} R_{00} &= -(\Gamma^1_{00})_{,1} + 2\Gamma^0_{10}\Gamma^1_{00} - \Gamma^1_{00}\frac{\partial}{\partial x^1} \ln \sqrt{-g(x)} \\ &= -\frac{1}{2}(\nu'e^{\nu-\lambda})' + \frac{1}{2}\nu'^2 e^{\nu-\lambda} - \frac{1}{2}(\nu'e^{\nu-\lambda})\left(\frac{\nu'+\lambda'}{2} + \frac{2}{r}\right) = 0, \end{aligned}$$

or

$$-\frac{1}{2}\nu''e^{\nu-\lambda} - \frac{1}{2}\nu'^2 e^{\nu-\lambda} + \frac{1}{2}\nu'\lambda'e^{\nu-\lambda} + \frac{1}{2}\nu'^2 e^{\nu-\lambda} - \frac{1}{4}\nu'^2 e^{\nu-\lambda}$$

$$-\frac{1}{4}\nu'\lambda'e^{\nu-\lambda} - \frac{1}{r}\nu'e^{\nu-\lambda} = 0,$$

$$\nu'' + \frac{1}{2}\nu'^2 - \frac{1}{2}\nu'\lambda' + \frac{2\nu'}{r} = 0. \quad (16.46)$$

(b) Similarly the equation $R_{11} = 0$ gives

$$\frac{\partial^2}{(\partial x^1)^2} \ln \sqrt{-g(x)} - (\Gamma^\alpha_{11})_{,\alpha} + \Gamma^\beta_{\tau 1}\Gamma^\tau_{\beta 1} - \Gamma^\tau_{11}(\ln \sqrt{-g(x)})_{,\tau} = 0,$$

$$(\ln \sqrt{-g})_{,1,1} - (\Gamma^1_{11})_{,1} + \Gamma^0_{10}\Gamma^0_{10} + \Gamma^1_{11}\Gamma^1_{11} + \Gamma^2_{21}\Gamma^2_{21} + \Gamma^3_{31}\Gamma^3_{31} - \Gamma^1_{11}(\ln \sqrt{-g})_{,1} = 0,$$

$$\frac{1}{2}(\nu'' + \lambda'') - \frac{2}{r^2} - \frac{1}{2}\lambda'' + \frac{1}{4}\nu'^2 + \frac{1}{4}\lambda'^2 + \frac{2}{r^2} - \frac{1}{2}\lambda'\left(\frac{\nu' + \lambda'}{2} + \frac{2}{r}\right) = 0.$$

Rearranging terms the last equation now reduces to

$$\nu'' + \frac{1}{2}\nu'^2 - \frac{1}{2}\nu'\lambda' - \frac{2\lambda'}{r} = 0. \tag{16.47}$$

Thus we have two coupled equations of the second order in $\nu(r)$ and $\lambda(r)$.

Example 16.3: Solution of $R_{00} = 0, R_{11} = 0, R_{22} = 0, R_{33} = 0$

Obtain the solutions of the equations $R_{00} = 0$ and $R_{11} = 0$ together with the equations $R_{22} = 0, R_{33} = 0$.

Solution: In Example 16.2 we obtained two equations of the second order for the functions $\nu(r)$ and $\lambda(r)$, Eqs. (16.46) and (16.47). Subtracting one from the other we obtain

$$\lambda'(r) + \nu'(r) = 0, \quad i.e. \quad \lambda(r) + \nu(r) = \text{const.} \tag{16.48}$$

Since asymptotically $\lambda(r), \nu(r) \to 0$ for $r \to \infty$, the constant must be zero, and hence

$$\lambda(r) = -\nu(r). \tag{16.49}$$

We substitute this result for $\nu(r)$ in Eq. (16.46) and obtain:

$$-\lambda'' + \lambda'^2 - \frac{2\lambda'}{r} = 0, \quad i.e. \quad (re^{-\lambda})'' = 0, \quad \text{or} \quad (re^{-\lambda})' = \text{const.} \tag{16.50}$$

In order to determine the constant, we consider the equation $R_{22} = 0$. From Eq. (16.41) we deduce that this equation is

$$(\ln \sqrt{-g})_{,2,2} - (\Gamma^\alpha_{22})_{,\alpha} + \Gamma^\beta_{\tau 2}\Gamma^\tau_{\beta 2} - \Gamma^\tau_{22}(\ln \sqrt{-g})_{,\tau} = 0,$$

or

$$(\ln \sqrt{-g})_{,2,2} - (\Gamma^1_{22})_{,1} + \Gamma^1_{22}\Gamma^2_{21} + \Gamma^2_{21}\Gamma^1_{22} + \Gamma^3_{23}\Gamma^3_{23} - \Gamma^1_{22}(\ln \sqrt{-g})_{,1} = 0. \tag{16.51}$$

This is, using Eqs. (16.44b), (16.33), (16.36) and (16.39),

$$\frac{\partial^2}{\partial\theta^2}(\ln|\sin\theta|) + (e^{-\lambda}r)' + 2(-e^{-\lambda}) + \cot^2\theta + e^{-\lambda}r\left(\frac{\lambda' + \nu'}{2} + \frac{2}{r}\right) = 0.$$

Since $\lambda' + \nu' = 0$, this is

$$\frac{\partial^2}{\partial\theta^2}(\ln|\sin\theta|) + \cot^2\theta = -(re^{-\lambda(r)})'. \tag{16.52}$$

But

$$\frac{\partial^2}{\partial\theta^2}\ln|\sin\theta| + \cot^2\theta = -\frac{1}{\sin^2\theta} + \frac{\cos^2\theta}{\sin^2\theta} = -\frac{1}{\sin^2\theta} + \frac{1 - \sin^2\theta}{\sin^2\theta} = -1,$$

so that

$$(e^{-\lambda(r)}r)' = 1. \tag{16.53}$$

Integrating this equation we obtain

$$re^{-\lambda(r)} = r - 2l,$$

(16.54)

where $-2l$ is a constant of integration. Hence

$$e^{\nu(r)} = e^{-\lambda(r)} = 1 - \frac{2l}{r}.$$

(16.55)

Finally we consider $R_{33} = 0$ and obtain parallel to the considerations above an equivalent result. The nondiagonal elements of $R_{\mu\nu}$ are identically zero. Thus we have determined all equations resulting from the vanishing of $R_{\mu\nu}$.

Inserting the results (16.55) into the metric of Eq. (16.42) we obtain the *Schwarzschild solution*

$$
\begin{aligned}
ds^2 &= \left(1 - \frac{2l}{r}\right)c^2 dt^2 - \left(1 - \frac{2l}{r}\right)^{-1} dr^2 - r^2(d\theta^2 + \sin^2\theta d\phi^2) \\
&\equiv g_{00}(t,r)c^2 dt^2 + g_{11}(t,r)dr^2 - r^2(d\theta^2 + \sin^2\theta d\phi^2) \\
&= dx^\mu g_{\mu\nu}(x)dx^\nu \equiv dx^\mu dx_\mu,
\end{aligned}
$$

(16.56a)

so that

$$
g_{00}(t,r) = \left(1 - \frac{2l}{r}\right), \qquad g_{11}(t,r) = -\left(1 - \frac{2l}{r}\right)^{-1},
$$

$$
g_{22}(t,r) = -r^2, \qquad g_{33}(t,r) = -r^2 \sin^2\theta.
$$

(16.56b)

The corresponding contravariant components of the metric are obtained from $g^{\mu\nu}g_{\nu\rho} = \delta^\mu_\rho$, and one has

$$
g^{00}(t,r) = \left(1 - \frac{2l}{r}\right)^{-1}, \qquad g^{11}(t,r) = -\left(1 - \frac{2l}{r}\right),
$$

$$
g^{22}(t,r) = -\frac{1}{r^2}, \qquad g^{33}(t,r) = -\frac{1}{r^2 \sin^2\theta}.
$$

(16.56c)

The Schwarzschild metric is seen to satisfy the asymptotic conditions at $r = \infty$. The singularity at $r = 2l$ (called the *apparent event horizon*) is not a singularity of the metric as one might presume, but of the system of coordinates — in fact, the singularity can be eliminated by changing to a different system of coordinates as we demonstrated in Sec. 12.4. A recognizable defect of the Schwarzschild coordinates and hence of (16.56a) is that at $r = 2l$ the roles of ct and r are reversed: In the region $r > 2l$ the t-direction is timelike (*i.e.* $(1 - 2l/r) > 0$ and for t, θ, ϕ constant $d\tau^2 = g_{rr}dr^2 > 0$) and the r-direction is spacelike; in the region $r < 2l$ these roles are reversed.

It remains to determine the constant l which has the dimensionality of a length. We recall that in the case of a weak gravitational field we had obtained (*cf.* Eq. (13.25)) the expression

$$g_{00}(\mathbf{x}) \simeq 1 + \frac{2\phi(\mathbf{x})}{c^2},$$

(16.57)

where $\phi(\mathbf{x})$ is the potential (*cf.* Eq. (13.25) and Eqs. (15.114), (15.126))

$$\phi(r) = -\frac{GM}{r}, \tag{16.58a}$$

the quantity M being the mass of the source and G Netwon's universal gravitational constant $G = 6.67 \times 10^{-8} \text{cm}^3 \text{g}^{-1} \text{s}^{-2} > 0$. Thus

$$g_{00} = 1 - \frac{2GM}{c^2 r}, \tag{16.58b}$$

which is to be compared with the Schwarzschild g_{00}, *i.e.*

$$1 - \frac{2GM}{c^2 r} = 1 - \frac{2l}{r}, \quad i.e. \quad l = \frac{GM}{c^2}. \tag{16.59}$$

The *Schwarzschild radius* is the quantity

$$r_S = 2l = \frac{2GM}{c^2}. \tag{16.60}$$

We recall from Example 2.3 that in the context there the quantity $2GM/R$ (R the radius of the Earth) was v^2, the square of the velocity of escape from the Earth. This factor also appeared in the trick calculation of the redshift in Sec. 11.3.1. Thus with $R = 2GM/v^2$ this parameter re-appears with R replaced by r_S and v^2 by c^2. In the case of a macroscopic body the radius r_S is that of the interior of the body, *e.g.* the Schwarzschild radius of the sun — which is where the above relativistic equations are *invalid* (*cf.* the discussion at the beginning of Sec. 16.2).

16.4 The Schwarzschild Solution for $\Lambda \neq 0$

We now wish to solve the following equation with $\Lambda \neq 0$ and the asymptotic conditions as before. We have, now with $\Lambda \neq 0$, *Einstein's equation*:

$$R_{\mu\nu} - \frac{1}{2} g_{\mu\nu} R + \Lambda g_{\mu\nu} = 0. \tag{16.61}$$

Multiplying the equation by $g^{\mu\nu}$ we obtain (since $R = g^{\mu\nu} R_{\mu\nu}, g^{\mu\nu} g_{\mu\nu} = 4$):

$$R - 2R + 4\Lambda = 0, \quad i.e. \quad R = 4\Lambda. \tag{16.62}$$

Inserting this into Eq. (16.61) we obtain

$$R_{\mu\nu} = \Lambda g_{\mu\nu}. \tag{16.63}$$

We now proceed as before and obtain

$$R_{00} = \Lambda g_{00}, \quad \text{and} \quad R_{kl} = \Lambda g_{kl},$$

and hence for the first case (cf. Eq. (16.46))

$$-e^{\nu-\lambda}\left[\frac{\nu''}{2} + \frac{\nu'}{r} + \frac{1}{4}\nu'(\nu'-\lambda')\right] = e^{\nu}\Lambda,$$

$$\nu'' + \frac{1}{2}\nu'(\nu'-\lambda') + \frac{2}{r}\nu' = -2e^{\lambda}\Lambda. \tag{16.64}$$

Considering $R_{kl} = \Lambda g_{kl}$, we obtain (for $k, l = 1, 2, 3$ — we cite this result here without presenting the calculation which proceeds parallel to that in Example 16.2 —

$$\left[\frac{1}{2}\nu'' - \frac{\lambda'}{r} + \frac{1}{4}\nu'(\nu'-\lambda')\right]\frac{x_k x_l}{r^2}$$

$$+ \left[-\frac{1}{r^2} + e^{-\lambda}\left(\frac{1}{r^2} + \frac{\nu'-\lambda'}{2r}\right)\right]\left(\delta_{kl} - \frac{x_k x_l}{r^2}\right)$$

$$= \Lambda\left[-\delta_{kl} + (1 - e^{\lambda})\frac{x_k x_l}{r^2}\right],$$

and hence for $k \neq l$:

$$\frac{1}{2}\nu'' - \frac{\lambda'}{r} + \frac{1}{4}\nu'(\nu'-\lambda') + \frac{1}{r^2} - \frac{1}{r^2}e^{-\lambda} - \frac{\nu'}{2r}e^{-\lambda} + \frac{\lambda'}{2r}e^{-\lambda} = \Lambda(1 - e^{\lambda}), \tag{16.65}$$

and for $k = l$ (note that $x_k x_k = \sum x_k^2 = r^2, \delta_{kk} = 3$):

$$\frac{1}{2}\nu'' - \frac{\lambda'}{r} + \frac{1}{4}\nu'(\nu'-\lambda') - \frac{2}{r^2} + \frac{2}{r^2}e^{-\lambda} + \frac{\nu'}{r}e^{-\lambda} - \frac{\lambda'}{r}e^{-\lambda} = \Lambda(-2 - e^{\lambda}). \tag{16.66}$$

Subtracting Eq. (16.66) from Eq. (16.65) we obtain

$$\frac{3}{r^2} - \frac{3}{r^2}e^{-\lambda} - \frac{3\nu'}{2r}e^{-\lambda} + \frac{3\lambda'}{2r}e^{-\lambda} = 3\Lambda,$$

which can be rewritten as

$$e^{-\lambda}\left(\frac{\lambda'-\nu'}{2r} - \frac{1}{r^2}\right) = \left(\Lambda - \frac{1}{r^2}\right). \tag{16.67}$$

On the other hand, adding Eqs. (16.65) and (16.66), we obtain

$$\nu'' - \frac{2\lambda'}{r} + \frac{1}{2}\nu'(\nu'-\lambda') - \underbrace{e^{-\lambda}\left(\frac{\lambda'-\nu'}{2r} - \frac{1}{r^2}\right)} = -\underbrace{\left(\Lambda - \frac{1}{r^2}\right)} - 2\Lambda e^{\lambda}.$$

Here the underbraced terms cancel out in view of Eq. (16.67), so that we obtain

$$\nu'' - \frac{2\lambda'}{r} + \frac{1}{2}\nu'(\nu' - \lambda') = -2\Lambda e^\lambda. \tag{16.68}$$

Equations (16.64) and (16.67) now imply by subtraction

$$\nu' + \lambda' = 0, \quad \nu + \lambda = \text{const.} \equiv \ln k. \tag{16.69}$$

Substituting $\lambda = \ln k - \nu$ in Eq. (16.64), we obtain

$$\nu'' + \frac{1}{2}\nu'^2 - \frac{1}{2}\nu'(\ln k - \nu)' + \frac{2}{r}\nu' = -2e^{\ln k - \nu}\Lambda, \quad \nu'' + \nu'^2 + \frac{2}{r}\nu' = -2ke^{-\nu}\Lambda,$$

or

$$\underbrace{e^\nu \nu'' + e^\nu \nu'^2} + \frac{2}{r}\nu' e^\nu = -2k\Lambda. \tag{16.70}$$

With (where $\nu = \nu(r)$)

$$(e^\nu)' = e^\nu \nu', \quad (e^\nu)'' = \underbrace{e^\nu \nu'' + \nu'^2 e^\nu},$$

we obtain

$$(e^\nu)'' + \frac{2}{r}(e^\nu)' = -2k\Lambda, \quad \text{or} \quad \frac{1}{r}(re^\nu)'' = -2k\Lambda, \tag{16.71}$$

and hence, with constants α and β,

$$(re^\nu)' = -k\Lambda r^2 + \beta, \quad \text{or} \quad re^\nu = \alpha + \beta r - k\Lambda \frac{r^3}{3}. \tag{16.72}$$

Inserting $\lambda' = -\nu'$ into Eq. (16.67), we have:

$$-e^{-\lambda}\left(\frac{\nu'}{r} + \frac{1}{r^2}\right) = \Lambda - \frac{1}{r^2}, \quad \text{or} \quad e^{-\lambda}(\nu'r + 1) = 1 - \Lambda r^2.$$

With

$$(e^{-\lambda}r)' = e^{-\lambda} - \lambda' e^{-\lambda}r = e^{-\lambda}(\nu'r + 1),$$

we obtain

$$(e^{-\lambda}r)' = 1 - \Lambda r^2. \tag{16.73}$$

But from Eq. (16.68) $\lambda = \ln k - \nu$, so that

$$e^{-\lambda} = e^{\ln(1/k)}e^\nu = \frac{1}{k}e^\nu,$$

and hence, multiplying by r,

$$(e^\nu r)' = k(e^{-\lambda} r)' = k(1 - \Lambda r^2). \tag{16.74}$$

Comparing this equation with Eq. (16.71), we obtain

$$k = \beta, \tag{16.75}$$

and hence from Eq. (16.71), i.e.

$$re^\nu = \alpha + \beta r - k\Lambda \frac{r^3}{3}, \quad \therefore \quad e^\nu = k\left(1 + \frac{\alpha}{kr} - \Lambda \frac{r^2}{3}\right). \tag{16.76}$$

Now,

$$\lambda = \ln k - \nu, \quad e^{-\lambda} = e^{\ln(1/k)} e^\nu = \frac{1}{k} e^\nu, \quad e^{-\lambda} = 1 + \frac{\alpha}{kr} - \Lambda \frac{r^2}{3}. \tag{16.77}$$

Inserting these results into the coefficients of the metric, we obtain

$$
\begin{aligned}
ds^2 &= e^{\nu(r)} c^2 dt^2 - e^{\lambda(r)} dr^2 - r^2(d\theta^2 + \sin^2\theta d\phi^2) \\
&= k\left(1 + \frac{\alpha}{kr} - \Lambda \frac{r^2}{3}\right) c^2 dt^2 - \left(1 + \frac{\alpha}{kr} - \Lambda \frac{r^2}{3}\right)^{-1} dr^2 \\
&\quad - r^2(d\theta^2 + \sin^2\theta d\phi^2).
\end{aligned}
\tag{16.78}
$$

Choosing the unit of time such that

$$(dt')^2 = k(dt)^2, \quad k = 1, \quad \text{then} \quad e^\nu = 1 + \frac{\alpha}{r} - \Lambda \frac{r^2}{3} = e^{-\lambda}, \tag{16.79}$$

so that (with $k = 1$)

$$ds^2 = \left(1 + \frac{\alpha}{r} - \Lambda \frac{r^2}{3}\right) c^2 dt^2 - \left(1 + \frac{\alpha}{r} - \Lambda \frac{r^2}{3}\right)^{-1} dr^2 - r^2(d\theta^2 + \sin^2\theta d\phi^2). \tag{16.80}$$

For $\alpha = -2l, \Lambda = 0$, one regains the Schwarzschild solution. In the approximation of the new potential defined by

$$g_{00} := 1 + \frac{2\phi(r)}{c^2}, \quad 1 + \frac{2\phi(r)}{c^2} = 1 - \frac{2l}{r} - \Lambda \frac{r^2}{3}, \tag{16.81}$$

one obtains

$$\phi(r) = -\frac{lc^2}{r} - \frac{\Lambda c^2}{6} r^2 = -G\frac{M}{r} - \frac{1}{6}\Lambda c^2 r^2. \tag{16.82}$$

Thus the constant Λ leads to a modification of Newton's potential — but with no apparent observational necessity.

16.5   The Relativistic Kepler Problem

The Kepler problem has already been treated in Chapter 7. In particular
the equation of motion of a planet around the sun was derived from the
Schwarzschild metric (16.56a) in Example 3.13 as an illustration of the variational principle. Thus with the derivation of the Schwarzschild metric from
Einstein's equation, we have also obtained the full Kepler equation with
both Newton's gravitational potential and Einstein's relativistic correction,
the latter explaining the perihelion anomaly, *i.e.* the nonclosure of the elliptic orbit of the planet Mercury, the planet nearest to the sun in the solar
system. In Newton's celestial mechanics based on his equation of motion
and the inverse square law of his gravitational force the orbit of a planet is
a closed ellipse. In actual fact, as we saw in Chapter 7, the point on the
orbit closest to the sun at a focus, called the *perihelion*, is found to precess slightly up to an amount of 43 seconds of arc per century. We now see
that Einstein's general relativity accounts for this difference. This result is
considered to be a test of Einstein's General Theory of Relativity, and its
confirmation by experiment a triumph of his theory. Since the equation has
been derived from the Schwarzschild metric in Example 3.13, and all other
aspects of the Kepler problem, including Einstein's relativistic effects, have
been discussed in Chapter 7, there is no need to repeat the considerations
here, and we refer back to these earlier chapters. In the following we therefore
first relate the present relativistic calculations to those of Chapter 7 and, in
particular, obtain Eq. (7.93) of the Kepler problem from the Schwarzschild
metric, and then concentrate on some aspects not dealt with previously. We
begin with Example 16.4 in which we show that the Schwarzschild metric
together with the two conservation laws (integration constants) for energy E
and angular momentum L — derived in Example 3.13 from the Schwarzschild
metric — permit us to re-express the relativistic energy-momentum formula
$p^\mu p_\mu = m^2 c^2$ (which corresponds to that in Minkowski space, Eq. (12.47)) in
terms of E and L, m being the mass of the particle.

Example 16.4: The Schwarzschild energy-momentum relation
Show that the momenta conjugate to the cyclic Schwarzschild coordinates t and ϕ, *i.e.* $p_0 = E/c$
and $p_\phi = \pm L, L \geq 0$, when inserted into $p^\mu p_\mu = m^2 c^2$ yield the relation

$$m^2 c^2 \left(\frac{dr}{d\tau}\right)^2 = E^2 - \left(m^2 c^4 + \frac{L^2 c^2}{r^2}\right)\left(1 - \frac{2l}{r}\right) \equiv E^2 - V_{\text{eff}}^2(r). \qquad (16.83)$$

This relation defines the *Schwarzschild effective potential* $V_{\text{eff}}(r)$.

Solution: In Example 3.13 we derived the Euler–Lagrange equations from the Schwarzschild
metric and observed two constants of integration (resulting from cyclic coordinates t, ϕ), which
— as in the Kepler problem — turn out to be the energy E and the angular momentum L (in
Example 3.13 respectively per unit mass called kc and hc). In terms of the proper time τ measured

in the rest frame of a particle in Schwarzschild space (hence, as remarked in Sec. 12.8, the mass m here is that which is frequently called "rest mass", e.g. in the book of Raine and Thomas [40], pp. 21, 23), we have from Eq. (16.56a) (effectively replacing ds by $cd\tau$ with the reasoning of Sec. 12.8)

$$1 = \left(1 - \frac{2l}{r}\right)\left(\frac{dt}{d\tau}\right)^2 - \left(1 - \frac{2l}{r}\right)^{-1}\left(\frac{dr}{cd\tau}\right)^2 - r^2\left[\left(\frac{d\theta}{cd\tau}\right)^2 + \sin^2\theta\left(\frac{d\phi}{cd\tau}\right)^2\right]. \tag{16.84}$$

The two conservation laws we obtained are (cf. after Eq. (3.98) in Example 3.13)

$$\left(1 - \frac{2l}{r}\right)\frac{dt}{d\tau} = kc, \qquad r^2\frac{d\phi}{d\tau} = hc. \tag{16.85}$$

Replacing in Eq. (16.84) $dt/d\tau$ by the expression in terms of k and $d\phi/d\tau$ by the expression in terms of h, and recalling from Example 3.13 that $\theta = \pi/2$, we obtain

$$1 = \left(1 - \frac{2l}{r}\right)^{-1}c^2k^2 - \left(1 - \frac{2l}{r}\right)^{-1}\left(\frac{dr}{cd\tau}\right)^2 - \frac{h^2}{r^2}. \tag{16.86}$$

Multiplying this relation by the square of the mass of the particle, m^2, we obtain, with $E/c^2 = mkc, L = mhc$, the relation

$$\begin{aligned}
m^2 &= \left(1 - \frac{2l}{r}\right)^{-1}\frac{E^2}{c^4} - m^2\left(1 - \frac{2l}{r}\right)^{-1}\left(\frac{dr}{cd\tau}\right)^2 - \frac{L^2}{c^2r^2}, \\
m^2\left(\frac{dr}{cd\tau}\right)^2 &= \frac{E^2}{c^4} - \left(m^2 + \frac{L^2}{c^2r^2}\right)\left(1 - \frac{2l}{r}\right), \\
m^2c^2\left(\frac{dr}{d\tau}\right)^2 &= E^2 - \left(m^2c^4 + \frac{L^2c^2}{r^2}\right)\left(1 - \frac{2l}{r}\right).
\end{aligned} \tag{16.87}$$

Hence, by comparison with Eq. (16.83), we identify the square of the effective potential as

$$V_{\text{eff}}^2 = \left(m^2 + \frac{L^2}{c^2r^2}\right)\left(1 - \frac{2l}{r}\right)c^4. \tag{16.88}$$

Next we consider the energy-momentum relation $p^\mu p_\mu = m^2c^2$. In the present case this is $g^{\mu\nu}p_\mu p_\nu = m^2c^2$. We consider first the distinction between p^μ and p_μ. We have, using the expressions of Eq. (16.56c) for $g^{\mu\nu}$ (with $\theta = \pi/2$ and $p_0 = p_t, p_1 = p_r, p_2 = p_\theta, p_3 = p_\phi$):

$$\begin{aligned}
m^2c^2 &= p^\mu p_\mu = p_\mu g^{\mu\nu}p_\nu|_{\theta=\pi/2} \\
&= (p_0, p_1, p_2, p_3)\begin{pmatrix} (1 - 2l/r)^{-1} & 0 & 0 & 0 \\ 0 & -(1 - 2l/r) & 0 & 0 \\ 0 & 0 & -1/r^2 & 0 \\ 0 & 0 & 0 & -1/r^2 \end{pmatrix}\begin{pmatrix} p_0 \\ p_1 \\ p_2 \\ p_3 \end{pmatrix} \\
&= \left(1 - \frac{2l}{r}\right)^{-1}p_0^2 - \left(1 - \frac{2l}{r}\right)p_1^2 - \frac{1}{r^2}p_2^2 - \frac{1}{r^2}p_3^2,
\end{aligned} \tag{16.89}$$

where in the present case $p_2 = p_\theta = 0$. Next we recall that, with m the mass of the particle under consideration,

$$(p^\mu) = (p^0, p^1, p^2, p^3) = m\left(\frac{cdt}{d\tau}, \frac{dr}{d\tau}, \frac{d\theta}{d\tau}, \frac{d\phi}{d\tau}\right). \tag{16.90}$$

Hence the momenta with lower indices are given by the following relation, in which we now use the coefficients of Eq. (16.56b), and as before $\theta = \pi/2$,

$$
\begin{aligned}
(p_\alpha) &= (p_0, p_1, p_2, p_3) = (g_{\alpha\rho})(p^\rho) \\
&= \begin{pmatrix} (1 - 2l/r) & 0 & 0 & 0 \\ 0 & -(1-2l/r)^{-1} & 0 & 0 \\ 0 & 0 & -r^2 & 0 \\ 0 & 0 & 0 & -r^2 \end{pmatrix} \begin{pmatrix} p^0 \\ p^1 \\ p^2 \\ p^3 \end{pmatrix} \\
&= \left(\left(1 - \frac{2l}{r}\right)p^0, -\left(1 - \frac{2l}{r}\right)^{-1} p^1, -r^2 p^2, -r^2 p^3 \right) \\
&= m\left(c\left(1 - \frac{2l}{r}\right)\frac{dt}{d\tau}, -\left(1 - \frac{2l}{r}\right)^{-1}\frac{dr}{d\tau}, -r^2\frac{d\theta}{d\tau}, -r^2\frac{d\phi}{d\tau} \right).
\end{aligned}
\tag{16.91}
$$

Comparing the first and the last expressions on the right hand side of this equation we obtain

$$
p_0 = mc\left(1 - \frac{2l}{r}\right)\frac{dt}{d\tau}, \quad p_1 = -m\left(1 - \frac{2l}{r}\right)^{-1}\frac{dr}{d\tau}, \quad p_2 = 0 \ (\theta = \pi/2), \quad p_3 = -mr^2\frac{d\phi}{d\tau}. \tag{16.92}
$$

We now insert these expressions into Eq. (16.89) and obtain

$$
m^2 c^2 = m^2 c^2 \left(1 - \frac{2l}{r}\right)\left(\frac{dt}{d\tau}\right)^2 - m^2\left(1 - \frac{2l}{r}\right)^{-1}\left(\frac{dr}{d\tau}\right)^2 - m^2 r^2\left(\frac{d\phi}{d\tau}\right)^2. \tag{16.93}
$$

Next we use the two conditions (16.85) for the replacement of the outer two derivative expressions. Then

$$
m^2 c^2 = m^2 k^2 c^4 \left(1 - \frac{2l}{r}\right)^{-1} - m^2\left(1 - \frac{2l}{r}\right)^{-1}\left(\frac{dr}{d\tau}\right)^2 - m^2 c^2 \frac{h^2}{r^2}. \tag{16.94}
$$

Dividing this equation by $m^2 c^2$ we obtain

$$
1 = c^2 k^2 \left(1 - \frac{2l}{r}\right)^{-1} - \left(1 - \frac{2l}{r}\right)^{-1}\left(\frac{dr}{cd\tau}\right)^2 - \frac{h^2}{r^2}. \tag{16.95}
$$

This equation is seen to be identical with Eq. (16.86), thus proving the equivalence of the latter with the relativistic Schwarzschild-space energy-momentum relation $p^\mu p_\mu = m^2 c^2$.

We collect from Example 16.4 the orbit equations which allow us to obtain also $dr/d\phi$ as a function of r, and hence to derive the paths of particles in space. The relevant equations are first Eq. (16.85) with hc there replaced by L/m and with $\theta = \pi/2$ (*cf.* Example 16.4), *i.e.*

$$
mr^2\frac{d\phi}{d\tau} = L, \tag{16.96}
$$

and then Eq. (16.87), *i.e.*

$$
m^2 c^2 \left(\frac{dr}{d\tau}\right)^2 = E^2 - \left(m^2 c^4 + \frac{L^2 c^2}{r^2}\right)\left(1 - \frac{2l}{r}\right) \equiv E^2 - V_{\text{eff}}^2(r).
$$

In Fig. 16.1 we sketch the effective potential. For the plots depicted there (redrawn MAPLE plots) we rewrote the potential

$$
V_{\text{eff}} = \sqrt{\left(m^2 c^4 + \frac{L^2 c^2}{r^2}\right)\left(1 - \frac{2l}{r}\right)} \tag{16.97a}
$$

as

$$V = \frac{V_{\text{eff}}(R)}{mc^2} = \sqrt{\left(1 + \frac{\gamma^2}{R^2}\right)\left(1 - \frac{1}{R}\right)}, \tag{16.97b}$$

where

$$\gamma = \frac{L}{2lmc}, \qquad R = \frac{r}{2l}. \tag{16.97c}$$

One can observe a considerable difference with the potential in the Kepler problem. However, there is a very small domain around the parameter value $\gamma = 2$ in which a binding of the particle is possible.[‡]

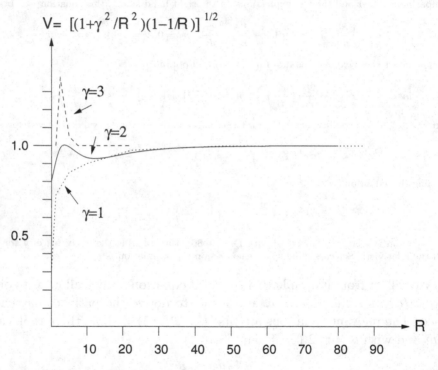

Fig. 16.1 The Schwarzschild effective potential $V = V_{\text{eff}}(R)/mc^2$ for $\gamma = 1, 2, 3$.

In the case of $m^2 = 0$, the photon, the effective potential is

$$V_{\text{eff}}(r) = \frac{Lc}{r}\sqrt{1 - \frac{2l}{r}}, \qquad V_{\text{eff}}^{\max}(r_0) = \frac{Lc}{\sqrt{27l}}, \tag{16.97d}$$

with a clear maximum at $r \to r_0 = 3l$. The potential is shown (resketched from a MAPLE plot) in Fig. 16.2.

[‡]This aspect is also discussed by D. Raine and E. Thomas [40], pp. 22–23.

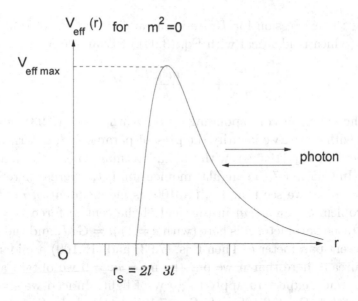

Fig. 16.2 The Schwarzschild effective potential $V_{\text{eff}}(r)$ for $m^2 = 0$.
For capture of the photon by the black hole, its energy must exceed $V_{\text{eff max}}$.

Again considering $m^2 \neq 0$, and setting

$$A^2 = \frac{E^2}{2m^2c^2}, \qquad B^2 = \frac{L^2}{m^2},$$

we can rewrite Eq. (16.87) (multiplied by $1/2$) as

$$\frac{1}{2}\left(\frac{dr}{d\tau}\right)^2 = A^2 - \frac{c^2}{2}\left(1 + \frac{B^2}{c^2 r^2}\right)\left(1 - \frac{2l}{r}\right). \tag{16.98}$$

Differentiating this equation with respect to r we obtain

$$\frac{d^2 r}{d\tau^2} = \frac{d}{dr}\left[\frac{1}{2}\left(\frac{dr}{d\tau}\right)^2\right] = -\frac{lc^2}{r^2} + \frac{B^2}{r^3} - 3\frac{B^2 l}{r^4}. \tag{16.99}$$

We can now use the calculations preceding Eq. (7.25) in order to change from
r, τ to $u = 1/r, \phi$. Then

$$\frac{d^2 r}{d\tau^2} = -\frac{L^2}{m^2}u^2\frac{d^2 u}{d\phi^2} = -lc^2 u^2 + B^2 u^3 - 3B^2 l u^4,$$

or

$$\frac{d^2 u}{d\phi^2} = \frac{lc^2 m^2}{L^2} - \frac{B^2 m^2}{L^2}u + 3\frac{B^2 l m^2}{L^2}u^2.$$

Replacing the expression for B^2 and rearranging terms, we thus obtain the equation (which is identical with Eq. (3.99) of Example 3.13)

$$\frac{d^2u}{d\phi^2} + u = \frac{m^2c^2l}{L^2} + 3lu^2. \tag{16.100}$$

This is the equation corresponding to our earlier Eq. (7.93) of the Kepler problem; with $c^2 = 1$ we identify the present parameter l, determined here as $l = GM/c^2$ (see Eq. (16.60)), with the undetermined perturbation parameter μ there. (In Chapter 7 the angular momentum is l, whereas here it is L, and since $L = mr^2\dot{\phi}$, we see that Eq. (16.100) is independent of m. Considered as the problem of a mass m in the field of the central force $f = -m\mu u^2$ of Eq. (7.25), the parameter μ is here (with $c = 1$) $\mu = GM$, and therefore equal to the present parameter l. Then Eqs. (7.93) and (16.100) are identical). We observe already here that if we put the mass $m = 0$ we obtain an equation which we would expect to apply to a ray of light. Indeed we shall find this in Sec. 16.6 (*cf.* Eq. (16.108)). In Sec. 7.10 we solved the Scharzschild orbit equation (16.100) perturbatively. In Appendix A we consider Eq. (16.100) in more detail and derive the solution in closed form in terms of Jacobian elliptic functions.

16.6   The Light Ray in the Schwarzschild Field

We now consider the path of a ray of light as it travels through a gravitational field as given by the Schwarzschild metric. Light or photons differ from ordinary particles in having no mass. Thus whereas the world line of a massive particle with velocity $v < c$ necessarily lies within the light cones, *cf.* Fig. 16.3, that of a photon has constant velocity, *i.e.* always has "*null light cones*" given by $\int ds = 0$. Since the photon has no rest frame, it "experiences" no time, as can be seen from the following. The *eigentime* τ measured by a clock fixed in the particle, *i.e.* in its rest frame, is obtained from the relation*

$$ds^2 = c^2dt^2 - d\mathbf{x}^2 = c^2dt^2\left[1 - \frac{v^2}{c^2}\right] = c^2d\tau^2.$$

Thus $d\tau = dt\sqrt{1 - v^2/c^2}$. But for $v = c$ one has $ds = 0$ and no eigentime τ. We postulate therefore that its trajectory is a geodesic defined by the null condition at a fixed velocity, *i.e.*

$$ds^2 = g_{\mu\nu}(x)dx^\mu dx^\nu = 0. \tag{16.101}$$

---

*See Sec. 12.8, Eq. (12.32a).

In terms of a parameter p, the variational principle we start from is now, like that before of Eq. (13.3),

$$\delta \int dp\, g_{\mu\nu}(x) \frac{dx^\mu}{dp} \frac{dx^\nu}{dp} = 0, \tag{16.102}$$

and the Euler–Lagrange equation obtained from this is that of the geodesic given by Eqs. (13.13) and (16.24), *i.e.*

$$\frac{d}{dp}\left(\frac{dx^\mu}{dp}\right) + \Gamma^\mu_{\nu\gamma} \frac{dx^\nu}{dp} \frac{dx^\gamma}{dp} = 0. \tag{16.103}$$

The parameter p which varies along the world-line of a light ray is called an "*affine parameter*", because it maintains for light rays the usual form of the

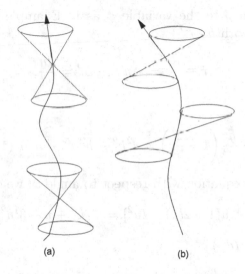

(a)   (b)

Fig. 16.3 Light cones on the world-line of
(a) a massive particle (world-line always within the cone),
and (b) of a photon (world-line on surface of the cone).

geodesic equation.[†] In the case of the Schwarzschild metric (16.55), *i.e.*

$$\left(\frac{ds}{dp}\right)^2 = \left(1 - \frac{2l}{r}\right)c^2 \left(\frac{dt}{dp}\right)^2 - \left(1 - \frac{2l}{r}\right)^{-1}\left(\frac{dr}{dp}\right)^2 - r^2\left(\frac{d\theta}{dp}\right)^2 - r^2 \sin^2\theta \left(\frac{d\phi}{dp}\right)^2,$$

we found in Example 3.13 for $\theta = \pi/2$ that (*cf.* after Eq. (3.98))

$$r^2 \frac{d\phi}{dp} = \text{const.} = \tilde{h} \quad \text{(previously } h\text{)},$$

$$\left(1 - \frac{2l}{r}\right)\frac{dt}{dp} = \text{const.} = \tilde{k}. \quad \text{(previously } k\text{)}. \tag{16.104}$$

[†]D. Raine and E. Thomas [40], p. 11.

With these conditions one obtains from the Schwarzschild metric (16.55) with $ds/dp = 0$:

$$0 = \left(1 - \frac{2l}{r}\right)c^2 \frac{\tilde{k}^2}{(1 - 2l/r)^2} - \left(1 - \frac{2l}{r}\right)^{-1}\left(\frac{dr}{dp}\right)^2 - r^2 \underbrace{\sin^2\theta}_{=1} \left(\frac{\tilde{h}}{r^2}\right)^2, \quad (16.105)$$

i.e.

$$0 = \left(1 - \frac{2l}{r}\right)^{-1}c^2\tilde{k}^2 - \left(1 - \frac{2l}{r}\right)^{-1}\left(\frac{dr}{dp}\right)^2 - \frac{\tilde{h}^2}{r^2}. \quad (16.106)$$

Setting

$$u(p) = \frac{1}{r(p)},$$

and changing from p to the variable ϕ as in Example 3.13, where it was shown that (here with $h \to \tilde{h}$)

$$\dot{r} \equiv \frac{dr}{dp} = -u'\tilde{h}, \quad u' = \frac{du}{d\phi},$$

we obtain

$$0 = c^2\tilde{k}^2 - \dot{r}^2 - \frac{\tilde{h}^2}{r^2}\left(1 - \frac{2l}{r}\right) = c^2\tilde{k}^2 - \tilde{h}^2u'^2 - \tilde{h}^2u^2(1 - 2lu). \quad (16.107)$$

Differentiating the equation with respect to angle ϕ, we obtain

$$u'[u'' + u(1 - 2lu) - lu^2] = u'(u'' + u - 3lu^2) = 0,$$

and hence $(u' = du/d\phi)$

$$u'' + u - 3lu^2 = 0, \quad \text{or} \quad u' = 0, \quad (16.108)$$

which agrees with Eq. (16.100) for $m^2 = 0$. What is the significance of the second equation? In the general equation (16.89) $u' = 0$ implies $u = $ const. $= u_0$, and hence $r_0 = $ const., which means that there is only this one physically significant value. The equation of the light ray must, however, be of the second order so that rays are possible from any point and any direction. In Example 16.5 we derive a specific solution of the differential equation (16.108) and show how the general solution may be obtained, but here we proceed as follows.[‡]

[‡]Recall that the Schwarzschild solution employs polar coordinates. Here we wish to explore the deviation of a ray of light from a straight line path. In planar Cartesian coordinates (x, y) a straight line is given by $y = mx + c$. In 2-dimensional polar coordinates (r, ϕ) this equation becomes $r\sin\phi = mr\cos\phi + c$ or $c/r = \sin\phi - m\cos\phi$ — a normally unfamiliar form. The solution of Eq. (16.108) (see Example 16.5) does not have this form, which tells us already that the path of the ray is not a straight line.

One can convince oneself that $3lu^2$ is small. Consider $3lu^2/u = 3lu$, or since $r_S = 2l$ (*cf.* Eq. (16.60)):

$$3lu = \frac{3}{2}\frac{2l}{r} = \frac{3}{2}\frac{r_S}{r}. \tag{16.109}$$

In the case of the sun $r_S \simeq 1\,\mathrm{km}$, so that for a trajectory outside the sun the term is small. We set

$$3l = \epsilon. \tag{16.110}$$

The equation of the trajectory of the light ray is then

$$u''(\phi) + u(\phi) = \epsilon u^2(\phi). \tag{16.111}$$

We assume a solution of the form

$$u(\phi) = u_0(\phi) + \epsilon v(\phi) + O(\epsilon^2), \tag{16.112}$$

so that

$$u_0'' + u_0 + \epsilon v'' + \epsilon v = \epsilon u_0^2 + O(\epsilon^2), \tag{16.113}$$

or in lowest order

$$u_0''(\phi) + u_0(\phi) = 0 \quad \text{and} \quad u_0 = A\cos(\phi + \delta). \tag{16.114}$$

If the axes are such that for $\delta = 0$, we have $u_0 = A\cos\phi$, then to first order

$$r \to r_0 = \frac{1}{u_0} = \frac{1}{A\cos\phi}, \quad \text{or} \quad r_0\cos\phi = \frac{1}{A} \equiv \tilde{r}, \quad u_0 = \frac{\cos\phi}{\tilde{r}}. \tag{16.115}$$

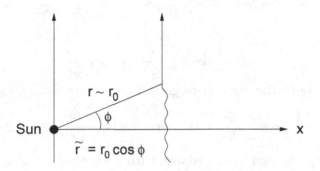

Fig. 16.4 A ray of light in the field of the sun in dominant order.

Since $\tilde{r} = r_0\cos\phi \equiv x_0$, this represents the straight line $x_0 = 1/A$ parallel to the y-axis, as indicated in Fig. 16.4. But this is only the lowest order approximation which does not involve a deviation due to the gravitational field of the sun. Thus without the term $3lu^2$ in Eq. (16.108) the path of the ray of light is a straight line.

More generally we have, as in Eq. (16.114),

$$r_0 \cos(\phi + \delta) = \frac{1}{A}, \quad i.e. \quad r_0 \cos\phi \cos\delta - r_0 \sin\phi \sin\delta = \frac{1}{A}, \qquad (16.116)$$

so that with $x_0 = r_0 \cos\phi$, $y_0 = r_0 \sin\phi$:

$$x_0 \cos\delta - y_0 \sin\delta = \frac{1}{A}, \qquad (16.117)$$

which is a straight line with gradient $\cos\delta/\sin\delta$.

Considering now the next equation obtained from Eq. (16.113) by equating coefficients of ϵ on both sides, we obtain

$$v''(\phi) + v(\phi) = u_0^2 = \frac{1}{\tilde{r}^2}\cos^2\phi = \frac{1}{2\tilde{r}^2}(1 + \cos 2\phi). \qquad (16.118)$$

We set

$$v = \alpha + \beta\cos 2\phi, \quad v' = -2\beta\sin 2\phi, \quad v'' = -4\beta\cos 2\phi,$$

so that Eq. (16.118) becomes

$$\alpha - 3\beta\cos 2\phi = \frac{1}{2\tilde{r}^2}(1 + \cos 2\phi), \qquad (16.119)$$

and hence

$$\alpha = \frac{1}{2\tilde{r}^2}, \quad \beta = -\frac{1}{6\tilde{r}^2}. \qquad (16.120)$$

Inserting these values into the solution v, we have

$$v = \frac{1}{2\tilde{r}^2} - \frac{1}{6\tilde{r}^2}\cos 2\phi = \frac{3 - (2\cos^2\phi - 1)}{6\tilde{r}^2} = \frac{2}{3\tilde{r}^2} - \frac{1}{3\tilde{r}^2}\cos^2\phi. \qquad (16.121)$$

Thus finally

$$u = \frac{1}{\tilde{r}}\cos\phi - \frac{\epsilon}{3\tilde{r}^2}\cos^2\phi + \frac{2\epsilon}{3\tilde{r}^2}. \qquad (16.122)$$

The asymptote to the ray is here given by $u \to 0$ $(r \to \infty)$, i.e. by

$$\frac{\epsilon}{3\tilde{r}^2}\cos^2\phi - \frac{1}{\tilde{r}}\cos\phi - \frac{2\epsilon}{3\tilde{r}^2} = 0, \quad \text{or} \quad \cos^2\phi - \frac{3\tilde{r}}{\epsilon}\cos\phi - 2 = 0, \quad (16.123)$$

and (choosing the sign in accordance with $\cos\phi \leq 1$)

$$\begin{aligned} \cos\phi &= \frac{3\tilde{r}}{2\epsilon} \pm \sqrt{\frac{9\tilde{r}^2}{4\epsilon^2} + 2} = \frac{3\tilde{r}}{2\epsilon}\left[1 \overset{(+)}{-} \sqrt{1 + \frac{8\epsilon^2}{9\tilde{r}^2}}\right] \\ &= \frac{3\tilde{r}}{2\epsilon}\left[1 - 1 - \frac{4\epsilon^2}{9\tilde{r}^2}\right] = -\frac{2}{3}\frac{\epsilon}{\tilde{r}} = -\frac{2l}{\tilde{r}}. \end{aligned} \qquad (16.124)$$

Since $l/\tilde{r}$ is small, $\phi \simeq \pm\pi/2$, which is possible for the asymptotes. Setting now for a ray of light passing the sun as in Fig. 16.5

$$\phi = \pm\frac{\pi}{2} + \delta, \qquad (16.125)$$

we have for the asymptotic directions:

$$\cos\left(\pm\frac{\pi}{2} + \delta\right) = -\frac{2l}{\tilde{r}}, \qquad \sin\delta = \pm\frac{2l}{\tilde{r}}, \qquad \delta \simeq \pm\frac{2l}{\tilde{r}}. \qquad (16.126)$$

The deviation of the ray from the ingoing asymptote to the outgoing asymptote is therefore

$$\Delta = 2\delta = \frac{4l}{\tilde{r}} = \frac{4GM}{c^2\tilde{r}}, \qquad \left(\text{since } l \stackrel{(16.59)}{=} \frac{GM}{c^2}\right). \qquad (16.127)$$

For a ray of light tangential to the sun, *i.e.* just grazing the sun, as in Fig. 16.5, one has $\Delta = 1.75''$. This is the result of the standard relativistic formula

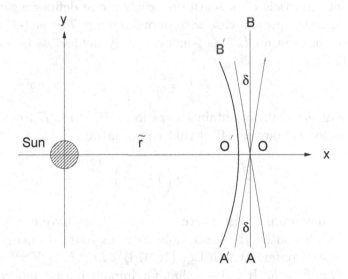

Fig. 16.5 The deflection of a ray of light in the field of the sun.

(16.127), also cited *e.g.* by Carmeli [10].[§] All experiments up to around 1972 examined the case of grazing passage.[¶] Every early such experiment measured the deflection of the light of a star during a total eclipse of the sun.[‖]

[§] M. Carmeli [10], p. 120.

[¶] See in particular W. Misner, K. S. Thorne and J. A. Wheeler [33], p. 1103.

[‖] For a summary of these experiments up to around 1972 see W. Misner, K. S. Thorne and J. A. Wheeler [33], p. 1105.

The observed value of van Biesbroek [5] is $\triangle = 1.7'' \pm 0.1''$. In Fig. 16.5 the straight line $AB, x = $ const. is the path of the ray without the gravitational deflection. In actual fact the ray coming from infinity at A' in direction $A'O$ ($\phi = -\pi/2 - \delta$) is deflected to $A'O'B'$ and then again for $r \to \infty$ follows the direction of OB' ($\delta = \pi/2 + \delta$). Thus altogether the deflection is through angle 2δ. Of course, one can also calculate the deflection of the light of a star at other cases of incidence of the ray. The deflection angle is zero when the ray comes in opposite to the sun's direction.** Thus the bending of light rays by gravity implies that non-radial rays can be captured by a black hole. An unspecified grazing angle measurement quoted by Carmeli (2006, reference above) as one of the latest measurements is $1.75 \pm 0.10''$. Only general relativity and its variant cosmological general relativity predict the correct factor, as Carmeli emphasizes.

In Sec. 16.5 we wrote out the Schwarzschild effective potential for a massless particle, i.e. with $m^2 = 0$, and obtained its maximum at $r_0 = 3l$ outside the horizon at $r_S = 2l$. For the capture of a light ray by the black hole we require a criterion to decide for which rays this will ensue. In analogy to the scattering of a particle off a scattering centre, one defines again an *impact parameter* b. In the present case with photon energy $T = pc$ (*cf.* Eq. (12.48)) and angular momentum L, this is conveniently defined as (since $L = pb$ as in Eq. (7.112))

$$b = \frac{cL}{T}.$$

In the case of $m^2 = 0$, we obtain the ratio $cL/T \equiv cL/E$ from Eq. (16.83) (note the factor m^2 on the left of that equation) as

$$\left. \frac{cL}{E} \right|_{m^2=0} = r\left(1 - \frac{2l}{r}\right)^{-1/2}.$$

Thus at the maximum of V_{eff} where $r \to r_0 = 3l$, we have $b = 3l\sqrt{3} = \sqrt{27}l$. Capture of a photon by the black hole requires that its energy exceed the maximum of the potential (see Eq. (16.97d)), *i.e.* $E > V_{\text{eff}}^{\max} = Lc/\sqrt{27}l$, or $\sqrt{27}l > Lc/E = b$. It follows that for impact parameters $b < \sqrt{27}l$ an incoming photon is captured by the black hole. For $b > \sqrt{27}l$ the photon is scattered by the potential back to infinity.††

We have seen therefore that matter bends space and time. This finding raises the question as to whether one can generate this bending also in the laboratory and bend a time-like world-line even back to itself into a closed loop. This idea would open the possibility to travel back in time, and this,

---

**W. Misner, K. S. Thorne and J. A. Wheeler [33], p. 1103.

††For additional discussion of these and related aspects see D. Raine and E. Thomas [40], p. 29.

of course, stimulates the human imagination. The simplest closed loops are circles and hence suggest a doughnut-shaped vacuum in which spacetime could get bent upon itself using focussed gravitational fields to form the closed time-like curve. To go back in time a traveller would race around the inside of the doughnut, thereby going further back into the past with each lap. This type of travel machine would be spacetime itself. However, the machine could not be used to travel back into a time before the machine was constructed. For a reference on this topic we refer to a recent publication of Ori [37].

Example 16.5: Specific solution of the Schwarzschild ray equation
Show that a specific but not the most general solution of Eq. (16.108) is given by

$$r = \frac{1}{u} = l(1 + \cos \phi), \tag{16.128}$$

and that the general solution involves an elliptic integral.

Solution: We use the relation

$$\frac{d}{du}\left[\frac{1}{2}\left(\frac{du}{d\phi}\right)^2\right] = \frac{d^2u}{d\phi^2} = 3lu^2 - u.$$

Hence, taking first the constant of integration c_0 as zero,

$$\frac{1}{2}\left(\frac{du}{d\phi}\right)^2 = \int^u du(3lu^2 - u) = lu^3 - \frac{u^2}{2},$$

so that

$$\frac{du}{d\phi} = \sqrt{2lu^3 - u^2}, \quad \int^\phi d\phi = \int^u \frac{du}{\sqrt{2lu^3 - u^2}} = \int^u \frac{du}{u\sqrt{2lu - 1}}.$$

The integral in u can be looked up in Tables of Integrals.[‡‡] One obtains

$$\phi = 2\tan^{-1}\sqrt{2lu - 1},$$

so that with $\tan^2\phi = \sec^2\phi - 1$, one has

$$\frac{1}{2lu} = \cos^2\frac{\phi}{2} = \frac{1}{2}[1 + \cos\phi], \quad \therefore \frac{1}{u} = l(1 + \cos\phi).$$

Differentiation of the result with respect to ϕ reproduces Eq. (16.108). But since for $l \to 0$ we obtain $r = 1/u = 0$, this is not the solution above. Thus one has to consider the case $c_0 \neq 0$, so that

$$\int^\phi d\phi = \int \frac{du}{\sqrt{2lu^3 - u^2 + 2c_0}}.$$

The integral in u is now of elliptic type and can be integrated and evaluated by the method outlined in Appendix A. However, for $l \to 0$ one obtains (then setting $\phi_0 = -\pi/2$)

$$\phi - \phi_0 = \sin^{-1}\frac{u_0}{\sqrt{2c_0}}, \quad u_0 = \sqrt{2c_0}\sin(\phi - \phi_0) = \sqrt{2c_0}\cos\phi,$$

and one achieves contact with Eq. (16.115).

[‡‡]H. B. Dwight [13], formula 192.11, p. 42.

Appendix A: Schwarzschild Orbit Solution

A.1 Introductory remarks

In the preface to his *"Jacobian Elliptic Function Tables"* Milne–Thomson [30] begins with the remark: "The widespread belief that calculations involving elliptic functions are difficult is due not to the nature of the calculations themselves but to the lack of suitable numerical tables wherewith to perform them. Calculations of sines and cosines would present analogous difficulties in the absence of tables of trigonometric functions. Elliptic functions are but natural generalizations of trigonometric functions." In many cases the first handicap in earlier days may have been a helplessness in dealing with *e.g.* cubic equations or other expressions in casting them into a standard elliptic form, or in dealing with elliptic integrals. However, today extensive tables of integrals like those of Byrd and Friedman [9] exist, and complicated elliptic expressions can be handled and plotted *e.g.* with MAPLE on computers, so that computational obstacles to the use of elliptic functions and integrals can be overcome with reasonable ease. In this appendix we show that the equation of the orbit resulting from Schwarzschild's solution can be solved in terms of Jacobian elliptic functions. The most immediate but manageable obstacle one encounters is the necessity to solve a cubic equation resulting from the nonlinear Einstein term in the orbit equation.

In Sec. 16.5 we obtained from the Schwarzschild solution to Einstein's equation the equation of the orbit, *i.e.* Eq. (16.100). In Chapter 7 we arrived at the same equation by adding to Kepler's equation (for Newton's law of gravitation) a perturbation term, with an arbitrary perturbation parameter. With Eq. (16.100) this perturbation parameter has been uniquely determined (see comments after Eq. (16.100)). Since we solved the equation perturbatively in Chapter 7, and hence determined the perihelion anomaly, it is an interesting exercise to show that the equation can, in fact, be solved exactly in terms of Jacobian elliptic functions.

A.2   The elliptic integral

Thus our starting point here is Eq. (16.100),

$$\frac{d^2u}{d\phi^2} = \frac{d}{du}\left[\frac{1}{2}\left(\frac{du}{d\phi}\right)^2\right] = \frac{m^2c^2l}{L^2} - u + 3lu^2, \tag{A.1}$$

and we convert this equation into a standard elliptic integral. First by integration we obtain, including an integration constant c_0,

$$\left(\frac{du}{d\phi}\right)^2_u = \frac{2m^2c^2l}{L^2}u - u^2 + 2lu^3 + 2c_0 \equiv f(u), \tag{A.2}$$

or

$$\frac{du}{d\phi} = \pm\sqrt{f(u)}, \quad \int d\phi = (\phi - \phi_0) = \pm\int\frac{du}{\sqrt{f(u)}}. \tag{A.3}$$

In the limit $m^2 \to 0$ which we wish to rederive later as a verification, we have

$$\int d\phi = \pm\int\frac{du}{\sqrt{2lu^3 - u^2}} = \pm\int\frac{du}{u\sqrt{2lu - 1}}. \tag{A.4}$$

The first step in integrating Eq. (A.3) is the determination of the zeros of $f(u) = 0$. To this end one first removes the quadratic term u^2 in $f(u)$ by setting

$$y = u - \frac{1}{6l}, \quad u = y + \frac{1}{6l}. \tag{A.5}$$

With this transformation we obtain

$$f(u) = \frac{2m^2c^2l}{L^2}u - u^2 + 2lu^3 + 2c_0 \equiv 2l[y^3 - Py + Q], \tag{A.6}$$

where

$$P = \frac{1}{12l^2} - \frac{m^2c^2}{L^2} > 0, \quad Q = \frac{m^2c^2}{6lL^2} - \frac{1}{3\times6^2l^3} + \frac{c_0}{l}. \tag{A.7}$$

In the following it is convenient to think of both l and m^2 as being small, so that $1/l$ is large, P is positive and Q is negative. It is also necessary to make some binomial expansions, even to second order, since the first term vanishes in some cases, so that these straight-forward expansions cannot be reproduced here in detail. Thus we consider — here for convenience for $c_0 = 0$ — the ratio (the size of this determines whether the 3 roots are real*)

$$\frac{27Q^2}{4P^3} = (1-3\alpha)^2(1-2\alpha)^{-3} = 1 - 3\alpha^2 + O(\alpha^3) < 1, \quad \alpha = \frac{6l^2m^2c^2}{L^2}. \tag{A.8}$$

---

*L. M. Milne–Thomson [30], p. 37.

Since this ratio is less than 1, it follows — see Milne–Thomson [30] — that there must be three real roots. One now sets

$$\cos 3\theta = -\left(\frac{27Q^2}{4P^3}\right)^{1/2} \simeq -[1 - 3\alpha^2 + O(\alpha^3)]^{1/2} \simeq -1 + \frac{3}{2}\alpha^2. \qquad (A.9)$$

We obtain by similar expansion

$$\gamma := \sqrt{\frac{4P}{3}} \simeq \frac{1}{3l}(1 - \alpha). \qquad (A.10)$$

Next we consider the three roots of the *cubic equation* (*cf.* Eq. (A.6))

$$y^3 - Py \pm Q = 0, \quad P > 0. \qquad (A.11)$$

The three real roots are given by

$$\begin{aligned}
y_1 &= \mp\gamma \cos\theta, \\[4pt]
y_2 &= \mp\gamma \cos(\theta + 2\pi/3) = \mp\gamma\left[-\frac{1}{2}\cos\theta - \frac{\sqrt{3}}{2}\sin\theta\right], \\[4pt]
y_3 &= \mp\gamma \cos(\theta + 4\pi/3) = \mp\gamma\left[-\frac{1}{2}\cos\theta + \frac{\sqrt{3}}{2}\sin\theta\right]. \qquad (A.12)
\end{aligned}$$

It is now necessary to determine the relative order of the three roots, and for this it is useful to consider the following expansions for $\epsilon > 0$ and small. We obtain[†] by setting $\theta = 60° + \epsilon$:

$$\begin{aligned}
\cos\theta &= \cos(60° + \epsilon) = \frac{1}{2}\cos\epsilon - \frac{\sqrt{3}}{2}\sin\epsilon \simeq \frac{1}{2}(1 - \sqrt{3}\epsilon), \\[4pt]
\cos(\theta + 120°) &= -\frac{1}{2}\cos\theta - \frac{\sqrt{3}}{2}\sin\theta = \cdots = -1 + O(\epsilon^2), \\[4pt]
\cos(\theta + 240°) &= -\frac{1}{2}\cos\theta + \frac{\sqrt{3}}{2}\sin\theta = \cdots = \frac{1}{2}(1 + \sqrt{3}\epsilon). \qquad (A.13)
\end{aligned}$$

Further:

$$\cos\theta = \cos(60° + \epsilon) \simeq \frac{1}{2}\left[1 - \sqrt{3}\epsilon - \frac{1}{2}\epsilon^2\right],$$

$$\cos 3\theta = 4\cos^3\theta - 3\cos\theta = \cdots \simeq -1 + \frac{9}{2}\epsilon^2 + O(\epsilon^3) \overset{(A.9)}{\equiv} -1 + \frac{3}{2}\alpha^2,$$

$$(A.14)$$

[†]H. B. Dwight [13], p. 78.

so that $\epsilon^2 \simeq \alpha^2/3$. The three roots (A.12) are in the corresponding approximations

$$y_1 \simeq \overset{-}{(+)} \frac{1}{2}\gamma(1 - \sqrt{3}\epsilon),$$

$$y_2 \simeq \overset{-}{(+)} \gamma(-1),$$

$$y_3 \simeq \overset{-}{(+)} \frac{1}{2}\gamma(1 + \sqrt{3}\epsilon). \tag{A.15}$$

We take the minus signs, so that for $\epsilon \to 0$ we have $y_1 = y_3 = -\gamma/2$, and $u = y_{1,3} + 1/6l = 0$. The order of these zeros,

$$y_3 < y_1 < y_2$$

is illustrated in Fig. A.1. Notice there that $f(u) = 2l(y - y_1)(y - y_2)(y - y_3)$ tends to $+\infty$ for $y \to +\infty$. We have therefore:

$$\int d\phi = \pm \int \frac{du}{\sqrt{f(u)}} = \pm \int \frac{dy}{\sqrt{2l}\sqrt{(y - y_1)(y - y_2)(y - y_3)}}. \tag{A.16}$$

Thus we consider

$$\pm\sqrt{2l} \int d\phi = \int \frac{dy}{\sqrt{(y - y_1)(y - y_2)(y - y_3)}}, \quad \text{with} \quad y_3 < y_1 < y_2. \tag{A.17}$$

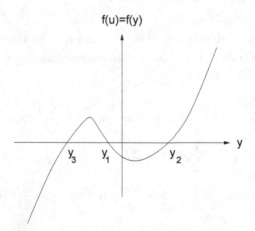

Fig. A.1 Ordering of the roots of the function $f(u)$ in terms of $y = u - 1/6l$.

A.3   Evaluating the elliptic integral

We evaluate the integral with the help of Tables of Elliptic Integrals. Thus we obtain[‡] ($sn[x, k^2]$ being a Jacobian elliptic function of argument x and

---

[‡]We use the first integral given by L. M. Milne–Thomson [30], p. 29.

elliptic modulus k)

$$I := \sqrt{2l}(\phi - \phi_0) \;=\; \int_y^\infty \frac{dy}{\sqrt{(y - y_1)(y - y_2)(y - y_3)}}$$

$$= \frac{2}{\sqrt{y_2 - y_3}} \mathrm{sn}^{-1}\left[\sqrt{\frac{y_2 - y_3}{y - y_3}}, k^2\right], \qquad \text{(A.18)}$$

where the elliptic modulus k is given by

$$k^2 = \frac{y_1 - y_3}{y_2 - y_3} = \frac{1}{2}(1 - \sqrt{3}\cot\theta) \simeq \cdots \simeq \frac{2\epsilon}{\epsilon + \sqrt{3}}. \qquad \text{(A.19)}$$

Here we inserted already the (approximate) expressions from above. Evaluating the expressions appearing here with the approximations developed above, we obtain

$$y_2 - y_3 = \gamma\sqrt{3}\sin\theta \simeq \frac{\sqrt{3}}{2}\gamma(\sqrt{3} + \epsilon). \qquad \text{(A.20)}$$

We obtain therefore with

$$I = \int_y^\infty \frac{dy}{\sqrt{(y - y_1)(y - y_2)(y - y_3)}}, \qquad \lambda := \frac{2}{\sqrt{y_2 - y_3}}$$

and using the relation $\mathrm{sn}^2[x, k^2] + \mathrm{cn}^2[x, k^2] = 1$, by inversion of Eq. (A.18)

$$(y_2 - y_3) = (y - y_3)\mathrm{sn}^2[I/\lambda, k^2] = (y - y_3)[1 - \mathrm{cn}^2[I/\lambda, k^2]], \qquad \text{(A.21)}$$

so that with a reordering of terms

$$(y - y_2) \;=\; (y - y_3)\mathrm{cn}^2[I/\lambda, k^2],$$

$$I/\lambda = \frac{\sqrt{2l}}{\lambda}(\phi - \phi_0) \;=\; \frac{1}{2}(\phi - \phi_0)\rho, \qquad \rho = \sqrt{2\sqrt{3}l\gamma}\sin\theta. \qquad \text{(A.22)}$$

Thus the exact solution in terms of y is

$$y = \frac{y_2 - y_3\mathrm{cn}^2[I/\lambda, k^2]}{1 - \mathrm{cn}^2[I/\lambda, k^2]}. \qquad \text{(A.23)}$$

It is now interesting to go to the limit of $\epsilon \to 0$, which corresponds to $m_0^2 \to 0$. We expect, of course, to obtain the ray solution (16.128). Since $\epsilon \to 0$ implies $k^2 \to 0$, and in this limit the Jacobian elliptic function $\mathrm{cn}[x, k^2]$ becomes $\cos(x)$ we obtain from Eq. (A.22) together with Eq. (A.15):

$$(y - \gamma) \;=\; \left(y + \frac{\gamma}{2}\right)\cos^2[I/\lambda],$$

$$y \;=\; \left(\frac{\gamma}{2}\right)\frac{2 + \cos^2[I/\lambda]}{1 - \cos^2[I/\lambda]} = \left(\frac{\gamma}{2}\right)\frac{2 + \cos^2[I/\lambda]}{\sin^2[I/\lambda]}. \qquad \text{(A.24)}$$

In the limit we are interested in we have:

$$\gamma \overset{(A.10)}{\longrightarrow} \frac{1}{3l}, \quad \lambda \to \frac{2}{\sqrt{3\gamma/2}},$$

$$\therefore \frac{I}{\lambda} \longrightarrow \frac{\sqrt{2l}(\phi - \phi_0)}{2}\sqrt{\frac{3\gamma}{2}} = \frac{1}{2}(\phi - \phi_0). \tag{A.25}$$

Hence, since $y = u - 1/6l$, we obtain

$$u = \frac{1}{6l} + \frac{1}{6l}\frac{2 + \cos^2[I/\lambda]}{\sin^2[I/\lambda]} = \frac{3}{6l\sin^2[I/\lambda]},$$

$$u = \frac{1}{2l\sin^2[(\phi - \phi_0)/2]}. \tag{A.26}$$

With the relation $\cos(\theta) = 1 - 2\sin^2(\theta/2)$ we obtain

$$ul = \frac{1}{1 - \cos(\phi - \phi_0)}, \quad \frac{1}{ul} = 1 - \cos(\phi - \phi_0), \tag{A.27}$$

and with the choice of $\phi_0 = \pi$, we obtain

$$\frac{1}{ul} = 1 + \cos\phi, \tag{A.28}$$

and have thus regained the special case of the equation for the light ray, Eq. (16.128). Rewriting the solution (A.23) in terms of u we obtain

$$\begin{aligned} u(\phi) &= \frac{1}{6l} + \frac{y_2 - y_3 \text{cn}^2[I/\lambda \equiv \rho(\phi - \phi_0)/2, k^2]}{1 - \text{cn}^2[I/\lambda \equiv \rho(\phi - \phi_0)/2, k^2]} \\ &= \frac{1}{6l} \mp \gamma \frac{\cos(\theta + 2\pi/3) - \cos(\theta + 4\pi/3)\text{cn}^2[I/\lambda \equiv \rho(\phi - \phi_0)/2, k^2]}{1 - \text{cn}^2[I/\lambda \equiv \rho(\phi - \phi_0)/2, k^2]}. \\ &\equiv \frac{1}{r(\phi)}. \end{aligned} \tag{A.29}$$

This solution could now be investigated in more detail numerically or graphically, *e.g.* with MAPLE.

Appendix B:
Reissner–Nordstrom Metric

B.1 Introductory remarks

It is an instructive calculation to derive the metric for a charged particle in the gravitational field. This solution of *Einstein's equation* is frequently cited, but its derivation is rarely given in detail. Therefore we present in some detail the calculations for the case of an electrostatic field in the presence of the gravitational field, which is again assumed to be static, and in the 3-dimensional space part spherically symmetric, although in the first place we shall not assume invariance under time inversion, which means we do not immediately consider $g_{0k} = 0$. In the following we make use of detailed calculations contained in the article of Leite–Lopes [28].

B.2 The metric

We consider first a static metric (no time dependence in the coefficients) with spherical symmetry, *i.e.* we consider $(g_{0k} = g_{k0})$

$$ds^2 = g_{\mu\nu}(x)dx^\mu dx^\nu = g_{00}(dx^0)^2 + 2g_{0k}dx^0dx^k + g_{ik}dx^i dx^k. \qquad \text{(B.1)}$$

We demand that for $r \to \infty$ (superscript L for Lorentz):

$$g_{\mu\nu}(x) \overset{r\to\infty}{\longrightarrow} g^L_{\mu\nu}(x), \quad ds^2_\infty = (dx^0)^2 - (d\mathbf{x})^2. \qquad \text{(B.2)}$$

We set (observe the coefficient functions are assumed to be functions of $r, r^2 = \sum_i (x^i)^2$, and the definitions involve onlx x-components with upper indices):

$$g_{00}(x) := A(r), \quad g_{0k}(x) := B(r)\frac{x^k}{r}, \quad g_{ik}(x) := -C(r)\delta^{ik} + D(r)\frac{x^i x^k}{r^2}.$$

$$\text{(B.3)}$$

Hence

$$ds^2 := A(r)(dx^0)^2 + 2B(r)\frac{x^k}{r}dx^0 dx^k - C(r)\delta^{ik}dx^i dx^k$$

$$+ D(r)\frac{x^i x^k}{r^2}dx^i dx^k. \tag{B.4}$$

In terms of some transformed coordintes $(x \to x')$ this last expression is

$$ds^2 = A'(r')(dx'^0)^2 + 2B'(r')\frac{x'^k}{r'}dx'^0 dx'^k - C'(r')\delta^{ik}dx'^i dx'^k$$

$$+ D'(r')\frac{x'^i x'^k}{r'^2}dx'^i dx'^k. \tag{B.5}$$

In the first place we demonstrate — in two separate steps (a) and (b) — that one can choose coordinates in such a way that $g_{0k} = 0$ and $C(r) = 1$.

(a)   We first show that we can choose coordinates such that $g_{0k} = 0$. To this end we consider the following transformation from coordinates x to x', and the inversion from x' to x:

$$x'^0 = x^0 + F(r), \quad \text{or} \quad x^0 = x'^0 - f(r'),$$
$$x'^k = x^k, \quad r' = r, \quad \text{or} \quad x^k = x'^k, \quad r = r'. \tag{B.6}$$

Then (*cf.* the transformation of tensors, Eq. (14.1)), since $\partial x^j / \partial x'^0 = 0$,

$$g'_{\mu\nu}(x') = \frac{\partial x^\alpha}{\partial x'^\mu}\frac{\partial x^\beta}{\partial x'^\nu}g_{\alpha\beta}(x)$$

$$\therefore g'_{0k}(x') = \frac{\partial x^0}{\partial x'^0}\frac{\partial x^0}{\partial x'^k}g_{00}(x) + \frac{\partial x^0}{\partial x'^0}\frac{\partial x^j}{\partial x'^k}g_{0j}(x) + 0 + 0$$

$$= \frac{\partial x^0}{\partial x'^k}g_{00}(x) + g_{0k}(x) = g_{0k}(x) - \frac{\partial f(r')}{\partial x'^k}g_{00}(x). \tag{B.7}$$

With the definitions of Eqs. (B.3) to (B.5) we obtain from the last equation

$$B'(r')\frac{x'^k}{r'} \overset{(B.5)}{\equiv} g'_{0k}(x') \overset{(B.7)}{\equiv} B(r)\frac{x^k}{r} - \frac{df(r')}{dr'}\frac{\partial r'}{\partial x'^k}g_{00}(x)$$

$$\overset{(B.6)}{\equiv} B(r')\frac{x'^k}{r'} - \frac{df(r')}{dr'}\frac{x'^k}{r'}g_{00}(x)$$

$$= \left(B(r') - \frac{df(r')}{dr'}A(r') \right)\frac{x'^k}{r'},$$

i.e.

$$B'(r') = B(r') - \frac{df(r')}{dr'}A(r'). \tag{B.8}$$

Hence if we choose the function $f(r')$ in the coordinate transformation (B.6) such that

$$\frac{df(r')}{dr'} = \frac{B(r')}{A(r')}, \tag{B.9}$$

we have

$$g'_{00}(x') = A'(r'), \quad g'_{0k}(x') = 0, \quad g'_{ik}(x') = -C'(r')\delta^{ik} + D'(r')\frac{x'^i x'^k}{r'^2}. \tag{B.10}$$

(b) Next we show that in addition we can choose the system of co-ordinates, *i.e.* transform to a new system, such that in the new frame $C(r) = 1$. We consider now the following transformation of the spatial coordinates $(k = 1, 2, 3)$:

$$x''^k = \phi(r)x^k, \quad x^k = \psi(r'')x''^k. \tag{B.11}$$

With these equations we have:

$$r''^2 = \sum_k (x''^k)^2 = \phi^2(r)\sum_k (x^k)^2 \equiv \phi^2(r)r^2,$$

$$r^2 \equiv \sum_k (x^k)^2 = \psi^2(r'')\sum_k (x''^k)^2 = \psi^2(r'')\phi^2(r)\sum_k (x^k)^2,$$

so that

$$\psi^2(r'')\phi^2(r) = 1, \quad \psi(r'')\phi(r) = +1 \tag{B.12}$$

($+1$ for the so-called "*proper*" transformation). Hence

$$r''^2 = \frac{r^2}{\psi^2(r'')}, \quad r = \psi(r'')r'', \quad r'' = \phi(r)r,$$

$$\frac{x''^k}{r''} = \frac{\phi(r)\,x^k}{\phi(r)\,r} = \frac{x^k}{r}, \tag{B.13}$$

and

$$\frac{r''}{r} = \frac{1}{\psi(r'')}, \quad r = \psi(r'')r''.$$

Now $(i, k, m, n, \ldots = 1, 2, 3$ and $\partial r/\partial x^i = \partial\sqrt{\sum_j (x^j)^2}/\partial x^i = x^i/r)$

$$\frac{\partial x^m}{\partial x''^i} \overset{(B.11)}{=} \frac{\partial}{\partial x''^i}(x''^m \psi(r'')) = \psi(r'')\delta^{im} + x''^m \frac{\partial\psi(r'')}{\partial r''}\frac{\partial r''}{\partial x''^i}$$

$$= \psi(r'')\delta^{im} + x''^m \frac{\partial\psi(r'')}{\partial r''}\frac{x''^i}{r''}. \tag{B.14}$$

Hence

$$g''_{ik}(x'') = \frac{\partial x^m}{\partial x''^i}\frac{\partial x^n}{\partial x''^k}g_{mn}(x) = \left(\psi(r'')\delta^{im} + x''^m\frac{\partial\psi(r'')}{\partial r''}\frac{x''^i}{r''}\right)$$

$$\times\left(\psi(r'')\delta^{kn} + x''^n\frac{\partial\psi(r'')}{\partial r''}\frac{x''^k}{r''}\right)g_{mn}(x). \qquad (B.15)$$

Here

$$\frac{x''^m x''^i}{r''} = \frac{\phi^2(r)x^m x^i}{r}\psi(r'') = \frac{x^m x^i}{r\psi(r'')} = \frac{r''x^m x^i}{r^2}, \qquad (B.16)$$

so that, with $g_{nm}(x)$ replaced by the expression in Eq. (B.3) and x''^k by $x^k r''/r$ of Eq. (B.13),

$$g''_{ik}(x'')$$

$$\overset{(B.14)}{=} \left(\psi(r'')\delta^{im} + r''\frac{x^m x^i}{r^2}\frac{\partial\psi(r'')}{\partial r''}\right)\left(\psi(r'')\delta^{kn} + r''\frac{x^n x^k}{r^2}\frac{\partial\psi(r'')}{\partial r''}\right)$$

$$\times\left(-C(r)\delta^{mn} + D(r)\frac{x^m x^n}{r^2}\right),$$

and multiplied out

$$g''_{ik}(x'') = -\psi^2(r'')C(r)\delta^{ik} + \left[\psi^2(r'')\frac{x^i x^k}{r^2} + 2\psi(r'')r''\frac{x^i x^k}{r^2}\frac{\partial\psi(r'')}{\partial r''}\right.$$

$$\left. + r''^2\frac{x^i x^k}{r^2}\left(\frac{\partial\psi(r'')}{\partial r''}\right)^2\right]D(r)$$

$$-2r''\frac{\partial\psi(r'')}{\partial r''}\frac{x^i x^k}{r^2}\psi(r'')C(r) - r''^2\left(\frac{\partial\psi(r'')}{\partial r''}\right)^2 C(r)\frac{x^i x^k}{r^2},$$

or

$$g''_{ik}(x'') = -\psi^2(r'')C(r)\delta^{ik} + \left[\left(\psi(r'') + r''\frac{d\psi(r'')}{dr''}\right)^2 D(r)\right.$$

$$\left. -r''\frac{d\psi(r'')}{dr''}\left(2\psi(r'') + r''\frac{d\psi(r'')}{dr''}\right)C(r)\right]\frac{x^i x^k}{r^2}. \qquad (B.17)$$

We can now show that the expression contained in $[\cdots]$ is a function $\mathcal{H}(r)$. This can be shown by differentiating with respect to r'' the two equations $r = \psi(r'')r''$, $\psi(r'')\phi(r) = 1$ (cf. Eqs. (B.12) and (B.13)), and using $r'' = r\phi(r)$. Equating from both cases dr/dr'', one obtains an equation which yields

$$\frac{d\psi(r'')}{dr''} = -[\phi^2(r)\{r + \phi(r)/(d\phi/dr)\}]^{-1}.$$

Thus if we choose $\psi(r'') = 1/\phi(r) = C^{-1/2}(r)$ we have:

$$g''_{ik}(x'') = -\delta^{ik} + \mathcal{H}(r)\frac{x^i x^k}{r^2} \tag{B.18}$$

(which corresponds to the choice $C(r) = 1$ in Eq. (B.3)) with

$$\begin{aligned}
g_{00} &= A(r) = e^{\nu(r)}, \quad y_{0k} = 0, \\
g_{ik} &= -\delta^{ik} + \mathcal{H}(r)\frac{x^i x^k}{r^2} \equiv -\delta^{ik} + (1 - e^{\lambda(r)})\frac{x^i x^k}{r^2}. \tag{B.19}
\end{aligned}$$

If we now go to spherical coordinates,

$$x = r\cos\phi\sin\theta, \quad y = r\sin\phi\sin\theta, \quad z = r\cos\theta,$$

implying

$$(dx)^2 + (dy)^2 + (dz)^2 = (d\mathbf{x})^2 = r^2(\sin^2\theta d\phi^2 + d\theta^2) + dr^2,$$

we have (note that $x^i dx^i/r = d(x^i)^2/2r = dr^2/2r = dr$)

$$\begin{aligned}
ds^2 &= e^{\nu(r)}(dx^0)^2 + \left[-\delta^{ik} + (1 - e^{\lambda(r)})\frac{x^i x^k}{r^2}\right]dx^i dx^k \\
&= e^{\nu(r)}(dx^0)^2 + [-d\mathbf{x}^2 + dr^2 - e^{\lambda(r)}dr^2] \\
&= e^{\nu(r)}(dx^0)^2 - e^{\lambda(r)}dr^2 - r^2(\sin^2\theta d\phi^2 + d\theta^2). \tag{B.20}
\end{aligned}$$

We see that the general form of ds^2 corresponds to that of the Schwarzschild solution (16.23).

B.3   The energy-momentum tensor

For the solution of Einstein's equation in the present case we also need the energy-momentum tensor of the electromagnetic field of a static point charge. We derived the energy-momentum tensor of the electromagnetic field in Sec. 15.4.1 . We recall the following points. The energy-momentum tensor $T_{\mu\nu}$ is effectively the 4-dimensional generalization of the Legendre transform, i.e. (cf. Eq. (6.70))

$$T_{\mu\nu} = \frac{\partial\mathcal{L}}{\partial(\partial A^\lambda/\partial x_\mu)}\frac{\partial A^\lambda}{\partial x^\nu} - \mathcal{L}g_{\mu\nu}. \tag{B.21}$$

In the case of the electromagnetic field the Lagrangian density (actually that of the free field, which we consider as that not at the point charge) is given by

$$\mathcal{L} = -\frac{1}{4}F_{\mu\nu}F^{\mu\nu}, \quad F_{\mu\nu} = \partial_\mu A_\nu - \partial_\nu A_\mu. \tag{B.22a}$$

Then:

$$\frac{\partial \mathcal{L}}{\partial(\partial A^\lambda/\partial x_\mu)} \equiv \frac{\partial \mathcal{L}}{\partial(\partial^\mu A^\lambda)} = F_{\lambda\mu},$$

and

$$T_{\mu\nu} = F_{\lambda\mu}\frac{\partial A^\lambda}{\partial x^\nu} + \frac{1}{4}g_{\mu\nu}F_{\alpha\beta}F^{\alpha\beta}. \tag{B.22b}$$

We symmetrize $T_{\mu\nu}$ by addition of the quantity $-F_{\lambda\mu}\partial A_\nu/\partial x_\lambda$ which does not contribute when integrated over since

$$F_{\lambda\mu}\frac{\partial A_\nu}{\partial x_\lambda} = \frac{\partial}{\partial x_\lambda}(F_{\lambda\mu}A_\nu) - \frac{\partial F_{\lambda\mu}}{\partial x_\lambda}A_\nu.$$

Here the first term on the right is a divergence which when integrated over gives a surface integral at infinity where the field is zero,[§] and the second term is zero in the absence of a current j_μ, *i.e.*

$$\partial^\lambda F_{\lambda\mu} = j_\mu. \tag{B.23}$$

Hence

$$T_{\mu\nu} = F_{\lambda\mu}\left(\frac{\partial A^\lambda}{\partial x^\nu} - \frac{\partial A_\nu}{\partial x_\lambda}\right) + \frac{1}{4}g_{\mu\nu}F_{\alpha\beta}F^{\alpha\beta},$$

i.e.

$$T_{\mu\nu} = F_{\mu\lambda}F^\lambda{}_\nu + \frac{1}{4}g_{\mu\nu}F_{\alpha\beta}F^{\alpha\beta} \tag{B.24}$$

in agreement with Eqs. (15.66), (15.76). That $T_{\mu\nu} = T_{\nu\mu}$ can be seen as follows. We have for the first part of $T_{\mu\nu}$ with $\mu, \nu \to \nu, \mu$:

$$T_{\nu\mu}^{(1)} = F_{\nu\lambda}F^\lambda{}_\mu = g^{\lambda\alpha}F_{\alpha\mu}F_{\nu\lambda} = g^{\alpha\lambda}F_{\lambda\nu}F_{\mu\alpha} = g^{\lambda\alpha}F_{\alpha\nu}F_{\mu\lambda} = F_{\mu\lambda}F^\lambda{}_\nu = T_{\mu\nu}^{(1)},$$

which had to be shown. Multiplying $T_{\nu\mu}$ by $g^{\mu\nu}$, we obtain (note the minus sign occurring in the shift of the upper index λ in $F^\lambda_\mu = -F_\mu{}^\lambda$):

$$\begin{aligned}
g^{\mu\nu}T_{\nu\mu} &= T^\mu{}_\mu = g^{\mu\nu}\left(F_{\nu\lambda}F^\lambda{}_\mu + \frac{1}{4}g_{\nu\mu}F_{\alpha\beta}F^{\alpha\beta}\right) = F^\mu{}_\lambda F^\lambda{}_\mu + \frac{1}{4}\delta^\mu_\mu F_{\alpha\beta}F^{\alpha\beta} \\
&= F^\mu{}_\lambda F^\lambda{}_\mu + F_{\alpha\beta}F^{\alpha\beta} = g^{\mu\alpha}F_{\alpha\lambda}g^{\lambda\beta}F_{\beta\mu} + F_{\alpha\beta}F^{\alpha\beta} \\
&= F_{\alpha\lambda}F^{\lambda\alpha} + F_{\alpha\beta}F^{\alpha\beta} = -F_{\alpha\lambda}F^{\alpha\lambda} + F_{\alpha\beta}F^{\alpha\beta} = 0, \\
\therefore \quad T^\mu{}_\mu &= \mathrm{Tr}(T^\mu{}_\nu) = 0. \tag{B.25}
\end{aligned}$$

[§]These are standard assumptions which are physically sensible although in the specific case of the free electromagnetic field of infinite range not so easy to justify. Such a justification is beyond the scope of the present text.

B.4 The energy-momentum tensor for an electrostatic field

In the case of an *electrostatic field* (A^0 is the scalar field usually written $\phi(r)/c$, where c is the velocity of light)

$$A^k = 0, \quad k = 1, 2, 3; \quad A^0 = A^0(r). \tag{B.26}$$

In this case we have

$$T_{00} \overset{(B.24)}{=} F_{0\lambda} F^\lambda{}_0 + \frac{1}{4} g_{00} F_{\alpha\beta} F^{\alpha\beta}. \tag{B.27}$$

Here we have only the electric field (one index of $F_{\mu\nu}$ is zero)

$$F_{k0} = \frac{\partial A_0}{\partial x^k} - \frac{\partial A_k}{\partial x^0} = \frac{\partial A_0}{\partial x^k} = \frac{dA_0}{dr}\frac{x^k}{r} = -F_{0k}, \quad F_{ik} = 0,$$

$$F^k{}_0 = g^{k\alpha} F_{\alpha 0} = g^{kl} F_{l0}. \tag{B.28}$$

Thus

$$F_{0\lambda} F^\lambda{}_0 = F_{0k} F^k{}_0 = g^{kl} F_{0k} F_{l0} = -g^{kl} F_{0k} F_{0l},$$
$$F_{\alpha\beta} F^{\alpha\beta} = F_{0k} F^{0k} + F_{k0} F^{k0} = 2g^{k\beta} g^{0\alpha} F_{0k} F_{\alpha\beta} = 2g^{kl} g^{00} F_{0k} F_{0l}, \tag{B.29}$$

and

$$T_{00} = -g^{kl} F_{0k} F_{0l} + \frac{1}{2} g^{kl} F_{0k} F_{0l} g^{00} g_{00}. \tag{B.30}$$

Now, $g^{\mu\lambda} g_{\lambda\nu} = \delta^\mu_\nu$, so that $g^{0\lambda} g_{\lambda 0} = 1$ and $g^{00} g_{00} + g^{0k} g_{k0} = 1$, but $g_{0k} = 0$. Hence $g_{00} g^{00} = 1$. Since we have (*cf.* Eq. (B.19)) $g_{00} = e^\nu$, we have

$$g^{00} = e^{-\nu(r)}, \quad g_{00} = e^{+\nu(r)}. \tag{B.31}$$

Then

$$T_{00} = -\frac{1}{2} g^{kl} F_{0k} F_{0l}. \tag{B.32}$$

From Eq. (B.19) we deduce that

$$g_{0k} = 0, \quad g_{ik} = -\delta^{ik} + (1 - e^{\lambda(r)}) \frac{x^i x^k}{r^2}. \tag{B.33}$$

We now need g^{ik}, *i.e.* with indices "upstairs". By checking that $\sum_\mu g^{i\mu} g_{\mu l} = g^{i0} g_{0l} + \sum_k g^{ik} g_{kl} = \delta^i_l$, one can verify that[¶]

$$g^{ik} = -\delta^{ik} + (1 - e^{-\lambda(r)}) \frac{x^i x^k}{r^2}. \tag{B.34}$$

[¶] The product can be seen to be (ignoring indices): $1 + (1-e^\lambda)(1-e^{-\lambda}) - (1-e^{-\lambda}) - (1-e^\lambda) = 1$.

Hence, from Eqs. (B.32) and (B.34):

$$
\begin{aligned}
T_{00} &= -\frac{1}{2}\left[-\delta^{ik} + (1 - e^{-\lambda(r)})\frac{x^i x^k}{r^2}\right]F_{0i}F_{0k} \\
&= \frac{1}{2}(F_{0k})^2 - \frac{1}{2}(1 - e^{-\lambda(r)})\frac{x^i}{r}F_{0i}\frac{x^k}{r}F_{0k}.
\end{aligned}
\tag{B.35}
$$

But from Eqs. (B.27) to (B.28) we have:

$$
F_{0k} = -\frac{dA_0}{dr}\frac{x^k}{r} \equiv -A_0'(r)\frac{x^k}{r},
\tag{B.36}
$$

so that

$$
T_{00} = \frac{1}{2}(A_0')^2 - \frac{1}{2}(A_0')^2 + \frac{1}{2}e^{-\lambda(r)}(A_0')^2 = \frac{1}{2}e^{-\lambda(r)}(A_0')^2.
\tag{B.37a}
$$

Also, since $g_{0k} = 0$:

$$
T_{k0} \overset{(B.24)}{=} F_{k\lambda}F^\lambda{}_0 = F_{k0}F^0{}_0 = 0.
\tag{B.37b}
$$

Now,

$$
\begin{aligned}
T_{kl} &\overset{(B.24)}{=} F_{k\lambda}F^\lambda{}_l + \frac{1}{4}g_{kl}F_{\alpha\beta}F^{\alpha\beta} = F_{k0}F^0{}_l + \frac{1}{4}g_{kl}F_{m0}F^{m0} \\
&\overset{(B.29)}{=} F_{k0}F^0{}_l + \frac{1}{4}g_{kl}2g^{mn}g^{00}F_{0m}F_{0n}.
\end{aligned}
$$

Here

$$
F_{k0}F^0{}_l = F_{k0}g^{0\alpha}F_{\alpha l} = F_{k0}g^{00}F_{0l}.
\tag{B.38}
$$

Hence

$$
T_{kl} = g^{00}\left(F_{k0}F_{0l} + \frac{1}{2}g_{kl}g^{mn}F_{0m}F_{0n}\right).
\tag{B.39}
$$

Inserting g^{00} of Eq. (B.31), F_{0k} and g_{ik} and g^{ik} of Eqs. (B.33) and (B.34), one obtains:

$$
\begin{aligned}
T_{kl} &= g^{00}\left(F_{k0}F_{0l} + \frac{1}{2}g_{kl}g^{mn}F_{0m}F_{0n}\right) \\
&= e^{-\nu(r)}\Bigg[-(A_0')^2\frac{x^k x^l}{r^2} + \frac{1}{2}\left\{-\delta^{kl} + (1 - e^{\lambda(r)})\frac{x^k x^l}{r^2}\right\} \\
&\quad \times \left\{-\delta^{mn} + (1 - e^{-\lambda(r)})\frac{x^m x^n}{r^2}\right\}(A_0')^2\frac{x^m x^n}{r^2}\Bigg].
\end{aligned}
$$

Performing the multiplications here, we have

$$T_{kl} = e^{-\nu(r)}(A'_0)^2 \left[-\frac{x^k x^l}{r^2} + \frac{1}{2}\delta^{kl} - \frac{1}{2}(1 - e^{+\lambda(r)})\frac{x^k x^l}{r^2}\delta^{mn}\frac{x^m x^n}{r^2} \right.$$

$$\left. -\frac{1}{2}\delta^{kl}(1 - e^{-\lambda(r)}) + \frac{1}{2}(1 - e^{+\lambda(r)})(1 - e^{-\lambda(r)})\frac{x^k x^l}{r^2}\frac{x^{m2} x^{n2}}{r^2 r^2} \right]$$

$$= e^{-\nu(r)}(A'_0)^2 \left[\frac{1}{2}\delta^{kl} e^{-\lambda(r)} - \frac{1}{2}\frac{x^k x^l}{r^2} + \frac{1}{2}e^{-\lambda(r)}\frac{x^k x^l}{r^2} \right].$$

Thus finally

$$T_{kl} = \frac{1}{2}e^{-(\nu(r)+\lambda(r))}(A'_0)^2 \left[-e^{\lambda(r)}\frac{x^k x^l}{r^2} + \left(\delta^{kl} - \frac{x^k x^l}{r^2} \right) \right]. \qquad (B.40)$$

B.5 Christoffel symbols and Riemann tensor

Having obtained the explicit expressions of the components of the metric tensor $g_{\mu\nu}$, we can proceed and use these to evaluate Christoffel symbols and elements of the Riemann tensor. The *Einstein equation* is (*cf.* Eq. (15.86))

$$R_{\mu\nu} - \frac{1}{2}g_{\mu\nu}R = -\kappa T_{\mu\nu}. \qquad (B.41)$$

Multiplying the equation by $g^{\mu\nu}$, we obtain in terms of the Ricci tensor R of Eq. (15.79):

$$R - 2R = -\kappa T^{\mu}{}_{\mu}, \quad \therefore \ R = \kappa T^{\mu}{}_{\mu}. \qquad (B.42)$$

Inserting this result into Eq. (B.41), we have

$$R_{\mu\nu} = -\kappa\left(T_{\mu\nu} - \frac{1}{2}g_{\mu\nu}T^{\alpha}{}_{\alpha} \right). \qquad (B.43)$$

In the present case with Eq. (B.25) we have

$$R_{\mu\nu} = -\kappa T_{\mu\nu}. \qquad (B.44)$$

Since $T_{k0} = 0$ (*cf.* Eq. (B.37b)) we have the case of

$$R_{00} = -\kappa T_{00}, \quad R_{k0} = 0, \quad R_{kl} = -\kappa T_{kl}. \qquad (B.45)$$

We now consider the *Christoffel symbols*

$$\Gamma^{\alpha}_{\mu\nu} = \frac{1}{2}g^{\alpha\lambda}\left(\frac{\partial g_{\mu\lambda}}{\partial x^{\nu}} + \frac{\partial g_{\nu\lambda}}{\partial x^{\mu}} - \frac{\partial g_{\mu\nu}}{\partial x^{\lambda}} \right), \qquad (B.46)$$

and collect first from Eqs. (B.31) to (B.34) the results for for $g_{\mu\nu}$ and $g^{\mu\nu}$. We have

$$g_{00} = e^{\nu(r)}, \quad g_{0k} = 0, \quad g_{kl} = -\delta^{kl} + (1 - e^{\lambda(r)})\frac{x^k x^l}{r^2},$$

$$g^{00} = e^{-\nu(r)}, \quad g^{k0} = 0, \quad g^{kl} = -\delta^{kl} + (1 - e^{-\lambda(r)})\frac{x^k x^l}{r^2}. \quad \text{(B.47)}$$

Thus it is now necessary to insert these expressions into Eq. (B.46). We do not give all the calculational details below and so cite the results along with some typical calculational steps as examples. Consider

$$
\begin{aligned}
\Gamma^k_{00} &= \frac{1}{2}g^{kl}\left(\frac{\partial g_{0l}}{\partial x^0} + \frac{\partial g_{0l}}{\partial x^0} - \frac{\partial g_{00}}{\partial x^l}\right) = -\frac{1}{2}g^{kl}\frac{\partial g_{00}}{\partial x^l} \\
&= -\frac{1}{2}\nu'(r)\frac{x^l}{r}\left[-\delta^{kl} + (1 - e^{-\lambda(r)})\frac{x^k x^l}{r^2}\right]e^{\nu(r)} \\
&= \frac{1}{2}\nu'(r)\left[\frac{x^k}{r} - (1 - e^{-\lambda(r)})\frac{x^k}{r}\right]e^{\nu(r)} \\
&= \frac{1}{2}\nu'(r)e^{(\nu(r)-\lambda(r))}\frac{x^k}{r}. \quad\quad\quad\quad\quad\quad\quad\quad\quad\quad \text{(B.48)}
\end{aligned}
$$

Analogously one obtains:

$$\Gamma^0_{0k} = \frac{1}{2}\nu'(r)\frac{x^k}{r}, \quad\quad\quad \text{(B.49)}$$

and after a lengthier calculation:

$$\Gamma^j_{kl} = \frac{x^j}{r}\left[\frac{1}{2}\frac{x^k x^l}{r^2}\lambda'(r) + \frac{1 - e^{-\lambda(r)}}{r}\left(\delta^{kl} - \frac{x^k x^l}{r^2}\right)\right], \quad \Gamma^l_{kl} = \frac{x^k}{2r}\lambda'(r). \quad \text{(B.50)}$$

Also:

$$\Gamma^0_{kl} = 0, \quad \Gamma^0_{00} = 0. \quad\quad\quad \text{(B.51)}$$

We now have to insert these expressions into $R_{\mu\nu}$, i.e.

$$R_{\mu\nu} \overset{(15.28)}{=} (\Gamma^\alpha_{\alpha\nu})_{,\mu} - (\Gamma^\alpha_{\mu\nu})_{,\alpha} + \Gamma^\alpha_{\eta\mu}\Gamma^\eta_{\alpha\nu} - \Gamma^\alpha_{\eta\alpha}\Gamma^\eta_{\mu\nu}. \quad \text{(B.52)}$$

Again we evaluate only a typical example and cite the results for others. Consider

$$
\begin{aligned}
R_{00} &= (\Gamma^\alpha_{\alpha 0})_{,0} - (\Gamma^\alpha_{00})_{,\alpha} + \Gamma^\alpha_{\eta 0}\Gamma^\eta_{\alpha 0} - \Gamma^\alpha_{\eta\alpha}\Gamma^\eta_{00} \\
&\overset{(B.51)}{=} 0 - (\Gamma^k_{00})_{,k} + \Gamma^0_{k0}\Gamma^k_{00} + \Gamma^k_{00}\Gamma^0_{k0} - \Gamma^0_{k0}\Gamma^k_{00} - \Gamma^l_{kl}\Gamma^k_{00}.
\end{aligned}
$$

Here the first contribution vanishes since this is the time derivative of a time-independent quantity. For the other contributions we obtain:

$$(\Gamma_{00}^k)_{,k} \overset{(B.48)}{=} \frac{1}{2}\nu'(r)e^{(\nu(r)-\lambda(r))}\frac{3}{r} + \frac{1}{2}\nu'(r)e^{(\nu(r)-\lambda(r))}\left(-\frac{1}{r^2}\right)\frac{x^k x^k}{r}$$

$$+ \frac{1}{2}\nu''(r)e^{(\nu(r)-\lambda(r))}\frac{x^{k2}}{r^2} + \frac{1}{2}\nu'(r)(\nu'(r)$$

$$- \lambda'(r))e^{(\nu(r)-\lambda(r))}\frac{x^{k2}}{r^2}$$

$$= e^{(\nu(r)-\lambda(r))}\left[\frac{\nu'(r)}{r} + \frac{1}{2}\nu'(r)(\nu'(r) - \lambda'(r)) + \frac{\nu''(r)}{2}\right],$$

$$\Gamma_{k0}^0\Gamma_{00}^k = \frac{1}{2}\nu'(r)\frac{x^k}{r}\frac{1}{2}\nu'(r)e^{(\nu(r)-\lambda(r))}\frac{x_k}{r} = \frac{1}{4}(\nu'(r))^2 e^{(\nu(r)-\lambda(r))},$$

$$\Gamma_{lol}^l\Gamma_{00}^k = \frac{1}{r}\left[\frac{1}{2}x^k\lambda'(r) + \frac{1-e^{-\lambda(r)}}{r}(x^{lo} - x^k)\right]\frac{1}{2}\nu'(r)e^{(\nu(r)-\lambda(r))}\frac{x^k}{r}$$

$$= \frac{1}{4}\nu'(r)\lambda'(r)e^{\nu(r)-\lambda(r)}.$$

Hence

$$R_{00} = -e^{(\nu(r)-\lambda(r))}\left[\frac{\nu''(r)}{2} + \frac{\nu'(r)}{r} + \frac{1}{4}\nu'(r)(\nu'(r) - \lambda'(r))\right].$$

We cite the results for the rest and have altogether:

$$R_{0k} = 0,$$

$$R_{00} = -e^{(\nu(r)-\lambda(r))}\left[\frac{\nu''(r)}{r} + \frac{\nu'(r)}{r} + \frac{1}{4}\nu'(r)(\nu'(r) - \lambda'(r))\right],$$

$$R_{kl} = \left[\frac{1}{2}\nu''(r) - \frac{\lambda'(r)}{r} + \frac{1}{4}(\nu'(r))^2 - \frac{1}{4}\nu'(r)\lambda'(r)\right]\frac{x^k x^l}{r^2}$$

$$+ \left[-\frac{1}{r^2} + e^{-\lambda(r)}\left\{\frac{1}{r^2} + \frac{\nu'(r) - \lambda'(r)}{2r}\right\}\right]\left(\delta^{kl} - \frac{x^k x^l}{r^2}\right),$$

$$T_{0k} \overset{(B.37b)}{=} 0, \qquad T_{00} \overset{(B.37a)}{=} \frac{1}{2}e^{-\lambda(r)}(A_0'(r))^2,$$

$$T_{kl} \overset{(B.40)}{=} \frac{1}{2}e^{-(\nu(r)+\lambda(r))}(A_0'(r))^2\left[-e^{\lambda(r)}\frac{x^k x^l}{r^2}\right.$$

$$\left. + \left(\delta^{kl} - \frac{x^k x^l}{r^2}\right)\right]. \tag{B.53}$$

B.6   The Einstein equation

The component equations of *Einstein's equation* are (*cf.* Eq. (B.45)):

$$(a)\quad R_{00} = -\kappa T_{00}, \quad (b)\quad R_{kl} = -\kappa T_{kl}. \tag{B.54}$$

In our case these equations are with insertion of R_{00} and R_{kl}:

$$(a)\quad e^{(\nu-\lambda)}\left[\frac{\nu''}{2} + \frac{\nu'}{r} + \frac{1}{4}\nu'(\nu'-\lambda')\right] = \frac{\kappa}{2}e^{-\lambda}(A_0')^2 = Ge^{-\lambda}(A_0')^2,$$

$$(b)\quad \left[\frac{\nu''}{2} - \frac{\lambda'}{r} + \frac{\nu'(\nu'-\lambda')}{4}\right]\frac{x^k x^l}{r^2}$$

$$+ \left[-\frac{1}{r^2} + e^{-\lambda}\left\{\frac{1}{r^2} + \frac{\nu'-\lambda'}{2r}\right\}\right]\left(\delta^{kl} - \frac{x^k x^l}{r^2}\right)$$

$$= -\frac{\kappa}{2}e^{-(\nu+\lambda)}(A_0')^2\left[-e^{\lambda}\frac{x^k x^l}{r^2} + \left(\delta^{kl} - \frac{x^k x^l}{r^2}\right)\right], \tag{B.55}$$

since $G = \kappa/2$. One can verify that the trace

$$T^{\mu}{}_{\mu} = g^{00}T_{00} + g^{kl}T_{kl}$$

is zero as claimed by (B.25). The equation $R = 0$ implies

$$R = g^{00}R_{00} + g^{kl}R_{kl} = 0. \tag{B.56}$$

Substituting the expressions in Eq. (B.53), one obtains the equation

$$\frac{1}{2}\nu''(r) + \frac{1}{r}(\nu'(r) - \lambda'(r)) + \frac{1}{4}\nu'(r)(\nu'(r) - \lambda'(r)) + \frac{1}{r^2} = \frac{1}{r^2}e^{\lambda(r)}. \tag{B.57}$$

We can rewrite this equation as

$$\frac{1}{2}\nu''(r) + \frac{1}{r}\nu'(r) + \frac{1}{4}\nu'(r)(\nu'(r) - \lambda'(r)) = \frac{1}{r^2}e^{\lambda(r)} - \frac{1}{r^2} + \frac{1}{r}\lambda'(r). \tag{B.58}$$

The left hand side of this equation is contained in the first of Eqs. (B.55), which can therefore be written as (with $G = \kappa/2$)

$$e^{(\nu(r)-\lambda(r))}\left(\frac{e^{\lambda(r)}}{r^2} - \frac{1}{r^2} + \frac{\lambda'(r)}{r}\right) = Ge^{-\lambda(r)}(A_0'(r))^2,$$

or

$$e^{\nu(r)}\left[e^{-\lambda(r)}\left(\frac{1}{r^2} - \frac{\lambda'(r)}{r}\right) - \frac{1}{r^2}\right] + Ge^{-\lambda(r)}(A_0'(r))^2 = 0. \tag{B.59}$$

We can also insert Eq. (B.58) into the second of Einstein's equation (B.55), *i.e.* we rewrite Eq. (B.58) as

$$\frac{1}{r^2}e^{\lambda(r)} = \frac{1}{2}\nu''(r) + \frac{1}{r}(\nu'(r) - \lambda'(r)) + \frac{1}{4}\nu'(r)(\nu'(r) - \lambda'(r)) + \frac{1}{r^2}, \quad \text{(B.60)}$$

and insert this for $e^{\lambda(r)}/r^2$ in the following expression contained in the second of Eqs. (B.55):

$$\left[-\frac{1}{r^2} + e^{-\lambda}\left\{ \frac{1}{r^2} + \frac{\nu' - \lambda'}{2r} \right\} \right] = -e^{-\lambda}\left[\frac{e^{\lambda}}{r^2} - \left\{ \frac{1}{r^2} + \frac{\nu' - \lambda'}{2r} \right\} \right]$$

$$= -\frac{e^{-\lambda}}{2}\left[\nu'' + \frac{\nu' - \lambda'}{r} + \frac{1}{2}\nu'(\nu' - \lambda') \right],$$

so that the second Einstein equation becomes:

$$\left[\frac{1}{2}\nu'' - \frac{\lambda'}{r} + \frac{1}{4}\nu'(\nu' - \lambda') \right] \frac{x^k x^l}{r^2}$$

$$- \frac{e^{-\lambda}}{2}\left[\nu'' + \frac{\nu' - \lambda'}{r} + \frac{1}{2}\nu'(\nu' - \lambda') \right]\left(\delta^{kl} - \frac{x^k x^l}{r^2} \right)$$

$$= -Ge^{-(\nu+\lambda)}(A_o')^2\left[-e^{\lambda}\frac{x^k x^l}{r^2} + \left(\delta^{kl} - \frac{x^k x^l}{r^2} \right) \right]. \quad \text{(B.61)}$$

For $k \neq l$ we obtain one equation from this, and for $k = l$ another. We subtract these and again use Eq. (B.60) but rewritten as:

$$\nu'' + \frac{\nu' - \lambda'}{r} + \frac{1}{2}\nu'(\nu' - \lambda') = 2\frac{e^{\lambda} - 1}{r^2} - \frac{\nu' - \lambda'}{r}. \quad \text{(B.62)}$$

We then obtain the equation

$$\frac{\lambda' - \nu'}{2r} + \frac{e^{\lambda} - 1}{r^2} = \frac{\kappa}{2}e^{-\nu}(A_0')^2. \quad \text{(B.63)}$$

In order to obtain equations in a somewhat different form we can proceed in another way. We consider again the equations (a) and (b) of Eq. (B.54) contained in the Einstein equation,

$$R_{\mu\nu} - \frac{1}{2}g_{\mu\nu}R = -\kappa T_{\mu\nu}, \quad R = 0. \quad \text{(B.64)}$$

Thus we consider first

$$R_{00} - \frac{1}{2}g_{00}R = -\kappa T_{00}, \quad \text{with} \quad R = 0, \quad \text{(B.65)}$$

which is simply Eq. (B.59),

$$-\left(\frac{\lambda'}{r} - \frac{1 - e^\lambda}{r^2}\right) = -\frac{\kappa}{2}e^{-\nu}(A_0')^2. \tag{B.66}$$

The other equation,

$$R_{kl} - \frac{1}{2}g_{kl}R = -\kappa T_{kl}, \quad R = 0, \tag{B.67}$$

is more involved. But equating coefficients of

$$\frac{x^k x^l}{r^2} \quad \text{and} \quad \left(\delta^{kl} - \frac{x^k x^l}{r^2}\right)$$

on both sides, one arrives at the following two equations:

$$-\left(\frac{\nu'}{r} + \frac{1 - e^\lambda}{r^2}\right) = \frac{\kappa}{2}e^{-\nu}(A_0')^2, \tag{B.68}$$

and

$$\left[\frac{\nu''}{2} + \frac{1}{2r}(\nu' - \lambda') + \frac{1}{4}\nu'(\nu' - \lambda')\right] = \frac{\kappa}{2}e^{-\nu}(A_0')^2. \tag{B.69}$$

The usefulness of the derivation of these additional equations can be seen by adding Eqs. (B.66) and (B.68), which yield immediately

$$\lambda'(r) + \nu'(r) = 0, \quad \lambda(r) + \nu(r) = \text{const.} \equiv \ln k. \tag{B.70}$$

This relation also follows by comparison of Eqs. (B.63) and (B.66). We can therefore concentrate from now on on these latter equations. In order to be able to proceed we need to evaluate A_0'. This is obtained in the following section.

B.7   Evaluating the electrostatic and gravitational fields

The equations (B.65) and (B.67) above for the gravitational field $g_{\mu\nu}$ have to be supplemented by those of the electromagnetic field $A_0(r)$, since we are considering a theory involving both fields. It is clear from our earlier considerations that the Euler–Lagrange equation of the electromagnetic field in the present case is simply the usual free-field equation with the ordinary derivative replaced by the covariant derivative. Hence we have

$$F^{\mu\nu}_{;\nu} = 0 \tag{B.71}$$

with (in the present case of the electrostatic field) $F_{ik} = 0$ (no magnetic component). One then obtains from Eq. (B.71) with the covariant derivative (14.67b) of the field tensor $F^{\mu\nu}$ the equation

$$0 = F^{\mu\nu}_{\ \ ;\nu} = \partial_\nu F^{\mu\nu} + \underbrace{\Gamma^\mu_{\lambda\nu} F^{\lambda\nu}}_{0} + \Gamma^\nu_{\nu\lambda} F^{\mu\lambda},$$

and for $\mu = 0$:

$$0 = F^{0\nu}_{\ \ ;\nu} = \partial_k F^{0k} + \Gamma^\nu_{\nu k} F^{0k},$$

or

$$F^{0k}_{\ \ ,k} + \Gamma^\eta_{\ k\eta} F^{0k} = 0. \tag{B.72}$$

The term $\Gamma^\mu_{\lambda\nu} F^{\lambda\nu}$ vanishes because Γ is symmetric in the indices λ and ν, and F antisymmetric. Here (see Eq. (B.36)):

$$F_{0k} = -A'_0(r) \frac{x^k}{r}, \qquad F^{0l} = g^{0\alpha} g^{k\beta} F_{\alpha\beta}. \tag{B.73}$$

Hence

$$
\begin{aligned}
-F^{0k} &\underset{(B.31),(B.34)}{=} g^{00} g^{kl} F_{0l} \\
&= e^{-\nu(r)} \left[-\delta^{kl} + \left(1 - e^{-\lambda(r)}\right) \frac{x^k x^l}{r^2} \right] A'_0(r) \frac{x^l}{r} \\
&= e^{-\nu(r)} \left(-\frac{x^k}{r} + \frac{x^k}{r} - e^{-\lambda(r)} \frac{x^k}{r} \right) A'_0(r) \\
&= -e^{-(\nu(r)+\lambda(r))} A'_0(r) \frac{x^k}{r}. \tag{B.74}
\end{aligned}
$$

Then

$$\Gamma^\eta_{\ k\eta} = \Gamma^0_{k0} + \Gamma^l_{kl} \underset{(B.49),(B.50)}{=} \frac{1}{2}(\nu'(r) + \lambda'(r)) \frac{x^k}{r}, \tag{B.75}$$

and Eq. (B.72) implies (with $\partial r/\partial x^k = x^k/r$ and $\partial(x^k/r)/\partial x^k = 2/r$)

$$
\begin{aligned}
&-(F^{0k}_{\ \ ,k} + \Gamma^\eta_{\ k\eta} F^{0k}) \\
&= \partial_k \left[-e^{-(\nu+\lambda)} A'_0 \frac{x^k}{r} \right] + \frac{x^k}{2r} \lambda'(r) F^{0k} \\
&= -e^{-(\nu+\lambda)} \left[A''_0 + \frac{2}{r} A'_0 - (\nu' + \lambda') A'_0 + \frac{1}{2}(\nu' + \lambda') A'_0 \right],
\end{aligned}
$$

and since this is to be zero (Eq. (B.72)), we obtain

$$A''_0 + \frac{2}{r} A'_0 - \frac{1}{2}(\nu' + \lambda') A'_0 = 0. \tag{B.76}$$

A first integral of this equation is

$$\left[r^2 e^{-(\nu+\lambda)/2} \right]^{-1} \frac{d}{dr} \left(r^2 e^{-(\nu+\lambda)/2} A_0' \right) = 0,$$

and thus we obtain for the derivative of the electromagnetic field A_0

$$A_0' = -\frac{\epsilon}{r^2} e^{(\nu+\lambda)/2}, \quad \epsilon = \text{const.} \tag{B.77}$$

We can now insert this result into our Eqs. (B.66) to (B.69). We set

$$\chi = e^{-\lambda(r)}, \quad \lambda(r) = -\ln \chi, \quad \chi' = -\lambda'(r) e^{-\lambda(r)}. \tag{B.78}$$

Equation (B.66) now becomes

$$-\left(\frac{\lambda'}{r} - \frac{1 - e^\lambda}{r^2} \right) = -\frac{\kappa \, \epsilon^2}{2 \, r^4} e^\lambda, \quad \frac{d\chi}{dr} + \frac{\chi - 1}{r} + \frac{\kappa \, \epsilon^2}{2 \, r^3} = 0. \tag{B.79}$$

This is a differential equation of the first order, hence we expect one constant of integration. The equation is solved with the help of an integration factor $\exp(-\int^r dr/r) = 1/r$, in which the integrand of the integral is the coefficient of χ.$^{\|}$ Then

$$\chi = e^{-\ln r} \int^r dr \left\{ \frac{1}{r} - \frac{\kappa \epsilon^2}{2r^3} \right\} e^{+\ln r} = \frac{1}{r} \left[r + \frac{\kappa \epsilon^2}{2r} + \text{const.} \right],$$

and hence, with integration constant l,

$$\chi(r) = 1 - \frac{\kappa l}{r} + \frac{\kappa \, \epsilon^2}{2 \, r^2}, \quad l = \text{const.} \tag{B.80}$$

$^{\|}$ Consider the following first-order differential equation in $y(x)$ and set $y(x) = y_0(x) f(x)$ with

$$\frac{dy}{dx} + P(x)y = Q(x), \quad \frac{dy_0}{dx} + P(x)y_0 = 0.$$

Here $y_0(x)$ supplies the integration factor, and we set

$$y(x) = f(x) e^{-\int^x P(x)dx}.$$

Inserting this into the original equation, one obtains

$$\frac{df(x)}{dx} = Q(x) e^{+\int^x P(x)dx}, \quad \therefore f(x) = \int^x dx Q(x) e^{+\int^x P(x)dx}.$$

It follows that

$$y(x) = e^{-\int^x P(x)dx} \int^x dx Q(x) e^{+\int^x P(x)dx}.$$

The remaining integration now allows one constant of integration.

From Eq. (B.70) we obtain ($\ln k = $ const.)

$$\nu + \lambda = \ln k.$$

Choosing $k = 1$, we have $\nu = -\lambda$. Thus, with $e^{\lambda(r)} = 1/e^{\nu(r)}$,

$$\chi \quad = \quad e^{\nu(r)} \overset{(B.80)}{=} g_{00} = 1 - \frac{\kappa l}{r} + \frac{\kappa \epsilon^2}{2r^2}, \quad g_{0k} = 0,$$

$$g_{kl} \overset{(B.19)}{=} -\delta^{kl} + \left[1 - \left\{1 - \frac{\kappa l}{r} + \frac{\kappa \epsilon^2}{2r^2}\right\}^{-1}\right] \frac{x^l x^k}{r^2}. \tag{B.81}$$

We have thus determined the gravitational field $g_{\mu\nu}$ in the presence of a static electric charge. Hence finally:

$$ds^2 = (dx^0)^2 \left(1 - \frac{\kappa l}{r} + \frac{\kappa \epsilon^2}{2r^2}\right) - r^2 d\Omega^2 - (dr)^2 \left(1 - \frac{\kappa l}{r} + \frac{\kappa \epsilon^2}{2r^2}\right)^{-1}, \tag{B.82}$$

with

$$\frac{x^l x^k}{r^2} dx^l dx^k = \left[\frac{d(x^l)^2}{2r}\right]^2 = \left(\frac{dr^2}{2r}\right)^2 = (dr)^2.$$

The result (B.82) is the *Reissner–Nordstrom metric*. For charge $\epsilon \to 0$ the result reduces to the *Schwarzschild metric*.

Bibliography

[1] A. C. Aitken, *Determinants and Matrices* (Oliver and Boyd, 1954).

[2] H. Aratyn and C. Rasinariu, *A Short Course in Mathematical Methods with Maple* (World Scientific, 2006).

[3] C. M. Bender, D. D. Holm and D. W. Hook, *Complex Rotation of a Rigid Body*, arXiv: 0705.3893, hep-th (May 2007).

[4] A. Ben–Naim, *Entropy Demystified* (World Scientific, 2007).

[5] van Biesbroek, probably cited by Leite–Lopes.

[6] E. H. Booth and P. M. Nicol, *Physics, Fundamental Laws and Principles*, 12th ed. (Australian Medical Publ. Co., 1952).

[7] H. Bucerius and M. Schneider, *Himmelsmechanik I, II*, BI 143a, 144a.

[8] K. E. Bullen, *An Introduction to the Theory of Mechanics*, 2nd ed. (Science Press Sydney, 1951).

[9] P. F. Byrd and M. D. Friedman, *Handbook of Elliptic Integrals for Engineers and Scientists* (Springer, 1971).

[10] M. Carmeli, *Cosmological Relativity* (World Scientific, 2006).

[11] S. M. Carroll, *Lecture Notes on General Relativity*, only available to the author's knowledge from http://itp.ucsb.edu/ carroll/notes (December 1997).

[12] R. Courant, *Differential and Integral Calculus*, Vols. I, II (Blackie & Son, 1951).

[13] H. B. Dwight, *Tables of Integrals and other Mathematical Data* (Macmillan Co., 1957).

[14] A. Einstein, *Ann. Physik* **17** (1905) 891.

[15] R. V. Eötvös, *Über die Anziehung der Erde auf verschiedene Substanzen* (On the attraction of the Earth of different substances), *Math. Naturw. Berichte aus Ungarn* **8** (1889) 65; R. V. Eötvös, D. Pekár and E. Fekete, *Ann. d. Physik* **68** (1922) 11.

[16] B. Felsager, *Geometry, Particles and Fields* (Odense University Press, 1981).

[17] G. W. Gibbons and N. S. Manton, *Nucl. Phys.* **B 274** 183; see p. 201.

[18] R. H. Giese, *Weltraumforschung I*, BI 107/107a.

[19] H. Goldstein, *Classical Mechanics*, 6th ed. (Addison–Wesley, 1959); *Klassische Mechanik*, 2nd ed. (Akad. Verlagsgesellschaft, 1972).

[20] I. S. Gradshteyn and I. M. Ryzhik, *Table of Integrals, Sums and Products* (Academic Press, 1965).

[21] W. Gröbner and N. Hofreiter, *Integraltafel I, Unbestimmte Integrale* (Springer, 1958).

[22] W. Gröbner and N. Hofreiter, *Integraltafel II, Bestimmte Integrale* (Springer, 1958).

[23] L. Iorio, *On the possibility of testing the Weak Equivalence Principle with artificial Earth satellites*, *Gen. Rel. Grav.* **36** (2004) 361, gr-qc/0309105.

[24] J. D. Jackson, *Classical Electrodynamics* (Wiley, 1975).

[25] T. W. B. Kibble and F. H. Berkshire, *Classical Mechanics*, 5th ed. (Imperial College Press, 2004).

[26] H. Koppe, *Quantenmechanik*, mimeographed lecture notes, University of Munich (1964).

[27] D. F. Lawden, *Tensor Calculus and Relativity* (Chapman and Hall, 1975).

[28] J. Leite–Lopes, *Notions de Relativité Generale*, CBPF Report 003/86, Rio de Janeiro.

[29] L. R. Lieber, *The Einstein Theory of Relativity* (Dobson, 1949).

[30] L. M. Milne–Thomson, *Jacobian Elliptic Function Tables* (Dover, 1950).

[31] L. M. Milne–Thomson, *Quart. J. Mech. Applied Maths.* **2** (1949) 479.

[32] C. W. Misner and P. Putnam, *Phys. Rev.* **116** (1959) 1045.

[33] C. W. Misner, K. S. Thorne and J. A. Wheeler, *Gravitation* (W. H. Freeman and Co., 1973).

[34] H. J. W. Müller–Kirsten, *Electrodynamics: An Introduction Including Quantum Effects* (World Scientific, 2004).

[35] H. J. W. Müller–Kirsten, *Introduction to Quantum Mechanics: Schrö-dinger Equation and Path Integral* (World Scientific, 2006).

[36] P. W. Norris and W. S. Legge, *Mechanics via the Calculus* (Cleaver–Hume Press, 1953). This book treats a large number of problems on a more elementary basis than the book of A. S. Ramsey.

[37] A. Ori, *Phys. Rev.* **D**, according to Yahoo, August 3 (2007).

[38] R. Penrose, *The Road to Reality, A complete guide to the laws of the universe* (Vintage Books, 2004).

[39] R. Penrose *Proc. Camb. Phil. Soc.* **55** (1959) 137.

[40] D. Raine and E. Thomas, *Black Holes, An Introduction* (Imperial College Press, 2005). This text of 155 pages assumes a first acquaintance with general relativity.

[41] A. S. Ramsey, *Dynamics* (Cambridge University Press, 1954). Although this book does not use the Lagrangian formalism to obtain equations of motion (which are derived from Newton's equation or conservation laws), it treats a lot of central force problems and makes considerable use of the geometry of conics. The text of P. W. Norris and W. S. Legge is on a more elementary basis.

[42] R. Regge, *An Elementary Course on General Relativity*, CERN Yellow Report 83-09 (1983) (this report has only 25 pages and very few formulas).

[43] F. Reif, *Fundamentals of Statistical and Thermal Physics* (McGraw–Hill, 1965).

[44] P. G. Roll, R. Krotkov and R. H. Dicke, *Ann. Phys. (N.Y.)* **26** (1964) 442.

[45] P. Rowlands, *Phys. Educ.* **32** (1997) 49.

[46] D. M. Y. Sommerville, *Analytical Conics* (G. Bell and Sons, 1951).

[47] J. Terrell, *Phys. Rev.* **116** (1959) 1041.

[48] The Physics Coaching Class, University of Science and Technology of China, *Problems and Solutions on Mechanics* (World Scientific, 1994). This very useful and readable book contains 272 problems on Newtonian mechanics, 84 problems on Lagrangian and Hamiltonian mechanics, and 54 problems on Special Relativity.

[49] E. Vogt, *Elementary Derivation of Kepler's Laws, Am. J. Phys.* (1995), TRIUMF report TRI-PP 95/3.

[50] S. Weinberg, *Gravitation and Cosmology* (Wiley, 1972).

[51] D. A. Wells, *Lagrangian Dynamics* (McGraw–Hill, 1967).

[52] F. Wilczek, *Fantastic Realities* (World Scientific, 2006).

[53] R. Williams, *An Introduction to Curved Space–Times*, CERN Report 91-06 (1991).

Index

a priori probability, 169, 197
a priori probability, definition of, 168
semi-latus rectum, 212

aberration formula, 430
aberration law, 420, 423, 429
aberration, definition of, 421
accelerated frame,
    noninertial frame, 130
acceleration,
    absolute, 312
    centrifugal, 442
    centripetal, 312
    Coriolis, 313, 442
    definition of, 5
    due to gravity, 316
    radial, 109
    tangential, 210
    tidal, 374
    transversal, 109, 202
action, 42, 103, 183
action functional, 485
addition of velocities, 419
affine connection, 436, 459
affine connection,
    Newtonian, 438
affine parameter, 517
affinity,
    definition, 459
    metric affinity, 463
    symmetry of, 467
    transformation law of, 460

air drifts, 317
Aitken, 270
Ampére's law, 192
angular deflection, 357
angular momentum, 5, 11, 96, 156, 158
aphelion, 226
apsis, definition of, 228
atmosphere, limit, 20
auxiliary circle, 219
axial vector, 128

bending of light ray, 361, 516, 522
Bianchi identity, 469, 476, 478
    first, 472
Bianchi symmetry, 472, 474
binding, 207
black hole, 404, 499
blueshift, 363
Bohr, Bohr radius, 256
boomerang, 331
boost, 185, 400
boost in electrodynamics, 410
Bucerius and Schneider, 233
Byrd and Friedman, 525

canonical Hamilton equations, 79
canonical momentum, 51
canonical variable, 77, 96, 141
cardioid, 218
Carmeli, 233, 353, 501, 521, 522
Carroll, 353, 498
Cayley–Klein parameters, 120, 122

and Euler angles, 124
centre of gravity, 258
centre of mass, 10, 200, 258
centrifugal acceleration, 356
centrifugal barrier, 206
centripetal acceleration, 312
characteristic equation, 272
charge density, 482
Christoffel symbol, 435, 438, 459,
        488, 501, 539
    for spherical coordinates, 439
    from geodesic equation, 502
    of second kind, 464
clock paradox, 413, 431, 432
clock, fast, slow, 406
clock, stationary, moving, 405, 432
closed system, 107
cofactor, alien, 447
cofactor, definition, 446
colatitude angle, 319, 356, 381
collective coordinate, 10
comet, 236
compact, noncompact, 184
cone, 369, 370
cone, null, see null cone, 401
cone, past, future, 401
configuration space, 30, 32, 42, 43
conical section,
    definition, 211
    geometry, 211
    polar equation, 211
conjugate diameter, 219
conjugate momentum, 51
conjugate point, 219
connection quantity,
    see Christoffel symbol, 489
connection, definition of, 463
conservation
    of angular momentum, 9
    of linear momentum, 9
conservation law, 138

conservation of 4-momentum, 415,
        427, 428
conservative system, 8
conserved quantity, 152
constraint, 30
    holonomic, 30, 39, 73
    nonholonomic, 31, 48, 73
    rheonomic, 32
    scleronomic, 32
continuity equation, 194
contraction of indices, 187
contravariant tensor, 409
    other definition, 447
contravariant vector, 395
coordinate,
    curvilinear, 88
    cyclic, 79, 201
Coriolis force, 364, 442
cosine theorem, 36, 242
cosmic velocity,
    first, 215
    second, 215
    third, 215
cosmological constant, 497
couple, 7
covariance, 105
covariant derivative, 76, 455, 461
    in electrodynamics, 467
    of tensor, 471
    of vector, 471
covariant metric tensor, 450
covariant tensor, 409
covariant vector, 395
cross section, 236
    differential, 236
    total, 240
cubic equation, solution of, 527
current density, 236, 482, 484
    of wind, 318
current of momentum, 482
current, steady, stationary, 179

curvature, 365, 375

    and Lorentz contraction, 411

curvature, Gaussian, 368

curvature, positive, negative, 369

cusp, 59

cusp, definition of, 62

cyclic coordinate, 51, 79, 141

cyclic variable, 51

cycloid, 59

cyclone, 319

cylinder, 53, 369, 370

D'Alembert principle, 37

D'Alembertian

    in Riemann space, 465, 468

decay of nucleus, 426

deflection of light, 361, 387, 498, 516, 518, 522

degeneracy, 169, 351

degree of freedom, 109

degree of freedom,

    definition, 30

delta function,

    see Dirac's delta function, 349

derivative, covariant, 76, 455

derived quantity, 96

determinant, 446

    expansion of, 446

deterministic, 168, 236

diagonalization, 265

diffeomorphism, 196

dilation, 387, 413, 441

dipole approximation, 26

Dirac's delta function, 349, 495

direction cosine, 262, 267, 290

    definition, 266

direction cosine,

    advantage of, 267

distribution, 495

divergence, 466

Doppler effect, 421

dual tensor, 192

dual, dual map, 450, 451

Eötvös experiment, 356, 359

eccentric angle, 219

eccentricity, 212, 214, 215, 219, 238

Eddington–Finkelstein coordinate, 403

eigentime, 400, 412, 413, 516

eigenvalue, 269

eigenvalue problem, 269

eigenvector, 269

Einstein, 186, 395, 421, 430, 488, 498

Einstein tensor, 471

Einstein's equation, 56, 487, 488, 491, 497, 499, 507, 539

Einstein's formula, 386, 444, 493

Einstein's formula,

    derivation of, 416

Einstein's gravitational law, 491

Einstein-Hilbert action, 489, 497

Einstein-Maxwell

    field equation, 492

electrodynamics, 391

electromagnetic field tensor, 410

electromagnetic field, free, 492

electron-positron

    annihilation, 427

ellipse, 213, 214

    polar equation, 26

ellipse, area of, 379

ellipsoid, 23, 267

ellipsoid of inertia, 265, 268, 286, 339

ellipsoid,

    momental, 265

    normal form, 267

    of revolution, 267

elliptic integral of first kind, 59

elliptic modulus, 59, 529

energy,
  conservation of, 9
  free, 81
  total, 415
energy-momentum
  tensor, 194, 481, 498, 535
energy-momentum tensor,
  in electrostatics, 537
ensemble, 170, 175
enthalpy, 81
entropy, 81, 348
equilibrium, 20, 21, 33, 179
equipotentials, 26
escape velocity, 19, 387, 507
ether theory, 392
Euclid, 11th axiom of, 366, 368
Euler angles, 117, 122, 275, 322
Euler differential equation, 46
Euler equations, 275, 278, 293
Euler-Lagrange equation, 30, 45,
      47, 434, 486, 502, 517
Euler-Lagrange equation,
  in field theory, 191
event, 359, 396, 417
event horizon, 404, 499
  apparent, 506
expanding universe, 402
expectation value, 179
extremum, 42
extremum principle, 434

Faraday's law, 192
field, 183
  density, 183
FitzGerald-Lorentz contrac-
    tion, 386, 394, 411, 422, 423
FitzGerald-Lorentz contraction,
  visibility, 394, 422
flat space, 359, 370, 371
fluctuation equation, 234
force,

applied, 33, 276
central, 202, 248
centrifugal, 21, 311, 312, 364,
    374
centripetal, 51, 312, 313
constraining, 33
Coriolis, 311, 313, 316, 317, 364
definition of central, 202
effective, 312
external, 9
fictitious, 359, 364, 365
field, 8
generalized, 38, 41, 72, 276
gravitational, 104, 470
internal, 9
Minkowski, 413
repelling Kepler, 238
tidal, 375, 405
virtual, 107
weight is a, 8
form invariance, 106, 141
frame, inertial, local, 358
free energy, 81
free enthalpy, 81
free fall, 4, 355, 361, 363, 436
frequency, 252
friction, 40
functional determinant
  see Jacobi, 98
fundamental mixed tensor, 462
fundamental postulate of relativ-
      ity, 363
future cone, 401

Galilean relativity, 4
Galilei, 5, 15, 355
  transformation, 105
Galilei transformation,
  general, 107
Gauss theorem, 468
Gauss' law, 192

Gaussian curvature, 368
generalized coordinate, 30, 32, 34
generating function, 142
generator, 154
geodesic, 366, 370, 375, 379, 380,
        433, 516
    equation, 501
    in Riemann space, 436
geodesic equation
    and covariant derivative, 470
geodesic, determination of, 434
Gibbs function, 81
glories, forward, backward, 241
Goldstein, xii, 79, 167, 267, 333,
        337, 342
gravimeter, 21
gravitational
    acceleration, 355
    blueshift, 387
    constant, 355
    deflection of light, 387
    field, weak, time-independent,
        492
    force, 355
    mass, 354, 355, 362
    potential, 389
    redshift, 387
    time dilation, 406
gravitational constant, 15
gravitational field, 258, 363
graviton, 363, 387, 492
grazing angle, 522
Green's function, 495
group,
    $O(2)$, 114
    $O(3)$, 116
    $O(n)$, 112
    $SO(2)$, 114, 136, 397, 399
    $SO(3)$, 116, 184, 260, 267
    $SO(3,1)$, 185, 397, 399
    $SU(2)$, 119

$SU(n)$, 119
$U(n)$, 119
    axioms, 110
    element, 110
    property, 109
    special, 114
group, Abelian, 116
group, commutative, 116
group, Poincaré, 186

Hamilton density, 183
Hamilton function, 52, 78
Hamilton principle, 42
Hamilton principle,
        extended, 46
Hamilton's action, 145
Hamilton's equations, 79, 88
Hamilton-Jacobi equation, 145, 165
harmonic oscillator, 165
harmonic oscillators,
        coupled, 162
Heisenberg picture, 182
helix, 370
Hermitian, 119, 265, 273
Hilbert space, 96
Hilbert-Einstein action, 489, 497
homomorphism, 111
Hooke's law, 41, 340, 351, 352
horizon, 404
Hubble, 498
hurrying physicist, 429
Huygens, 59, 75
hyperbola, 213, 214, 398

impact parameter, 236, 522
inclined plane, 16
incompressibility, 173
incompressibility condition, 170
indices, raising, lowering, 464
inert, inertia, meaning of, 4
inertia,

coefficient of moment of, 260
ellipsoid of, 265
moment of, 257, 260, 261
inertia, principal moment of, 265
inertial
    frame, 354, 358, 392
    mass, 354, 362
    motion, 358
    system, 363
    tensor, 260
inertial frames,
    clocks in, 395
    equivalent frames, 392
    observation of rotation in, 440
inertial tensor,
    diagonalization of, 269
inertial, definition of, 4
intensity, 236
invariable plane, 294
invariance, 104, 105
inversion, 115
isomorphism, 111, 122

Jacobi determinant, 90
Jacobi identity, 98, 152, 192
Jacobian elliptic functions, 293, 301,
    525, 528

Kepler problem, 210, 389, 498, 514
    relativistic, 511
Kepler, first (orbit) law, 214
Kepler, second (area) law, 223, 227
Kepler, third (period) law, 227
kernel, 111
Kibble and Berkshire, xii, 167, 320
Killing vector, 196
Kronecker delta, 462

Lagrange bracket, 91
Lagrange brackets, fundamental, 92
Lagrange density, 183

Lagrange multiplier, 48, 55, 67, 70,
    72, 379
Lagrangian, 40, 434, 436
Lagrangian equation, 39
latitude, 32, 203, 313, 315, 319,
    371, 380
Lawden, 353, 455
Legendre polynomial, 24, 203
Legendre polynomial,
    associated, 203
Legendre transform, 77, 78, 80
Leibnitz rule, 466, 490
Leite-Lopes, 353, 360, 498, 531
Levi-Civita form, 451
Levi-Civita symbol, 6, 136, 451
light cone, 373
light cone, future, 397, 413
line element, 358
Liouville equation, 170, 175
Liouville theorem, 168, 174
longitude, 32, 203, 371, 380
Lorentz
    boost, 407
    contraction, *see* FitzGerald-
        Lorentz, 393
    force, 412, 416, 484
    geometry, 365
    group, 185
    indices, 406
    invariance, 358
    invariant, 187
    transformation, 184, 186, 394,
        407
lune, 367

Mössbauer effect, 444
mapping, 111
Mars, 223
mass,
    geometric, 402
    gravitational, 5, 15, 354, 362

gravitational, active, passive, 15

inertial, 5, 15, 354, 362

reduced, 201

relativistic, 413

solar, 405

Maxwell energy-momentum tensor, 486

Maxwell equations, 465, 468, 481

in flat space, 469

in Riemann space, 469

Maxwell relation, first, 81

Maxwell stress tensor, 482

metric, 186, 358, 418

coefficents, 418

metric field, 363, 492

Michelson and Morley experiment, 392

Milne-Thomson, 525, 527

minimal coupling, 467

Minkowski

force, 413

metric, 450, 499

metric, flat, 365

space, 184, 359, 397, 406, 488

spacetime tensor, 443

Misner, Thorne and Wheeler, 353

mixed tensor, 409

moment of inertia, 74, 257

moment of inertia, principal, 267

momentum conservation, 52

momentum,

canonical, 76

conjugate, 76

kinematical, 76

mechanical, 76

Newton's

gravitational law, 211, 316

gravitational potential, 235, 493, 496, 499

law of gravitation, 14, 20, 21, 354

second law, 278, 398

Newton's equation, 439, 481

covariantized, 412

in gravitational field, 433

reduction to, 437

Newton's laws, 106, 317, 354

Noether current, 194

Noether's theorem, 138, 192

normal form, 267, 339

normal modes, 350, 352

normal modes of vibration, 334

null cone, 400, 401, 516

nutations, 307, 322

nutations, damped, 309

observable, 97

observing, seeing, difference between these, 422

oil field, 21

orthogonality, 116

orthogonality relation, 111

parabola, 16, 18, 213, 214

parallel transport, 445, 455, 457

past cone, 401

Pauli matrices, 121

Pauli matrices, properties of, 124

pendulum,

centrifugal, 67

compound, 281

cycloidal, 59, 74

double, 334

Foucault, 319

rigid body, 281

rubber band, 40

simple, 58

spherical, 64
under gravity, 315
Penrose, 29, 256, 333, 353, 422, 423, 432, 472, 486
perihelion, 226, 228, 291, 511
of Mercury, 388
perihelion anomaly, 211
secular, 232
period, 227, 252
phase space, 43, 78, 176
coordinates, 484
photon, 357, 363, 392, 416, 514, 516
physical space, 32
Poincaré group, 408
Poincaré transformation, 456
point charge, 535
Poisson algebra, 98
Poisson bracket, 91, 92, 98, 147
fundamental, 95
Poisson equation, 488, 491, 495, 496
Poisson equation, solution of, 495
Poisson theorem, 153
pole, 202
postulates, of Special Relativity, 392
potential, 8
centrifugal, 205
Coulomb, 236, 237, 240
effective, 205
gravitational, 23, 443
thermodynamical, 81
velocity-dependent, 40, 47, 51
power, 415, 417
precession, 228, 295, 307, 313, 331
precession of Mercury's perihelion, 388
principal axis, 257, 265, 267, 269, 339
principal axis,
transformation to, 257

principal moment of inertia, 26
probability density, 169
probability, classical a priori, 168
probability, quantal a priori, 169
proper time, 400, 405, 412, 441
pseudo-tensor, 450
pseudoscalar, 454

quantization, Bohr-Sommerfeld
-Wilson, 254
quantum mechanics, 158

rainbow singularity, 241
Raine and Thomas, 512
rapidity, 185
redshift, 433, 443, 444
reflection, 114, 115
refractive index, 388
Regge, 353
Reissner-Nordstrom
metric, 406, 531, 547
relativistic mass, 386, 415
relativity principle, 104
relaxation time, 179
rescaling, 436
residual acceleration, 374
resonance frequency, 340
rest frame, 413
rest mass, 386, 415
Ricci curvature scalar, 371, 471, 487, 497
definition, 478
Ricci flat, 492
Ricci tensor, 471, 487
definition, 476
symmetry, 479
Ricci-Einstein tensor, 471, 475, 479, 487, 488
definition, 478
Riemann curvature tensor, 371, 375, 471

definition, 472
symmetry, 474
Riemann space, 406, 433, 488
rigid body, 13, 31, 109
rocket, 19
rolling, 69, 72, 284
rotation,
    operator, 466
    active, objective, 135
    passive, subjective, 135
    proper, improper, 116, 118
Routh function, 79
Rutherford formula, 240

saddle configuration, 343
scalar, 408
scalar field, 139
scalar function, definition of, 134
scalar triple product, 159, 259
scattering, 235
scattering angle, 239
Schrödinger picture, 182
Schwarzschild
    effective potential, 522
    orbit solution, 525
    ray equation, 523
    solution, 535
Schwarzschild coordinates, 499, 506
Schwarzschild energy-momentum re-
    lation, 511
Schwarzschild metric, 402, 506, 517
Schwarzschild radius, 507
Schwarzschild solution, 56, 498, 499,
    506
    nonzero cosmological constant,
    507
Schwarzschild space,
    light ray in, 516
screening, 240
secular equation, 272, 303, 335, 341
seeing, observing, difference

between these, 422
selfadjoint, Hermitian, 121, 265
SO(1,3), 407
Sommerville, 200
space, definition when flat, 473
space, flat, nonflat, 456
spacelike direction, 400
spaceship, 374
spacetime, 184, 359, 392
special, what makes the theory spe-
    cial, 391
spherical triangle, 366
spherical triangle, area of a, 368
spinning top, 302
spinor, 119
spring equinox, 247
stability, 253, 342, 349
    condition, 235
stability equation, 350
stability of an orbit,
    definition, 233
state, 97
statics, 32
Steiner's theorem, 263, 281, 287,
    330
Stokes' theorem, 8, 468, 490, 496
strong equivalence principle, 359,
    361, 362
symmetry, 104
symmetry axis, 268
symmetry condition, 193
symplectic matrix, 88

tangent vector, 417
tangential equation, 217
tangential plane, 294
tangential vector, 413
tangential velocity,
    acceleration, 210
tensor, 409
    fundamental mixed, 462

mixed, 445
rank 2, 260
symmetric, antisymmetric, 449
tensor calculus, 445
tensor, inertial, 260, 265
tensor, rank 2, definition, 138
Terrell, 422, 425
thermodynamics,
second law, 348
thought experiment, 360
tidal acceleration, 374
tidal effect, 376
tidal forces, 405
tides, 365
tilting light cones, 403
time dilation, 413, 441
timelike direction, 400
torque, 7, 40, 277
torsion balance experiment, 357
torsion-free, 455
torus, 332
total energy, 52
transformation,
active Lorentz, 400
active, passive, 395
canonical, 141, 154
generalized coordinate, 195
generator of, 136
identity, 114
Lorentz, 397
point, 141
similarity, 112, 117
unitary, 136
transversal velocity,
acceleration, 108, 202, 203, 230
trick calculations, 385
turning points, 306
twin problem, 431

umbrella, 15
unit tangent, 435

unitarity, 116, 119, 270, 273

van Biesbroek, 522
variational principle, 42
vector function, definition of, 135
vector product, 259
vector,
axial, 452
contravariant, 184, 407
covariant, 184, 407
polar, 452
velocity addition, 106
velocity of escape, 21, 215, 387, 507
velocity, transversal, 203
virial theorem, 483
virtual displacement, 33
virtual work, principle of, 32
virtual, definition of, 33
Voigt, 394
volume element, 88
volume form, 452
volume of parallelopiped, 453

weak equivalence principle, 355, 357,
362, 363, 374, 376
weighting, classical *a priori*, 168
Weinberg, 353
Wilczek, 353, 498
worldline, 359, 372, 401, 434
parameter, 434

zero mode, 342, 349, 351